MULTIVARIATE ANALYSIS, DESIGN OF EXPERIMENTS, AND SURVEY SAMPLING

STATISTICS: Textbooks and Monographs

A Series Edited by

D. B. Owen, Founding Editor, 1972–1991

1. The Generalized Jackknife Statistic, *H. L. Gray and W. R. Schucany*
2. Multivariate Analysis, *Anant M. Kshirsagar*
3. Statistics and Society, *Walter T. Federer*
4. Multivariate Analysis: A Selected and Abstracted Bibliography, 1957–1972, *Kocherlakota Subrahmaniam and Kathleen Subrahmaniam*
5. Design of Experiments: A Realistic Approach, *Virgil L. Anderson and Robert A. McLean*
6. Statistical and Mathematical Aspects of Pollution Problems, *John W. Pratt*
7. Introduction to Probability and Statistics (in two parts), Part I: Probability; Part II: Statistics, *Narayan C. Giri*
8. Statistical Theory of the Analysis of Experimental Designs, *J. Ogawa*
9. Statistical Techniques in Simulation (in two parts), *Jack P. C. Kleijnen*
10. Data Quality Control and Editing, *Joseph I. Naus*
11. Cost of Living Index Numbers: Practice, Precision, and Theory, *Kali S. Banerjee*
12. Weighing Designs: For Chemistry, Medicine, Economics, Operations Research, Statistics, *Kali S. Banerjee*
13. The Search for Oil: Some Statistical Methods and Techniques, *edited by D. B. Owen*
14. Sample Size Choice: Charts for Experiments with Linear Models, *Robert E. Odeh and Martin Fox*
15. Statistical Methods for Engineers and Scientists, *Robert M. Bethea, Benjamin S. Duran, and Thomas L. Boullion*
16. Statistical Quality Control Methods, *Irving W. Burr*
17. On the History of Statistics and Probability, *edited by D. B. Owen*
18. Econometrics, *Peter Schmidt*
19. Sufficient Statistics: Selected Contributions, *Vasant S. Huzurbazar (edited by Anant M. Kshirsagar)*
20. Handbook of Statistical Distributions, *Jagdish K. Patel, C. H. Kapadia, and D. B. Owen*
21. Case Studies in Sample Design, *A. C. Rosander*

22. Pocket Book of Statistical Tables, *compiled by R. E. Odeh, D. B. Owen, Z. W. Birnbaum, and L. Fisher*
23. The Information in Contingency Tables, *D. V. Gokhale and Solomon Kullback*
24. Statistical Analysis of Reliability and Life-Testing Models: Theory and Methods, *Lee J. Bain*
25. Elementary Statistical Quality Control, *Irving W. Burr*
26. An Introduction to Probability and Statistics Using BASIC, *Richard A. Groeneveld*
27. Basic Applied Statistics, *B. L. Raktoe and J. J. Hubert*
28. A Primer in Probability, *Kathleen Subrahmaniam*
29. Random Processes: A First Look, *R. Syski*
30. Regression Methods: A Tool for Data Analysis, *Rudolf J. Freund and Paul D. Minton*
31. Randomization Tests, *Eugene S. Edgington*
32. Tables for Normal Tolerance Limits, Sampling Plans and Screening, *Robert E. Odeh and D. B. Owen*
33. Statistical Computing, *William J. Kennedy, Jr., and James E. Gentle*
34. Regression Analysis and Its Application: A Data-Oriented Approach, *Richard F. Gunst and Robert L. Mason*
35. Scientific Strategies to Save Your Life, *I. D. J. Bross*
36. Statistics in the Pharmaceutical Industry, *edited by C. Ralph Buncher and Jia-Yeong Tsay*
37. Sampling from a Finite Population, *J. Hajek*
38. Statistical Modeling Techniques, *S. S. Shapiro and A. J. Gross*
39. Statistical Theory and Inference in Research, *T. A. Bancroft and C.-P. Han*
40. Handbook of the Normal Distribution, *Jagdish K. Patel and Campbell B. Read*
41. Recent Advances in Regression Methods, *Hrishikesh D. Vinod and Aman Ullah*
42. Acceptance Sampling in Quality Control, *Edward G. Schilling*
43. The Randomized Clinical Trial and Therapeutic Decisions, *edited by Niels Tygstrup, John M Lachin, and Erik Juhl*
44. Regression Analysis of Survival Data in Cancer Chemotherapy, *Walter H. Carter, Jr., Galen L. Wampler, and Donald M. Stablein*
45. A Course in Linear Models, *Anant M. Kshirsagar*
46. Clinical Trials: Issues and Approaches, *edited by Stanley H. Shapiro and Thomas H. Louis*
47. Statistical Analysis of DNA Sequence Data, *edited by B. S. Weir*
48. Nonlinear Regression Modeling: A Unified Practical Approach, *David A. Ratkowsky*
49. Attribute Sampling Plans, Tables of Tests and Confidence Limits for Proportions, *Robert E. Odeh and D. B. Owen*
50. Experimental Design, Statistical Models, and Genetic Statistics, *edited by Klaus Hinkelmann*
51. Statistical Methods for Cancer Studies, *edited by Richard G. Cornell*
52. Practical Statistical Sampling for Auditors, *Arthur J. Wilburn*
53. Statistical Methods for Cancer Studies, *edited by Edward J. Wegman and James G. Smith*
54. Self-Organizing Methods in Modeling: GMDH Type Algorithms, *edited by Stanley J. Farlow*
55. Applied Factorial and Fractional Designs, *Robert A. McLean and Virgil L. Anderson*
56. Design of Experiments: Ranking and Selection, *edited by Thomas J. Santner and Ajit C. Tamhane*
57. Statistical Methods for Engineers and Scientists: Second Edition, Revised and Expanded, *Robert M. Bethea, Benjamin S. Duran, and Thomas L. Boullion*
58. Ensemble Modeling: Inference from Small-Scale Properties to Large-Scale Systems, *Alan E. Gelfand and Crayton C. Walker*
59. Computer Modeling for Business and Industry, *Bruce L. Bowerman and Richard T. O'Connell*
60. Bayesian Analysis of Linear Models, *Lyle D. Broemeling*
61. Methodological Issues for Health Care Surveys, *Brenda Cox and Steven Cohen*

62. Applied Regression Analysis and Experimental Design, *Richard J. Brook and Gregory C. Arnold*
63. Statpal: A Statistical Package for Microcomputers—PC-DOS Version for the IBM PC and Compatibles, *Bruce J. Chalmer and David G. Whitmore*
64. Statpal: A Statistical Package for Microcomputers—Apple Version for the II, II+, and IIe, *David G. Whitmore and Bruce J. Chalmer*
65. Nonparametric Statistical Inference: Second Edition, Revised and Expanded, *Jean Dickinson Gibbons*
66. Design and Analysis of Experiments, *Roger G. Petersen*
67. Statistical Methods for Pharmaceutical Research Planning, *Sten W. Bergman and John C. Gittins*
68. Goodness-of-Fit Techniques, *edited by Ralph B. D'Agostino and Michael A. Stephens*
69. Statistical Methods in Discrimination Litigation, *edited by D. H. Kaye and Mikel Aickin*
70. Truncated and Censored Samples from Normal Populations, *Helmut Schneider*
71. Robust Inference, *M. L. Tiku, W. Y. Tan, and N. Balakrishnan*
72. Statistical Image Processing and Graphics, *edited by Edward J. Wegman and Douglas J. DePriest*
73. Assignment Methods in Combinatorial Data Analysis, *Lawrence J. Hubert*
74. Econometrics and Structural Change, *Lyle D. Broemeling and Hiroki Tsurumi*
75. Multivariate Interpretation of Clinical Laboratory Data, *Adelin Albert and Eugene K. Harris*
76. Statistical Tools for Simulation Practitioners, *Jack P. C. Kleijnen*
77. Randomization Tests: Second Edition, *Eugene S. Edgington*
78. A Folio of Distributions: A Collection of Theoretical Quantile-Quantile Plots, *Edward B. Fowlkes*
79. Applied Categorical Data Analysis, *Daniel H. Freeman, Jr.*
80. Seemingly Unrelated Regression Equations Models: Estimation and Inference, *Virendra K. Srivastava and David E. A. Giles*
81. Response Surfaces: Designs and Analyses, *Andre I. Khuri and John A. Cornell*
82. Nonlinear Parameter Estimation: An Integrated System in BASIC, *John C. Nash and Mary Walker-Smith*
83. Cancer Modeling, *edited by James R. Thompson and Barry W. Brown*
84. Mixture Models: Inference and Applications to Clustering, *Geoffrey J. McLachlan and Kaye E. Basford*
85. Randomized Response: Theory and Techniques, *Arijit Chaudhuri and Rahul Mukerjee*
86. Biopharmaceutical Statistics for Drug Development, *edited by Karl E. Peace*
87. Parts per Million Values for Estimating Quality Levels, *Robert E. Odeh and D. B. Owen*
88. Lognormal Distributions: Theory and Applications, *edited by Edwin L. Crow and Kunio Shimizu*
89. Properties of Estimators for the Gamma Distribution, *K. O. Bowman and L. R. Shenton*
90. Spline Smoothing and Nonparametric Regression, *Randall L. Eubank*
91. Linear Least Squares Computations, *R. W. Farebrother*
92. Exploring Statistics, *Damaraju Raghavarao*
93. Applied Time Series Analysis for Business and Economic Forecasting, *Sufi M. Nazem*
94. Bayesian Analysis of Time Series and Dynamic Models, *edited by James C. Spall*
95. The Inverse Gaussian Distribution: Theory, Methodology, and Applications, *Raj S. Chhikara and J. Leroy Folks*
96. Parameter Estimation in Reliability and Life Span Models, *A. Clifford Cohen and Betty Jones Whitten*
97. Pooled Cross-Sectional and Time Series Data Analysis, *Terry E. Dielman*
98. Random Processes: A First Look, Second Edition, Revised and Expanded, *R. Syski*
99. Generalized Poisson Distributions: Properties and Applications, *P. C. Consul*
100. Nonlinear L_p-Norm Estimation, *Rene Gonin and Arthur H. Money*
101. Model Discrimination for Nonlinear Regression Models, *Dale S. Borowiak*
102. Applied Regression Analysis in Econometrics, *Howard E. Doran*
103. Continued Fractions in Statistical Applications, *K. O. Bowman and L. R. Shenton*

104. Statistical Methodology in the Pharmaceutical Sciences, *Donald A. Berry*
105. Experimental Design in Biotechnology, *Perry D. Haaland*
106. Statistical Issues in Drug Research and Development, *edited by Karl E. Peace*
107. Handbook of Nonlinear Regression Models, *David A. Ratkowsky*
108. Robust Regression: Analysis and Applications, *edited by Kenneth D. Lawrence and Jeffrey L. Arthur*
109. Statistical Design and Analysis of Industrial Experiments, *edited by Subir Ghosh*
110. *U*-Statistics: Theory and Practice, *A. J. Lee*
111. A Primer in Probability: Second Edition, Revised and Expanded, *Kathleen Subrahmaniam*
112. Data Quality Control: Theory and Pragmatics, *edited by Gunar E. Liepins and V. R. R. Uppuluri*
113. Engineering Quality by Design: Interpreting the Taguchi Approach, *Thomas B. Barker*
114. Survivorship Analysis for Clinical Studies, *Eugene K. Harris and Adelin Albert*
115. Statistical Analysis of Reliability and Life-Testing Models: Second Edition, *Lee J. Bain and Max Engelhardt*
116. Stochastic Models of Carcinogenesis, *Wai-Yuan Tan*
117. Statistics and Society: Data Collection and Interpretation, Second Edition, Revised and Expanded, *Walter T. Federer*
118. Handbook of Sequential Analysis, *B. K. Ghosh and P. K. Sen*
119. Truncated and Censored Samples: Theory and Applications, *A. Clifford Cohen*
120. Survey Sampling Principles, *E. K. Foreman*
121. Applied Engineering Statistics, *Robert M. Bethea and R. Russell Rhinehart*
122. Sample Size Choice: Charts for Experiments with Linear Models: Second Edition, *Robert E. Odeh and Martin Fox*
123. Handbook of the Logistic Distribution, *edited by N. Balakrishnan*
124. Fundamentals of Biostatistical Inference, *Chap T. Le*
125. Correspondence Analysis Handbook, *J.-P. Benzécri*
126. Quadratic Forms in Random Variables: Theory and Applications, *A. M. Mathai and Serge B. Provost*
127. Confidence Intervals on Variance Components, *Richard K. Burdick and Franklin A. Graybill*
128. Biopharmaceutical Sequential Statistical Applications, *edited by Karl E. Peace*
129. Item Response Theory: Parameter Estimation Techniques, *Frank B. Baker*
130. Survey Sampling: Theory and Methods, *Arijit Chaudhuri and Horst Stenger*
131. Nonparametric Statistical Inference: Third Edition, Revised and Expanded, *Jean Dickinson Gibbons and Subhabrata Chakraborti*
132. Bivariate Discrete Distribution, *Subrahmaniam Kocherlakota and Kathleen Kocherlakota*
133. Design and Analysis of Bioavailability and Bioequivalence Studies, *Shein-Chung Chow and Jen-pei Liu*
134. Multiple Comparisons, Selection, and Applications in Biometry, *edited by Fred M. Hoppe*
135. Cross-Over Experiments: Design, Analysis, and Application, *David A. Ratkowsky, Marc A. Evans, and J. Richard Alldredge*
136. Introduction to Probability and Statistics: Second Edition, Revised and Expanded, *Narayan C. Giri*
137. Applied Analysis of Variance in Behavioral Science, *edited by Lynne K. Edwards*
138. Drug Safety Assessment in Clinical Trials, *edited by Gene S. Gilbert*
139. Design of Experiments: A No-Name Approach, *Thomas J. Lorenzen and Virgil L. Anderson*
140. Statistics in the Pharmaceutical Industry: Second Edition, Revised and Expanded, *edited by C. Ralph Buncher and Jia-Yeong Tsay*
141. Advanced Linear Models: Theory and Applications, *Song-Gui Wang and Shein-Chung Chow*

142. Multistage Selection and Ranking Procedures: Second-Order Asymptotics, *Nitis Mukhopadhyay and Tumulesh K. S. Solanky*
143. Statistical Design and Analysis in Pharmaceutical Science: Validation, Process Controls, and Stability, *Shein-Chung Chow and Jen-pei Liu*
144. Statistical Methods for Engineers and Scientists: Third Edition, Revised and Expanded, *Robert M. Bethea, Benjamin S. Duran, and Thomas L. Boullion*
145. Growth Curves, *Anant M. Kshirsagar and William Boyce Smith*
146. Statistical Bases of Reference Values in Laboratory Medicine, *Eugene K. Harris and James C. Boyd*
147. Randomization Tests: Third Edition, Revised and Expanded, *Eugene S. Edgington*
148. Practical Sampling Techniques: Second Edition, Revised and Expanded, *Ranjan K. Som*
149. Multivariate Statistical Analysis, *Narayan C. Giri*
150. Handbook of the Normal Distribution: Second Edition, Revised and Expanded, *Jagdish K. Patel and Campbell B. Read*
151. Bayesian Biostatistics, *edited by Donald A. Berry and Dalene K. Stangl*
152. Response Surfaces: Designs and Analyses, Second Edition, Revised and Expanded, *André I. Khuri and John A. Cornell*
153. Statistics of Quality, *edited by Subir Ghosh, William R. Schucany, and William B. Smith*
154. Linear and Nonlinear Models for the Analysis of Repeated Measurements, *Edward F. Vonesh and Vernon M. Chinchilli*
155. Handbook of Applied Economic Statistics, *Aman Ullah and David E. A. Giles*
156. Improving Efficiency by Shrinkage: The James-Stein and Ridge Regression Estimators, *Marvin H. J. Gruber*
157. Nonparametric Regression and Spline Smoothing: Second Edition, *Randall L. Eubank*
158. Asymptotics, Nonparametrics, and Time Series, *edited by Subir Ghosh*
159. Multivariate Analysis, Design of Experiments, and Survey Sampling, *edited by Subir Ghosh*

Additional Volumes in Preparation

Statistical Process Monitoring and Control, *S. Park and G. Vining*

MULTIVARIATE ANALYSIS, DESIGN OF EXPERIMENTS, AND SURVEY SAMPLING

edited by
SUBIR GHOSH
University of California, Riverside
Riverside, California

A Tribute to Jagdish N. Srivastava

CRC Press
Taylor & Francis Group
Boca Raton London New York

CRC Press is an imprint of the
Taylor & Francis Group, an **informa** business

First published 1999 by Marcel Dekker, Inc.

Published 2018 by CRC Press
Taylor & Francis Group
6000 Broken Sound Parkway NW, Suite 300
Boca Raton, FL 33487-2742

First issued in paperback 2020

ISBN-13: 978-0-367-57918-0 (pbk)
ISBN-13: 978-0-8247-0052-2 (hbk)

Visit the Taylor & Francis Web site at
http://www.taylorandfrancis.com

and the CRC Press Web site at
http://www.crcpress.com

Jagdish N. Srivastava

Preface

Multivariate analysis, design of experiments, and survey sampling are three widely used areas of statistics. Many researchers have contributed to develop these areas in ways that we see today and many others are working toward further development. As a result, we observe a profusion of research. This reference book is a collection of articles describing some of the recent developments and surveying some topics.

This book is a tribute to Professor Jagdish N. Srivastava, who has contributed vigorously to three areas of statistics, namely, multivariate analysis, design of experiments, and survey sampling. He is the founder of the *Journal of Statistical Planning and Inference*. This collection of articles is a present to Professor Srivastava for his 65th birthday to celebrate his contribution, leadership, and dedication to our profession. This is a collection not just by his friends but by the world leaders in their special research areas. The topics covered are broader than the title describes. Parametric, nonparametric, frequentist, Bayesian, and bootstrap methods, spatial processes, point processes, Markov models, ranking and selection, robust regression, calibration, model selection, survival analysis, queuing and networks, and many others are also discussed in this book. All the articles have been refereed and are in general expository. This book should be of value to students, instructors, and researchers at colleges and universities as well as in business, industry, and government organizations.

The following individuals were truly outstanding for their cooperation and help in reviewing the articles: David Allen, Rebecca Betensky, Derek Bond, Hamparsum Bozdogan, Jay Briedt, Yasuko Chikuse, Vernon Chinchilli, Philip David, Jeffrey Eisele, David Findley, Robin Flowerdew, Philip Hougaard, John Hubert, Clifford Hurvich, Robert Keener, John Klein, Steffen Lauritzen, TaChen Liang, Brenda MacGibbon, J.M. (anonymous), Susan Murphy, Hans-George Muller, Jukka Nyblom (two articles), Hannu Oja, Stan Openshaw, Scott Overton, Judea Pearl, Don Poskitt, Ronald Randles, Jennifer Seberry, Michael Sherman, Nozer Singpurwalla, Adrian Smith, Tom Stroud, Rolf Sundberg, Lori Thombs, Blaza Toman, Peter van der Heijden, Nanny Wermuth, Dean Wichern, Zhiliang Ying, Mai Zhou. I am grateful to all our distinguished reviewers.

My deep appreciation and heartfelt thanks go to our renowned contributors who I hope forgive me for not telling them in advance about some details regarding this book. But then, a surprise for Professor Jagdish N. Srivastava and our contributors will uplift our spirits and encourage us to contribute more to our society.

My sincere thanks go to Russell Dekker, Maria Allegra, and others at Marcel Dekker, Inc. I would like to thank my wife, Susnata, and our daughter, Malancha, for their support and understanding of my efforts in completing this project.

Subir Ghosh

Contents

Contributors

Marilyn Agin Central Research Division, Pfizer, Inc., Groton, Connecticut

Jeremy Aldworth Lilly Research Laboratories, Eli Lilly and Company, Indianapolis, Indiana

Steen A. Andersson Department of Mathematics, Indiana University, Bloomington, Indiana

C. Armero Department of Statistics and Operations Research, University of Valencia, Valencia, Spain

M. J. Bayarri Department of Statistics and Operations Research, University of Valencia, Valencia, Spain

José M. Bernardo Department of Statistics and Operations Research, University of Valencia, Valencia, Spain

Biman Chakraborty Department of Statistics and Applied Probability, National University of Singapore, Republic of Singapore, and Indian Statistical Institute, Calcutta, India

Kathryn Chaloner School of Statistics, University of Minnesota, St. Paul, Minnesota

Probal Chaudhuri Indian Statistical Institute, Calcutta, India

D. Stephen Coad School of Mathematical Sciences, University of Sussex, Brighton, England

Noel Cressie Department of Statistics, The Ohio State University, Columbus, Ohio

Jan de Leeuw Department of Statistics, University of California, Los Angeles, California

Holger Dette Department of Mathematics, Ruhr University at Bochum, Bochum, Germany

Malay Ghosh Department of Statistics, University of Florida, Gainesville, Florida

Takesi Hayakawa Graduate School of Economics, Hitotsubashi University, Tokyo, Japan

Richard A. Johnson Department of Statistics, University of Wisconsin, Madison, Wisconsin

Chul-Ki Kim Department of Statistics, Ewha Woman's University, Seoul, Korea

Sadanori Konishi Graduate School of Mathematics, Kyushu University, Fukuoka, Japan

Sergio G. Koreisha Department of Decision Sciences, University of Oregon, Eugene, Oregon

Tze Leung Lai Department of Statistics, Stanford University, Stanford, California

David Madigan Department of Statistics, University of Washington, Seattle, Washington

John I. Marden Department of Statistics, University of Illinois at Urbana-Champaign, Champaign, Illinois

George Michailidis Department of Statistics, University of Michigan, Ann Arbor, Michigan

Klaus J. Miescke Department of Mathematics, Statistics, and Computer Science, University of Illinois at Chicago, Chicago, Illinois

Axel Munk Department of Mathematics, Ruhr University of Bochum, Bochum, Germany

Wayne Myers School of Forest Resources and Environmental Resources Research Institute, The Pennsylvania State University, University Park, Pennsylvania

Kannan Natarajan Pharmaceutical Research Institute, Bristol-Myers Squibb, Princeton, New Jersey

Hiroyuki Ohmori Department of Mathematics, Ehime University, Matsuyama, Japan

Samuel D. Oman Department of Statistics, Hebrew University, Jerusalem, Israel

Yasuhiro Omori Department of Economics, Tokyo Metropolitan University, Tokyo, Japan

Efstathios Paparoditis Department of Mathematics and Statistics, University of Cyprus, Nicosia, Cyprus

G. P. Patil Center for Statistical Ecology and Environmental Statistics, The Pennsylvania State University, University Park, Pennsylvania

Michael D. Perlman Department of Statistics, University of Washington, Seattle, Washington

Dimitris N. Politis Department of Mathematics, University of California, San Diego, La Jolla, California

Tarmo Pukkila Insurance Department, Ministry of Social Affairs and Health, Helsinki, Finland

Thomas S. Richardson Department of Statistics, University of Washington, Seattle, Washington

Joseph P. Romano Department of Statistics, Stanford University, Stanford, California

Teruhiro Shirakura Department of Mathematics and Informatics, Faculty of Human Development, Kobe University, Kobe, Japan

Muni S. Srivastava Department of Statistics, University of Toronto, Toronto, Ontario, Canada

Charles Taillie Center for Statistical Ecology and Environmental Statistics, The Pennsylvania State University, University Park, Pennsylvania

Dietrich von Rosen Department of Mathematics, Uppsala University, Uppsala, Sweden

Thorsten Wagner Department of Mathematics, Ruhr University at Bochum, Bochum, Germany

Deborah Y. Wang Department of Statistics, University of California, Los Angeles, California

Michael Woodroofe Department of Statistics, University of Michigan, Ann Arbor, Michigan

Jagdish N. Srivastava: Life and Contributions of a Statistician, a Combinatorial Mathematician, and a Philosopher

In 1959, a newly arrived Ph.D. student at the University of North Carolina, Chapel Hill, was asked by his renowned adviser, Professor R. C. Bose, to create three mutually orthogonal Latin squares of size 10×10 for his research work. At that time, Professor Bose, in collaboration with Professors S. S. Shrikhande and E. T. Parker, made his famous discovery on the falsity of Euler's conjecture and produced two mutually orthogonal Latin squares of size 10×10. The student argued, by quoting R. A. Fisher and F. Yates, that large row–column designs are not very desirable from the statistical perspective. Did the student lose his job?

No, the great professor discovered a bright student by agreeing with him and appreciated his brave spirit. In two years, the student finished his Ph.D. degree, working on a different topic. (It is interesting to note that three mutually orthogonal squares have still not been made, nor is it proven that they cannot be made.)

During 1961–63, that student—J. N. Srivastava—worked as a Research Associate with two giants, Professor S. N. Roy in multivariate analysis and Professor Bose in design theory. He also had two more giants, Professors H.

Hotelling and W. Hoeffding, as his teachers and advisers. In 1963, Professor Bose asked him to try to improve BCH codes, discovered by Professor Bose in collaboration with his student Ray Chaudhuri and independently by Hocquengham. This time the bright student accepted the challenge. This resulted in the work that in 1968 Professor E. Berlekamp named *Srivastava Codes*.

In multivariate analysis, Professor Srivastava's research includes MANOVA with complete as well as incomplete data. For incomplete data, his work is in estimation and hypothesis testing, classification problems, and meta-analysis. His monograph, jointly with R. Gnanadesikan and S. N. Roy, *Analysis and Design of Certain Quantitative Multiresponse Experiments*, appeared in 1971 and was published by Pergamon Press in New York.

In design theory, Professor Srivastava developed optimum balanced designs for fractional factorial experiments, introducing and studying concepts such as balanced arrays and multidimensional partially balanced association schemes, leading to the noncommutative algebra of Bose and Srivastava, which is a multiset generalization of the Bose–Mesner algebra.

In the 1970s, Professor Srivastava realized that the "optimality" of a design is fictitious, unless the model assumed is correct. He discussed this with Professor Jack Kiefer. Professor Srivastava directed his efforts toward the problem of identifying the model and created the new and influential fields of search linear models and search designs. Recently, he developed multistage designs for the same purpose. All these fields emphasize the concept of the *revealing power of a design*, which measures a design's ability to identify the nonnegligible parameters. In the last few years, he did some pioneering research on nonadditivity in row-column designs, which confirms the remarks he made to Professor Bose in 1959.

In 1985, Professor Srivastava introduced a new class of designs for comparative experiments in the field of reliability, called self-relocating designs (SRD). He established the superiority of SRD over individual censoring of type II in the sense of having higher accuracy owing to a smaller value of the "trace of the asymptotic variance-covariance matrix" for the maximum likelihood estimates of the hazard rates, and simultaneously, having lesser experimental costs because of a great reduction in the "total expected time under experiment."

In 1985, Professor Srivastava introduced a very general class of estimators, which is considered preeminent in sampling theory. This has been called the Srivastava estimation. Almost all of the well-known estimators are special cases of this general class. The theory shows the inadequacy of many earlier procedures and suggests potential techniques for further improvement.

An important contribution of Professor Srivastava to the statistics profession is his founding of the *Journal of Statistical Planning and Inference* (JSPI). He has been its Editor-in-Chief and Advisory Board Chair, and for several terms he has also been its Executive Editor. His dedication in developing the JSPI is remarkable. Because of the hard work of the very distinguished scientists who have been appointed to its Editorial Board by Professor Srivastava, the journal quality has been high, and its volume of papers is among the largest.

Professor Srivastava organized international conferences in 1971, 1973, 1978, and 1995. The 1973 conference was particularly significant because of the participation of a large number of the greatest living statisticians of the century. He brought Professor R. C. Bose to Colorado in 1971, where Professor Bose passed away in 1987. In 1995, Professor Srivastava arranged a highly distinguished conference in honor of Professor Bose.

Professor Srivastava was elected Fellow of the American Statistical Association in 1971, and of the Institute of Mathematical Statistics in 1972, and became an elected member of the International Statistical Institute in 1972. In 1991, he was made a Foundation Fellow of the Institute of Combinatorics and its Applications, and in 1992, a Fellow of the Third World Academy of Science.

Professor Srivastava served as the Session President of the Indian Society of Agricultural Statistics in 1977, and as the President of the Forum for Interdisciplinary Mathematics during 1993-95, and also as the President of the International Indian Statistical Association during 1994-97.

Professor Srivastava has also studied quantum mechanics and mathematical logic. Gödel's theorem made a deep impression on him, and he began to realize the limitations of science. Slowly he turned toward spirituality and studied the great religions of the world. Because of his nonsectarian outlook, he was particularly drawn to the *Bhagavad-Gita*. This interest led him to obtain a joint appointment in 1991 in the Philosophy Department at Colorado State University in Fort Collins, where he has been a Professor of Statistics and Mathematics since 1966.

J. N. Srivastava was born in Lucknow, India, on June 20, 1933. His mother, Madhuri Devi, passed away when he was only two years old. He was raised by his stepmother, Lila. His father, Mahabir Prasad, coached him, emphasizing English and mathematics, and enrolled him in the fifth grade at the age of 8. He completed the master's degree in mathematical statistics in 1954 from Lucknow University and the Statistician's Diploma from the Indian Statistical Institute in Calcutta in 1958. In the master's examination, he stood first with the highest score among students of all departments in the College of Science, and was admitted to the

Governor's Camp, a group organized by the State Governor for bright students.

During 1954-59, he worked at institutions in Lucknow and Delhi in agricultural statistics. In 1958, he was asked to work on a theoretical design problem needed for an experiment in Kerala, in the course of which he generalized the theory of confounding in factorial experiments from the symmetrical to the asymmetrical case. He also worked at the University of Nebraska, Lincoln, from 1963 till 1966. Since then he has been at Colorado State University in Fort Collins.

Professor Srivastava has mentored many doctoral students, who have benefited from his demand for quality in their research work. He does not look at the time when he works. This results in long hours of work but shows his dedication to what he does.

Truly an international academician and scholar, Professor Srivastava is interested in philosophical as well as fundamental issues in science and religion. He likes to travel for professional, spiritual, and other reasons. He and his wife, Usha, have been married for nearly fifty years. Her devotion, patience, and resilience are truly remarkable. Mrs. Srivastava has been of immense help to many Ph.D. students of Professor Srivastava in surviving the stress of intense academic demand. Her assistance to the *Journal of Statistical Planning and Inference* deserves commendation. Both Professor and Mrs. Srivastava share a common spiritual interest. Their two sons, Arvind and Ashok, and daughter, Gita, are all professionals in medicine and law, electrical and computer engineering, and library science, respectively.

Professor Srivastava has suffered a series of severe medical problems because of birth defects, but fortunately he survived through all of them. In the process he developed a great familiarity with health-related issues, particularly in alternative medicine. He gives free advice to his family and friends. He even uses design of experiments, particularly factorial experiments, in his everyday life for determining effective medication.

It is with great pleasure, pride, and admiration that we dedicate this book to Professor Jagdish N. Srivastava in honor of his 65th birthday. We wish him and his family the very best on this happy occasion and in years to come.

Subir Ghosh

1

Sampling Designs and Prediction Methods for Gaussian Spatial Processes

JEREMY ALDWORTH Lilly Research Laboratories, Eli Lilly and Company, Indianapolis, Indiana

NOEL CRESSIE The Ohio State University, Columbus, Ohio

1. INTRODUCTION

For a phenomenon that varies over a continuous (or even a large finite) spatial domain, it is seldom feasible, or even possible, to observe every potential datum of some study variable associated with that phenomenon. Thus, an important part of statistics is statistical sampling theory, where inference about the study variable may be made from a subset, or sample, of the population of potential data.

Spatial sampling refers to the sampling of georeferenced or spatially labeled phenomena. In the spatial context, interest is usually in the *prediction* of (some function of) the study variable at multiple unsampled sites, and it is in this sense that the prediction problem is multivariate. Given some predictand together with its predictor, a *best* sampling plan or network refers to the choice of locations at which to sample the phenomenon in order to achieve optimality according to a given criterion (e.g., minimize average mean squared prediction error, where the average is taken over multiple prediction locations). In practice, optimal sampling plans may be extremely difficult to achieve, but good, although suboptimal, sampling plans may be relatively easy to obtain and these designs, at least, should be sought.

A commonly chosen predictand in survey sampling is the total (or mean) of the study variable over a specified spatial domain. In this article, we shall also consider predictands defined over some "local" subregion of the domain, and predictands that are nonlinear functions of the study variable at multiple spatial locations.

The objective of this paper is to gauge, through a carefully designed simulation experiment, the performance of different prediction methods under different sampling designs, over several realizations of a spatial process whose strength of spatial dependence varies from zero to very strong. Included are both "spatial" and "nonspatial" analyses and designs. Our emphasis is on prediction of spatial statistics defined on both "local" and "global" regions, based on data obtained from a global network of sampling sites. A brief review of geostatistical theory, survey-sampling theory, and of the spatial-sampling literature are given in Section 2. Based on this knowledge, we designed a simulation experiment whose details are described in Section 3. Section 4 analyzes the results of the experiment and conclusions are given in Section 5.

2. BRIEF REVIEW OF GEOSTATISTICAL THEORY, SURVEY-SAMPLING THEORY, AND SPATIAL SAMPLING

2.1 Geostatistical Theory

Suppose some phenomenon of interest is indexed by spatial location in a domain $D \subset \mathbb{R}^2$. We wish to choose a sample size n and sample locations $\{\mathbf{s}_1, \ldots, \mathbf{s}_n\} \subset D$ so that "good" inferences may be made about the phenomenon from the sample data. Such spatially labeled data often exhibit dependence in the sense that observations closer together tend to be more similar than observations farther apart, which should be exploited in the search for an optimal (or good) network of sites. A brief synopsis of the geostatistical theory characterizing this spatial dependence follows (see Cressie, 1993a, part I, for more details).

2.1.1 The spatial process

A spatial planar process is a real- (or vector-) valued stochastic process $\{Z(\mathbf{s}) : \mathbf{s} \in D\}$ where $D \subset \mathbb{R}^2$. In all that is to follow, the case of real-valued $Z(\cdot)$ will be considered; inference is desired on unobserved parts of the process at multiple locations.

In studies where spatially labeled data exhibit spatial dependence, the following model is useful:

$$Z(\mathbf{s}) = \mu(\mathbf{s}) + \delta(\mathbf{s}); \qquad \mathbf{s} \in D \tag{1}$$

where $\mu(\cdot)$ is the large-scale, deterministic, mean structure of the process (i.e., trend) and $\delta(\cdot)$ is the small-scale stochastic structure that models the spatial dependence among the data. That is,

$$E[\delta(\mathbf{s})] = 0; \qquad \mathbf{s} \in D \tag{2}$$

$$\text{cov}[\delta(\mathbf{s}), \delta(\mathbf{u})] \equiv C(\mathbf{s}, \mathbf{u}); \qquad \mathbf{s}, \mathbf{u} \in D \tag{3}$$

Hence $E[Z(\mathbf{s})] = \mu(\mathbf{s})$; $\mathbf{s} \in D$, and $\text{cov}[Z(\mathbf{s}), Z(\mathbf{u})] = C(\mathbf{s}, \mathbf{u})$; $\mathbf{s}, \mathbf{u} \in D$.

Another useful measure of spatial dependence is the *variogram*:

$$\begin{aligned} 2\gamma(\mathbf{s}, \mathbf{u}) &\equiv \text{var}[Z(\mathbf{s}) - Z(\mathbf{u})] \\ &= C(\mathbf{s}, \mathbf{s}) + C(\mathbf{u}, \mathbf{u}) - 2C(\mathbf{s}, \mathbf{u}); \qquad \mathbf{s}, \mathbf{u} \in D \end{aligned} \tag{4}$$

The quantity $\gamma(\cdot, \cdot)$ is called the *semivariogram*.

2.1.2 Stationarity

The process $Z(\cdot)$ is *first-order stationary* if the following condition holds:

$$E[Z(\mathbf{s})] = \mu; \qquad \mathbf{s} \subset D \tag{5}$$

If $Z(\cdot)$ is first-order stationary and if it satisfies

$$\text{var}[Z(\mathbf{s}) - Z(\mathbf{u})] \equiv 2\gamma(\mathbf{s}, \mathbf{u}) = 2\gamma^o(\mathbf{s} - \mathbf{u}); \qquad \mathbf{s}, \mathbf{u} \in D \tag{6}$$

then $Z(\cdot)$ is said to be *intrinsically stationary*. Note that the variogram $2\gamma^o(\cdot)$ is a function only of the vector difference $\mathbf{s} - \mathbf{u}$. Intrinsically stationary processes are more general than *second-order stationary* processes, for which equation (5) is assumed and assumption (6) is replaced with

$$\text{cov}[Z(\mathbf{s}), Z(\mathbf{u})] \equiv C(\mathbf{s}, \mathbf{u}) = C^o(\mathbf{s} - \mathbf{u}) \tag{7}$$

where $C^o(\cdot)$ is a function only of the vector difference $\mathbf{s} - \mathbf{u}$.

2.1.3 Ordinary kriging

Assume that $Z(\cdot)$ is first-order stationary (i.e., assume equation (5)). Suppose that the data consist of observations $Z(\mathbf{s}_1), \ldots, Z(\mathbf{s}_n)$ of the process at locations $\{\mathbf{s}_1, \ldots, \mathbf{s}_n\} \subset D$. Let $\mathbf{s}_0 \in D$ be some unsampled location and suppose we wish to predict $Z(\mathbf{s}_0)$. Or, more generally, suppose we wish to predict

$$Z(B) \equiv \int_B Z(\mathbf{u})\, d\mathbf{u}/|B|; \qquad B \subset D \tag{8}$$

where $|B| \equiv \int_B d\mathbf{u}$, the area of B. Note that B may or may not contain sample locations. The spatial best linear unbiased predictor (BLUP), also known as the *ordinary (block) kriging* predictor, of $Z(B)$ is

$$\hat{Z}(B) = \sum_{i=1}^{n} \lambda_i Z(\mathbf{s}_i) \tag{9}$$

where $\lambda_1, \ldots, \lambda_n$ are chosen such that

$$E[\hat{Z}(B)] = E[Z(B)] \equiv \mu \tag{10}$$

and they minimize

$$E[Z(B) - \sum_{i=1}^{n} l_i Z(\mathbf{s}_i)]^2 \tag{11}$$

with respect to $l_1, \ldots, l_n$.

By expressing $\hat{Z}(B)$ as $\boldsymbol{\lambda}_{ok}(B)'\mathbf{Z}$, where $\boldsymbol{\lambda}_{ok}(B)' = (\lambda_1, \ldots, \lambda_n)$ and $\mathbf{Z} = (Z(\mathbf{s}_1), \ldots, Z(\mathbf{s}_n))'$, it is not difficult to show (e.g., Cressie, 1993a, p. 142) that

$$\boldsymbol{\lambda}_{ok}(B)'\mathbf{Z} = \left(\gamma(B) + \left\{\frac{1 - \mathbf{1}'\Gamma^{-1}\gamma(B)}{\mathbf{1}'\Gamma^{-1}\mathbf{1}}\right\}\mathbf{1}\right)' \Gamma^{-1}\mathbf{Z} \tag{12}$$

where $\mathbf{1}$ is an $n \times 1$ vector of n ones, Γ is an $n \times n$ matrix whose (i, j)th element is $\gamma(\mathbf{s}_i, \mathbf{s}_j)$, $\gamma(B) \equiv (\gamma(B, \mathbf{s}_1), \ldots, \gamma(B, \mathbf{s}_n))'$, and $\gamma(B, \mathbf{s}_i) \equiv \int_B \gamma(\mathbf{u}, \mathbf{s}_i) d\mathbf{u}/|B|$; $i = 1, \ldots, n$. The subscript "ok" on $\boldsymbol{\lambda}_{ok}(B)$ emphasizes that we are considering the ordinary kriging vector of coefficients.

The ordinary kriging predictor can also be expressed in terms of the covariance function $C(\cdot, \cdot)$. Assuming equations (2), (3), and (5), it can be shown (e.g., Cressie, 1993a, p. 143) that

$$\hat{Z}(B) \equiv \boldsymbol{\lambda}_{ok}(B)'\mathbf{Z} = \left(\mathbf{c}(B) + \left\{\frac{1 - \mathbf{1}'\Sigma^{-1}\mathbf{c}(B)}{\mathbf{1}'\Sigma^{-1}\mathbf{1}}\right\}\mathbf{1}\right)' \Sigma^{-1}\mathbf{Z} \tag{13}$$

where $\Sigma \equiv \text{var}(\mathbf{Z})$, an $n \times n$ matrix whose (i, j)th element is $C(\mathbf{s}_i, \mathbf{s}_j)$, $\mathbf{c}(B) \equiv (C(B, \mathbf{s}_1), \ldots, C(B, \mathbf{s}_n))'$, and $C(B, \mathbf{s}_i) \equiv \int_B C(\mathbf{u}, \mathbf{s}_i) d\mathbf{u}/|B|$; $i = 1, \ldots, n$. Note that equation (13) can be written as

$$\hat{Z}(B) = \hat{\mu}_{gls} + \mathbf{c}(B)'\Sigma^{-1}(\mathbf{Z} - \hat{\mu}_{gls}\mathbf{1}) \tag{14}$$

where $\hat{\mu}_{gls} \equiv \mathbf{1}'\Sigma^{-1}\mathbf{Z}/\mathbf{1}'\Sigma^{-1}\mathbf{1}$ is the generalized least-squares estimator of μ, and is also the best linear unbiased estimator (BLUE) of μ. In the case where μ is known, optimal linear prediction has been discussed inter alia by Graybill (1976, pp. 429–439), who shows that the best linear predictor $p_{opt}(\mathbf{Z})$ has the form

$$p_{opt}(\mathbf{Z}) = \mu + \mathbf{c}(B)'\Sigma^{-1}(\mathbf{Z} - \mu\mathbf{1}) \tag{15}$$

Further, if $Z(\cdot)$ is a Gaussian process, then equation (15) is the best (minimum mean-squared error) predictor, namely $E(Z(B)|\mathbf{Z})$. In geostatistics, equation (15) is also known as the *simple kriging* predictor. If μ is unknown, then $p_{opt}(\mathbf{Z})$ is not a statistic, in which case the spatial BLUP can be obtained by replacing μ in equation (15) with its BLUE, $\hat{\mu}_{gls}$ (Goldberger, 1962).

The *ordinary (point) kriging* predictor of $Z(\mathbf{s}_0)$ is $\hat{Z}(\mathbf{s}_0) = \boldsymbol{\lambda}_{ok}(\mathbf{s}_0)'\mathbf{Z}$, where $\mathbf{s}_0 \in D$ is typically some unsampled location and has the same form as equation (12), but with $\gamma(B)$ replaced by $\gamma(\mathbf{s}_0) \equiv (\gamma(\mathbf{s}_0, \mathbf{s}_1), \ldots, \gamma(\mathbf{s}_0, \mathbf{s}_n))'$. Written in terms of the covariance function, it has the same form as equation (13), but with $\mathbf{c}(B)$ replaced by $\mathbf{c}(\mathbf{s}_0) \equiv (C(\mathbf{s}_0, \mathbf{s}_1), \ldots, C(\mathbf{s}_0, \mathbf{s}_n))'$.

Define the *ordinary (block) kriging variance*

$$\sigma^2_{ok}(B) \equiv E[Z(B) - \sum_{i=1}^{n} \lambda_i Z(\mathbf{s}_i)]^2 \tag{16}$$

which is the minimized mean-squared prediction error. Note once again that the subscript "ok" on $\sigma^2_{ok}(B)$ emphasizes that we are considering the ordinary kriging variance. This can be expressed more explicitly as

$$\sigma^2_{ok}(B) = -\gamma(B, B) + \gamma(B)'\Gamma^{-1}\gamma(B) - (\mathbf{1}'\Gamma^{-1}\gamma(B) - 1)^2/(\mathbf{1}'\Gamma^{-1}\mathbf{1}) \tag{17}$$

where $\gamma(B, B) \equiv \int_B \int_B \gamma(\mathbf{u}, \mathbf{v})\, d\mathbf{u}\, d\mathbf{v}/|B|^2$ and the other terms are as defined in equation (12).

The *ordinary (point) kriging variance* is defined as

$$\sigma^2_{ok}(\mathbf{s}_0) - \gamma(\mathbf{s}_0)'\Gamma^{-1}\gamma(\mathbf{s}_0) - (\mathbf{1}'\Gamma^{-1}\gamma(\mathbf{s}_0) - 1)^2/(\mathbf{1}'\Gamma^{-1}\mathbf{1}) \tag{18}$$

where $\gamma(\mathbf{s}_0) \equiv (\gamma(\mathbf{s}_0, \mathbf{s}_1), \ldots, \gamma(\mathbf{s}_0, \mathbf{s}_n))'$.

If we assume that $Z(\cdot)$ is intrinsically stationary, then $\gamma(\mathbf{s}, \mathbf{u}) \equiv \gamma^o(\mathbf{s} - \mathbf{u})$ in equations (12), (17), and (18). If we assume the stricter condition of second-order stationarity, then $C(\mathbf{s}, \mathbf{u}) \equiv C^o(\mathbf{s} - \mathbf{u})$ in equations (13) and (14).

Note that $\sigma^2_{ok}(B)$ (or $\sigma^2_{ok}(\mathbf{s}_0)$) does not depend on the data $\mathbf{Z} = (Z(\mathbf{s}_1), \ldots, Z(\mathbf{s}_n))'$, but only on the sample locations $\{\mathbf{s}_1, \ldots, \mathbf{s}_n\}$, the prediction region B (or prediction location $\mathbf{s}_0$), the number, n, of locations sampled, and the semivariogram γ. This property makes kriging very useful for designing spatial sampling plans (e.g., Cressie et al., 1990), since the data play no role in the search for a good sampling plan; all that is required is an accurately modeled variogram of the spatial process.

2.1.4 Constrained kriging

Suppose we wish to predict $g(Z(B))$, where $g(\cdot)$ is some nonlinear function. Could we not use $g(\boldsymbol{\lambda}_{ok}(B)'\mathbf{Z})$? Cressie (1993b) concludes that $\boldsymbol{\lambda}_{ok}(B)'\mathbf{Z}$ is "too smooth," resulting in an often unacceptable bias for the predictor $g(\boldsymbol{\lambda}_{ok}(B)'\mathbf{Z})$. A predictor of $g(Z(B))$ with better bias properties, called *constrained kriging* (Cressie, 1993b), follows.

Assume only that $Z(\cdot)$ satisfies equations (2), (3), and (5). Suppose that $g(\cdot)$ is sufficiently smooth to possess at least two continuous derivatives. Then, by the δ-method, we have

$$E[g(Z(B))] \simeq g\{E(Z(B))\} + g''\{E(Z(B))\}\text{var}(Z(B))/2$$
$$= g(\mu) + g''(\mu)\text{var}(Z(B))/2 \tag{19}$$

Let the form of the predictor of $g(Z(B))$ be $g(\boldsymbol{\alpha}'\mathbf{Z})$ satisfying at least the unbiasedness condition on the original scale,

$$E[\boldsymbol{\alpha}'\mathbf{Z}] = E[Z(B)] \equiv \mu \tag{20}$$

Using the δ-method, we obtain

$$E[g(\boldsymbol{\alpha}'\mathbf{Z})] \simeq g\{E(\boldsymbol{\alpha}'\mathbf{Z})\} + g''\{E(\boldsymbol{\alpha}'\mathbf{Z})\}\text{var}(\boldsymbol{\alpha}'\mathbf{Z})/2$$
$$= g(\mu) + g''(\mu)\text{var}(\boldsymbol{\alpha}'\mathbf{Z})/2 \tag{21}$$

Thus, as a predictor of $g(Z(B))$,

$$Bias(g(\boldsymbol{\alpha}'\mathbf{Z})) = E[g(Z(B))] - E[g(\boldsymbol{\alpha}'\mathbf{Z})]$$
$$\simeq g''(\mu)\{\text{var}(\boldsymbol{\alpha}'\mathbf{Z}) - \text{var}(Z(B))\}/2 \tag{22}$$

and

$$MSE(g(\boldsymbol{\alpha}'\mathbf{Z})) = E[g(Z(B)) - g(\boldsymbol{\alpha}'\mathbf{Z})]^2$$
$$\simeq (g'(\mu))^2 E\{\boldsymbol{\alpha}'\mathbf{Z} - Z(B)\}^2 \tag{23}$$

Note that if $g(\cdot)$ is linear in $\mathbf{Z}$, then $g''(\mu) \equiv 0$, and $Bias(g(\boldsymbol{\alpha}'\mathbf{Z})) = 0$.

Equation (22) indicates that for nonlinear (and sufficiently smooth) $g(\cdot)$, in order to obtain an (approximately) unbiased predictor of $g(Z(B))$ of the form $g(\boldsymbol{\alpha}'\mathbf{Z})$, where $\boldsymbol{\alpha} = (\alpha_1, \ldots, \alpha_n)'$ is chosen to minimize equation (23), we need to minimize $E[\sum_{i=1}^n \alpha_i Z(\mathbf{s}_i) - Z(B)]^2$ with respect to $\alpha_1, \ldots, \alpha_n$, subject to the unbiasedness constraint equation (20) *and* the variance constraint

$$\text{var}(\boldsymbol{\alpha}'\mathbf{Z}) = C(B, B) \equiv \int_B \int_B C(\mathbf{u}, \mathbf{v})\, d\mathbf{u}\, d\mathbf{v}/|B|^2 = \text{var}(Z(B)) \tag{24}$$

Note that if $Z(\cdot)$ is a Gaussian process, then $g(\boldsymbol{\alpha}'\mathbf{Z})$ is an unbiased predictor of $g(Z(B))$ for *any* measurable function g, provided that equations (20) and (24) are satisfied (Cressie, 1993b).

On most occasions (see below) this constrained minimization can be carried out, yielding the *constrained (block) kriging* predictor $\boldsymbol{\lambda}_{ck}(B)'\mathbf{Z}$. Cressie (1993b) shows that

$$\boldsymbol{\lambda}_{ck}(B)' = \frac{1}{m_2}(\mathbf{c}(B) + m_1\mathbf{1})'\Sigma^{-1} \tag{25}$$

$$m_1 = \frac{m_2 - \mathbf{1}'\Sigma^{-1}\mathbf{c}(B)}{\mathbf{1}'\Sigma^{-1}\mathbf{1}} \tag{26}$$

$$m_2 = \left\{\frac{(\mathbf{c}(B)'\Sigma^{-1}\mathbf{c}(B))(\mathbf{1}'\Sigma^{-1}\mathbf{1}) - (\mathbf{1}'\Sigma^{-1}\mathbf{c}(B))^2}{(\mathbf{1}'\Sigma^{-1}\mathbf{1})\mathrm{var}(Z(B)) - 1}\right\}^{1/2} \tag{27}$$

where $\mathbf{c}(B)$ and Σ are as defined in equation (13). The numerator of equation (27) is well defined by the Cauchy–Schwarz inequality, and the denominator is well defined if

$$\mathrm{var}(Z(B)) > (\mathbf{1}'\Sigma^{-1}\mathbf{1})^{-1} = \mathrm{var}(\hat{\mu}_{gls}) \tag{28}$$

where $\hat{\mu}_{gls}$ is the generalized least-squares estimator of μ.

The constrained kriging predictor can also be expressed as

$$\boldsymbol{\lambda}_{ck}(B)'\mathbf{Z} = \hat{\mu}_{gls} + \left\{\frac{C(B, B) - \mathrm{var}(\hat{\mu}_{gls})}{\mathrm{var}(\mathbf{c}(B)'\Sigma^{-1}(\mathbf{Z} - \hat{\mu}_{gls}\mathbf{1}))}\right\}^{1/2} \mathbf{c}(B)'\Sigma^{-1}(\mathbf{Z} - \hat{\mu}_{gls}\mathbf{1}) \tag{29}$$

and it is the best predictor in the class of linear, unbiased, and variance-matching predictors. The *constrained (block) kriging variance* is

$$\sigma^2_{ck}(B) = 2C(B, B) - 2\boldsymbol{\lambda}_{ck}(B)'\mathbf{c}(B) \tag{30}$$

where $C(B, B)$ is given in equation (24).

The *constrained (point) kriging* predictor of $Z(\mathbf{s}_0)$ is $\hat{Z}(\mathbf{s}_0) = \boldsymbol{\lambda}_{ck}(\mathbf{s}_0)'\mathbf{Z}$, for some $\mathbf{s}_0 \in D$, and it has the same form as equation (25), but with $\mathbf{c}(B)$ replaced by $\mathbf{c}(\mathbf{s}_0)$ and $C(B, B)$ replaced by $C(\mathbf{s}_0, \mathbf{s}_0)$. Similarly, the constrained (point) kriging variance is

$$\sigma^2_{ck}(\mathbf{s}_0) = 2C(\mathbf{s}_0, \mathbf{s}_0) - 2\boldsymbol{\lambda}_{ck}(\mathbf{s}_0)'\mathbf{c}(\mathbf{s}_0) \tag{31}$$

Note that the constrained kriging predictor is unlikely to perform as well as the kriging predictor if $g(\cdot)$ is linear (e.g., if we wish to predict $g(Z(B)) \equiv Z(B)$), especially if $\mathrm{var}(Z(B))$ and $\mathrm{var}(\boldsymbol{\lambda}_{ok}(B)'\mathbf{Z})$ are substantially different.

2.1.5 Trend

Suppose the data $\mathbf{Z}$ are generated by a spatial model with trend, that is, nonconstant $\mu(\cdot)$ in equation (1). When $\mu(\cdot)$ is linear in explanatory variables, the ordinary kriging predictor may be generalized to yield the *universal kriging* predictor (see Cressie, 1993a, Section 3.4). Alternatively, $\mu(\cdot)$ may be estimated nonparametrically (e.g., by median polish; see Cressie, 1993a, Section 3.5), subtracted from the data, and ordinary kriging can then be applied to the residuals. However, the two components $\mu(\cdot)$ and $S(\cdot)$ are not observed individually, so it can happen that a part of $\mu(\cdot)$ is inadvertently included as part of the small-scale stochastic component $S(\cdot)$. In that case, there may be "leakage" of part of the trend into the (estimated) covariance function in equation (3) or the (estimated) variogram function in equation (4). An example of this is given in Section 3.4.

2.1.6 Measurement error

Data from a spatial process are usually contaminated with measurement error, for which the following model is useful

$$Z(\mathbf{s}) = S(\mathbf{s}) + \epsilon(\mathbf{s}); \quad \mathbf{s} \in D \tag{32}$$

where $\epsilon(\cdot)$ represents a zero-mean, white-noise measurement-error process, and

$$S(\mathbf{s}) = \mu(\mathbf{s}) + \delta(\mathbf{s}); \quad \mathbf{s} \in D \tag{33}$$

where $\mu(\cdot)$ and $\delta(\cdot)$ are defined as in equation (1); the ϵ and S processes are assumed to be independent. Note that if two observations are taken at a single location, that is, if $Z_1(\mathbf{s})$ and $Z_2(\mathbf{s})$ are observed, they differ from one another only in their error terms, $\epsilon_1(\mathbf{s})$ and $\epsilon_2(\mathbf{s})$, respectively.

The S-process is sometimes referred to as the "state" process or the "signal," to which measurement errors are added yielding the "noisy" Z-process. It is very important to realize that now we are interested in predicting the "noiseless" S-process over D, but what we actually measure are noisy $\{Z(\mathbf{s}_1), \ldots, Z(\mathbf{s}_n)\}$.

The form of the ordinary kriging and constrained kriging predictors given by equations (13) and (25), respectively, do not change under the measurement error model in equation (32), except when predicting back at a data location (Cressie, 1993a, p. 128). Further, note that under the model in equation (32), we have

$$\Sigma \equiv \text{var}(\mathbf{Z}) = \text{var}((S(\mathbf{s}_1), \ldots, S(\mathbf{s}_n))') + \tau^2\mathbf{I} \tag{34}$$

where $\text{var}(\epsilon(\mathbf{s})) \equiv \tau^2$ and I is the $n \times n$ identity matrix. Thus, equation (34) allows the predictors in equations (13) and (25) to "filter out" the measurement error from the data.

2.1.7 Estimating and modeling the variogram

In practice, the variogram 2γ is seldom known and is usually estimated by some nonparametric estimator such as

$$2\hat{\gamma}(\mathbf{h}) = \frac{1}{|N(\mathbf{h})|}\sum_{N(\mathbf{h})}(Z(\mathbf{s}_i) - Z(\mathbf{s}_j))^2 \tag{35}$$

where $N(\mathbf{h}) \equiv \{(\mathbf{s}_i, \mathbf{s}_j) : \mathbf{s}_i - \mathbf{s}_j = \mathbf{h}\}$ and $|N(\mathbf{h})|$ is the number of distinct ordered pairs in the set $N(\mathbf{h})$. A robust alternative estimator (Cressie, 1993a, p. 75) is

$$2\bar{\gamma}(\mathbf{h}) = \frac{1}{|N(\mathbf{h})|}\left\{\sum_{N(\mathbf{h})}|Z(\mathbf{s}_i) - Z(\mathbf{s}_j)|^{1/2}\right\}^4 /\{0.457 + 0.494/|N(\mathbf{h})|\} \tag{36}$$

Note that these variogram estimators are functions only of the vector difference, $\mathbf{h} = \mathbf{u} - \mathbf{v}$, implicitly assuming that equation (6) holds. If $Z(\cdot)$ is intrinsically stationary, (i.e., both equation (5) and equation (6) hold), then $2\hat{\gamma}$ is an unbiased estimator of $2\gamma^o$.

The distinction we made between $\gamma(\cdot,\cdot)$ as a function of two vectors in equation (4) and $\gamma^o(\cdot)$ as a function of vector differences in equation (6) should now be clear. Intrinsic stationarity is required for (unbiased) estimation of the variogram, and not for the kriging equations to be valid. Hereafter we shall be concerned only with variograms and covariances as functions of vector differences, and we shall notate them simply as $2\gamma(\cdot)$ and $C(\cdot)$, respectively.

The nonparametrically estimated variogram cannot be used satisfactorily in equations (12) or (13), or to obtain the kriging variance σ^2_{ok}, because, among other reasons covered by Cressie (1993a), the estimates are not conditionally negative definite. Moreover, 2γ is estimated only at the lags corresponding to the set of all pairs among the data locations, and these may or may not coincide with the lags required to predict $Z(B)$. Hence, the usual practice is to fit a model $2\gamma(\mathbf{h}; \boldsymbol{\theta})$, whose form is known (apart from a few parameters $\boldsymbol{\theta}$), to $2\hat{\gamma}(\mathbf{h})$ (or $2\bar{\gamma}(\mathbf{h})$). Thus, we use $2\gamma(\cdot; \hat{\boldsymbol{\theta}})$ in place of $2\gamma(\cdot)$ to obtain the kriging predictor and σ^2_{ok}. (For further discussion on candidate models for $2\gamma(\cdot; \boldsymbol{\theta})$, see Cressie, 1993a, p. 66.) In the case of second-order stationarity, given by equations (5) and (7), the stationary covariance function can be obtained from $C(\mathbf{h}) = C(\mathbf{0}) - \gamma(\mathbf{h})$.

Assuming the measurement-error model in equation (32), the measurement error variance τ^2 can be estimated from multiple samples at selected sites, if they are available. However, for spatial phenomena, this may not always be possible (e.g., once a soil core has been taken from the ground, it is gone!) However, it may be possible to take extra samples sufficiently close together to avoid contamination by a "microscale" process. If we assume that $\delta(\cdot)$ is L_2-continuous (i.e., $E[(\delta(\mathbf{s}+\mathbf{h}) - \delta(\mathbf{s}))^2] \to 0$ as $||\mathbf{h}|| \to 0$), then τ^2 can be estimated by the "nugget effect" of the modeled variogram, where the nugget effect $\hat{c}_0$ is defined as

$$\hat{c}_0 = \lim_{||\mathbf{h}|| \to 0} \gamma(\mathbf{h}, \hat{\boldsymbol{\theta}}) \tag{37}$$

2.2 Survey-Sampling Theory

We now present a very brief summary of some of the elements of survey-sampling theory. For more details, the reader is referred to Särndal et al. (1992) and Cochran (1977).

2.2.1 Finite-population sampling

Consider a population of N *labels* which, for convenience, will be represented by the *finite* set $D_f = \{\mathbf{s}_1, \mathbf{s}_2, \ldots, \mathbf{s}_N\}$. Associated with each label $\mathbf{s}_j$ is a real number $Z(\mathbf{s}_j)$, a value of the *study variable* Z corresponding to that label. We assume for the moment that all the elements of the *parameter vector* $\mathbf{Z} = (Z(\mathbf{s}_1), \ldots, Z(\mathbf{s}_N))'$ are fixed and can be obtained without error. The parameter vector $\mathbf{Z}$ may also be referred to as the *target population*.

We are usually interested in making inferences about some numerical summary of $\mathbf{Z}$, in the form of a *finite population parameter* $\theta(\mathbf{Z})$. Common examples of $\theta(\mathbf{Z})$ include: the population size, $N \equiv \sum_{\mathbf{s}_j \in D_f} 1$; the population total, $T = \sum_{\mathbf{s}_j \in D_f} Z(\mathbf{s}_j)$; the population mean, $Z(D_f) \equiv N^{-1}T$; and the population variance, $S_N^2 = (N-1)^{-1} \sum_{\mathbf{s}_j \in D_f} (Z(\mathbf{s}_j) - Z(D_f))^2$. Other examples include the population cumulative distribution function (CDF), defined as

$$F(z) \equiv N^{-1} \sum_{\mathbf{s}_j \in D_f} I(Z(\mathbf{s}_j) \le z); \qquad z \in \mathbb{R} \tag{38}$$

where $I(\cdot)$ is the indicator function (i.e., $I(A) = 1$ if A is true and $I(A) = 0$ if A is not true), and the inverse function of the CDF, the quantile function, defined as

$$q(\alpha) \equiv \inf\{z : \ F(z) \ge \alpha\}; \qquad \alpha \in [0, 1] \tag{39}$$

Suppose that a subset of labels is selected from D_f, randomly or otherwise. This set, $A \subset D_f$, is called a *sample*; $\mathbf{s} \in A$ is called a *sampling unit*. The process of drawing the sample and obtaining the corresponding Z-values is referred to as a *survey sample*. The *data* collected in a survey sample consist of both the labels and their corresponding measurements, written as

$$X = \{(\mathbf{s}_j, Z(\mathbf{s}_j)) : \mathbf{s}_j \in A\} \tag{40}$$

A *sampling design* (or *design*) is a probability mass function $p(\cdot)$ defined on subsets of D_f, such that $\Pr(A = a) = p(a)$. This defines the probability that the sample a is selected. If $p(a) = 1$, for some $a \in D_f$, then the design has no randomization and is said to be *purposive*; if $p(D_f) = 1$, then the design is a *census*.

Define the random indicator function

$$I_j = \begin{cases} 1 & if\ \mathbf{s}_j \in A \\ 0 & if\ \mathbf{s}_j \notin A \end{cases}$$

This is called the *sample membership indicator* of element $\mathbf{s}_j$. The probability that element $\mathbf{s}_j$ is included in a sample is given by the *first-order inclusion probability* of element $\mathbf{s}_j$ as follows:

$$\pi_j = Pr(\mathbf{s}_j \in A) = Pr(I_j = 1) = \sum_{a:\ \mathbf{s}_j \in a} p(a) \tag{41}$$

The probability that both elements $\mathbf{s}_i$ and $\mathbf{s}_j$ are included in a sample is given by the *second-order inclusion probability* of $\mathbf{s}_i$ and $\mathbf{s}_j$ as follows:

$$\pi_{ij} = Pr(\mathbf{s}_i \in A\ and\ \mathbf{s}_j \in A) = Pr(I_i I_j = 1) = \sum_{a:\ \mathbf{s}_i \& \mathbf{s}_j \in a} p(a) \tag{42}$$

Note that $\pi_{jj} = \pi_j$.

A *probability sampling design* is sometimes defined (e.g., Särndal et al., 1992, p. 32) as a design for which

$$\pi_j > 0, \qquad \text{for all } \mathbf{s}_j \in D_f \tag{43}$$

However, Overton (1993) suggests that a probability sampling design should be defined as a design for which equation (43) holds, *and* for which

$$\pi_j \text{ is known, for all } \mathbf{s}_j \in \mathrm{A} \tag{44}$$

In this chapter, we shall use Overton's (1993) "stronger" definition of a probability sampling design, because it explicitly (and correctly) demands knowledge of the inclusion probabilities for the sample.

A sample a realized by a probability sampling design is called a *probability sample* (or *p-sample*). If for some reason the inclusion probabilities are separated from the sample a, then by condition (44) a no longer qualifies to be called a p-sample. A probability sampling design p has three desirable properties: (i) it eliminates selection bias; (ii) it is objective (as opposed to the purposive selection of "representative" elements or the haphazard selection of convenient elements); and, in particular, (iii) statistical inferences can be made about $\theta(\mathbf{Z})$ based on the probability structure provided by p, without having to appeal to any statistical model from which $\mathbf{Z}$ is assumed to be a realization. Such inferences are referred to as *design-based* inferences.

If p is a probability sampling design, we can obtain an unbiased estimator of the population total T (see equation (47)). Further, a *measurable probability sampling design* is sometimes defined (e.g., Särndal et al., 1992, p. 33) as a probability sampling design for which

$$\pi_{ij} > 0, \qquad \text{for all } \mathbf{s}_i, \mathbf{s}_j \in D_f \tag{45}$$

However, for similar reasons to those given above, we shall use the "stronger" definition that a measurable probability sampling design requires, in addition to equation (45), that

$$\pi_{ij} \text{ is known, for all } \mathbf{s}_i, \mathbf{s}_j \in A \tag{46}$$

For such designs, we can obtain from the sample an unbiased estimator of the sampling variance of equation (47) (see equation (49)).

If p is a probability sampling design, then the *Horvitz–Thompson estimator* (Horvitz and Thompson, 1952), also called the *π-estimator*, of the total T, is defined as

$$\hat{T}_{ht} \equiv \sum_{\mathbf{s}_j \in A} \frac{Z(\mathbf{s}_j)}{\pi_j} = \sum_{\mathbf{s}_j \in D_f} \frac{Z(\mathbf{s}_j) I_j}{\pi_j} \tag{47}$$

This is an unbiased estimator of T and its *sampling variance* is given by

$$\text{var}(\hat{T}_{ht}) = \sum_{\mathbf{s}_i \in D_f} \sum_{\mathbf{s}_j \in D_f} \frac{(\pi_{ij} - \pi_i \pi_j)}{\pi_i \pi_j} Z(\mathbf{s}_i) Z(\mathbf{s}_j) \tag{48}$$

If p is a measurable probability sampling design, an unbiased estimator of this variance is

$$\hat{V}_{ht} = \sum_{\mathbf{s}_i \in A} \sum_{\mathbf{s}_j \in A} \frac{(\pi_{ij} - \pi_i \pi_j)}{\pi_{ij} \pi_i \pi_j} Z(\mathbf{s}_i) Z(\mathbf{s}_j) \tag{49}$$

The Horvitz–Thompson estimator can be used to estimate several other "totals" of interest, such as the population size, mean, and variance. For example, the Horvitz–Thompson estimator of the population CDF by equation (38) is

$$\hat{F}_{ht}(z) = \frac{1}{N}\sum_{\mathbf{s}_j \in A} \pi_j^{-1} I(Z(\mathbf{s}_j) \leq z); \qquad z \in \mathbb{R} \tag{50}$$

2.2.2 Horvitz–Thompson estimation for a continuous population

Cordy (1993) extends the formulation of the Horvitz–Thompson estimator in equation (47), its variance in equation (48), and its variance estimator in equation (49) to the case where the population (of labels) is continuous and the sampling units are points (e.g., in Euclidean space).

Let $D \subset \mathbb{R}^d$; $d \geq 1$, be a continuous population of labels. Assuming a fixed sample size of n, define the *sample space* $\mathcal{A}$ as

$$\mathcal{A} = \{a \equiv (\mathbf{s}_1, \ldots, \mathbf{s}_n) : \mathbf{s}_i \in D; \; i = 1, \ldots, n\}$$

that is, the sample $A \equiv (\boldsymbol{S}_1, \ldots, \boldsymbol{S}_n) \in \mathcal{A}$ is an ordered n-tuple of random locations, and A has values in D^n.

A (continuous-population) sampling design is a joint probability density function of A with support in D^n, denoted as

$$f(a); \qquad \mathbf{s}_i \in D, \; i = 1, \ldots, n$$

The first-order inclusion probabilities are defined as

$$\pi_j^* = \sum_{i=1}^{n} f_i(\mathbf{s}_j); \qquad \mathbf{s}_j \in D, \; j = 1, \ldots, n$$

where $f_i(\cdot)$ is the marginal probability density function of $\boldsymbol{S}_i$, the ith element of A; $i = 1, \ldots, n$.

The second-order inclusion probabilities are defined as

$$\pi_{ij}^* = \sum_{k=1}^{n} \sum_{l \neq k}^{n} f_{kl}(\mathbf{s}_i, \mathbf{s}_j); \qquad \mathbf{s}_i, \mathbf{s}_j \in D, \; i, j = 1, \ldots, n$$

where $f_{kl}(\cdot, \cdot)$ is the joint marginal probability density function of $(\boldsymbol{S}_k, \boldsymbol{S}_l)$, the kth and lth elements of A; $k, l = 1, \ldots, n, \; k \neq l$.

Suppose we wish to estimate the (continuous-population) total

$$T^* = \int_D Z(\mathbf{s})\, \mathrm{d}(\mathbf{s})$$

Cordy (1993) proves that if $\pi_j^* > 0$, for all $\mathbf{s}_j \in D$, and $\int_D \frac{1}{\pi_j^*}\, \mathrm{d}\mathbf{s}_j < \infty$, then an unbiased estimator of T^* is given by

$$\hat{T}^*_{ht} \equiv \sum_{\mathbf{s}_j \in A^*} \frac{Z(\mathbf{s}_j)}{\pi^*_j}$$

where A^* is the *set* whose elements comprise the elements of the n-tuple A. Cordy (1993) shows that the *sampling variance* of $\hat{T}^*_{ht}$ is given by

$$\text{var}(\hat{T}^*_{ht}) = \int_D \frac{(Z(\mathbf{s}_j))^2}{\pi^*_j}\, d\mathbf{s}_j + \int_D \int_D \frac{(\pi^*_{ij} - \pi^*_i \pi^*_j)}{\pi^*_i \pi^*_j} Z(\mathbf{s}_i) Z(\mathbf{s}_j)\, d\mathbf{s}_i\, d\mathbf{s}_j$$

Finally, Cordy (1993) shows that if, in addition, $\pi^*_{ij} > 0$, for all $\mathbf{s}_i, \mathbf{s}_j \in D$, then an unbiased estimator of $\text{var}(\hat{T}^*_{ht})$ is given by

$$\hat{V}^*_{ht} = \sum_{\mathbf{s}_j \in A^*} \left(\frac{Z(\mathbf{s}_j)}{\pi^*_j} \right)^2 + \sum_{\mathbf{s}_i \in A^*} \sum_{\mathbf{s}_j \in A^*;\, \mathbf{s}_j \neq \mathbf{s}_i} \frac{(\pi^*_{ij} - \pi^*_i \pi^*_j)}{\pi^*_{ij} \pi^*_i \pi^*_j} Z(\mathbf{s}_i) Z(\mathbf{s}_j)$$

In the rest of the chapter, we shall use instead the finite-population formulation, and compare it with a geostatistical approach adapted to deal with a finite number of units in the domain of interest.

2.2.3 Superpopulation models in survey sampling

Given that the study variable Z can be measured without error, we have so far assumed that $\mathbf{Z}$ consists of N fixed elements and inference is design-based, that is, based on the randomization scheme imposed on the population of labels D_f. Thus, the probability structure that supports design-based inference is an exogenous or externally imposed one.

Suppose we now assume that our target population is a single realization of the random N-vector $\mathbf{Z}$, and that the joint distribution of $\mathbf{Z}$ can be described by some model ξ, sometimes called the *superpopulation* model. Superpopulation models are used to extend the basis of inference and to formulate estimators with better properties than purely design-based ones.

For example, assume that a sample A has been drawn. Then inference based on the probability structure provided by the superpopulation model ξ conditional on A is called *model-based* inference. Clearly, model-based inference requires that the model be well specified, that is, that the (model-based) inferences be consistent (in the sense of Fisher consistency; Fisher, 1956, p. 143) with the target population (Overton, 1993).

Superpopulation models are also invoked to formulate estimation methods that may perform substantially better than purely design-based estimation methods if the model is well specified, and no worse if the model is not well specified (e.g., the regression estimator discussed by Särndal et al., 1992, p. 225). Such methods are said to be *model-assisted*, but not model-dependent. For unconditional inference, the probability structures of both p and ξ are used.

2.2.4 Measurement error

The assumption that the elements of the observable vector $\mathbf{Z} = (Z(\mathbf{s}_1), \ldots, Z(\mathbf{s}_N))'$ are free of measurement error may be unrealistic. Suppose, more realistically, that

$$Z(\mathbf{s}_i) = S(\mathbf{s}_i) + \epsilon(\mathbf{s}_i); \qquad i = 1, \ldots, N \tag{51}$$

where $S(\mathbf{s}_i)$ is the "true" (fixed) value of the ith element of the study population and $\epsilon(\mathbf{s}_i) \equiv Z(\mathbf{s}_i) - S(\mathbf{s}_i)$ represents the ith observational error.

Assuming the measurement error model in equation (51), we need to place stochastic structure on the error term in equation (51) if we wish to make any statistical statements about estimators of $\theta(\mathbf{S})$, where $\mathbf{S} \equiv (S(\mathbf{s}_1), \ldots, S(\mathbf{s}_N))'$ is the target population.

Let $\hat{\theta}$ be an estimator of $\theta(\mathbf{S})$ and assume some stochastic model for the error process $\epsilon(\cdot)$. Then the estimation error $\hat{\theta} - \theta(\mathbf{S})$ is a random variable whose probability distribution is determined jointly by the sampling design p and the error process $\epsilon(\cdot)$.

2.2.5 Estimation in the presence of measurement error

For simple error-process models, population-total and population-mean estimation is straightforward. For example, assume the model in equation (51) and suppose ϵ is a zero-mean, white-noise process with $\text{var}(\epsilon) \equiv \tau^2$. Consider the "true" population total $T_S \equiv \sum_{\mathbf{s}_j \in D_f} S(\mathbf{s}_j)$ and the Horvitz–Thompson estimator $\hat{T}_{ht} = \sum_{\mathbf{s}_j \in A} \pi_j^{-1} Z(\mathbf{s}_j)$. Assume $\epsilon(\cdot)$ and the design p are independent and define the joint expectation

$$E_{p\epsilon}(\cdot) \equiv E_p[E_\epsilon(\cdot|A)]$$

where $E_p(\cdot)$ denotes expectation with respect to p, and $E_\epsilon(\cdot|A)$ is the expectation with respect to ϵ, conditional on the sample A. Then

$$\begin{aligned} E_{p\epsilon}(\hat{T}_{ht}) &= E_p[E_\epsilon(\sum_{\mathbf{s}_j \in A} \pi_j^{-1} Z(\mathbf{s}_j)|A)] \\ &= E_p[\sum_{\mathbf{s}_j \in A} \pi_j^{-1} S(\mathbf{s}_j)] = T_S \end{aligned}$$

and it can be shown (Särndal et al., 1992, Chap. 16) that

$$MSE_{p\epsilon}(\hat{T}_{ht}) = V_1 + V_2 \tag{52}$$

where $V_1 = \sum_{\mathbf{s}_i \in D_f} \sum_{\mathbf{s}_j \in D_f} (\pi_{ij} - \pi_i \pi_j) \frac{S(\mathbf{s}_i)S(\mathbf{s}_j)}{\pi_i \pi_j}$, and $V_2 = \tau^2 \sum_{i=1}^{N} \pi_i^{-1}$. Thus, the Horvitz–Thompson estimator $\hat{T}_{ht}$ is unbiased for T_S and its variance can be simply partitioned into sampling error and measurement error components.

By contrast, CDF estimation is not so straightforward, even if the simple error model given above is assumed. Consider the CDF of S,

$$F_S(z) \equiv \frac{1}{N} \sum_{\mathbf{s}_j \in D_f} I(S(\mathbf{s}_j) \leq z); \qquad z \in \mathbb{R}, \tag{53}$$

and consider the Horvitz–Thompson estimator of $F_S(z)$ given by equation (50). It is easy to show that $E_p(\hat{F}_{ht}(z)) = F_Z(z) \equiv \frac{1}{N} \sum_{\mathbf{s}_j \in D_f} I(Z(\mathbf{s}_j) \leq z)$. Assuming that $\epsilon(\cdot)$ is a zero-mean, white-noise process with $\text{var}(\epsilon) \equiv \tau^2$, we have

$$E_{p\epsilon}(\hat{F}_{ht}(z)) = E_\epsilon(F_Z(z))) = \frac{1}{N} \sum_{\mathbf{s}_j \in D_f} G_\epsilon(z - S(\mathbf{s}_j))$$

$$\equiv (G_\epsilon * F_S)(z) \neq F_S(z)j \qquad \text{if } \tau^2 > 0$$

where $G_\epsilon(z) \equiv \Pr(\epsilon(\mathbf{s}_j) \leq z)$, and "$*$" denotes convolution.

Stefanski and Bay (1996) use a simulation–extrapolation deconvolution argument to provide a bias-adjusted CDF estimator. They assume the model in equation (51) where S is fixed and $(\epsilon(\mathbf{s}_1), \ldots, \epsilon(\mathbf{s}_N)) \sim NI(0, \tau^2)$ (i.e., each $\epsilon(\mathbf{s}_i)$ is independently distributed as a $N(0, \tau^2)$ random variable), independent of the sampling design p. Denote the CDF of $\epsilon(\mathbf{s})$ as $\Phi_\tau(\cdot)$. They show that $E_{p\epsilon}[\hat{F}_{ht}(z)] = (\Phi_\tau * F_S)(z)$, which is a special case of the previous result. In order to obtain a deconvoluted CDF estimator, the simulation–extrapolation method of Cook and Stefanski (1994) was followed, where: (i) additional pseudo random errors of known variance are added to the original data, creating error-inflated "pseudo data" sets; (ii) "pseudo CDF estimators" are recomputed from the pseudo data; and (iii) a trend is established, of the pseudo estimators as a function of the variance of the added errors, and extrapolated backwards to the case of no measurement error.

More specifically, suppose $\boldsymbol{\eta} = (\eta_1, \ldots, \eta_m)'$, where the error variance of the ith pseudo random variable is $\tau^2(1 + \eta_i)$, with $\eta_1 < \cdots < \eta_m$. Stefanski and Bay (1996) obtain the generic pseudo CDF estimator

$$\hat{F}_{Z,\eta}(z) = \frac{1}{N} \sum_{i=1}^{n} \frac{1}{\pi_i} \Phi\left(\frac{z - Z(\mathbf{s}_i)}{\tau(1 + \eta)^{1/2}} \right) \tag{54}$$

where $\Phi(\cdot)$ is the standard normal CDF. The method they propose depends on the fact that the expectation of equation (54) can be well approximated by a quadratic function in η, that is, by $\beta_0 + \beta_1 \eta + \beta_2 \eta^2$, where $\{\beta_0, \beta_1, \beta_2\}$

are unknown but estimable using linear-model theory. Fix a z in $F_S(z)$. Extrapolating backwards to the case of zero measurement error is equivalent to letting $\eta \to -1$, resulting in the measurement-error-free estimator $\hat{F}_S(z) = \hat{\beta}_0 - \hat{\beta}_1 + \hat{\beta}_2$. From linear-model theory, this can be expressed as

$$\hat{F}_S(z) = \mathbf{a}'(D'D)^{-1}D'\mathbf{v} \tag{55}$$

where $\mathbf{a}' = (1, -1, 1)$, $D = (\mathbf{1}, \boldsymbol{\eta}, \boldsymbol{\eta}^2)$, $\mathbf{1} = (1, \ldots, 1)'$, $\boldsymbol{\eta} = (\eta_1, \ldots, \eta_m)'$, $\boldsymbol{\eta}^2 = (\eta_1^2, \ldots, \eta_m^2)$, and $\mathbf{v} = (\hat{F}_{Z,\eta_1}(z), \ldots, \hat{F}_{Z,\eta_m}(z))'$ whose elements are given by equation (54).

The authors note that when applying this procedure to the range $z_1 < z_2 < \cdots < z_k$, $\hat{F}_S(z)$ may not be monotonic in z, and it is possible that $\hat{F}_S(z) \notin [0, 1]$. As a solution to these problems, they propose fitting an unweighted isotonic regression model to the point estimates $\hat{F}_S(z_i)$; $i = 1, \ldots, k$, and truncating the lower and upper ends of the fitted model at 0 and 1, respectively, if necessary.

It is suggested that the η-values be taken equally spaced over the interval [0.05, 2], and that $m \geq 5$. The authors note that the theory underlying simulation–extrapolation assumes that τ^2 is "small." Even so, they suggest that this method should significantly reduce bias when τ^2 is moderate or even large.

Fuller (1995) invokes a superpopulation model to obtain an estimator of the quantile function in equation (39). Using the same superpopulation model, we can derive a CDF estimator of $F_S(z)$ with better bias properties than equation (50). Fuller (1995) assumes the measurement error model in equation (51), where $(\epsilon(\mathbf{s}_1), \ldots, \epsilon(\mathbf{s}_N)) \sim NI(0, \tau^2)$, $(S(\mathbf{s}_1), \ldots, S(\mathbf{s}_N)) \sim NI(\mu, \sigma^2)$, and the ϵ and S processes are independent. By invoking this model we get

$$F_S(z) = \Phi\left(\frac{z - \mu}{\sigma}\right) \tag{56}$$

where $\Phi(\cdot)$ is the standard normal CDF. Suppose a sample of size n is taken, such that the labels $(\mathbf{s}_1, \ldots, \mathbf{s}_k)$, with $k < n$, are selected, and n_j replicate samples are taken at label $\mathbf{s}_j$; $j = 1, \ldots, k$, where $\sum_{j=1}^{k} n_j = n$. Define

$$\hat{\mu} \equiv \frac{1}{k}\sum_{i=1}^{n}\frac{1}{n_i}\sum_{j=1}^{n_i} Z_j(\mathbf{s}_i)$$

From the analysis of variance of the random-effects model, Fuller (1995) obtains the following variance estimator:

$$\hat{\sigma}^2 \equiv \frac{1}{r}\left(\frac{1}{k-1}\sum_{i=1}^{k} n_i(\bar{Z}(\mathbf{s}_i) - \hat{\mu})^2 - \hat{\tau}^2\right)$$

where $\bar{Z}(\mathbf{s}_i) = \frac{1}{n_i}\sum_{j=1}^{n_i} Z_j(\mathbf{s}_i)$, $r = (k-1)^{-1}(n - n^{-1}\sum_{i=1}^{k} n_i^2)$, and $\hat{\tau}^2 = (\sum_{i=1}^{k}(n_i - 1))^{-1}\sum_{i=1}^{k}\sum_{j=1}^{n_i}(Z_j(\mathbf{s}_i) - \bar{Z}(\mathbf{s}_i))^2$. It is straightforward to show that $E(\hat{\mu}) = \mu$ and $E(\hat{\sigma}^2) = \sigma^2$. Substituting $\hat{\mu}$ for μ and $\hat{\sigma}$ for σ in equation 56) we obtain the following CDF estimator:

$$\hat{F}_S(z) = \Phi\left(\frac{z - \hat{\mu}}{\hat{\sigma}}\right) \tag{57}$$

Nusser et al. (1996) extend this approach to non-Gaussian S and ϵ by assuming that a transformation exists such that both S and ϵ, suitably transformed, are normally distributed.

2.2.6 Inference in spatial sampling

Consider the spatial-process model in equation (1), and consider the set

$$X = \{(\mathbf{s}, Z(\mathbf{s})) : \ \mathbf{s} \in A\} \tag{58}$$

where A is a finite (possibly random) subset of $D \in \mathbb{R}^2$. (In practice, D is bounded and is often discretized to a finite grid of locations D_f.) Spatial model-based inference is based upon the probability structure ξ defined by equation (1), conditional on the sample A of locations. If A is selected by a probability sampling design p, then spatial design-based inference is supported by the probability structure defined by p, conditional on $Z(\cdot)$.

Now assume the spatial model in equation (32) that includes measurement error. Spatialmodel-based inference is similar to that for the model in equation (1), but now ξ also includes the measurement-error component $\epsilon(\cdot)$. Further, for equation (32), spatial design-based inference is supported by the design p *and* the error process $\epsilon(\cdot)$ (see Sections 2.2.4 and 2.2.5), conditional on $S(\cdot)$. Here, both model-based and design-based inferences depend on the probability structure defined by $\epsilon(\cdot)$.

To avoid confusion, we wish to clarify some terminology. In the spatial-model context, interest is usually in the "prediction" of quantities assumed to be random (i.e., functions of $S(\cdot)$); in the survey-sampling context, those same quantities (i.e., functions of $(S(\mathbf{s}_1), \ldots, S(\mathbf{s}_N))'$) may be assumed fixed and consequently the corresponding inference is termed "estimation." In our work, we shall generally use the term "prediction" for making inference on $S(\cdot)$ or $(S(\mathbf{s}_1), \ldots, S(\mathbf{s}_N))'$, regardless of whether it is model-based or design-based.

2.3 Review of the Spatial Sampling Literature

The past literature on spatial sampling has been concerned with a number of issues related to choosing $A \equiv \{\mathbf{s}_1, \ldots, \mathbf{s}_n\}$ and $n = |A|$ in a spatial domain of interest. The first issue is that of model-based versus design-based sampling approaches. The second issue deals with design criteria by which the performance of a spatial design is assessed, for a given predictor. The third is one of comparing various popular, although not necessarily optimal, designs, and the fourth relates to adaptive sampling procedures.

2.3.1 Model-based and design-based sampling approaches

Model-based and design-based approaches to spatial-sampling theory each depend on a different source of probability structure upon which inferences may be made (Sections 2.2.3 and 2.2.6). What emerges from the literature is that design-based inference may be more robust than model-based inference, but that an appropriate model-based analysis may perform substantially better, provided that the model-based inferences are (Fisher) consistent with the target population (Overton, 1993).

The central problem with model-based inference is that if the model is not consistent with the target population, then a purely model-based analysis may yield substantially biased estimates of population parameters and very misleading estimates of sampling precision (McArthur, 1987; Overton, 1993). This suggests that the design-based approach may be more appropriate if a consistent spatial model cannot be identified reliably. De Gruijter and ter Braak (1990) argue that, although design-based efficiency may be less than the optimal model-based efficiency, such a loss may be a worthwhile premium to pay for robustness against model errors and for achieving p-unbiasedness (i.e., design-unbiasedness).

Some model-based methods incorporate design-based approaches (Cox et al., 1997). Overton (1993) says that model-based inference can improve the precision of a p-sample, often greatly, and is likely to be consistent with the target population. Cressie and Ver Hoef (1993) and Stevens (1994) suggest that model-based inference may be strengthened by characteristics of a p-sample (e.g., a systematic design with random starting point), by eliminating selection bias in the choice of the sample locations, and by providing a mechanism for inferences free of the assumed spatial model. However, Cressie and Ver Hoef (1993) also suggest that purely design-based inferences, ignoring models describing small-scale spatial dependence structures, are severely limited in the range of questions they can address. Thus, armed with flexible spatial models and good model diagnostics, the use of spatial models can greatly enhance the science of spatial sampling.

2.3.2 Design criteria

A central problem in sampling theory is the search for an optimal sampling *strategy*, that is, the search for a design/predictor combination that best achieves our objectives. Prediction objectives commonly include point and block prediction (see Section 2.1) and Cox et al. (1997) list several other prediction objectives in the spatial context. These include the prediction of the average of a nonlinear function of the spatial process $S(\cdot)$, the prediction of the maximum of $S(\cdot)$, and the prediction of the subregion for which the spatial process exceeds a given threshold.

In much of the spatial sampling literature, the major concern seems to be about design optimization for a specified predictor. That is, for a given predictor, *design* criteria are usually considered. This discussion will concentrate largely on model-based design criteria, mainly because of their particular applicability to the spatial sampling situation. Design-based design criteria (e.g., Horvitz–Thompson sampling variances or MSEs with respect to the sampling design) are, as Cox et al. (1997) note, elementary from a statistical point of view and can be obtained from any reasonable text on survey sampling (e.g., Cochran, 1977; Särndal et al., 1992).

Thus, given some measurement-error spatial-process model and the accompanying spatial analyses discussed in Section 2.1, various model-based design criteria can be formulated to assess the performance of different sampling designs for given predictors. Considering only linear unbiased predictors $\{\hat{S}(\mathbf{u}; \mathbf{Z}) : \mathbf{u} \in A\}$, which are functions of $X = \{(\mathbf{s}_1, Z(\mathbf{s}_1)), \ldots, (\mathbf{s}_n, Z(\mathbf{s}_n))\}$, Cox et al. (1997) enumerate three sampling-design criteria:

(D-1) The integrated mean squared error (IMSE) criterion: Minimize with respect to sampling locations $\{\mathbf{s}_1, \ldots, \mathbf{s}_n\}$,

$$IMSE(\mathbf{s}_1, \ldots, \mathbf{s}_n) \equiv \int_D E[\hat{S}(\mathbf{u}; \mathbf{Z}) - S(\mathbf{u})]^2 \, d\mathbf{u}$$

(D-2) The maximum MSE (MMSE) criterion: Minimize with respect to $\{\mathbf{s}_1, \ldots, \mathbf{s}_n\}$,

$$MMSE(\mathbf{s}_1, \ldots, \mathbf{s}_n) \equiv \sup_{\mathbf{u} \in D} E[\hat{S}(\mathbf{u}; \mathbf{Z}) - S(\mathbf{u})]^2$$

(D-3) The entropy criterion: Maximize with respect to $\{\mathbf{s}_1, \ldots, \mathbf{s}_n\}$,

$$H(\mathbf{s}_1, \ldots, \mathbf{s}_n) \equiv E[-\log\{f(Z(\mathbf{s}_1), \ldots, Z(\mathbf{s}_n))\}]$$

where f denotes the probability density function of $(Z(\mathbf{s}_1), \ldots, Z(\mathbf{s}_n))$.

Note that, for ordinary kriging (constrained kriging) predictors, criteria (D-1) and (D-2) simplify to $\int_D \sigma^2_{ok(ck)}(\mathbf{u})\, d\mathbf{u}$ and $\sup_{\mathbf{u}\in D}\{\sigma^2_{ok(ck)}(\mathbf{u})\}$, respectively.

Cressie (1993a) modifies (D-1) and (D-2) by including weight functions in the integrals. He gives as an example of weights, indicator functions that focus attention on subregions whose mean and variance exceed some given threshold.

Clearly, it makes sense to choose criteria that relate to the major objectives in the study. Thus, (D-1) is an appropriate criterion if a design is needed to perform "best on average," and (D-2) is useful if one seeks to minimize the worst case. The entropy criterion (D-3) is claimed by Caselton and Zidek (1984) and Guttorp et al. (1993) to be useful in studies with multiple objectives. It is usually considered only in cases where finitely many potential locations are available, but this need not be a limitation, since a very fine grid of possible locations is still finite in number. A feature of (D-3) is that if $Z(\cdot)$ is a Gaussian process, the maximization of H is equivalent to the maximization of the determinant of the covariance matrix of $(Z(\mathbf{s}_1), \ldots, Z(\mathbf{s}_n))$, and hence it is also known as the D-optimality criterion (Cox et al., 1997).

It has been well documented (Olea, 1984; Cressie et al., 1990) that regular, particularly triangular, networks do well in terms of minimizing maximum σ_{ok}^2. However, Haas (1992) points out that, in practice, the variogram in equation (6) typically has to be estimated and then modeled (as described in Section 2.1.7), so that some clustering of the design points is necessary for good variogram estimation at short lags, where accuracy is usually most important.

Haas (1992) addresses the problem of redesigning a continental-scale monitoring network by providing a method for optimally adding sites to a subregion of the continent, using a bivariate criterion. He defines the relative error estimate (REE) of a region B as $\hat{\sigma}_{ok}(B)/\hat{Z}(B)$, where the (estimated) kriging standard error is obtained from equation (17), but with $\gamma(\cdot)$ replaced by $\gamma(\cdot; \hat{\boldsymbol{\theta}})$. The criterion he proposes is to minimize both the REE over the subregion in question and the standard deviation of the REE at the subregion's center; the latter, he suggests, could be estimated by the sample standard deviation of a simulated sampling distribution of the REE at the subregion's center.

Guttorp et al. (1993) make use of the entropy criterion (D-3), which they propose as a generic objective designed to meet some quality requirement for all objectives, even though it may not be best for any given objective. Caselton and Zidek (1984) explore the usefulness of the entropy criterion for long-term studies where all possible uses of data are unlikely to be unique or foreseen. They view network optimization as a problem of selecting a set M of monitored site locations so that the increase in information about a set U of unmonitored sites, after observing $\{Z(\mathbf{s}) : \mathbf{s} \in M\}$, is maximized. Here $D_f = M \cup U$. Thus, if $\mathbf{Z}_M$, $\mathbf{Z}_U$ are random variable vectors of the process at the monitored and unmonitored sites, respectively, let $f(\mathbf{Z}_M, \mathbf{Z}_U)$ be the joint distribution of the vectors and $f(\mathbf{Z}_M)$, $f(\mathbf{Z}_U)$ be the marginal distributions. If $f(\mathbf{Z}_U)$ is interpreted as a prior density function, then Caselton and Zidek (1984) propose that the locations M should be chosen to maximize what Pérez-Abreu and Rodríguez (1996) refer to as the *Shannon information index*,

$$I(U, M) = \iint log\left\{\frac{f(\mathbf{Z}_M, \mathbf{Z}_U)}{f(\mathbf{Z}_M)f(\mathbf{Z}_U)}\right\} f(\mathbf{Z}_M, \mathbf{Z}_U)\,\mathrm{d}\mathbf{Z}_U\,\mathrm{d}\mathbf{Z}_M$$

which is the information about $\mathbf{Z}_U$ contained in $\mathbf{Z}_M$. Note that if $f(\mathbf{Z}_M)$ and $f(\mathbf{Z}_U)$ are independent, then $I(U, M) = 0$ (i.e., the monitored sites provide no information about the unmonitored sites). Pérez-Abreu and Rodríguez (1996) extend this to the multivariate $\mathbf{Z}(\cdot)$ case.

Haas (1992) observes some problems with this approach: (i) the prior distribution $f(\mathbf{Z}_U)$ may be difficult to specify in practice; (ii) a parametric multivariate distribution needs to be specified for all of the sites, monitored and unmonitored; (iii) the mean and covariance structure of the multivariate distribution must be estimated in practice, thereby introducing error into the computation minimizing $I(U, M)$, and this has not been taken into account in the current version of the theory.

So far, we have been concerned mainly with the precision aspects of model-based design criteria, whether or not the invoked models relate meaningfully to the target population. If a superpopulation model is assumed, we must be clear whether our target population is the "real-world" population (i.e., a single realization of the invoked model), or if it is the superpopulation itself. For example, given the spatial model in equation (1), are we interested in predicting the real-world mean $Z(B)$, or is our interest in estimating the model mean $E(Z(B)) \equiv \mu(B)$? Superpopulation-model parameter estimation may be appropriate if we believe that the model describes some causal or mechanistic behavior, but in the spatial context, interest is more likely to be in the real-world population.

Overton (1993) strongly suggests that if our interest is in the real-world population, then our primary statistical criterion should be that (model-based) inferences be (Fisher) consistent with the real-world population. Sampling designs, which provide precise predictors according to the models but which are not consistent with the real-world population, are simply unacceptable. He further suggests that probability sampling designs best ensure real-world-population consistency.

Certain criteria may be useful either for selecting a good design or for selecting a good predictor, depending on how the optimization is done. For example, suppose inferences about some real-world population are desired, ξ is some assumed superpopulation model and p is a probability sampling design. Then Särndal et al. (1992, p. 516) suggest that the predictor that minimizes the unconditional MSE be selected. For example, suppose we wish to predict $S(B)$ with $\hat{S}(B)$. The criterion to minimize, with respect to choice of predictor $\hat{S}(B)$, is

$$E_{p\xi}[(\hat{S}(B) - S(B))^2] \tag{59}$$

where $E_{p\xi}(\cdot)$ is the expectation taken with respect to both p and ξ. If $E_{p\xi}[\hat{S}(B) - S(B)] = 0$, then equation (59) is called the *anticipated variance* (Isaki and Fuller, 1982), and this is commonly used as a design criterion; that is, it is minimized with respect to p for a given predictor and a given model ξ (e.g., Breidt, 1995a).

In cases where design-based inference is to be emphasized, Overton and Stehman (1993) state that, while (design-based) precision should be a primary design criterion, an important secondary criterion is the ability to obtain adequate variance estimators. Unbiased variance estimators of the form of equation (49) exist for Horvitz–Thompson estimators if p is a measurable probability sampling design, but equation (49) may behave badly under certain conditions (e.g., it may be negative; Särndal et al., 1992, p. 47). However, for those designs that have some pairwise inclusion probabilities that are zero (e.g., systematic sampling designs), adequate variance estimators are not so obviously obtainable, and models may have to be invoked to derive them.

The sampling optimality criteria presented so far have all been based on statistical considerations. Frequently, cost or economic considerations provide a very important limitation, which should be introduced into the objective function.

Bras and Rodriguez-Iturbe (1976) propose the objective function

$$\delta(n, A) + \beta\kappa(n, A) \tag{60}$$

where $\delta(n, A)$ is a measure of statistical accuracy, $\kappa(n, A)$ is the cost of sampling, β is a measure of accuracy obtained from a unit increase in cost, n is the sample size, and $A \equiv \{\mathbf{s}_1, \ldots, \mathbf{s}_n\}$ is the set of sample locations. Bras and Rodriguez-Iturbe (1976) suggest how equation (60) can be optimized numerically over sampling locations A and sample size n. Bogardi et al. (1985) consider the problem of optimal spatial sampling design as one of multicriterion decision making. A composite objective function, measuring statistical and economic tradeoffs, is proposed. The optimal *rectangular* network is achieved through compromise programming (Zeleny, 1982). Englund and Heravi (1992) propose the use of conditional simulation (i.e., simulation from a spatial model whose realizations always go through the observations at the sampling locations $\{\mathbf{s}_1, \ldots, \mathbf{s}_n\}$) as a powerful tool for the optimization of economic objective functions.

2.3.3 Comparison of sampling designs

The computational problem of obtaining optimal IMSE designs is difficult (Cressie, 1993a, pp. 319, 320; Cox et al., 1997), but it is often relatively easy to obtain "good" designs, and standard iterative algorithms usually make substantial improvements over the initial design after only a few iterations (Cox et al., 1997). Cox et al. (1997) further state that in their experience, good designs tend to spread points uniformly in the design region, echoing the results of Dalenius et al. (1960), Olea (1984), and Currin et al. (1991).

Using model-based design criteria, Olea (1984) and Cressie et al. (1990) compare various popular, but not necessarily optimal, designs. Olea (1984) shows that the geometrical configuration of a network is the most important consideration in optimal network design and, in particular, regular triangular grids minimize the maximum σ_{ok}^2 over the spatial domain D. Cressie et al. (1990) illustrate that for spatial processes with increasing spatial dependence (up to a point), systematic sampling designs (SYS) are better than stratified random sampling designs (STS), which are in turn better than simple random sampling designs (SRS), with respect to a risk function that depends on σ_{ok}^2. Olea (1984) reaches conclusions that agree with this ranking, adding that clustered designs are by far the worst.

Overton and Stehman (1993) compare designs with respect to two criteria, viz., the relative precision of the designs with respect to design-based criteria as the primary consideration, and the capability for adequate variance estimation of the designs as a second important consideration. They used three types of surface $Z(\cdot)$: planar, quadratic, and sinusoidal. The designs considered were SYS, SRS, and another design called tessellation-stratified sampling (TSS), where the hexagons of a triangular-grid tessellation were used as strata and one random point per stratum was selected. They note that TSS has the advantages of a systematic design (evenly distributed samples over the domain of interest) without its disadvantage (an inability to handle certain periodicities in the surface). In almost all cases, TSS outperformed SYS with respect to both criteria under consideration, and in some cases greatly so. The sampling design SRS performed worst in all cases.

A study by McArthur (1987) involved sampling over a peaked surface corresponding to environmental contaminants emerging from a point source, where the pollutants are more concentrated near the source. In this case, a preferential grid (stratification of two systematic grids, with one of higher sampling density centered on the location of the peak) is the most accurate and precise by far, but estimation of the design-based sampling variance from one realization of the design is not possible. It was found that STS and importance sampling (IS), a Monte Carlo method for computing integrals (see, e.g., Hammersley and Handscomb, 1964), were very useful for predicting $Z(B)$ if good prior knowledge of the surface $Z(\cdot)$ were available. If not, SYS or SRS is suggested, with preference given to the former if no periodicities are likely to occur in the surface. Estimation of the design-based sampling variance from one realization of a SYS design is also not possible, but Wolter (1984) suggests approximate methods to get around this.

Markov chain (MC) designs for one-per-stratum sampling are presented by Breidt (1995a) for finite populations, and by Breidt (1995b) for continuous populations. For simplicity, consider for the moment a one-dimensional sampling domain partitioned into equally sized strata. After initially selecting a sampling location from an "end" stratum by means of some probability function, the location-selection procedure moves sequentially along the strata of the domain to the other end by means of a stochastic process satisfying the Markov property (i.e., the probability of a future selection depends only on the present selection and not on past selections). Thus, the probability of a location being selected depends only on the selection in the immediately preceding stratum. This procedure is easily extended to two dimensions by considering two independent stochastic processes, with one operating on the "latitude" coordinate, and the other on the "longitude" coordinate. Given the stochastic processes specified by Breidt (1995b) in the continuous population case, what results is a range of designs that include as special cases SYS, STS with one sampling unit per stratum, and balanced systematic sampling designs (BAS). A balanced design is one in which the sampling location within one stratum and the sampling location within an adjacent stratum are equidistant from the stratum boundary that separates them. Breidt (1995a,b) evaluates MC designs under a variety of superpopulation models, using the *anticipated variance* criterion (see equation (59)). He demonstrates that new designs from within the broader MC class are competitive with the standard designs, SYS, STS, and BAS, under a variety of models. In particular, he shows that for models with a dominant trend term, the optimal design is close to BAS, but if an autocorrelation (spatial dependence) term dominates, then the optimal design shifts closer to SYS.

It is well known that systematic designs are potentially disastrous in the face of periodicities occurring within the sampling domain. It has been noted that while few periodicities occur in nature, human effects on the landscape are often systematic. Much of the midwest of the USA has imposed upon it a one-mile square grid of gravel roads and, as Breidt (1995a) remarks, a survey sampler using a systematic design with a one-mile interval and an unlucky random start might conclude that Iowa is covered by gravel roads! Such unfortunate occurrences may be avoided by selecting a nonsystematic MC design.

2.3.4 Adaptive sampling

For spatial phenomena that are rare or clustered or both, adaptive sampling methods (Thompson, 1992) may be far more useful than traditional sampling methods. Suppose that some variable of interest, Z, is mostly zero, but that its nonzero values are spatially clustered. We wish to (i) estimate the average $Z(D_f)$ or total T of Z, and (ii) locate the "pockets" where Z is nonzero. Traditional sampling methods provide unbiased estimates of $Z(D_f)$ and T, but with high variance, and the maps they provide detailing the occurrence of pockets of Z are usually highly inaccurate (Seber and Thompson, 1994). The adaptive-sampling procedure goes as follows: (i) select an initial sample by some conventional design (e.g., SRS); (ii) add to the initial sample any units that satisfy some specified condition (e.g., add neighboring units to all initial sample units where Z occurs). Thompson (1992) provides a theory to modify selection probabilities and to obtain consistent estimators, which he shows can be much more efficient than SRS estimators. Cox et al. (1997) suggest that this may be useful for "hot-spot" identification in environmental problems.

3. COMPUTER SIMULATION EXPERIMENT

Consider a spatial domain over which a Gaussian spatial process (superpopulation) model is defined. The aim of this study is to use design-based criteria to compare different analyses (i.e., prediction methods) and different sampling designs under a variety of conditions. In much of the spatial sampling literature, the emphasis is on gauging sampling designs for given predictors; in this study the emphasis is on the choice of prediction method under different sampling designs.

A computer experiment was devised, complete with "factors" and "responses." The factors of the experiment include the sampling designs, the prediction methods, and the different conditions under which the designs and analyses were conducted. Performance criteria constitute the responses. The details of the spatial-process model and of the computer experiment are now presented.

3.1 Gaussian Spatial Models

The spatial domain D considered in this experiment is a square region in $\mathbb{R}^2$ which is discretized into $D_f = \{(x, y) : x = 1, \ldots, 10;\ y = 1, \ldots, 10\}$, a 10×10 grid of 100 locations. Over this domain, the following measurement-error spatial model was invoked:

$$Z(\mathbf{s}) \equiv S(\mathbf{s}) + \epsilon(\mathbf{s}); \qquad \mathbf{s} = (x, y) \in D_f \tag{61}$$

where $\epsilon(\cdot)$ is a zero-mean, white-noise measurement-error process, and we define the state process

$$S(\mathbf{s}) = \mu(\mathbf{s}) + \delta(\mathbf{s}); \qquad \mathbf{s} = (x, y) \in D_f \tag{62}$$

where $\mu(x, y) = \beta(x - 4.5);\ x = 1, \ldots, 10;\ y = 1, \ldots, 10$, with β to be specified in Section 3.2.3. The set of 100 δ-values $\{\delta(x, y):\ x = 1, \ldots, 10;\ y = 1, \ldots, 10\}$ corresponding to the 100 grid locations $\{(x, y):\ x = 1, \ldots, 10;\ y = 1, \ldots, 10\}$ were generated according to a zero-mean, Gaussian spatial process model with covariance function

$$\operatorname{cov}(\delta(i, j), \delta(k, l)) = \sigma^2 \begin{cases} \rho^h & \text{for } \rho \in \{0.1, 0.2, \ldots, 0.9\} \\ 1 & \text{if } \rho = 0 \text{ and } (i, j) = (k, l) \\ 0 & \text{if } \rho = 0 \text{ and } (i, j) \neq (k, l) \end{cases} \tag{63}$$

where $h \equiv \{(i - k)^2 + (j - l)^2\}^{1/2}$ and, without loss of generality, $\sigma^2 = 1$ was chosen.

Note that $\delta(\cdot)$ is a second-order stationary process and, for $\rho = 0$, $(\delta(1, 1), \ldots, \delta(10, 10)) \sim NI(0, 1)$. As $\rho \to 1$ (although $\rho > 0.9$ is out of the range of values used in this study), the δ-process tends towards a common value, that value being distributed as $N(0, 1)$.

The Z-process in equation (61) was obtained by defining the measurement-error process through $(\epsilon(1, 1), \ldots, \epsilon(10, 10)) \sim NI(0, \tau^2)$, where τ^2 is specified in Section 3.2.4.

3.2 Factors of the Computer Experiment

3.2.1 Regions

Let the entire domain of the 10×10 locations be called the *global* region G. Demarcate a 3×3 subregion of 9 locations $\{(x, y):\ x = 1, 2, 3;\ y = 1, 2, 3\}$ of G, and call this the *local* region L. These are the two regions over which statistical inference of various forms will be made.

3.2.2 Strength of spatial dependence

The strength of spatial dependence in the δ-process, characterized by equation (63), ranges from zero to very strong and is indexed by $\rho \in \{0, 0.1, \ldots, 0.9\}$.

3.2.3 Trend

Two values were given to the trend term in equation (62), viz., $\beta \in \{0, (8.25)^{-1/2}\}$. Note that for $\beta = 0$, $S(\cdot)$ and $Z(\cdot)$ are both (second-order) stationary Gaussian processes. The choice of the other value of β is discussed below.

3.2.4 Noise

Two levels of measurement error were considered: $\tau^2 = 0.1$ and 2. These two levels will be referred to as *low* noise and *high* noise, respectively. Their choice is discussed below.

The trend and noise parameters were chosen according to a square-root signal-to-noise ratio scale. Define the "signal" variance, σ_s^2, as the sum of the model variance of the δ-process and the sample variance of the trend process; that is, $\sigma_s^2 = \text{var}_\xi(\delta(\mathbf{s})) + \frac{1}{10}\sum_{x=1}^{10} \beta^2(x - 4.5)^2 = 1 + 8.25\beta^2$. Thus, $\beta \in \{0, (8.25)^{-1/2}\}$ yields $\sigma_s^2 \in \{1, 2\}$. Combining both levels of β and both levels of τ^2 we obtain the following four ratios: $\sigma_s/\tau \in \{0.71, 1, 3.16, 4.47\}$.

These ratios do not take into account the spatial "ρ-effect," (i.e., strength of spatial correlation). As ρ increases, the δ "realization" becomes much smoother, even though $\text{var}_\xi(\delta(\mathbf{s})) \equiv 1$. For large ρ, $\text{var}_\xi(\delta(\mathbf{s}))$ accurately represents the variability due to $\delta(\cdot)$ over a very large spatial domain, but over small spatial domains (such as 10×10), where $\delta(\cdot)$ is likely to be much smoother, the signal variance may appear to be substantially less than 1. Thus the apparent ratio from any given realization may be much smaller than the prespecified σ_s/τ.

3.2.5 Realizations of the state process

The S-process in equation (62) is held fixed over the randomization component in the design. A vector $\mathbf{Y}$ of 100 $NI(0, 1)$ values were generated, and the 100 values of $\delta(\cdot)$ (i.e., δ-realizations) were obtained from $V(\rho)^{1/2}\mathbf{Y}$ where, from equation (63), the matrix $V(\rho)$ with elements $\text{cov}(\delta(\mathbf{u}), \delta(\mathbf{s}))$; $\mathbf{u}, \mathbf{s} \in G$ is a function of $\rho \in \{0, 0.1, \ldots, 0.9\}$. Thus, a set of *ten* δ-realizations were generated corresponding to the set of *ten* ρ values, using the *same* $\mathbf{Y}$ vector, and two S-realizations, corresponding to the two values of the trend parameter β, were created from each δ-realization by adding the appropriate trend term given in equation (62) (see Figure 1, where S-realizations are shown for the two β-values in question, and for $\rho \in \{0, 0.5, 0.9\}$).

Three different $\mathbf{Y}$ vectors of 100 $NI(0, 1)$ values were used to generate three realizations of the S-process (i.e., S-realizations) for each value of ρ and β.

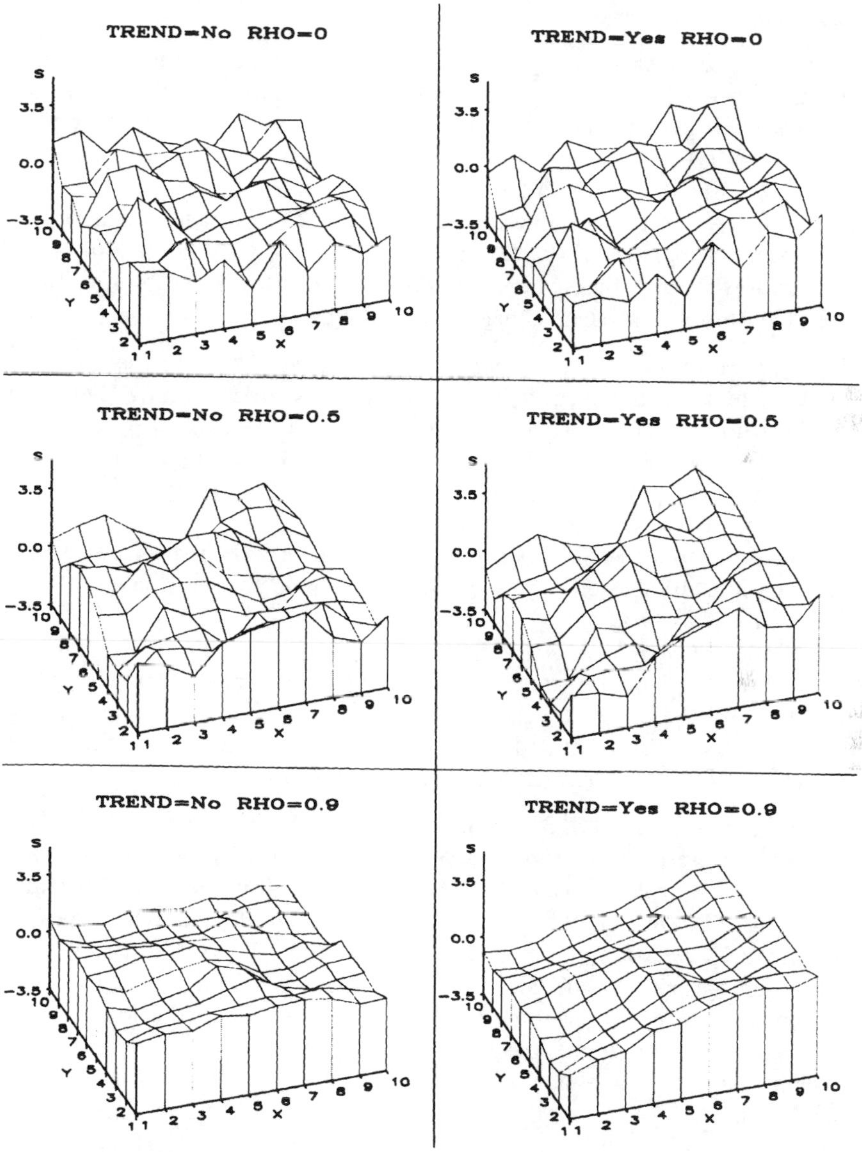

Figure 1. Generated S-realizations over global region $G = \{(x, y) : x = 1, \ldots, 10;\ y = 1, \ldots, 10\}$, for trend parameter $\beta \in \{0, (8.25)^{-1/2}\}$ and spatial-correlation parameter $\rho \in \{0, 0.5, 0.9\}$

3.2.6 Sampling designs

A subset of 20 locations $\{\mathbf{s}_1, \ldots, \mathbf{s}_{20}\}$ was selected according to various sampling designs from among the 100 grid locations, and the corresponding Z-values were generated by adding 20 ϵ-values, independently generated from a $N(0, \tau^2)$ distribution, to the corresponding values from the state process, $\{S(\mathbf{s}_j) : j = 1, \ldots, 20\}$. A different random sample of locations yields a different set of S-values and a different set of measurement errors, but the underlying 100 values $\{S(x, y) : x = 1, \ldots, 10;\ y = 1, \ldots, 10\}$ of the S-realization remain the same over randomization of the sampling locations. The sampling designs considered are below.

1. *Systematic random sampling* (SYS, notated "Y" on the figures). Consider the design configuration where the $\mathbf{s}_j$-th column has sampling location at $(1, j) + (2j - 2, 0)$ and $(6, j) + (2j - 2, 0)$; $j = 1, \ldots, 10$, and componentwise addition is modulo 10. Randomly "start" the 20 samples by choosing the first location to be $(1, 1) + (0, k)$, where k is uniformly distributed on $\{0, 1, 2, 3, 4\}$.
2. *Stratified random sampling* (STS, notated "T" on the figures). Two locations are chosen randomly (without replacement) from each "column" of the grid.
3. *Simple random sampling* (SRS, notated "R" on the figures). Twenty locations are chosen randomly (without replacement) from the entire grid.
4. *Clustered* (CLU, notated "C" on the figures). Ten observations are chosen from location (2,2) and ten from location (9,9).

The three designs, SYS, STS, and SRS, are probability sampling designs, and CLU is a purposive design. The S-process is stationary within the strata defined by STS, irrespective of the value of the trend parameter β. Because of its desirable properties with respect to spatial model-based criteria (see Section 2.3.2), SYS may be regarded as a "spatial" design.

The sample size was fixed at $n = 20$ and thus is not a factor in the experiment. However, for some prediction methods over the local region (see Section 3.2.7), the sample size is a random variable. Note that it is possible that the 20 sampled locations may all fall outside L for STS and SRS designs.

3.2.7 Prediction methods

Prediction presupposes some target. In this study, interest is in two predictands, the spatial mean and the spatial cumulative distribution function (SCDF), which are defined below.

The spatial mean

The spatial mean of $S(\cdot)$ over a region B is defined as

$$S(B) \equiv \sum_{\mathbf{u} \in B} S(\mathbf{u})/|B| \tag{64}$$

where $|B|$ is the number of S-values whose locations fall in B. Thus,

$$S(G) = \frac{1}{100} \sum_{x=1}^{10} \sum_{y=1}^{10} S(x, y) \tag{65}$$

$$S(L) = \frac{1}{9} \sum_{x=1}^{3} \sum_{y=1}^{3} S(x, y) \tag{66}$$

We will consider four predictors of these two quantities, all of which are functions of $\mathbf{Z} \equiv (Z(\mathbf{s}_1), \ldots, Z(\mathbf{s}_{20}))'$, the data vector corresponding to the 20 sampled locations.

1. *Ordinary kriging* (OK). Define

$$\hat{S}_{ok}(B) \equiv \boldsymbol{\lambda}_{ok}(B)'\mathbf{Z}; \qquad B = G \text{ or } L \tag{67}$$

where $\boldsymbol{\lambda}_{ok}(B)'\mathbf{Z}$ is given by equation (13) with $n = 20$. Recall that $\hat{S}_{ok}(B)$ is the *best linear unbiased predictor* (BLUP) of $S(B)$.

2. *Constrained kriging* (CK). Define

$$\hat{S}_{ck}(B) \equiv \begin{cases} \boldsymbol{\lambda}_{ck}(B)'\mathbf{Z} & \text{if } \rho > 0, \quad B = G \text{ or } L \\ \bar{Z} + \left\{ \frac{n - |B|(1+\tau^2)}{|B|(1+\tau^2)\{n \sum_{i=1}^{m} |B_i|^2 - N_m^2\}} \right\}^{\frac{1}{2}} & \\ \quad \times \left(\sum_{i=1}^{m} |B_i| Z(\mathbf{v}_i) - N_m \bar{Z} \right) & \text{if } \rho = 0, \ B = L \\ \bar{Z} & \text{if } \rho = 0, \ B = G \end{cases} \tag{68}$$

where $\boldsymbol{\lambda}_{ck}(B)'$ is given by equation (25) with $n = 20$. The solution for the case where $\rho = 0$, $B = L$, and an explanation of the terms in equation (68) are given in Appendix 1. No solution to the constrained kriging equations exists for $\rho = 0$, $B = G$, so $\bar{Z}$ was used instead.

3. *Regional poststratification* (RP). Define

$$\hat{S}_{rp}(G) \equiv \begin{cases} \frac{|L|}{|G|}(\bar{Z}_L) + \frac{|G|-|L|}{|G|}(\bar{Z}_{G-L}) & \text{if } n_L > 0 \\ \bar{Z} & \text{if } n_L = 0 \end{cases} \tag{69}$$

where $\bar{Z}_L$, $\bar{Z}_{G-L}$, and $\bar{Z}$ are averages of the Z-values contained in

L, $G - L$, and G respectively, and n_L is the number of sample locations in L. The local-region predictor is defined as

$$\hat{S}_{rp}(L) \equiv \begin{cases} \bar{Z}_L & \text{if } n_L > 0 \\ \bar{Z} & \text{if } n_L = 0 \end{cases} \tag{70}$$

Note that the sample size specified in equation (70) is a random variable.

4 *Arithmetic mean* (AM). Define

$$\hat{S}_{am}(B) \equiv \bar{Z}; \qquad B = G \text{ or } L \tag{71}$$

It should be noted that $\hat{S}_{am}(G)$ is a Horvitz–Thompson estimator for $S(G)$ for designs with equal first-order inclusion probabilities, that is, for SYS, STS, and SRS. (All pairwise inclusion probabilities for the designs STS and SRS are positive, and hence $\hat{V}_{ht}$ given by equation (49) can be obtained for those designs. This is not the case for SYS.)

Observe that the first two predictors, ordinary kriging and constrained kriging, are model-based "spatial" predictors and the last two predictors, regional poststratification and arithmetic mean, are design-based "non-spatial" predictors. The model assumptions of ordinary and constrained kriging are that the spatial covariance parameters are known.

The spatial cumulative distribution function

Let F_B denote the SCDF of $\{S(\mathbf{s}) : \mathbf{s} \in B\}$; specifically,

$$F_B(z) \equiv \frac{1}{|B|} \sum_{\mathbf{u} \in B} I(S(\mathbf{u}) \leq z); \qquad z \in \mathbb{R} \tag{72}$$

where $|B| \equiv \sum_{\mathbf{u} \in B} 1$. The SCDF is discussed more fully in Majure et al. (1996). Now define the quantile function of the SCDF F_B as follows:

$$q(\alpha) \equiv \inf\{z : F_B(z) \geq \alpha\}; \qquad \alpha \in [0, 1] \tag{73}$$

In our study, we restrict $\alpha \in \{0.1, 0.25, 0.5, 0.75, 0.9\}$ and $B = G$ or L. Note that if $B = G$, then equations (72) and (73) are equivalent to the finite-population CDF in equation (38) and the finite-population quantile function in equation (39), respectively, in Section 2.2.1.

Thus, $F_G(q(\alpha)) = \alpha$, for all α in question, but this is not the case for F_L, since L contains only nine locations (e.g., $F_L(q(0.1)) = 0.111$). For consistency in our experiment, the quantity to be predicted is always $F_B(q(\alpha))$, rather than α.

Consider the region $B \subset G$, containing $|B|$ locations $\{\mathbf{u}_1, \ldots, \mathbf{u}_{|B|}\}$. Six SCDF predictors are considered in this study.

1. *Ordinary kriging* (OK). Define

$$\hat{F}_{B;ok}(z) = \frac{1}{|B|}\sum_{i=1}^{|B|} I(\hat{S}_{ok}(\mathbf{u}_i) \le z); \qquad z \in \mathbb{R},\ B = G \text{ or } L \tag{74}$$

where $\hat{S}_{ok}(\mathbf{u}_i) = \boldsymbol{\lambda}_{ok}(\mathbf{u}_i)'\mathbf{Z}$ is defined as in equation (13) except that $\mathbf{c}(B)$ is replaced by $\mathbf{c}(\mathbf{u}_i) \equiv (C(\mathbf{u}_i, \mathbf{s}_1), \ldots, C(\mathbf{u}_i, \mathbf{s}_n))'$; $i = 1, \ldots, |B|$.

2. *Constrained kriging* (CK). Define

$$\hat{F}_{B;ck}(z) = \frac{1}{|B|}\sum_{i=1}^{|B|} I(\hat{S}_{ck}(\mathbf{u}_i) \le z); \qquad z \in \mathbb{R},\ B = G \text{ or } L \tag{75}$$

where

$$\hat{S}_{ck}(\mathbf{u}_i) = \begin{cases} \boldsymbol{\lambda}_{ck}(\mathbf{u}_i)'\mathbf{Z} & \text{if } \rho > 0 \\ \bar{Z} + \left\{\frac{n-(1+\tau^2)}{(1+\tau^2)m(n-m)}\right\}^{\frac{1}{2}}(\sum_{i=1}^{m} Z(\mathbf{v}_i) - m\bar{Z}) & \text{if } \rho = 0 \end{cases} \tag{76}$$

and $\boldsymbol{\lambda}_{ck}(\mathbf{u}_i)'\mathbf{Z}$ is defined as in equation (25) except that $\mathbf{c}(B)$ is replaced by $\mathbf{c}(\mathbf{u}_i) \equiv (C(\mathbf{u}_i, \mathbf{s}_1), \ldots, C(\mathbf{u}_i, \mathbf{s}_n))'$; $i = 1, \ldots, |B|$. The solution for $\rho = 0$ and an explanation of the terms in equation (76) are given in Appendix 1.

3. *Best predictor* (BP). From Bayesian decision theory (e.g., Cressie, 1993a, p. 107), the optimal predictor of $F_B(z)$ is $E[F_B(z)|\mathbf{Z}]$. Since we are dealing with Gaussian processes, it is easy to show that

$$E[F_B(z)|\mathbf{Z}] = \frac{1}{|B|}\sum_{i=1}^{|B|} \Phi\left(\frac{z - \hat{S}_{sk}(\mathbf{u}_i)}{\sqrt{\sigma^2 - \mathbf{c}(\mathbf{u}_i)'\Sigma^{-1}\mathbf{c}(\mathbf{u}_i)}}\right); \quad z \in \mathbb{R},\ B = G \text{ or } L \tag{77}$$

where $\hat{S}_{sk}(\mathbf{u}_i)$ is the simple kriging predictor in equation (15) of $S(\mathbf{u}_i)$ and $\Phi(\cdot)$ is the standard normal CDF. Thus we define the best ("plug-in") predictor as

$$\hat{F}_{B;bp}(z) = \frac{1}{|B|}\sum_{i=1}^{|B|} \Phi\left(\frac{z - \hat{S}_{ok}(\mathbf{u}_i)}{\sqrt{\sigma^2 - \mathbf{c}(\mathbf{u}_i)'\Sigma^{-1}\mathbf{c}(\mathbf{u}_i)}}\right); \quad z \in \mathbb{R},\ B = G \text{ or } L \tag{78}$$

where $\hat{S}_{ok}$ is the ordinary kriging predictor.

4. *Horvitz–Thompson* (HT). Define

$$\hat{F}_{G;ht}(z) = \frac{1}{n}\sum_{i=1}^{n} I(Z(\mathbf{s}_i) \le z); \qquad z \in \mathbb{R} \tag{79}$$

where $n = 20$ is the sample size, and

$$\hat{F}_{L;ht}(z) = \begin{cases} \frac{1}{n_L}\sum_{s\in A_L} I(Z(\mathbf{s}) \le z); \quad z \in \mathbb{R}, & \text{if } n_L > 0 \\ \hat{F}_{G;ht}(z); \quad z \in \mathbb{R} & \text{if } n_L = 0 \end{cases} \tag{80}$$

where $A_L = A \cap L$ and $n_L = |A_L|$.
Note that $\hat{F}_{G;ht}(z)$ is an unbiased estimator of $\sum_{\mathbf{u}\in G} I(Z(\mathbf{u}) \le z)/\sum_{\mathbf{u}\in G} 1;\ z \in \mathbb{R}$, although this does not account for the measurement error associated with the study (see Section 2.2.5).

5. *Simplified model* (SM). Following Fuller (1995) (see Section 2.2.5), we define

$$\hat{F}_{G;sm}(z) \equiv \Phi\left(\frac{z-\hat{\mu}_G}{\hat{\sigma}_G}\right); \qquad z \in \mathbb{R}, \tag{81}$$

where $\hat{\mu}_G \equiv \bar{Z}$ and $\hat{\sigma}_G^2 \equiv \frac{1}{n-1}\sum_{i=1}^{n}(Z(\mathbf{s}_i) - \bar{Z})^2 - \tau^2$. Also, define

$$\hat{F}_{L;sm}(z) \equiv \begin{cases} \Phi\left(\frac{z-\hat{\mu}_L}{\hat{\sigma}_L}\right); & z \in \mathbb{R}, \quad 1 \text{ if } n_L > 1 \\ \hat{F}_{G;sm}(z); & z \in \mathbb{R}, \quad 1 \text{ if } n_L \le 1 \end{cases} \tag{82}$$

where $\hat{\mu}_L \equiv \bar{Z}_L$, and $\hat{\sigma}_L^2 \equiv \frac{1}{n_L-1}\sum_{u\in A_L}(Z(\mathbf{u}) - \bar{Z}_L)^2 - \tau^2$, if $n_L > 1$. In cases where $\hat{\sigma}_B^2$ was negative, a small positive number (viz., $\hat{\sigma}_B^2 = 0.00001$) was substituted for $B = G$ or L. This predictor is "simplified" in the sense that the assumed model ξ only requires that the marginal distribution of each $S(\mathbf{s})$ be $N(\mu_B, \sigma_B^2)$, $B = G$ or L; no attempt is made to model the spatial dependence structure.

6. *Simulation extrapolation deconvolution* (DC). Following Stefanski and Bay (1996) (see Section 2.2.5), we define

$$\hat{F}_{G;dc}(z) \equiv \frac{1}{n}\sum_{i=1}^{n} \mathbf{a}'(D'D)^{-1}D'\mathbf{v}_i; \qquad y \in \mathbb{R} \tag{83}$$

where $\mathbf{a}' = (1, -1, 1)$, $D = (\mathbf{1}, \boldsymbol{\eta}, \boldsymbol{\eta}^2)$, $\boldsymbol{\eta} = (\eta_1, \ldots, \eta_m)'$, $\boldsymbol{\eta}^2 = (\eta_1^2, \ldots, \eta_m^2)$, and $\mathbf{v}_i = (\Phi(\frac{z-Z(\mathbf{s}_i)}{\tau\sqrt{\eta_1}}), \ldots, \Phi(\frac{z-Z(\mathbf{s}_i)}{\tau\sqrt{\eta_m}}))'$; $i = 1, \ldots, n$, which is equivalent to equation (55). Following the authors' recommendations, we set $\boldsymbol{\eta} = (0.05, 0.2, 0.4, \ldots, 2)'$ with $m = 11$.
For the local region L, define

$$\hat{F}_{L;dc}(z) \equiv \frac{1}{n_L}\sum_{i=1}^{n} \mathbf{a}'(D'D)^{-1}D'\mathbf{v}_i I(\mathbf{s}_i \in L); \qquad z \in \mathbb{R} \tag{84}$$

provided $n_L \ge 1$. Otherwise $\hat{F}_{L;dc} = \hat{F}_{G;dc}$.

Note that the first three predictors, ordinary kriging, constrained kriging, and the best predictor are "spatial" predictors in the sense that the invoked model relates directly to the spatial process. The three remaining predictors ignore the spatial stucture and can be called "nonspatial" predictors. Observe also that for the three spatial predictors and the SM predictor, the filtering out of measurement error is a straightforward procedure (see Sections 2.1.6 and 2.2.5). In addition, within the geostatistical methodology there exist techniques for estimating the variance of the error process in cases where it is not possible to replicate observations of the study variable at a site (see the discussion about the nugget effect in equation (37)). On the other hand, filtering out measurement error adequately for SCDF prediction purposes is a nontrivial problem if no model is invoked for the state variable S.

The best predictor, ordinary kriging, and constrained kriging all require that the spatial covariance parameters be known. The best predictor also requires that $S(\cdot)$ and $\mathbf{Z}$ be jointly normal. For the simplified model predictor, it is assumed that each $S(s)$ is $N(\mu, \sigma^2)$, that each $\epsilon(s)$ is $NI(0, \tau^2)$, that $S(\cdot)$ and $\epsilon(\cdot)$ are independent, and that the parameters of these distributions are known. The only modeling assumption of the deconvolution predictor is for the error process to be a zero-mean, white-noise Gaussian process whose parameter is known. The Horvitz–Thompson estimator has no model assumptions but it does assume that the first-order inclusion probabilities can be calculated; for the probability sampling designs we considered, they are all equal.

3.3 Responses of the Computer Simulation Experiment

The "responses" of this computer simulation experiment are performance criteria of the predictors of the spatial mean and the SCDF. The *design-based* prediction MSE was taken to be the primary criterion, and the design-based prediction bias constituted a secondary criterion. Comparison is in terms of the MSE, subject to the bias not being too large, noting Hájek's (1971, p. 236) dictum that greatly biased estimators are poor no matter what other properties they have.

Why did we not choose model-based criteria? The spatial models specified in Section 3.1 were used to ascertain how different analyses, in particular "spatial" vs. "nonspatial" analyses, perform under different sampling designs, under different conditions. To use this same spatial model to obtain model-based performance criteria as well could be perceived as unfairly favoring the "spatial" analyses, so we chose not to do it.

In many cases, expressions for the design-based criteria were unavailable. Consequently, a computer-simulation experiment was conducted so that for all combinations of the experimental factors, 400 realizations of each of the sampling designs were generated, and spatial means and SCDFs were predicted from the sampled values in each case. When estimating proportions, such as for the SCDF, 400 realizations guarantees accuracy to the first decimal place.

3.3.1 Spatial-mean responses

Suppose $\hat{S}(B; \mathbf{Z})$ is a predictor of $S(B)$, and define $\mathbf{S} \equiv (S(\mathbf{u}_1), \ldots, S(\mathbf{u}_{|B|}))'$; $B = G$ or L. Then

$$Bias_{p\epsilon}(\hat{S}(B; \mathbf{Z})) \equiv E_{p\epsilon}[\hat{S}(B; \mathbf{Z}) - S(B)|\mathbf{S}] \tag{85}$$

$$\begin{aligned} MSE_{p\epsilon}(\hat{S}(B; \mathbf{Z})) &\equiv E_{p\epsilon}[(\hat{S}(B; \mathbf{Z}) - S(B))^2|\mathbf{S}] \\ &= \mathrm{var}_{p\epsilon}(\hat{S}(B; \mathbf{Z})|\mathbf{S}) + [Bias_{p\epsilon}(\hat{S}(B; \mathbf{Z}))]^2 \end{aligned} \tag{86}$$

where $E_{p\epsilon}(\cdot|\mathbf{S})$ is the design expectation for the measurement-error model in equation (32), (conditional on $\mathbf{S}$).

The estimators of equations (85) and (86) are

$$\widehat{Bias}_{p\epsilon}(\hat{S}(B; \mathbf{Z})) = \frac{1}{400} \sum_{i=1}^{400} [\hat{S}(B; \mathbf{Z}^{(i)}) - S(B)] \tag{87}$$

$$\widehat{MSE}_{p\epsilon}(\hat{S}(B; \mathbf{Z})) = \frac{1}{400} \sum_{i=1}^{400} [\hat{S}(B; \mathbf{Z}^{(i)}) - S(B)]^2 \tag{88}$$

where $\mathbf{Z}^{(i)}$ is the ith random sample; $i = 1, \ldots, 400$. Note that $\mathbf{Z}^{(i)}$ will be different from $\mathbf{Z}^{(i')}$ because of both the randomness in the sampling design and the randomness in the measurement error, and that $S(B)$ remains fixed over the 400 $p\epsilon$-realizations.

3.3.2 SCDF responses

Define F_B as the SCDF of $\{S(\mathbf{s}) : \mathbf{s} \in B\}$; $B = G$ or L. Let $\hat{F}_B(q(\alpha); \mathbf{Z})$ be a predictor of $F_B(q(\alpha))$; $\alpha \in \{0.1, 0.25, 0.5, 0.75, 0.9\}$. Then

$$Bias_{p\epsilon}(\hat{F}_B(q(\alpha); \mathbf{Z})) = E_{p\epsilon}[\hat{F}_B(q(\alpha); \mathbf{Z}) - F_B(q(\alpha))|\mathbf{S}] \tag{89}$$

$$MSE_{p\epsilon}(\hat{F}_B(q(\alpha); \mathbf{Z})) = E_{p\epsilon}[(\hat{F}_B(q(\alpha); \mathbf{Z}) - F_B(q(\alpha)))^2|\mathbf{S}] \tag{90}$$

The estimators of these quantities are

$$\widehat{Bias}_{p\epsilon}(\hat{F}_B(q(\alpha); \mathbf{Z})) = \frac{1}{400} \sum_{i=1}^{400} [\hat{F}_B(q(\alpha); \mathbf{Z}^{(i)}) - F_B(q(\alpha))] \tag{91}$$

$$\widehat{MSE}_{p\epsilon}(\hat{F}_B(q(\alpha); \mathbf{Z})) = \frac{1}{400} \sum_{i=1}^{400} [\hat{F}_B(q(\alpha); \mathbf{Z}^{(i)}) - F_B(q(\alpha))]^2 \tag{92}$$

where $\mathbf{Z}^{(i)}$ is the ith random sample; $i = 1, \ldots, 400$.

3.4 Comments

The simplifications in this study should be noted. First, the variances and covariances of S and Z were completely specified, for the practical reason that variogram-parameter estimation over the 400 p-realizations would have been prohibitive. Thus, a source of variability, which would occur in practice, has been removed from the spatial predictors $\hat{S}_{ok}$ and $\hat{S}_{ck}$. This also has obvious sampling-design implications. First, there is no longer any need to follow Laslett's (1994) "geostatistical credo" of supplementing a basic grid design with extra clustered points in order to estimate the variogram accurately at short lags. Second, knowledge of the components of $Z(\cdot)$, viz., $S(\cdot)$ and $\epsilon(\cdot)$, easily allowed the kriging equations to be modified so that the measurement-error process $\epsilon(\cdot)$ could be filtered out. In practice, $S(\cdot)$ and $\epsilon(\cdot)$ are usually unobservable individually, and τ^2 is estimated either from replicated observations (preferably) or from the nugget effect of the variogram if some assumptions are made (see Section 2.1.7). The specified error variance τ^2 was also used in the simplified-model and deconvolution SCDF predictors.

Consider the nonparametric estimator in equation (35). In our experiment $E[2\hat{\gamma}(\mathbf{u} - \mathbf{v})] = 2E[(Z(\mathbf{u}) - Z(\mathbf{v}))^2] = 2\gamma\ (\mathbf{u} - \mathbf{v}) + 2\beta^2(x_u - x_v)^2$; $\mathbf{u} = (x_u, y_u)' \in D_f$, $\mathbf{v} = (x_v, y_v)' \in D_f$, $\beta \in \{0, 1/(8.25)^{-1/2}\}$. Thus, for nonzero β, $2\hat{\gamma}(\mathbf{u} - \mathbf{v})$ has a "leakage" term, $2\beta^2(x_u - x_v)^2$, which will inflate the estimated variogram quadratically as distances increase in the east–west direction. In our simulation experiment, the leakage term is included in the covariance function when $\beta \neq 0$ (see Section 2.1.5). That is, we formally used the "variogram"

$$2\gamma^*(\mathbf{u} - \mathbf{v}) \equiv 2(1 - \rho^{\|\mathbf{u}-\mathbf{v}\|}) + 2\beta^2(x_u - x_v)^2; \quad \mathbf{u} = (x_u, y_u)', \mathbf{v} = (x_v, y_v)' \in D_f$$

from which we obtained the "covariance function"

$$C^*(\mathbf{u} - \mathbf{v}) \equiv 1 - \gamma^*(\mathbf{u} - \mathbf{v}); \qquad \mathbf{u}, \mathbf{v} \in D_f$$

used for spatial prediction.

The averages of the spatial-mean and the SCDF $p\epsilon$-MSEs (see Section 3.3) over the three S-realizations (see Section 3.2.5) serve as a crude approximation to equation (59), and the three values themselves give some indication of how the $p\epsilon$-MSE *varies* over S-realizations.

4. RESULTS OF THE EXPERIMENT

Not unexpectedly, the design CLU performed extremely poorly throughout the experiment. We did not expect CLU to perform well in this spatial context, but included it because such a design may be used for ecological studies in which a regional process is sampled repeatedly at, or very close to, one (or a few) prespecified spatial location, in the belief that this location is "representative." This belief can only be supported under the following rather restrictive assumptions on the parameters: β is near zero and ρ is near one. This might occur if the phenomenon being studied "mixes" well (such as the composition of the atmosphere, after several years), but most ecological processes (e.g., timber on forested lands) do not mix well, even after decades. With regard to the study of the nation's ecological resources, the message from this simulation experiment is clear: no matter how many measurements are taken from so-called "representative sites," their skill in predicting national and regional ecological resources is extremely low. This is best illustrated by the plots in Figure 2, where the spatial mean of the global region is predicted using constrained kriging for varying values of ρ in the low noise and zero trend case. Shown on the plots are mean-squared prediction error (MSE) and absolute bias for the four designs described in Section 3.2.6, including CLU. The performance of CLU was so poor that one must henceforth doubt the ability of representative sites to say anything about the *regional* behavior of an ecological phenomenon.

In order to present the results of the other three designs on a comparable scale, it was decided to exclude CLU from the rest of this discussion. Not surprisingly, with 20 out of a possible 100 values sampled, spatial-mean prediction over the global region was uniformly good, irrespective of predictor or design, when the response was averaged over all other factors. On the other hand, the results for SCDF prediction over the local region were inconclusive, again not surprisingly given the small size of the region. Therefore, it was decided to include in this discussion only the results of spatial-mean prediction over the local region, and SCDF prediction over the global region.

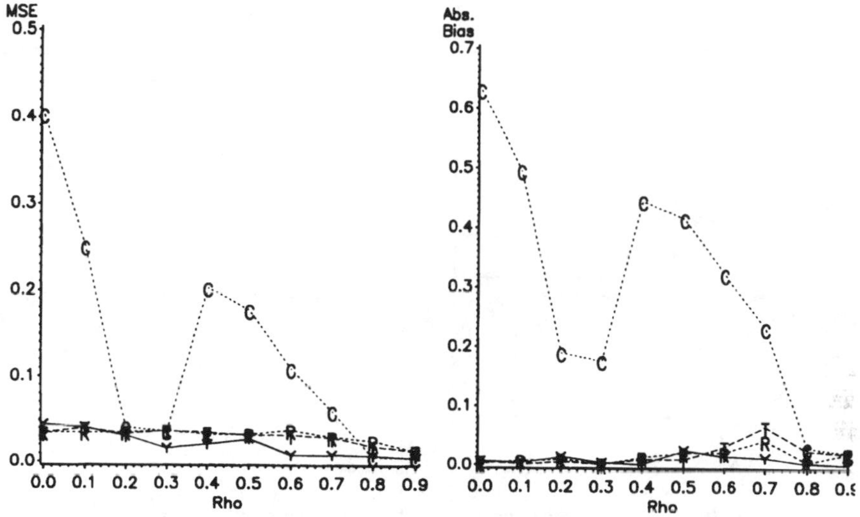

Figure 2. Spatial-mean prediction. Mean-squared error (MSE) and absolute bias (Abs. Bias) versus spatial-correlation parameter (RHO). Plots are for constrained kriging (CK), no trend, and low noise over the global region. Sampling designs: Y, systematic random; T, stratified random; R, simple random; C, clustered

4.1 Spatial-Mean Prediction over the Local Region

The most important features of spatial-mean prediction over the local region L can be summarized from the results presented in Figure 3 and Table 1. Figure 3 shows the MSE and absolute bias of the predictors and designs in a series of plots corresponding to the levels of the trend and noise factors. All plots are conditioned on the "region" factor being $B = L$ and averaged over those factors, which are not shown. Table 1 displays an analysis of variance (ANOVA) of the MSEs of the local (i.e., $B = L$), spatial-mean prediction part of the experiment. The ANOVA here serves merely as an arithmetic partition of selected sums of squares and corresponding variance ratios; no distributional assumptions nor strict statistical inferences are made. The names of the factors in the ANOVA are self-explanatory (e.g., REALN refers to S-realization). Factors with relatively large variance ratios (i.e., indicating large effects) are highlighted in bold script. The variance ratio, marked VR in the table, is the ratio between the "treatment" mean-squared error (marked MS in the table) and the residual mean-squared error.

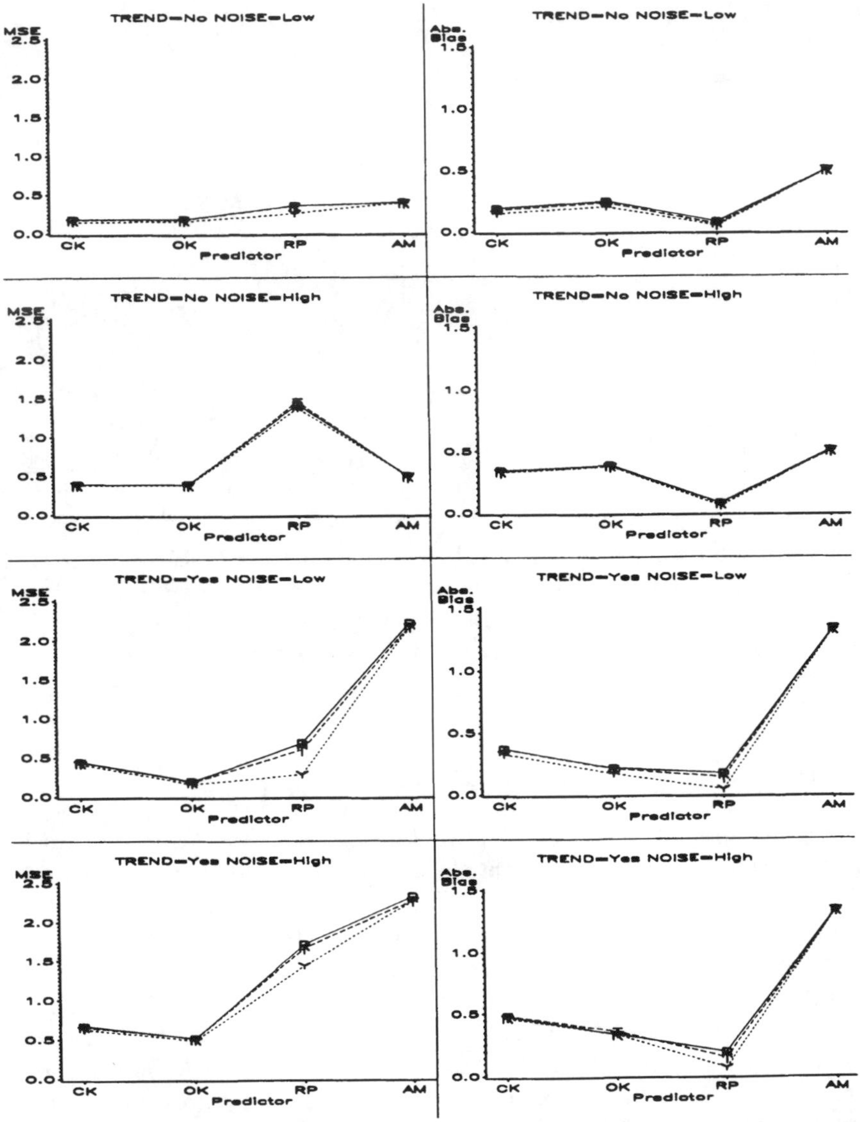

Figure 3. Spatial-mean prediction. Mean-squared error (MSE) and absolute bias (Abs. Bias) plots for all levels of trend and noise over the local region. On the horizontal axis are the four spatial-mean predictors, constrained kriging (CK), ordinary kriging (OK), regional poststratification (RP), and arithmetic mean (AM). Sampling designs: Y, systematic random; T, stratified random; and R, simple random

Table 1. Spatial-mean prediction. Analysis of variance (ANOVA) of mean-squared error (MSE) over the local region.

Source	DF	SS	MS	VR
TREND	**1**	**121.379031**	**121.379031**	**2021.03**
REALN	**2**	**23.749024**	**11.874512**	**197.72**
RHO	9	9.474361	1.052707	17.53
NOISE	**1**	**63.136746**	**63.136746**	**1051.26**
DESIGN	2	1.456829	0.728415	12.13
PRED	**3**	**251.920640**	**83.973547**	**1398.21**
TREND*REALN	**2**	**33.274012**	**16.637006**	**277.02**
TREND*RHO	9	8.080051	0.897783	14.95
TREND*NOISE	1	0.063514	0.063514	1.06
TREND*DESIGN	2	0.446606	0.223303	3.72
TREND*PRED	**3**	**178.165490**	**59.388497**	**988.85**
REALN*RHO	18	28.758647	1.597703	26.60
REALN*NOISE	2	0.108080	0.054040	0.90
REALN*DESIGN	4	0.190963	0.047741	0.79
REALN*PRED	**6**	**74.958407**	**12.493068**	**208.02**
RHO*NOISE	9	1.415973	0.157330	2.62
RHO*DESIGN	18	0.567555	0.031531	0.53
RHO*PRED	27	42.102472	1.559351	25.96
NOISE*DESIGN	2	0.042704	0.021352	0.36
NOISE*PRED	**3**	**55.602956**	**18.534319**	**308.61**
DESIGN*PRED	6	1.793600	0.298933	4.98
TREND*REALN*PRED	**6**	**88.174307**	**14.695718**	**244.69**
TREND*RHO*PRED	27	31.142113	1.153412	19.20
TREND*NOISE*PRED	3	0.187600	0.062533	1.04
REALN*RHO*PRED	54	44.180658	0.818160	13.62
REALN*NOISE*PRED	6	0.097971	0.016329	0.27
Residual	1213	72.85037	0.06006	
Total	1439	1133.32068		

Relatively large VR values are highlighted in bold script. DF, degrees of freedom; SS, sums of squares; MS, mean squares; VR, variance ratio, i.e., MS/(residual MS).

The most striking feature in this part of the experiment is the very large predictor effect and the almost negligible design effect (see Figure 3 and Table 1). We should warn that the small design effect may in part be due to the small size of the global region and the relatively large sample size (i.e., within the constraints of this experiment, the spatial configurations of the three nonclustered designs considered here may not differ too markedly). Nevertheless, in what follows we see that a spatial analysis is always preferred, regardless of the design. It is also very clear that trend and noise are highly significant factors in this experiment.

Consider the stationary (i.e., no trend) case. Figure 3 indicates that OK and CK perform best with respect to the MSE criterion, and their bias properties seem to be reasonably good at both noise levels. The AM also performs reasonably well with respect to both criteria, at both noise levels. On the other hand, the RP MSE explodes if the noise level is high.

In the presence of trend, AM performs terribly with respect to MSE and bias. The relative precision of OK, CK, and RP do not change much, except that in the presence of trend, OK outperforms CK, but not by much. Note the slight design effect for RP; here, SYS outperforms the other two designs.

The S-realization effect is not displayed here, but for OK, CK, and RP it is negligible. However, for AM this effect is large in the presence of trend, accounting for the fairly large variance ratios of the factor REALN and the interactions REALN*PRED and TREND*REALN*PRED in Table 1.

The ρ-effect is surprisingly small (see Table 1), and a referee has suggested that because of the geometric decay of the correlation function with distance, values of ρ exceeding 0.9 would also be interesting to look at. Trend and noise appear to be far more important than strength of spatial correlation in the local spatial-mean prediction part of this study.

In conclusion, OK, the spatial BLUP, is the preferred predictor, with CK as a competitive alternative, especially for stationary processes. The arithmetic mean performs very badly in the presence of trend, and RP performs very badly if the data are noisy. This demonstrates that by using a spatial model describing both large-scale spatial structure (if it exists) and small-scale spatial structure, observations from *outside* the local region can be used effectively for local spatial-mean prediction.

4.2 SCDF Prediction over the Global Region

The reader is directed to Figures 4–6, and Table 2 for the following discussion of the results of SCDF prediction over the global region G. Figure 4 displays, on the left, MSE plots of the six predictors for the five α-levels (corresponding to the five quantile predictands) over both noise levels, for $\beta = 0$. On the right, the plots display the bias indirectly by showing how the predictors (joined by lines) track their respective predictands (represented by stars). For example, take OK in the top right-hand plot of Figure 4: nearly 80% of the probability of the predicted distribution lies within the inter-

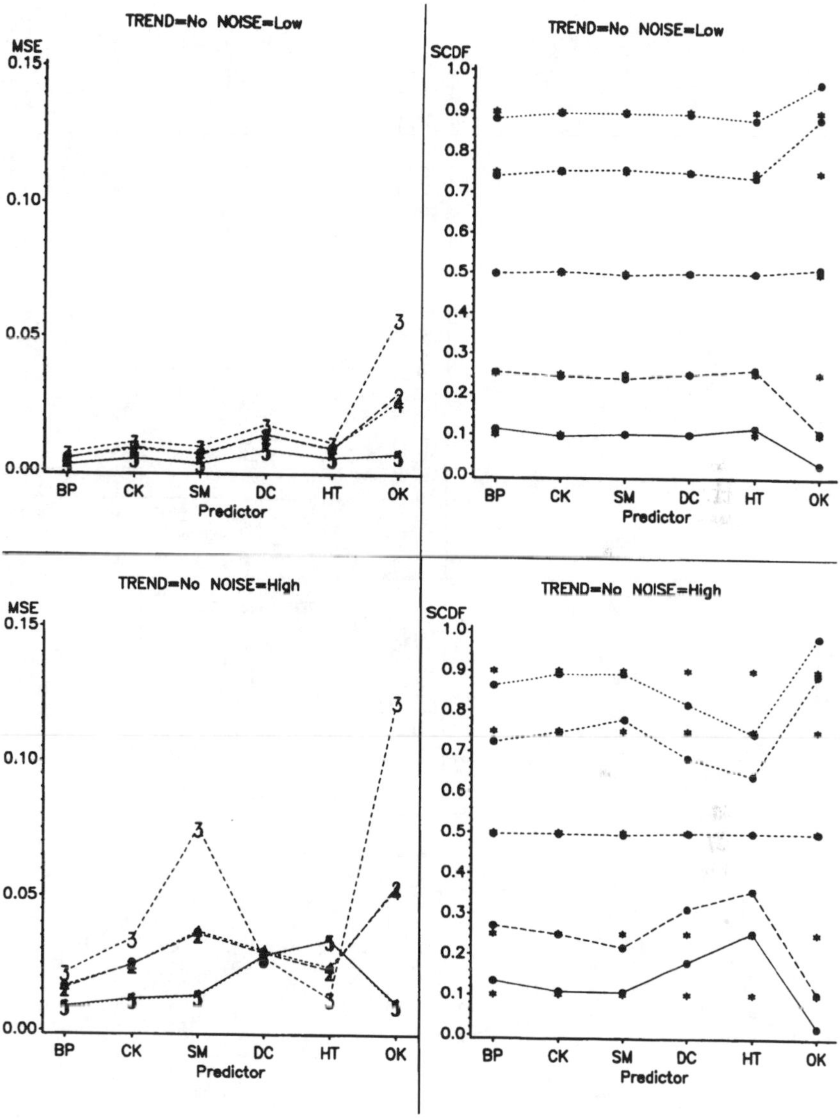

Figure 4. Spatial cumulative distribution function (SCDF) prediction. Mean-squared error (MSE) and SCDF-prediction plots for both levels of noise and no trend over the global region. On the horizontal axis are the six SCDF predictors, best predictor (BP), constrained kriging (CK), simplified model (SM), deconvolution (DC), Horvitz–Thompson (HT), and ordinary kriging (OK). The numbers 1, ..., 5 in the MSE plots on the left, represent $\alpha \in \{0.1, 0.25, 0.5, 0.75, 0.9\}$, respectively (e.g., 3 represents median-prediction). In the plots on the right * denotes $F_G(q(\alpha))$; $\alpha \in \{0.1, 0.25, 0.5, 0.75, 0.9\}$, and the dots represent the corresponding predictors

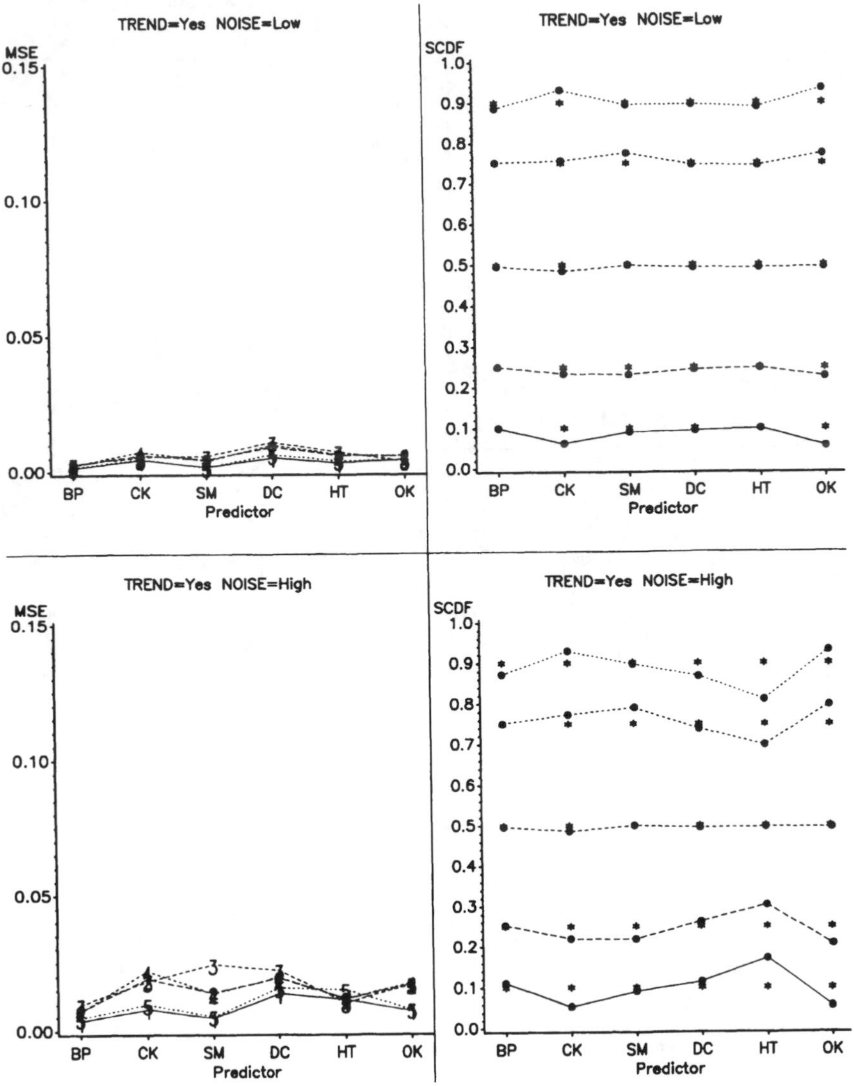

Figure 5. Spatial cumulative distribution function (SCDF) prediction. Mean-squared error (MSE) and SCDF-prediction plots for both levels of noise and with trend over the global region. On the horizontal axis are the six SCDF predictors, best predictor (BP), constrained kriging (CK), simplified model (SM), deconvolution (DC), Horvitz–Thompson (HT), and ordinary kriging (OK). The numbers 1, ..., 5 in the MSE plots on the left, represent $\alpha \in \{0.1, 0.25, 0.5, 0.75, 0.9\}$, respectively (e.g., 3 represents median-prediction). In the plots on the right, $*$ denotes $F_G(q(\alpha))$; $\alpha \in \{0.1, 0.25, 0.5, 0.75, 0.9\}$, and the dots represent the corresponding predictors

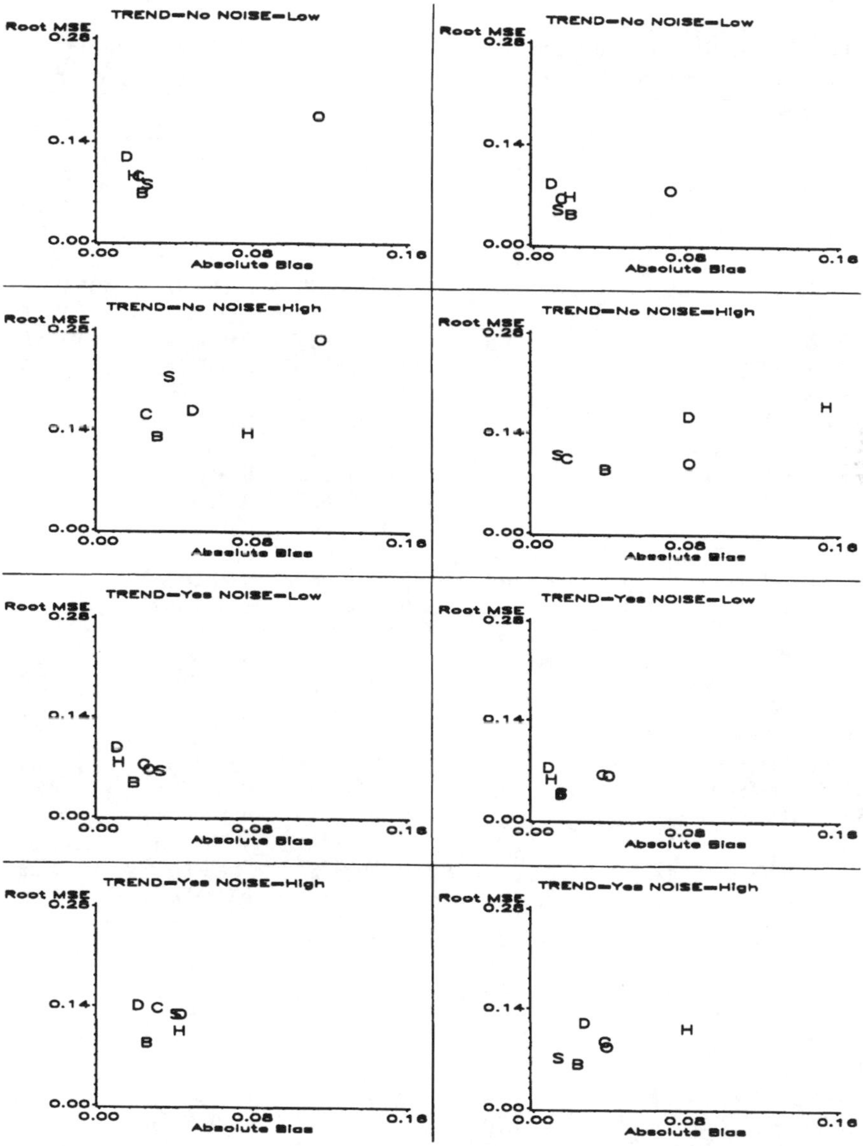

Figure 6. Spatial cumulative distribution function (SCDF) prediction. Plots of root-mean-squared error (root MSE) versus absolute bias for all levels of trend and noise over the global region. On the left, root MSE and absolute bias are averaged over the three inner quantiles; on the right, root MSE and absolute bias are averaged over the two outer quantiles. SCDF predictors: B, best predictor; C, constrained kriging; S, simplified model; D, deconvolution; H, Horvitz–Thompson; and O, ordinary kriging

Table 2. Spatial cumulative distribution function (SCDF) prediction. Analysis of variance (ANOVA) of mean-squared error (MSE) over the global region.

Source	DF	SS	MS	VR
TREND	**1**	**0.06014255**	**0.06014255**	**10756.48**
REALN	2	0.00181021	0.00090510	161.88
RHO	9	0.00334629	0.00037181	66.50
NOISE	**1**	**0.09870255**	**0.09870255**	**17652.92**
DESIGN	2	0.00104630	0.00052315	93.56
PRED	**5**	**0.05353429**	**0.01070686**	**1914.92**
TREND*REALN	2	0.00065443	0.00032722	58.52
TREND*RHO	9	0.00242916	0.00026991	48.27
TREND*NOISE	**1**	**0.01533655**	**0.01533655**	**2742.94**
TREND*DESIGN	2	0.00009405	0.00004702	8.41
TREND*PRED	**5**	**0.03771380**	**0.00754276**	**1349.02**
REALN*RHO	18	0.00186621	0.00010368	18.54
REALN*NOISE	2	0.00150047	0.00075024	134.18
REALN*DESIGN	4	0.00050638	0.00012659	22.64
REALN*PRED	10	0.00035548	0.00003555	6.36
RHO*NOISE	9	0.00441639	0.00049071	87.76
RHO*DESIGN	18	0.00067312	0.00003740	6.69
RHO*PRED	45	0.01821994	0.00040489	72.41
NOISE*DESIGN	2	0.00000527	0.00000263	0.47
NOISE*PRED	5	0.00765350	0.00153070	273.77
DESIGN*PRED	10	0.00018881	0.00001888	3.38
TREND*REALN*PRED	10	0.00019198	0.00001920	3.43
TREND*RHO*PRED	45	0.01246132	0.00027692	49.53
TREND*NOISE*PRED	5	0.00473842	0.00094768	169.49
REALN*RHO*PRED	90	0.00019311	0.00000215	0.38
REALN*NOISE*PRED	10	0.00039291	0.00003929	7.03
Residual (a)	1837	0.01027120	0.00000559	
Whole-plot Total	2159	0.33844468		
ALPHA	**4**	**0.29125668**	**0.07281417**	**1436.54**
TREND*ALPHA	**5**	**0.40984807**	**0.08196961**	**1617.12**
REALN*ALPHA	10	0.00960998	0.00096100	18.96
RHO*ALPHA	45	0.05281559	0.00117368	23.15
NOISE*ALPHA	**5**	**0.54215456**	**0.10843091**	**2139.15**
ALPHA*DESIGN	10	0.00675377	0.00067538	13.32
ALPHA*PRED	25	0.60924425	0.02436977	480.77
TREND*ALPHA*PRED	25	0.48131501	0.01925260	379.82
REALN*ALPHA*PRED	50	0.00589376	0.00011788	2.33
NOISE*ALPHA*PRED	25	0.17655751	0.00706230	139.33
Residual (b)	8436	0.42759653	0.00005069	
Total	10799	3.35149040		

Relatively large VR values are highlighted in bold script. DF, degrees of freedom; SS, sums of squares; MS, mean squares; VR, variance ratio, i.e., MS/(residual MS).

quartile range of the "true" distribution. This means that the distribution predicted by OK is too "peaked," that is, OK yields a surface that is too "smooth" for effective SCDF prediction. All results displayed in Figure 4 are for the global (i.e., $B = G$) SCDF part of the experiment, averaged over those factors not shown. Figure 5 displays plots similar to those of Figure 4 except that here the trend component is present.

In Figure 6, the square root of the MSE is plotted against absolute bias for each predictor, again in a series of plots corresponding to the levels of trend and noise. Those on the left show results averaged over values corresponding to the three inner-quantile predictions, and those on the right show results averaged over values corresponding to the two outer-quantile predictions.

Table 2 is an ANOVA of the prediction MSEs of SCDF predictors over the global region. Its construction is similar to that of Table 1 except for the extra subplot factor ALPHA, which represents the five quantile predictands under consideration.

Table 2 and Figures 4 and 5 demonstrate the large predictor effect, the small design effect, and the very large effects of the trend and noise factors, echoing the results for the local spatial-mean prediction part of this experiment. In addition, there appears to be a sizable α-effect.

Figures 4–6 show the effect of trend on SCDF prediction. In the case of positive trend, all predictors perform fairly comparably. This is not surprising, because if a "mountain" (i.e., a trend) in the data dominates, it will be picked up, irrespective of prediction method! It is in the stationary case, particularly when small-scale variation is dominated by noise, that the merits of the different predictors will likely be demonstrated most clearly.

Thus, considering only the stationary case, a large prediction effect is immediately discernable and, although not displayed in the figures, the design effect is negligible (see Table 2). By invoking Hájek's (1971) dictum that badly biased predictors are unacceptable, it seems that OK fails the bias test (see Figure 4). When the noise level is high, the HT predictor, and possibly the DC predictor, also fail the bias test (see Figure 4). Among the others, the BP appears to perform best with respect to the MSE. The CK predictor performs nearly as well and has excellent bias properties. The SM predictor does not perform well when predicting the middle portion of the distribution when the noise level is high. However, it does predict the tails of the distribution well, irrespective of noise level, accounting in large part for the large variance ratio of the factor ALPHA and the interaction ALPHA*NOISE in Table 2.

All predictors behaved consistently over S-realizations, and exhibited little ρ-effect. Therefore, as in the case of local spatial-mean prediction, trend and noise appear to be more important factors than the strength of spatial correlation in the SCDF-prediction (over G) part of this study (while noting the comment about looking at larger ρ-values given in Section 4.1).

Clearly, the preferred predictor in all cases is BP (see Figures 4–6), although it demands the strongest model assumptions (see the discussion on model assumptions, Section 3.2.7), and it may be sensitive to departures from those assumptions. Constrained kriging may be a good alternative if those assumptions cannot be verified as holding at least approximately. Constrained kriging does not require Gaussianity to perform well (Cressie, 1993b); however, it does require that all spatial covariance parameters be known or well estimated. In the low-noise case only, the other predictors, except OK, could be considered acceptable, depending on the number of modeling assumptions one is willing to make.

5 CONCLUSIONS

Several conclusions can be drawn from this study, but before we do that, we wish to emphasize that these conclusions pertain to the characteristics of the spatial phenomenon described in Section 3.1. In particular, this spatial phenomenon does not contain values of interest that are either clustered or rare. The conclusions are listed below.

1. Clustered designs, which correspond to so-called "representative-site" selection, should be avoided.
2. The choice of sampling design from among SYS, STS, or SRS designs appears to be unimportant for both spatial-mean and SCDF prediction.
3. For spatial-mean prediction over the local region, the spatial BLUP (i.e., ordinary kriging) is the preferred predictor, although constrained kriging performs competitively, especially for stationary processes. Both predictors require that the spatial covariance parameters be known or well estimated. The regional poststratification predictor should be avoided if the measurement error is large, and the arithmetic mean should be avoided in the presence of a trend component.
4. For SCDF prediction over the global region, the so-called "best predictor" performs best, but requires the strongest model assumptions. Constrained kriging performs well and requires fewer model assumptions. The simplified model, deconvolution, and Horvitz–Thompson predictors perform well only if the measurement-error component is small. Ordinary kriging should be avoided.
5. The effects of different factors/levels on SCDF prediction are only discernible for larger sample sizes, in comparison with those for

spatial-mean prediction. In those cases, Conclusions 3 and 4 tell us that constrained kriging is a superior predictor.

6. The conclusions stated above were generally consistent across the three *S*-realizations generated, with minor exceptions as noted in Section 4.

ACKNOWLEDGMENTS

The research presented in this article was supported by the US Environmental Protection Agency under cooperative agreement No. CR822919-01-0. The article has not been subjected to the review of the EPA and thus does not reflect the view of the agency and no official endorsement should be inferred. The authors would like to express their appreciation to Scott Overton and an anonymous referee for their helpful comments.

REFERENCES

Bogardi I, Bardossy A, Duckstein L. Multicriterion network design using geostatistics. *Water Resour. Res.*, 21:199–208, 1985.

Bras RL, Rodriguez-Iturbe I. Network design for the estimation of areal mean of rainfall events. *Water Resour. Res.*, 12:1185–1195, 1976.

Breidt FJ. Markov chain designs for one-per-stratum sampling. *Survey Methodol.*, 21:63–70, 1995a.

Breidt FJ. Markov chain designs for one-per-stratum spatial sampling. *Proceedings of the Section on Survey Research Methods*, Vol. 1, ASA, Orlando, FL, pp 356–361, 1995b.

Caselton WF, Zidek JV. Optimal monitoring network designs. *Stat. Probab. Lett.*, 2:223–227, 1984.

Cochran WG. *Sampling Techniques*, 3rd edn. New York: Wiley, 1977.

Cook JR, Stefanski LA. Simulation–extrapolation estimation in parametric measurement error models. *J. Am. Stat. Assoc.*, 89:1314–1328, 1994.

Cordy CB. An extension of the Horvitz–Thompson theorem to point sampling from a continuous universe. *Stat. Probab. Lett.*, 18:353–362, 1993.

Cox DD, Cox LH, Ensor KB. Spatial sampling and the environment: some issues and directions. *Environ. Ecol. Stat.*, 4: 219–233, 1997.

Cressie N. *Statistics for Spatial Data*, rev. ed. New York: Wiley, 1993a.

Cressie N. Aggregation in geostatistical problems. In: A Soares, ed. *Geostatistics Tróia '92*, Vol. 1. Dordrecht: Reidel, pp. 25–36, 1993b.

Cressie N, Ver Hoef JM. Spatial statistical analysis of environmental and ecological data. In: Goodchild MF, Parks BO, Steyaert LT, eds. *Environmental Modeling with GIS*. New York: Oxford University Press, pp. 404–413, 1993.

Cressie N, Gotway CA, Grondona MO. Spatial prediction from networks. *Chemometrics Intell. Lab. Syst.*, 7:251–271, 1990.

Currin C, Mitchell T, Morris M, Ylvisaker D. Bayesian prediction of deterministic functions with applications to the design and analysis of computer experiments. *J. Am. Stat. Assoc.*, 86:953–963, 1991.

Dalenius T, Hájek J, Zubrzycki S. On plane sampling and related geometrical problems. *Proceedings of the 4th Berkeley Symposium on Mathematical Statistics and Probability*, Vol. 1. Berkeley, CA: University of California Press, pp. 125–150, 1960.

de Gruijter JJ, ter Braak CJF. Model-free estimation from spatial samples: A reappraisal of classical sampling theory. *Math. Geol.*, 22:407–415, 1990.

Englund EJ, Heravi N. Conditional simulation: Practical application for sampling design optimization. In: Soares A, ed. *Geostatistics Tróia '92*, Vol. 2. Dordrecht: Reidel, pp. 612–624, 1993.

Fisher RA. *Statistical Methods and Scientific Inference*. Edinburgh: Oliver & Boyd, 1956.

Fuller WA. Estimation in the presence of measurement error. *Int. Stat. Rev.*, 6:121–147, 1995.

Goldberger AS. Best linear unbiased prediction in the generalized linear regression model. *J. Am. Stat. Assoc.*, 57:369–375, 1962.

Graybill FA. *Theory and Application of the Linear Model*. Pacific Grove, CA: Wadsworth & Brooks/Cole, 1976.

Guttorp P, Le ND, Sampson PD, Zidek JV. Using entropy in the redesign of an environmental monitoring network. In: Patil GP, Rao CR, eds. *Multivariate Environmental Statistics*. Amsterdam: North Holland, pp. 347–386, 1993.

Haas TC. Redesigning continental-scale monitoring networks. *Atmos. Environ.*, 26A:3323–3333, 1992.

Hájek J. Comment on paper by D Basu. In: Godambe VP, Sprott DA, eds. *Foundations of Statistical Inference*. Toronto: Holt, Rinehart and Winston, 1971.

Hammersley JM, Handscomb DC. *Monte Carlo Methods*. London: Methuen, 1964.

Horvitz DG, Thompson DF. A generalization of sampling without replacement from a finite universe. *J. Am. Stat. Assoc.*, 47:663–685, 1952.

Isaki CT, Fuller WA. Survey design under the regression superpopulation model. *J. Am. Stat. Assoc.*, 77:89–96, 1982.

Laslett GM. Kriging and splines: An empirical comparison of their predictive performance in some applications. *J. Am. Stat. Assoc.*, 89:391–409, 1994.

Majure JJ, Cook D, Cressie N, Kaiser MS, Lahiri SN, Symanzik J. Spatial CDF estimation and visualization with applications to forest health monitoring. *Comput. Sci. Stat.*, 27:93–101, 1996.

McArthur RD. An evaluation of sample designs for estimating a locally concentrated pollutant. *Commun. Stat. Simulation Comput.*, 16:735–759, 1987.

Nusser SM, Carriquiry AL, Dodd KW, Fuller WA. A semiparametric transformation approach to estimating usual daily intake distributions. *J. Am. Stat. Assoc.*, 91:1440–1449, 1996.

Olea RA. Sampling design optimization for spatial functions. *Math. Geol.*, 16:369–392, 1984.

Overton WS. Probability sampling and population inference in monitoring programs. In: Goodchild MF, Parks BO, Steyaert CT, eds. *Environmental Modeling with GIS*. New York: Oxford University Press, pp. 470–480, 1993.

Overton WS, Stehman S. Properties of designs for sampling continuous spatial resources from a triangular grid. *Commun. Stat. Theory Methods*, 22:2641–2660, 1993.

Pérez-Abreu V, Rodríguez JE. Index of effectiveness of a multivariate environmental monitoring network. *Environmetrics*, 7:489–501, 1996.

Särndal C-E, Swensson B, Wretman J. *Model Assisted Survey Sampling*. New York: Springer, 1992.

Seber GAF, Thompson SK. Environmental adaptive sampling. In: Patil GP, Rao CR, eds. *Handbook of Statistics*, Vol. 12. Amsterdam: Elsevier, pp. 201–220, 1994.

Stefanski LA, Bay JM. Simulation extrapolation deconvolution of finite population cumulative distribution function estimators. *Biometrika*, 83:406–417, 1996.

Stevens DL. Implementation of a national monitoring program. *J. Environ. Manage.*, 42:1–29, 1994.

Thompson SK. *Sampling*. New York: Wiley, 1992.

Wolter KM. An investigation of some estimators of variance for systematic sampling. *J. Am. Stat. Assoc.*, 79:781–790, 1984.

Zeleny M. *Multiple Criteria Decision Making*. New York: McGraw-Hill, 1982.

APPENDIX 1

The constrained kriging predictor in equation (25) is not defined if $m_2 = 0$, and this occurs when $\rho = 0$. We use the limiting result as $\rho \to 0$ to provide a solution.

Assume that $S(\cdot)$ is a spatial process with constant mean μ and covariance function $C(\mathbf{s}, \mathbf{u})$, and that $Z(\cdot) = S(\cdot) + \epsilon(\cdot)$, where $\epsilon(\cdot)$ is a zero-mean, white-noise process with variance τ^2. Then from equation (29), we define the constrained (point) kriging predictor of $S(\mathbf{s}_0)$ as

$$\boldsymbol{\lambda}_{ck}(\mathbf{s}_0)'\mathbf{Z} = \hat{\mu}_{gls} + \left\{\frac{C(\mathbf{s}_0, \mathbf{s}_0) - \mathrm{var}(\hat{\mu}_{gls})}{\mathrm{var}(\mathbf{c}(\mathbf{s}_0)'\Sigma^{-1}(\mathbf{Z} - \hat{\mu}_{gls}\mathbf{1}))}\right\}^{1/2} \mathbf{c}(\mathbf{s}_0)'\Sigma^{-1}(\mathbf{Z} - \hat{\mu}_{gls}\mathbf{1})$$

Suppose that exactly $m < n$ sampling locations are equally closest to $\mathbf{s}_0$ and assume these to be $\{\mathbf{v}_1, \ldots, \mathbf{v}_m\}$; that is, we have $||\mathbf{v}_1 - \mathbf{s}_0|| = \ldots = ||\mathbf{v}_m - \mathbf{s}_0|| = \min_{j=1,\ldots,n} ||\mathbf{s}_j - \mathbf{s}_0||$.

Thus, for $C(\mathbf{s}, \mathbf{u})$ given by equation (29),

$$\begin{aligned}\mathbf{c}(\mathbf{s}_0) &= \sigma^2\left(\rho^{||\mathbf{v}_1 - \mathbf{s}_0||}, \ldots, \rho^{||\mathbf{v}_1 - \mathbf{s}_0||}, \rho^{||\mathbf{v}_{m+1} - \mathbf{s}_0||}, \ldots, \rho^{||\mathbf{v}_n - \mathbf{s}_0||}\right)' \\ &= \sigma^2\rho^{||\mathbf{v}_1 - \mathbf{s}_0||}\mathbf{r}'\end{aligned}$$

where $\mathbf{r}' = (1, \ldots, 1, \rho^{||\mathbf{v}_{m+1} - \mathbf{s}_0|| - ||\mathbf{v}_1 - \mathbf{s}_0||}, \ldots, \rho^{||\mathbf{v}_n - \mathbf{s}_0|| - ||\mathbf{v}_1 - \mathbf{s}_0||})$. Consequently,

$$\lim_{\rho\to 0}\boldsymbol{\lambda}_{ck}(\mathbf{s}_0)'\mathbf{Z} = \lim_{\rho\to 0}\left\{\hat{\mu}_{gls} + \left\{\frac{C(\mathbf{s}_0,\mathbf{s}_0) - \mathrm{var}(\hat{\mu}_{gls})}{\mathrm{var}(\mathbf{c}(\mathbf{s}_0)'\Sigma^{-1}(\mathbf{Z}-\hat{\mu}_{gls}\mathbf{1}))}\right\}^{1/2}\right.$$

$$\left.\times\, \mathbf{c}(\mathbf{s}_0)'\Sigma^{-1}(\mathbf{Z}-\hat{\mu}_{gls}\mathbf{1})\right\}1$$

$$= \bar{Z} + \sqrt{\sigma^2 - \mathrm{var}(\bar{Z})}\left(\lim_{\rho\to 0}\left\{\frac{\mathbf{r}'\Sigma^{-1}(\mathbf{Z}-\hat{\mu}_{gls}\mathbf{1})}{\sqrt{\mathrm{var}(\mathbf{r}'\Sigma^{-1}(\mathbf{Z}-\hat{\mu}_{gls}\mathbf{1}))}}\right\}\right)$$

Now, $\lim_{\rho\to 0}\rho^{||\mathbf{v}_j-\mathbf{s}_0||-||\mathbf{v}_1-\mathbf{s}_0||} = 0$; $j = m+1,\ldots,n$, and so $\lim_{\rho\to 0}\mathbf{r}' = (\mathbf{1}_m', 0,\ldots,0)'$. This yields the final result

$$\lim_{\rho\to 0}\boldsymbol{\lambda}_{ck}(\mathbf{s}_0)'\mathbf{Z} = \bar{Z} + \left\{\frac{(n-1)\sigma^2-\tau^2}{m(n-m)(\sigma^2+\tau^2)}\right\}^{\frac{1}{2}}\left(\sum_{i=1}^{m} Z(\mathbf{v}_i) - m\bar{Z}\right) \tag{93}$$

Several consequences of equation (93) should be noted.

1. There is no solution if $m = n$. This case is rare and only occurs when $\mathbf{s}_1,\ldots,\mathbf{s}_n$ are on the circumference of a circle and $\mathbf{s}_0$ is at its center. The constrained kriging equations with $\rho > 0$ also break down when this occurs.
2. If $m = 1$ and $\tau^2 = 0$, then $\lim_{\rho\to 0}\boldsymbol{\lambda}_{ck}(\mathbf{s}_0)'\mathbf{Z} = Z(\mathbf{v}_j)$, where $||\mathbf{v}_j - \mathbf{s}_0|| = \min_{i=1,\ldots,n}||\mathbf{s}_i - \mathbf{s}_0||$. That is, constrained kriging yields a piecewise constant prediction surface, constant on Voronoi polygons.

The extension of this result to constrained kriging of blocks is not difficult. Assume that exactly $m \le n$ sampling locations are equally closest to *any* location in B and assume these to be $\{\mathbf{v}_1,\ldots,\mathbf{v}_m\}$. Take the ith of these, $\mathbf{v}_i$, and define $B_i \equiv \{\mathbf{u}\in B: \; ||\mathbf{u}-\mathbf{v}_i|| = \min_{j=1,\ldots,|B|}||\mathbf{u}_j - \mathbf{v}_i||\}$, $i = 1,\ldots,m$. Using similar arguments to the point-prediction case, we obtain

$$\lim_{\rho\to 0}\boldsymbol{\lambda}_{ck}(B)'\mathbf{Z} = \bar{Z} + \left\{\frac{n\sigma^2 - |B|(\sigma^2+\tau^2)}{|B|(\sigma^2+\tau^2)\{n\sum_{i=1}^{m}|B_i|^2 - N_m^2\}}\right\}^{\frac{1}{2}} \times\left(\sum_{i=1}^{m}|B_i|Z(\mathbf{v}_i) - N_m\bar{Z}\right) \tag{94}$$

where $N_m = \sum_{i=1}^{m}|B_i|$.

Several consequences of equation (94) should be noted.

1. Let A be the set of sampling locations $\{\mathbf{s}_1, \ldots, \mathbf{s}_n\}$. If $A \subset B$, then $|B_i| = 1$; $i = 1, \ldots, m$, and $m = n$. This means $n\sum_{i=1}^{m} |B_i|^2 - N_m^2 = n^2 - n^2 = 0$. Hence no solution exists when $B = G$.
2. Equation (94) is not defined if $\{n\sigma^2 - |B|(\sigma^2 + \tau^2)\}/\{n\sum_{i=1}^{m} |B_i|^2 - N_m^2\} < 0$.
3. If $B \equiv \mathbf{s}_0$, then equation (94) reduces to equation (93).

2

Design Techniques for Probabilistic Sampling of Items with Variable Monetary Value

JOSÉ M. BERNARDO University of València, València, Spain

1. THE PROBLEM

To ensure the quality of their output, organizations often face the problem of sampling items from a collection of different, possibly related, categories (for instance, different suppliers, or alternative production methods) with possibly very different monetary value within each category (for instance, the same supplier may produce simple electronic components and expensive chips). In this chapter, we will consider situations where each of the items may be classified either as correct or as *nonconforming* (that is, failing to meet whatever requirements are established to qualify as correct). We will further assume that all items selected for inspection are correctly classified (so that any nonconforming items in the sample will be found), and that nonconforming items which are not sampled remain undetected. Under these conditions, we use Bayesian decision theory and hierarchical modelling to propose an inspection strategy designed to *minimize the probable monetary value of undetected nonconforming items.*

Formally, let $\boldsymbol{v} = \{v_1, \ldots, v_N\}$ be the known monetary values of the N *items* $\{i_1, \ldots, i_N\}$ which comprise a shipment provided by a supplier, let

$\boldsymbol{\theta} = \{\theta_1, \ldots, \theta_N\}$, $0 \geq \theta_j \geq 1$, be the *unknown* probabilities of each of them being *nonconforming*, and let us assume that an inspection strategy is required to minimize the probable total monetary value of undetected nonconforming items. If N is large, the cost of a complete inspection will typically be prohibitive, and hence a sampling procedure is necessary. It is also required that the items to be inspected could not be anticipated by the supplier, and hence a *probabilistic* sampling procedure is essential.

Any *sample* $\boldsymbol{z}$ from the shipment may be represented by a vector of zeros and ones, so that $\boldsymbol{z} = \{z_1, \ldots, z_N\}$, $z_j \in \{0, 1\}$, where $z_j = 1$ if, and only if, the j-th item has been chosen for inspection. Any probabilistic sampling *strategy* is determined by (i) the required *sample size n* and (ii) the *sampling mechanism* (without replacement), $\boldsymbol{s}$, specified by the functions

$$p(z_j = 1 | z_{j_1} = 1, \ldots, z_{j_k} = 1), \qquad j \notin \{j_1, \ldots, j_k\} \tag{1}$$

which describe, for each possible case, the *probability* of selecting for inspection the jth item, given that items $\{j_1, \ldots, j_k\}$ have already been selected.

We shall assume that any nonconforming item is detected if inspected. Hence, with the notation established, the *expected loss* due to undetected nonconforming items, given a sample $\boldsymbol{z} = \{z_1, \ldots, z_N\}$ of size n, is given by

$$l_n(\boldsymbol{\theta}, \boldsymbol{z}) = \sum_{j=1}^{N} (1 - z_j)\, v_j\, \theta_j, \quad \sum_{j=1}^{N} z_j = n \tag{2}$$

that is, by the sum, for all noninspected items, of the products $v_j \theta_j$ of their monetary values and their probabilities of being nonconforming. Consequently, the corresponding *relative expected loss* ρ_n is given by

$$\rho_n(\boldsymbol{\theta}, \boldsymbol{z}) = \frac{1}{V} \sum_{j=1}^{N} (1 - z_j)\, v_j\, \theta_j = \sum_{j=1}^{N} (1 - z_j)\, w_j\, \theta_j \tag{3}$$

where $V = \sum v_j$ is the total monetary value of the shipment, and $w_j = v_j / V$ is the relative value of the jth item.

Since the objective is to minimize the relative expected loss in equation (3), which is simply the sum of the products $w_j \theta_j$ which correspond to noninspected items, the sample should be selected by *preferably* choosing for inspection those items for which the product $w_j \theta_j$ is maximum. This is precisely achieved by selecting for inspection available items with a probability proportional to $w_j \theta_j$. Formally, the sampling mechanism $\boldsymbol{s}$ should therefore be

$$p(z_j = 1 | z_{j_1} = 1, \ldots, z_{j_k} = 1) = \frac{w_j \theta_j}{\sum_{j \notin \{j_1, \ldots, j_k\}} w_j \theta_j}, \qquad j \notin \{j_1, \ldots, j_k\} \tag{4}$$

For such a sampling mechanism, the expected loss due to undetected nonconforming parts will clearly be a decreasing function of the sample size n, and will converge to zero as the sample size n approaches the shipment size N. On the other hand, the *expected cost* c_n of inspecting n items will obviously increase with n. It follows that the total loss to be expected if a sample of size n is inspected will be

$$l_n(\boldsymbol{\theta}, z) + c_n = V\rho_n(\boldsymbol{\theta}, z) + c_n, \qquad 0 \leq n \leq N \tag{5}$$

Since the expected value of ρ_n decreases with n while c_n increases with n, there will be some value n^* of the sample size which minimizes the expected value of equation (5), and this will be the *optimal* sample size.

If the probabilities $\boldsymbol{\theta} = \{\theta_1, \ldots, \theta_N\}$ of the items being nonconforming were known, the expected relative loss $\rho_n(\boldsymbol{\theta}, z)$ which corresponds to any given sample $z = \{z_1, \ldots, z_N\}$ could be computed, and the optimal sample size could easily be determined. Since the θ_js are *not* known, a probability model must be established relating the available data D to the unknown θ_js. Within the Bayesian framework (see, e.g. Bernardo and Smith, 1994, for an introduction), such a model may then be used to obtain the *posterior distribution* $p(\boldsymbol{\theta}|D)$ which describes the information provided by the data on the possible values of the θ_js, and this may be used to obtain the *predictive distribution* $p(\rho_n|z, D)$ which probabilistically describes the possible loss values as a function of the selected sample z and the available information D. Finally, the predictive distribution of ρ_n may be used to derive the optimal sample size n^*.

In Section 2, we describe the data structure, propose a flexible hierarchical logistic model for the behaviour of the θ_js, and describe the statistical analysis of the model proposed. In Section 3, we analyse the computing aspects involved in the implementation of the proposed solution. In Section 4, we describe an example and evaluate the results obtained. Section 5 contains a final discussion.

2. DATA STRUCTURE AND STATISTICAL ANALYSIS

We shall assume that each item may be classified into one of k possible categories, denoted by $\{\delta_1, \ldots, \delta_k\}$, and that the available past data

$$D = \{(n_i, r_i), \qquad i = 1, \ldots, k\} \tag{6}$$

consist of the number of items n_i which have been inspected in the past within each category, and the number r_i of nonconforming items which have been found among them. Obviously, one must have $0 \leq r_i \leq n_i$.

We shall further assume that the probability of an item being nonconforming may appropriately be described by a hierarchical logistic model of the form

$$\begin{aligned} p(x_{ij}|\delta_i, \theta_i) &= \theta_i^{x_{ij}}(1-\theta_i)^{1-x_{ij}}, \quad x_{ij} \in \{0, 1\}, \quad j = 1, \ldots, n_i, \quad i = 1, \ldots, k \\ \theta_i &= \frac{\exp[\beta_i]}{1+\exp[\beta_i]}, \quad \beta_i \in \Re, \quad i = 1, \ldots, k \\ p(\beta_i|\mu, \sigma) &= \mathrm{N}(\beta_i|\mu, \sigma), \quad i = 1, \ldots, k \\ p(\mu, \sigma) &= \pi_\beta(\mu, \sigma) \end{aligned} \tag{7}$$

where $x_{ij} \in \{0, 1\}$ is the unknown random quantity which identifies whether the jth item in the i-th category is nonconforming ($x_{ij} = 1$) or is correct ($x_{ij} = 0$). Thus, we are assuming that the β_is, that is the logodds which correspond to the event that an item which belongs to the i-th category is nonconforming, namely,

$$\boldsymbol{\beta} = \{\beta_1, \ldots, \beta_k\}, \qquad \beta_i = \log\frac{\phi_i}{1-\phi_i}, \qquad i = 1, \ldots, k \tag{8}$$

may be considered as a normal random sample with mean μ and standard deviation σ. This would be appropriate when the items in the different categories belong to the same general population, as would be the case, for instance, when the categories identify different suppliers producing similar products. Moreover, no prior information is assumed beyond that provided by the data D, and hence (Bernardo, 1979, 1998; Bernardo and Ramón, 1998) the prior distribution for $p(\mu, \sigma)$ is assumed to be the corresponding nonsubjective *reference* prior $\pi_\beta(\mu, \sigma)$, when the β_is are the parameters of interest.

Using standard hierarchical normal analysis (Good 1965, 1980; Lindley, 1971, 1972; Lindley and Smith, 1972; Smith, 1973; Goel and DeGroot, 1981; Berger and Bernardo, 1992) on the full exponential Laplace approximations to the marginal distributions of the β_is, it may be established that the marginal reference *posterior* distribution of each of the β_is is approximately given by

$$\pi(\beta_i|D) \approx \mathrm{N}(\beta_i|m_i, s_i), \qquad i = 1, \ldots, k \tag{9}$$

where

$$
\begin{aligned}
m_i &= \gamma_i\,[\psi(r_i+1/2)-\psi(n_i-r_i+1/2)]+(1-\gamma_i)\,m, \qquad i=1,\ldots,k\\
s_i &= (h_i+h)^{-1/2}, \qquad i=1,\ldots,k\\
h_i &= (n_i+2)\,q_i(1-q_i), \qquad q_i=\frac{r_i+1/2}{n_i+1}, \qquad i=1,\ldots,k\\
m &= \frac{1}{k}\sum\nolimits_{i=1}^{k} m_i, \quad h=\frac{k-1}{k\,s^2}, \qquad s^2=\frac{1}{k}\sum\nolimits_{i=1}^{k}(m_i-m)^2\\
\gamma_i &= \frac{h_i}{h+h_i}, \qquad 0\le\gamma_i\le 1
\end{aligned}
\tag{10}
$$

where $\psi(x)=\Gamma'(x)/\Gamma(x)$ is the *digamma* function that, except for very small values of x, may be well approximated by $\log(x)-1/(2x)$.

It should be noted that if only one class exists, so that $k=1$, one has $\boldsymbol{\beta}=\beta$, $h=0$, $\gamma=1$, and equation (9) reduces to

$$\pi(\beta|D)\approx \mathrm{N}(\beta|m,\ s) \tag{11}$$

where

$$
\begin{aligned}
m &= \psi(r+1/2)-\psi(n-r+1/2)\\
s &= \frac{1}{\sqrt{(n+1)q(1-q)}}, \qquad q=\frac{r+1/2}{n+1}
\end{aligned}
\tag{12}
$$

Consider now a shipment which consists of items $\{i_1,\ldots,i_N\}$, and let a_j be a real constant, measured in standard deviations of the logodds and to be specified by an expert, which *describes an assessment of the marginal propensity of the jth item to be either more prone* ($a_j>0$) *or less prone* ($a_j<0$) *to being nonconforming than the average item in its class*. Thus, the structure of the shipment may be described by the matrix

Item	Class	Monetary value	Propensity	Probability	Sample	Result
i_1	$\delta(i_1)$	v_1	a_1	θ_1	z_1	x_1
i_2	$\delta(i_2)$	v_2	a_2	θ_2	z_2	x_2
...	...	...	...	...	...	...
i_j	$\delta(i_j)$	v_j	a_j	θ_j	z_j	x_j
...	...	...	...	...	...	...
i_N	$\delta(i_N)$	v_N	a_N	θ_N	z_N	x_N

where, for each item i_j, we know its class, $\delta(i_j)$, its monetary value, v_j, and its marginal propensity, a_j, to being nonconforming, but *not* its probability, θ_j, of being nonconforming. For completion, we have also included in that matrix the z_j values which identify whether or not the jth item is selected for inspection (to be decided), and the x_j values which denote whether or not the jth item is nonconforming (which may be determined without error if that item is selected for inspection).

Our main task is to use all available information, including any past relevant data, to "estimate" the probabilities θ_j of each of the items being nonconforming. Within a Bayesian framework (see Bernardo and Smith, 1994, Chap. 5 for details), this is done by deriving the joint posterior distribution, $p(\boldsymbol{\theta}|\boldsymbol{\delta}, \boldsymbol{a}, D)$ given the classes $\{\delta(i_1), \ldots, \delta(i_N)\}$, the marginal propensities $\{a_1, \ldots, a_N\}$, and the available relevant past data D. This will allow the use of the sampling mechanism in equation (4) to select an appropriate sample $\boldsymbol{z} = \{z_1, \ldots, z_N\}$, and to "estimate" the expected relative loss ρ_n by determining its predictive distribution

$$p(\rho_n|\boldsymbol{z}, \boldsymbol{\delta}, \boldsymbol{a}, D) = \int_{\Theta} p(\rho_n|\boldsymbol{\theta}, \boldsymbol{z})\, p(\boldsymbol{\theta}|\boldsymbol{\delta}, \boldsymbol{a}, D)\, d\boldsymbol{\theta} \tag{13}$$

To incorporate the *possibility* of using an expert's judgement on the probability of a particular item being nonconforming, we shall assume that the behaviour of such probability is described by the *extended* hierarchical logistic model

$$\begin{aligned} p(x_j|\delta_i, \theta_{ij}) &= \theta_{ij}^{x_j}(1-\theta_{ij})^{1-x_j}, \qquad i = 1, , k, \qquad j = 1, \ldots, N \\ \theta_{ij} &= \frac{\exp[\beta_i + a_j\sigma]}{1 + \exp[\beta_i + a_j\sigma]}, \qquad \beta \in \Re, \qquad a_j \in \Re \\ p(\beta_i|\mu, \sigma) &= \mathrm{N}(\beta_i|\mu, \sigma), \qquad i = 1, \ldots, k \\ p(\mu, \sigma) &= \pi_\beta(\mu, \sigma) \end{aligned} \tag{14}$$

Thus, we are extending the model in equation (6) by assuming that the logodds β_{ij} which correspond to the event that an item which belongs to category δ_i, and has marginal propensity a_j, is nonconforming and has a linear structure

$$\beta_{ij} = \log\frac{\theta_{ij}}{1-\theta_{ij}} = \beta_i + a_j\sigma, \qquad j = 1, \ldots, n_i, \qquad i = 1, , k \tag{15}$$

with an element β_i which is common to all items in the same class, and an element a_j which depends on the item, and models, in terms of standard deviations of the β_is, the marginal propensity of the item to being nonconforming as compared with the average in its class. Note that if no information about the a_js is available, they may simply be set to their default value, namely $a_j = 0$.

It follows from equation (9), (10), and (14), and the logistic transformation in equation (8), that the marginal reference posterior distribution of the probability θ_{ij} of an item to be nonconforming, given that it belongs to class i and has marginal propensity a_j, is approximately given by

$$\pi(\theta_{ij}|\delta_i, a_j, D) \approx \theta_{ij}^{-1}(1-\theta_{ij})^{-1}\, \mathrm{N}\left(\frac{\exp(\theta_{ij})}{1+\exp(\theta_{ij})} \,\middle|\, m_{ij}, s_i\right) \tag{16}$$

where

$$\begin{aligned}
m_{ij} &= \gamma_i\, m_i + (1-\gamma_i)\, m + a_j\, s_i, \qquad i = 1, \ldots, k, \qquad j = 1, \ldots, N\\
s_i &= (h_i + h)^{-1/2}, \qquad i = 1, \ldots, k\\
m_i &= \psi(r_i + 1/2) - \psi(n_i - r_i + 1/2), \qquad i = 1, \ldots, k\\
h_i &= (n_i + 2)\, q_i(1 - q_i), \qquad q_i = \frac{r_i + 1/2}{n_i + 1}, \qquad i = 1, \ldots, k\\
m &= \frac{1}{k}\sum\nolimits_{i=1}^{k} m_i, \qquad h = \frac{k-1}{k\, s^2}, \qquad s^2 = \frac{1}{k}\sum\nolimits_{i=1}^{k}(m_i - m)^2\\
\gamma_i &= \frac{h_i}{h + h_i}, \qquad 0 \le \gamma_i \le 1
\end{aligned} \tag{17}$$

where, again, $\psi(x)$ is the digamma function.

Since, given z, $\rho_n(\boldsymbol{\theta}, z)$ is a one-to-one function of $\boldsymbol{\theta}$, equations (3), (13) and (16) may be used to derive the reference posterior predictive distribution of the relative loss

$$\pi(\rho_n|z, \boldsymbol{\delta}, \boldsymbol{a}, D) = \int_\Theta p(\rho_n|\boldsymbol{\theta}, z)\, \pi(\boldsymbol{\theta}|\boldsymbol{\delta}, \boldsymbol{a}, D)\, d\boldsymbol{\theta} \tag{18}$$

with expected value

$$\begin{aligned}
\mathbf{E}[\rho_n|z, \boldsymbol{\delta}, \boldsymbol{a}, D] &= \sum\nolimits_{j=1}^{N}(1 - z_j)\, w_j\, \mathbf{E}[\theta_j|\delta, a_j, D]\\
&\approx \sum\nolimits_{\{j;\, z_j = 0\}} w_j\, \frac{\exp(m_{ij})}{1 + \exp(m_{ij})}
\end{aligned} \tag{19}$$

3. IMPLEMENTATION

As described in Section 1, the main practical objective of this analysis is to estimate the expected loss associated with a probabilistic inspection due to undetected nonconforming items, and this is conveniently encapsulated by the relative loss ρ_n. Using the formulae listed in equation (17), equation (19) offers a simple derivation of the posterior expected value of ρ_n; this may easily be computed with a hand calculator, and may provide a useful first approximation. However, only the full posterior predictive distribution $\pi(\rho_n|z, \boldsymbol{\delta}, \boldsymbol{a}, D)$ provides the necessary information for a complete understanding of the risk involved, and hence for a correct determination of the appropriate sample size. In this section, we outline a numerical procedure designed to obtain $\pi(\rho_n|z, \boldsymbol{\delta}, \boldsymbol{a}, D)$ by probabilistic simulation. For general accounts of simulation techniques see, for example, Gelfand and Smith (1990), Tanner (1991), Casella and George (1992), or Smith and Roberts (1993).

3.1 Predictive Distribution of ρ_n and Optimal Sample Size

To determine the predictive distribution $\pi(\rho_n|z, \boldsymbol{\delta}, \boldsymbol{a}, D)$ of the relative loss for any set of possibly interesting sample sizes, we proceed as follows:

1. *Compute past sufficient statistics.* Using past available data $D = \{(n_i, r_i),\ i = 1, \ldots, k\}$ and the formulae in equation (10), compute the means and standard deviations (m_i, s_i), $i = 1, \ldots, k$, of the marginal reference posterior distributions of the logodds which correspond to the event that an item is nonconforming.
2. *Get new shipment data.* Read and store the relevant information $\{i_j, \delta(i_j), a_j, v_j\}$, for each of the N items in the shipment to be analysed. Compute the total monetary value V of the shipment, and the relative value w_j of each item.
3. *Compute new sufficient statistics.* Using the new shipment data and the formulae in equation (17), compute the mean and standard deviations $(m_{ij}, s_i), j = 1, \ldots, N$, for the marginal reference posterior distributions of the logodds which correspond to the event that each item in the shipment is nonconforming.

Repeat for different sample sizes:

4. *Generate relative losses.* For any desired sample size n, implement a double loop, generate a number l_s of samples of size n with the appropriate probabilistic structure and, for each of those samples, generate a number l_r of simulated relative losses. More specifically:

   Repeat l_s times:

   a. For each item in the shipment, generate a standard normal random variate t, such that $p(t) = \mathrm{N}(t|0, 1)$, and simulate the probability θ_j that item j known to belong to category i is nonconforming by

$$p_j = \frac{\exp(m_{ij} + t\,s_i)}{1 + \exp(m_{ij} + t\,s_i)}, \qquad j = 1, \ldots, N \tag{20}$$

and store the results.

b. For each item in the shipment, simulate the expected relative loss $w_j\theta_j$ by $w_j p_j$ and, using random numbers on [0, 1], generate a sample of size n, $\{z_1, \ldots, z_N\}$, $\sum_j z_j = n$, by implementing the sampling mechanism in equation (4).

Repeat l_r times.

c. For each item which is *not* in the sample obtained, generate a random number u on [0, 1], simulate this as nonconforming ($x_j = 1$) if, and only if, $p_j \leq u$, compute and store the corresponding relative loss of the sample as

$$\rho_n = \sum_{i=1}^{N} (1 - z_j) x_j w_j \tag{21}$$

d. Use the stored ρ_n values of the relative loss to construct the predictive distribution of ρ_n by directly computing its moments, range, and quantiles.

5. *Derive the estimated probabilities.* Estimate the probabilities θ_j of an item being nonconforming by the average $\bar{p}_j$ of the p_j values generated in Step 4.a.

An appropriate sample size may usually be determined by inspection of the predictive distributions of the relative loss which are induced from different sample sizes. As mentioned in Section 1, if the cost structure of the sampling mechanism is specified, then the optimal sample size may easily be found by minimizing equation (5).

3.2 Final Sample

Once an appropriate sample size n^* has been determined:

1. *Generate the final sample.* For each item in the shipment, use the estimated probabilities $\bar{p}_j$ to simulate the expected relative loss $w_j\theta_j$ by $w_j\bar{p}_j$. Using random numbers on [0, 1], generate one sample of size n^* by implementing the sampling mechanism in equation (4) with $\theta_j = \bar{p}_j$, and print the identifiers of the resulting sample. Alternatively, a manually selected sample may be read in, in order to analyze its probable risk.
2. **Repeat** l_f times:
   a. For each item in the shipment i_j, use equation (20) to simulate its probability θ_j of being nonconforming.

b. For each item which is *not* in the final sample, generate a random number u on [0, 1], simulate this as nonconforming ($x_j = 1$) if, and only if $p_j \leq u$, use equation (21) to compute the corresponding relative loss of the sample, and store the result.

3. *Generate the final predictive distribution.* Use the stored values of the relative loss to construct the final predictive distribution of ρ_{n^*} by directly computing its moments, range, and quantiles. If desired, use those values to plot the corresponding probability density function.

4. EXAMPLE

In the problem which motivated this research, shipments contained between 100 and 500 items of very different values, and items belonged to one of eight possible categories. Actually, shipments were homogeneous, in that all items in the same shipment belonged to the same category, although this is certainly not required to use the model proposed.

Table 1 reproduces the available past data, together with their corresponding sufficient statistics. Merging all categories, there were a total of 381 inspected items, 11 of which (2.89%) turned out to be nonconforming. It can be observed that predicted probabilities $\theta_i = \exp(m_i)/(1 + \exp(m_i))$ of nonconforming for each category smooth the corresponding raw proportions by pooling them towards the overall proportion of nonconforming. It can also be observed that category δ_4 is special in having a considerably larger probability of containing nonconforming items than the rest.

Table 2 reproduces the information which corresponds to the first 15 items in a particular shipment which only contained items from category δ_8 with a total monetary value of $V = 28\,894.252$ monetary units. This

Table 1. Past available data and sufficient statistics

Category	n_i	r_i	% NC	m_i	s_i	$100 * \phi_i$
δ_1	60	0	0.00	−5.64	1.22	0.35
δ_2	75	0	0.00	−5.81	1.22	0.30
δ_3	42	3	7.14	−2.66	0.54	6.53
δ_4	20	5	25.00	−1.24	0.49	22.45
δ_5	41	0	0.00	−5.36	1.22	0.47
δ_6	71	0	0.00	−5.77	1.22	0.31
δ_7	39	3	7.69	−2.59	0.54	7.0
δ_8	33	0	0.00	−5.20	1.22	0.55

Table 2. Partial data of a new shipment

Number	a_j	Monetary value	w_j
1	0	1 187	0.00
2	0	223 533	0.77
3	0	6 690	0.02
4	0	13 660	0.05
5	0	42 896	0.15
6	0	35 210	0.12
7	0	234 444	0.81
8	0	34 578	0.12
9	0	58 674	0.20
10	0	2 463	0.01
11	1	253 487	0.88
12	0	23 694	0.08
13	0	13 695	0.05
14	0	52 690	0.18
15	1	78 924	0.27

include their marginal propensity a_j, monetary value v_j, and relative value, as a percentage, $w_j = 100 * v_j/V$.

Table 3 reproduces, for several sample sizes, the expected loss $E[l_n]$, the expected relative loss $E[\rho_n]$, and the posterior probability that the relative loss ρ_n is smaller than 1.0% of the total monetary value V. It should be

Table 3. Expected losses as a function of the sample size

n	$E[l_n]$	$E[\rho_n]$	$P(\rho_n < 1)$
10	158.253	0.548	0.856
20	73.822	0.255	0.934
30	45.623	0.158	0.967
40	27.775	0.096	0.989
50	17.909	0.062	0.997
60	11.441	0.040	0.999
70	7.855	0.027	1.000
80	4.994	0.017	1.000
90	2.853	0.010	1.000
100	1.153	0.004	1.000

appreciated, for example, that with a relatively small sample of size 20, the expected loss is 73.822 monetary units, which is 0.25% of the total monetary value of the shipment, and, with that sample size, the probability of a relative loss smaller than 1.0% is 0.934.

Table 4 reproduces the optimal sample of size 15. For each selected item, this table describes the number in the original list, its associated propensity, its monetary value, and the estimated probability that such an item is nonconforming, which only depends on the class δ_j to which the item belongs (in this shipment, all belong to class δ_8), and on its propensity a_j.

The final computation shows that the total monetary value of this sample is 21 888.147 monetary units, which accounts for 75.75% of the total monetary value of the shipment. The *expected* loss in this inspection sample is only 93.834 units, or 0.325% of the total value. The probability that the relative loss is actually less than 0.5% is 0.814, and the probability that it is less than 1.0% is 0.906.

5. DISCUSSION

We have provided a sophisticated risk prediction model which may be used to optimally select inspection samples from shipments which contain items of *highly variable monetary value* to minimize the probable risk associated

Table 4. Optimal sample of size 15

Number	a_j	v_j	$P[x_j = 1]$
8	0.00	34.578	0.011
21	1.00	1063.230	0.034
28	1.00	13220.868	0.034
31	0.00	1235.250	0.011
32	0.00	1256.112	0.011
33	1.00	1033.340	0.034
64	1.00	976.497	0.034
71	1.00	635.255	0.034
72	1.00	219.809	0.034
77	1.00	547.813	0.034
82	0.00	1014.213	0.011
84	0.00	13.265	0.011
90	0.00	53.218	0.011
91	1.00	367.991	0.034
111	1.00	216.709	0.034

with accepting nonconforming items, and to estimate their corresponding value. The computations have been performed with specifically written Pascal code; typical examples take only a few minutes on a conventional Mac or PC.

The use of a hierarchical structure makes it possible vastly to improve the precision of the predictions computed by allowing the incorporation of information relative to the failure rate of items which, although *related* to those being considered, belong to a different category.

The use of reference analysis makes it possible to use the powerful Bayesian hierarchical machinery without *necessarily* introducing subjective input on the parameter values; however, the algorithm is built to facilitate the incorporation of subjective information on the individual propensity of an item to be defective, if so desired.

The formal Bayesian solution to the main problem posed, namely to estimate the risk ρ_n involved in accepting a particular sampling design of, say, size n, is its posterior predictive distribution $p(\rho_n | \mathbf{z}, \boldsymbol{\delta}, \boldsymbol{a}, D)$. This cannot be analytically expressed in a closed form but, using the approach described, it is easily approximated by standard simulation techniques.

ACKNOWLEDGMENTS

The author is grateful to the anonymous referees for their helpful comments. This research was partially funded with grant PB96-0754 of the DGICYT, Madrid, Spain.

REFERENCES

Berger, J. O. and Bernardo, J. M. (1992). Reference priors in a variance components problem. *Bayesian Analysis in Statistics and Econometrics* (P. K. Goel and N. S. Iyengar, eds.). Berlin: Springer, pp. 323–340.

Bernardo, J. M. (1979). Reference posterior distributions for Bayesian inference. *J. Roy. Stat. Soc. B*, **41**, 113–147 (with discussion). Reprinted in *Bayesian Inference* (N. G. Polson and G. C. Tiao, eds.), Brookfield, VT: Edward Elgar, 1995, pp. 229–263.

Bernardo, J. M. (1998). Noninformative priors do not exist. *J. Stat. Plann. Inference*, **65**, 159–189 (with discussion).

Bernardo, J. M. and Ramón, J. M. (1998). An introduction to Bayesian reference analysis: Inference on the ratio of multinomial parameters. *Statistician*, **47**, 101–135.

Bernardo, J. M. and Smith, A. F. M. (1994). *Bayesian Theory*. Chichester: Wiley.

Casella, G. and George, E. I. (1992). Explaining the Gibbs sampler. *Am. Stat.* **46**, 167–174.

Gelfand, A. E. and Smith, A. F. M. (1990). Sampling based approaches to calculating marginal densities. *J. Am. Stat. Assoc.*, **85**, 398–409.

Goel, P. K. and DeGroot, M. H. (1981). Information about hyperparameters in hierarchical models. *J. Am. Stat. Assoc.*, **76**, 140–147.

Good, I. J. (1965). *The Estimation of Probabilities. An Essay on Modern Bayesian Methods*. Cambridge, MA: MIT Press.

Good, I. J. (1980). Some history of the hierarchical Bayesian methodology. *Bayesian Statistics* (J. M. Bernardo, M. H. DeGroot, D. V. Lindley, and A. F. M. Smith, eds.). Valencia: University Press, pp. 489–519 (with discussion).

Lindley, D. V. (1971). The estimation of many parameters. *Foundations of Statistical Inference* (V. P. Godambe and D. A. Sprott, eds.), Toronto: Holt, Rinehart and Winston, pp. 435–453 (with discussion).

Lindley, D. V. (1972). *Bayesian Statistics, A Review*. Philadelphia, PA: SIAM.

Lindley, D. V. and Smith, A. F. M. (1972). Bayes estimates for the linear model. *J. R. Stat. Soc. B*, **34**, 1–41 (with discussion)

Smith, A. F. M. (1973). A general Bayesian linear model. *J. R. Stat. Soc. B*, **35**, 67–75.

Smith, A. F. M. and Roberts, G. O. (1993). Bayesian computation via the Gibbs sampler and related Markov chain Monte Carlo methods. *J. R. Stat. Soc. B*, **55**, 3–23 (with discussion).

Tanner, M. A. (1991). *Tools for Statistical Inference: Observed Data and Data Augmentation Methods*. Berlin: Springer.

3

Small Area Estimation: A Bayesian Perspective

MALAY GHOSH University of Florida, Gainesville, Florida

KANNAN NATARAJAN Pharmaceutical Research Institute, Bristol-Myers Squibb, Princeton, New Jersey

1. INTRODUCTION

Small area estimation is becoming a topic of growing importance in recent years. The terms "small area" and "local area" are commonly used to denote a small geographical area such as a county, municipality, or a census division. They may also describe a "small domain," that is a small subpopulation such as a specific age–sex–race group of people in a large geographical area.

Of late, there is a growing demand for small area statistics from both the public and private sectors. As a recent example, we may cite Bill H.R. 1645 passed by the House of Representatives on November 21, 1993. This bill requires the Secretary of Commerce to produce and publish at least every 2 years, beginning in 1996, current data related to the incidence of poverty in the United States. Specifically, the legislation states that "to the extent feasible," the Secretary shall produce estimates of poverty for states, counties, and local jurisdictions of governments and school districts. For school districts, estimates are to be made of the number of poor children aged 5–17 years. It also specifies production of state and county estimates of the number of poor persons aged 65 and over. These statistics for small areas will be used by a broad range of customers, including policy-makers at the state and local level as well as the private sector. This includes the allocation of Federal and State funds, Federal funds being estimated as more than $30 billion.

Ghosh and Rao (1994) have provided an extensive discussion of small area estimation methods from both the frequentist and the Bayesian points of view, their primary emphasis being the discussion of continuous data. Many examples of small area estimation are cited in their paper. This chapter discusses small area estimation exclusively from a Bayesian angle, with more emphasis on generalized linear models which encompass both discrete and continuous data.

The need for Bayesian techniques in the context of small area estimation has long been recognized. Particularly effective in this regard are the hierarchical Bayes (HB) and empirical Bayes (EB) methods which connect the local areas through models. For the theory and applications of small area estimation based on normal models, the reader is referred to Fay and Herriot (1979), Ghosh and Meeden (1986), Ghosh and Lahiri (1987, 1992), Stroud (1987), Prasad and Rao (1990), and Datta and Ghosh (1991), among others. Empirical Bayes analyses of binary survey data began with Dempster and Tomberlin (1980), and have subsequently appeared in MacGibbon and Tomberlin (1989) and Farrell et al. (1997). Stroud (1991) discusses hierarchical Bayes analysis of the binomial distribution. More recently, Ghosh, et al. (1998) discussed small area estimation in the context of generalized linear models which include binary data and count data as well as continuous data.

In Section 2 of this review we discuss some synthetic and composite estimators, and Bayesian interpretations of the same. In Section 3, we discuss some specific simple normal models that have been used in the context of small area estimation. Bayesian inference based on hierarchical Bayes generalized linear models is discussed in Section 4. Finally, Section 5 introduces Bayesian model averaging in the context of small area estimation, and this idea is illustrated with an example.

2. SYNTHETIC AND COMPOSITE ESTIMATION

In this section, we present a few simple examples of the so-called "synthetic" and "composite" estimators with or without auxiliary information. We shall also examine how some of the composite estimators can be viewed as model-based as well as Bayes estimators.

To begin with, let y_{ij} denote the characteristic of interest for the jth unit in the ith local area, where $j = 1, \ldots, N_i$, $i = 1, \ldots, m$. For notational simplicity, let y_{ij} $(j = 1, \ldots, n_i, i = 1, \ldots, m)$ denote the characteristics corresponding to the sampled units in these local areas. We shall write $y(s) = (y_{11}, \ldots, y_{1n_1}, \ldots, y_{m1}, \ldots, y_{mn_m})$ and $y(\bar{s}) = (y_{1,n_1+1}, \ldots, y_{1N_1}, \ldots, y_{m,n_m+1}, \ldots, y_{mN_m})$.

The population mean for the ith local area is denoted by $\bar{y}_i = N_i^{-1} \sum_{j=1}^{N_i} y_{ij}$, $i = 1, \ldots, m$. The direct estimators of $\bar{y}_i$ are the corresponding sample means, namely $\bar{y}_{is} = n_i^{-1} \sum_{j=1}^{n_i} y_{ij}$, $i = 1, \ldots, m$. A synthetic estimator common to all these local areas is the overall mean, namely $\bar{y}_s = \sum_{i=1}^{m} n_i \bar{y}_{is} / \sum_{i=1}^{m} n_i$. A composite estimator for the ith local area is a weighted average of the local area mean and the overall mean, that is, it is of the form $w_i \bar{y}_{is} + (1 - w_i)\bar{y}_s$, with w_i being the weight attached to the ith local area mean.

A natural question is how to choose the weights w_i. Both design- and model-based approaches are proposed towards this aim. A simple-minded choice is to use $w_i = n_i/N_i$, $i = 1, \ldots, m$, or $w_i = n_T/N_T$, where $n_T = \sum_{i=1}^{m} n_i$ and $N_T = \sum_{i=1}^{m} N_i$. The former choice has a simple and yet interesting model-based justification which is given below.

Suppose y_{ij} $(j = 1, \ldots, N_i, i = 1, \ldots, m)$ are iid with common mean θ and variance 1. Then

$$E(\bar{y}_i | y(s)) = (n_i/N_i)\bar{y}_{is} + (1 - n_i/N_i)\theta. \tag{1}$$

However, the best linear unbiased estimator of θ is $\bar{y}_s$. This leads to the best linear unbiased predictor $(n_i/N_i)\bar{y}_{is} + (1 - n_i/N_i)\bar{y}_s$ of $\bar{y}_i$.

Interestingly, this estimator has a simple Bayesian interpretation as well. Assume that conditional on θ, y_{ij} $(j = 1, \ldots, N_i, i = 1, \ldots, m)$ are iid $N(\theta, 1)$. Suppose also that $\theta \sim$ uniform $(-\infty, \infty)$. Then the posterior distribution of θ given $y(s)$ is $N(\bar{y}_s, n_T^{-1})$. This implies that $y(\bar{s})$ is multivariate normal with common mean $E(\theta | y(s)) = \bar{y}_s$, common variance $V(\theta | y(s)) = n_T^{-1} + 1$, and common covariance $V(\theta | y(s)) = n_T^{-1}$. Consequently, conditional on $y(s)$, $\bar{y}_i$ is normal with mean the same as the one given in equation (1), and variance given by

$$\begin{aligned}
V[N_i^{-1} \sum_{j=1}^{N_i} y_{ij} | y(s)] &= V[N_i^{-1} \sum_{j=n_i+1}^{N_i} y_{ij} | y(s)] \\
&= N_i^{-2}[\sum_{j=n_i+1}^{N_i} V(y_{ij} | y(s)) \\
&\quad + \sum \sum_{n_i+1 \le j \ne j' \le N_i} Cov(y_{ij}, y_{ij'} | y(s))] \\
&= N_i^{-2}[(N_i - n_i)(n_T^{-1} + 1) + (N_i - n_i)(N_i - n_i - 1)n_T^{-1}] \\
&= N_i^{-2}(N_i - n_i) + N_i^{-2}(N_i - n_i)^{-2} n_T^{-1}
\end{aligned} \tag{2}$$

One may question the utility of the Bayes procedure in this context since the model-based approach yields the same estimator of the finite population mean without any distributional assumption. However, the variance of the ith local area population mean based only on the assumption of the first two moments is $N_i^{-2}(N_i - n_i)$. This is clearly an underestimate, since it does not incorporate the uncertainty involved in estimating θ. This is rectified in equation (2), where the second term accounts for the variability due to not knowing θ. Second, without any distributional assumption, it is virtually impossible to construct any meaningful confidence or credible set. Thus, the Bayesian approach has its merits, and this becomes more apparent in the context of more complex models.

Next, we discuss synthetic and composite estimation in the presence of auxiliary information. As in the previous section, we denote by y_{ij} the value of the jth unit in the ith local area, $j = 1, \ldots, N_i$, $i = 1, \ldots, m$. The corresponding vectors of auxiliary characteristics are denoted by x_{ij}, $j = 1, \ldots, N_i$, $i = 1, \ldots, m$. For simplicity, we shall consider in this section only scalar x_{ij}.

The direct estimator of the finite population mean $\bar{y}_i$ is the ratio estimator $(\bar{y}_{is}/\bar{x}_{is})\bar{X}_i$, where $\bar{y}_{is}$ is as defined above, while $\bar{x}_{is} = n_i^{-1}\sum_{j=1}^{n_i} x_{ij}$, and $\bar{X}_i = N_i^{-1}\sum_{j=1}^{N_i} x_{ij}$. The ratio synthetic estimator of $\bar{y}_i$ is given by $\hat{\bar{y}}_i^{RS} = (\bar{y}_s/\bar{x}_s)\bar{X}_i$, where $\bar{y}_s$ is as defined above, while $\bar{x}_s = \sum_{i=1}^{m} n_i\bar{x}_{is}/n_T$. A composite estimator of $\bar{y}_i$ is given by $\hat{\bar{y}}_i^{*} = (n_i/N_i)\bar{y}_{is} + (1 - n_i/N_i)\hat{\bar{y}}_i^{RS}$. An alternative composite estimator of $\bar{y}_i$ is given by $\hat{\bar{y}}_i^{C} = (n_i/N_i)\bar{y}_{is} + (1 - n_i/N_i)(\bar{y}_s/\bar{x}_s)\bar{X}'_i$, where $\bar{X}'_i = (N_i - n_i)^{-1}\sum_{j=n_i+1}^{N_i} x_{ij}$.

A model-based justification of $\hat{\bar{y}}_i^{C}$ is now given.

Suppose y_{ij} are independent with mean bx_{ij}, and variance $\sigma^2 x_{ij}$. Then the best linear unbiased estimator of b is obtained by minimizing $\sum_{i=1}^{m}\sum_{j=1}^{n_i}(y_{ij} - bx_{ij})^2/(\sigma^2 x_{ij})$ with respect to b. The resulting estimator of b is given by

$$\hat{b} = \sum_{i=1}^{m}\sum_{j=1}^{n_i} y_{ij} / \sum_{i=1}^{m}\sum_{j=1}^{n_i} x_{ij} = \bar{y}_s/\bar{x}_s \tag{3}$$

Now, $E(\bar{y}_i|y(s)) = (n_i/N_i)\bar{y}_{is} + b(N_i - n_i)\bar{X}'_i$. Substitution of $\hat{b}$ for b leads to the estimator $\hat{\bar{y}}_i^{C}$ of $\bar{y}_i$. This estimator is proposed in Holt et al. (1979). Equally interesting is the Bayesian interpretation of $\hat{\bar{y}}_i^{C}$. Assume that conditional on b, y_{ij} are independent $N(bx_{ij}, \sigma^2 x_{ij})$, while marginally b is uniform $(-\infty, \infty)$. Then the posterior pdf of b given $y(s)$ is

$$f(b|y(s)) \propto \exp[-\sum_{i=1}^{m}\sum_{j=1}^{n_i}(y_{ij} - bx_{ij})^2/(2\sigma^2)] \tag{4}$$

that is, $b|y(s) \sim N(\bar{y}_s/\bar{x}_s, \sigma^2/(n_T\bar{x}_s))$. Thus the joint posterior of y_{ij} $(j = 1, \cdots, N_i, i = 1, \cdots, m)$ given $y(s)$ is multivariate normal with

$$E(y_{ij}|y(s)) = E(b|y(s))x_{ij} = (\bar{y}_s/\bar{x}_s)x_{ij} \tag{5}$$

$$V(y_{ij}|y(s)) = V(b|y(s))x_{ij}^2 + \sigma^2 x_{ij} \tag{6}$$

$$Cov(y_{ij}, y_{ij'}|y(s)) = V(b|y(s))x_{ij}^2 \tag{7}$$

This implies

$$E[\bar{y}_i|y(s)] = N_i^{-1}[n_i\bar{y}_{is} + (\bar{y}_s/\bar{x}_s) \sum_{j=n_i+1}^{N_i} E(y_{ij}|y(s)]$$

$$= N_i^{-1}[n_i\bar{y}_{is} + (\bar{y}_s/\bar{x}_s)(N_i - n_i)\bar{X}_i'] \tag{8}$$

Note that the above estimator is the same as the one proposed in Holt et al. (1979). Note, however that

$$V(\bar{y}_i|y(s)) = V[N_i^{-1} \sum_{j=n_i+1}^{N_i} y_{ij}|y(s)]$$

$$= V(b|y(s))N_i^{-2}(N_i - n_i)^2(\bar{X}_i')^2$$

$$+ \sigma^2 N_i^{-2}(N_i - n_i)\bar{X}_i' \tag{9}$$

In contrast, the model-based procedure does not incorporate the uncertainty in b, and yields the estimator $\sigma^2 N_i^{-2}(N_i - n_i)\bar{X}_i'$ for $V(\bar{y}_i|y(s))$.

3. NORMAL THEORY SMALL AREA ESTIMATION

This section illustrates normal theory small area estimation. In particular, we introduce a simple model where only area-specific auxiliary data $\boldsymbol{x}_i = (x_{i1}, \ldots, x_{ip})^T$ are available, and the parameters of interest, θ_i, are related to the $\boldsymbol{x}_i$s. In particular, it is assumed that

$$\theta_i = \boldsymbol{x}_i^T \boldsymbol{b} + u_i \ (i = 1, \ldots, m) \tag{10}$$

where $\boldsymbol{b}(p \times 1)$ is the vector of regression coefficients, and the u_i are iid $N(0, \sigma^2)$. We shall write $\boldsymbol{\theta} = (\theta_1, \ldots, \theta_m)^T$, $\boldsymbol{u} = (u_1, \ldots, u_m)^T$, $\boldsymbol{X}^T = (\boldsymbol{x}_1, \ldots, \boldsymbol{x}_p)$, and assume $r(\boldsymbol{X}) = p < m$.

For making inferences about the θ_i under the model in equation (10), we assume that direct estimators y_i are available and $y_i \overset{ind}{\sim} N(\theta_i, V_i)$, $i = 1, \ldots, m$. Write $\boldsymbol{y} = (y_1, \ldots, y_m)^T$. In addition, it is assumed that there exist independent statistics $\hat{V}_i$ such that $\hat{V}_i^{-1} \overset{ind}{\sim} V_i \frac{\chi^2_{n_i}}{n_i} (i = 1, \ldots, m)$. We shall assign prior distributions to $\boldsymbol{b}, V_1, \ldots, V_m$ and σ^2.

The full hierarchical model is described below.

1. $\boldsymbol{y} \mid \boldsymbol{\theta}, V_1, \ldots, V_m, \boldsymbol{b}, \sigma^2 \sim \mathrm{N}(\boldsymbol{\theta}, \boldsymbol{\Lambda}^{-1})$, where $\boldsymbol{\Lambda}^{-1} = \mathrm{Diag}(V_1, \ldots, V_m)$.
2. $\boldsymbol{\theta} \mid V_1, \ldots, V_m, \boldsymbol{b}, \sigma^2 \sim \mathrm{N}(\boldsymbol{X b}, \sigma^2 \boldsymbol{I}_m)$.
3. $\boldsymbol{b}, \sigma^2, V_1, \ldots, V_m$ are marginally independent with $\boldsymbol{b} \sim \mathrm{Uniform}(\boldsymbol{R}^p)$, $\sigma^{-2} \sim \mathrm{Gamma}(\frac{1}{2}\mathrm{c}, \frac{1}{2}\mathrm{d})$, and $V_i^{-1} \overset{ind}{\sim} \mathrm{Gamma}(\frac{1}{2}\mathrm{a}, \frac{1}{2}\mathrm{g})$. By a $\mathrm{Gamma}(\alpha, \beta)$ random variable Z, we mean one with pdf $\mathrm{f}(z) \propto \exp(-\alpha z) z^{\beta-1}$.

This method is a generalization of the one proposed by Datta *et al.* (1992), where the V_is are assumed to be known. The posterior distribution of $\boldsymbol{\theta}$ given $\boldsymbol{y}$ is analytically intractable. Instead, it is found by Markov chain Monte Carlo (MCMC) numerical integration techniques. To this end, we find the full conditionals:

(a) $\boldsymbol{b} \mid \boldsymbol{y}, \hat{v}_1, \ldots, \hat{v}_m, \boldsymbol{\theta}, \sigma^2 \sim N_p\big((\boldsymbol{X}^T\boldsymbol{X})^{-1}\boldsymbol{X}^T\boldsymbol{\theta}, \sigma^2(\boldsymbol{X}^T\boldsymbol{X})^{-1}\big)$;

(b) $\sigma^{-2} \mid \boldsymbol{y}, \hat{v}_1, \ldots, \hat{v}_m, \boldsymbol{b}, \boldsymbol{\theta} \sim \mathrm{Gamma}\big(\frac{1}{2}(c + \sum_{i=1}^m (\theta_i - \boldsymbol{x}_i^T\boldsymbol{b})^2), \frac{1}{2}(d + m)\big)$;

(c) $V_1^{-1}, \cdots, V_m^{-1} \mid \boldsymbol{y}, \theta, \boldsymbol{b}, \sigma^2, \hat{v}_1, \ldots, \hat{v}_m \overset{ind}{\sim} \mathrm{Gamma}(\frac{1}{2}(a + (y_i - \theta_i)^2 + n_i\hat{v}_i), \frac{1}{2}(n_i + g + 1))$;

(d) $\theta_1, \ldots, \theta_m \mid \boldsymbol{y}, \hat{v}_1, \ldots, \hat{v}_m, \boldsymbol{b}, \sigma^2 \overset{ind}{\sim} N\big((1 - B_i)y_i + B_i\boldsymbol{x}_i^T\boldsymbol{b}, \sigma^2(1 - B_i)\big)$, where $B_i = V_i/(\sigma^2 + V_i)$.

We now illustrate the above theory with the adjustment of 1990 census counts.

Adjustment of census counts has been a topic of heated debate for nearly a decade. The 1980 counts were never officially adjusted owing to a decision of the then Commerce Secretary. However, in several lawsuits brought against the Bureau of the Census by different states and cities who demanded revision of the reported counts, the topic of adjustment came up repeatedly in the courtroom testimony of statisticians appearing as expert witnesses on both sides. The issue was again hotly discussed and debated in subsequent scientific publications (see Ericksen and Kadane, 1985; Freedman and Navidi, 1986). This issue came up again with the appearance of the 1990 census data. Once again the Commerce Secretary announced on July 15, 1991, that the results of the 1990 census would not be adjusted, thus overturning the Census Bureau recommendation to use adjusted census data. Almost immediately after this, the city of New York and others brought a lawsuit seeking to overturn the decision of the Commerce Secretary. The case was tried in the courtroom of Federal Judge Joseph M. McLaughlin of Manhattan during May, 1992, and the verdict was in favor of not adjusting the census counts. There were eminent statisticians on both sides arguing for and against the adjustment of census counts.

The objective here is not to deal with the pros and cons of adjustment but show that the HB methodology can be used for adjustment of the 1990 census if needed.

Like other proponents of adjustment, we agree that the 1990 post enumeration survey (PES) data collected in August, 1990, forms the basis of adjustment. The 1990 PES is a sample of 170,000 housing units in 5400 sample block clusters, each cluster being either one block or a collection of several small blocks. To be useful, the PES results must be extended to nonsampled blocks. To this end, the population is divided into several groups or poststrata. The census count is known for each such poststratum, while the PES estimates the corresponding true population. The ratio of the PES estimate of the true population to the census count is known as the adjustment factor. The construction of poststrata has undergone several revisions, with the original proposal of 1392 poststrata being now replaced by 357 poststrata.

These poststrata are constructed as follows. First, for non-Hispanic white and other owners, there are four geographic areas under consideration: (i) northeast, (ii) south, (iii) midwest, and (iv) west. Each geographic area is then divided into (a) urbanized areas with population 250,000+, (b) other urban areas, and (c) nonurban areas. This leads to 12 strata. Similarly for non-Hispanic white and other non-owners (renters), there are 12 such strata. Next, black owners in urbanized areas with population 250,000+ are classified into four strata according to four geographic areas. However, black owners in other urban areas are collapsed into one stratum, as are black owners in nonurban areas. This leads to six strata for black owners. Similarly each category of black nonowners, nonblack Hispanic owners, and nonblack Hispanic nonowners is divided into six strata following the same pattern used in the construction of strata for the black owners.

So far we have reached a total of 12 + 12 + 6 + 6 + 6 + 6 = 48 strata. Added to these are three strata containing (i) Asian and Pacific-Islander owners, (ii) Asian and Pacific-Islander nonowners, and (iii) American Indians on reservations. This leads to a total of 51 strata. Each such stratum is now cross-classified with seven age–sex categories: (a) 0–17 (males and females), (b) 18–29 (males), (c) 18–29 (females), (d) 30–49 (males), (e) 30–49 (females), (f) 50+ (males), and (g) 50+ (females). This leads to a total of 51 x 7 = 357 poststrata.

The set of adjustment factors and the sample variances are available for all the 357 poststrata. However, for performing the HB analysis, we have not taken into account the last three categories of (i) Asian and Pacific-Islander owners, (ii) Asian and Pacific-Islander nonowners, and (iii)

American Indians on reservations, as it is generally felt that these categories should not be merged with the rest, and an HB analysis combines information from all the sources in calculating the smoothed adjustment factors. This leads to a HB analysis based on 336 poststrata. We do not report that analysis here, but discuss instead the results of a simpler analysis based on 48 poststrata where all seven age–sex categories are pooled into one. Even with this simplification, the main messages of this chapter, namely the need for (i) smoothing the adjustment factors and (ii) providing more reliable estimates of the associated standard errors, are clearly conveyed in our analysis.

We consider the hierarchical model as given above with a = g = c = d = 0 in point 3, above, to ensure some form of diffuse gamma priors for the inverse of the variance components in our model. The results, however, are not very sensitive to the choice of a, g, c, d as long as some version of diffuse prior is used. This means that the posterior inference does not change much when these parameters are close to zeroes. Next, the n_i, the degrees of freedom for the χ^2 distribution associated with $\hat{V}_i$ in the ith poststratum, represent the P-sample (the number of persons counted in the PES) in the ith poststratum divided by some factor, here 300. We admit that the number 300 is ad hoc, but feel that division by some such factor is essential to perform some meaningful analysis. Indeed, this number was suggested to us from conversations with people at the Bureau of the Census. The design matrix $\boldsymbol{X}$ provided to us from the Bureau of the Census was obtained via best subsets regression and is of the form

$$\boldsymbol{X}^T = (\boldsymbol{x}_1, \cdots, \boldsymbol{x}_{48}) \,,$$

where each $\boldsymbol{x}_i$ is a nine-component column-vector with the first element equal to 1, the second element equal to the indicator for nonowner, the third and fourth elements equal to the indicators for black and Hispanic, respectively, the fifth and the sixth elements denoting the indicators for an urbanized area with a population of 250,000+ and a nonurbanized area, respectively, and finally the seventh, eighth, and ninth elements denoting, the indicator or proportion in northeast, south, and west, respectively.

The HB analysis was performed by using the Gibbs sampler. In performing the analysis, we have taken t (the number of iterations needed to generate a sample) equal to 50, while the number of samples is taken as 2500. The stability in the point estimates of the adjustment factors is achieved once a sample of 1500 is generated, while stability in the associated standard errors is achieved once a sample of 2500 is generated.

The results of the HB analysis are reported in Table 1, which provides the adjustment factors (y), the corresponding standard errors (SD.y), the smoothed adjustment factors using the hierarchical model (HB1), the associated standard errors (SD.HB1), the smoothed adjustment factors using the model of Datta et al. (1992) (HB2) and the associated standard errors (SD.HB2) for all the 48 poststrata.

It is clear from Table 1 that both the present method and the one of Datta et al. essentially lead to the same point estimates of the adjustment factors and both methods lead to a substantial reduction in the standard errors. However, in most of the 48 poststrata, the estimated standard errors obtained by the present method (SD.HB1) are 1.5–2 times (sometimes even more) bigger than the ones of Datta et al. (SD.HB2). A few exceptions are poststrata 12, 25, 27, 28, and 31–34 where the estimated standard errors using the present method are lower than the ones using the model of Datta et al. This is somewhat surprising, although we should point out that there is no algebraic inequality between the two HB standard errors.

We conclude with the assertion that a model-based approach for smoothing the adjustment factors is strongly recommended. Also, hierarchical modeling is particularly well suited to meet this need.

4. GENERALIZED LINEAR MODELS

The results of the previous section are now extended to generalized linear models (GLMs) where the data are not necessarily continuous. Important special cases are the binomial logit and Poisson log–linear models. The hierarchical Bayes GLM that we consider is as follows:

1. conditional on $\theta_1, \ldots, \theta_m, y_1, \ldots, y_m$ are independent with probability density function (pdf)

$$f(y_i \mid \theta_i) = \exp\left[\phi_i^{-1}(y_i\theta_i - \psi(\theta_i)) + \rho(y_i; \phi_i)\right], \qquad 1 \le i \le m; \qquad (11)$$

   where the ϕ_is are assumed to be known;
2. $\theta_i \mid \boldsymbol{b}, \sigma^2 \overset{ind}{\sim} \mathrm{N}(\boldsymbol{x}_i^T\boldsymbol{b}, \sigma^2)$;
3. marginally $\boldsymbol{b}$ and σ^2 are mutually independent with $\boldsymbol{b} \sim \mathrm{Uniform}(\boldsymbol{R}^p)$, $\sigma^{-2} \sim \mathrm{Gamma}(\frac{1}{2}a, \frac{1}{2}g)$.

We first prove a result which provides sufficient conditions under which the joint posterior of $\boldsymbol{\theta} = (\theta_1, \ldots, \theta_m)^T$, $\boldsymbol{b}$ and σ^2 given $\boldsymbol{y} = (y_1, \ldots, y_m)^T$ is proper. This theorem generalizes the corresponding result of Ghosh (1998).

THEOREM 1: Assume that $f(y_i \mid \theta_i)$ is bounded for all i and

$$I_i = \int_{-\infty}^{\infty} \exp\left[\phi_i^{-1}(y_i\theta_i - \psi(\theta_i))\right] d\theta_i < \infty$$

Table 1. Raw adjustment factors, HB estimators and standard errors

I	Y	SD.Y	HB1	SD.HB1	HB2	SD.HB2
1	0.9792	0.0104	0.9902	0.0038	0.9897	0.0027
2	1.0069	0.0072	1.0038	0.0042	1.0030	0.0020
3	0.9974	0.0039	0.9948	0.0023	0.9949	0.0015
4	0.9966	0.0064	1.0027	0.0034	1.0054	0.0024
5	0.9893	0.0048	0.9908	0.0030	0.9909	0.0021
6	1.0052	0.0043	1.0044	0.0034	1.0041	0.0023
7	0.9990	0.0040	0.9954	0.0024	0.9961	0.0020
8	1.0063	0.0058	1.0035	0.0026	1.0067	0.0029
9	0.9947	0.0069	0.9937	0.0046	0.9926	0.0025
10	1.0018	0.0069	1.0072	0.0043	1.0057	0.0032
11	0.9930	0.0116	0.9981	0.0048	0.9975	0.0032
12	1.0029	0.0069	1.0063	0.0034	1.0083	0.0037
13	1.0117	0.0143	1.0238	0.0062	1.0272	0.0036
14	1.0262	0.0156	1.0374	0.0045	1.0404	0.0025
15	1.0239	0.0170	1.0283	0.0046	1.0322	0.0027
16	1.0328	0.0172	1.0365	0.0055	1.0430	0.0025
17	1.0353	0.0162	1.0245	0.0060	1.0284	0.0029
18	1.0330	0.0186	1.0380	0.0040	1.0415	0.0025
19	1.0124	0.0113	1.0288	0.0048	1.0332	0.0028
20	1.0470	0.0147	1.0371	0.0052	1.0442	0.0026
21	1.0697	0.0467	1.0274	0.0079	1.0301	0.0029
22	1.0665	0.0193	1.0409	0.0060	1.0433	0.0029
23	1.0293	0.0160	1.0318	0.0073	1.0350	0.0030
24	1.0648	0.0206	1.0400	0.0068	1.0459	0.0033
25	1.0165	0.0196	1.0108	0.0057	1.0071	0.0084
26	1.0221	0.0094	1.0242	0.0065	1.0202	0.0065
27	1.0082	0.0088	1.0151	0.0058	1.0120	0.0066
28	1.0649	0.0216	1.0234	0.0058	1.0230	0.0063
29	1.0136	0.0101	1.0230	0.0067	1.0198	0.0055
30	1.0364	0.0203	1.0271	0.0084	1.0226	0.0051
31	1.0913	0.0193	1.0445	0.0075	1.0448	0.0095
32	1.0669	0.0217	1.0579	0.0067	1.0578	0.0075
33	1.0638	0.0191	1.0489	0.0070	1.0496	0.0078
34	1.1106	0.0335	1.0570	0.0071	1.0604	0.0072
35	1.0433	0.0128	1.0561	0.0071	1.0568	0.0065
36	1.0484	0.0595	1.0605	0.0094	1.0598	0.0059
37	1.0068	0.0444	1.0132	0.0052	1.0107	0.0047
38	1.0259	0.0095	1.0267	0.0046	1.0239	0.0030

Table 1. (continued)

I	Y	SD.Y	HB1	SD.HB1	HB2	SD.HB2
39	0.9585	0.0238	1.0176	0.0049	1.0156	0.0036
40	1.0298	0.0092	1.0258	0.0046	1.0265	0.0030
41	1.0095	0.0170	1.0255	0.0044	1.0249	0.0029
42	1.0280	0.0283	1.0282	0.0053	1.0262	0.0030
43	1.0721	0.0404	1.0469	0.0075	1.0483	0.0055
44	1.1030	0.0311	1.0604	0.0054	1.0614	0.0039
45	1.0711	0.0374	1.0513	0.0066	1.0532	0.0043
46	1.0629	0.0209	1.0595	0.0067	1.0640	0.0036
47	1.0707	0.0310	1.0584	0.0062	1.0618	0.0032
48	1.1876	0.0724	1.0621	0.0077	1.0644	0.0032

is finite for at least one i. Let $S = \{i : I_i < \infty\}$ and s = cardinality of S. Then $\pi(\boldsymbol{\theta}, \boldsymbol{b}, \sigma^2 \mid \boldsymbol{y})$ is proper if $s + g > p,\ a > 0$.

PROOF: The joint posterior pdf of $\boldsymbol{\theta}$, $\boldsymbol{b}$, and σ^2 given $\boldsymbol{y}$ is

$$\pi(\boldsymbol{\theta}, \boldsymbol{b}, \sigma^2 \mid \boldsymbol{y}) \propto \prod_{i=1}^{m} f(y_i \mid \theta_i)(\sigma^2)^{-\frac{1}{2}m} \exp\left[-\frac{1}{2\sigma^2}\sum_{i=1}^{m}(\theta_i - \boldsymbol{x}_i^T\boldsymbol{b})^2\right] \exp\left(-\frac{a}{2\sigma^2}\right)(\sigma^2)^{-\frac{1}{2}g-1} \tag{12}$$

First integrate out those θ_i for which I_i is infinite. Denote by $\boldsymbol{\theta}^*$ the vector of remaining θ_is. Then the joint pdf of $\boldsymbol{\theta}^*$, $\boldsymbol{b}$, and σ^2 is given by

$$\pi(\boldsymbol{\theta}^*, \boldsymbol{b}, \sigma^2 \mid \boldsymbol{y}) \propto \prod_{i \in S} f(y_i \mid \theta_i)(\sigma^2)^{-\frac{1}{2}s} \exp\left[-\frac{1}{2\sigma^2}\sum_{i \in S}(\theta_i - \boldsymbol{x}_i^T\boldsymbol{b})^2\right] \exp\left(-\frac{a}{2\sigma^2}\right)(\sigma^2)^{-\frac{1}{2}g-1}. \tag{13}$$

Now first integrating with respect to $\boldsymbol{b}$, and then with respect to σ^2, it follows that the posterior of $\boldsymbol{\theta}^*$ is given by

$$\pi(\boldsymbol{\theta}^* \mid \boldsymbol{y}) \leq K \prod_{i \in S} f(y_i \mid \theta_i)$$

where $K(> 0)$ is a constant which does not depend on θ. From the assumptions of the theorem,

$$\int_{-\infty}^{\infty} f(y_i \mid \theta_i)\,\mathrm{d}\theta_i < \infty \qquad \text{for } i \in S$$

This proves the theorem.

Sometimes, however, element-specific auxiliary data $\boldsymbol{x}_{ij}$ are available. As in Ghosh et al. (1998), then we also have data y_{ij}, where y_{ij} denotes the observation for the jth unit in the ith local area ($j = 1, \ldots, n_i$; $i = 1, \ldots, m$). Now we have the HB generalized linear model given below:

1. Conditional on the θ_{ij}, y_{ij} are independent with

$$f(y_{ij} \mid \theta_{ij}) = \exp\left[\phi_{ij}^{-1}(y_{ij}\theta_{ij} - \psi(\theta_{ij})) + \rho(y_{ij}; \phi_{ij})\right]$$

2. $\theta_{ij} = \boldsymbol{x}_{ij}^T \boldsymbol{b} + u_i + \epsilon_{ij}$, where the u_i and the ϵ_{ij} are mutually independent with u_i iid $\mathrm{N}(0, \sigma^2)$ and ϵ_{ij} are iid $\mathrm{N}(0, \tau^2)$.
3. $\boldsymbol{b}, \sigma^2$ and τ^2 are mutually independent with $\boldsymbol{b} \sim \mathrm{Uniform}(\boldsymbol{R}^p)$, $\sigma^{-2} \sim \mathrm{Gamma}(\frac{1}{2}a, \frac{1}{2}g)$, $\tau^{-2} \sim \mathrm{Gamma}(\frac{1}{2}c, \frac{1}{2}d)$.

Such HB models are analyzed in Ghosh et al. (1998).

5. MULTIPLE MODELS

The discussion so far has concentrated on small area estimation based on specific HB generalized linear models. Often, there are situations where there is no clearcut choice among several models. In such situations, one can find the posterior probabilities of the different models, pick the one with the highest posterior probability, and find small area estimates and standard errors based on that model. Another option is not to report estimates and standard errors based on a single model, but report estimates which are weighted averages of estimates based on the different contemplated models, the respective weights being proportional to the posterior probabilities of these models. Similar views are expressed, for example, in Raftery (1996), and the same basic concept was used in Rubin and Stroud (1987). This method, being adaptive in nature, has intrinsic appeal, especially in situations when one particular model does not outperform the rest. Moreover, in finding the standard errors associated with the small area estimates, there is an extra layer of uncertainty due to the choice of models. This results in larger standard errors associated with the estimates, but the procedure seems worthwhile, especially when none of the contemplated models emerges as a clearcut winner.

To be specific, suppose there are K contemplated models labeled $M_1, \ldots, M_K$. Suppose the data is $\boldsymbol{y}$ and the parameter of interest is Δ. Then

$$P(\Delta|\boldsymbol{y}) = \sum_{k=1}^{K} P(\Delta|\boldsymbol{y}, M_k)P(M = M_k|\boldsymbol{y}). \tag{14}$$

This leads to

$$E(\Delta|\boldsymbol{y}) = \sum_{k=1}^{K} E(\Delta|\boldsymbol{y}, M_k)P(M = M_k|\boldsymbol{y}) \tag{15}$$

$$\begin{aligned} V(\Delta|\boldsymbol{y}) &= E(\Delta^2|\boldsymbol{y}) - (E(\Delta|\boldsymbol{y}))^2 \\ &= \sum_{k=1}^{K} E(\Delta^2|\boldsymbol{y}, M_k)P(M = M_k|\boldsymbol{y}) - [E(\Delta|\boldsymbol{y})]^2 \\ &= \sum_{k=1}^{K} V(\Delta|\boldsymbol{y}, M_k)P(M = M_k|\boldsymbol{y}) \\ &\quad + \sum_{k=1}^{K}(E(\Delta|\boldsymbol{y}, M_k))^2 P(M = M_k|\boldsymbol{y}) - E(\Delta|\boldsymbol{y})^2. \end{aligned} \tag{16}$$

Clearly, the first term in the right-hand side of equation (16) represents the expectation of the conditional variance of Δ given the data and the model, while the second and third terms combined represent the variance of the conditional expectation of Δ given the data and the model. Raftery (1993) gives a discussion of multiple models.

We present here the idea of mixing the models using the general description of the HB GLMs and later illustrate this idea with an example. For simplicity of discussion, we assume there are two possible models labelled M_1 and M_2.

Let Y_{ik} denote the minimal sufficient statistic (discrete or continuous) for the kth unit within the ith stratum ($k = 1, \ldots, n_i$; $i = 1, \ldots, m$), and Y_{ik} are assumed to be conditionally independent with pdf

$$f(y_{ik} \mid \theta_{ik}) = \exp\left[\phi_{ik}^{-1}(y_{ik}\theta_{ik} - \psi(\theta_{ik})) + \rho(y_{ik}; \phi_{ik})\right] \tag{17}$$

($k = 1, \ldots, n_i$; $i = 1, \ldots, m$). Under model M_1, the canonical parameter θ_{ik} is modeled as

$$\theta_{ik} = \boldsymbol{x}_{ik}^{(1)^T}\boldsymbol{b}_1^{(M_1)} + \boldsymbol{x}_{ik}^{(2)^T}\boldsymbol{b}_2^{(M_1)} + u_i + \epsilon_{ik} \tag{18}$$

and under model M_2, θ_{ik} is modeled as

$$\theta_{ik} = \boldsymbol{x}_{ik}^{(1)^T}\boldsymbol{b}_1^{(M_2)} + u_i + \epsilon_{ik}, \tag{19}$$

where u_is and ϵ_{ik}s are mutually independent, with u_is iid $N(0, r_u^{-1})$, while ϵ_{ik}s iid $N(0, r^{-1})$.

Note that it is important to distinguish between $\boldsymbol{b}_1^{(M_1)}$ and $\boldsymbol{b}_1^{(M_2)}$ in the two models, as they carry different interpretations. However, either model will cause estimates of the θ_{ik}s to borrow strength from other strata, as well as other cells within a given stratum.

Since this case involves finding posterior probabilities or the Bayes factor of the two models along with small area estimates, if one assigns a diffuse prior to the regression coefficients, then the posterior distribution of M (the indicator variable for the model) given the data becomes improper. Hence, one needs to assign a proper prior to $\boldsymbol{b}$.

The full hierarchical model is described below. For model M_1:

1. conditional on θ, $\boldsymbol{b}_1^{(M_1)}$, $\boldsymbol{b}_2^{(M_1)}$, $\boldsymbol{u}$, $R_u = r_u$ and $R = r$, Y_{ik} are independent with pdf given in equation (17);
2. conditional on $\boldsymbol{b}_1^{(M_1)}$, $\boldsymbol{b}_2^{(M_1)}$, $\boldsymbol{u}$, $R_u = r_u$ and R = r,

$$\theta_{ik} \overset{ind}{\sim} N(\boldsymbol{x}_{ik}^{(1)^T} \boldsymbol{b}_1^{(M_1)} + \boldsymbol{x}_{ik}^{(2)^T} \boldsymbol{b}_2^{(M_1)} + u_i, r^{-1})$$

3. conditional on $\boldsymbol{b}_1^{(M_1)}$, $\boldsymbol{b}_2^{(M_1)}$, $R_u = r_u$ and $R = r$, $u_i \overset{ind}{\sim} N(0, r_u^{-1})$;
4. $\boldsymbol{b}_1^{(M_1)}$, $\boldsymbol{b}_2^{(M_1)}$, $R_u = r_u$ and R = r are mutually independent with

$$\boldsymbol{b}^{(M_1)} = \begin{pmatrix} \boldsymbol{b}_1^{(M_1)} \\ \boldsymbol{b}_2^{(M_1)} \end{pmatrix} \sim N\left(\begin{pmatrix} \boldsymbol{0} \\ \boldsymbol{0} \end{pmatrix}, \quad \begin{pmatrix} \eta_1^2 \boldsymbol{I} & \boldsymbol{0} \\ \boldsymbol{0} & \eta_2^2 \boldsymbol{I} \end{pmatrix}\right)$$

$$R_u \sim \text{Gamma } (\tfrac{1}{2}a, \tfrac{1}{2}g) \text{ and R} \sim \text{Gamma } (\tfrac{1}{2}c, \tfrac{1}{2}d).$$

We shall choose η_1^2 to be large and η_2^2 small, to reflect the strong belief that $\boldsymbol{b}_2^{(M_1)}$ is close to 0 but attach a small amount of uncertainity to this belief. The near diffuseness of the prior on $\boldsymbol{b}_1^{(M_1)}$ reflects the vagueness in its choice in conformity with earlier models.

Model M_2 sets $\boldsymbol{b}_2^{(M_2)} = \boldsymbol{0}$. Other than that, the remainder of model M_2, remains the same as in model M_1. Also, we assign $P(M = M_1) = \pi$.

On notations, superscripts M_j, $j = 1, 2$ for $\boldsymbol{u}, r_u$ and r indicates that samples were observed from their respective full conditional distributions based on model $M_j, j = 1, 2$.

The full conditional distributions based on model M_1 are given by:

1. $\boldsymbol{b}^{(M_1)} | \boldsymbol{\theta}^{(M_1)}, \boldsymbol{u}^{(M_1)}, r_u^{(M_1)}, r^{(M_1)}, \boldsymbol{y} \sim$

$$N\left(\left(r^{(M_1)}\boldsymbol{X}^T\boldsymbol{X} + \begin{pmatrix} \eta_1^{-2}\boldsymbol{I} & \boldsymbol{0} \\ \boldsymbol{0} & \eta_2^{-2}\boldsymbol{I} \end{pmatrix}\right)^{-1} r^{(M_1)}\boldsymbol{X}^T(\boldsymbol{\theta}^{(M_1)} - \Sigma_i u_i^{(M_1)} \Sigma \boldsymbol{x}_{ik}),\right.$$
$$\left.\left(r^{(M_1)}\boldsymbol{X}^T\boldsymbol{X} + \begin{pmatrix} \eta_1^{-2}\boldsymbol{I} & \boldsymbol{0} \\ \boldsymbol{0} & \eta_2^{-2}\boldsymbol{I} \end{pmatrix}\right)^{-1}\right)$$

2. $u_i^{(M_1)}|\boldsymbol{\theta}^{(M_1)}, \boldsymbol{b}^{(M_1)}, r_u^{(M_1)}, r^{(M_1)}, \boldsymbol{y} \sim$
   $\mathrm{N}\left((r^{(M_1)}n_i + r_u^{(M_1)})^{-1} r^{(M_1)} \sum_k(\theta_{ik}^{(M_1)} - \boldsymbol{x}_{ik}^T\boldsymbol{b}^{(M_1)}), (r^{(M_1)}n_i + r_u^{(M_1)})^{-1}\right)$
3. $R^{(M_1)}|\boldsymbol{\theta}^{(M_1)}, \boldsymbol{b}^{(M_1)}, \boldsymbol{u}^{(M_1)}, r_u^{(M_1)}, \boldsymbol{y} \sim$
   Gamma $\left(\frac{1}{2}(c + \sum_i \sum_k(\theta_{ik}^{(M_1)} - \mathrm{x}_{ik}^T\boldsymbol{b}^{(M_1)} - u_i^{(M_1)})^2), \frac{1}{2}(d + \sum_1^m n_i)\right)$
4. $R_u^{(M_1)}|\boldsymbol{\theta}^{(M_1)}, \boldsymbol{b}^{(M_1)}, \boldsymbol{u}^{(M_1)}, r^{(M_1)}, \boldsymbol{y} \sim$
   Gamma $\left(\frac{1}{2}(a + \boldsymbol{u}^{(M_1)^T}\boldsymbol{u}^{(M_1)}), \frac{1}{2}(g + \sum_1^m n_i)\right)$
5. $\theta_{ik}^{(M_1)}|\boldsymbol{b}^{(M_1)}, \boldsymbol{u}^{(M_1)}, r_u^{(M_1)}, r^{(M_1)}, \boldsymbol{y} \overset{ind}{\sim}$
   $\Pi_{i,k}(\theta_{ik}^{(M_1)}|\boldsymbol{b}^{(M_1)}, \boldsymbol{u}^{(M_1)}, r_u^{(M_1)}, r^{(M_1)}, \boldsymbol{y}) \propto \exp[\{y_{ik}\theta_{ik}^{(M_1)} - \psi(\theta_{ik}^{(M_1)})\}\phi_{ik}^{-1}$
   $-\frac{1}{2}r^{(M_1)}(\theta_{ik}^{(M_1)} - \boldsymbol{x}_{ik}^T\boldsymbol{b}^{(M_1)} - u_i^{(M_1)})^2]$

For the full conditionals based on M_2, replace M_1 by M_2, $\boldsymbol{X}$ by $\boldsymbol{X}^{(1)}$, x_{ik} by $\boldsymbol{x}_{ik}^{(1)}$, and $\begin{pmatrix} \eta_1^{-2}\boldsymbol{I} & \boldsymbol{0} \\ \boldsymbol{0} & \eta_2^{-2}\boldsymbol{I} \end{pmatrix}$ by $\eta_1^{-2}\boldsymbol{I}$.

Finally, the full conditional for the model indicator variable M is given by

$$P\Big(M = M_1|\boldsymbol{\theta}^{(M_1)}, \boldsymbol{\theta}^{(M_2)}, \boldsymbol{b}_1^{(M_1)}, \boldsymbol{b}_2^{(M_1)}, \boldsymbol{b}_1^{(M_2)}, \boldsymbol{u}^{(M_1)}, r_u^{(M_1)}, r^{(M_1)},$$
$$\boldsymbol{u}^{(M_2)}, r_u^{(M_2)}, r^{(M_2)}, \boldsymbol{y}\Big)$$
$$= \pi \exp\left[-\frac{r^{(M_1)}}{2}\sum_i\sum_k(\theta_{ik}^{(M_1)} - \boldsymbol{x}_{ik}^T\boldsymbol{b}^{(M_1)} - u_i^{(M_1)})^2\right.$$
$$\left. - \frac{\eta_1^{-2}}{2}||\boldsymbol{b}_1^{(M_1)}||^2 - \frac{\eta_2^{-2}}{2}||\boldsymbol{b}_2^{(M_1)}||^2\right]$$
$$\div\left\{\pi \exp\left[-\frac{r^{(M_1)}}{2}\sum_i\sum_k(\theta_{ik}^{(M_1)} - \boldsymbol{x}_{ik}^T\boldsymbol{b}^{(M_1)} - u_i^{(M_1)})^2\right.\right.$$
$$\left. - \frac{\eta_1^{-2}}{2}||\boldsymbol{b}_1^{(M_1)}||^2 - \frac{\eta_2^{-2}}{2}||\boldsymbol{b}_2^{(M_1)}||^2\right]$$
$$+(1-\pi)\exp\left[-\frac{r^{(M_2)}}{2}\sum_i\sum_k(\theta_{ik}^{(M_2)} - \boldsymbol{x}_{ik}^{(1)T}\boldsymbol{b}_1^{(M_2)} - u_i^{(M_2)})^2\right.$$
$$\left.\left. - \frac{\eta_1^{-2}}{2}||\boldsymbol{b}_1^{(M_2)}||^2\right]\right\}$$

This idea is illustrated with a dataset given in Table 2, p. 613 of Fowlkes et al. (1988). The data relate to job satisfaction based on a 1981 survey of employees of a large national corporation. These employees were asked the question "Are you satisfied with your job?" The responses were "satisfied" and "not satisfied." The data give a breakdown of employees in these two categories according to seven geographic regions, namely, Northeast, Mid-Atlantic, Southern, Midwest, Northwest, Southwest and Pacific, as well as demographic categories such as age (less than 35, 35–44, and greater than 44), sex (male and female), and race (white and others). The sample size in some of these $7 \times 3 \times 2 \times 2 = 84$ cells is indeed small, for example non-white females in the 35–44 age group in the Mid-Atlantic region is only 3, while the size is only 2 for nonwhite females in the greater than 44 age group in the same region. Also, in the Southwest, nonwhite females in the greater than 44 age group is only 2.

The target is to estimate the true proportion of employees satisfied with their jobs in the 84 cells described in the preceding paragraph. In addition, one needs to produce standard errors associated with these estimates. Clearly, there is a need to borrow strength, especially for those cells where the sample size is very small. Moreover, since some of the cells with small sample sizes provide sample proportion 1 or 0 of employees satisfied with their job, reporting these estimates is not very meaningful. Also, reporting zero standard errors for those cells with 1 or 0 sample proportion is even more suspicious. The region variable reflects not only possible geographic diversity, but also differences due to local company management and policies. We mention, though, that no attempt is made to examine the question of possible bias due to nonresponse.

We shall use HB models which can overcome many of the existing problems. As possible covariates in this case, one may include the main effects due to sex and race as well as the different first- and second-order interactions. A preliminary analysis of the data, however, found the second-order interaction to be highly insignificant, and the interactions age–sex and age–race were also slightly nonsignificant. Consequently, two models are contemplated, one using as covariates all the main effects as well as the first-order interactions, while the second retains the main effects, and only the sex–race interaction as covariates.

Let Y_{ik} denote the number of people who responded "satisfied" with their jobs within the kth cell of the ith region, $i = 1, \ldots, 7$ and $k = 1, \ldots, 12$, where the cells represent a particular category of the variables age, sex, and race. The Y_{ik}s are then distributed as

$$Y_{ik} \overset{ind}{\sim} Bin(n_{ik}, p_{ik})$$

where n_{ik} represents the number of people sampled within the kth cell of the ith region, and p_{ik} represents the true proportion of people who are satisfied with their jobs within the kth cell of the ith region.

For the model involving all the main effects as well as all the first-order interactions, referred to as model M_1, the natural parameter $\theta_{ik} = \log \frac{p_{ik}}{1-p_{ik}}$ is modelled as

$$\theta_{ik}^{(M_1)} = \boldsymbol{x}_{ik}^{(1)^T} \boldsymbol{b}_1^{(M_1)} + \boldsymbol{x}_{ik}^{(2)^T} \boldsymbol{b}_2^{(M_1)} + u_i + \epsilon_{ik} \tag{20}$$

where u_i and ϵ_{ik} are mutually independent, with $u_i \overset{ind}{\sim} \mathrm{N}(0, r_u^{-1})$ and $\epsilon_{ik} \overset{ind}{\sim} \mathrm{N}(0, r^{-1})$, and

$$\boldsymbol{x}_{ik}^T \boldsymbol{b} = \mu^{(M_1)} + \alpha_a^{(M_1)} + \gamma_s^{(M_1)} + \eta_r^{(M_1)} + (\gamma\eta)_{sr}^{(M_1)} + (\alpha\gamma)_{as}^{(M_1)} + (\alpha\eta)_{ar}^{(M_1)} \tag{21}$$

In the above model, $\mu^{(M_1)}$ is the general effect, $\alpha_a^{(M_1)}$ is the main effect of the ath age group, $\gamma_s^{(M_1)}$ is the main effect of the sth sex, $\eta_r^{(M_1)}$ is the main effect of the rth race, $(\alpha\gamma)_{as}^{(M_1)}$ is the interaction effect of the ath age and sth sex, $(\alpha\eta)_{ar}^{(M_1)}$ is the interaction effect of the ath age and rth race, and $(\gamma\eta)_{sr}^{(M_1)}$ is the interaction effect of the sth sex and rth race. To avoid redundancy, we impose the corner-point restriction, namely,

$$\begin{aligned}\alpha_1^{(M_1)} &= \gamma_1^{(M_1)} = \eta_1^{(M_1)} = (\alpha\gamma)_{a1}^{(M_1)} = (\alpha\gamma)_{1s}^{(M_1)} \\ &= (\alpha\eta)_{a1}^{(M_1)} = (\alpha\eta)_{1r}^{(M_1)} = (\gamma\eta)_{s1}^{(M_1)} = (\gamma\eta)_{1r}^{(M_1)} = 0\end{aligned} \tag{22}$$

for all a, r, and s.

Under the reduced model,

$$\theta_{ik}^{(M_2)} = \boldsymbol{x}_{ik}^{(1)^T} \boldsymbol{b}_1^{(M_2)} + u_i + \epsilon_{ik} \tag{23}$$

where u_i and ϵ_{ik} are mutually independent with $u_i \overset{ind}{\sim} \mathrm{N}(0, r_u^{-1})$ and $\epsilon_{ik} \overset{ind}{\sim} \mathrm{N}(0, r^{-1})$, and

$$\boldsymbol{x}_{ik}^{(1)^T} \boldsymbol{b}_1^{(M_2)} = \mu^{(M_2)} + \alpha_a^{(M_2)} + \gamma_s^{(M_2)} + \eta_r^{(M_2)} + (\gamma\eta)_{sr}^{(M_2)} \tag{24}$$

where the notations α, γ, η, etc. have similar significance as in model M_1. Note that in the reduced model, the age–sex and age–race interactions are absent. Once again, corner-point restrictions similar to equation (22), but this time for this reduced model, are imposed. Several priors for η_1^2 and η_2^2 [$(\eta_1^2 = 10^4, \eta_2^2 = 0.001), (\eta_1^2 = 10^6, \eta_2^2 = 0.00001)$ etc.] with a combination of priors with near zero values for a, g, c, and d[$a = g = c = d = 5 \times 10^{-6}$, $a = g = c = d = 2 \times 10^{-5}$, and $a = g = c = d = 0.2$ etc.] were tried. A neutral prior assigning 1/2 probability to each of these models resulted in 0.69 posterior probability for the reduced model. However, rather than using only the reduced model, we present estimates which are weighted averages of estimates from the two models, the weights being proportional to the corresponding posterior probabilities of these models. The HB estimates and the associated standard errors of the proportion of people who would response "satisfied" if sampled remained virtually insensitive to the choice of the hyperparameters a, g, c, d.

For model diagnostics, in addition to the Bayes factor, we also computed the quantity

$$D^2 = \frac{\text{Dev}(\text{M}_1) - \text{Dev}(\text{M}_0)}{\text{Dev}(\text{M}_2) - \text{Dev}(\text{M}_0)}$$

where M_2 denotes the saturated model which includes the intercept as well as the first- and second-order interactions; M_0 is the intercept model. As M_1 we use first the model with the intercept as well as all the first-order interactions, while subsequently we use as M_1 the model without the age–sex and age-race interactions. The quantity D^2 was also used in Malec et al. (1997) as a tool for model diagnostics in a related small area estimation problem. For the present example, the quantity D^2 turned out to be 0.9448 and 0.8706, respectively in these two cases. This also suggests from a frequentist standpoint that it is preferable to use multiple models in the present case.

Table 2 reports (i) the raw estimates, the associated standard errors, (ii) the HB estimates for the model with the intercept, main effects, and first-order interaction, and the associated posterior s.ds, (iii) the HB estimates for the model without the age–sex and age–race interactions and the associated posterior s.ds, and (iv) finally HB estimates which are weighted averages of HB estimates based on the two models, and the associated posterior s.ds. Again, for brevity, Table 3.2 provides estimates only for regions northeast, southwest, and Pacific. Clearly, the weighted estimates lean more towards the estimates based on the reduced model than the ones for the full model (i.e., the model with all main effects and all first-order interactions), but one can clearly notice that there is a nonnegligible difference between the reduced model estimates and the weighted average estimates. Also, as anticipated, the posterior s.ds associated with the weighted average estimates are typically larger than the posterior s.ds based on either the full or reduced model estimates. As described earlier, this phenomenon is anticipated because the latter also incorporates uncertainty due to the model. However, all the posterior s.ds are much smaller than the raw standard errors.

Finally, in Table 3 we provide the summary estimates and the associated standard errors for all the seven regions where all the demographic categories are combined. The findings are very similar to those given in Table 2.

Table 2. Job Satisfaction Survey

Region	Category	N	n	Sample		H. Bayes					
						M_2		M_1		Mixed	
				Prop.	SD	Prop.	SD	Prop.	SD	Prop.	SD
Northwest	White, <35, M	450	270	.60000	.02309	.60637	.01948	.60715	.01953	.60661	.01950
	White, <35, F	327	176	.53823	.02757	.56173	.02358	.55731	.02354	.56037	.02357
	White, 35–44, M	323	215	.66563	.02625	.66034	.02088	.66018	.02144	.66029	.02105
	White, 35–44, F	120	80	.66667	.04303	.64494	.02862	.65097	.03064	.64680	.02926
	White, >44, M	405	269	.66420	.02347	.67927	.01950	.67857	.01953	.67906	.01951
	White, >44, F	150	110	.73333	.03611	.70277	.02608	.70651	.02727	.70392	.02646
	Other, <35, M	56	36	.64286	.06403	.65808	.03317	.66403	.03432	.65991	.03353
	Other, <35, F	41	25	.60976	.07618	.54935	.03766	.55125	.03890	.54994	.03805
	Other, 35–44, M	16	9	.56250	.12402	.68463	.03697	.67006	.04355	.68014	.03912
	Other, 35–44, F	16	11	.68750	.11588	.58219	.04145	.57982	.04590	.58146	.04287
	Other, >44, M	19	16	.84211	.08365	.74755	.03265	.74441	.04197	.74659	.03578
	Other, >44, F	9	4	.44444	.16563	.62012	.04103	.62628	.05038	.62202	.04412
Southwest	White, <35, M	378	252	.66667	.02425	.64866	.02073	.64990	.02085	.64904	.02077
	White, <35, F	158	97	.61392	.03873	.60420	.02720	.59893	.02846	.60258	.02759
	White, 35–44, M	234	162	.69231	.03017	.67305	.02322	.67322	.02406	.67311	.02348
	White, 35–44, F	74	47	.63514	.05596	.63273	.03115	.63957	.03392	.63484	.03203
	White, >44, M	292	199	.68151	.02726	.69175	.02056	.69091	.02103	.69149	.02071
	White, >44, F	86	62	.72093	.04837	.69445	.02865	.69853	.03078	.69571	.02932

Table 2. (continued)

						H. Bayes					
				Sample		M_2		M_1		Mixed	
Region	Category	N	n	Prop.	SD	Prop.	SD	Prop.	SD	Prop.	SD
	Other, <35, M	96	69	.71875	.04589	.67857	.02973	.68508	.03086	.68057	.03008
	Other, <35, F	81	45	.55556	.05521	.54233	.03373	.54430	.03421	.54293	.03388
	Other, 35–44, M	21	14	.66667	.10287	.69148	.03542	.67696	.04245	.68701	.03773
	Other, 35–44, F	12	8	.66667	.13608	.57853	.04177	.57630	.04615	.57785	.04317
	Other, >44, M	19	14	.73684	.10102	.73987	.03283	.73625	.04208	.73875	.03593
	Other, >44, F	2	2	1.00000	.00000	.63211	.04129	.63916	.05166	.63428	.04474
Pacific	White, <35, M	177	119	.67232	.03528	.64270	.02584	.64461	.02574	.64329	.02581
	White, <35, F	95	62	.65263	.04885	.61452	.03063	.60845	.03208	.61265	.03109
	White, 35–44, M	86	66	.76744	.04556	.68489	.03031	.68619	.03171	.68529	.03074
	White, 35–44, F	30	20	.66667	.08607	.63694	.03446	.64475	.03786	.63935	.03554
	White, >44, M	88	67	.76136	.04544	.71948	.02732	.71876	.02820	.71926	.02760
	White, >44, F	35	25	.71429	.07636	.68816	.03193	.69270	.03451	.68956	.03275
2-12	Other, <35, M	61	45	.73770	.05632	.67764	.03261	.68498	.03292	.67990	.03271
	Other, <35, F	37	22	.59459	.08072	.54541	.03775	.54723	.03930	.54597	.03824
	Other, 35–44, M	25	15	.60000	.09798	.68413	.03590	.67037	.04220	.67989	.03795
	Other, 35–44, F	18	10	.55556	.11712	.57194	.04062	.56898	.04541	.57103	.04215
	Other, >44, M	14	8	.57143	.13226	.73094	.03464	.72706	.04296	.72974	.03740
	Other, >44, F	8	6	.75000	.15309	.63345	.04057	.63995	.05053	.63545	.04388

Table 3. Job satisfaction survey overall region estimates

Region	Sample		Model M_2 H. Bayes		Modek M_1 H. Bayes		Mixed Model H. Bayes	
	Prop.	SD	Prop.	SD	Prop.	SD	Prop.	SD
Northeast	.6114	.0112	.6315	.0096	.6313	.0097	.6314	.0096
Mid-Atlantic	.7098	.0190	.6713	.0132	.6718	.0139	.6715	.0134
Southern	.6406	.0127	.6392	.0093	.6395	.0094	.6393	.0094
Midwest	.6234	.0109	.6341	.0086	.6341	.0088	.6341	.0087
Northwest	.6320	.0110	.6352	.0088	.6352	.0088	.6352	.0088
Southwest	.6683	.0124	.6556	.0097	.6560	.0100	.6557	.0098
Pacific	.6899	.0178	.6554	.0124	.6559	.0128	.6556	.0125

6. CONCLUDING REMARKS

This chapter discusses some Bayesian methods that have so far been used for small area estimation. There are many other applications, for example time series modeling, modeling of spatial data, etc., which are not included in this chapter. The following general comments are from a referee. Bayesian small area estimation may be performed either using the usual infinite population model or finite population theory. In the present chapter, the latter approach is used only in the section on synthetic and composite estimation, but finite populations may be incorporated into HB theory by imposing a Bayesian model on the finite population members, of which the sample is a subset. This is used, for example in Datta and Ghosh (1991). Whether this is desirable in small area estimation depends on whether the populations are small enough for the finite population correction to make any difference. This will not be so when the small area is "large," e.g., if it is a state, but the finiteness of the population may be relevant when the small area is a county, municipality, or census division, depending on the number of individuals being studied in the small area. This can be done along the lines of Section 2, where a predictive approach to small area estimation is presented. The method is clearly applicable for complex hierarchical models as well.

ACKNOWLEDGMENTS

Thanks are due to two referees for their very careful reading of the manuscript and for many helpful suggestions. Research was partially supported by NSF Grant Number SBR-9423996.

REFERENCES

Datta, G. S. and Ghosh, M. (1991). Bayesian prediction in linear models: Applications to small area estimation. *Ann. Stat.*, **19**, 1748–1770.

Datta, G. S., Ghosh, M., Huang, E. T., Isaki, C. T., Schultz, L. K. and Tsay, J. H. (1992). Hierarchical and empirical Bayes methods for adjustment of census undercount: The 1988 Missouri dress rehearsal data. *Survey Methodol.*, **18**, 95–108.

Dempster, A. P. and Tomberlin, T. J. (1980). The analysis of census undercount from a post enumeration survey. *Proceedings of the Conference on Census Undercount*, Bureau of the Census, pp. 88–94.

Ericksen, E.P. and Kadane, J.B. (1985). Estimating the population in a census year (with discussion). *J. Am. Stat. Assoc.*, **80**, 98–131.

Farrell, P., MacGibbon, B., and Tomberlin, T. J. (1997). Empirical Bayes estimators of small area proportions in multistage designs. *Statistica Sinica*, in press.

Fay, R. E. and Herriot, R. (1979). Estimates of income for small places: An application of James–Stein procedures to census data. *J. Am. Stat. Assoc.*, **74**, 269–277.

Fowlkes, E. B., Freeny, A. E., and Landwehr, J. M. (1988). Evaluating logistic models for large contingency tables. *J. Am. Stat. Assoc.*, **83**, 611–622.

Freedman, D. A. and Navidi, W. C. (1986). Regression models for adjusting the 1980 census (with discussion). *Stat. Sci.*, **1**, 3–39.

Ghosh, M. and Lahiri, P. (1987). Robust empirical Bayes estimation of means from stratified samples. *J. Am. Stat. Assoc.*, **82**, 1153–1162.

Ghosh, M. and Lahiri, P. (1992). A hierarchical Bayes approach to small area estimation with auxillary information. In:. P. K. Goel and N. S. Iyengar, eds. *Bayesian Analysis in Statistics and Econometrics*, Lecture Notes in Statistics, **75**. New York: Springer, pp. 107–125.

Ghosh, M. and Meeden, G. (1986). Empirical Bayes estimation in finite population sampling. *J. Am. Stat. Assoc.*, **74**, 269–277.

Ghosh, M. and Rao, J.N.K. (1994). Small area estimation: An appraisal (with discussion). *Stat. Sci.*, **9**, 65–93.

Ghosh, M., Natarajan, K., Stroud, T. W. F., and Carlin, B. P. (1998). Generalized linear models for small area estimation. *J. Am. Stat. Assoc.*, in press.

Holt, D., Smith, T. M. F., and Tomberlin, T. J. (1979). A model-based approach to estimation for small subgroups of a population. *J. Am. Stat. Assoc.*, **74**, 405–410.

MacGibbon, B. and Tomberlin, T. J. (1989). Small area estimates of proportions via empirical Bayes techniques. *Survey Methodol.*, **15**, 237–252.

Malec, D., Sedransk, J., Moriarity, C. L., and Leclere, F. B. (1997). Small area inference for binary variables in the National Health Interview Survey. *J. Am. Stat. Assoc.*, **92**, 815–826.

Prasad, N.G.N. and Rao, J.N.K. (1990). The estimation of mean squared errors of small-area estimators. *J. Am. Stat. Assoc.*, **85**, 163–171.

Raftery, A.E. (1996). Approximate Bayes factors and accounting for model uncertainty in generalized linear models. *Biometrika*, **83**, 251–266.

Rubin, D. B. and Stroud, T. W. F. (1987). Bayesian break-point forecasting in parallel time series, with application to university admissions. *Can. J. Stat.*, **15**, 1–19.

Stroud, T. W. F. (1987). Bayes and empirical Bayes approaches to small area estimation. R. Platek, J. N. K. Rao, C. E. Saerndal, and M. P. Singh, eds. *Small Area Statistics. An International Symposium*. Wiley, pp. 124–140.

Stroud, T. W. F. (1991). Hierarchical Bayes predictive means and variances with application to sample survey inference. *Commun. Stat. Theory Methods.*, **20**, 13–36.

4

Bayes Sampling Designs for Selection Procedures

KLAUS J. MIESCKE University of Illinois at Chicago, Chicago, Illinois

1. INTRODUCTION

Let $P_1, \ldots, P_k$ be k populations which belong to a given one-parameter exponential family $\{F_\theta\}, \theta \in \Omega \subseteq \mathfrak{R}$, where P_i is associated with a certain parameter $\theta_i \in \Omega$, $i = 1, \ldots, k$, but the values of $\theta_1, \ldots, \theta_k$ are unknown. Suppose one wants to select, based on independent random samples of respective sizes $m_1, \ldots, m_k$, that population which has the largest θ-value, $\theta_{[k]}$ say. In the decision theoretic approach, let $L(\boldsymbol{\theta}, i)$ be a given loss for selecting population P_i at any $\boldsymbol{\theta} = (\theta_1, \ldots, \theta_k) \in \Omega^k$, $i = 1, \ldots, k$. Two special types of loss functions, which will be considered later on, are the so-called *"0-1" loss* $L(\boldsymbol{\theta}, i) = 0$ *if* $\theta_i = \theta_{[k]}$, and 1 otherwise, and the *linear loss* $L(\boldsymbol{\theta}, i) = \theta_{[k]} - \theta_i$, $i = 1, \ldots, k$. The performance of a selection rule d, i.e., a measurable function from the sampling space into the set $\{1, \ldots, k\}$, can then be measured by its expected loss, i.e., its frequentist risk, at each parameter configuration $\boldsymbol{\theta} \in \Omega^k$. Extending this framework to the Bayes approach, it is assumed that the parameters $\boldsymbol{\Theta} = (\Theta_1, \ldots, \Theta_k)$, say, are a priori random and follow a known prior density $\pi(\boldsymbol{\theta})$, $\boldsymbol{\theta} \in \Omega^k$. The purpose of this chapter is to provide an introduction to Bayes selection procedures, a brief review of multistage selection procedures, and a thorough discussion of recent results on Bayes look-ahead sampling designs for selection procedures.

The history of selection procedures dates back to the 1950s. The first such procedures considered were based on k independent samples of equal sample sizes. An overview and thorough discussions of these early procedures, and of the development of numerous branches of the theory of selection thereafter until 1979, is provided in the by now classical monograph by Gupta and Panchapakesan (1979). In celebration of "40 Years of Statistical Selection Theory," and its pioneers Robert E. Bechhofer, Shanti S. Gupta, and Milton Sobel, a conference was held on September 5–10, 1993, at Bad Doberan in Germany, and the proceedings were included in a special journal issue, edited by Miescke and Rasch (1996).

To reduce the effort or cost of sampling, without losing too much power in the decisions, selection procedures which incorporate combinations of various types of sampling, stopping, and selection components have been proposed and studied in the literature over the past three decades. In their fundamental monograph, Bechhofer et al. (1968) have derived optimum sequential selection rules for exponential families in the frequentist approach. These are based on vector-at-a-time sampling, i.e., sampling of the same number of observations from each population at a time or stage, the natural terminal selection decision, and an optimum stopping rule. Elimination of certain populations from further sampling, because they emerged as apparently inferior populations during the sampling process, is not allowed as an option at intermediate stages of the sampling process. At this point it should be mentioned that in a case where elimination of populations from further sampling was allowed, the option exists of extending this elimination to the pool of populations available for selection at the end, or not. An overview of sequential ranking and selection procedures is provided by Gupta and Panchapakesan (1991).

A simple selection procedure, based on vector-at-a-time sampling, which incorporates elimination from sampling and selection is a two-stage selection procedure of the following type. After $t_1$observations have been drawn from each population at Stage 1, a suitable set of populations is eliminated from further sampling. Then t_2 observations are drawn from each population that has not been screened out, and a final selection is then made from the latter, using all of the data observed from them. Here an option exists regarding the number of populations retained for Stage 2: it can be chosen as random or pre-determined. The former case has been studied by Gupta and Miescke (1984a) using the Bayes approach. Both cases and their extensions to multistage selection procedures have been treated, by Gupta and Miescke (1984b), also with the Bayes approach, using backward optimization. An overview of this class of procedures, which select the best population efficiently in terms of risk and sampling costs, is provided by Miescke (1984). The results derived depend heavily on the assumption that at each sampling stage an equal number of observations, which may, however, vary from stage to stage, is drawn from each population still in the running. This assumption allows the use of permutation symmetry in the posterior risks in connection with permutation invariant priors, which simplifies the analysis of such procedures.

Whenever a population is eliminated from the pool of populations retained for a final selection, a conflict of the following type may arise. The data collected from such a population prior to its elimination could make it look better again, relative to the other populations, after further sampling from the latter have turned out to be not so favorable. The Bayes approach clearly calls for the use of information from all observations that have been drawn, since more observations cannot increase the Bayes risk. Using this approach, it may in fact occur that such an eliminated population emerges, in terms of the posterior risk, as the population that appears to be the best. Elimination of populations from final selections is thus unreasonable from a Bayesian point of view. Elimination, or just temporary elimination, of populations from sampling at some stages of the sampling process can be incorporated in a natural way into the Bayes approach. The advantage of inherent permutation symmetry, however, is then lost. In conclusion, the allocation of a possibly unequal number of observations, where some of them may actually be zero, to the k populations at various stages seems to be more appropriate. Temporarily drawing no new observation from certain populations, but retaining all k populations in the pool for the final selection decision, may be called a soft elimination. Such a soft elimination also invites the use of priors which are not permutation symmetric, since updated priors of this type occur anyway in a natural manner at the various stages due to soft elimination.

In many statistical experiments, the sampling process extends over a substantial length of time. One of the advantages of the Bayesian approach in a statistical analysis is that it allows conclusions to be drawn at intermediate time points. In particular, such conclusions can be made toward modifications of the future sampling and decision process. Such types of adaptive sampling or sampling allocation schemes will be the main topic of discussions later on, when Bayes look-ahead selection procedures are considered. Dealing with information from samples of possibly unequal sizes from the k populations may occur quite naturally. A test person may not always come on schedule, or may drop out of the study, a test object may break under stress, a budget cut may force some test runs to be abandoned, or random time spans are under study which are subject to some form of censoring.

The main type of problem considered in this chapter is how to allocate observations to the k populations in a stepwise manner, where the goal is to select the best population at the end of the sampling process. More precisely, assume that k independent samples of respective sizes $n_1, \ldots, n_k$ have already been observed at a first stage from populations $P_1, \ldots, P_k$, which may be the combined outcomes of several previous stages, and that m additional observations can be taken at a future second stage. One interesting problem that is considered later is how to allocate $m_1, \ldots, m_k$ observations, subject to $m_1 + \cdots + m_k = m$, in an optimum way among the k populations, given all the information, prior and first-stage observations, gathered so far. Looking ahead with the expected posterior Bayes risk, given the information presently at hand, and then minimizing it, does not only provide an optimum allocation of observations in the future. It also allows us to assess how much better the final decision can be expected to be after further sampling has been done, following this optimum allocation. In marketing research such as direct marketing, medical research such as clinical trials (Whitehead, 1991), and social research such as survey sampling (Govindarajulu and Katehakis, 1991), interim analyses are very often performed at certain stages to decide if sampling should be continued, and if so, how to allocate new observations. Such Bayes designs have been studied in the binomial case, under various loss functions, by Gupta and Miescke (1993) for the more general problem of simultaneous selection and estimation, including cost of sampling. For the sake of simplicity of presentation, simultaneous estimation with selection and cost of sampling will only be considered briefly. The former would require the use of more involved loss functions, and the latter the incorporation of stopping rules. Modifications of the allocations considered later to such extended features are straightforward, but technically more involved.

Allocating m new observations at a second stage, using the expected posterior risk, can only be done after the terminal selection rule is known. The latter is the Bayes selection which is based on all observations drawn in the complete sampling process. Thus, the first step toward a Bayes design for the second stage is to determine the optimum (Bayes) single-stage selection rules for various sample sizes $m_1, \ldots, m_k$. This has been done, under both the 0–1 loss and the linear loss, for the binomial case by Abughalous and Miescke (1989), including extensions to a larger class of loss functions, and for the normal case by Gupta and Miescke (1988).

After the Bayes terminal selection decisions are known, one can proceed as described above, looking ahead for all possible sampling allocations, which in the present setting are restricted by $m_1 + \cdots + m_k = m$, compare the associated expected posterior risk, find its minimum value, and then implement a design associated with it. At this point one may wonder why all m observations are allocated at once, rather than allocating only a few (or just one), learning more through them (it), and then initiating a new allocation optimization process for the remaining allocations. As will be shown later, such a breakdown of allocation of m observations, if done properly, cannot increase the Bayes risk and may in fact be better than allocating all m observations at once. The best possible allocation scheme is to allocate one observation at a time, in m consecutive steps, which are altogether determined by backward optimization, starting at the end with the Bayes terminal selection for every possible allocation $m_1, \ldots, m_k$ with $m_1 + \cdots + m_k = m$, and then optimizing successively every single previous allocation. However, this appears to be feasible only for discrete distributions, and the only Bayes sequential design of this type that has been treated up to now is for the binomial case (Miescke and Park, 1997a). Alternative allocation schemes, which appear to perform close to the best based on backward optimization, are studied and discussed for the normal case by Gupta and Miescke (1994, 1996a), and for the binomial case by Gupta and Miescke (1996b) and Miescke and Park (1997a). Another type of adaptive sampling and selection for Bernoulli populations, which is in the frequentist approach, can be found in Bechhofer and Kulkarni (1982).

One reasonable procedure is to allocate one observation at a time in an optimum way, pretending that it is the last one to be drawn before final selection, and then to iterate this process until all m observations have been taken. This will be considered later in this chapter. Other procedures, which may allocate more than one observation at a time will also be considered. However, they appear to be less appealing, since with each new observation more can be learnt about the unknown parameters, which in turn can improve the basis for further decisions. Look-ahead procedures, which have been used previously by Govindarajulu and Katehakis (1991) in survey sampling, and which are described and discussed in various other settings in Berger (1985), will be discussed thoroughly later.

Selecting in terms of the largest sample mean is called in the literature the natural selection rule. It is the uniformly best permutation invariant selection procedure, in the frequentist sense, for a general class of loss functions, as long as the sample sizes are equal. However, for unequal sample sizes, the natural selection rule appears to be less powerful, although it still remains intuitively appealing. In view of this fact, optimum sample size allocations for the natural selection rule have been considered in the frequentist approach by Bechhofer (1969), Dudewicz and

Dalal (1975), and Bechhofer et al. (1991, 1995). Bayes selection rules under unequal sample sizes can have complicated forms which may not be represented in closed form, as has been shown by Abughalous and Miescke (1989) and Gupta and Miescke (1988). Bayes rules for more involved normal models have been studied by Berger and Deely (1988) and Fong and Berger (1993). Earlier ideas and results on sampling allocations for Bayes rules under normality and linear loss are from Dunnett (1960).

An introduction to Bayes selection procedure is provided in Section 2. Here, as well as in the remaining sections, special emphasis is given to two specific models under the 0–1 loss and the linear loss. The first is the normal case with independent normal priors for the k parameters, which is called the *normal–normal* model, and the second is the binomial case with independent beta priors, which is called the *binomial–beta* model. As a first step toward Bayes look-ahead sequential sampling designs, Bayes one- and two-stage sampling designs are studied in Section 3. Finally, Bayes look-ahead sequential sampling designs are treated in Section 4.

2. BAYES SELECTION PROCEDURES

Let $P_1, \ldots, P_k$ belong to a one-parameter exponential family $\{F_\theta\}$, $\theta \in \Omega \subseteq \Re$, where P_i is associated with a certain parameter $\theta_i \in \Omega$, $i = 1, \ldots, k$, but where the values of $\theta_1, \ldots, \theta_k$ are unknown. Let the goal be to find that population which has the largest parameter value. Special emphasis will be given to the normal family $\{N(\theta, \sigma^2)\}$, $\theta \in \Omega = \Re$, with $\sigma^2 > 0$ known, and to the binomial family $\{B(n, \theta)\}$, $\theta \in \Omega = [0, 1]$. Let $X_1, \ldots, X_k$ denote sufficient statistics from independent random samples of sizes $n_1, \ldots, n_k$ from $P_1, \ldots, P_k$, respectively. Since Bayes selection procedures are the topic, only nonrandomized decision rules need to be considered. These can be represented as measurable functions $\mathrm{d}(\boldsymbol{x})$ with values in $\{1, \ldots, k\}$, where $\boldsymbol{x} = (x_1, \ldots, x_k)$ are the observed values of $X = (X_1, \ldots, X_k)$. In the decision theoretic approach, let $L(\boldsymbol{\theta}, i)$ be the loss for selecting population P_i at $\boldsymbol{\theta} = (\theta_1, \ldots, \theta_k)$, with $L(\boldsymbol{\theta}, i) \leq L(\boldsymbol{\theta}, j)$ for $\theta_i \geq \theta_j$, $i, j = 1, \ldots, k$. Later, emphasis will be given to two special loss functions, the 0–1 loss and the linear loss, which are defined by

$$
\begin{aligned}
L(\boldsymbol{\theta}, i) &= 1 - I_{\{\theta_{[k]}\}}(\theta_i) && 0-1 \text{ loss} \\
L(\boldsymbol{\theta}, i) &= \theta_{[k]} - \theta_i && \text{linear loss}
\end{aligned}
\tag{1}
$$

where $\theta_{[k]} = \max\{\theta_1, \ldots, \theta_k\}, \mathrm{i} = 1, \ldots, k$.

Finally, the Bayesian component is added to the problem. Here the parameters $\boldsymbol{\Theta} = (\Theta_1, \ldots, \Theta_k)$ are assumed to be a priori independent random variables which follow a prior distribution with a known density $\pi(\boldsymbol{\theta}) = \pi_1(\theta_1) \times \cdots \times \pi_k(\theta_k)$. For the normal family, normal priors $\Theta_i \sim N(\mu_i, \nu_i^{-1})$ with $\mu_i \in \Re$, $\nu_i > 0$ will be assumed, and for the binomial family, beta priors $\Theta_i \sim \text{Beta}(\alpha_i, \beta_i)$, with $\alpha_i, \beta_i > 0$, $i = 1, \ldots, k$. The frequentist risk of a selection rule d at $\boldsymbol{\Theta} = \boldsymbol{\theta}$ is given by $R(\boldsymbol{\theta}, d) = E_{\boldsymbol{\theta}}(L(\boldsymbol{\theta}, d(\mathbf{X}))$, and its Bayes risk by $r(\pi, d) = E^{\pi}(R(\boldsymbol{\Theta}, d))$. The latter is minimized by every Bayes rule, and this minimum $r(\pi)$, say, is called the Bayes risk of the problem. The Bayes risk of a rule d can be represented in two ways as follows:

$$\begin{aligned} r(\pi, d) &= E(L(\boldsymbol{\Theta}, d(\mathbf{X}))) \\ &= E^{\pi}(\, E\{L(\boldsymbol{\Theta}, d(\mathbf{X})) \mid \boldsymbol{\Theta}\}) \\ &= E^{m}(\, E\{L(\boldsymbol{\Theta}, d(\mathbf{X})) \mid \mathbf{X}\}) \end{aligned} \tag{2}$$

where m denotes the marginal density or discrete probability function of $\mathbf{X}$. The standard way of determining a Bayes rule d^B, say, is to minimize, at every $\mathbf{X} = \mathbf{x}$, the posterior expected loss, i.e., the posterior Bayes risk $E\{L(\boldsymbol{\Theta}, d(\mathbf{x})) \mid \mathbf{X} = \mathbf{x}\}$. Depending on the type of loss function, one arrives at the following criteria:

$$\begin{aligned} E\{L(\boldsymbol{\Theta}, d^B(\mathbf{x})) \mid \mathbf{X} = \mathbf{x}\} &= \min_{i=1,\ldots,k} E\{L(\boldsymbol{\Theta}, i) \mid \mathbf{X} = \mathbf{x}\} \\ &\quad \text{in general} \\ &= 1 - \max_{i=1,\ldots,k} P\{\Theta_i = \Theta_{[k]} \mid \mathbf{X} = \mathbf{x}\} \\ &\quad \text{for 0–1 loss} \\ &= E\{\Theta_{[k]} \mid \mathbf{X} = \mathbf{x}\} - \max_{i=1,\ldots,k} E\{\Theta_i \mid \mathbf{X} = \mathbf{x}\} \\ &\quad \text{for linear loss} \end{aligned} \tag{3}$$

Apparently, under the linear loss, $E\{\boldsymbol{\Theta}_{[k]} \mid \mathbf{X} = \mathbf{x}\}$ does not need to be considered for finding a Bayes rule. However, it is relevant for the evaluation of the Bayes risk $r(\pi)$, and thus it will be relevant for the Bayes designs to be considered in the subsequent sections. For the two special loss functions, the quantities to be minimized or maximized in equation (3) can be represented by

$$E\{L(\boldsymbol{\Theta}, i)|\mathbf{X} = \mathbf{x}\} = \int_{\Re^k} L(\boldsymbol{\theta}, i)\ \pi(\boldsymbol{\theta}\,|\mathbf{x})\ \mathrm{d}\boldsymbol{\theta}$$

in general

$$P\{\Theta_i = \Theta_{[k]}\,|\,\mathbf{X} = \mathbf{x}\} = \int_{\Re} \prod_{j \neq i} \left(\int_{-\infty}^{\theta_i} \pi_j(\theta_j|x_j)\ \mathrm{d}\theta_j \right) \pi_i(\theta_i|x_i)\ \mathrm{d}\theta_i \tag{4}$$

for 0–1 loss

$$E\{\Theta_i\,|\,\mathbf{X} = \mathbf{x}\} = \int_{\Re} \theta_i\ \pi_i(\theta_i|x_i)\ \mathrm{d}\theta_i$$

for linear loss

where $\pi(\boldsymbol{\theta}\,|\,\mathbf{x})$ is the posterior density of $\boldsymbol{\Theta}$, given $\mathbf{X} = \mathbf{x}$, and $\pi_r(\theta_r|x_r)$ is the posterior marginal density of Θ_r, given $\mathbf{X} = \mathbf{x}$, $r = 1, \ldots, k$. In the second and third part of equation (4), the fact that $\pi(\boldsymbol{\theta}\,|\,\mathbf{x}) = \pi_1(\theta_1|x_1) \times \cdots \times \pi_k(\theta_k|x_k)$ is utilized. This means that a posteriori, $\Theta_1, \ldots, \Theta_k$ are not only independent, but that the posterior distribution of Θ_r, given $\mathbf{X} = \mathbf{x}$, depends on $\mathbf{x}$ only through x_r, $r = 1, \ldots, k$.

In the remainder of this section, Bayes rules for the normal–normal and the binomial–beta model will be studied under the 0–1 loss and the linear loss. Because of the inherent independence, only the distributions associated with each individual ith of the k populations, $i \in \{1, \ldots, k\}$, have to be specified. For the normal–normal model one has

$$\begin{aligned} \text{Normal–Normal}: \quad & X_i|\boldsymbol{\Theta} = \boldsymbol{\theta} \sim N\left(\theta_i, p_i^{-1}\right), \qquad \Theta_i \sim N\left(\mu_i, \nu_i^{-1}\right) \\ & \Theta_i|\mathbf{X} = \mathbf{x} \sim N\left(\frac{p_i x_i + \nu_i \mu_i}{p_i + \nu_i}, (p_i + \nu_i)^{-1}\right) \\ & X_i \sim N\left(\mu_i, p_i^{-1} + \nu_i^{-1}\right) \end{aligned} \tag{5}$$

where X_i is the sample mean of the ith population, and $p_i = n_i \sigma^{-2}$ is its precision. The Bayes selection rules under 0–1 loss and linear loss can now be seen, in view of equations (3)–(5), to be $d^B(\mathbf{x}) = i_0$, if for $i = 1, \ldots, k$, the following respective quantity is maximized at $i = i_0$.

$$\begin{aligned} P\{\Theta_i = \Theta_{[k]}\,|\,\mathbf{X} = \mathbf{x}\} &= \int_{\Re} \prod_{j \neq i} \Phi_j(\theta\,|x_j)\ d\Phi_i(\theta\,|x_i) && \text{for } 0-1 \text{ loss} \\ E\{\Theta_i\,|\,\mathbf{X} = \mathbf{x}\} &= \frac{p_i x_i + \nu_i \mu_i}{p_i + \nu_i} && \text{for linear loss} \end{aligned} \tag{6}$$

where $\Phi_r(\cdot\,|\,x_r)$ is the c.d.f. of the conditional distribution of Θ_r, given $\mathbf{X} = \mathbf{x}$, $r = 1, \ldots, k$, which is represented in equation (5).

At this point it should be mentioned that there is a natural selection rule d^N, say, which selects in terms of the largest of the sample means $\bar{X}_1, \ldots, \bar{X}_k$. In terms of the frequentist risk, and for a large class of loss functions, including the 0–1 loss and the linear loss, it is the uniformly best permutation invariant selection procedure if the sample sizes $n_1, \ldots, n_k$ are all equal. More generally, an analogous fact holds for monotone likelihood ratio families. The history of its proofs, one of which is in the Bayesian approach utilizing permutation-invariant priors, can be found in Gupta and Miescke (1984b). For the present situation where $n_1, \ldots, n_k$ may not be equal, no optimum properties of d^N under the 0–1 loss are known. Investigations toward possible admissibility have been made only recently by Miescke and Park (1997b). Gupta and Miescke (1988) have shown that d^N is minimax under the 0–1 loss if and only if $n_1 = \ldots = n_k$. Here the minimax value of the problem is $1 - 1/k$, which can be proved with a suitable sequence of independent normal priors.

An undesirable property of d^N under the 0–1 loss was first discovered (Lam and Chiu, 1976; Tong and Wetzell, 1979) by noting that the frequentist risk of d^N is not always increasing in each of the sample sizes $n_1, \ldots, n_k$. On the other hand, under the linear loss, d^N is a proper Bayes rule and, because of its uniqueness, also admissible (Berger, 1985). This can be readily seen (Gupta and Miescke, 1988) from equation (6), by letting $\mu_r = \mu$ and $\nu_r = c\, p_r$, $r = 1, \ldots, k$, for some fixed real μ and a positive c. Properties of Bayes rules for other priors are also discussed there, as well as in Berger and Deely (1988) and Fong and Berger (1993).

Turning now to the binomial–beta model, the situation regarding the distributions presents itself as follows:

$$\begin{aligned} \text{Binomial–Beta}: X_i|\boldsymbol{\Theta} = \boldsymbol{\theta} &\sim B(n_i, \theta_i), \qquad \Theta_i \sim \text{Beta}(\alpha_i, \beta_i) \\ \Theta_i|\mathbf{X} = \mathbf{x} &\sim \text{Beta}(\alpha_i + x_i, \beta_i + n_i - x_i) \\ X_i &\sim \text{PE}(n_i, \alpha_i, \beta_i, 1) \end{aligned} \tag{7}$$

where $\alpha_i, \beta_i > 0$, $i = 1, \ldots, k$. Here the unconditional marginal distribution of X_i, $i = 1, \ldots, k$, is a Pólya–Eggenberger-type distribution (Johnson and Kotz, 1969), sometimes called beta–binomial distribution, which in the present situation turns out to be

$$P\{X_i = x_i\} = \binom{n_i}{x_i} \frac{\Gamma(\alpha_i + \beta_i)}{\Gamma(\alpha_i)\Gamma(\beta_i)} \frac{\Gamma(\alpha_i + x_i)\Gamma(\beta_i + n_i - x_i)}{\Gamma(\alpha_i + \beta_i + n_i)}, \quad x_i = 0, 1, \ldots, n_i. \tag{8}$$

The Bayes selection rules under 0–1 loss and linear loss can be seen, in view of equations (3), (4), and (7), to be $d^B(\mathbf{x}) = i_0$, if for $i = 1, \ldots, k$, the following is maximized at $i = i_0$.

$$P\{\Theta_i = \Theta_{[k]} \mid \mathbf{X} = \mathbf{x}\} = \int_{[0,1]} \prod_{j \neq i} F_j(\theta \mid x_j)\, dF_i(\theta \mid x_i), \qquad \text{for 0–1 loss}$$

$$E\{\Theta_i \mid \mathbf{X} = \mathbf{x}\} = \frac{\alpha_i + x_i}{\alpha_i + \beta_i + n_i}, \qquad \text{for linear loss} \tag{9}$$

where $F_r(\cdot \mid x_r)$ is the c.d.f. of the conditional distribution of Θ_r, given $\mathbf{X} = \mathbf{x}$, $r = 1, \ldots, k$, which is represented in equation (7).

At this point, again, the natural selection rule d^N, which selects in terms of the largest of the sample means $X_1/n_1, \ldots, X_k/n_k$, has to be discussed. Since the binomial family is also an exponential family, d^N is also the uniformly best invariant selection rule if and only if the sample sizes are equal. Likewise, Abughalous and Miescke (1989) have shown that under the 0–1 loss, minimaxity of d^N holds if and only if $n_1 = \ldots = n_k$. As in the normal case, the minimax value of the problem is $1 - 1/k$, which can be proved with a suitable sequence of independent beta priors. Other results regarding d^N turn out to be different from their counterparts in the normal case. For example, d^N is not a proper Bayes rule here. Properties of Bayes rules, for various priors, under the 0–1 loss and the linear loss have been studied in the same paper. Questions regarding the optimality of d^N for unequal sample sizes $n_1, \ldots, n_k$, in the present setting, have been addressed in Bratcher and Bland (1975), Risko (1985), and Abughalous and Miescke (1989).

There are two other types of Bayes selection procedures which are based on different goals or philosophies. These will be described briefly at the end of this section. The first is within the subset selection approach, which is due to Gupta (1956, 1965). Here the goal is to select a nonempty subset $s \subseteq \{1, \ldots, k\}$, preferably of small size, which contains the best population. This goal can be represented in various ways by means of the loss function. The first paper within this framework was by Deely and Gupta (1968), and deals with the linear loss function $L(\boldsymbol{\theta}, s) = \sum_{i \in s} \alpha_{s,i}(\theta_{[k]} - \theta_i)$. The loss function $L(\boldsymbol{\theta}, s) = \sum_{i \in s}(a - b\, I_{\{\theta_{[k]}\}}(\theta_i))$ has been used by Bratcher and Bhalla (1974) and Gupta and Hsu (1977), the loss function $L(\boldsymbol{\theta}, s) = c\,|s| + \theta_{[k]} - \max_{i \in s} \theta_i$ has been used by Goel and Rubin (1977), and the additive loss function $L(\boldsymbol{\theta}, s) = \sum_{i \in s} l_i(\boldsymbol{\theta})$, with emphasis on the special case $L(\boldsymbol{\theta}, s) = \sum_{i \in s} (\theta_{[k]} - \theta_i - \varepsilon)$, has been used by Miescke (1979). Another, nonadditive loss function has been used by Chernoff and Yahav (1977). Further references in this regard can be found in Gupta and Panchapakesan (1979).

The other type of selection procedures combines selection of the best population with the estimation of the parameter of the selected population. The decision rules are now of the form $\delta(\mathbf{x}) = (d(\mathbf{x}), e_{d(\mathbf{x})}(\mathbf{x}))$, where $d(\mathbf{x}) \in \{1, \ldots, k\}$ is the selection rule, and $e_i(\mathbf{x})$, $i = 1, \ldots, k$, is a collection of estimates for θ_i, $i = 1, \ldots, k$, which are available after selection. As is shown in Cohen and Sackrowitz (1988) and Gupta and Miescke (1990), the decision theoretic treatment of the combined selection and estimation problem consists of two steps of optimization. First the possible estimates are determined, which turn out to be the usual Bayes estimates of the related problem of estimation without considering selection. Then, after knowing the available estimates, the optimum selection is made. Detailed results for the normal case, under the additive loss function $L(\boldsymbol{\theta}, \delta) = A(\boldsymbol{\theta}, d) + B(\theta_d, e_d)$, and various special cases, are presented in Gupta and Miescke (1990). Here $A(\boldsymbol{\theta}, d)$ is the loss due to selecting population P_d, and $B(\theta_d, e_d)$ is the loss of estimating θ_d with e_d, $d = 1, \ldots, k$. Similar work for the binomial case has been done by Gupta and Miescke (1993). In both papers, overviews of work in this direction and further references can be found.

3. BAYES ONE- AND TWO-STAGE SAMPLING DESIGNS

Starting with the situation described in the previous section up to equation (3), let us now consider a fixed total sample size allocation problem. Suppose that in the planning stage of the experiment, a total of $n_1 + \cdots + n_k = n$ observations are planned to be drawn from the respective populations. Since, after every observed $\mathbf{X} = \mathbf{x}$, the posterior Bayes risk will be equal to $\min_{i=1,\ldots,k} E\{L(\boldsymbol{\Theta}, i)|\mathbf{X} = \mathbf{x}\}$, the optimum allocation of $n_1, \ldots, n_k$, i.e., the Bayes design, is given by the following criterion:

$$\begin{aligned}
&\min_{n_1+\cdots+n_k=n} E\big(\min_{i=1,\ldots,k} E\{L(\boldsymbol{\Theta}, i)|\mathbf{X}\}\big) && \text{in general} \\
&\max_{n_1+\cdots+n_k=n} E\big(\max_{i=1,\ldots,k} P\{\Theta_i = \Theta_{[k]}|\mathbf{X}\}\big), && \text{for 0–1 loss} \\
&\max_{n_1+\cdots+n_k=n} E\big(\max_{i=1,\ldots,k} E\{\Theta_i|\mathbf{X}\}\big), && \text{for linear loss}
\end{aligned} \tag{10}$$

where the outer expectation is with respect to the unconditional marginal distribution of $\mathbf{X}$. For the normal–normal model and the binomial–beta model, its representation is given in equation (5) and in equation (7) and (8), respectively. The inner conditional expectations and probabilities in equation (10) are represented in general form in equation (4), and in their special forms for the normal–normal model and the binomial–beta model in equations (6) and (9), respectively.

In the first part of this section, Bayes designs for the two special models will be studied under the 0–1 loss and the linear loss. For the normal–normal model, the Bayes designs consist of those sample sizes $n_1, \ldots, n_k$ which achieve the following maximum:

$$\max_{n_1+\cdots+n_k=n} E\left(\max_{i=1,\ldots,k} \int_{\Re} \prod_{j\neq i} \Phi\left(\frac{\delta_i\theta + \mu_i(X_i) - \mu_j(X_j)}{\delta_j}\right)\varphi(\theta)\mathrm{d}\theta\right), \quad \text{for 0–1 loss}$$

$$\max_{n_1+\cdots+n_k=n} E\left(\max_{i=1,\ldots,k} \frac{p_i X_i + \nu_i \mu_i}{p_i + \nu_i}\right), \quad \text{for linear loss} \tag{11}$$

where in the first criterion, Φ and φ denote the c.d.f. and density, respectively, of N(0,1), and where $\delta_r = (p_r + \nu_r)^{-1/2}$, $\mu_r(X_r) = (p_r X_r + \nu_r \mu_r)/(p_r + \nu_r)$, and $p_r = n_r \sigma^{-2}$, $r = 1 \ldots k$, for brevity. The outer expectations are with respect to the marginally independent random variable $X_r \sim N(\mu_r, p_r^{-1} + \nu_r^{-1})$, $r = 1, \ldots k$.

The special case of $k = 2$ populations has been completely analyzed in Gupta and Miescke (1994). For both loss functions it has been shown there, using different techniques, that the Bayes design is determined by minimizing $|p_1 + \nu_1 - p_2 - \nu_2|$, the absolute difference between the two posterior precisions. It is zero if and only if the joint posterior distribution of Θ_1 and Θ_2 is decreasing in transposition, a situation in which the Bayes selection rules are of simple forms (Gupta and Miescke, 1988). The results for $k = 2$, mentioned above, carry over to $k \geq 3$ populations to some extent. This has been shown in Gupta and Miescke (1996a) for the linear loss, mainly for the special case of $n = 1$. The latter plays an important role in optimum sequential allocations, which will be discussed in more detail in the next section.

For the binomial–beta model, the Bayes designs consist of those sample sizes $n_1, \ldots, n_k$ which achieve the following maximum:

$$\max_{n_1+\cdots+n_k=n} E\left(\max_{i=1,\ldots,k} \int_{\Re} \prod_{j\neq i} H_{j,X_j}(\theta)\, h_{i,X_i}(\theta)\, \mathrm{d}\theta\right), \quad \text{for 0–1 loss}$$

$$\max_{n_1+\cdots+n_k=n} E\left(\max_{i=1,\ldots,k} \frac{\alpha_i + X_i}{\alpha_i + \beta_i + n_i}\right), \quad \text{for linear loss} \tag{12}$$

where in the first criterion, H_{r,x_r} and h_{r,x_r} denote the c.d.f. and the density, respectively, of Beta$(\alpha_r + x_r, \beta_r + n_r - x_r)$, $r = 1, \ldots, k$, for brevity. The outer expectations are with respect to the marginally independent random variables $X_r \sim PE(n_r, \alpha_r, \beta_r, 1)$, the Pólya–Eggenberger distribution given by equation (8), $r = 1, \ldots, k$.

In the second part of this section, the one-stage model considered above will be extended to a two-stage model, which can be summarized, after a standard reduction by sufficiency, as follows. At $\boldsymbol{\theta} \in \Omega^k$, let X_i and Y_i be sufficient statistics of samples of sizes n_i and m_i from population P_i at Stage

1 and Stage 2, respectively, which altogether are independent. It is assumed that sampling at Stage 1 has been completed already, where $\mathbf{X} = (x_1, \ldots, x_k)$ has been observed, and that it is planned to allocate observations $\mathbf{Y} = (Y_1, \ldots, Y_k)$ for Stage 2, subject to $m_1 + \cdots + m_k = m$, where m is fixed given.

First, let us consider the situation at the end of Stage 2, where both, $\mathbf{X} = \mathbf{x}$ and $\mathbf{Y} = \mathbf{y}$ have been observed. From equations (2) and (3), it follows that every Bayes selection rule $d^B(\mathbf{x}, \mathbf{y})$, say, is determined by

$$E\{L(\mathbf{\Theta}, d^B(\mathbf{x}, \mathbf{y})) \mid \mathbf{X} = \mathbf{x}, \mathbf{Y} = \mathbf{y}\} = \min_{i=1,\ldots,k} E\{L(\mathbf{\Theta}, i) \mid \mathbf{X} = \mathbf{x}, \mathbf{Y} = \mathbf{y}\} \quad (13)$$

Here $\mathbf{X}$ and $\mathbf{Y}$ are not combined into an overall sufficient statistic, since the situation at the end of Stage 1 will now be studied. The criterion for allocating observations $\mathbf{Y}$ for Stage 2, after having observed $\mathbf{X} = \mathbf{x}$ at Stage 1, is to find $m_1, \ldots, m_k$, subject to the side condition $m_1 + \cdots + m_k = m$, for which the following minimum is achieved:

$$\begin{aligned} &\min_{m_1+\cdots+m_k=m} E\{E\{L(\mathbf{\Theta}, d^B(\mathbf{x}, \mathbf{Y})) \mid \mathbf{X} = \mathbf{x}, \mathbf{Y}\} \mid \mathbf{X} = \mathbf{x}\} \quad (14) \\ &= \min_{m_1+\cdots+m_k=m} E\{\min_{i=1,\ldots,k} E\{L(\mathbf{\Theta}, i) \mid \mathbf{X} = \mathbf{x}, \mathbf{Y}\} \mid \mathbf{X} = \mathbf{x}\} \end{aligned}$$

where the outer expectation is with respect to the conditional distribution of $\mathbf{Y}$, given $\mathbf{X} = \mathbf{x}$. It should be pointed out that in equations (13) and (14), d^Bdoes not only depend on $n_1, \ldots, n_k$, which are fixed here, but also on $m_1, \ldots, m_k$, which are varying, since every design of specific ns and ms has its own Bayes selection rules.

From now on it is assumed that Stage 1 has been completed already, i.e., that $\mathbf{X} = \mathbf{x}$ has been observed, and that a Bayes design for Stage 2 with $m_1 + \cdots + m_k = m$ has to be determined. In this situation, it is convenient to update the prior with the information provided by $\mathbf{X} = \mathbf{x}$ (Berger, 1985), i.e. to treat Stage 2 with observations $\mathbf{Y}$ as a first stage and to use the updated prior density $\pi(\boldsymbol{\theta} \mid \mathbf{x})$ as a prior density. The Bayes designs are then all sampling allocations $m_1, \ldots, m_k$ for which the following minimum or maximum, respectively, is achieved.

$$\begin{aligned} &\min_{m_1+\cdots+m_k=m} E_x(\min_{i=1,\ldots,k} E_{\mathbf{x}}\{L(\mathbf{\Theta}, i) \mid \mathbf{Y}\}) && \text{in general} \quad (15) \\ &\max_{m_1+\cdots+m_k=m} E_x(\max_{i=1,\ldots,k} P_{\mathbf{x}}\{\Theta_i = \Theta_{[k]} \mid \mathbf{Y}\}) && \text{for 0–1 loss} \\ &\max_{m_1+\cdots+m_k=m} E_x(\max_{i=1,\ldots,k} E_{\mathbf{x}}\{\Theta_i \mid \mathbf{Y}\}) && \text{for linear loss} \end{aligned}$$

where here and in the following, the subscript $\mathbf{x}$ at all probabilities and expectations indicates that the updated prior, based on $\mathbf{X} = \mathbf{x}$, is used as prior. Comparing equation (15) with equation (10), one can see that the design problem for Stage 2, posed at the end of Stage 1, can be considered as a one-stage design problem and treated as such.

On the other hand, if one prefers not to update the prior, then in equation (15) the inner operations $E_{\mathbf{x}}$ and $P_{\mathbf{x}}$ are with respect to the conditional distribution of $\boldsymbol{\Theta}$, given $\mathbf{X} = \mathbf{x}$ and $\mathbf{Y}$, and the outer operations $E_{\mathbf{x}}$ are with respect to the conditional distribution of $\mathbf{Y}$, given $\mathbf{X} = \mathbf{x}$. Both approaches are valid, equivalent, and lead to the same results.

To conclude this section, Bayes designs for the two special models will be studied under the 0–1 loss and the linear loss. For the normal–normal model, it can be shown (Gupta and Miescke, 1994) that the Bayes designs consist of those sample sizes $m_1, \ldots, m_k$ which achieve the following respective maximum:

$$\max_{m_1+\cdots+m_k=m} E\left(\max_{i=1,\ldots,k} \int \prod_{j\neq i} \Phi(\vartheta_j^{-1}[\vartheta_i z + \mu_i(x_i) - \mu_j(x_j) + \gamma_i N_i - \gamma_j N_j]) \varphi(z)\mathrm{d}z \right) \quad \text{for 0–1 loss}$$

$$\max_{m_1+\cdots+m_k=m} E\left(\max_{i=1,\ldots,k} [\mu_i(x_i) + \gamma_i N_i] \right), \quad \text{for linear loss} \tag{16}$$

where $\vartheta_r = (p_r + q_r + \nu_r)^{-1/2}$, $\gamma_r = \vartheta_r(p_r + \nu_r)^{-1/2} q_r^{1/2}$, $\mu_r(x_r) = (p_r x_r + \nu_r \mu_r)/(p_r + \nu_r)$, with $p_r = n_r \sigma^{-2}$ and $q_r = m_r \sigma^{-2}$, $r = 1, \ldots, k$, and $N_1, \ldots, N_k$ are generic independent $N(0, 1)$ random variables.

The results for the special case of $k = 2$ are analogous to the one-stage Bayes design mentioned earlier. Results for the case of $k \geq 3$ are only known for the linear loss. They are based, in view of equation (16), on the properties of $E(\max_{i=1,\ldots,k}\{\lambda_i + \tau_i N_i\})$ as a function of $\lambda_i \in \Re$ and $\tau_i > 0$, $i = 1, \ldots, k$. These properties were derived by Gupta and Miescke (1996a) using the auxiliary function

$$T(w) = w\Phi(w) + \varphi(w) = \int_{-\infty}^{w} \Phi(v)\,\mathrm{d}v, \qquad w \in \Re \tag{17}$$

Here the special case of $m = 1$ has been worked out in detail, which is relevant for the Bayes sequential allocations considered in Section 4. Discussion of further results in this respect will therefore be postponed until Section 4.

For the binomial–beta model, the Bayes designs for Stage 2 consist of those sample sizes $m_1, \ldots, m_k$ for which, subject to $m_1 + \cdots + m_k = m$, the following maximums are achieved:

$$\max_{m_1+\cdots+m_k=m} \sum_{\mathbf{y}} \left(\max_{i=1,\ldots,k} \int_{\Re} \prod_{j \neq i} H_{j,x_j y_j}(\theta)\, h_{i,x_i y_i}(\theta)\, d\theta \right) P_x\{\mathbf{Y}=\mathbf{y}\}, \quad \text{for 0–1 loss,} \tag{18}$$

$$\max_{m_1+\cdots+m_k=m} \sum_{\mathbf{y}} \left(\max_{i=1,\ldots,k} \frac{a_i + y_i}{a_i + b_i + m_i} \right) P_x\{\mathbf{Y}=\mathbf{y}\}, \quad \text{for linear loss}$$

where in the first criterion, H_{r,x_r,y_r} and h_{r,x_r,y_r} denote the c.d.f. and the density, respectively, of Beta$(a_r + y_r,\ b_r + m_r - y_r)$, with $a_r = \alpha_r + x_r$ and $b_r = \beta_r + n_r - x_r$, $r = 1, \ldots, k$, for brevity. The sums in equation (18) are expectations with respect to the conditional distribution of $\mathbf{Y}$, given $\mathbf{X} = \mathbf{x}$, which, analogously to equation (8), are given by

$$P_{\mathbf{x}}\{\mathbf{Y}=\mathbf{y}\} = \prod_{i=1}^{k} \binom{m_i}{y_i} \frac{\Gamma(a_i+b_i)}{\Gamma(a_i)\,\Gamma(b_i)} \frac{\Gamma(a_i+y_i)\,\Gamma(b_i+m_i-y_i)}{\Gamma(a_i+b_i+m_i)} \tag{19}$$

where $y_i = 0, 1, \ldots, m_i$, $i = 1, \ldots, k$.

No further results are known in this situation under the 0–1 loss. However, under the linear loss, interesting theoretical as well as numerical results have been found by Gupta and Miescke (1996b) and Miescke and Park (1997a). These are relevant for the Bayes sequential allocations and will be discussed in the next section.

To conclude this section, some comments will be made regarding cost of sampling. Suppose that every single observation costs a certain amount λ, say. Then equations (10) and (15) have to be adjusted for the respective cost of sampling $n\lambda$ and $m\lambda$. If the cost of sampling turns out to be larger than the gain in maximum posterior expectation given in equations (10) or (15), respectively, then apparently it is not worth taking all of these observations. This approach has been treated by Gupta and Miescke (1993) within the problem of combined selection and estimation, in the binomial–beta model. It leads, among other considerations, to finding the largest sample size, subject to its given upper bound, which is worth allocating to incur a gain. In the next section, where observations are allocated and taken in a sequential fashion, cost of sampling would require the incorporation of a stopping rule. However, this will not be done to keep the presentation of basic ideas simple. Modifications of these sequential allocation rules to this more general setting are straightforward, but more involved. Therefore, cost of sampling will not be considered any further.

4. BAYES LOOK-AHEAD SEQUENTIAL SAMPLING DESIGNS

From now on it is assumed that Stage 1 has been completed, i.e., that $\mathbf{X} = \mathbf{x}$ has been observed already, and that m additional observations $\mathbf{Y} = (Y_1, \ldots, Y_k)$ are planned to be drawn at Stage 2. The optimum allocations of sample sizes $m_1, \ldots, m_k$, i.e., the Bayes designs, are determined by the criterion in equation (15), i.e., by

$$\min_{m_1+\cdots+m_k=m} E_x(\min_{i=1,\ldots,k} E_x\{L(\boldsymbol{\Theta}, i) \mid \mathbf{Y}\}) \tag{20}$$

The first step toward sequential Bayes designs is to consider an intermediate step of Stage 2 sampling, where so far only $\tilde{\mathbf{Y}} = \tilde{\mathbf{y}}$ has been observed, with $\tilde{m}_i$ observations from population P_i, $i = 1, \ldots, k$, where $\tilde{m}_1 + \cdots + \tilde{m}_k = \tilde{m}$, and where $\tilde{m}$ with $1 \leq \tilde{m} < m$ is fixed. The best allocation of the remaining $\hat{m} = m - \tilde{m}$ observations with $\hat{m}_i = m_i - \tilde{m}_i \geq 0$, $i = 1, \ldots, k$, achieves

$$\min_{\hat{m}_1+\cdots+\hat{m}_k=\hat{m}} E_x\left\{ \min_{i=1,\ldots,k} E_{\mathbf{x}}\{L(\boldsymbol{\Theta}, i) \mid \tilde{\mathbf{Y}} = \tilde{\mathbf{y}}, \hat{\mathbf{Y}}\} \,\middle|\, \tilde{\mathbf{Y}} = \tilde{\mathbf{y}} \right\} \tag{21}$$

where the outer expectation is with respect to the conditional distribution of the new observations $\hat{\mathbf{Y}}$, say, given $\tilde{\mathbf{Y}} = \tilde{\mathbf{y}}$ (and $\mathbf{X} = \mathbf{x}$).

Returning now to the end of Stage 1, the optimum two-step allocation for drawing first $\tilde{m}$ and then $\hat{m} = m - \tilde{m}$ observations at Stage 2 is found by backward optimization. First one has to consider every possible sample size configuration $\tilde{m}_1, \ldots, \tilde{m}_k$ and every possible outcome $\tilde{\mathbf{Y}} = \tilde{\mathbf{y}}$. For each such setting, one allocation $\hat{m}_i(\tilde{\mathbf{y}}, \tilde{m}_1, \ldots, \tilde{m}_k)$, $i = 1, \ldots, k$, has to be found which achieves equation (21). Then one has to find an allocation $\tilde{m}_1, \ldots, \tilde{m}_k$ which achieves

$$\min_{\tilde{m}_1+\cdots+\tilde{m}_1=\tilde{m}} E_x\left(\min_{\hat{m}_1+\cdots+\hat{m}_1=\hat{m}} E_{\mathbf{x}}\left\{ \min_{i=1,\ldots,k} E_{\mathbf{x}}\{L(\boldsymbol{\Theta}, i) \mid \tilde{\mathbf{Y}}, \hat{\mathbf{Y}}\} \,\middle|\, \tilde{\mathbf{Y}} \right\} \right) \tag{22}$$

Here one should be aware of the fact that the information contained in $\mathbf{Y}$ is the combined information gained from $\tilde{\mathbf{Y}}$ and $\hat{\mathbf{Y}}$. It should also be pointed out clearly that in the middle minimization operation of equation (22), $\hat{m}_1, \ldots, \hat{m}_k$ depend on $\tilde{m}_1, \ldots, \tilde{m}_k$ and on $\tilde{\mathbf{Y}}$, and thus they are random variables themselves! This is the very reason why equation (22) can be handled numerically and in computer simulations in the binomial–beta model, where $\mathbf{Y}$ is discrete and assumes only finitely many values, but not in the normal–normal model.

Now comparing equation (20) with equation (22), one can show that the latter must be less than or equal to the former. If one deletes the minimum to the right of the first expectation in equation (22), and inserts a minimum to the left of that term, subject to $\hat{m}_i = m_i - \tilde{m}_i \geq 0$, and subject to

$\hat{m}_1 + \cdots + \hat{m}_k = \hat{m}$, then the resulting value cannot be smaller. Combining the two iterated minimization operations into one leads to equation (20). To summarize, one can state (Miescke and Park, 1997a) the following result.

THEOREM 1: *For fixed m and $\tilde{m} < m$, the best allocation for drawing first $\tilde{m}$ and then $\hat{m} = m - \tilde{m}$ observations at Stage 2 is at least as good as the best allocation of all m observations in one step, in the sense that the posterior Bayes risk in equation (22) of the former is not larger than that of the latter, given by equation (20). This process of stepwise optimum allocation can be iterated for further improvements. The overall best allocation scheme is to draw, in m steps, one observation at a time, which are determined by backward optimization.*

In view of this theorem, several reasonable sampling allocation schemes can be constructed which utilize information gained from observations at previous steps. Let R_t, for $t \leq m$, denote the allocation of t observations determined by equation (15), with m replaced by t there. Moreover, let $R_{t,1}$ allocate any single observation to one of the populations sampled by R_t. In a similar way let $R^*_{t,1}$ allocate one observation to one of the populations to which R_t assigns the largest allocation. Finally, denote by B_1 the optimum allocation of one observation, knowing all future allocation strategies. It should be pointed out that, unlike the other allocations considered above, B_1 is not a stand-alone procedure, since it requires knowledge of what will be done after it has been applied.

Using these three types of intermediate allocation rules, the following schemes of allocating m observations are possible. (R_m) allocates all m observations at once, using equation (15). This fixed sample size m Bayes design will be denoted by **OPT** in what follows. A better allocation scheme, in terms of the Bayes risk, is $(R_{m,1}, R_{m-1})$, which uses $R_{m,1}$ for the first allocation and then R_{m-1} for the rest. Better than $(R_{m,1}, R_{m-1})$, of course, is (B_1, R_{m-1}), which uses backward optimization B_1 for the first allocation, knowing that R_{m-1} will be used for allocating the remaining $m - 1$ observations in one step. In this fashion, similar and also more complicated allocation schemes can be constructed (Gupta and Miescke 1996a), which are linked through a partial ordering in terms of their Bayes risks. Such constructions are motivated by the fact that the overall optimum allocation scheme $(B_1, B_1, \ldots, B_1, R_1)$, denoted by **BCK**, is not practicable, except for small m and k, up to about $m = 20$ for $k = 3$, in the binomial–beta model. For this model, the allocation scheme **APP**, say, which is $(R^*_{m,1}, R^*_{m-1,1}, \ldots, R^*_{2,1} R_1)$, appears to be a very good approximation to **BCK** under the linear loss. This will be justified at the end of this section.

The allocation scheme $(R_1, R_1, \ldots, R_1)$ allocates in m steps one observation at a time, using R_1, pretending that it would be the last one before making the final (selection) decision. It looks ahead one observation at a time (Amster, 1963; Berger, 1985) and will henceforth be denoted by **LAH**. It should not be confused with the allocation scheme **SOA**, say, which allocates in m steps one observation at a time, using the "state of the art." To be more specific, suppose that $\tilde{\mathbf{Y}} = \tilde{\mathbf{y}}$ has been drawn so far. Then **SOA** allocates the next observation to any one of those populations which are associated with the minimum of the k values of $E_{\mathbf{x}}\{L(\boldsymbol{\Theta}, i) \mid \tilde{\mathbf{Y}} = \tilde{\mathbf{y}}\}$, $i = 1, \ldots, k$. Two other allocation schemes, which will be considered later in the simulation study of the binomial–beta model, should be mentioned here. The first assigns one observation at a time, each purely at random, regardless of the previous observations, and is denoted by **RAN**. The second assigns m/k observations to each population, provided that m is divisible by k, and is denoted by **EQL**.

Theoretical results for allocation scheme **LAH** in the normal–normal model and under the linear loss, which are presented in Gupta and Miescke (1994,1996a), will now be discussed. Here, it is sufficient to consider the first allocation R_1 in $(R_1, R_1, \ldots, R_1)$, which is based on $\mathbf{X} = \mathbf{x}$. All consecutive allocations R_1 are decided analogously, based on $\mathbf{X} = \mathbf{x}$ and the observations $\tilde{\mathbf{Y}} = \tilde{\mathbf{y}}$ that have been taken so far at Stage 2. Starting with criterion in equation (16) for the linear loss with $m = 1$, where exactly one of the sample sizes $m_1, \ldots, m_k$ is equal to one and all others are zero, i.e., where exactly one of $q_1, \ldots, q_k$ is equal to σ^{-2} and all others are zero, this first observation is taken from one of the populations which yields

$$\max_{i=1,\ldots,k} E\left(\max\left\{\mu_i(x_i) + \sigma_i N_i\,,\ \max_{j\neq i}\{\mu_j(x_j)\}\right\}\right) \tag{23}$$
$$= \max_{i=1,\ldots,k}\left\{\mu_i(x_i) + \sigma_i T([\max_{j\neq i}\{\mu_j(x_j)\} - \mu_i(x_i)]/\sigma_i)\right\}$$

where $\sigma_r = (p_r + \sigma^{-2} + \nu_r)^{-1/2}(p_r + \nu_r)^{-1/2}\sigma^{-1}$, and $\mu_r(x_r)$, $r = 1, \ldots, k$, are defined just after equation (16), and the function T is given by equation (17).

To describe the properties of the first allocation R_1 in $(R_1, R_1, \ldots, R_1)$, it proves useful to consider the ordered values $\mu_{[1]}(\mathbf{x}) < \mu_{[2]}(\mathbf{x}) < \ldots < \mu_{[k]}(\mathbf{x})$ of $\mu_r(\mathbf{x}) = \mu_r(x_r)$, $r = 1, \ldots, k$. Let $P_{(i)}$ be the population, and let $\sigma_{(i)}$ be the standard deviation, which is associated with $\mu_{[i]}(\mathbf{x})$, $i = 1, \ldots, k$. Then one can state the following result.

THEOREM 2: *After $\mathbf{X} = \mathbf{x}$ has been observed, the preferences of the first allocation R_1 in $(R_1, R_1, \ldots, R_1)$ are as follows.*

1. *If* $\sigma_{(k-1)} < (=, >)\ \sigma_{(k)}$, *then allocating to* $P_{(k-1)}$ *is worse than (equivalent to, better than) allocating to* $P_{(k)}$.
2. *If for* $1 \le i < j \le k-2$, $\sigma_{(i)} \le \sigma_{(j)}$, *then allocating to* $P_{(i)}$ *is worse than allocating to* $P_{(j)}$.
3. *If for* $1 \le i < j \le k-2$, $\sigma_{(i)} > \sigma_{(j)}$, *and* $\sigma_{(i)} < (=, >)\ \sigma_{ij}$, *then allocating to* $P_{(i)}$ *is worse than (equivalent to, better than) allocating to* $P_{(j)}$, *where* σ_{ij} *is determined by*

$$\mu_{[j]}(\mathbf{x}) + \sigma_{(j)} T\big((\mu_{[k]}(\mathbf{x}) - \mu_{[j]}(\mathbf{x}))/\sigma_{(j)}\big)$$
$$= \mu_{[i]}(\mathbf{x}) + \sigma_{ij} T\big((\mu_{[k]}(\mathbf{x}) - \mu_{[i]}(\mathbf{x}))/\sigma_{ij}\big). \tag{24}$$

4. *Let* $P_{(*)}$ *be a best allocation to either* $P_{(k-1)}$ *or* $P_{(k)}$ *according to step 1. Likewise, let* $P_{(\bullet)}$ *be a best allocation to* $P_{(1)}, \ldots, P_{(k-2)}$ *according to steps 2 and 3. Then an overall best allocation is found by using steps 2 and 3 with* (i), (j), σ_{ij}, $\mu_{[i]}(\mathbf{x})$, *and* $\mu_{[j]}(\mathbf{x})$ *replaced by* $(\bullet)$, $(*)$, $\sigma_{\bullet *}$, $\mu_{[\bullet]}(\mathbf{x})$, *and* $\mu_{[k-1]}(\mathbf{x})$, *respectively.*

The proof can be found in Gupta and Miescke (1996a), along with further comments. Moreover, a numerical example with real-life data is presented there, in which comparisons are also made with respect to standard multiple comparison procedures, such as Scheffé's and Tukey's methods.

Theoretical and numerical simulation results for the binomial–beta model under the linear loss are presented in Gupta and Miescke (1996b) and Miescke and Park (1997a). These will be discussed in the remainder of this section. First, properties of **LAH**, the allocation scheme $(R_1, R_1, \ldots, R_1)$, will be studied. As has been justified above, it suffices to consider the first allocation R_1. Starting with the criterion in equation (18) for the linear loss with $m = 1$, where exactly one of the sample sizes $m_1, \ldots, m_k$ is equal to one, and all others are zero, this first observation is taken from one of the populations which yield the following maximum:

$$\max_{i=1,\ldots,k} E_X\left\{\max\left\{\frac{a_i + Y_i}{a_i + b_i + 1}, \max_{j \ne i}\left\{\frac{a_j}{a_j + b_j}\right\}\right\}\right\}$$
$$= \max_{i=1,\ldots,k}\left\{\max\left\{\frac{a_i + 1}{a_i + b_i + 1}, \max_{j \ne i}\left\{\frac{a_j}{a_j + b_j}\right\}\right\} \frac{a_i}{a_i + b_i}\right.$$
$$\left. + \max\left\{\frac{a_i + 0}{a_i + b_i + 1}, \max_{j \ne i}\left\{\frac{a_j}{a_j + b_j}\right\}\right\} \frac{b_i}{a_i + b_i}\right\} \tag{25}$$

where $a_r = \alpha_r + x_r$ and $b_r = \beta_r + n_r - x_r$, $r = 1, \ldots, k$, for brevity. To summarize, one can state here the following result.

THEOREM 3: *The first allocation R_1 in $(R_1, R_1, \ldots, R_1)$ is made with respect to one of the populations P_i, $i = 1, \ldots, k$, for which the maximum in equation (25) is achieved. All consecutive allocations of the type R_1 are made analogously, with a_r and b_r updated to $\alpha_r + x_r + \tilde{y}_r$ and $\beta_r + n_r - x_r + \tilde{m}_r - \tilde{y}_r$, respectively, with respect to the $\tilde{m}_r$ observations, represented by $\tilde{y}_r$, $r = 1, \ldots, k$, which have been made so far at Stage 2.*

The proof is given in Gupta and Miescke (1996b), along with further details of the behavior of this allocation scheme. One interesting point worth mentioning is related to the fact that under the linear loss (with no cost of sampling), the Bayes look-ahead risk cannot increase when more future observations are included. This fact implies that the maximum in equation (25) is always greater than or equal to $\max_{i=1,\ldots,k}\{a_i/(a_i + b_i)\}$. However, equality may occur, in which case one additional observation from any of the populations would not be worth taking from the Bayesian point of view. A similar situation may arise at any allocation of the type R_1 in $(R_1, R_1, \ldots, R_1)$, and especially at the last allocation. In the latter case, one can accelerate the process by stopping one observation short of m. In the context of sequentially allocating m observations at Stage 2, however, this is of minor concern and will not be considered any further.

Numerical results for allocation schemes **EQL** and **OPT**, and computer simulation results for **RAN**, **SOA**, and **LAH** have been presented for $k = 3$ populations by Gupta and Miescke (1996b). The computer programs were written in Microsoft Quick Basic Version 4.5, using subroutines from Sprott (1991). These results have been extended, using Microsoft Visual Basic Version 4.0, to numerical results for the overall optimum allocation scheme **BCK** in Miescke and Park (1997a), where it has also been found that **APP** appears to be a very good approximation to **BCK**. In summary, the performances of the following allocation schemes have been studied, which have been explained in more detail earlier in this section.

RAN	Assign one observation at a time, each purely at random.
EQL	Assign $m/3$ observations to each population P_1, P_2, P_3.
SOA	Assign one observation at a time, following the state of the art.
LAH	Assign one observation at a time, using $(R_1, R_1, \ldots, R_1)$.
OPT	Assign m_t observations to P_t, $t = 1, 2, 3$, using (R_m).
APP	Assign one observation at a time, using $(R^*_{m,1}, R^*_{m-1,1}, \ldots, R^*_{2,1}, R_1)$.
BCK	Assign one observation at a time, using backward optimization.

The performances of these allocation schemes have been compared in three examples, with suitably chosen values for $a_r = \alpha_r + x_r$ and $b_r = \beta_r + n_r - x_r$,

$r = 1, 2, 3$, to cover various interesting settings, The values for m considered have been 1, 3, 9, and 15. For $m = 1$, **EQL** has been set to take its observation from population P_1, rather than leaving the respective spaces empty in the tables. As to ties, **RAN**, **SOA**, **LAH**, and **APP** have been used, and are recommended, with ties broken purely at random, with equal probabilities, whenever they occur. This recommendation is corroborated by findings in the numerical studies.

Comparing the expected posterior gains of the first five allocation schemes, it turns out that overall, **LAH** and **OPT** are performing similarly well, each sometimes better than the other, but clearly better than **RAN**, **EQL**, and **SOA**. The latter effect is found to be increasing in m. That **LAH** is not always as good as **OPT** proves that it cannot be any version, i.e. with any type of breaking ties, of the allocation scheme $(R_{m,1}, R_{m-1,1}, \ldots, R_{2,1} R_1)$, and thus in particular it cannot be equal to $(R^*_{m,1}, R^*_{m-1,1}, \ldots, R^*_{2,1} R_1)$, i.e., **APP**, since the latter two are always at least as good as **OPT**. One advantage of **LAH**, besides its easy implementation, is that each of its individual allocations is self-contained. Thus, if in an ongoing experiment the total number of m observations has to be changed, this has only minor effects on its usage. The numerical results for **BCK** in Miescke and Park (1997a) became feasible with the release of Microsoft Visual Basic Version 4.0, which can handle, on a typical IBM-type Pentium computer, a 6-dimensional array (for the a_is and b_is) with a common subscript range of $1, 2, \ldots, 15$ (for the $\tilde{m}_i$s), i.e., more than 10^7 variables. As anticipated, **LAH** and **OPT** turn out to be good approximations to **BCK**.

One striking fact has been observed by comparing the first allocation of **BCK** with the allocation m_1, m_2, m_3 of **OPT**. In all but one of the 96 parameter settings considered in the three examples, the population to which **OPT** allocates the largest sample size is one of those which **BCK** would allow to start with. This clearly indicates that the allocation scheme $(R^*_{m,1}, R^*_{m-1,1}, \ldots, R^*_{2,1} R_1)$, i.e., **APP**, should be considered as a good approximation to **BCK**, and thus be used in practice. A study of the performance of **APP** within the framework of the three examples does not appear to be feasible at the moment because of the length of computing time required for such a task. It would be a combination of calculating the individual steps $R^*_{t,1}$, randomizing tied populations, and simulating the m outcomes.

ACKNOWLEDGMENTS

The author is grateful to a referee for carefully reading the manuscript and for suggested changes. This research was supported in part by US Army

Research Office Grant DAAH04-95-1-0165 at Purdue University under the direction of Professor Shanti S. Gupta.

REFERENCES

Abughalous, M.M., Miescke, K.J. On selecting the largest success probability under unequal sample sizes. *J. Stat. Plann. Inference*, **21**:53–68, 1989.

Amster, S.J. A modified Bayesian stopping rule. *Ann. Math. Stat.* **34**:1404–1413, 1963.

Bechhofer, R.E. Optimal allocation of observations when comparing several treatments with a control. In: Krishnaiah P.R., ed. *Multivariate Analysis*. New York: Academic Press, 1969, pp. 465–473.

Bechhofer, R.E., Kulkarni, R.V. Closed adaptive sequential procedures for selecting the best of $k \geq 2$ Bernoulli populations. In: Gupta, S.S., Berger, J.O., eds. *Statistical Decision Theory and Related Topics—III*, Vol. 1. New York: Academic Press, 1982, pp. 61–108.

Bechhofer, R.E., Kiefer, J., Sobel, M. *Sequential Identification and Ranking Procedures*. Chicago, IL: University of Chicago Press, 1968.

Bechhofer, R.E., Hayter, A.J., Tamhane, A.C. Designing experiments for selecting the largest normal mean when the variances are known and unequal: Optimal sample size allocation. *J. Stat. Plann. Inference*, **28**:271–289, 1991.

Bechhofer, R.E., Santner, T.J., Goldsman, D.M. *Design and Analysis of Experiments for Statistical Selection, Screening, and Multiple Comparisons*. New York: Wiley, 1995.

Berger, J.O. *Statistical Decision Theory and Bayesian Analysis*. 2nd edn. New York: Springer, 1985.

Berger, J.O., Deely, J. A Bayesian approach to ranking and selection of related means with alternatives to AOV methodology. *J. Am. Stat. Assoc.*, **83**:364–373, 1988.

Bratcher, T.L., Bhalla, P. On the properties of an optimal selection procedure. *Commun. Stat.*, **3**:191–196, 1974.

Bratcher, T.L., Bland, R.P. On comparing binomial probabilities from a Bayesian viewpoint. *Commun. Stat.*, **4**:975–985, 1975.

Chernoff, H., Yahav, J.A. A subset selection problem employing a new criterion. In: Gupta S.S., Moore, D.S., eds. *Statistical Decision Theory and Related Topics—II*, New York: Academic Press, 1977, pp. 93–119.

Cohen, A., Sackrowitz, H.B. A decision theoretic formulation for population selection followed by estimating the mean of the selected population. In: Gupta, S.S., Berger, J.O., eds. *Statistical Decision Theory and Related Topics—IV*, Vol. 2, New York, Springer, 1988, pp. 33–36.

Deely, J.J., Gupta, S.S. On the properties of subset selection procedures. *Sankhyā*, **A-30**:37–50, 1968.

Dudewicz, E.J., Dalal, S.R. Allocation of observations in ranking and selection with unequal variances. *Sankhyā*, **B-37**:28–78, 1975.

Dunnett, C.W. On selecting the largest of *k* normal populations means. With discussion. *J. R. Stat. Soc.*, **B-22**:1–40, 1960.

Fong, D.K.H., Berger, J.O. Ranking, estimation, and hypothesis testing in unbalanced two-way additive models—a Bayesian approach. *Stat. Dec.*, **11**:1–24, 1993.

Goel, P.K., Rubin, H. On selecting a subset containing the best population—a Bayesian approach. *Ann. Stat.*, **5**:969–983, 1977.

Govindarajulu, Z., Katehakis, M.N. Dynamic allocation in survey sampling. *Am. J. Math. Manage. Sci.*, **11**: 199–221, 1991.

Gupta, S.S. On a decision rule for a problem in ranking means. Mimeo Series No. 150, Institute of Statistics, University of North Carolina, Chapel Hill, NC, 1956.

Gupta, S.S. On some multiple decision (selection and ranking) rules. *Technometrics*, **7**:225–245, 1965.

Gupta, S.S., Hsu, J.C. On the monotonicity of Bayes subset selection procedures. In: *Proceedings of the 41st Session of the ISI*, Book 4, pp. 208–211, 1977.

Gupta, S.S., Miescke, K.J. On two-stage Bayes selection procedures. *Sankhyā*, **B-46**:123–134, 1984a.

Gupta, S.S., Miescke, K.J. Sequential selection procedures—a decision theoretic approach. *Ann. Stat.*, **12**:336–350, 1984b.

Gupta, S.S., Miescke, K.J. On the problem of finding the largest normal mean under heteroscedasticity. In: Gupta, S.S., Berger, J.O., eds.

Statistical Decision Theory and Related Topics—IV, Vol. 2. New York: Springer, 1988, pp. 37–49.

Gupta, S.S., Miescke, K.J. On finding the largest normal mean and estimating the selected mean. *Sankhyā*, **B-52**:144–157, 1990.

Gupta, S.S., Miescke, K.J. On combining selection and estimation in the search of the largest binomial parameter. *J. Stat. Plann. Inference*, **36**:129–140, 1993.

Gupta, S.S., Miescke, K.J. Bayesian look ahead one-stage sampling allocations for selecting the largest normal mean. *Stat. Pap.*, **35**:169–177, 1994.

Gupta, S.S., Miescke, K.J. Bayesian look ahead one-stage sampling allocations for selection of the best population. *J. Stat. Plann. Inference*, **54**:229–244, 1996a.

Gupta, S.S., Miescke, K.J. Bayesian look ahead one-stage sampling allocations for selection of the best Bernoulli population. In: Brunner, E., Denker, M., eds. *Research Developments in Probability and Statistics*, Festschrift in Honor of M.L. Puri., Zeist: VSP Interntional, 1996b, pp. 353–369.

Gupta, S.S., Panchapakesan, S. *Multiple Decision Procedures: Theory and Methodology of Selecting and Ranking Populations*. New York: Wiley, 1979.

Gupta, S., Panchapakesan, S. Sequential ranking and selection procedures. In: Ghosh, B.K., Sen, P.K., eds. *Handbook of Sequential Analysis*. New York: Marcel Dekker, 1991, pp. 363–380.

Johnson, N.L., Kotz, S. *Discrete Distributions*. Boston, MA: Houghton Mifflin, 1969.

Lam, K., Chiu, W.K. On the probability of correctly selecting the best of several normal populations. *Biometrika*, **63**:410–411, 1976.

Miescke, K.J. Bayesian subset selection for additive and linear loss functions. *Commun. Stat.*, **A8-12**:1205–1226, 1979.

Miescke, K.J. Recent results on multi-stage selection procedures. In: Iosifescu, M., ed. *Proceedings of the 7th Conference on Probability Theory*. Bucharest: Editura Academiei Republicii Socialiste Romania, 1984, pp. 259–268.

Miescke, K.J., Park, H. Bayes *m*-truncated sampling allocations for selecting the best Bernoulli population. In: Panchapakesan, S., Balakrishnan,

N., eds. *Advances in Statistical Decision Theory and Applications*, in Honor of S.S. Gupta. Boston, MA: Birkhäuser, 1997a, pp. 31–47.

Miescke, K.J., Park, H. On the natural selection rule under normality. Discussion paper, 16 pp. in MS Word 7, 1997b.

Miescke, K.J., Rasch, D. (eds.): 40 Years of Statistical Selection Theory, Parts I & II. Special Issue, *J. Stat. Plann. Inference*, **54**(2/3), 1996.

Risko, K.R. Selecting the better binomial population with unequal sample sizes. *Commun. Stat. Theory Methodol.*, **14**:123–158, 1985.

Sprot, J.C. *Numerical Recipes: Routines and Examples in BASIC*, Cambridge: Cambridge University Press, 1991.

Tong, Y.L., Wetzell, D.E. On the behavior of the probability function for selecting the best normal population. *Biometrika*, **66**:174 176, 1979.

Whitehead, J. Sequential methods in clinical trials. In: Ghosh, B.K., Sen, P.K., eds. *Handbook of Sequential Analysis*. New York: Marcel Dekker, 1991, pp. 593–611.

5

Cluster Coordinated Composites of Diverse Datasets on Several Spatial Scales for Designing Extensive Environmental Sample Surveys: Prospectus on Promising Protocols

WAYNE MYERS School of Forest Resources and Environmental Resources Research Institute, Pennsylvania State University, University Park, Pennsylvania

G. P. PATIL and CHARLES TAILLIE Center for Statistical Ecology and Environmental Statistics, Pennsylvania State University, University Park, Pennsylvania

1. INTRODUCTION

Geographic information systems (GIS) have become obligatory for many agencies, organizations, and corporations concerned with the environment. GIS provides for dynamic mapping, overlay, and spatial analysis of geographically referenced variables in large multilayer (multivariate) databases that typically occupy many gigabytes of computer storage media. GIS accommodates both synoptic and localized information. GIS layers (variables) pertaining to points, plots, and linears are localized in the sense that they ascribe values only in certain places within a geographic extent of coverage. We use the term site data here in referring to information stored in a GIS that has been obtained for points or plots by field (ground-based)

Prepared with partial support from the NSF/EPA Water and Watersheds Program, National Science Foundation Cooperative Agreement Number DEB-9524722. The contents have not been subjected to Agency review and therefore do not necessarily reflect the views of the Agency and no official endorsement should be inferred.

data collection. Layers tessellated as grids or polygons ascribe a value to all locations within populated tessera. We use the term synoptic to indicate that all tessera within the scope of coverage for a layer are populated (as opposed to being holes), so that a value is available for all locations therein.

Landscape ecologists use the term scale to indicate relative level of detail in data according to size (or grain) of grid cells and polygons. A multitude of small (fine-grained) grid cells or polygons are considered as being larger scale than fewer more expansive (coarse-grained) ones. Larger scale data thus provide more detail under map enlargement than smaller scale data. The coarser grain of smaller scale data usually entails aggregating, averaging, or interpolating over larger spans of distance and area. Smaller scale data are therefore of lower resolution and tend to have less fidelity with local (site) measurements of the same variable than larger scale data. This sense of scale is compatible with, but different from, the cartographic concept of proportional relation between dimensions of terrain features and their representations on maps.

Remote sensing uses radiant electromagnetic energy as a signal for conveying environmental information to recording devices aboard satellites and aircraft. The selective sensitivity of recording devices to particular wavelengths allows simultaneous data acquisition for several spectral wavelength intervals (bands). Radiant energy is necessarily integrated over an instantaneous field of view for the sensor, with the responses of detectors to the integrated signals being recorded as a rectangular array of picture elements (pixels). An array of pixels comprising a remotely sensed scene is thus the counterpart of a cellular grid in GIS, with each spectral band being a layer (or variate). A large instantaneous field of view produces coarse resolution (small-scale) images, whereas a small field of view produces finer detail (larger-scale) images. Resampling by aggregating or splitting pixels permits small-scale and large-scale images to be spatially superimposed (registered), but does not add detail to coarse-grained imagery. Much of the effort in analyzing remotely sensed images has typically been devoted to pattern recognition, whereby algorithmic classification translates multiband image data into categorical (thematic) map layers of environmental interest to be incorporated in GIS.

Recent interest on the part of the US Environmental Protection Agency (EPA), as reflected in solicitations for research, is turning toward synthesis of disparate multiscale information from GIS and remote sensing to help focus the design of sample surveys for environmental monitoring, and to the formulation of indicators for environmental status. Allied agencies with environmental responsibilities are lending impetus to this emphasis on coordinating multi-tiered data.

2. CLUSTER-BASED APPROACH TO MULTIVARIATE REGIONALIZATION

Cluster-based multivariate regionalization is central to our strategy for coordinating a collage of coverages to facilitate the design of sample surveys for environmental assessment and monitoring. The landscape ecological rationale rests on mosaic patterns of landscapes formed by recurring assemblages of environmental elements that are characteristic of ecosystems (Forman, 1995). Such assemblages occur as more or less distinct patches having pronounced internal consistency and with much the same assemblage comprising kindred patch occurrences across the landscape. Human activity is likewise patterned in a manner that is substantially conditioned by the prevailing ecological pattern. The landscape pattern must necessarily be expressed in patterning of environmental measurements that have some temporal stability, regardless of what is measured. GIS layers and spectral image variables are acquired largely because they do exhibit appreciable temporal stability. Well-conceived multivariate clustering should thus be consistent with the underlying landscape pattern and provide a more or less "natural" regionalization for generic stratification.

Our method of clustering is an offshoot of work on image data compression that we call PHASES, which is an acronym for "pixel hyperclusters as segmented environmental signals." The PHASES dataform is a hybrid between image information and a GIS layer that connects these two analytical realms and is usable in either domain (Myers, 1998). Its intermediary nature leads to some confusion regarding terminology because of conflicts in jargon between remote sensing and GIS. The processes of generating and manipulating PHASES for the present purposes are rooted in remote sensing and artificial intelligence, but do not conform to the conventions of either.

Clustering has been a mainstay of environmental signal processing, with its usual use being for statistical pattern recognition to make categorical thematic maps by so-called "unsupervised" analysis (Lillesand and Kiefer, 1994). Our major motive for clustering has instead been to compress multiband image data so that it can be stored as if consisting of only a single band. This manner of collapsing multiband data is contrary to past practice in remote sensing, which would have taken the first principal component as an optimal variance preserving linear combination. The problem with principal component compression in this regard is its propensity to relegate areally restricted but important elements of the landscape such as roads and riparian belts to low-order components, thus effectively "squeezing them out" of the first principal component that is being taken as a distillation of scene information.

The precedent for our compressive clustering lies in the work of Kelly and White (1993) on computer-assisted photointerpretive mapping from image data that extends the ideas of unsupervised analysis. The essence of unsupervised analysis is clustering to a level that is commensurate with thematic interest, and then empirically investigating instances of clusters to determine their thematic correspondence. Clusters typically outnumber thematic classes by two to three times in performing unsupervised analysis. In order to make the landscape pattern more evident to an interpreter, Kelly and White extracted a much larger number of clusters that they termed hyperclusters. They used a K-means clustering tactic (Hartigan, 1975) in their SPECTRUM implementation of the mapping technique under the auspices of the Khoros group at the Los Alamos National Laboratory. The SPECTRUM hyperclusters captured a surprising amount of detail in multiband image data from NASA's Landsat Thematic Mapper (TM) sensor, which inspired the PHASES cluster-based compression effort apart from any particular pattern recognition scenario.

In a compression context there is an obvious answer to the otherwise bothersome question of how many clusters. Decide how many bits of file storage will be allotted to each pixel (grid cell), and then extract as many clusters as can be accommodated. Since the most common mode of storage for image data is byte binary, this implies 255 clusters, with zero being reserved as a missing data code. There are, however, two special concerns relative to compression. One is that distinctive but diminutive landscape elements should not be dominated by those that occur more profusely, as happens with principal component compression. The second concern is that the availability of fast computer memory should not constrain the size of dataset that can be processed. A collateral concern is that the computations should be amenable to parallelization when that facet of computer technology eventually becomes commonplace. These concerns and considerable experimentation led to a weightable modification of the ISODATA clustering tactic (Tou and Gonzalez, 1974). The name ISODATA derives from "iterative self-organizing data analysis" technique. The basic ISODATA protocol begins with a prespecified number of variously chosen cluster "seeds" which become cluster mean vectors (centroids) after the first iteration. Each data vector becomes associated with the closest seed in a Euclidean distance sense. At the end of a pass, the seed moves to the centroid of its associates. Seeds thus migrate through multidimensional variate space toward the more densely populated sectors.

Although the eventual ISODATA endpoint is supposedly independent of initial values for seed vectors, this is certainly not true for reasonable numbers of cycles through large multivariate geospatial datasets. Accordingly, a special preclustering algorithm has been added to disperse the seeds through occupied sectors of multivariate space in a manner that mitigates against having seeds close together. This helps to ensure that there are clusters representing less common features. The format of the initial seed file is the same as that of the cluster centroid file, so that clustering can be conducted a few cycles at a time with interim visual inspection of cluster images.

A way of controlling minimum cluster size without reducing the number of clusters has been provided by introducing a dynamic cluster splitting capability to augment the basic ISODATA protocol. Miniscule clusters have their seeds replanted in a manner that splits clusters having a larger sum of squared distances from the centroid. The latter is accomplished by tracking the migration of cluster seeds through multidimensional space.

The PHASE combination of tactics has some parallels with neural networks, but is more easily constrained and less of a "black box." The PHASE approach also shares the computationally intensive nature of neural networks, but to a lesser degree. Compressing a 4-band dataset having 3000–4000 rows and columns in the grid will require several hours on 300+ megahertz PC-level computers. Operability and efficacy has been well established by experience, with a variety of image data on both PC computers and Unix workstations.

The PHASES compressed dataform consists of a cellular grid (raster) file containing cluster numbers for pixels, accompanied by relational files containing tables of cluster characteristics by cluster number. One tabular file has means and frequencies for clusters, and another has information on variability among clusters. Additional auxiliary files such as cluster edge relations are optional according to purpose. The cluster numbers are ordered according to increasing length of cluster centroid vector. This ordering permits the grid file of cluster numbers to be rendered for direct viewing as a sort of intensity image. Renderings of the clusters as images can likewise be colorized according the cluster mean values for variates. Such pseudocolor images can be made to mimic the appearance of original image displays or to innovatively enhance things of special interest.

These file structures can be used directly with GIS as image backdrops or imported to grid-based (raster) GIS for spatial analysis (Myers et al., 1997a). To the degree that clusters replicate the landscape patch mosaic, so also should the PHASES compression serve as a surrogate for the original dataset. Aside from statistical summaries, however, original details within clusters will have been sacrificed to parsimony. In this regard, some inability to restore the original data exactly can actually be an advantage relative to copyright concerns for redistribution. Clusters are also advantageous in making some of the spatial structure that resides in an image more explicit. The fragmentation statistics of McGarigal and

Marks (1995) thus become usable for quantifying some properties of the spatial pattern (Myers et al., 1997b).

PHASES software is distributed with a demonstration clip from a 4-band Pennsylvania mosaic of image data from the Landsat MSS (multispectral scanner) satellite sensor. Band 1 consists of green light, and band 2 of red light. Bands 3 and 4 are both in the photographic (near) infrared wavelengths, with band 4 covering a wider range of wavelengths. This subscene includes a small man-made lake that provides a good illustration of the PHASES data structure. Deeper water is notable for absorbing most of the radiation in these wavelengths and thus reflecting very little to the sensor. The ordering of PHASES clusters by length of centroid vector corresponds to increasing overall brightness. Surfaces having low reflectance thus have low cluster numbers. Table 1 is an excerpt from the PHASES grid encompassing the lake, in which open water is represented by cluster No. 1. Table 2 is the corresponding extract from the centroid table showing the first 10 rows pertaining to clusters 1–10, respectively.

3. QUANTIFIED CATEGORIES AND ENVIRONMENTAL INDICATORS AS SIGNALS

Clustering fits the remote sensing context in a fairly intuitive manner, since it is not difficult to imagine the eye working in similar fashion when viewing the landscape from an aircraft. The fit may not be as obvious, however, when considering how categorical themes from GIS layers might be included in the scenario. This becomes clearer when the relational character of many GIS themes is taken into account. GIS thematic categories are often typologies wherein each type jointly determines a whole suite of relational variables, many of which are quantitative. Soils and geology are prime examples for which great effort has been expended in developing characterization tables for the classes that appear on a map. The categories of these themes are effectively already in a cluster-coded form, and must be decoded in terms of their (relational) cluster variables before doing further quantitative analysis. Physical and chemical properties of soils are fundamentally quantitative, even though the characterization level for mapped soil types may be determined on only an ordinal scale. This is parallel to much of the available remotely sensed image data for which idiosyncracies of sensor detectors or "preprocessing" operations often render the data more ordinal than interval. The point here is that relational variables for GIS themes can serve as (pseudo)signals for use in PHASE clustering.

Table 1 Excerpt from sample PHASES grid showing open lake water as cluster No. 1

37	95	155	164	145	88	50	41	31	38	83	95	25	5	11	40
49	124	141	151	139	64	40	63	82	83	57	54	25	12	12	40
49	124	141	151	140	127	81	82	117	117	88	74	41	25	30	38
59	83	138	151	140	127	81	82	117	97	97	38	8	12	23	50
83	72	93	145	138	127	93	83	97	83	72	6	2	12	38	41
83	72	97	148	132	95	16	4	12	5	2	2	2	2	46	17
97	148	154	156	132	77	4	1	1	1	1	1	2	2	2	4
112	156	156	148	117	97	4	1	1	1	1	1	1	1	2	24
119	124	119	119	95	29	2	1	1	1	1	1	1	1	6	63
124	124	106	91	30	3	1	1	1	1	1	1	2	6	19	46
124	120	131	122	23	3	1	1	1	1	1	2	3	13	23	38
151	151	106	91	30	3	1	1	1	1	1	4	6	8	11	46
142	140	69	30	2	1	1	1	1	1	2	8	6	6	9	59
11	4	23	4	1	1	1	2	4	4	5	8	8	8	11	46
43	2	2	2	1	1	1	2	4	8	9	12	24	29	23	23
32	1	1	1	1	1	1	2	6	16	12	24	40	41	30	15
48	2	1	1	1	1	1	4	25	38	46	63	83	54	48	26
46	4	1	1	1	1	2	17	63	83	97	97	46	64	93	63
81	49	2	1	1	2	4	38	82	97	132	148	120	72	93	59
101	90	37	15	2	4	41	72	112	132	148	148	112	72	81	72
151	115	48	8	4	12	59	117	148	145	132	93	72	93	93	72
88	57	41	9	12	46	83	117	148	120	46	38	59	97	117	97
54	17	12	38	59	82	120	154	138	81	41	38	63	97	112	112
31	16	12	38	72	117	135	148	117	81	38	40	72	112	112	112
64	31	24	38	83	120	130	132	102	83	46	63	112	130	120	83

So-called "environmental indicators" constitute another major source of (pseudo)signals for incorporating GIS information composities into PHASE analysis. Environmental indicators are essentially recipes of informational ingredients for "cooking up" a rating on the condition of the environment in much the same manner that body temperature serves as an index of human health. Expertise is required to formulate such indicators, but environmental indicator "cookbooks" and "premixes" are becoming more generally available. O'Neill et al. (1997) offer a synopsis regarding indicators for monitoring environmental quality at the landscape scale. EPA (1997) provides an index of watershed indicators that is an impressive compendium. Having environmental indicators incorporated in clustering also helps with the interpretation of cluster groupings.

Table 2 First 10 rows of the cluster centroid table for a PHASES sample. The first field is the cluster number, the second field is the frequency, and the remaining fields are the respective band means

1	70	13.214286	9.614285	10.928572	5.671429
2	32	14.843750	11.875000	24.281250	21.562500
3	22	17.272728	16.454546	32.454544	25.045454
4	84	14.690476	11.892858	34.035713	33.511906
5	120	14.000000	11.525000	35.841667	36.125000
6	116	14.534483	11.982759	36.491379	38.301723
7	66	17.969696	21.681818	35.530304	38.060608
8	101	15.039604	12.386139	38.950497	36.306931
9	151	14.377483	11.761589	38.715233	38.516556
10	29	19.241379	25.827587	37.172413	33.793102

4. GROUPED ANALYSIS

Clustering in the PHASES fashion produces what amounts to a multivariate frequency table with unequal class intervals. This is most apparent in the structure of the centroid table (Table 2), but also holds for other auxiliary tables not illustrated here. In the same manner that one can do statistical computations from a frequency table, there are also PHASE counterparts to most of the analytical undertakings that could be conducted on the parent dataset. Instead of being done on a cell (pixel) basis, the analyses are done in a weighted manner with cluster frequencies serving as weights and cluster means substituting for original data values. The centroid table and a companion table of variability among clusters have formats that facilitate importing into spreadsheets for performing computations that are not directly supported by software modules in PHASES.

Analytical results obtained in this way are based entirely on variation between clusters, and may therefore differ somewhat from corresponding analysis of raw data.

Post-clustering principal components analysis can be helpful in understanding the multidimensional configuration of clusters. The analysis begins with computation of a weighted covariance matrix or correlation matrix from the centroid data in a spreadsheet, which is then passed to standard statistical software for computation of eigenvalues and eigenvectors. There is also a PHASES software facility that computes overall and within-cluster covariance matrices when other clustering operations are completed. Comparison of size and orientation of compressed components relative to original components can be instructive regarding the nature of information that has been suppressed in clustering. Loadings of high-order compressed components also indicate which variates had most influence on cluster formation.

5. RECLUSTERING

For most survey situations, the multiplicity of hyperclusters goes well beyond the number of strata that are logistically tractable in a sampling design. Grouping of hyperclusters will therefore be required. The most expedient way of accomplishing a major reduction is by clustering the clusters. The PHASES clustering facility has a weighting option for this purpose. For weighted reclustering, the hypercluster mean vectors are treated as data vectors and the cluster frequencies serve as weights. The weights give larger hyperclusters proportionately more influence on the multidimensional migration of seeds.

PHASE formulations also afford ample opportunity for less generic reclustering. One of the auxiliary tables contains information on maximum and minimum standard deviations for variates in each cluster to give an indication of departure from sphericity. This table also gives distance from each centroid to nearest centroid, distance from each centroid to farthest data vector (cluster radius), and distance from each centroid to nearest data vector for another cluster. This information can provide a basis for considering clusters as proximity networks in multidimensional space, from which hierarchical reductions in number of clusters can be strategically configured.

Since clusters induce a spatial patchwork, the incorporation of spatial relations in cluster reduction can likewise be considered. Clusters that participate in the formation of localized complexes are candidates for unification into a stratum. Information on edge apportionment among clusters can serve to indicate which sets of clusters are preferentially juxtaposed, and therefore appropriately unified as a stratum.

PHASE facilities also provide for rendering views in which clusters are selectively colorized on a graytone background of cluster numbers as intensity values. Such a display allows judgment to be exercised in deciding which clusters should be grouped as strata. This often proves useful for refining preliminary groupings from reclustering.

6. SPATIAL SIMPLIFICATION

It bears emphasis that much of the underutilization for GIS data in environmental sampling design is due to information overload. There are too many layers with disparate quantification, varying resolution, partial coverage and temporal inconsistency to be incorporated into the sampling design individually. Thus our concern is with compositing, compressing and simplifying to the point of manageability. In a GIS context, there is a quantal/spatial duality to simplification. So far we have focused mostly on the quantal side of the duality. Quantal simplification will also induce spatial simplification due to the fact that some of the patch boundaries are not distinguished. The most "intelligent" simplification comes from merging landscape pattern elements that have both quantial similarity and preferential geographic juxtaposition, since this combination creates low contrast edges which can be eliminated with minimal impact on visual perception of pattern (Myers et al., 1997b).

Even after quantal amalgamation has been accomplished, there will almost certainly remain "flecks and flakes" that lend some "salt and pepper" character to the pattern. We use fleck to mean an isolated cell of a raster grid that is not touched by any like neighbor. A flake is a patch that is tiny relative to the average patch size. Exactly what constitutes a flake will depend on the spatial resolution that is of concern in a particular situation. Flecks tend to be inconsequential, and thus distracting for many practical purposes. Some degree of flake suppression is often a practical necessity for converting from cell to polygon form so that boundaries of strata can be plotted as line maps. It may be desirable to suppress some types of flakes and not others.

PHASE software provides for suppressing both flecks and flakes. Fleck suppression is a fairly straightforward matter of changing the cluster number for a cell to that of its most similar neighbor. Flake suppression is a more complicated and computationally intensive task which involves doing topological annexation of small patches by larger neighboring patches.

7. COORDINATING SYNOPTIC SCALES AND QUILTING COVERAGES

Coordinating two synoptic datasets is largely a matter of finding the counterpart(s) in one dataset for the elements of another. This is true whether or not the two datasets have the same scale. A typical context is multiple scenes of partially overlapping image data, with some or all being collected on different days from a satellite in a progressive (polar) orbit. The time lapse may be years instead of days, and the images may even come from different sensors at different scales. Complete coverage of a region requires "quilting" scenes together using correspondence in overlap areas for linkage. Spatial rectification and registration are prerequisite for undertaking this sort of linkage. In sampling terms, the idea is to map zones that could serve as strata in a seamless manner across several scenes. Such zones may be only vaguely characterized at the time of mapping.

GIS provides for visually assessing the spatial integrity of zones, which allows judgment to substitute for characterization in many respects.

The coordination scenario begins with separate (hyper)clustering for each of the datasets. The task is thus reduced to matching clusters as counterparts among the scenes. The coarsest scale (least detailed) coverage becomes the base or "common denominator" of the zonal coordination. The base is reclustered as necessary to obtain a reasonable number of zones having coherent spatial patterns. The LNKPHASE facility of PHASES next produces a list of base-cluster counterparts using most frequent (modal) occurrence as a matching criterion. The counterparts then become de facto "training sets" for doing a classification in a "supervised" manner whereby similarity to training sets is the membership criterion. Classification is rapid, since it is done using cluster centroids. The base zones are thus carried into the match scene, which is subsequently matched to another scene for zonal carryover, and so on. Such zones are effectively unnamed thematic categories.

8. SYNOPTIC LINKAGE AMONG SITES

The preceding protocol pertains in the absence of site-level data to particularize the process. When site data are available, initiation and propagation of zones is keyed to sites or "site guided." The most distinctive or most important site gives rise to the first training set. A synoptic excerpt like Table 1 is obtain for the vicinity of the site via the BLKPHASE facility. Clusters that are proximate to the site become training sets for zonal classification. If known instances of like sites occur elsewhere, then they also contribute training sets for the same zone. Other important and/or contrasting sites likewise become progenitors for additional zones. When zones have been established for an initial scene, matching of scenes and determination of counterparts is conducted as outlined earlier. Counterpart clusters again provide training to transfer zones into the next scene. If additional sites exist in the neigboring coverage, then they augment the training and/or serve as consistency checks for propagating zones. It may also be helpful to perform preliminary clustering of sites before embarking on zonation.

The strategy of doing supervised classification entirely at a cluster level is one the innovations of PHASES relative to image analysis. The similarity metric for the adaptive classifier is Euclidian distance, with thresholding on the basis of intersecting hyperspheres having radii that are specified multiples of maximum or minimum standard deviations for cluster variables. In analogy with the usual maximum likelihood spectral classifier, clusterwise classification could also be based upon the Kullback–Liebler information metric instead of Euclidean distance.

9. CAUTIONARY CONSIDERATIONS

Incorporating generic GIS data into sample designs is essentially informational opportunism, and opportunism often tends to be untidy. In other words, you take what you can get when the opportunity arises without being certain of achieving substantial gains in any given case. Gains are likely to be uneven across an extent of sampling, since the strength of linkage between the sampling variable(s) of interest and the data layers usually depends on the immediate environmental setting. To the degree that the GIS mirrors environmental settings, there should be overall statistical advantage in attempting the linkages.

When (zonal) spatial partitions are formed from clusters, it is likely that some partitions will contain quite consistent site data, whereas others may have such variability in site data as to be almost noninformative. Likewise, regressions of site variables on cluster mean variables may exhibit residuals that are suggestive of outliers. It then becomes prudent to determine which clusters are responsible for the outliers and segregate them for recomputation of regression on the remainder. Site-guided partitioning is based on the ASSUMPTION that similar clusters should be similar with regard to site data. Whether or not that assumption is borne out in sampling remains to be seen, and may even not be easily determined in retrospect.

GIS linkages are mostly based on hypotheses drawn from the data. Such data mining may uncover substantial veins of informational ore, or the traces of ore may dwindle rapidly upon pursuit. In the course of prospecting, it is unwise to build costly superstructures for extraction on the basis of preliminary and perhaps tenuous evidence. Sampling policy should therefore favor designs that are capable of exploiting suspected linkages, but are not informationally impaired if the suspicions prove unfounded. At the most conservative extreme, the spatial partitions might only be used for poststratification.

10. STABLE SURVEY DESIGNS

Suggestions for candidate sampling designs are circumspect in light of the aforegoing cautionary discussion. Although a wide variety of sampling designs could involve GIS information, concern here is mostly with those that would incur only a less desirable sample allocation if GIS data should turn out to be locally or even globally misleading. The most obvious design technique is stratification, which has been alluded to throughout the presentation. Spatial partitions derived from GIS become strata, and any site data are used to help determine allocations among strata and how many total samples to obtain. If GIS data are particularly suspect, then allocation can be decided on other grounds and GIS-based partitions used to post-stratify for estimation.

The GIS-based spatial partitions also fit readily into multistage or sub-sampling schemes whereby a sample of secondary units is drawn from sampled primary units, a sample of tertiary units is drawn from sampled secondary units, etc. This has been a popular mode of using remotely sensed imagery for sampling of forests, and should apply equally well to multiscale data in GIS, which is often derived from remotely sensed imagery. If it is desirable to avoid the complications of quilting, then the respective zonal coverages can serve directly as primary units, with clusters being secondary units.

Multiphase designs such as double sampling tend to be more strongly dependent on form and consistency of relationships between variables, as might be the case for the association of cluster means with site variables. If relationships between such variables appear strong, then ranked-set sampling may offer a less rigid alternative to double sampling. Myers et al. (1995) consider some of the possibilities of ranked-set sampling with spatially based concommitant variables as the ranking mechanism. The basic strategy is to make preliminary draws of sets of m samples, with each being internally ranked via the concommitant variable, and then one of each set being selected on a rotation of ranks. The efficiency of the design comes from spreading the sample over the distribution of the target variable. Errors of ranking reduce the efficiency but do not otherwise impact the estimates.

There are some interesting possibilities for using adaptive sampling (Thompson, 1992) in a reverse sort of mode for detecting and investigating situations where the presumed linkage between the GIS information and site data weakens. The synoptic GIS information is available and readily mapped for carrying by field crews. When sampling sites are encountered during field work where the observations depart from the GIS context in a specified degree, then supplementary samples can be taken in that region to determine the cause and scope of the GIS inconsistency. Each sampling operation thus contributes to the validation or correction of the GIS as well as providing global estimates of the current target variable(s).

11. PERSPECTIVE

GIS is one of the fastest growing information technologies and has become well entrenched in most facets of environmental work. Satellite-based remote sensing is one of the more prolific sources of spatial environmental information, with a massive archive of image data and several new higher resolution systems on the technological horizon. Despite their broad usage in cartographic and geometric modes, both have been underutilized in a statistical sense. (Hyper)clustering can facilitate compositing of layer variates and their organization as spatial partitions relevant to sampling, as well as allow linkage between scales. Rather than offering a particular template for incorporating such information in sampling designs, we have attempted a preliminary consideration of strategies by which it might contribute to different designs without placing the survey effort in jeopardy if apparent relationships prove tenuous.

REFERENCES

EPA. The index of watershed indicators. EPA-841-R-97-010, Office of Water, United States Environmental Protection Agency, Washington, DC, 56 pp., 1997.

Forman, R.T.T. *Land Mosaics: The Ecology of Landscapes and Regions*. Cambridge: Cambridge University Press, 1995.

Hartigan, J.A. *Clustering Algorithms*. New York: Wiley, 1975.

Kelly, P., White, J. Preprocessing remotely sensed data for efficient analysis and classification. *Applications of Artificial Intelligence 1993: Knowledge-Based Systems in Aerospace and Industry*. Proceedings SPIE 1993, pp. 24–30.

Lillesand, T., Kiefer, R.W. *Remote Sensing and Image Interpretation*. 3rd edn. New York: Wiley, 1994.

McGarigal, K., Marks, B. FRAGSTATS: Spatial pattern analysis program for quantifying landscape structure. USDA, Forest Service, Pacific Northwest Research Station, General Technical Report PNW-GTR-351, 1995.

Myers, W. Remote Sensing and Quantitative Geogrids in PHASES [Pixel Hyperclusters as Segmented Environmental Signals]. Report ER93803, Environmental Resources Research Institute, Pennyslvania State University, University Park, PA, 1998.

Myers, W., Johnson, G.D., Patil, G.P. Rapid mobilization of spatial/temporal information in the context of natural catastrophes. *COENOSES*, **10**(2–3):89–94, 1995.

Myers, W., Patil, G.P., Taillie, C. PHASE formulation of synoptic multivariate landscape data. Technical Report Number 97-1102, Center for Statistical Ecology and Environmental Statistics, Department of Statistics, Pennsylvania State University, University park, PA, 1997a.

Myers, W., Patil, G.P., Taillie, C. Exploring landscape scaling properties through constrictive analysis. Technical Report Number 97-1101, Center for Statistical Ecology and Environmental Statistics, Department of Statistics, Pennsylvania State University Park, PA, 1997b.

O'Neill, R.V., Hunsaker, C.T., Jones, K.B., Ritters, K.H., Wickham, J.D., Schwartz, P.M., Goodman, I.A., Jackson, B.L., Baillargeon, W.W. Monitoring environmental quality at the landscape scale: Using landscape indicators to assess bioteic diversity, watershed integrity, and landscape stability. *Bioscience*, **47**:513–519, 1997.

Thompson, S.K. *Sampling*. New York: Wiley, 1992.

Tou, J.T., Gonzalez, R.C. *Pattern Recognition Principles*. Reading, MA: Addison-Wesley, 1974.

6

Corrected Confidence Sets for Sequentially Designed Experiments: Examples

MICHAEL WOODROOFE University of Michigan, Ann Arbor, Michigan

D. STEPHEN COAD University of Sussex, Falmer, Brighton, England

1. INTRODUCTION

Ford and Silvey (1980) proposed an adaptive design for estimating the point at which a regression function attains its minimum. In their model, there are potential observations of the form

$$y_k = \theta_1 x_k + \theta_2 x_k^2 + \epsilon_k, \qquad k = 1, 2, \ldots, \tag{1}$$

where x_k are design points chosen from the interval $-1 \leq x \leq 1$, θ_1 and $\theta_2 > 0$ are unknown parameters, and $\epsilon_1, \epsilon_2, \ldots$ are i.i.d. standard normal random variables. Interest centered on estimating the value $-\theta_1/2\theta_2$ at which the regression function attains its minimum. After examining the asymptotic variance of the maximum likelihood estimator, Ford and Silvey proposed the following *adaptive design*: first take observations at $x_1 = -1$ and $x_2 = +1$; thereafter, if (x_k, y_k), $k = 1, \ldots, n$, have been determined, let y_n^- or y_n^+ denote the sum of y_k for which $k \leq n$ and $x_k = -1$ or $x_k = +1$, respectively; take the next observation at $x_{n+1} = +1$ if $|y_n^+| < |y_n^-|$ and take the next observation at $x_{n+1} = -1$ otherwise. After the experiment is run, investigators may want confidence intervals, and a problem arises here. Owing to the adaptive nature of the design, it is not the case that the maximum likelihood estimators, say $\hat{\theta}_{n,1}$ and $\hat{\theta}_{n,2}$, have a bivariate normal distribution. This problem was addressed by Ford et al. (1985) and Wu (1985). The former paper proposed an exact solution which seems overly conservative. The latter proposed an asymptotic solution which, at a practical level, ignores the adaptive nature of the design. This paper contains an asymptotic solution which does not ignore the adaptive nature of the design.

The Ford–Silvey example fits nicely into the following more general model: letting $'$ denote transpose, suppose that there are potential observations of the form

$$y_k = x_k'\theta + \sigma\epsilon_k, \quad k = 1, 2, \ldots \tag{2}$$

where $x_k = (x_{k,1}, \ldots, x_{k,p})'$ are design variables which may be chosen from a subset $\mathcal{X} \subseteq \Re^p$, $\theta = (\theta_1, \ldots, \theta_p)'$ is a vector of unknown parameters, $\sigma > 0$ may be known or unknown, and $\epsilon_1, \epsilon_2, \ldots$ are i.i.d. standard normal random variables. The design vectors x_k, $k = 1, 2, \ldots$, may be chosen *adaptively*; that is, each x_k may be a (measurable) function of previous responses and auxiliary randomization, say

$$x_k = x_k(u_1, \ldots, u_k, y_1, \ldots, y_{k-1}), \qquad k = 1, 2, \ldots, \tag{3}$$

where $u_1, u_2 \ldots$ are independent of $\epsilon_1, \epsilon_2, \ldots$ and have a known distribution. Letting $\mathbf{y}_n = (y_1, \ldots, y_n)'$, $X_n = (x_1, \ldots, x_n)'$, and $\mathbf{e}_n = (\epsilon_1, \ldots, \epsilon_n)'$, the model in equation (2) may be written in the familiar form

$$\mathbf{y}_n = X_n\theta + \sigma\mathbf{e}_n, \qquad n = 1, 2, \ldots.$$

The model is very general. It includes adaptive procedures for finding the maximum or minimum of a regression function, as in Ford and Silvey (1980), models for sequential clinical trials, as in Coad (1995), adaptive biased coin designs, as in Eisele (1994), and time series and controlled time series, as in Lai and Wei (1982), among others.

The likelihood function is not affected by the adaptive nature of the design, or by the optional stopping introduced later (see, for example, Berger and Wolpert, 1984). So the maximum likelihood estimator of θ has the familiar form

$$\hat{\theta}_n = (X_n'X_n)^{-1}X_n'\mathbf{y}_n \tag{4}$$

provided that $X_n'X_n$ is positive definite. The maximum likelihood estimator of σ^2 also has a familiar form. The usual estimator of σ^2 for a nonadaptive design is

$$\hat{\sigma}_n^2 = \frac{\|\mathbf{y}_n - X_n\hat{\theta}_n\|^2}{n-p} \tag{5}$$

when $X_n'X_n > 0$ and $n > p$. This estimator is unbiased for nonadaptive designs and is nearly unbiased in the absence of optional stopping. See Section 2 for the bias when there is a stopping time.

The sampling distributions of these estimators may be affected by the adaptive design, however, and it is the purpose of this paper to explain how approximate expressions for the sampling distributions of $\hat{\theta}_n$ may be obtained. Approximations to the sampling distributions are presented in Sections 2–4. These are illustrated by simple examples in Sections 2–4, and by more complicated examples in Sections 5 and 6. The accuracy of the approximations is assessed by simulation. The presentation in Sections 2–4 is informal. Precise statements, with conditions, are deferred to Section 7. Proofs are discussed briefly in Section 8.

1.1 Remark on Notation

Below, $P_{\sigma,\theta}$ denotes a probability model under which equation (2) holds, and $E_{\sigma,\theta}$ denotes expectation with respect to $P_{\sigma,\theta}$. When σ is known and fixed, it may be omitted from the notation. For example, probability and expectation may be denoted by P_θ and E_θ.

2. BIASES

The effect of the adaptive design may be illustrated by computing the biases of $\hat{\theta}_n$ and $\hat{\sigma}_n^2$. With a view towards Example 3 below, suppose that the model in equation (2) is observed for $k = 1, \dots, N$, where N is a stopping time with respect to (u_k, y_k), $k = 1, 2 \dots$. That is, the event $\{N = n\}$ can depend only on (u_k, y_k), $k = 1, \dots, n$. The approximate biases are derived from asymptotic expansions. Thus, suppose that N depends on a parameter a, say $N = N_a$, and that $N_a \to \infty$ in probability as $a \to \infty$. The case of nonrandom sample size $N = n \to \infty$ is not excluded here. Suppose that $X_N'X_N > 0$ *w.p.*1 for all a, and that

$$\eta(\sigma, \theta) := \lim_{a\to\infty} a(X_N'X_N)^{-1} \tag{6}$$

and

$$\rho(\sigma^2, \theta) := \lim_{a\to\infty} \frac{a}{N} \tag{7}$$

exist in $P_{\sigma,\theta}$-probability for *a.e.* θ for each $\sigma > 0$. Actually, more is required (see Section 8). Write $\eta(\sigma, \theta) = [\eta_{ij}(\sigma, \theta) : i, j = 1, \dots, p]$ and let $\mathbf{1} = (1, \dots, 1)'$. Then

$$E_{\sigma,\theta}(\hat{\theta}_N - \theta) \approx \frac{\sigma^2}{a}\eta^{\#}(\sigma, \theta)\mathbf{1} \tag{8}$$

and

$$E_{\sigma,\theta}(\hat{\sigma}_N^2 - \sigma^2) \approx \frac{2\sigma^4}{a}\rho'(\sigma^2, \theta) \tag{9}$$

in the very weak sense of Woodroofe (1986, 1989), where $'$ denotes differentiation with respect to σ^2 and

$$\eta_{ij}^{\#}(\sigma, \theta) = \frac{\partial}{\partial \theta_j}\eta_{ij}(\sigma, \theta) \tag{10}$$

These are among the main findings of Coad and Woodroofe (1998), who also obtain approximations to variances and covariances. Observe that the bias of $\hat{\sigma}_N^2$ is determined primarily by the optional stopping; that is, if $N = n = a$, then $\rho(\sigma^2, \theta) = 1$ for all σ^2 and θ, so that $\rho'(\sigma^2, \theta) = 0$ and, therefore, $E_{\sigma,\theta}(\hat{\sigma}_n^2 - \sigma^2) = o(1/n)$.

EXAMPLE 1: *An Autoregressive Process.* To illustrate the use of equations (8) and (9), consider the simple autoregressive process $y_k = \theta y_{k-1} + \sigma\epsilon_k$, $k = 1, 2, \ldots$, with $y_0 = 0$. Suppose that the process is observed for n time units, so that $N = n$. Then $p = 1$, $X_n'X_n = y_0^2 + y_1^2 + \cdots + y_{n-1}^2$, and

$$\lim_{n\to\infty} n(X_n'X_n)^{-1} = \lim_{n\to\infty} \frac{n}{y_0^2 + y_1^2 + \cdots + y_{n-1}^2} = \frac{1-\theta^2}{\sigma^2}$$

$w.p.1$ for all $|\theta| < 1$ and $\sigma > 0$. It then follows from equation (8) that

$$E_{\sigma,\theta}(\hat{\theta}_n - \theta) \approx -\frac{2\theta}{n} \tag{11}$$

for large n, and $E_{\sigma,\theta}(\hat{\sigma}_n^2 - \sigma^2) = o(1/n)$ as $n \to \infty$, and equation (11) agrees well with Coad and Woodroofe's (1998) simulations for n as small as 25. □

EXAMPLE 2: *The Ford–Silvey Example.* In the Ford–Silvey example in equation (1), let $N = n = a$ since there is no stopping time, and write

$$X_n'X_n = \begin{pmatrix} \sum x_k^2 & \sum x_k^3 \\ \sum x_k^3 & \sum x_k^4 \end{pmatrix} = \begin{pmatrix} n & s_n \\ s_n & n \end{pmatrix}$$

where $s_n = x_1 + \cdots + x_n$. Ford and Silvey (1980) showed that $w.p.1$, $s_n/n \to -\kappa(\theta)$, where $\kappa(\theta) = \theta_1/\theta_2$, if $|\theta_1| \leq |\theta_2|$, and $\kappa(\theta) = \theta_2/\theta_1$ otherwise. It follows that if $|\theta_1| \neq |\theta_2|$, then

$$\eta(\theta) = \lim_{n\to\infty} n(X_n'X_n)^{-1} = \frac{1}{1-\kappa^2}\begin{pmatrix} 1 & \kappa \\ \kappa & 1 \end{pmatrix}$$

where κ has been written for $\kappa(\theta)$. The expression for $\eta^{\#}(\theta)$ is complicated and will not be detailed. After some algebra, however, it is easily seen that

$$E_\theta(\hat{\theta}_n - \theta) \approx \frac{1}{n}\eta^{\#}(\theta)\mathbf{1} = \frac{1}{n\beta(\theta)}\begin{pmatrix} \theta_1 \\ \theta_2 \end{pmatrix} \tag{12}$$

where $\beta(\theta) = \max[\theta_1^2, \theta_2^2] - \min[\theta_1^2, \theta_2^2]$. Again, the approximation in equation (12) agrees well with Coad and Woodroofe's (1998) simulations for n as small as 25. □

3. SAMPLING DISTRIBUTIONS

In this section, σ is assumed to be known, and probability is denoted by P_θ. Let m be an integer, $1 \leq m \leq p$, and let A_n be an $m \times p$ matrix and B_n be a $p \times p$ matrix for which

$$A_n A_n' = I_m \qquad \text{and} \qquad X_n'X_n = B_n B_n', \tag{13}$$

where I_m denotes the $m \times m$ identity matrix. There are many possible choices for A_n and B_n, and some advantages of using a Cholesky decomposition for $X_n'X_n$ are described by Woodroofe and Coad (1997). The only requirements, however, are equation (13), equation (16) below, and that A_n and B_n depend measurably on X_n and $\mathbf{y}_n$. If $X_n'X_n > 0$, let

$$Z_n^o = \frac{1}{\sigma} B_n'(\theta - \hat{\theta}_n) \tag{14}$$

and

$$W_n^o = A_n Z_n^o = \frac{1}{\sigma} A_n B_n'(\theta - \hat{\theta}_n) \tag{15}$$

Here W_n^o and Z_n^o are first approximations to approximately pivotal quantities. Of course, Z_n^o would have an (exactly) standard p-variate normal distribution in the absence of an adaptive design and optional stopping.

The problem is to find an approximation to the distribution of W_N^o. Let

$$Q^{(a)} = \sqrt{a} A_N B_N'^{-1}, \qquad a \geq 1$$

where $N = N_a$ denotes the stopping time. The conditions for the expansions require that $Q^{(a)}$ have a limit as $a \to \infty$, say

$$\lim_{a\to\infty} Q^{(a)} = Q(\theta) \tag{16}$$

in probability, where $Q(\theta) = [q_{ij}(\theta) : i = 1, \ldots, m, \ j = 1, \ldots, p]$. Again, a stronger form of convergence is required (see Section 7). Suppose that the entries $q_{ij}(\theta)$ are differentiable with respect to θ, and let

$$q^{\#}_{ij}(\theta) = \frac{\partial}{\partial \theta_j} q_{ij}(\theta)$$

$$m_{ij}(\theta) = \sum_{k=1}^{p} \sum_{\ell=1}^{p} \frac{\partial^2}{\partial \theta_k \partial \theta_\ell} [q_{ik}(\theta) q_{j\ell}(\theta)]$$

$Q^{\#}(\theta) = [q^{\#}_{ij}(\theta) : i = 1, \ldots, m, \ j = 1, \ldots, p]$ and $M(\theta) = [m_{ij}(\theta) : i, j = 1, \ldots, m]$. Next, let Φ^m denote the standard m-variate normal distribution and write

$$\Phi^m h = \int_{\Re^m} h(w) \Phi^m \{dw\}$$

$$\Phi^m_1 h = \int_{\Re^m} w h(w) \Phi^m \{dw\}, \qquad (m \times 1) \tag{17}$$

and

$$\Phi^m_2 h = \frac{1}{2} \int_{\Re^m} (ww' - I_m) h(w) \Phi^m \{dw\}, \ (m \times m)$$

whenever the integrals are meaningful. If $h : \Re^m \to \Re$ is a function of quadratic growth (that is, $|h(w)| \leq C(1 + \|w\|^2)$ for all $w \in \Re^m$ for some constant $0 < C < \infty$), then

$$E_\theta[h(W^o_N)] \approx \Phi^m h - \frac{\sigma}{\sqrt{a}} (\Phi^m_1 h)' Q^{\#}(\theta) 1 + \frac{\sigma^2}{a} \mathrm{tr}[(\Phi^m_2 h) M(\theta)] \tag{18}$$

for large a in the very weak sense. This is Theorem 1 of Woodroofe and Coad (1997). See Section 7 for a precise statement with conditions. Specializing equation (18) to the cases $h(w) = w_i$ and $h(w) = w_i w_j$ gives an approximate mean and covariance matrix of W^o_N; that is,

$$E_\theta(W^o_N) \approx -\frac{\sigma}{\sqrt{a}} Q^{\#}(\theta) 1 = \mu_a(\theta), \qquad \text{say} \tag{19}$$

and

$$E_\theta(W^o_N W^{o\prime}_N) \approx I_m + \frac{\sigma^2}{a} M(\theta) \tag{20}$$

Relation (18) may be summarized as W^o_N is approximately normal with mean $\mu_a(\theta)$ and covariance matrix $I_m + a^{-1}\sigma^2 M(\theta) - \mu_a(\theta)\mu_a(\theta)'$ because, letting Ψ_a denote the latter distribution, a three-term Taylor series expansion of $\int_{\Re^m} h d\Psi_a$ agrees with the right-hand side of equation (18). This is a remarkably simple description, given the generality of the basic model in equation (2).

The magnitude of the effect and the accuracy of the approximation are illustrated in the following example.

EXAMPLE 3: Hayre and Gittins (1981) considered a problem in which two treatments, A and B say, produce normally distributed responses with unknown means μ and ν and variance σ^2, say. Such a model may be written in the form of equation (2) with $\theta_1 = \mu$, $\theta_2 = \nu - \mu$, and $x = (1, z)'$, where $z = 0$ or 1 accordingly as an observation is made on A or B. Motivated by clinical trials in which there was an ethical cost for giving an inferior treatment, Hayre and Gittins (1981) suggested the sampling rule $z_1 = 0$, $z_2 = 1$, and

$$z_{n+1} = 1 \quad \text{iff} \quad \frac{s_n}{n - s_n} \leq w(\hat{\theta}_{n,2}), \qquad n \geq 2$$

where $s_n = z_1 + \cdots + z_n$ and w is a positive function on $\Re$. For example, the function $w(\delta) = \sqrt{1 + 10\delta}$ for $\delta > 0$ and $w(\delta) = 1/\sqrt{1 + 10|\delta|}$ for $\delta \leq 0$ is used in the simulations below. For the case of known σ^2, the sequential design was to be used with the stopping time

$$N = \inf\{n \geq 3 : |\imath_n^2 \hat{\theta}_{n,2}| > a\sigma^2\}$$

where

$$\imath_n^2 = \frac{s_n(n - s_n)}{n}$$

and $a > 0$ is a design parameter, chosen to control the error probabilities of a sequential probability ratio test. In this example,

$$X_n' X_n = \begin{pmatrix} n & s_n \\ s_n & s_n \end{pmatrix}, \qquad n \geq 1$$

Further, there is special interest in $\theta_2 = \nu - \mu$, the difference of the two means, and it is natural to let $A_n B_n' = c_n(0, 1)$, where c_n is a constant. This may be accomplished by letting

$$A_n = \left[-\sqrt{\frac{s_n}{n}}, \sqrt{\frac{n - s_n}{n}} \right]$$

and

$$B_n = \begin{pmatrix} \sqrt{n - s_n} & \sqrt{s_n} \\ 0 & \sqrt{s_n} \end{pmatrix}$$

in which case $c_n = i_n$ and

$$Q^{(a)} = \sqrt{a} A_N B_N^{-1} = \frac{\sqrt{a}}{i_N}\left[-\frac{s_N}{N}, 1\right]$$

To apply equation (18), it is necessary to find the limiting matrix $Q(\theta)$. Hayre and Gittins (1981) showed that

$$\frac{a}{i_N^2} \to \frac{|\theta_2|}{\sigma^2}, \qquad \frac{s_N}{N} \to \frac{w(\theta_2)}{1 + w(\theta_2)}$$

and

$$\frac{a}{N} \to \frac{|\theta_2| w(\theta_2)}{\sigma^2[1 + w(\theta_2)]^2}$$

$w.p.1$. It follows that

$$Q^{(a)} \to \frac{\sqrt{|\theta_2|}}{\sigma}\left[-\frac{w(\theta_2)}{1 + w(\theta_2)}, 1\right] = Q(\theta)$$

$$Q^{\#}(\theta) = \frac{1}{\sigma}\left[0, \frac{\text{sign}(\theta_2)}{2\sqrt{|\theta_2|}}\right]$$

and $m_{11}(\theta) = 0$ for $\theta_2 \neq 0$. Thus, $\mu_a(\theta) = -\text{sign}(\theta_2)/(2\sqrt{a|\theta_2|})$ in equation (19). When specialized to indicator functions, equation (18) asserts

$$P_\theta\{W_N^o \leq c\} \approx \Phi(c) + \frac{1}{\sqrt{a}}\varphi(c)\frac{\text{sign}(\theta_2)}{2\sqrt{|\theta_2|}} \tag{21}$$

where Φ and φ denote the standard normal univariate distribution and density functions.

It is clear from equation (21) that the correction term $\varphi(c)\text{sign}(\theta_2)/(2\sqrt{a|\theta_2|})$ can be significant, and the approximation in equation (21) effectively predicts the simulated values in Table 1 for the cases $\theta_1 = 0$, $\theta_2 = 0.5$, $\theta_2 = 1$, and $a = 6$. The accuracy of the approximation deteriorates as $|\theta_2|$ decreases. For $a = 6$, the approximation overcorrects when $\theta_2 = 0.25$. Woodroofe (1989) has reported simulations of a similar nature for a closely related example, due to Robbins and Siegmund (1974). In this example, the approximation does not depend on the function w that determines the allocation rule. Coad (1991) has shown that this independence holds more generally, for essentially different allocation rules, and has reported simulations which indicate that the approximation in equation (21) works well for other allocation rules too. □

Table 1 $P_\theta\{W_N^o \le c\}$ in the Hayre–Gittins example

		$\theta_2 = 0.25$		$\theta_2 = 0.5$		$\theta_2 = 1$	
c	Normal	MC	Approx.	MC	Approx.	MC	Approx.
−2.00	.0228	.0381	.0448	.0369	.0383	.0324	.0338
−1.75	.0401	.0623	.0753	.0622	.0650	.0536	.0577
−1.50	.0668	.0994	.1197	.0998	.1042	.0911	.0932
−1.25	.1056	.1525	.1802	.1572	.1584	.1443	.1429
−1.00	.1587	.2264	.2574	.2272	.2285	.2148	.2081
−0.75	.2266	.3199	.3496	.3084	.3186	.2748	.2881
−0.50	.3085	.4245	.4523	.4082	.4102	.3631	.3804
−0.25	.4013	.5325	.5593	.5092	.5129	.4754	.4802
0.00	.5000	.6338	.6629	.6126	.6152	.5780	.5814
0.25	.5987	.7242	.7566	.7054	.7103	.6740	.6776
0.50	.6915	.8003	.8352	.7863	.7931	.7593	.7633
0.75	.7734	.8619	.8963	.8525	.8603	.8307	.8348
1.00	.8413	.9048	.9401	.9021	.9112	.8826	.8907
1.25	.8944	.9314	.9689	.9409	.9471	.9288	.9316
1.50	.9332	.9451	.9861	.9649	.9706	.9548	.9596
1.75	.9599	.9513	.9952	.9787	.9848	.9758	.9776
2.00	.9772	.9534	.9993	.9869	.9928	.9870	.9883

Note: Based on 10,000 replications with $a = 6, \sigma = 1, \theta_1 = 0, w(\delta) = \sqrt{1 + 10\delta}$ for $\delta > 0$, and $w(\delta) = 1/\sqrt{1 + 10|\delta|}$ for $\delta \le 0$.

3.1 Corrected Confidence Sets

The main application of equation (18) is to form corrected confidence sets, confidence sets whose actual coverage probability differs from the nominal by $o(1/a)$. To do this, it is convenient to standardize W_N^o. For the case of known σ^2, let $\hat{\mu}_a^o$ denote an estimator of $\mu_a(\theta)$, for example, $\hat{\mu}_a^o = \mu_a(\hat{\theta}_N)$. Then it may be shown that

$$E_\theta[(W_N^o - \hat{\mu}_a^o)(W_N^o - \hat{\mu}_a^o)'] \approx I_m + \frac{1}{a}\Delta^o(\theta)$$

where

$$\Delta_{ij}^o(\theta) = \sigma^2 \sum_{k=1}^{p} \sum_{\ell=1}^{p} \frac{\partial}{\partial \theta_\ell} q_{ik}(\theta) \frac{\partial}{\partial \theta_k} q_{j\ell}(\theta)$$

for $i, j = 1, \ldots, m$. Let $\hat{\Delta}_a^o$ denote an estimator of $\Delta^o(\theta)$, for example, $\hat{\Delta}_a^o = \Delta^o(\hat{\theta}_N)$, and let

$$W_N^{o*} = \left(I_m + \frac{\hat{\Delta}_a^o}{2a}\right)^{-1} (W_N^o - \hat{\mu}_a^o)$$

Then W_N^{o*} is approximately standard m-variate normal to a high order; that is

$$E_\theta[h(W_N^{o*})] \approx \Phi^m h \tag{22}$$

for functions h of quadratic growth. Here $\approx$ means equality up to $o(1/a)$ in the very weak sense of Woodroofe (1986,1989). This is Theorem 2 of Woodroofe and Coad (1997). See Section 7 for precise statements with conditions.

It is easy to use equation (22) to form corrected confidence sets for θ. Given a subset $C \subseteq \Re^m$, let

$$\mathcal{C} = \{\theta : W_N^{o*} \in C\} \tag{23}$$

Then

$$P_\theta\{\theta \in \mathcal{C}\} = P_\theta\{ W_N^{o*} \in C\} \approx \Phi^m(C)$$

That is, $\mathcal{C}$ is an approximate confidence set of level $\Phi^m(C)$ to order $o(1/a)$. In applications, C might be of the form $C = \{\theta : |\theta| \leq c\}$, where $|\cdot|$ is a norm on $\Re^m$ and $c > 0$. There may be interest in a given set of linear functions, say $\Gamma\theta$, where Γ is an $m \times p$ matrix. For example, the rows of Γ might be a set of contrasts. In such cases, it is easy to construct A_n, so that $\mathcal{C}$ is, in fact, a confidence set for $\Gamma\theta$. To do so, let G_n be an $m \times m$ matrix for which $G_n\Gamma(X_n'X_n)^{-1}\Gamma' G_n' = I_m$, and let $A_n = G_n\Gamma B_n'^{-1}$. Then $A_nA_n' = I_m$ and $A_nB_n' = G_n\Gamma$, so that

$$\mathcal{C} = \{\theta : \Gamma\theta \in G_n^{-1}C\}$$

There is further simplification if $m = 1$. Then $G_n = 1/\sqrt{\Gamma(X_n'X_n)^{-1}\Gamma'}$, $A_nB_n^{-1} = G_n\Gamma(X_n'X_n)^{-1}$, and

$$Q^{(a)} \to \frac{\Gamma\eta(\theta)}{\sqrt{\Gamma\eta(\theta)\Gamma'}} = Q(\theta) \tag{24}$$

if equation (6) holds.

4. UNKNOWN σ

The procedure is similar for the case of unknown σ^2, but the answers are more complicated. For unknown σ^2, let

$$Z_n = \frac{1}{\hat{\sigma}_n} B_n'(\theta - \hat{\theta}_n) \tag{25}$$

and

$$W_n = A_n Z_n = \frac{1}{\hat{\sigma}_n} A_n B_n'(\theta - \hat{\theta}_n) \tag{26}$$

where A_n and B_n are as in equation (13). Then the analogue of (18) takes the following form. If h is a function of quadratic growth, then

$$\begin{aligned} E_{\sigma,\theta}[h(W_N)] \approx \Phi^m h &- \frac{\sigma}{\sqrt{a}}(\Phi_1^m h)' Q^{\#}(\sigma, \theta)\mathbf{1} \\ &+ \frac{\sigma^2}{a}\operatorname{tr}\{(\Phi_2^m h)[M(\sigma,\theta) - \sigma^2 \rho'(\sigma^2, \theta) I_m]\} \\ &+ \frac{1}{a}(\Phi_4^m h)\rho(\sigma^2, \theta) \end{aligned} \tag{27}$$

where

$$\Phi_4^m h = \frac{1}{4}\int_{\Re^m} h(w)[\|w\|^4 - 2m\|w\|^2 + m(m-2)]\Phi^m\{dw\}$$

and ρ is as in equation (7). The final two terms on the right-hand side of equation (27) arise from the variability in $\hat{\sigma}_N^2$. In the absence of optional stopping, $\rho(\sigma^2, \theta) = 1$ and $\rho'(\sigma^2, \theta) = 0$, and the right-hand side of equation (27) simplifies.

To form corrected confidence sets, write $\mu_a(\theta) = \mu_a(\sigma, \theta)$ and $\Delta^o(\theta) = \Delta^o(\sigma, \theta)$ to emphasize the dependence on σ. Further, let

$$\Delta(\sigma, \theta) = \Delta^o(\sigma, \theta) - \sigma^2 \rho'(\sigma^2, \theta) I_m$$

let $\hat{\mu}_a$ and $\hat{\Delta}_a$ denote estimators of $\mu_a(\sigma, \theta)$ and $\Delta(\sigma, \theta)$, for example $\hat{\mu}_a = \mu_a(\hat{\sigma}_N, \hat{\theta}_N)$ and $\hat{\Delta}_a = \Delta(\hat{\sigma}_N, \hat{\theta}_N)$, and let

$$W_N^* = \left(I_m + \frac{\hat{\Delta}_a}{2a}\right)^{-1}(W_N - \hat{\mu}_a) \tag{28}$$

If h is a function of quadratic growth, then

$$E_{\sigma,\theta}[h(W_N^*)] \approx \Phi^m h + (\Phi_4^m h)\frac{\rho(\sigma^2, \theta)}{a} \tag{29}$$

The dependence of the right-hand side of equation (29) on the parameters is more apparent than real, since $\rho(\sigma^2, \theta)/a$ may be estimated by $1/N$ (see equation (7)). Let $\mathcal{T}_n^m$ denote the standard m-variate t-distribution on n degrees of freedom, and write $\mathcal{T}_n^m h = \int_{\Re^m} h(y)\mathcal{T}_n^m\{dy\}$ for suitable functions h. Then it is not difficult to see that the right-hand side of equation (29) is an approximation to $\mathcal{T}_{a/\rho}^m h$ which may be estimated by $\mathcal{T}_N^m h$ or $\mathcal{T}_{N-p}^m h$.

The construction of corrected confidence sets now proceeds as in the previous section. If $C_a \subseteq \Re^m$ are measurable and

$$\mathcal{C}_a = \{\theta : W_N^* \in C_a\}$$

then

$$P_{\sigma,\theta}\{\theta \in \mathcal{C}_a\} \approx \mathcal{T}^m_{a/\rho}(C_a)$$

EXAMPLE 1 REVISITED: In the autoregressive example with $N = n = a$,

$$Q(\sigma, \theta) = \frac{\sqrt{1-\theta^2}}{\sigma}$$

$$\mu_n(\sigma, \theta) = \frac{\theta}{\sqrt{n(1-\theta^2)}}$$

and

$$\Delta(\sigma, \theta) = \frac{\theta^2}{1-\theta^2}$$

Both μ_n and Δ are independent of σ. Let $\tilde{\theta}_n = \hat{\theta}_n$ if $|\hat{\theta}_n| \leq 1 - 1/n$ and let $\tilde{\theta}_n = \pm(1 - 1/n)$ otherwise. Further, let $\hat{\mu}_n = \mu_n(1, \tilde{\theta}_n)$, let $\hat{\Delta}_n = \Delta(1, \tilde{\theta}_n)$, and define W_n^* by equation (28). Let $i_n^2 = X_n' X_n = y_0^2 + \cdots + y_{n-1}^2$. If c_n is the upper $\alpha/2$ quantile of the t-distribution on $n-1$ degrees of freedom, then

$$P_{\sigma,\theta}\left\{-\frac{\hat{\sigma}_n}{i_n}\left[c_n\left(1+\frac{\hat{\Delta}_n}{2n}\right) - \hat{\mu}_n\right] \leq \theta - \hat{\theta}_n \leq \frac{\hat{\sigma}_n}{i_n}\left[c_n\left(1+\frac{\hat{\Delta}_n}{2n}\right) + \hat{\mu}_n\right]\right\}$$
$$\approx 1 - \alpha \tag{30}$$

for $|\theta| < 1$ and $\sigma > 0$.

To assess the accuracy of equation (30), a simulation experiment was conducted in which 10,000 samples of size n were generated for selected values of α, n, and θ with $\sigma = 1$. For each choice of α, n, and θ, the coverage probability on the left-hand side of equation (30) was estimated by Monte Carlo. The results are reported in Table 2. They show that the approximation in equation (30) is excellent, even for θ near to ± 1. □

EXAMPLE 3 REVISITED: If σ^2 is unknown in Example 3 (the Hayre–Gittins example), then the stopping time N is modified as follows. Five observations are taken from each population at the beginning, and

$$N = \inf\left\{n \geq 11 : |i_n^2 \hat{\theta}_{n,2}| > a\left(\frac{n+6}{n-6}\right)\hat{\sigma}_n^2\right\}$$

Table 2 Coverage probabilities for autoregressive processes

θ	$n = 25$		$n = 50$	
	$\alpha = 0.05$	$\alpha = 0.10$	$\alpha = 0.05$	$\alpha = 0.10$
0.0	.949	.899	.948	.899
0.2	.947	.896	.949	.896
0.4	.946	.894	.946	.896
0.6	.945	.895	.945	.896
0.8	.953	.894	.947	.897
0.9	.954	.904	.953	.899
±	.0044	.006	.0044	.006

Note: Based on 10,000 replications with $\sigma = 1$; ± is two standard deviations.

Then, $Q^{\#}(\sigma, \theta) = [0, \text{sign}(\theta_2)/2\sqrt{|\theta_2|}]/\sigma$, $\mu_a(\sigma, \theta) = -\text{sign}(\theta_2)/(2\sqrt{a|\theta_2|})$, and $\Delta^o(\sigma, \theta) = 1/4|\theta_2|$, as above. Further,

$$\rho(\sigma^2, \theta) = \frac{|\theta_2| w(\theta_2)}{[1 + w(\theta_2)]^2 \sigma^2}$$

and

$$\rho'(\sigma^2, \theta) = -\frac{|\theta_2| w(\theta_2)}{[1 + w(\theta_2)]^2 \sigma^4}$$

In view of the singularity at $\theta_2 = 0$, it seems prudent to smooth the estimators near $\hat{\theta}_{N,2} = 0$. There are many ways to do this. Here we take

$$\hat{\mu}_a = -\frac{\text{sign}(\hat{\theta}_{N,2})}{2\sqrt{a(\frac{1}{N} + |\hat{\theta}_{N,2}|)}}$$

and

$$\hat{\Delta}_a = \frac{1}{4[\frac{1}{N} + |\hat{\theta}_{N,2}|]} - \hat{\sigma}_N^2 \rho'(\hat{\sigma}_N^2, \hat{\theta}_N)$$

Then equation (29) holds. In this example, $W_N = i_N(\theta_2 - \hat{\theta}_{N,2})$. So, if c_n is the upper $100\alpha/2$ percentile of the t-distribution on $n - 2$ degrees of freedom, then an approximate $100(1 - \alpha)$ percent confidence interval for θ_2 is

$$C = \left\{\theta : -\left[c_N\left(1+\frac{\hat{\Delta}_a}{2a}\right) - \hat{\mu}_a\right]\frac{\hat{\sigma}_N}{i_N} \le \theta_2 - \hat{\theta}_{N,2}\right.$$
$$\left. \le \left[c_N\left(1+\frac{\hat{\Delta}_a}{2a}\right) + \hat{\mu}_a\right]\frac{\hat{\sigma}_N}{i_N}\right\}$$

As above, the accuracy of this interval may be assessed by simulation. Simulated coverage probabilities are reported in Table 3 for selected values of α, θ_2, σ, and a, and $\theta_1 = 0$. These show excellent agreement with the nominal values for $a = 9$ and very good agreement for $a = 6$. It is interesting that the corrected confidence intervals are valid, even for small values of θ_2. □

Table 3 Coverage probabilities for the Hayre–Gittins example

$a = 6$									
σ		0.75			1.0			1.5	
α		0.05	0.10		0.05	0.10		0.05	0.10
θ_2	$E(N)$	CP	CP	$E(N)$	CP	CP	$E(N)$	CP	CP
0.100	102.5	.955	.908	165.7	.949	.905	338.2	.947	.899
0.250	69.8	.948	.898	113.0	.943	.894	234.3	.944	.892
0.375	52.9	.957	.907	85.5	.949	.901	176.0	.948	.896
0.500	43.2	.954	.902	68.9	.949	.898	141.7	.946	.891
0.750	32.8	.954	.906	51.9	.955	.907	104.6	.949	.900
1.000	27.5	.958	.905	42.9	.954	.904	85.0	.945	.898
±		.0044	.006		.0044	.006		.0044	.006
$a = 9$									
σ		0.75			1.0			1.5	
α		0.05	0.10		0.05	0.10		0.05	0.10
θ_2	$E(N)$	CP	CP	$E(N)$	CP	CP	$E(N)$	CP	CP
0.100	178.6	.951	.905	299.3	.952	.903	635.0	.948	.897
0.250	100.5	.954	.904	169.8	.954	.906	362.7	.951	.903
0.375	74.2	.951	.901	123.4	.948	.899	261.9	.953	.904
0.500	59.6	.951	.899	98.1	.950	.898	206.9	.952	.903
0.750	45.0	.953	.902	73.3	.948	.898	152.0	.946	.895
1.000	37.6	.952	.897	59.8	.949	.903	122.7	.949	.901
±		.0044	.006		.0044	.006		.0044	.006

Note: Based on 10,000 replications with $\theta_1 = 0$, $w(\delta) = \sqrt{1 + 10\delta}$ for $\delta > 0$, $w(\delta) = 1/\sqrt{1 + 10|\delta|}$ for $\delta \le 0$; ± is two standard deviations. The t-percentiles c_n were approximated using (26.7.8) of Abramowitz and Stegun (1964).

REMARK: There are other applications to clinical trials. Woodroofe and Coad (1997) have shown how equation (22) can be used to form simultaneous confidence intervals in Siegmund's (1993) and Betensky's (1996) procedures for comparing three treatments. In addition, Coad and Woodroofe (1997) have used similar ideas to construct confidence intervals for the ratio of two hazard rates, following a sequential test.

5. THE FORD–SILVEY EXAMPLE

In the Ford–Silvey example, equation (1) and Example 2 of Section 2, $N = a = n$,

$$X_n'X_n = \begin{pmatrix} n & s_n \\ s_n & n \end{pmatrix}$$

$$B_n = \frac{1}{\sqrt{n}}\begin{pmatrix} \sqrt{n^2 - s_n^2} & s_n \\ 0 & n \end{pmatrix}$$

and

$$\eta(\theta) = \lim_{n\to\infty} n(X_n'X_n)^{-1} - \frac{1}{1-\kappa^2}\begin{pmatrix} 1 & \kappa \\ \kappa & 1 \end{pmatrix}$$

if $\kappa \neq 1$, where $s_n = x_1 + \cdots + x_n$, $\kappa = \theta_1/\theta_2$ if $|\theta_1| \leq \theta_2$, and $\kappa = \theta_2/\theta_1$ if $0 < \theta_2 < |\theta_1|$.

5.1 Confidence Intervals for θ_1 and θ_2

To find a confidence interval for θ_1, let $\Gamma = (1, 0)$ in equation (24). Then $A_n = (1, 0)$,

$$W_n^o = \sqrt{\frac{n^2 - s_n^2}{n}}\,(\theta_1 - \hat{\theta}_{n,1})$$

and

$$Q(\theta) = \frac{\Gamma\eta(\theta)}{\sqrt{\Gamma\eta(\theta)\Gamma'}} = \frac{1}{\sqrt{1-\kappa^2}}\,[1, \kappa]$$

In the region $|\theta_1| < \theta_2$, $\kappa = \theta_1/\theta_2$, $Q(\theta) = (\theta_2, \theta_1)/\sqrt{\theta_2^2 - \theta_1^2}$, and, therefore, $Q^{\#}(\theta) = \theta_1\theta_2(1, -1)/\sqrt{(\theta_2^2 - \theta_1^2)^3}$, so that

$$\mu_n(\theta) = -\frac{1}{\sqrt{n}}\,Q^{\#}(\theta)\mathbf{1} = 0$$

and

$$\Delta^o(\theta) = 0$$

Similarly, in the region $0 < \theta_2 < |\theta_1|$, $\kappa = \theta_2/\theta_1$, $Q(\theta) = (\theta_1, \theta_2)/\sqrt{\theta_1^2 - \theta_2^2}$, and $Q^{\#}(\theta) = (-\theta_2^2, \theta_1^2)/\sqrt{(\theta_1^2 - \theta_2^2)^3}$, so that

$$\mu_n(\theta) = -\frac{1}{\sqrt{n(\theta_1^2 - \theta_2^2)}}$$

and

$$\Delta^o(\theta) = \frac{1}{\theta_1^2 - \theta_2^2}$$

Thus, if c is the upper $100\alpha/2$ percentile of the standard normal distribution and if $C = [-c, c]$ in equation (23), then

$$\mathcal{C} = \left\{\theta : -\sqrt{\frac{n}{n^2 - s_n^2}}\left[c\left(1 + \frac{\hat{\Delta}_n^o}{2n}\right) - \hat{\mu}_n^o\right] \le \theta_1 - \hat{\theta}_{n,1} \le \sqrt{\frac{n}{n^2 - s_n^2}} \times \left[c\left(1 + \frac{\hat{\Delta}_n^o}{2n}\right) + \hat{\mu}_n^o\right]\right\}$$

where $\hat{\mu}_n^o$ and $\hat{\Delta}_n^o$ denote estimators of $\mu_n(\theta)$ and $\Delta^o(\theta)$, possibly smoothed as in Example 3 of Section 4.

A similar procedure may be used to find approximate confidence intervals for θ_2. Letting $\Gamma = (0, 1)$ in equation (24), leads to

$$W_n^o = \sqrt{\frac{n^2 - s_n^2}{n}}\,(\theta_2 - \hat{\theta}_{n,2})$$

$$\mu_n(\theta) = \begin{cases} -1/\sqrt{n(\theta_2^2 - \theta_1^2)} & \text{if } |\theta_1| < \theta_2 \\ 0 & \text{if } 0 < \theta_2 < |\theta_1| \end{cases}$$

$$\Delta^o(\theta) = \begin{cases} 1/(\theta_2^2 - \theta_1^2) & \text{if } |\theta_1| < \theta_2 \\ 0 & \text{if } 0 < \theta_2 < |\theta_1| \end{cases}$$

and

$$\mathcal{C} = \left\{\theta : -\sqrt{\frac{n}{n^2 - s_n^2}}\left[c\left(1 + \frac{\hat{\Delta}_n^o}{2n}\right) - \hat{\mu}_n^o\right] \le \theta_2 - \hat{\theta}_{n,2} \le \sqrt{\frac{n}{n^2 - s_n^2}} \times \left[c\left(1 + \frac{\hat{\Delta}_n^o}{2n}\right) + \hat{\mu}_n^o\right]\right\}$$

Monte Carlo estimates of the actual coverage probabilities of these confidence intervals are reported in Table 4. It appears that the approximation is excellent, except on the lines $|\theta_1| = \theta_2$, where it is conservative. The theoretical justification for the procedure fails on these lines, as is evident from the infinite discontinuities in $\mu_n^o(\theta)$ and $\Delta^o(\theta)$.

5.2 Confidence Intervals for θ_1/θ_2

Since the Ford–Silvey example was motivated by the problem of finding a good estimator for $\upsilon = \theta_1/\theta_2$, it is natural to consider $\hat{\upsilon}_n = \hat{\theta}_{n,1}/\hat{\theta}_{n,2}$ in some detail. In some ways, approximations to the distribution of $\hat{\upsilon}_n$ are nicer than those for $\hat{\theta}_{n,1}$ and $\hat{\theta}_{n,2}$. The bias of $\hat{\upsilon}_n$ may be computed, as in Section 2, and

Table 4 Coverage probabilities for the Ford–Silvey example

		$n = 25$		$n = 50$	
θ_1	θ_2	$\alpha = 0.05$	$\alpha = 0.10$	$\alpha = 0.05$	$\alpha = 0.10$
1.0	1.0	.958	.912	.956	.909
		.956	.912	.953	.907
1.0	1.5	.954	.901	.952	.901
		.948	.897	.950	.899
1.0	2.0	.948	.896	.949	.898
		.948	.902	.951	.901
1.0	4.0	.950	.902	.950	.900
		.949	.899	.952	.899
1.5	1.0	.950	.902	.948	.898
		.949	.897	.947	.900
2.0	1.0	.948	.896	.948	.897
		.946	.897	.951	.903
2.0	2.0	.957	.911	.952	.901
		.952	.905	.952	.903
4.0	1.0	.950	.901	.949	.898
		.948	.896	.948	.898
4.0	4.0	.953	.904	.949	.902
		.953	.904	.949	.900
$\pm$		.0044	.006	.0044	.006

Note: Based on 10,000 replications; $\pm$ is two standard deviations. The upper figure is the coverage probability for θ_1, and the lower is for θ_2.

$$E_\theta(\hat{\upsilon}_n - \upsilon) \approx \begin{cases} 0 & \text{if } |\theta_1| < \theta_2 \\ \theta_1/n\theta_2^3 & \text{if } 0 < \theta_2 < |\theta_1| \end{cases} \tag{31}$$

Again the approximation is discontinuous when $|\theta_1| = \theta_2$, and the theoretical justification fails in this case.

To form confidence intervals, it is necessary to compute the probability that $\upsilon - \hat{\upsilon}_n \le b_n$, where b_n may depend on the data. Assuming that $\theta_2 > 0$, the inequality may be rewritten as

$$\theta_1 - (b_n + \hat{\upsilon}_n)\theta_2 \le 0 \tag{32}$$

Letting $\kappa_n = -s_n/n$, equation (32) may be written in the form

$$a_n' Z_n \le c \tag{33}$$

where

$$a_{n,1} = \frac{1 - (b_n + \hat{\upsilon}_n)\kappa_n}{\sqrt{1 - 2(b_n + \hat{\upsilon}_n)\kappa_n + (b_n + \hat{\upsilon}_n)^2}}$$

$$a_{n,2} = -\frac{\sqrt{(1 - \kappa_n^2)}(b_n + \upsilon_n)}{\sqrt{1 - 2(b_n + \hat{\upsilon}_n)\kappa_n + (b_n + \hat{\upsilon}_n)^2}}$$

and

$$c = \frac{\sqrt{n(1 - \kappa_n^2)}b_n\hat{\theta}_{n,2}}{\sqrt{1 - 2\kappa_n(b_n + \hat{\upsilon}_n) + (b_n + \hat{\upsilon}_n)^2}} \tag{34}$$

The probability of equation (33) may be approximated using equation (18). To do so, it is necessary to find the limit of $Q^{(n)} = \sqrt{n}a_n'B_n^{-1}$. First observe that if c remains bounded in equation (34), then $b_n \to 0$ in probability as $n \to \infty$. It follows easily that

$$\begin{aligned} Q^{(n)} &\to \frac{[1 - \upsilon\kappa, -\sqrt{(1 - \kappa^2)}\upsilon]}{\sqrt{(1 - 2\kappa\upsilon + \upsilon^2)(1 - \kappa^2)}} \begin{pmatrix} 1 & \kappa \\ 0 & \sqrt{1 - \kappa^2} \end{pmatrix} \\ &= \frac{[1 - \upsilon\kappa, \kappa - \upsilon]}{\sqrt{(1 - 2\kappa\upsilon + \upsilon^2)(1 - \kappa^2)}} = Q(\theta), \qquad \text{say} \end{aligned}$$

in probability, if $\upsilon \neq 1$. If $|\upsilon| < 1$, then $\kappa = \upsilon$ and $Q(\theta) = [1, 0]$, and if $|\upsilon| > 1$, then $\kappa = 1/\upsilon$ and $Q(\theta) = [0, -1]$. In either case, $Q(\theta)$ does not depend on θ, and therefore the probability of equation (33) is

$$P_\theta\{a_n' Z_n \le c\} = \Phi(c) + o(1/n)$$

in the very weak sense. Let c be the upper $100\alpha/2$ percentile of the standard normal distribution, let b_n^+ be the positive solution to equation (34), and let b_n^- be the negative solution to equation (34) with c replaced by $-c$. Then

$$P_\theta\{b_n^- \leq v - \hat{v}_n \leq b_n^+\} \approx 1 - \alpha \tag{35}$$

in the very weak sense.

As above, the theoretical justification for equation (35) fails on the lines $|\theta_1| = \theta_2$, but now the approximations compare well with simulated values, even on these lines. Monte Carlo estimates of the bias and coverage probabilities are provided in Table 5.

Table 5 Bias of $\hat{v}_n$ in the Ford–Silvey example

θ_1	θ_2	$n = 25$		$n = 50$	
		Approx.	M.C.	Approx.	M.C.
1.0	1.0	.040	.034	.020	.014
1.0	2.0	.000	−.002	.000	.000
1.0	4.0	.000	.000	.000	.000
2.0	1.0	.080	.098	.040	.044
2.0	2.0	.010	.088	.006	.004
4.0	1.0	.160	.196	.080	.088
4.0	4.0	.002	.002	.002	.000

Coverage Probabilities for v

θ_1	θ_2	$n = 25$		$n = 50$	
		$\alpha = .05$	$\alpha = .10$	$\alpha = .05$	$\alpha = .10$
1.0	1.0	.945	.888	.948	.896
1.0	1.5	.946	.899	.953	.904
1.0	2.0	.946	.895	.947	.900
1.0	4.0	.950	.899	.950	.900
1.5	1.0	.949	.894	.952	.903
2.0	1.0	.950	.895	.952	.901
2.0	2.0	.947	.894	.950	.899
4.0	1.0	.950	.895	.951	.900
4.0	4.0	.946	.892	.952	.904
±		.0044	.006	.0044	.006

Note: Based on 10,000 replications; ± is two standard deviations.

6. AUTOREGRESSIVE PROCESSES

Consider an autogressive process of order two, say

$$y_k = \theta_1 y_{k-1} + \theta_2 y_{k-2} + \sigma\epsilon_k, \quad k = 1, 2, \ldots,$$

with $y_{-1} = y_0 = 0$. Here $\sigma > 0$ and $\theta = (\theta_1, \theta_2)'$ is confined to a triangular region Ω in which $\theta_1 + \theta_2 < 1$, $\theta_1 - \theta_2 > -1$, and $|\theta_2| < 1$ (see, for example, Brockwell and Davis (1991, Chap. 8). Clearly, this model is of the form of equation (2), and

$$X_n' X_n = \begin{pmatrix} r_{n,0} & r_{n,1} \\ r_{n,1} & r_{n-1,0} \end{pmatrix}$$

where

$$r_{n,j} = \sum_{k=2}^{n} y_{k-1} y_{k-1-j}$$

for $n \geq 3$. It may be shown that

$$\eta(\sigma, \theta) = \lim_{n\to\infty} n(X_n' X_n)^{-1} = \frac{1}{\sigma^2}\begin{pmatrix} 1-\theta_2^2 & -\theta_1(1+\theta_2) \\ -\theta_1(1+\theta_2) & 1-\theta_2^2 \end{pmatrix}$$

$w.p.1$ for all $\sigma > 0$ and $\theta \in \Omega$. Using equation (8), it then follows that

$$E_{\sigma,\theta}(\hat{\theta}_n - \theta) \approx -\frac{1}{n}\begin{pmatrix} \theta_1 \\ 3\theta_2 + 1 \end{pmatrix} \tag{36}$$

and equation (36) agrees well with Coad and Woodroofe's (1998) simulations.

To form confidence intervals for θ_1, say, let $\Gamma = (1, 0)$ in equation (24). Then

$$G_n = \sqrt{r_{n,0} - \frac{r_{n,1}^2}{r_{n-1,0}}}$$

$$W_n = \frac{G_n}{\hat{\sigma}_n}(\theta_1 - \hat{\theta}_{n,1})$$

and

$$Q(\sigma, \theta) = \frac{\sqrt{1-\theta_2^2}}{\sigma}\left[1, -\frac{\theta_1}{1-\theta_2}\right]$$

Differentiation then yields

$$\mu_n(\sigma, \theta) = \frac{\theta_1}{(1 - \theta_2)\sqrt{n(1 - \theta_2^2)}}$$

and

$$\Delta(\sigma, \theta) = \frac{\theta_1^2 + 2\theta_2(1 - \theta_2)^2(1 + \theta_2)}{(1 - \theta_2)^3(1 + \theta_2)}$$

Observe that μ_n and Δ do not depend on σ. Let $\tilde{\theta}_{n,1} = \hat{\theta}_{n,1}$, $\tilde{\theta}_{n,2} = \hat{\theta}_{n,2}$ if $|\tilde{\theta}_{n,2}| \leq 1 - 1/n$, $\tilde{\theta}_{n,2} = \pm(1 - 1/n)$ otherwise, $\hat{\mu}_n = \mu_n(1, \tilde{\theta}_n)$, and $\hat{\Delta}_n = \Delta(1, \tilde{\theta}_n)$. Then confidence intervals for θ_1 may be formed as in equation (30), except that now there are $n - 2$ degrees of freedom.

The procedure is similar for θ_2. Letting $\Gamma = (0, 1)$ in equation (24), $W_n = G_n(\theta_2 - \hat{\theta}_{n,2})/\hat{\sigma}_n$, where

$$G_n = \sqrt{r_{n-1,0} - \frac{r_{n,1}^2}{r_{n,0}}}$$

Then

$$Q(\sigma, \theta) = \frac{\sqrt{1 - \theta_2^2}}{\sigma}\left[-\frac{\theta_1}{1 - \theta_2}, 1\right]$$

$$\mu_n(\sigma, \theta) = \frac{1 + 2\theta_2}{\sqrt{n(1 - \theta_2^2)}}$$

and

$$\Delta(\sigma, \theta) = \frac{1 + 2\theta_2 + 2\theta_2^2}{1 - \theta_2^2}$$

Estimators of μ_n and Δ may be constructed as above, and approximate confidence intervals of the form of equation (30) may also be constructed.

A simulation study was conducted to assess the accuracy of the approximation. For $n = 25$ and 50, the simulated and nominal values of the coverage probability agree well, except for θ_2 near one of the vertices of the triangle Ω. For $n = 100$, the agreement is still very good, but the nominal values may be slightly too conservative. Some representative values are included in Table 6.

Table 6 Coverage probabilities for autoregressive processes

		$n = 25$		$n = 50$		$n = 100$	
θ_1	θ_2	$\alpha = 0.05$	$\alpha = 0.10$	$\alpha = 0.05$	$\alpha = 0.10$	$\alpha = 0.05$	$\alpha = 0.10$
0.0	0.0	.949	.897	.949	.897	.957	.907
		.949	.897	.949	.902	.955	.906
0.0	0.5	.950	.900	.949	.897	.953	.906
		.952	.897	.954	.904	.953	.902
1.5	−0.6	.948	.895	.947	.898	.956	.905
		.946	.896	.948	.896	.957	.908
1.9	−0.9	.949	.895	.948	.891	.948	.899
		.942	.880	.942	.883	.948	.895
±		.0044	.006	.0044	.006	.0044	.006

Note: Based on 10,000 replications; ± is two standard deviations. The upper figure is the coverage probability for θ_1, and the lower is for θ_2.

7. FORMALITIES

The theoretical justification for the expansions in Sections 2–4 is in the very weak sense of Woodroofe (1986). Let $\Omega \subseteq \Re^p$ denote the parameter space, and suppose that Ω is a convex open set or a union of a countable collection of such sets, as in Coad and Woodroofe (1998). Further, let $\gamma_a(\theta)$ be a function of the unknown parameter, like a coverage probability or rescaled bias, and let $\gamma(\theta)$ be a candidate for a limiting function. If $q > 0$, then the notation

$$\gamma_a(\theta) = \gamma(\theta) + o(a^{-q}) \qquad \text{very weakly} \tag{37}$$

means that

$$\lim_{a\to\infty} a^q \int_\Omega [\gamma_a(\theta) - \gamma(\theta)]\xi(\theta)\mathrm{d}\theta = 0 \tag{38}$$

for all sufficiently smooth compactly supported prior densities ξ on Ω. The precise amount of smoothness required of ξ may depend on q. It is argued below that very weak approximation is strong enough to support the frequentist interpretation of confidence. An example in which $\gamma_a(\theta) = \gamma(\theta) + o(a^{-1})$ holds in the very weak sense, but not in the conventional sense, is also described below.

Letting $\|\cdot\|$ denote the trace norm for matrices, the conditions under which equation (18) holds may now be stated quite simply: *If there are matrices* $Q(\theta) = [q_{ij}(\theta) : i = 1, \ldots, m,\ j = 1, \ldots, p]$ *for which* q_{ij} *are twice continuously differentiable on* Ω, *if*

$$\lim_{a\to\infty} \int_K E_\theta \|Q^{(a)} - Q(\theta)\|^2 d\theta = 0 \tag{39}$$

for all compact $K \subseteq \Omega$, *and if* $\approx$ *is interpreted to mean equality up to* $o(a^{-1})$ *in the very weak sense, then equation (18) holds for all measurable, symmetric (sign invariant) functions* $h : \Re^m \to \Re$ *of quadratic growth; and if also*

$$\lim_{a\to\infty} \|\sqrt{a} \int_K E_\theta[Q^{(a)} - Q(\theta)] d\theta\| = 0 \tag{40}$$

for all compact $K \subseteq \Omega$, *then equation (18) holds for all measurable h of quadratic growth.* These assertions are the corollary to Theorem 1 in Woodroofe and Coad (1997), and there is substantial uniformity with respect to h in that theorem that is not reported here. The condition in equation (39) is not restrictive.

Surprisingly, less smoothness is required for the corrected confidence sets than for equation (18). With $Q(\theta)$ as in equation (16), suppose that q_{ij} are once differentiable on Ω and that

$$\int_K [\|Q^{\#}(\theta)\mathbf{1}\|^2 + \|\Delta^o(\theta)\|] d\theta < \infty \tag{41}$$

for all compact $K \subseteq \Omega$. In words, $\|Q^{\#}(\theta)\mathbf{1}\|^2$ and $\|\Delta^o(\theta)\|$ must be locally integrable in θ. Let $C = C_a$ be a confidence set of the form of equation (23), and let

$$\gamma_a(\theta) = P_\theta\{\theta \in C_a\} = P_\theta\{W_N^{o*} \in C\}$$

for $\theta \in \Omega$ and $a \geq a_0$. *If equation (39) and (41) hold, then there exist estimators* $\hat{\mu}_a^o$ *and* $\hat{\Delta}_a^o$ *for which*

$$\gamma_a(\theta) = \Phi^m(C) + o\left(\frac{1}{a}\right) \tag{42}$$

in the very weak sense for all measurable, symmetric (sign invariant) subsets $C \subseteq \Re^m$; *and if equation (40) holds too, then equation (42) holds for all measurable* $C \subseteq \Re^m$. These assertions follow from Theorem 2 and Proposition 3 of Woodroofe and Coad (1997). As stated, they are unsatisfactory in that only the existence of estimators $\hat{\mu}_a^o$ and $\hat{\Delta}_a^o$ for which equation (42) holds is claimed. However, if $Q^{\#}(\theta)$ and $\Delta^o(\theta)$ are bounded and continuous, then it is sufficient to use $\hat{\mu}_a^o = \mu_a(\hat{\theta}_N)$ and $\hat{\Delta}_a^o = \Delta^o(\hat{\theta}_N)$.

Similar conditions are sufficient for equations (8) and (9). These are included in Section 8, along with an outline of the proof of equation (8).

Very weak approximations are strong enough to support a frequentist interpretation of confidence. To see why, consider the case of known σ, let C_a denote a confidence set of the form of equation (23), and suppose that this procedure is put into routine use. If the procedure is used by a sequence of clients, then it seems reasonable to suppose that the values of θ will vary from client to client. If these values are drawn from a density ξ, say, then the long run relative frequency of coverage is

$$\bar{\gamma}_a(\xi) = \int_\Omega \gamma_a(\theta)\xi(\theta)\mathrm{d}\theta \tag{43}$$

Thus, in order to have a valid confidence procedure, it is enough to have $\bar{\gamma}_a(\xi)$ approximate a nominal value, and this is precisely the meaning of equation (37), assuming only that ξ is smooth and compactly supported. It is amusing to contrast the use of ξ here with conventional Bayesian uses. Here ξ has a clear frequentist interpretation. However, it is unknown to any given client and may be unknowable, since estimating ξ would require access to others' data sets and, even then, there is only indirect information about ξ.

Woodroofe and Keener (1987) developed an example in which equation (18) holds in a very weak sense, but not in a conventional one.

EXAMPLE 4: Let $y_1, y_2, \ldots$ be independent and normally distributed with unknown mean θ and unit variance (so that $p = 1$ and $x_k = 1$ for all $k = 1, 2, \ldots$). Suppose that θ is known to be positive and let

$$N = N_a = \inf\{n \geq 1 : y_1 + \cdots + y_n > a\}$$

for $a \geq 1$. This corresponds roughly to a one-sided sequential probability ratio test. Then $X_N' X_N = N$,

$$\lim_{a\to\infty} \frac{a}{N} = \theta$$

and equation (18) holds in the very weak sense with $Q(\theta) = \sqrt{\theta},\ \theta > 0$. However, equation (18) does not hold in the conventional sense. In fact,

$$P_\theta\{Z_N^o \leq c\} = \Phi(c) + \frac{1}{\sqrt{a\theta}} R_a(\theta, c) + o\left(\frac{1}{\sqrt{a}}\right)$$

for fixed $\theta > 0$, where $R_a(\theta, c)$ is bounded but oscillates wildly as a increases. The exact expression for R_a is complicated, owing to the presence of ladder variables, and is not reproduced here. □

8. OUTLINE OF A PROOF

Suppose that Ω is a convex open set or the union of a countable collection of convex open sets. If $\eta(\sigma, \theta)$ is continuously differentiable in θ for fixed σ and

$$\lim_{a\to\infty} \int_K E_{\sigma,\theta} \|a(X_N' X_N)^{-1} - \eta(\sigma, \theta)\| \mathrm{d}\theta = 0 \tag{44}$$

for all compact subsets $K \subseteq \Omega$, then equation (8) holds in the very weak sense. For equation (9) it is sufficient that $\rho(\sigma^2, \theta)$ be differentiable in σ^2, that $\rho'(\sigma^2, \theta)$ and the derivatives of η be continuous in (σ^2, θ), that equation (44) holds for each fixed σ, and that

$$\lim_{a\to\infty} \int_K E_{\sigma,\theta} \left| \frac{a}{N} - \rho(\sigma^2, \theta) \right| \mathrm{d}\sigma^2 \mathrm{d}\theta = 0$$

for all compact subsets $K \subseteq (0, \infty) \times \Omega$.

The proof of equation (8) is outlined next. For simplicity, σ is assumed to be known, say $\sigma = 1$, and is omitted from the notation. The meaning of equation (8) is that

$$\int_\Omega \left[E_\theta(\hat{\theta}_N - \theta) - \frac{1}{a} \eta^\#(\theta) \mathbf{1} \right] \xi(\theta) \mathrm{d}\theta = o\left(\frac{1}{a}\right) \tag{45}$$

as $a \to \infty$ for all continuously differentiable densities ξ with compact support. The first step in the proof is to write

$$\int_\Omega E_\theta(\hat{\theta}_N - \theta) \xi(\theta) \mathrm{d}\theta = E_\xi(\hat{\theta}_N - \theta)$$

where E_ξ denotes expectation in a Bayesian model in which θ has prior distribution ξ and equation (2) holds conditionally given θ. Then

$$E_\xi(\hat{\theta}_N - \theta) = E_\xi[E_\xi^N(\hat{\theta}_N - \theta)]$$

where E_ξ^N denotes conditional (posterior) expectation given the data. The next step is to approximate $E_\xi^N(\hat{\theta}_N - \theta)$. Let L_N denote the likelihood function,

$$L_N(\theta) = \exp[-\tfrac{1}{2}(\theta - \hat{\theta}_N)'(X_N' X_N)(\theta - \hat{\theta}_N)]$$

Then

$$E_\xi^N(\theta - \hat{\theta}_N) = \frac{1}{c} \int_\Omega (\theta - \hat{\theta}_N) L_N(\theta) \xi(\theta) \mathrm{d}\theta$$

where c is a normalizing constant, $c = \int_\Omega L_N(\theta)\xi(\theta)\mathrm{d}\theta$. Multiplying by $(X_N' X_N)$ and integrating by parts,

$$(X_N' X_N) E_\xi^N(\hat{\theta}_N - \theta) = \frac{1}{c} \int_\Omega \nabla L_N(\theta)\xi(\theta) \mathrm{d}\theta$$
$$= -\frac{1}{c} \int_\Omega L_N(\theta) \nabla \xi(\theta) \mathrm{d}\theta = -E_\xi^N \left[\frac{\nabla \xi}{\xi}(\theta) \right]$$

where ∇ denotes gradient with respect to θ. Thus, using equation (44),

$$a E_\xi^N(\hat{\theta}_N - \theta) = -a(X_N' X_N)^{-1} E_\xi^N \left[\frac{\nabla \xi}{\xi}(\theta) \right] \to -\eta(\theta) \frac{\nabla \xi}{\xi}(\theta)$$

and

$$a E_\xi(\hat{\theta}_N - \theta) \to \int_\Omega -\eta(\theta) \frac{\nabla \xi}{\xi}(\theta)\xi(\theta) \mathrm{d}\theta = \int_\Omega \eta^\#(\theta) \mathbf{1} \xi(\theta) \mathrm{d}\theta \tag{46}$$

where the final equality follows from another integration by parts. This completes the proof, because equation (46) is equivalent to equation (45).

□

The proofs of equations (9) and (18) use similar ideas, but the details of the integration by parts are more complicated. Justification for the corrected confidence sets combines these ideas with a Taylor series expansion.

REFERENCES

Abramowitz, M. and Stegun, I.A. (1964). *Handbook of Mathematical Functions*. US Department of Commerce.

Berger, J.O. and Wolpert, R. (1984). *The Likelihood Principle*. Haywood, CA.: Institute of Mathematical Statistics.

Betensky, R. (1996). An O'Brian–Fleming sequential trial for comparing three treatments. *Ann. Stat.*, **24**, 1765–1791.

Brockwell, P.J. and Davis, R.A. (1991). *Time Series: Theory and Methods*. New York: Springer.

Coad, D.S. (1991). Sequential allocation with data-dependent allocation and time trends. *Sequential Anal.*, **10**, 91–97.

Coad, D.S. (1995). Sequential allocation rules for multi-armed clinical trials. *J. Stat. Comput. Simulation*, **52**, 239–251.

Coad, D.S. and Woodroofe, M. (1997). Approximate confidence intervals after a sequential clinical trial comparing two exponential survival curves with censoring. *J. Stat. Plann. Inference*, **63**, 79–96.

Coad, D.S. and Woodroofe, M. (1998). Approximate bias calculations for sequentially designed experiments. *Sequential Anal.*, **17**, 1–31.

Eisele, J.R. (1994). The doubly adaptive biased coin design for sequential clinical trials. *J. Stat. Plann. Inference*, **38**, 249–262.

Ford, I. and Silvey, S.D. (1980). A sequentially constructed design for estimating a nonlinear parametric function *Biometrika*, **67**, 381–388.

Ford, I., Titterington, D.M. and Wu, C.F.J. (1985). Inference and sequential design. *Biometrika*, **72**, 545–551.

Hayre, L.S. and Gittins, J.C. (1981). Sequential selection of the larger of two normal means. *J. Am. Stat. Assoc.*, **76**, 696–700.

Lai, T.L. and Wei, C.Z. (1982). Least squares estimates in stochastic regression models with applications to identification and control of dynamic systems. *Ann. Stat.*, **10**, 154–166.

Robbins, H. and Siegmund, D. (1974). Sequential tests involving two populations. *J. Am. Stat. Assoc.*, **69**, 132–139.

Siegmund, D.O. (1993). A sequential clinical trial for comparing three treatments. *Ann. Stat.*, **21**, 464–483.

Woodroofe, M. (1986). Very weak expansions for sequential confidence levels. *Ann. Stat.*, **14**, 1049–1067.

Woodroofe, M. (1989). Very weak expansions for sequentially designed experiments: linear models. *Ann. Stat.*, **17**, 1087–1102.

Woodroofe, M. and Coad, D.S. (1997). Corrected confidence sets for sequentially designed experiments. *Stat. Sinica*, **7**, 53–74.

Woodroofe, M. and Keener, R. (1987). Asymptotic expansions in boundary crossing probabilities. *Ann. Probab.*, **15**, 102–114.

Wu, C.F.J. (1985). Asymptotic inference from a sequential design in a nonlinear situation. *Biometrika*, **72**, 553–558.

7

Resampling Marked Point Processes

DIMITRIS N. POLITIS University of California, San Diego, La Jolla, California

EFSTATHIOS PAPARODITIS University of Cyprus, Nicosia, Cyprus

JOSEPH P. ROMANO Stanford University, Stanford, California

1. INTRODUCTION AND NOTATION

Suppose $\{X(\mathbf{t}), \mathbf{t} \in \mathbf{R}^d\}$ is a homogeneous random field in d dimensions, with $d \in \mathbf{Z}^+$, that is, a collection of real-valued random variables $X(\mathbf{t})$ that are indexed by the continuous parameter $\mathbf{t} \in \mathbf{R}^d$. In the important special case where $d = 1$, the random field $\{X(\mathbf{t})\}$ is just a continuous time, stationary stochastic process. The probability law of the random field $\{X(\mathbf{t}), \mathbf{t} \in \mathbf{R}^d\}$ will be denoted by P_X. We will generally assume that $EX(\mathbf{t})^2 < \infty$, in which case homogeneity (i.e., strict stationarity) implies weak stationarity, namely that for any $\mathbf{t}, \mathbf{h} \in \mathbf{R}^d$, $EX(\mathbf{t}) = \mu$, and $Cov(X(\mathbf{t}), X(\mathbf{t}+\mathbf{h})) = R(\mathbf{h})$; in other words, $EX(\mathbf{t})$ and $Cov(X(\mathbf{t}), X(\mathbf{t}+\mathbf{h}))$ do not depend on $\mathbf{t}$ at all.

Our objective is statistical inference pertaining to features of the unknown probability law P_X on the basis of data; in particular, this paper will focus on estimation of the common mean μ. For the case where the data are of the form $\{X(\mathbf{t}), \mathbf{t} \in \mathbf{E}\}$, with $\mathbf{E}$ being a finite subset of the rectangular lattice $\mathbf{Z}^d$, different block-resampling techniques have been developed in the literature; see, for example, Hall (1985),

Carlstein (1986), Künsch (1989), Lahiri (1991), Liu and Singh (1992), Politis and Romano (1992a,b,c, 1993, 1994), Raïs (1992), and Sherman and Carlstein (1994, 1996). However, in many important cases, for example, queueing theory, spatial statistics, mining and geostatistics, meteorology, etc., the data correspond to observations of $X(\mathbf{t})$ at nonlattice, irregularly spaced points. For instance, if $d = 1$, $X(\mathbf{t})$ might represent the required service time for a customer arriving at a service station at time $\mathbf{t}$. If $d = 2$, $X(\mathbf{t})$ might represent a measurement of the quality or quantity of the ore found in location $\mathbf{t}$, or a measurement of precipitation at location $\mathbf{t}$ during a fixed time interval, etc. As a matter of fact, in case $d > 1$, irregularly spaced data seem to be the rule rather than the exception; see, for example, Cressie (1991), Karr (1991), and Ripley (1981).

A useful and parsimonious way to model the irregularly scattered $\mathbf{t}$-points is to assume they are generated by a homogeneous Poisson point process observable on a compact subset $K \in \mathbf{R}^d$, and assumed to be independent of the random field $\{X(\mathbf{t})\}$; see Karr (1986, 1991) for a thorough discussion on the plausibility of the Poisson assumption. So let N denote such a homogeneous Poisson process on $\mathbf{R}^d$, independent of $\{X(\mathbf{t})\}$, and possessing mean measure Λ, that is, $EN(A) = \Lambda(A)$ for any set $A \subset \mathbf{R}^d$; note that homogeneity of the process allows us to write $\Lambda(A) = \lambda|A|$, where λ is a positive constant signifying the "rate" of the process, and $|\cdot|$ denotes Lebesgue measure (volume). The point process N can then be expressed as $N = \sum_i \epsilon_{\mathbf{t}_i}$, where $\epsilon_{\mathbf{t}}$ is a point mass at $\mathbf{t}$, i.e., $\epsilon_{\mathbf{t}}(A)$ is 1 or 0 according to whether $\mathbf{t} \in A$ or not; in other words, N is a random (counting) measure on $\mathbf{R}^d$. The expected number of $\mathbf{t}$-points to be found in A is $\Lambda(A)$, whereas the actual number of $\mathbf{t}$-points found in set A is given by $N(A)$. The joint (product) probability law of the random field $\{X(\mathbf{t})\}$ and the point process N will be denoted by P. The observations are then described via the "marked point process" $\tilde{N} = \sum_i \epsilon_{\{\mathbf{t}_i, X(\mathbf{t}_i)\}}$, which is just the point process N with each $\mathbf{t}$-point being "marked" by the value of X at that point.

Hence, in this paper, our objective will be interval estimation of μ on the basis of measurements of the value of $X(\cdot)$ at a finite number of generally non lattice, irregularly spaced points $\mathbf{t} \in \mathbf{R}^d$. The observed marked point process is then defined as the collection of pairs $\{(\mathbf{t}_j, X(\mathbf{t}_j))$, $j = 1, \ldots, N(K)\}$, where $\{\mathbf{t}_j\}$ are the points at which the $\{X(\mathbf{t}_j)\}$ "marks" happen to be observed; see Daley and Vera-Jones (1988), Karr (1991), or Krickeberg (1982) for more details on marked point processes.

The paper is organized as follows: Section 2 contains some useful notions on mixing, and some necessary background on mean estimation, in Section 3 the marked point process "circular" bootstrap is introduced and studied, while in Section 4 the marked point process "block" bootstrap is introduced and studied; some concluding remarks are presented in Section 5, while all proofs are deferred to Section 6.

2. SOME BACKGROUND AND A USEFUL LEMMA ON MIXING

The continuous parameter random field $\{X(\mathbf{t}), \mathbf{t} \in \mathbf{R}^d\}$ will be assumed to satisfy a certain weak dependence condition that will be quantified in terms of mixing coefficients. Let $\rho(\cdot, \cdot)$ denote sup-distance (i.e., the distance arising from the l_∞ norm) on $\mathbf{R}^d$; the strong mixing coefficients of Rosenblatt (1985) are then defined as

$$\alpha_X(k) \equiv \sup_{E_1, E_2 \subset \mathbf{R}^d} \{|P(A_1 \cap A_2) - P(A_1)P(A_2)| : A_i \in \mathcal{F}(E_i), i = 1, 2,$$
$$\rho(E_1, E_2) \geq k\}$$

where $\mathcal{F}(E_i)$ is the σ-algebra generated by $\{X(\mathbf{t}), \mathbf{t} \in E_i\}$. Alternatively, in a random field set-up where $d > 1$, it is now customary to consider mixing coefficients that in general also depend on the size (volume) of the sets considered; see for example Doukhan (1994). Thus define $\alpha_X(k; l_1, l_2) \equiv$

$$\sup_{E_1, E_2 \subset \mathbf{R}^d} \{|P(A_1 \cap A_2) - P(A_1)P(A_2)| : A_i \in \mathcal{F}(E_i), |E_i| \leq l_i, i = 1, 2,$$
$$\rho(E_1, E_2) \geq k\}.$$

Note that $\alpha_X(k; l_1, l_2) \leq \alpha_X(k)$, and that in essence $\alpha_X(k) = \alpha_X(k; \infty, \infty)$. A random field is said to be strong mixing if $\lim_{k\to\infty} \alpha_X(k) = 0$. There are many interesting examples of strong mixing random fields; see Rosenblatt (1985). However, there is a big class of random fields of great interest in spatial statistics, namely Gibbs (Markov) random fields in $d > 1$ dimensions, that are *not* strong mixing, but instead satisfy a condition on the decay of the $\alpha_X(k; l_1, l_2)$ coefficients; see Doukhan (1994).

Nevertheless, for our results a yet weaker notion of mixing is required. So we define the coefficients

$$\bar{\alpha}_X(k; l) \equiv \sup\{|P(A_1 \cap A_2) - P(A_1)P(A_2)| : A_i \in \mathcal{F}(E_i), i = 1, 2,$$
$$E_2 = E_1 + \mathbf{t}, |E_1| = |E_2| \leq l, \rho(E_1, E_2) \geq k\}$$

where the supremum is now taken over all *compact and convex* sets $E_1 \subset \mathbf{R}^d$, and over all $\mathbf{t} \in \mathbf{R}^d$ such that $\rho(E_1, E_1 + \mathbf{t}) \geq k$. As before, we may also define $\bar{\alpha}_X(k) = \bar{\alpha}_X(k; \infty)$.

It is easy now to see that $\bar{\alpha}_X(k) \leq \alpha_X(k)$, so that if the random field is α-strong mixing, then it will necessarily be $\bar{\alpha}$-strong mixing as well, implying $\lim_{k\to\infty} \bar{\alpha}_X(k) = 0$. In other words, $\bar{\alpha}$-mixing is easier to satisfy than regular

α-mixing. Similarly, $\bar{\alpha}_X(k; l_1) \leq \alpha_X(k; l_1, l_1)$, and if the $\alpha_X(k; l_1, l_1)$ coefficients are small, then the same will be true for the $\bar{\alpha}_X(k; l_1)$ coefficients; our bootstrap results will consequently be based on this weaker notion. See Doukhan (1994), Roussas and Ioannides (1987), and Ivanov and Leonenko (1986, p. 34) for discussion and references on strong mixing coefficients.

We now give a useful lemma, its essence being that mixing properties of the continuous parameter random field are inherited by the observed marked point process; the lemma is a generalization of a result of Masry (1988), who considered renewal point processes on the real line (case $d = 1$).

LEMMA 1: *Let N_g be a general Poisson process (not necessarily homogeneous) on $\mathbf{R}^d$, possessing mean measure Λ_g, and assumed to be independent of the random field $\{X(\mathbf{t}), \mathbf{t} \in \mathbf{R}^d\}$. Let E_1, E_2 be two subsets of $\mathbf{R}^d$ such that $\rho(E_1, E_2) = k > 0$, and define $\tilde{Y}_i = N_g(E_i)^{-1} \int_{E_i} X(\mathbf{t})N(d\mathbf{t})$ and $\bar{Y}_i = (\Lambda_g(E_i))^{-1} \int_{E_i} X(\mathbf{t})N(d\mathbf{t})$ for $i = 1, 2$; also assume that $E|X(\mathbf{t})|^p = C_p < \infty$ for some $p > 2$. Then*

$$|Cov(\tilde{Y}_1, \tilde{Y}_2)| \leq 10C_p^{2/p}(\alpha_X(k; |E_1|, |E_2|))^{1-2/p}$$

and

$$|Cov(\bar{Y}_1, \bar{Y}_2)| \leq 10C_p^{2/p}(\alpha_X(k; |E_1|, |E_2|))^{1-2/p}$$

If E_1, E_2 are compact, convex, and are translates of one another, that is, if $E_1 = E_2 + \mathbf{t}$, then we also have

$$|Cov(\tilde{Y}_1, \tilde{Y}_2)| \leq 10C_p^{2/p}(\bar{\alpha}_X(k; |E_1|))^{1-2/p}$$

and

$$|Cov(\bar{Y}_1, \bar{Y}_2)| \leq 10C_p^{2/p}(\bar{\alpha}_X(k; |E_1|))^{1-2/p}$$

Although we will subsequently use Lemma 1 in the special case of a homogeneous Poisson process, the lemma's generality is noteworthy.

We now consider estimation of the mean of the random field $\mu = EX(\mathbf{t})$ on the basis of observing $\{X(\mathbf{t})\}$ for the $\mathbf{t}$-points generated by the homogeneous Poisson process N over the compact, convex set $K \subset \mathbf{R}^d$. It is natural to estimate μ by the sample mean which—as hinted at in Lemma 1—can be defined in two asymptotically equivalent ways:

$$\tilde{X}_K \equiv \frac{1}{\lambda|K|} \int_K X(\mathbf{t})N(d\mathbf{t})$$

and

$$\bar{X}_K \equiv \frac{1}{N(K)} \int_K X(\mathbf{t})N(d\mathbf{t})$$

the difference between the two being division by expected or actual sample size, respectively. Obviously, if λ is unknown, then our only practical choice is $\bar{X}_K$. Note that Karr (1986) presents some arguments in favor of using $\bar{X}_K$ even if λ is known; however, in this paper we will study both $\tilde{X}_K$ and $\bar{X}_K$ in the interest of completeness.

It is immediate that $\tilde{X}_K$ is unbiased for μ; $\bar{X}_K$ is also unbiased as a conditioning (on N) argument shows. See, for instance, Karr (1986) where it is also shown that under some regularity assumptions, as $|K| \to \infty$, $\tilde{X}_K$ and $\bar{X}_K$ are both consistent and asymptotically normal at rate $\sqrt{|K|}$ with the same asymptotic variance. We explicitly give Karr's (1986) theorem below as it will be useful for our bootstrap theory.

THEOREM 1 [Karr, 1986]: *Let $R(\mathbf{t}) = Cov(X(\mathbf{0}), X(\mathbf{t}))$, and assume that*

$$\int R(\mathbf{t})d\mathbf{t} < \infty \tag{1}$$

where $\int$ is short-hand for $\int_{\mathbf{R}^d}$. Also assume that

$$\frac{1}{\sqrt{|K|}} \int_K (X(\mathbf{t}) - \mu)d\mathbf{t} \overset{\mathcal{L}}{\Longrightarrow} N\left(0, \int R(\mathbf{t})d\mathbf{t}\right) \tag{2}$$

as $diam(K) \to \infty$, where $diam(K)$ denotes the supremum of the diameters of all l_∞ balls contained in K. Then, as $diam(K) \to \infty$, we have

$$\sqrt{|K|}(\tilde{X}_K - \mu) \overset{\mathcal{L}}{\Longrightarrow} N(0, \sigma^2), \qquad \sqrt{|K|}(\bar{X}_K - \mu) \overset{\mathcal{L}}{\Longrightarrow} N(0, \sigma^2)$$

and furthermore

$$\lim Var\left(\sqrt{|K|}\tilde{X}_K\right) = \lim Var\left(\sqrt{|K|}\bar{X}_K\right) = \sigma^2 \equiv \int R(\mathbf{t})d\mathbf{t} + \lambda^{-1}R(\mathbf{0})$$

Note that we have taken the liberty of correcting an obvious typo in the variance formula (3.10) of Karr (1986). To verify the assumptions of Karr's theorem, the following lemma may be used.

LEMMA 2: *If $E|X(\mathbf{t})|^{2+\delta} < \infty$ for some $\delta > 0$, and $\bar{\alpha}_X(k; l_1) \le const.(1 + l_1)^{\gamma d^{-1}} k^{-d-\epsilon}$ for some $\epsilon > 2d/\delta$, and some $\gamma < \frac{\delta(\epsilon\delta - 2d)}{2d\delta(1+\delta)}$, then equations (1) and (2) hold true.*

Different sufficient conditions for equations (1) and (2) are given in Yadrenko (1983). Nevertheless, to actually use the asymptotic normality of the sample mean to construct confidence intervals for the mean μ, the asymptotic variance must be explicitly estimated. While it is relatively easy to estimate $R(\mathbf{0})$, and λ is consistently estimable by $N(K)/|K|$, consistent estimation of $\int R(\mathbf{t})d\mathbf{t}$ is not a trivial matter, particularly in the case of the irregularly spaced data considered here.

The resampling methodology that is introduced in this paper is able to yield confidence intervals for the mean without explicit estimation of the asymptotic variance; alternatively, the resampling method may provide an estimate of the asymptotic variance to be used in connection with the asymptotic normality result of Karr (1986). The "circular" resampling methodology of the next section uses a blocking argument similar to the circular bootstrap of Politis and Romano (1992c, 1993), while the "block" resampling methodology of our Section 4 employs arguments similar to the block bootstrap of Künsch (1989) and Liu and Singh (1992).

3. "CIRCULAR" BOOTSTRAP FOR MARKED POINT PROCESSES

Our goal will be to construct bootstrap confidence intervals for μ on the basis of observing $\{X(\mathbf{t})\}$ for the $\mathbf{t}$-points generated by the Poisson process N over the compact, convex set $K \subset \mathbf{R}^d$. In this section we will further assume that K is a "rectangle," i.e., that $K = \{\mathbf{t} = (t_1, \ldots, t_d) : 0 \le t_i \le K_i, i = 1, \ldots, d\}$; in the next section the general case of K being possibly nonrectangular (but convex) will be addressed. The proposed circular bootstrap for marked point processes is described as follows.

1. Begin by imagining that K is "wrapped around" on a compact torus; in other words, we interpret the index $\mathbf{t}$ as being modulo K. If $\mathbf{t} \notin K$, we will redefine $\mathbf{t} = \mathbf{t}(\text{modulo } K)$, where the ith coordinate of vector $\mathbf{t}(\text{modulo } K)$ is $t_i(\text{modulo } K_i)$. With this redefinition, we have data $X(\mathbf{t})$ even if $\mathbf{t} \notin K$.
2. Let $c = c(K)$ be a number in $(0, 1)$ depending on K in a way that is made precise in Theorem 2, and define a scaled-down replica of K by $B = \{c\mathbf{t} : \mathbf{t} \in K\}$, where $\mathbf{t} = (t_1, \ldots, t_d)$ and $c\mathbf{t} = (ct_1, \ldots, ct_d)$; B has the same shape as K but smaller dimensions. Also define the displaced sets $B + \mathbf{y}$, and let $l = [1/c^d]$, where $[\cdot]$ denotes the integer part.
3. Generate random points $\mathbf{Y}_1, \mathbf{Y}_2, \ldots, \mathbf{Y}_l$ independent and identically distributed from a uniform distribution on K, and define

$$\bar{X}^* \equiv l^{-1} \sum_{i=1}^{l} \frac{1}{\lambda|B|} \int_{B+\mathbf{Y}_i} X(\mathbf{t}) N(d\mathbf{t})$$

and

$$\bar{X}^* \equiv l^{-1} \sum_{i=1}^{l} \frac{1}{N(B+\mathbf{Y}_i)} \int_{B+\mathbf{Y}_i} X(\mathbf{t}) N(\mathrm{dt})$$

4. This generation of the points $\mathbf{Y}_1, \mathbf{Y}_2, \ldots, \mathbf{Y}_l$ and subsequently of $\tilde{X}^*$ and $\bar{X}^*$ is governed by a probability mechanism which we will denote by P^*; note that the generation is performed conditionally on the marked point process data that were actually observed.
5. Let $P(\sqrt{|K|}(\bar{X}_K - \mu) \leq x)$ denote the distribution function of the sample mean (centered and normalized), and let $P^*(\sqrt{|K|}(\tilde{X}^* - E^*\tilde{X}^*) \leq x)$ and $P^*(\sqrt{|K|}(\bar{X}^* - E^*\bar{X}^*) \leq x)$ denote the conditional (given the marked point process data) distribution function of its bootstrap counterparts; E^* and Var^* denote expected value and variance, respectively, under the probability mechanism P^*.

An intuitive way of visualizing the construction of $\tilde{X}^*$ and $\bar{X}^*$ is to imagine a "re-tiling" of an area comparable to the rectangle K by putting side-by-side the small rectangles $B + \mathbf{Y}_i$, carrying along at the same time the **t**-points and their corresponding X-marks that the marked point process $\tilde{N}$ originally generated in $B + \mathbf{Y}_i$; as a final step, recalculate the sample mean of the re-tiled process to get $\tilde{X}^*$ and $\bar{X}^*$.

We are now ready to state our main results.

THEOREM 2 *Assume equations (1) and (2), let* $E|X(\mathbf{t})|^{6+\delta} < \infty$, *where* $\delta > 0$, *and assume that* $R(\mathbf{0}) = Var(X(\mathbf{t})) > 0$. *Also assume that* $\bar{\alpha}_X(k; l_1) \leq const.(1 + l_1)^{\bar{\gamma}} k^{-\bar{\beta}}$ *for some* $\bar{\beta} > 3d$, *and* $0 \leq \bar{\gamma} \leq \bar{\beta}/d$. *Let* $\min_i K_i \to \infty$, $\max_i K_i = O(\min_i K_i)$, *and let* $c = c(K) \to 0$, *but in such a way that* $c^d |K| \to \infty$. *Then*

$$E^*\tilde{X}^* = \tilde{X}_K, \qquad \frac{Var^*\tilde{X}^*}{Var\tilde{X}_K} \xrightarrow{P} 1$$

and

$$\sup_x |P^*(\sqrt{|K|}(\tilde{X}^* - \tilde{X}_K) \leq x) - P(\sqrt{|K|}(\tilde{X}_K - \mu) \leq x)| \xrightarrow{P} 0$$

REMARK 1: The result of Theorem 2 could be compactly expressed as

$$d_2\Big(P^*(\sqrt{|K|}(\tilde{X}^* - \tilde{X}_K) \leq x), P(\sqrt{|K|}(\tilde{X}_K - \mu) \leq x)\Big) \xrightarrow{P} 0$$

where $d_2(\cdot, \cdot)$ is Mallows metric between distributions—see Bickel and Freedman (1981) or Shao and Tu (1995, p. 73). Convergence in d_2 is stronger than weak convergence, as it also implies convergence of the first two moments. Nevertheless, this compact expression would obscure the fact that we have exact equality of the first moments here, as opposed to only approximate equality as in Theorem 4 of our next section.

REMARK 2: Also recall that in Lemma 2 we gave a sufficient mixing condition that—together with our moment condition—implies equations (1) and (2); however, since in general there exist different sufficient conditions, we followed Karr (1986) in placing equations (1) and (2) in the assumptions of the theorems.

THEOREM 3: *Under the conditions of Theorem 2 we have*

$$E^*\bar{X}^* = \bar{X}_K, \qquad \frac{Var^*\bar{X}^*}{Var\bar{X}_K} \xrightarrow{P} 1$$

and

$$\sup_x |P^*(\sqrt{|K|}(\bar{X}^* - \bar{X}_K) \le x) - P(\sqrt{|K|}(\bar{X}_K - \mu) \le x)| \xrightarrow{P} 0$$

REMARK 3: Although we have stated the results only for the sample mean (in its two forms), an application of the δ-method (see Bickel and Freedman, 1981) immediately shows that our resampling methodology is also valid for smooth functions of the sample mean. The circular bootstrap (and the block bootstrap discussed in Section 4) can be extended to other mean-like statistical functionals as well, for example, appropriately differentiable statistics (see Künsch, 1989; Liu and Singh, 1992), or general linear statistics (see Politis and Romano, 1993).

So far we have motivated the introduction of resampling for marked point processes in terms of by-passing the difficult problem of estimating the asymptotic variance σ^2 which is required in order to use the asymptotic normality of the sample mean for confidence intervals. Nevertheless, the bootstrap typically has a further advantage as compared with the asymptotic normal distribution, namely that it yields a more accurate distribution approximation; in other words, the bootstrap is typically "higher order accurate"—see, e.g., Efron and Tibshirani (1993) or Shao and Tu (1995) and references therein.

Nevertheless, the property of higher order accuracy will apply only to the bootstrap distribution of the standardized or studentized sample mean. For example, the block bootstrap in the case of stationary sequences in discrete time was recently shown to be higher order accurate; see Lahiri (1991) for the standardized sample mean, and Götze and Künsch (1996) for the studentized sample mean. The same higher order accuracy property characterizes the circular bootstrap; see Politis and Romano (1992c).

Although the necessary tools (for instance, Edgeworth expansions for the distribution of the sample mean) are not yet available to prove higher order accuracy of the bootstrap in a marked point process setting, we conjecture that this higher order accuracy of the standardized or studentized sample mean indeed obtains. With this in mind, we offer the following easy corollary of Theorems 2 and 3, that has to do with the bootstrap distribution of the standardized sample mean. A similar result would be true for the studentized sample mean, in which case a double (or iterated) bootstrap (cf. Hall, 1992) would be required if we intend to studentize $\bar{X}_K$ and $\tilde{X}_K$ using the bootstrap variance estimator.

COROLLARY 1: *Under the conditions of Theorem 2, we also have*

$$\sup_x \left| P^*\left(\frac{\tilde{X}^* - \tilde{X}_K}{\sqrt{Var^* \tilde{X}^*}} \le x \right) - P\left(\frac{\tilde{X}_K - \mu}{\sqrt{Var \tilde{X}_K}} \le x \right) \right| \xrightarrow{P} 0 \tag{3}$$

and

$$\sup_x \left| P^*\left(\frac{\bar{X}^* - \bar{X}_K}{\sqrt{Var^* \bar{X}^*}} \le x \right) - P\left(\frac{\bar{X}_K - \mu}{\sqrt{Var \bar{X}_K}} \le x \right) \right| \xrightarrow{P} 0 \tag{4}$$

4. "BLOCK" BOOTSTRAP FOR MARKED POINT PROCESSES

Now our observation region K can be *any* compact, convex subset of $\mathbf{R}^d$; for our asymptotic results, K will be assumed to expand uniformly in all directions, that is, we will assume that $diam(K) \to \infty$. As before, $diam(K)$ is the supremum of the diameters of all l_∞ balls contained in K, and we also define $Diam(K)$ to be the infimum of the diameters of all l_∞ balls that contain K. The block bootstrap for marked point processes is described as follows.

1. As in the previous section, let $c = c(K)$ be a number in $(0, 1)$ depending on K, and define a scaled-down replica of K by $B = \{c\mathbf{t} : \mathbf{t} \in K\}$, where $\mathbf{t} = (t_1, \ldots, t_d)$ and $c\mathbf{t} = (ct_1, \ldots, ct_d)$. However, since the wrap-around will not be used here, we define the set of "allowed" displacements $K_{1-c} = \{\mathbf{y} \in \mathbf{R}^d : B + \mathbf{y} \subset K\}$, and as before we let $l = [1/c^d]$.

2. Generate random points $\mathbf{Y}_1, \mathbf{Y}_2, \ldots, \mathbf{Y}_l$ independent and identically distributed from a uniform distribution on K_{1-c}, and let

$$\tilde{X}^* \equiv l^{-1} \sum_{i=1}^{l} \frac{1}{\lambda |B|} \int_{B+\mathbf{Y}_i} X(\mathbf{t}) N(d\mathbf{t})$$

and

$$\bar{X}^* \equiv l^{-1} \sum_{i=1}^{l} \frac{1}{N(B+\mathbf{Y}_i)} \int_{B+\mathbf{Y}_i} X(\mathbf{t}) N(d\mathbf{t})$$

This generation of the points $\mathbf{Y}_1, \mathbf{Y}_2, \ldots, \mathbf{Y}_l$ and subsequently of $\tilde{X}^*$ and $\bar{X}^*$ is governed by a probability mechanism which we will denote by P^*, with moments denoted by E^*, Var^*, etc. Note again that this generation is done conditionally on the marked point process data observed; thus P^* is really a conditional probability.
3. Let $P^*(\sqrt{|K|}(\tilde{X}^* - E^*\tilde{X}^*) \leq x)$ and $P^*(\sqrt{|K|}(\bar{X}^* - E^*\bar{X}^*) \leq x)$ denote the conditional (given the marked point process data) distribution functions of the bootstrap sample means.

We are now ready to state another set of consistency results.

THEOREM 4: *Assume equations (1) and (2), let* $E|X(\mathbf{t})|^{6+\delta} < \infty$, *where* $\delta > 0$, *and assume that* $R(\mathbf{0}) = Var(X(\mathbf{t})) > 0$. *Also assume that* $\bar{\alpha}_X(k; l_1) \leq const.(1 + l_1)^{\bar{\gamma}} k^{-\bar{\beta}}$ *for some* $\bar{\beta} > 3d$, *and* $0 \leq \bar{\gamma} \leq \bar{\beta}/d$. *Let* $diam(K) \to \infty$, $Diam(K) = O(diam(K))$, *and let* $c = c(K) \to 0$, *but in such a way that* $c^d |K| \to \infty$. *Then*

$$\frac{Var^* \tilde{X}^*}{Var \tilde{X}_K} \xrightarrow{P} 1$$

and

$$\sup_x |P^*(\sqrt{|K|}(\tilde{X}^* - E^*\tilde{X}^*) \leq x) - P(\sqrt{|K|}(\tilde{X}_K - \mu) \leq x)| \xrightarrow{P} 0$$

THEOREM 5: *Under the conditions of Theorem 4, we have*

$$\frac{Var^* \bar{X}^*}{Var \bar{X}_K} \xrightarrow{P} 1$$

and

$$\sup_x |P^*(\sqrt{|K|}(\bar{X}^* - E^*\bar{X}^*) \leq x) - P(\sqrt{|K|}(\bar{X}_K - \mu) \leq x)| \xrightarrow{P} 0$$

REMARK 4: Note that $E^*\bar{X}^* \neq \bar{X}_K$; instead, $E^*\bar{X}^* = \bar{X}_K + O_P(c)$, as the proof of Theorem 4 shows. Therefore, in order not to introduce bias in the bootstrap distribution, it is necessary to center the bootstrap distribution around its bootstrap mean (thus forcing it to have mean zero as in the true distribution of the sample mean); see Lahiri (1991) and Politis and Romano (1992c, 1993) for a similar discussion in the discrete-time case.

REMARK 5: In comparing the circular bootstrap of Section 3 with the block bootstrap of this section, we note that the block bootstrap is valid even for nonrectangular, convex observation regions K, at the expense of having explicitly to center the bootstrap distribution—see previous Remark—which the circular bootstrap does automatically.

Similarly to Corollary 1, we now offer a result on the block bootstrap distribution of the standardized sample mean in anticipation of its possible higher order accuracy properties.

COROLLARY 2: *Under the conditions of Theorem 4, we also have*

$$\sup_x \left| P^*\left(\frac{\tilde{X}^* - \tilde{X}_K}{\sqrt{Var^*\tilde{X}^*}} \leq x \right) - P\left(\frac{\tilde{X}_K - \mu}{\sqrt{Var\tilde{X}_K}} \leq x \right) \right| \xrightarrow{P} 0 \tag{5}$$

and

$$\sup_x \left| P^*\left(\frac{\bar{X}^* - \bar{X}_K}{\sqrt{Var^*\bar{X}^*}} \leq x \right) - P\left(\frac{\bar{X}_K - \mu}{\sqrt{Var\bar{X}_K}} \leq x \right) \right| \xrightarrow{P} 0 \tag{6}$$

5. CONCLUDING REMARKS

In this paper, we have introduced two different (but closely related) resampling techniques for marked point processes, and have shown that they both lead to consistent estimation of the sampling distribution of the sample mean; thus, the bootstrap estimate of sampling distribution can be used effectively for the construction of confidence intervals and hypothesis testing regarding the unknown true mean μ.

The proposed techniques may be viewed as variants of the well-known circular and block bootstrap methods originally designed for data observed over points on a rectangular integer lattice. Some further comments regarding the implementation of the methods are given below:

1. For practical implementation, note that since it is possible for us to generate as many pseudo-replicates of $\tilde{X}^*$, $\bar{X}^*$, $\tilde{X}^*$, and $\bar{X}^*$as we wish,

the aforementioned circular and block bootstrap distributions $P^{\star}$ and P^{*} (as well as their moments $E^{\star}$, E^{*}, $Var^{\star}$, and Var^{*}) can be evaluated approximately by a Monte Carlo procedure simply by looking at the empirical distribution (and moments) of the corresponding generated pseudo-replicates.

2. It is interesting to note that it is not necessary to take the uniform distribution on K and K_{1-c} as the distribution used to generate the i.i.d. points $\mathbf{Y}_1, \mathbf{Y}_2, \ldots, \mathbf{Y}_l$ that are central to the circular and block bootstrap procedures. For example, the (discrete) uniform distribution on $K \cap (h\mathbf{Z})^d$ and $K_{1-c} \cap (h\mathbf{Z})^d$ can instead be used respectively, for the construction of the circular and block bootstrap, for marked point processes *without* affecting the validity of our asymptotic results; here $h\mathbf{Z} = \{hk; k \in \mathbf{Z}\} = \{\ldots, -2h, -h, 0, h, 2h, 3h, \ldots\}$, and h is a positive real number that is either a constant, or in general may depend on K but in such a way that guarantees that the cardinality of the set $K \cap (h\mathbf{Z})^d$, and the cardinality of $K_{1-c} \cap (h\mathbf{Z})^d$ both tend to infinity as $diam(K) \to \infty$.
3. Our bootstrap results are in principle generalizable to the case where the **t**-points are generated by a Poisson process N_g that is not necessarily homogeneous; it is for this reason that our Lemma 1 is stated in such a general form. Nevertheless, since in effect the bootstrap for the sample mean will not work unless the sample mean is known to be asymptotically normal, a new version of Karr's theorem would also be required which will not rely on the homogeneity of the Poisson process involved; such a central limit theorem is expected to have rate $\sqrt{\Lambda_g(K)}$ as opposed to the $\sqrt{|K|}$ rate of Karr's theorem. Also required would be appropriate restrictions on the mean measure Λ_g of the general Poisson prossess N_g; for example, a natural requirement would be to assume that Λ_g is absolutely continuous with respect to Lebesgue measure on $\mathbf{R}^d$, with Radon–Nikodym derivative $\lambda_g(\mathbf{t})$ satisfying $\lambda_g(\mathbf{t}) > \epsilon$ for all $\mathbf{t}$, where ϵ is a positive constant.
4. In general, it is not even necessary to take $l = [1/c^d]$ for the methods to work. To be more specific, our Corollaries 1 and 2 are stated in a way that makes them remain true *verbatim* even if l is taken as being different from $[1/c^d]$, as long as $l \to \infty$ as $diam(K) \to \infty$; in contrast, an explicit renormalization is required for our theorems to remain true if we omit the $l = [1/c^d]$ (or even $l \sim [1/c^d]$) assumption. Nonetheless, we conjecture that—under some extra conditions—the simple choice $l = [1/c^d]$ will be required for higher order accuracy results, that is, in order to have the right-hand side of equations (3)–(6) of order $o_P(1/|K|)$.

5. Last but not least is the issue of "optimally" choosing the design parameter c in practice, again with the point of view of improving estimation accuracy. Although our first-order consistency results (for example Corollaries 1 and 2) remain true for any choice of $c = c(K)$ satisfying $c \to 0$ but $c^d|K| \to \infty$ as $diam(K) \to \infty$, it is highly plausible that there is an "optimal" choice of c as a function of K that will minimize the right-hand side of equations (3)–(6). Thus, the issues of higher order accuracy and of optimally choosing c are intertwined, and will be the subject of further research. Note, however, that the quest for the optimal c ultimately depends on the optimality criterion employed. For example, improving the accuracy of distribution estimation will generally entail a different optimal choice of c as compared with the one required for improving the accuracy of bootstrap variance estimation; see Hall et al. (1996) for a thorough discussion on the analogous problem of optimal block size in the time series case.

6. TECHNICAL PROOFS

PROOF OF LEMMA 1: Consider first the identity

$$E^{N_g(E_1),N_g(E_2)}\bar{Y}_1\bar{Y}_2 - E^{N_g(E_1),N_g(E_2)}\bar{Y}_1 E^{N_g(E_1),N_g(E_2)}\bar{Y}_2$$

$$= E^{N_g(E_1),N_g(E_2)}\{E(\bar{Y}_1\bar{Y}_2|N_g) - E(\bar{Y}_1|N_g)E(\bar{Y}_2|N_g)\}+$$

$$+E^{N_g(E_1),N_g(E_2)}\{E(\bar{Y}_1|N_g)E(\bar{Y}_2|N_g)\} - E^{N_g(E_1),N_g(E_2)}\bar{Y}_1 E^{N_g(E_1),N_g(E_2)}\bar{Y}_2$$

where $E^{N_g(E_1),N_g(E_2)}\bar{Y}$ is an alternative short-hand notation for the conditional expectation $E(\bar{Y}|N_g(E_1), N_g(E_2))$. Note that $E(\bar{Y}_i|N_g)$ is a function of just the $\mathbf{t}_1^{(i)}, \mathbf{t}_2^{(i)}, \ldots, \mathbf{t}_{N_g(E_i)}^{(i)}$ points that were generated by the point process N_g in the set E_i, for $i = 1, 2$. Also note that $\mathbf{t}_1^{(1)}, \mathbf{t}_2^{(1)}, \ldots, \mathbf{t}_{N_g(E_1)}^{(1)}$ and $\mathbf{t}_1^{(2)}, \mathbf{t}_2^{(2)}, \ldots, \mathbf{t}_{N_g(E_2)}^{(2)}$ are two *different* collections of t-points, i.e., there are no common t-points to both collections. Thus, by the Poisson assumption, $E(\bar{Y}_1|N_g)$ and $E(\bar{Y}_2|N_g)$ are conditionally (given $N_g(E_1)$, $N_g(E_2)$) independent (being functions of the distinct, independent sets of t-points $\mathbf{t}_1^{(1)}, \mathbf{t}_2^{(1)}, \ldots, \mathbf{t}_{N_g(E_1)}^{(1)}$ and $\mathbf{t}_1^{(2)}, \mathbf{t}_2^{(2)}, \ldots, \mathbf{t}_{N_g(E_2)}^{(2)}$, respectively). Hence,

$$E^{N_g(E_1),N_g(E_2)}\{E(\bar{Y}_1|N_g)E(\bar{Y}_2|N_g)\}$$

$$= E^{N_g(E_1),N_g(E_2)}\{E(\bar{Y}_1|N_g)\}E^{N_g(E_1),N_g(E_2))}\{E(\bar{Y}_2|N_g)\}$$

$$= E^{N_g(E_1),N_g(E_2)}\bar{Y}_1 E^{N_g(E_1),N_g(E_2)}\bar{Y}_2$$

where $E^{N_g(E_1),N_g(E_2)}\{E(\bar{Y}_i|N_g)\} = E^{N_g(E_1),N_g(E_2)}\bar{Y}_i$, since the σ-algebra generated by the random variables $N_g(E_1)$, $N_g(E_2)$ is coarser than that generated by the whole process N_g.

Thus we have shown that

$$E^{N_g(E_1),N_g(E_2)}\bar{Y}_1\bar{Y}_2 - E^{N_g(E_1),N_g(E_2)}\bar{Y}_1 E^{N_g(E_1),N_g(E_2))}\bar{Y}_2$$

$$= E^{N_g(E_1),N_g(E_2)}\{E(\bar{Y}_1\bar{Y}_2|N_g) - E(\bar{Y}_1|N_g)E(\bar{Y}_2|N_g)\} \tag{7}$$

Now note that $E^{N_g(E_1),N_g(E_2)}\bar{Y}_i = E^{N_g(E_i)}\bar{Y}_i$, and that $E^{N_g(E_i)}\bar{Y}_i$ is a function of the random variable $N_g(E_i)$ only. Also note that because E_1, E_2 are assumed disjoint (since $\rho(E_1, E_2) > 0$), the Poisson process properties imply that the random variables $N_g(E_1)$, $N_g(E_2)$ are independent. Hence,

$$E\{E^{N_g(E_1),N_g(E_2)}\bar{Y}_1 E^{N_g(E_1),N_g(E_2))}\bar{Y}_2\} = E\{E^{N_g(E_1)}\bar{Y}_1 E^{N_g(E_2))}\bar{Y}_2\}$$

$$= E\{E^{N_g(E_1)}\bar{Y}_1\}E\{E^{N_g(E_2)}\bar{Y}_2\} = E\bar{Y}_1 E\bar{Y}_2$$

Therefore, taking expectations on both sides of equation (7), we finally arrive at the relation

$$E\bar{Y}_1\bar{Y}_2 - E\bar{Y}_1 E\bar{Y}_2 = E\{E(\bar{Y}_1\bar{Y}_2|N_g) - E(\bar{Y}_1|N_g)E(\bar{Y}_2|N_g)\} \tag{8}$$

Conditionally on N_g, $\bar{Y}_i$ is just a sum of the $N_g(E_i)$ random variables $X(\mathbf{t})$ with $\mathbf{t}$-indices that happen to be in E_i. However, for any fixed (conditionally) set of points $\{\mathbf{t}_k, k = 1, 2, \ldots\}$, Minkowski's inequality coupled with the assumed homogeneity of the X-process yields

$$E^{N_g}|N_g(E_i)^{-1}\sum_{k=1}^{N_g(E_i)} X(\mathbf{t}_k)|^p \leq E^{N_g}|X(\mathbf{t})|^p = E|X(\mathbf{t})|^p \tag{9}$$

by the independence of the X-process to N_g. Now, by a well-known mixing inequality (see Roussas and Ioannides, 1987), we have that

$$|E(\bar{Y}_1\bar{Y}_2|N_g) - E(\bar{Y}_1|N_g)E(\bar{Y}_2|N_g)| \leq 10C_p^{2/p}(\alpha_X(k; |E_1|, |E_2|))^{1-2/p}$$

Taking expectations (with respect to N_g) on the above completes the first assertion of the lemma.

To prove the bound concerning the covariance of $\tilde{Y}_1$ and $\tilde{Y}_2$, exactly the same arguments apply only that equation (9) is replaced by

$$E^{N_g}|(\Lambda_g(E_i))^{-1}\sum_{k=1}^{N_g(E_i)} X(\mathbf{t}_k)|^p \leq \left(\frac{N_g(E_i)}{\Lambda_g(E_i)}\right)^p E|X(\mathbf{t})|^p$$

Therefore, using the above bound, we obtain

$$|E(\tilde{Y}_1\tilde{Y}_2|N_g) - E(\tilde{Y}_1|N_g)E(\tilde{Y}_2|N_g)| \\ \leq \frac{N_g(E_1)N_g(E_2)}{\Lambda_g(E_1)\Lambda_g(E_2)} 10C_p^{2/p}(\alpha_X(k; |E_1|, |E_2|))^{1-2/p}$$

Taking expectations (with respect to N_g), and using the independence of $N_g(E_1)$ and $N_g(E_2)$ and that $EN_g(E_i) = \Lambda_g(E_i)$ completes the second assertion of the lemma.

Finally, the last two assertions of the lemma involving $\bar{\alpha}$-mixing follow in a similar way to the ones we have proved involving α-mixing. □

PROOF OF LEMMA 2: Owing to Theorem 1.7.1 of Ivanov and Leonenko (1986)—that actually employs a yet weaker notion than our $\bar{\alpha}_X$-mixing, we just need to verify the finiteness of $\int R(\mathbf{t})d\mathbf{t}$ which by necessity will then be equal to the limiting variance of $\frac{1}{\sqrt{|K|}}\int_K X(\mathbf{t})d\mathbf{t}$; since a covariance kernel is nonnegative definite, the improper integral $\int R(\mathbf{t})d\mathbf{t}$ exists, it is nonnegative, but it may be infinite. Note also that, if we are not going to divide by $\int R(\mathbf{t})d\mathbf{t}$ to produce a standardized (or uniform) central limit theorem, it is not necessary to prove or assume that $\int R(\mathbf{t})d\mathbf{t} \neq 0$.

Note that by a well-known mixing inequality (see, for instance, Roussas and Ioannides, 1987) we have that $|Cov(X(\mathbf{0}), X(\mathbf{t}))| \leq$ const.$\bar{\alpha}_X(\max_i |t_i|; 0)^{1-2/(2+\delta)}$. Thus we have

$$\int |R(\mathbf{t})|d\mathbf{t} = O\left(\int \bar{\alpha}_X(\max_i |t_i|; 0)^{1-2/(2+\delta)}d\mathbf{t}\right)$$
$$= O\left(\int_0^\infty y^{d-1}\left(\frac{1}{y^{d+\epsilon}}\right)^{1-2/(2+\delta)} dy\right)$$
$$= O\left(\int_0^\infty \frac{1}{y^{1-2d(2+\delta)^{-1}+\epsilon(1-2/(2+\delta))}} dy\right) < \infty$$

where the assumed bound on the $\bar{\alpha}_X$-coefficients was used, together with the assumption $\epsilon > 2d/\delta$. □

PROOF OF THEOREM 2: Note first that $\tilde{X}^*$ is an average of l i.i.d. random variables, each one being distributed as $\frac{1}{\lambda|B|}\int_{B+\mathbf{Y}} X(\mathbf{t})N(d\mathbf{t})$, where $\mathbf{Y}$ has the uniform distribution on K.

Therefore,

$$E^*\tilde{X}^* = E^*\frac{1}{\lambda|B|}\int_{B+\mathbf{Y}} X(\mathbf{t})N(d\mathbf{t}) = \int_K\int_{B+\mathbf{Y}} \frac{1}{|K|\lambda|B|} X(\mathbf{t})N(d\mathbf{t})d\mathbf{Y} = \tilde{X}_K$$

Similarly,

$$Var^*\tilde{X}^* = l^{-1}Var^*\left(\frac{1}{\lambda|B|}\int_{B+\mathbf{Y}} X(\mathbf{t})N(\mathrm{d}\mathbf{t})\right)$$

and

$$Var^*\left(\frac{1}{\lambda|B|}\int_{B+\mathbf{Y}} X(\mathbf{t})N(\mathrm{d}\mathbf{t})\right) = \frac{1}{|K|}\int_K\left(\frac{1}{\lambda|B|}\int_{B+\mathbf{Y}} X(\mathbf{t})N(\mathrm{d}\mathbf{t}) - \tilde{X}_K\right)^2 \mathrm{d}\mathbf{Y}$$

$$= \frac{1}{|K|}\int_K\left(\frac{1}{\lambda|B|}\int_{B+\mathbf{Y}} X(\mathbf{t})N(\mathrm{d}\mathbf{t}) - \mu - (\tilde{X}_K - \mu)\right)^2 \mathrm{d}\mathbf{Y} = A_1 - A_2 + A_3,$$

where

$$A_1 = \frac{1}{|K|}\int_K\left(\frac{1}{\lambda|B|}\int_{B+\mathbf{Y}} X(\mathbf{t})N(\mathrm{d}\mathbf{t}) - \mu\right)^2 d\mathbf{Y}$$

$$A_2 = \frac{2}{|K|}\int_K\left(\frac{1}{\lambda|B|}\int_{B+\mathbf{Y}} X(\mathbf{t})N(\mathrm{d}\mathbf{t}) - \mu\right)(\tilde{X}_K - \mu)\mathrm{d}\mathbf{Y}$$

and

$$A_3 = \frac{1}{|K|}\int_K(\tilde{X}_K - \mu)^2\, d\mathbf{Y}$$

Now note that under the assumed conditions, Karr's (1986) theorem implies that $\sqrt{|K|}(\tilde{X}_K - \mu) \overset{\mathcal{L}}{\Longrightarrow} N(0, \sigma^2)$, where $\sigma^2 = \int R(\mathbf{t})\mathrm{d}\mathbf{t} + R(\mathbf{0})/\lambda$. Observe also that our assumption $R(\mathbf{0}) = Var(X(\mathbf{t})) > 0$ implies that $\sigma^2 > 0$, since $\int R(\mathbf{t})\mathrm{d}\mathbf{t} \geq 0$ by nonnegative definiteness of $\{R(\mathbf{t})\}$.

Hence, it follows that $\tilde{X}_K - \mu = O_P(1/\sqrt{|K|})$, and therefore $A_3 = O_P(1/|K|)$. Similarly, because (also by Karr's (1986) theorem) we have $\frac{1}{\lambda|B|}\int_{B+\mathbf{Y}} X(\mathbf{t})N(\mathrm{d}\mathbf{t}) - \mu = O_P(1/\sqrt{|B|})$, it follows that $A_2 = O_P(1/\sqrt{|B||K|})$. Since A_2 and A_3 are asymptotically negligible, we now focus on A_1.

Let $G(\mathbf{Y}) \equiv \left(\frac{1}{\lambda\sqrt{|B|}}\int_{B+\mathbf{Y}} X(\mathbf{t})N(\mathrm{d}\mathbf{t}) - \mu\sqrt{|B|}\right)^2$, and note that due to the homogeneity of the X-process and of the point process N, $G(\mathbf{Y})$ is itself a homogeneous random field with index $\mathbf{Y} \in \mathbf{R}^d$. Also note that since the X-process has more than four finite moments, $EG(\mathbf{Y})^2 < \infty$, and homogeneity yields $Cov(G(\mathbf{Y}_1), G(\mathbf{Y}_2)) = C(\mathbf{Y}_1 - \mathbf{Y}_2)$ for some covariance function $C(\cdot)$.

So we have

$$A_1 = \frac{1}{|K||B|}\int_K G(\mathbf{Y})\mathrm{d}\mathbf{Y}$$

and

$$EA_1 = \frac{1}{|K||B|}\int_K EG(\mathbf{Y})d\mathbf{Y} = \frac{1}{|B|}EG(\mathbf{0})$$

by homogeneity of the G-field. However, since $|B| \to \infty$, Karr's (1986) theorem implies that $EG(\mathbf{0}) \to \sigma^2$, and thus $EA_1 = \frac{\sigma^2}{|B|} + o(1/|B|)$.

Now look at

$$\begin{aligned} VarA_1 &= \frac{1}{|K|^2|B|^2}\int_K\int_K C(\mathbf{Y}_1 - \mathbf{Y}_2)d\mathbf{Y}_1 d\mathbf{Y}_2 \\ &= \frac{1}{|K||B|^2}\int_{\mathbf{R}^d} C(\mathbf{t})\frac{|K \cap (K - \mathbf{t})|}{|K|}d\mathbf{t} \end{aligned}$$

where again the homogeneity of G was used.

Recall that K is the rectangle $\{\mathbf{t} = (t_1, ..., t_d) : 0 \le t_i \le K_i, i = 1, ..., d\}$ and B is the rectangle $\{\mathbf{t} = (t_1, ..., t_d) : 0 \le t_i \le cK_i, i = 1, ..., d\}$. Define "symmetrized" versions of K and B by

$$K^* = \{\mathbf{t} = (t_1, ..., t_d) : |t_i| \le K_i, \quad i = 1, ..., d\}$$

and

$$B^* = \{\mathbf{t} = (t_1, ..., t_d) : |t_i| \le cK_i, \quad i = 1, ..., d\}$$

Now note that

$$\begin{aligned} VarA_1 &= \frac{1}{|K||B|^2}\int_{K^*} C(\mathbf{t})\frac{|K \cap (K - \mathbf{t})|}{|K|}d\mathbf{t} \\ &\le \frac{1}{|K||B|^2}\int_{B^*} |C(\mathbf{t})|d\mathbf{t} + \frac{1}{|K||B|^2}\int_{K^*-B^*} |C(\mathbf{t})|d\mathbf{t}. \end{aligned}$$

However, $\frac{1}{|K||B|^2}\int_{B^*} |C(\mathbf{t})|d\mathbf{t} = O(\frac{1}{|K||B|})$, since $|C(\mathbf{t})|$ is bounded by $|C(\mathbf{0})|$. Finally note that for $\mathbf{t} \in K^* - B^*$, our Lemma 1 affords us the possibility of using mixing bounds for the covariance $C(\mathbf{t})$. Letting $p = 3$ in Lemma 1, we obtain

$$|C(\mathbf{t})| \le 10(E|G(\mathbf{0})|^3)^{2/3}(\bar{\alpha}_X(\rho(\mathbf{t}, B^*); |B|))^{1/3};$$

but from Lemma 1.8.1 of Ivanov and Leonenko (1986) it follows that $E|G(\mathbf{0})|^3 \le \text{const.}(E|X(\mathbf{t}) - \mu|^{6+\delta})^{6/(6+\delta)}$, which is assumed to be finite. Putting all this together, and noting that

$$|K|^{-1}\int_{K^*-B^*} (\bar{\alpha}_X(\rho(\mathbf{t}, B^*); |B|))^{1/3}d\mathbf{t} \le |K|^{-1}\int_{K^*} (\bar{\alpha}_X(\rho(\mathbf{t}, \mathbf{0}); |B|))^{1/3}d\mathbf{t}$$

$$\leq |K|^{-1}\int_0^{\max_i K_i} y^{d-1}(\bar{\alpha}_X(y;|B|))^{1/3}\mathrm{d}y = O\left(\frac{|B|^{\bar{\gamma}/3}}{|K|}\int_0^{\max_i K_i} y^{d-1}y^{-\bar{\beta}/3}\mathrm{d}y\right)$$

$$= O\left(\frac{|B|^{\bar{\gamma}/3}}{|K|}(\max_i K_i)^{d-\bar{\beta}/3}\right) = O(c^{d\bar{\gamma}/3}|K|^{(\bar{\gamma}-\bar{\beta}d^{-1})/3}) = o(1)$$

because $c \to 0$ and it was assumed that $\bar{\beta} \geq d\bar{\gamma}$. In the above string of inequalities, the assumed bound on the $\bar{\alpha}_X$ coefficients was used, as well as the fact that $\max_i K_i = O(|K|^{1/d})$ by assumption. Therefore, it follows that

$$VarA_1 = O\left(\frac{1}{|K||B|}\right) + o\left(\frac{1}{|B|^2}\right) = o\left(\frac{1}{|B|^2}\right).$$

To recapitulate, we have shown that

$$lVar^{\star}\tilde{X}^{\star} = EA_1 + o_P\left(\frac{1}{|B|}\right) = \frac{\sigma^2}{|B|} + o_P\left(\frac{1}{|B|}\right)$$

Since $l|B|/|K| \to 1$ and $|K|Var\tilde{X}_K \to \sigma^2$, it follows that $\frac{Var^{\star}\tilde{X}^{\star}}{Var\tilde{X}_K} \xrightarrow{P} 1$ as claimed.

So far we have shown convergence of the first two bootstrap moments of $\tilde{X}^{\star}$ to the corresponding moments of $\tilde{X}_K$. Since $\tilde{X}_K$ is asymptotically normal, to complete the proof of Theorem 2 we need to show that $\tilde{X}^{\star}$ is also asymptotically normal, in $P^{\star}$ probability. However, recall that $\tilde{X}^{\star}$ is an average of l i.i.d. random variables, with $l \to \infty$. Therefore, to show that $\tilde{X}^{\star}$ is asymptotically normal in $P^{\star}$ (conditional) probability, it suffices (by the Lyapunov central limit theorem) to show that $E^{\star}|G(\mathbf{Y})|^{3/2} = O_P(1)$. However $E^{\star}|G(\mathbf{Y})|^{3/2} = \frac{1}{|K|}\int_K |G(\mathbf{Y})|^{3/2}\mathrm{d}\mathbf{Y} \equiv \xi$. Due to homogeneity and $\bar{\alpha}$-mixing, a calculation similar to the calculation of $VarA_1$ above yields $Var\xi \to 0$. Hence, a weak law of large numbers obtains, and $\xi \xrightarrow{P} E|G(\mathbf{Y})|^{3/2}$ which is finite by Lemma 1.8.1 of Ivanov and Leonenko (1986) and the assumed finiteness of $6+\delta$ moments of the X-process. Therefore, $E^{\star}|G(\mathbf{Y})|^{3/2} = O_P(1)$ and the proof is completed.

□

PROOF OF THEOREM 3: Note again that $E^{\star}\bar{X}^{\star}$ is an average of l i.i.d. random variables, each one being distributed as $\frac{1}{N(\mathbf{B}+\mathbf{Y})}\int_{\mathbf{B}+\mathbf{Y}} X(\mathbf{t})N(\mathrm{d}\mathbf{t})$, where $\mathbf{Y}$ has the uniform distribution on K.

Therefore,

$$E^{*}\bar{X}^{*} = E^{*}\frac{1}{N(B+\mathbf{Y})}\int_{B+\mathbf{Y}} X(\mathbf{t})N(\mathrm{d}\mathbf{t})$$
$$= \int_{K}\int_{B+\mathbf{Y}} \frac{1}{|K|N(B+\mathbf{Y})} X(\mathbf{t})N(\mathrm{d}\mathbf{t})\mathrm{d}\mathbf{Y} = \bar{X}_K$$

by a conditioning (on N) argument similar to the proof of the unbiasedness of $\bar{X}_K$ as an estimator of μ—see Karr (1986, 1991).

Furthermore, recall that under the assumed conditions, Karr's (1986) theorem implies that $\sqrt{|K|}(\bar{X}_K - \mu) \overset{\mathcal{L}}{\Longrightarrow} N(0, \sigma^2)$, where $\sigma^2 = \int R(\mathbf{t})d\mathbf{t} + R(\mathbf{0})/\lambda$.

Now due to the Poisson assumption, we have $N(K)/|K| \to \lambda$ almost surely. As a matter of fact, $EN(K) = VarN(K) = \lambda|K|$; therefore, $\frac{N(K)}{\lambda|K|} = 1 + O_P(\frac{1}{\sqrt{|K|}})$. By the aforegoing discussion, we have that

$$\sqrt{|K|}(\tilde{X}_K - \bar{X}_K) = \sqrt{|K|}(\tilde{X}_K - \mu + \mu - \bar{X}_K)$$
$$= \frac{N(K) - \lambda|K|}{N(K)\lambda\sqrt{|K|}}\int_K (X(\mathbf{t}) - \mu)N(\mathrm{d}\mathbf{t})$$
$$= O_P(N(K)^{-1}\int_K (X(\mathbf{t}) - \mu)N(\mathrm{d}\mathbf{t})) = O_P(\frac{1}{\sqrt{|K|}}) = o_P(1)$$

since $\bar{X}_K - \mu = O_P(\frac{1}{\sqrt{|K|}})$ by Karr's (1986) result.

Consequently, $\sqrt{|K|}(\tilde{X}_K - \mu)$ and $\sqrt{|K|}(\bar{X}_K - \mu)$ have the same asymptotic distribution, as well as the same asymptotic variance (by Karr's (1986) theorem); thus Theorem 3 follows from Theorem 2. □

PROOF OF THEOREM 4: The proof is similar to the proof of Theorem 2; below we point out only the differences and new elements.

Note again that $\tilde{X}^*$ is an average of l i.i.d. random variables, each one being distributed as $\frac{1}{\lambda|B|}\int_{B+\mathbf{Y}} X(\mathbf{t})N(\mathrm{d}\mathbf{t})$, where $\mathbf{Y}$ has the uniform distribution on K_{1-c}.

Therefore,

$$E^{*}\tilde{X}^{*} = E^{*}\frac{1}{\lambda|B|}\int_{B+\mathbf{Y}} X(\mathbf{t})N(\mathrm{d}\mathbf{t}) = \int_{K_{1-c}}\int_{B+\mathbf{Y}} \frac{1}{\lambda|B||K_{1-c}|} X(\mathbf{t})N(\mathrm{d}\mathbf{t})d\mathbf{Y}$$

Recall that

$$E^{*}\tilde{X}^{*} = \int_{K}\int_{B+\mathbf{Y}} \frac{1}{\lambda|B||K|} X(\mathbf{t})N(\mathrm{d}\mathbf{t})\mathrm{d}\mathbf{Y} = \tilde{X}_K$$

Hence,

$$|K|(\tilde{X}_K - \mu) - |K_{1-c}|(E^*\tilde{X}^* - \mu) = \int_{K-K_{1-c}} \int_{B+\mathbf{Y}} \frac{1}{\lambda|B|}(X(\mathbf{t}) - \mu)N(d\mathbf{t})d\mathbf{Y}$$
$$= \int_{K-K_{1-c}} O_P(1/\sqrt{|B|})d\mathbf{Y}$$
$$= O_P(|K - K_{1-c}|/\sqrt{|B|})$$

However $|K_{1-c}| = (1-c)^d|K|$, and $|K - K_{1-c}| = O(c|K|)$. Therefore,

$$E^*\tilde{X}^* = \frac{\tilde{X}_K - \mu}{(1-c)^d} + \mu + O_P(c/\sqrt{|B|}) = \mu + O_P(1/\sqrt{|K|}) + O_P(c/\sqrt{|B|})$$
$$= \mu + o_P(1/\sqrt{|B|})$$

Similarly,

$$lVar^*\tilde{X}^* = \frac{1}{|K_{1-c}|}\int_{K_{1-c}}\left(\frac{1}{\lambda|B|}\int_{B+\mathbf{Y}} X(\mathbf{t})N(d\mathbf{t}) - E^*\tilde{X}^*\right)^2 d\mathbf{Y}$$
$$= \frac{1}{|K_{1-c}|}\int_{K_{1-c}}\left(\frac{1}{\lambda|B|}\int_{B+\mathbf{Y}} X(\mathbf{t})N(d\mathbf{t}) - \mu - (E^*\tilde{X}^* - \mu)\right)^2 d\mathbf{Y}$$
$$= D_1 - D_2 + D_3,$$

where

$$D_1 = \frac{1}{|K_{1-c}|}\int_{K_{1-c}}\left(\frac{1}{\lambda|B|}\int_{B+\mathbf{Y}} X(\mathbf{t})N(d\mathbf{t}) - \mu\right)^2 d\mathbf{Y}$$
$$D_2 = \frac{2}{|K_{1-c}|}\int_{K_{1-c}}\left(\frac{1}{\lambda|B|}\int_{B+\mathbf{Y}} X(\mathbf{t})N(d\mathbf{t}) - \mu\right)(E^*\tilde{X}^* - \mu)d\mathbf{Y}$$

and

$$D_3 = \frac{1}{|K_{1-c}|}\int_{K_{1-c}} (E^*\tilde{X}^* - \mu)^2 d\mathbf{Y}$$

However, necessarily $D_3 = o_P(1/|B|)$, since we have shown that $E^*\tilde{X}^* = \mu + o_P(1/\sqrt{|B|})$. Similarly, $D_2 = o_P(1/|B|)$, since $\frac{1}{\lambda|B|}\int_{B+\mathbf{Y}} X(\mathbf{t})N(d\mathbf{t}) - \mu = O_P(1/\sqrt{|B|})$.

Note now that D_1 is of the same form as the quantity A_1 in the proof of Theorem 2; the only differences are: (a) integrating/averaging takes place over K_{1-c} as opposed to K, (b) K is not a rectangle anymore, but rather a general compact, convex set, and (c) we generally define "symmetrized" versions of the nonrectangular K_{1-c} and B by K^*_{1-c} = convex hull of $\{\mathbf{t} : |\mathbf{t}| \in K_{1-c}\}$, and B^* = convex hull of $\{\mathbf{t} : |\mathbf{t}| \in B\}$, where

$|\mathbf{t}| \equiv (|t_1|, \ldots, |t_d|)$. However, the same arguments used in the proof of Theorem 2 to show that $A_1 = \frac{\sigma^2}{|B|} + o_P(1/|B|)$ can be used to show $|B|D_1 \xrightarrow{P} \sigma^2$, which in turns implies that $\frac{Var^* \bar{X}^*}{Var \bar{X}_K} \xrightarrow{P} 1$. Finally, the proof of asymptotic normality follows *verbatim* the proof of Theorem 2. □

PROOF OF THEOREM 5: The proof follows from Theorem 4 using the same arguments used in the proof of Theorem 3. □

REFERENCES

Bickel, P. and Freedman, D. (1981), Some asymptotic theory for the bootstrap, *Ann. Stat.*, **9**, 1196–1217.

Carlstein, E. (1986), The use of subseries values for estimating the variance of a general statistic from a stationary sequence, *Ann. Stat.*, **14**, 1171–1179.

Cressie, N. (1991), *Statistics for Spatial Data*, New York: Wiley.

Daley, D.J. and Vera-Jones, D. (1988), *An Introduction to the Theory of Point Processes*, New York: Springer.

Doukhan, P. (1994), *Mixing: Properties and Examples*, Lecture Notes in Statistics No. 85, New York: Springer.

Efron, B. and Tibshirani, R. (1993), *An Introduction to the Bootstrap*, New York: Chapman & Hall.

Götze, F. and Künsch, H.R. (1996), Second order correctness of the blockwise bootstrap for stationary observations, *Ann. Stat.*, **24**, 1914–1933.

Hall, P. (1985), Resampling a coverage pattern, *Stochastic Process. Appl.*, **20**, 231–246.

Hall, P. (1992), *The Bootstrap and Edgeworth Expansion*, New York: Springer.

Hall, P., Horowitz, J.L., and Jing, B. (1996), On blocking rules for the block bootstrap with dependent data, *Biometrika*, **82**, 561–574.

Ivanov, A.V. and Leonenko, N.N. (1986), *Statistical Analysis of Random Fields*, Dordrecht: Kluwer.

Karr, A. F. (1986), Inference for stationary random fields given Poisson samples, *Adv. Appl. Prob.*, **18**, 406–422.

Karr, A. F. (1991), *Point Processes and their Statistical Inference*, 2nd edn., New York: Marcel Dekker.

Krickeberg, K. (1982), Processus ponctuels en statistique, École d' été de probabilités de Saint-Flour X—1980, In: P.L. Hennequin (ed.), *Lecture Notes in Mathematics No. 929*, Berlin: Springer, pp. 205–313.

Künsch, H.R. (1989), The jackknife and the bootstrap for general stationary observations, *Ann. Stat.*, **17**, 1217–1241.

Lahiri, S.N. (1991), Second order optimality of stationary bootstrap, *Stat. Prob. Lett.*, **11**, 335–341.

Liu, R.Y. and Singh, K. (1992), Moving blocks jackknife and bootstrap capture weak dependence, In: R. LePage and L. Billard (eds.), *Exploring the Limits of Bootstrap*, Wiley, New York, pp. 225–248.

Masry, E. (1988), Random sampling of continuous-parameter stationary processes: Statistical properties of joint density estimators, *J. Multivar. Anal.*, **36**, 133–165.

Politis, D.N. and Romano, J.P. (1992a), A general resampling scheme for triangular arrays of α-mixing random variables with application to the problem of spectral density estimation, *Ann. Stat.*, **20**, 1985–2007.

Politis, D.N. and Romano, J.P. (1992b), A nonparametric resampling procedure for multivariate confidence regions in time series analysis, In C. Page and R. LePage (eds.), *Computing Science and Statistics, Proceedings of the 22nd Symposium on the Interface*, Springer, New York, pp. 98–103.

Politis, D.N. and Romano, J.P. (1992c), A circular block-resampling procedure for stationary data, In: R. LePage and L. Billard (eds.), *Exploring the Limits of Bootstrap*, Wiley, New York, pp. 263–270.

Politis, D.N. and Romano, J.P. (1993). Nonparametric resampling for homogeneous strong mixing random fields, *J. Multivar. Anal.*, **47**, 301–328.

Politis, D.N. and Romano, J.P. (1994). The stationary bootstrap, *J. Am. Stat. Assoc.*, **89**, 1303–1313.

Raïs, N. (1992), Méthodes de reéchantillonage et de sous échantillonage dans le contexte spatial et pour des données dépendantes, PhD Thesis, Department of Mathematics and Statistics, University of Montreal, Montreal.

Ripley, B.D. (1981), *Spatial Statistics*, Wiley, New York.

Rosenblatt, M. (1985), *Stationary Sequences and Random Fields*, Boston, MA: Birkhäuser.

Roussas, G.G. and Ioannides, D. (1987), Moment inequalities for mixing sequences of random variables, *Stochastic Anal. Appl.*, **5**, 61–120.

Shao, J. and Tu, D. (1995), *The Jackknife and Bootstrap*, New York: Springer.

Sherman, M. and Carlstein, E. (1994), Nonparametric estimation of the moments of a general statistic computed from spatial data, *J. Am. Stat. Assoc.*, **89**, 496–500.

Sherman, M. and Carlstein, E. (1996), Replicate histograms, *J. Am. Stat. Assoc.*, **91**, 566–576.

Yadrenko, M.I. (1983), *Spectral Theory of Random Fields*, New York: Optimization Software.

8

Graphical Markov Models in Multivariate Analysis

STEEN A. ANDERSSON, Indiana University, Bloomington, Indiana

DAVID MADIGAN, MICHAEL D. PERLMAN, and THOMAS S. RICHARDSON University of Washington, Seattle, Washington

1. OVERVIEW

One of the most central ideas of statistical science is the assessment of dependencies among a set of random variates. The familiar concepts of correlation, regression, and prediction are manifestations of this idea, and many aspects of causal relationships ultimately rest on representations of multivariate dependence.

Graphical Markov models (GMM) and *structural equation models* (SEM) use *graphs* (either undirected, directed, or mixed) and *path diagrams*, respectively, to represent multivariate dependencies in an economical and computationally efficient manner. A GMM or SEM is constructed by specifying *local* dependencies for each variable $\equiv$ node of the graph in terms of its immediate neighbors, parents, or both, yet can represent a highly varied and complex system of multivariate dependencies by means of the *global* structure of the graph. The local specification permits efficiencies in modeling, inference, and probabilistic calculations.

GMMs and SEMs differ significantly in the *semantics* associated with the graph: a GMM uses its graph to represent Markov (≡ conditional independence) relations among the variables, while a SEM uses its graph (usually directed) to represent a set of equations that express each variable as a function (usually linear) of its parents in the graph and an error term. It is remarkable that often a GMM and a SEM with the same graph may be *Markov equivalent*, i.e., may represent the same statistical model, allowing results and techniques for each model to be applied to both. A main focus of current research is to find general necessary and sufficient conditions for the Markov equivalence of GMMs and SEMs.

Even within the classes of GMMs and SEMs themselves, questions of Markov equivalence can be theoretically nontrivial and highly significant for applications. It is known, for example, that *two or more different graphs or path diagrams can determine Markov-equivalent statistical models*, and hence cannot be distinguished on the basis of data regardless of sample size. Since the number of Markov-equivalent graphs can grow superexponentially with the number of vertices, such redundancy can render model search and selection computationally infeasible. For a wide variety of GMMs and SEMs, current research seeks to define, characterize, and efficiently construct *essential graphs* and *partial ancestral graphs*, which simultaneously represent all graphs or path diagrams in a given Markov equivalence class, thus overcoming these difficulties.

Despite the great interest in GMMs since 1980, only after the recent introduction of *chain graphs* (which have both directed and undirected edges but no fully or partly directed cycles) has it been noted that, conversely, *the same graph can admit more than one Markov interpretation*, and hence can simultaneously represent more than one statistical model. A third focus of current research is the systematic investigation of such *alternative Markov properties*, which reveal even more closely the links between GMMs and SEMs.

For multivariate normal distributions, Markov properties are equivalent to conditions on the covariance structure. Current research seeks *to incorporate linear regression structure* into GMMs. This allows, for example, the following nonstandard multivariate linear regression model to be analyzed by standard linear methods: a three-variate, two-way multivariate analysis of variance model with no interactions, no row effects for variable 1, no column effects for variable 2, and no row or column effects for variable 3.

Lattice conditional independence (LCI) models, a subclass of GMMs, are particularly well suited to the analysis of nonmonotone missing data patterns in continuous or categorical data, and of nonnested dependent linear regression models, including the *seemingly unrelated regressions model* in econometrics.

Among their many applications, GMMs have become prevalent in statistical science for the analysis of categorical data in contingency tables, for the modeling of spatially dependent processes such as the spread of epidemics in human and animal populations, and for the development of early warning systems for severe weather conditions; in computer science (as Bayesian networks) for information processing and retrieval, for robotics, computer vision, and pattern recognition, for the debugging of complex programs (such as Windows 95), and for the representation of expert systems for medical diagnosis; and in decision science (as influence diagrams) as models for information flow and control and for combining the opinions of many decision-makers. SEMs have long been used in fields such as genetics, sociology, econometrics, and psychometrics as networks for representing the structure of complex causal systems. A crucial feature of all these models is that they are designed for fast computational implementation, thereby facilitating the development of software that can "reason" about real-world problems.

2. INTRODUCTION: GRAPHICAL MARKOV MODELS AND STRUCTURAL EQUATION MODELS

The use of graphical Markov models (GMMs) to simplify complex statistical dependencies occurred early in probability theory (Markov chains) and continued with the application of Markov random fields, represented by undirected graphs (UG), in spatial statistics, image analysis, and related areas. Prominent early contributors in these areas included W. Gibbs, M. V. Averintsev, R. L. Dobrushin, F. Spitzer, J. Moussouris, J. Hammersley, P. Clifford, J. Besag, U. Grenander, D. Geman, and S. Geman. In this work, the vertices of the graphs represent sites or pixels, and the edges represent stochastic dependencies. More pertinently, missing edges represent conditional independences. For instance, in image analysis, edges might connect each pixel with its neighboring pixels in the image, thus representing the assumption that the value at a pixel (e.g., the pixel's intensity) is conditionally independent of all other pixels given its neighbors.

The use of structural equation models (SEM) to represent causal hypotheses originated in the work of Sewall Wright (1921) on path analysis in population genetics, and then expanded into psychometrics, econometrics, and many other fields. Prominent early contributors included L. Thurstone, H. Harman, P. Horst, K. Jöreskog, D. Lawley, and A. Maxwell (psychometrics), T. Haavelmo, H. Wold, and the Cowles Commission (econometrics), I. J. Good (philosophy), and H. T. Blalock and H. Costner (sociology). The discipline-specific models that arose, such as factor analysis, are all instances of SEMs and account for a large proportion of the models currently employed in the social sciences. The books by Bollen (1988) and Goldberger (1991) report the state of the art of structural equation modeling.

In statistics, the systematic development of GMMs to represent complex dependencies for both categorical and continuous data accelerated rapidly in the late 1970s, starting with work by Lauritzen, Darroch, Speed, Wermuth, and others on graphical log–linear and recursive linear models, and then was continued by Dawid, Spiegelhalter, Frydenberg, Cox, and others, with such applications as medical diagnosis and epidemiology. This development has been facilitated by the parallel development of Bayesian methodology and high-speed computational techniques. In applications, GMMs have simplified many standard statistical analyses, such as closed population estimation (Madigan and York, 1997) and double sampling (York et al., 1995). See Whittaker (1990), Edwards (1995), Cox and Wermuth [CW] (1996), and Lauritzen [L] (1996) for excellent overviews.

At the same time, separate but convergent developments of these ideas occurred in computer science, decision analysis, and philosophy, where GMMs have been called influence diagrams, belief networks, or Bayesian networks, and are used for the construction of expert systems, neural networks, and causal models. Directed graphs (DG), in particular, have proved to be well-suited for describing causal relations among random variates in complex systems: once a set of assumed local dependencies are depicted by arrows to construct the skeleton of the system, their global implications can be revealed and traced through the network structure by means of the *d-separation* criterion. The application of GMMs to expert systems has proved hugely successful—any recent Uncertainty in Artificial Intelligence (UAI) proceedings provides numerous real-world applications. The books by Pearl (1988), Neapolitan (1990), Oliver and Smith (1990), Heckerman (1991), Spirtes et al. (1993), Almond (1995), Jensen (1996), and Castillo et al. (1997) survey these areas.

GMMs determined by acyclic digraphs (ADG) (also called directed acyclic graphs ≡ DAG) admit especially simple statistical analysis. The joint probability density admits a recursive factorization (Lauritzen et al., 1990) that allows efficient updating algorithms for Bayesian analysis and, in expert system applications, simple causal interpretations (Lauritzen and Spiegelhalter, 1988; Spiegelhalter et al., 1993)—indeed, the vibrant UAI community focuses much of its effort on ADG models ≡ Bayesian networks. In the multinomial or multivariate normal cases, the likelihood function (i.e., both the joint density and the parameter space) factors into the product of likelihood functions (LF) of saturated multinomial models or normal linear regression models, respectively, and thus admits explicit maximum likelihood estimates (MLE) and likelihood ratio tests ([L], 1996; Andersson and Perlman [AP], 1998)—also see Section 3.3. The only UG models with these properties are the decomposable models, exactly those UG models that are Markov equivalent to some ADG models (Dawid and Lauritzen, 1993; Andersson et al., 1997a).

In the late 1980s, Lauritzen, Wermuth, and Frydenberg [LWF] generalized ADG models to adicyclic graphs ≡ *chain graphs* (CG), which include both ADGs and UGs as special cases. Roughly speaking, CGs attempt to simultaneously represent dependencies some of which are causal and some associative. Much current research on GMMs is focussed on CGs; e.g., Wermuth and Lauritzen (1990), [L] (1989, 1996), [CW] (1993, 1996), Højsgaard and Thiesson (1995), Bouckaert and Studený (1995), Buntine (1995), Studený (1996, 1997), Studený and Bouckaert (1998). We refer the reader to Lauritzen's 1996 book for a thorough treatment of the Markov properties associated with ADGs, UGs, and CGs.

It has been noted recently that a CG may admit Markov interpretations different than that of [LWF], hence may simultaneously represent different statistical models. [CW] (1993, 1996) introduced block-regression (full line, full arrow), multivariate regression (dashed line, dashed arrow) and concentration regression (full-line, dashed arrow) chain graphs. [CW] (1996) define corresponding models and present analyses for general distributions, in particular for generalized linear and nonlinear regressions. Andersson, Madigan, and Perlman [AMP] (1996, 1998a) introduced a related alternative Markov property for CGs which in some ways is a more direct generalization of the ADG Markov property than is the [LWF] Markov property for CGs. They show that for multivariate normal distributions, it is the [AMP], rather than the [LWF], Markov property that is satisfied by a block-recursive normal linear system naturally associated with the CG (see Sections 4.1 and 4.6). This model can be decomposed into a collection of conditional normal models, each of which combines the features of multivariate linear regression models and covariance selection models (Dempster, 1972; Speed and Kiiveri, 1986; Eriksen, 1997), facilitating the estimation of its parameters. The study of these competing Markov properties and corresponding *pathwise-separation* properties for CGs, together with their relation to the Markov properties of recursive and nonrecursive SEMs, is currently under intensive study (see Sections 4 and 5).

An interesting, although complicating, feature of ADG and CG models is the nonuniqueness of the graph associated with the model. Unlike UGs, two or more ADGs or CGs may determine the same GMM. For instance, an ADG with a single edge represents a unique statistical model, irrespective of the edge's orientation. This nonuniqueness can lead to computational inefficiency in model selection, and to inaccurate specification of prior distributions in Bayesian model averaging. Fundamental characterizations of Markov equivalence for ADGs and for [LWF] CGs (Wermuth and Lauritzen, 1983; Frydenberg [F], 1990; Verma and Pearl, 1990, 1992; [AMP], 1997a) allow one to determine whether two ADGs or two [LWF] CGs determine the same GMM. In [AMP] (1996, 1998a) a corresponding criterion for the equivalence of [AMP] CGs has been obtained; this new criterion is closer to that for ADGs than is the criterion for equivalence of [LWF] CGs (see Sections 3.1 and 4.4).

For an ADG D, [AMP] (1997b) have defined and characterized the *essential graph* D^*, a unique representative of the Markov equivalence class containing D, which overcomes the nonuniqueness difficulties noted above for model selection and model averaging (Chickering, 1995; Meek, 1995a; Madigan et al., 1996). Here we introduce and discuss the *essential graph* G^* associated with an [AMP] CG G (see Sections 3.1 and 4.4).

Given the prima facie similarity between SEMs and GMMs and the deep understanding of the Markov properties of GMMs, it is natural to seek analogous results for SEMs. Such questions have been considered by Kiiveri et al. (1984), Wermuth (1992), [CW] (1993), and [AP] (1998). Care must be taken, because the DGs associated with SEMs, also called path diagrams, allow certain features that do not occur in standard GMM models: directed cycles (representing feedback) and double-headed edges ($\leftrightarrow$) (indicating the presence of correlated errors). It is already known that univariate SEMs do satisfy the natural analogue to the global Markov property for the path diagram (Kiiveri et al., 1984; Geiger and Pearl, 1988; Spirtes et al., 1993, 1996; Spirtes, 1995; Koster, 1996). Thus, in these cases a simple extension of the path criterion (d-separation) used to read off conditional independences (CIs) from ADG models can be applied to path diagrams.

As for ADGs and CGs, the Markov equivalence of SEMs and the characterization of their Markov equivalence classes is of particular significance to the practice of structural equation modeling. What structural features (if any) are common to the path diagrams in a given Markov equivalence class? This question is complicated further by the fact that SEM models often postulate the existence of unmeasured (or latent) variables. It is therefore possible for two models containing different sets of unmeasured variables to be Markov equivalent on a set of observed variables which they have in common; this is the relevant notion of Markov equivalence, since we cannot directly test CIs among unobserved variables. Currently, model specification algorithms exist that use the results of CI tests to search among Markov equivalence classes of SEM models (Spirtes et al., 1993, 1995, 1996, 1998; Richardson [R], 1996a; Spirtes and Richardson, 1997).

Richardson (1996a) has introduced *partial ancestral graphs* (PAG) to represent graphical features and CIs common to a Markov equivalence class of SEM path diagrams (possibly with latent variables). The PAG formalism and related algorithms for constructing PAGs on the basis of CI tests or tests of d-separation relations on graphs, promise to greatly facilitate efficient SEM specification algorithms (see Section 5).

3. GRAPHICAL MARKOV MODELS BASED ON ACYCLIC DIGRAPHS (ADG)

We consider multivariate probability distributions P on a product probability space $\mathbf{X} \equiv \times(\mathbf{X}_v | v \in V)$, where V is a finite index set. Such a distribution will be represented by a random variate $X \equiv (X_v | v \in V) \in \mathbf{X}$. For any subset $A \subseteq V$, we define $X_A := (X_v | v \in A)$. We often abbreviate X_v and X_A by v and A, respectively. For three pairwise disjoint subsets $A, B, C \subseteq V$, write $A \perp\!\!\!\perp B \mid C\,[P]$ to indicate that X_A and X_B are conditionally independent (CI) given X_C under P. A graphical Markov model (GMM) uses a graph $G \equiv (V, E)$ with vertex set V and edge set E (either undirected, directed, or both) to specify a Markov property, i.e., a collection of conditional independences (CIs), among the random variates X_v, $v \in V$.

First we briefly review graph-theoretic terminology—see [F] (1990), [L] (1996), or [AMP] (1996, 1997a,b, 1998a) for more detail. For any subset $A \subseteq V$, the *induced subgraph* $G_A \equiv (A, E_A)$ is the graph with vertex set A and all edges involving only vertices in A.

A *path* of length $n \geq 1$ from v to w in G is a sequence $(v_0, v_1, \ldots, v_n)$ of $n+1$ distinct vertices such that $v_0 = v$, $v_n = w$, and $v_{i-1} \cdots v_i$ for all $i = 1, \ldots, n$, where "$\cdots$" indicates either — or $\rightarrow$. An *n-cycle* is a path of length $n \geq 3$ with the modification that $v_n = v_0$. If $v_{i-1} \rightarrow v_i \in G$ for (all) (at least one) (no) i, the path/cycle is called (*directed*), (*semi-directed*), (*undirected*). If G is a UG (digraph), all paths are undirected (directed). An n-cycle is *chordless* if no two nonconsecutive (mod n) vertices are linked in G.

A UG $G \equiv (V, E)$ is *chordal* if every n-cycle with $n \geq 4$ possesses a *chord*, that is, an edge between two nonconsecutive (mod n) vertices. A UG is chordal iff it is *decomposable*—cf. Lauritzen et al. (1984, Theorem 2), Whittaker (1990, Proposition 12.4.2), or [L] (1996, Proposition 2.5).

The UDG G is *connected* if, for every distinct $v, w \in V$, there is a path between v and w in G. A subset $A \subseteq V$ is *connected* in G if G_A is connected. The maximal connected subsets are called the *connected components* of G, and V can be uniquely partitioned into the disjoint union of the connected components of G. For pairwise disjoint subsets $A(\neq \emptyset)$, $B(\neq \emptyset)$, and S of V, A and B are *separated* by S in the UG G if all paths in G between A and B intersect S. Note that if $S = \emptyset$, then A and B and separated by S in G if and only if there are *no* paths connecting A and B in G. In this case, A and B are separated by *any* subset S disjoint from A and B.

A graph is called *adicyclic* if it contains no semidirected cycles. An adicyclic graph is commonly called a *chain graph* (CG). An *acyclic digraph* (ADG) is a digraph that contains no directed cycles. Thus, UGs and ADGs are special cases of CGs.

Next we recall the local and global Markov properties for ADGs and UGs, respectively. The following abbreviations for standard graph-theoretic terms are used: parents = pa, neighbors = nb, closure = cl, boundary = bd, nondescendants = nd (see [L], 1996, for definitions).

DEFINITION 1: (The local Markov property for ADGs.) Let $D \equiv (V, E)$ be an ADG. A probability measure P on $\mathbf{X}$ is said to be *local D-Markovian* if

$$v \perp\!\!\!\perp (\mathrm{nd}_D(v) \setminus \mathrm{pa}_D(v)) | \mathrm{pa}_D(v) \, [P] \quad \forall v \in V$$

DEFINITION 2: (The global Markov property for UGs.) Let $G \equiv (V, E)$ be a UG. A probability measure P on $\mathbf{X}$ is said to be *global G-Markovian* if $A \perp\!\!\!\perp B \mid S\,[P]$ whenever S separates A and B in G.

We have specified these two Markov properties because of their role in defining the Markov properties for CGs in Section 4.2. In fact, both local and global properties exist for ADGs and UGs (and for CGs). The local and global properties are equivalent for ADGs, but not in general for UGs unless further restrictions are imposed on P, e.g., if P admits a strictly positive density. Furthermore, for ADGs there exists an apparently weaker, but in fact equivalent, variant of the local property that depend on a total ordering (≡ "well-numbering") of the vertices rather than on descendant relationships. As a referee has noted, global properties are generally difficult to verify, while local properties are generally inadequate for inference—the complete arsenal of Markov properties is required for applications. We refer the reader to [L] (1996) for details.

In Sections 3.1–3.6 we survey current research on several overlapping topics involving ADG models and the related class of *lattice conditional independence* (LCI) models.

3.1 Bayesian Model Selection, Markov Equivalence, and Essential Graphs for ADG Models

As noted above, ADG models provide an elegant framework for Bayesian analysis (Spiegelhalter and Lauritzen, 1990). Much applied statistical work involving ADG models has adopted a Bayesian perspective: experts specify a prior distribution on competing ADG models, and then these prior distributions combine with likelihoods (typically integrated over parameters) to give posterior model probabilities. Model selection algorithms then seek out the ADG models with highest posterior probability, and subsequent inference proceeds conditionally on these selected models (Cooper and Herskovits, 1992; Buntine, 1994; Heckerman et al., 1994; Madigan and Raftery, 1994). Non-Bayesian model selection methods are similar, replacing posterior model probabilities by, for example, penalized maximum likelihoods (Chickering, 1995).

Heckerman et al. (1994) highlighted a fundamental problem with this general approach. Because several different ADGs may determine the same statistical model, i.e., may determine the same set of CI restrictions among a given set of random variates, the collection of all possible ADGs for these variates naturally coalesces into one or more classes of Markov-equivalent ADGs, where all ADGs within a Markov-equivalence class determine the same statistical model. For example, the first three graphs in Figure 1 are Markov equivalent: each specifies the single CI $X_b \perp\!\!\!\perp X_c | X_a$, abbreviated as $b \perp\!\!\!\perp c | a$. The fourth graph specifies the independence $b \perp\!\!\!\perp c$.

$b \rightarrow a \rightarrow c \qquad b \leftarrow a \rightarrow c \qquad b \leftarrow a \leftarrow c \qquad b \rightarrow a \leftarrow c$

Figure 1 Four acyclic digraphs with the vertex set $V \equiv \{a, b, c\}$

Model selection algorithms that ignore Markov equivalence face three basic difficulties.

1. Repeating analyses for equivalent ADGs leads to significant computational inefficiencies: in some cases, the Markov equivalence class $[D]$ containing a given ADG D may be superexponentially large.
2. Ensuring that equivalent ADGs have equal posterior probabilities severely constrains prior distributions.
3. Weighting individual ADGs in Bayesian model averaging procedures to achieve specified weights for all Markov-equivalence classes is impractical without an explicit representation of these classes.

Treating each Markov-equivalence class as a single model would overcome these difficulties.

Pearl (1988) and Verma and Pearl (1990, 1992) showed that two ADGs are Markov equivalent iff they have the same *skeleton* (≡ underlying UG) and same *immoral* configurations: $a \rightarrow c \leftarrow b$, a and b nonadjacent—see also [AMP], 1997b. [AMP] (1997b) have shown that for every ADG D, the equivalence class $[D]$ can be uniquely represented by a *single* Markov-equivalent chain graph $D^* := \cup\{D' | D' \in [D]\}$, the *essential graph*

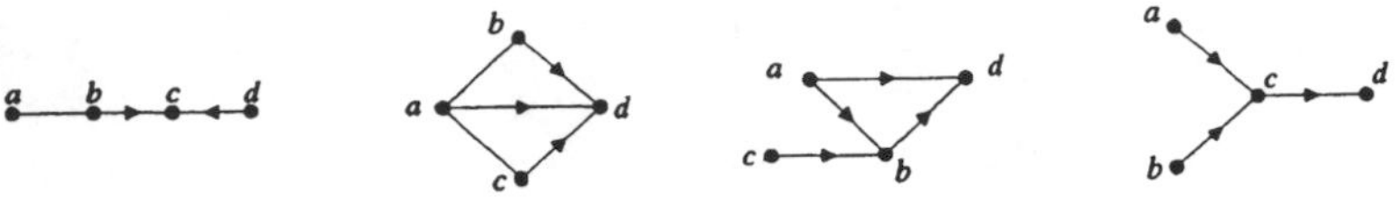

Figure 2 Four essential graphs D^* for 4 variables

associated with $[D]$. (The essential graph was called the *completed pattern* in Verma and Pearl, 1990.) Here, the *union* $\cup$ of ADGs with the same skeleton is again a graph with the same skeleton, but any directed edge ($\equiv$ arrow) that occurs with opposite orientations in two ADGs is replaced by an undirected edge ($\equiv$ line) in the union. Thus *an arrow occurs in the essential graph* D^* *iff it occurs with the same orientation in every* $D' \in [D]$. Clearly, every arrow involved in an immorality in D is essential, but other arrows may be essential as well, for example, the arrow $a \rightarrow d$ in the second graph in Figure 2. (Pearl and Verma, 1991; Spirtes et al., 1993; Chickering, 1995 noted that, under certain additional assumptions, the *essential arrows* of an ADG D, i.e., those directed edges of D that occur in D^*, may indicate causal influences.)

In principle, therefore, one can conduct model selection and model averaging over the reduced space of essential graphs, rather than over the larger space of individual ADGs. However, since $[D]$ can be superexponentially large (a complete graph admits $n!$ acyclic orderings) direct calculation of D^* is computationally infeasible. To overcome this obstacle, [AMP] (1997b) have given the following explicit characterization of essential graphs.

THEOREM 1: A graph G is D^* for some ADG D iff G satisfies the following four conditions:

1. G is a chain graph;
2. for each chain component $\tau \in \mathcal{T}(G)$ (Section 4.2), the UG G_τ is chordal ($\equiv$ decomposable);
3. the configuration $a \rightarrow b — c$ does not occur as an induced subgraph of G;
4. each arrow $a \rightarrow b$ in G occurs in at least one "strongly protected" configuration.

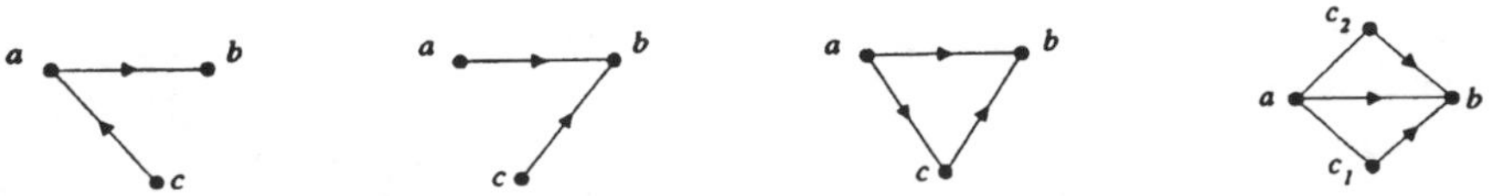

[AMP] (1997b) applied Theorem 1 to obtain a polynomial-time algorithm for constructing D^* from D and to establish thc irreducibility of certain Markov chains used for Monte Carlo search procedures over the space of essential graphs. (Computationally efficient algorithms for D^* appear in Chickering (1995) and Meek (1995a).) This approach yields more efficient model selection and model averaging procedures for ADG models, based on essential graphs. Such procedures were implemented by Madigan et al. (1996), who suggested that graphical modelers, both Bayesian and non-Bayesian, should focus their attention on essential graphs rather than on the larger but equivalent class of all ADGs (see also Chickering, 1996).

Four examples of essential graphs D^* for four variables are shown in Figure 2. In the first example, D can be taken to be the ADG $a \to b \to c \leftarrow d$; in the second example, D can be taken to be the ADG with additional arrows $a \to b$ and $a \to c$. In the third and fourth examples, $D = D^*$. [AMP] (1997b) gives all 185 essential graphs for four variables, compared with 543 ADGs; for five variables, these counts are 8782 and 29,281, respectively. The asymptotic behavior of these counts and their ratio as the number of variables $\to \infty$ is under investigation.

3.2 Hierarchical Partition Models for Bayesian Analysis of ADG Models

Spiegelhalter and Lauritzen (1990) introduced hyper-Dirichlet distributions as natural conjugate priors in ADG models for discrete data. Independently, Cooper and Herskovits (1992) and Buntine (1994) developed similar Bayesian frameworks. A critical assumption made implicitly or explicitly by these and other authors is that the various parameters that comprise the ADG model are independent a priori. While few would arguc that this assumption reflects any expert's beliefs, it has led to efficient learning algorithms.

Recently, however, Geiger and Heckerman (1995) showed that the assumption of parameter independence (in addition to other standard assumptions) is extraordinarily restrictive in that it is satisfied *only* by the hyper-Dirirchlet prior distribution and, furthermore, places severe constraints on the parameters of this prior. One possible approach to relaxing the parametric independence assumption is based on the hierarchical partition models of Consonni and Veronese (1995). Here the parameters are partitioned into a priori exchangeable subsets. A Markov chain Monte Carlo algorithm permits averaging across the set of possible partitions. Sung (1996) has suggested that this approach (which admits parametric independence as a special case) provides improved out-of-sample predictive performance (see also Golinolli et al. (1999)). This approach should be investigated more fully and evaluated for some larger-scale applications. In particular, hyper-Dirichlet prior distributions might be developed for essential graph models (cf. Section 3.1) and used to investigate hierarchical partition models.

3.3 Multivariate Linear Regression Models with ADG Covariance Structure

Early treatments of statistical inference for ADG models were given by Wermuth (1980), Kiiveri et al. (1984), and Shachter and Kenley (1989) for continuous multivariate distributions, by Wermuth and Lauritzen (1983) for discrete distributions, and by Lauritzen and Wermuth (1989) for distributions with both continuous and discrete components. [L] (1996) presents a thorough review.

Most recursive models studied to date have concentrated only on the ADG Markov condition, which takes the form of constraints on the covariance structure in the multivariate normal case, with only very simple structure (if any) allowed for the regression (≡ mean-value) subspace. [AP] (1998) have addressed the following question: under the Markov covariance constraints determined by an ADG D, what is the largest class of multivariate linear regression subspaces L, called *D-linear subspaces*, or simply *D-subspaces*, for which the likelihood function (LF) continues to admit a D-recursive factorization? Such *linear ADG models* will maintain the amenable properties of ADG models already noted. ([AP] (1998) have characterized the D-subspaces as those that remain invariant under the action of an algebra of block-triangular matrices with additional zero blocks determined by the structure of D.)

Conversely, for a given multivariate linear regression subspace L, [AP] (1998) have shown how to construct the most parsimonious ADG model $D(L)$ such that L remains a $D(L)$-subspace. (It turns out that $D(L)$ is a transitive ADG, and hence is equivalent to some LCI model—see Section 3.4.) For example, consider the following non-standard multivariate linear regression subspace L: a three-variate two-way MANOVA model with no interactions, no row effects for variable 1, no column effects for variable 2, and no row or column effects for variable 3. It can be shown that the ADG Markov model determined by $D(L)$ imposes the single CI condition $2 \perp\!\!\!\perp 3 \mid 1$, allowing this nonstandard MANOVA model to be analyzed by standard linear methods.

In general, this approach applies to the analysis of a family of *nonnested, dependent* univariate regression models, which includes Zellner's (1962) seemingly unrelated regressions (SUR) model as a special case. If L represents the combined multivariate regression subspace, then by imposing the minimal Markov constraints determined by the ADG $D(L)$, standard linear methods can be applied to the combined multivariate regression model, leading to explicit MLEs for the regression and covariance parameters. Perlman and Wu (1998c) have studied the efficiency of this approach compared with standard iterative methods for SUR models, which can be highly sensitive to the choice of starting point.

[AP] (1998) also derived the likelihood ratio test (LRT) for testing the validity of these ADG Markov assumptions. If this hypothesis is rejected, the constrained MLE obtained under the ADG model $D(L)$ may still provide an appropriate starting point for iterative algorithms to find the unconstrained MLE.

The class of D-subspaces can be enlarged to a suitably defined class of *D^*-subspaces*, where D^* is the essential graph (Section 3.1) associated with an ADG D. Since D^* is obtained by converting the nonessential arrows of D into lines, in general fewer constraints are imposed on D^*-subspaces than on D-subspaces, which may significantly extend the applicability of linear ADG models (see also Section 4.7).

3.4 ADG Models and Lattice Conditional Independence (LCI) Models

Lattice conditional independence (LCI) models were introduced by Andersson and Perlman [AP] (1988, 1993, 1995a,b) in the context of multivariate normal distributions, motivated by analogy with the lattice structure of balanced ANOVA designs (cf. Andersson (1990)). They showed that LCI models share the desirable statistical properties of ADG models—an LCI model admits a simple recursive density factorization, which in turn yields explicit MLEs and LRTs—and furthermore admits a transitive and proper group action leading to a classical integral representation of the density of the maximal invariant statistic. Furthermore, [AP] (1991, 1994) showed that LCI models are precisely suited to allow similar explicit likelihood analyses for nonmonotone missing data patterns and for nonnested dependent linear regressions (≡ generalized SUR models, cf. Section 3.3). Perlman and Wu (1998b) have investigated the efficiency of this approach compared with standard iterative methods for missing data, such as the EM algorithm.

A general LCI model $\mathbf{L}(\mathcal{K})$ specifies the CI relations $X_K \perp\!\!\!\perp X_L \mid X_{K\cap L}\ \forall\, K, L \in \mathcal{K}$, abbreviated as $K \perp\!\!\!\perp L \mid K \cap L$, among the variables $X_v, v \in V$, of a multivariate distribution, where $\mathcal{K}$ is a ring (≡ a finite distributive lattice) of subsets of the index set V. For example, in

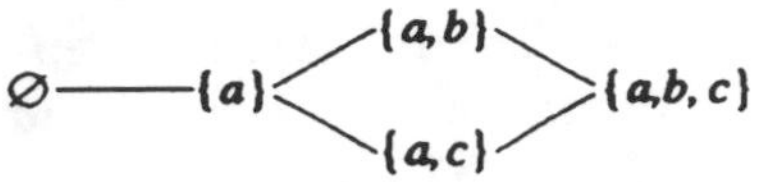

Figure 3   A ring $\mathcal{K}$ of subsets of $V \equiv \{a, b, c\}$

Figure 3 the LCI model $\mathbf{L}(\mathcal{K})$ specifies the CI $\{a, b\} \perp\!\!\!\perp \{a, c\} \mid \{a\}$, or equivalently, $b \perp\!\!\!\perp c \mid a$.

These similarities between LCI and ADG models prompted Andersson, Madigan, Perlman, and Triggs [AMPT] (1995, 1997c) to study the relation between these two classes. First they established that the *class* of LCI models coincides with the class of transitive ADG ($\equiv$ TADG) models. (An ADG D is *transitive* if $a \rightarrow b \rightarrow c \in D \Rightarrow a \rightarrow c \in D$.) Then they found the following theorem, which leads to a polynomial-time algorithm to determine whether a *specific* ADG D is Markov equivalent to some LCI model, that is, to determine whether the Markov equivalence class $[D]$ (possibly superexponentially large) contains at least one TADG.

THEOREM 2:   Let D be an ADG. Then D is Markov equivalent to some TADG iff none of the following five configurations occurs as an induced subgraph of the essential graph D^*.

$$
\begin{array}{l}
a \rightarrow b \\
| \quad\ \uparrow \qquad a \rightarrow b \rightarrow c \qquad a \leftarrow b\!-\!c \rightarrow d \qquad a\!-\!b\!-\!c \rightarrow d \qquad a\!-\!b\!-\!c\!-\!d. \\
c - d
\end{array}
$$

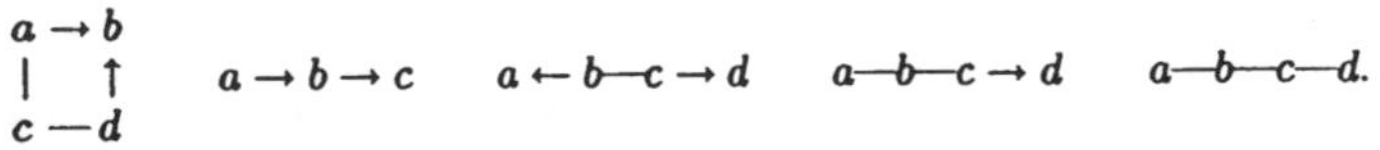

3.5   LCI Models for Nonnested Missing Data Patterns in Categorical Data

Perlman and Wu (1998a) have applied multinomial LCI models to the analysis of non-monotone missing data patterns in categorical data from an I-way contingency table. Consider a sample $X_1, \ldots, X_n \sim P$, a multinomial probability distribution on an I-way contingency table $C := \times(J_i \mid i \in I)$. Suppose that some or all $X_1, \ldots, X_n$ are observed incompletely, so that for each $j = 1, \ldots, n$, only a subvector $(X_j)_{K_j}$ is observed. The set $S := \{K_j \mid j = 1, \ldots, n\}$ of subsets of I is the *observed data pattern*. It is well known that if S is *monotone* (totally ordered under inclusion), then the likelihood function for the observed data factors recursively into a product of conditional multinomial LFs for a nested family of lower-dimensional tables. The MLEs of the conditional probabilities are obtained explicitly in terms of cell frequencies, and then used to reconstruct the MLEs of the unconditional cell probabilities (Little and Rubin, 1987).

If $\mathcal{S}$ is nonmonotone, however, then no such factorization of the LF is possible. In this case, it is possible to construct a parsimonious LCI model under which the LF does admit an amenable recursive factorization. Let $\mathcal{K} \equiv \mathcal{K}(\mathcal{S})$ be the ring of subsets of I generated by $\mathcal{S}$. The multinomial LCI model $M(\mathcal{K})$ is defined to be the set of all multinomial P on $\mathcal{C}$ such that, if $X \equiv (X_i | i \in I) \sim P$, then $X_K \perp\!\!\!\perp X_L \mid X_{K \cap L}\ \forall\ K, L \in \mathcal{K}$, where X_K denotes the projection of X onto the marginal table $\mathcal{C}_K := \times(J_i | i \in K)$. Similar to the multivariate normal case, if the LCI constraint $P \sim \mathbf{M}(\mathcal{K})$ is imposed, then the LF based on the observed data factors into the product of conditional saturated ($\equiv$ unconstrained) multinomial likelihoods for a (nonnested) family of lower-dimensional tables determined by $\mathcal{K}$. The MLEs of the (unconstrained) conditional cell probabilities are obtained explicitly in terms of ordinary cell frequencies, and then used to reconstruct the MLEs of the unconditional cell probabilities subject to the LCI constraint.

To assess the validity of the LCI model $\mathbf{M}(\mathcal{K})$, Perlman and Wu (1998a) have derived the LRT for testing a multinomial LCI model against the unconstrained alternative (or, more generally, for testing one such LCI model against another). Even if the LCI hypothesis is rejected, the constrained MLE obtained under $\mathbf{M}(\mathcal{K})$ promises to be a good starting point for the EM algorithm to compute the unconstrained MLE.

It may be possible to develop logit regression ADG models for categorical data analogous to the linear ADG models for continuous data discussed in Section 3.3. This will lead to a treatment of nonnested logit regression models for contingency tables analogous to that for continuous data in Section 3.3.

We refer the interested reader to related papers by Geng et al. (1996, 1997).

3.6 Covariance Models Combining Conditional Independence and Group Symmetry Constraints

Andersson, Brøns, and Jensen [ABJ] have developed an elegant general theory for the analysis of multivariate normal statistical models with covariance structure determined by group symmetry (GS), i.e., by invariance under a finite group of orthogonal transformations. They have shown that most of the classical patterned covariance structures (e.g., intraclass structure, compound symmetry, circular symmetry, the complex normal distribution) and testing problems (e.g., testing independence, testing equality of covariance matrices, cf. Anderson, 1984) are special cases of GS models and associated testing problems. At the same time they have exposed many new models (e.g., the quaternion normal distribution) and new testing problems (e.g., testing complex vs. real structure) covered by their general theory. Like ADG and LCI models, GS models allow explicit likelihood inference under normality; of course the forms of the MLE and LRT statistics are somewhat different. References in English include Andersson (1975, 1982), [ABJ] (1983), [AP] (1984), Jensen (1988), Perlman (1987), and Bertelsen (1989). Massam (1994) and Massam and Neher (1998) give interesting related results for distributions on symmetric cones.

Recently, Andersson and Madsen (1998) introduced the class of *GS–LCI models* for multivariate normal distributions which allow *both* GS and LCI restrictions on the covariance structure. This allows, for example, covariance models where certain blocks of variables satisfy CI restrictions across blocks and GS restrictions within blocks, or vice versa. Again, explicit likelihood inference is possible for a GS–LCI model, which again admits a transitive and proper group action yielding an integral representation for the distribution of the maximal invariant statistic. Massam and Neher (1998) have studied LCI models on symmetric cones, which are closely related to GS–LCI models.

It is possible to extend many of the results obtained for GS–LCI models to the larger class of GS–ADG models, although no longer will a transitive proper group action be present in general. It may also be possible to extend the classes of GS–LCI and GS–ADG models in such a way as to allow, for example, group symmetry that acts not only within blocks but also across blocks of variables. Hylleberg et al. (1993), Andersen et al. (1995), and Madsen (1995) treat several special cases of these generalizations.

4. GRAPHICAL MARKOV MODELS BASED ON ADICYCLIC GRAPHS ≡ CHAIN GRAPHS

4.1 An Alternative Markov Property for Chain Graphs

Lauritzen and Wermuth (1989) and Frydenberg [F] (1990) introduced a Markov property for CGs that generalized the classical Markov properties of both ADGs and UGs in a particular way. While this represents a significant and useful extension of the class of GMMs, the LWF interpretation of CGs has been criticized on several grounds, as described below. This led Andersson, Madigan, and Perlman [AMP] (1996, 1998a) to propose an alternative Markov property (AMP) for CGs that more nearly retains the desirable recursive nature of the Markov property for ADGs. Unlike the LWF property, the AMP property for a CG G can be shown to coincide with the Markov property satisfied by a certain block-recursive normal linear system naturally associated with G (a type of Cox–Wermuth (1996) concentration regression model (cf. Section 4.6), as illustrated in Figure 4.

To motivate the AMP Markov property, consider the simple CG in Figure 4, which represents a set of conditional independences satisfied by random variables $X_1, X_2, X_3, X_4 \equiv 1, 2, 3, 4$. The LWF Markov property for G reduces to the three conditions

$$1 \perp\!\!\!\perp 4 \mid 2, 3, \qquad 2 \perp\!\!\!\perp 3 \mid 1, 4, \qquad 1 \perp\!\!\!\perp 2 \tag{1}$$

whereas the AMP property can be expressed in terms of the three conditions

$$1 \perp\!\!\!\perp 4 \mid 2, \qquad 2 \perp\!\!\!\perp 3 \mid 1, \qquad 1 \perp\!\!\!\perp 2 \tag{2}$$

Although both Markov interpretations of the CG G may be useful for modeling (Cox and Wermuth [CW], 1993, p. 206), Cox (1993, p. 369) states that "While from one perspective this [condition (1)] is easily interpreted, it clearly does not satisfy the requirement of specifying a direct mode of data generation."

By contrast, a direct mode of data generation for the AMP is easily specified. Consider the recursive linear system

$$\begin{aligned} X_1 &= \epsilon_1 \\ X_2 &= \epsilon_2 \\ X_3 &= b_{31}X_1 + \epsilon_3 \\ X_4 &= b_{42}X_2 + \epsilon_4 \end{aligned} \tag{3}$$

where b_{31}, b_{42} are nonrandom scalars, and ϵ_1, ϵ_2, and (ϵ_3, ϵ_4) are mutually independent random errors with zero means. ϵ_1 and ϵ_2 have univariate normal distributions with arbitrary variances, and (ϵ_3, ϵ_4) has a *bivariate* normal distribution with arbitrary covariance matrix. Then (X_1, X_2, X_3, X_4) satisfies the AMP condition (2) for G, but not the LWF condition (1) when ϵ_3 and ϵ_4 have nonzero correlation.

Such a linear representation remains valid for a general CG G: the AMP for G is equivalent to the set of conditional independences (CIs) satisfied by

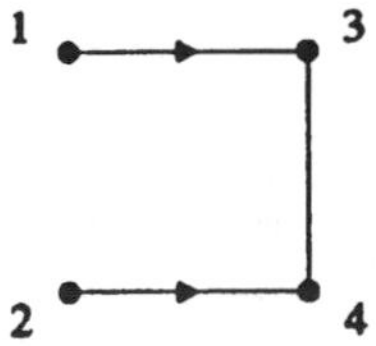

Figure 4 A simple chain graph G

the system of block-recursive linear equations naturally associated with G. (As noted in [L] (1996, p. 154), in general this does not hold for the LWF property; Theorem 5, below, describes those G for which LWF $\equiv$ AMP.) Each variate is given as a linear function of its parents in the chain graph, together with a normal error term—see equation (8). The errors are independent across blocks, while within each block the errors are (possibly) correlated according to the undirected edges connecting them, as in a Gaussian UG model $\equiv$ "covariance selection model" (Dempster, 1972). The linear model constructed in this way differs from a standard linear structural equation model (SEM): a SEM model usually specifies zeroes in the covariance matrix for the error terms, while the covariance selection model sets to zero elements of the *inverse* error covariance matrix.

For a general CG G under the assumption of multivariate normality, both the LWF and AMP Markov properties imply that the joint distribution factors into a product of conditional normal distributions (see equations (4) and (5)), where each conditional distribution involves a regression matrix β and a conditional covariance matrix Λ. Under the AMP interpretation, the Markov conditions take the form of zero restrictions on certain elements of β and of Λ^{-1}, equivalent to a CW concentration regression CG model ([CW], 1993, p. 205). By contrast, under the LWF Markov property, zero restrictions are imposed on certain elements of $\Lambda^{-1}\beta$ (the natural parameters when the normal distribution is expressed as an exponential family) and of Λ^{-1}, a CW block regression CG model. Since regression coefficients (β) are directly interpretable as measures of dependence, the AMP formulation may be preferable in this regard.

Yet another striking contrast between the LWF and AMP Markov properties for the CG G in Figure 4 appears when the marginal distribution of (X_1, X_3, X_4) is considered. Under the AMP property, the marginal distribution inherits the independence $1 \perp\!\!\!\perp 4$ implied by equation (2). Under the LWF property, however, no CI holds between X_1 and X_4 in the marginal distribution, despite the absence of an edge between 1 and 4 in G.

For these reasons it is surprising that, with the notable exception of [CW] (1993, 1996), the study of Markov models for CGs has been limited to the LWF interpretation.

[CW] (1996, p. 34) introduce *joint-response models* for CGs under the assumption of multivariate normality. Their CGs G may have both solid and dashed lines and arrows, with the restriction that the lines in any chain component of G must be either all solid or all dashed, and all arrows entering any chain component must be either all solid or all dashed. The CI assumptions associated with solid vs. dashed lines/arrows are different: the Markov property for a CG with solid lines and dashed arrows (a CW concentration regression model) is equivalent to the AMP block-recursive

Markov property, while the Markov property for a CG with solid lines and solid arrows (a CW block regression model) is equivalent to the LWF block-recursive Markov property.

[CW] (1996, p. 40) note that whereas the global Markov properties of block regression CG models are well known, "no similar results are currently available" for any other cases of their joint-response CG Markov models, including their concentration regression CG models.The AMP global Markov property gives the first such result (Theorem 4).

4.2 The LWF and AMP Block-Recursive and Global Markov Properties for Chain Graphs

Hereafter, $G \equiv (V, E)$ denotes a general chain graph (CG). We shall formulate a block-recursive version of the LWF Markov property associated with G, and then establish its equivalence to the original LWF global Markov property given by Frydenberg [F] (1990). It is this block-recursive property, not the global property, that suggests our new AMP block-recursive Markov property, which then leads to the new AMP global Markov property.

Define the following binary relations on V:

$$v \leq w \Longleftrightarrow \exists \text{ a path in } G \text{ from } v \text{ to } w \in G, \text{ or } v = w$$
$$v \ll w \Longleftrightarrow \exists \text{ a fully directed path in } G \text{ from } v \text{ to } w \in G, \text{ or } v = w$$
$$v \sim w \Longleftrightarrow \exists \text{ a fully undirected path in } G \text{ from } v \text{ to } w \in G, \text{ or } v = w.$$

A subset $A \subseteq V$ is *anterior* if $v \in A$ whenever $v \leq a$ for some $a \in A$. For any subset $A \subseteq V$, define $\mathrm{At}(A) :=$ the smallest anterior set containing A, so $\mathrm{At}(A) = \{v \in V \mid v \leq a \text{ for some } a \in A\}$. A subset $A \subseteq V$ is *ancestral* if $v \in A$ whenever $v \ll a$ for some $a \in A$. For $A \subseteq V$, define $\mathrm{An}(A) :=$ the smallest ancestral set containing A, so $\mathrm{An}(A) = \{v \in V \mid v \ll a \text{ for some } a \in A\}$.

As in [F] (1990), let $\mathcal{T} \equiv \mathcal{T}(G)$ denote the set of *chain components* of G, i.e., the equivalence classes in V induced by the equivalence relation $\sim$. Equivalently, if $\tau \subseteq V$, then $\tau \in \mathcal{T}$ iff the induced subgraph G_τ is a connected component of $G^\wedge$, the graph obtained from G by deleting all arrows (see Figure 5a,b). A subset $A \subseteq V$ is *coherent* if $v \in A$ whenever $v \sim a$ for some $a \in A$, that is, if A is a union of chain components of G. For any subset $A \subseteq V$, define $\mathrm{Co}(A) :=$ the smallest coherent set containing A, so $\mathrm{Co}(A) = \{v \in V \mid v \sim a \text{ for some } a \in A\} = \cup(\tau \in \mathcal{T} \mid \tau \cap A \neq \emptyset)$.

The chain components of G themselves comprise the vertices of the ADG $\mathcal{D}(G) \equiv (\mathcal{T}(G), \mathcal{E}(G))$, where an arrow connects two chain components in $\mathcal{D}$ iff at least one arrow connects vertices of these chain components in G (see

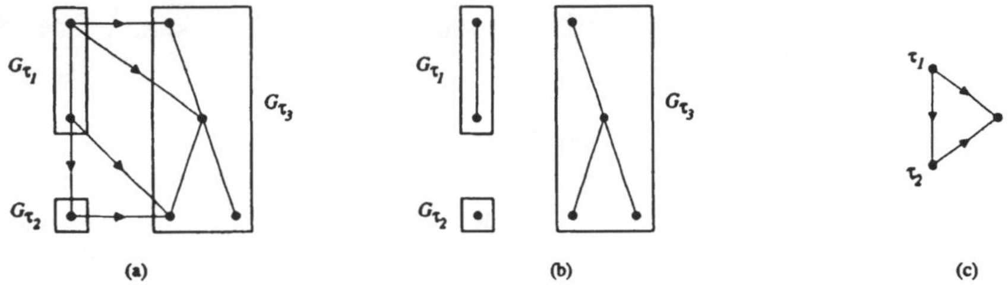

Figure 5 (a) A chain graph G with $\mathcal{T}(G) = \{\tau_1, \tau_2, \tau_3\}$. (b) The UG $G^{\wedge}$. (c) The ADG $\mathcal{D}(G)$

Figure 5c). Our block-recursive version of the LWF Markov property is naturally expressed in terms of the local ($\equiv$ global) Markov property of $\mathcal{D}$, the global Markov properties of the UGs G_τ, $\tau \in \mathcal{T}$, and a critical third property which, when suitably modified, yields the AMP block-recursive Markov property.

DEFINITION 3: Let $G \equiv (V, E)$ be a chain graph and let $\mathcal{D} = \mathcal{D}(G)$. A probability measure P on $\mathbf{X}$ is said to be *LWF block-recursive G-Markovian* if P satisfies the following three conditions:

1. $\forall \tau \in \mathcal{T},\ \tau \perp\!\!\!\perp (\mathrm{nd}_{\mathcal{D}}(\tau) \setminus \mathrm{cl}_{\mathcal{D}}(\tau)) | \mathrm{pa}_{\mathcal{D}}(\tau)\,[P]$, i.e., P is local $\mathcal{D}$-Markovian on $\mathbf{X}$;
2. $\forall \tau \in \mathcal{T}$, the conditional distribution $P_{\tau|\mathrm{pa}_{\mathcal{D}}(\tau)}$ is global G_τ-Markovian on $\mathbf{X}_\tau$;
3. $\forall \tau \in \mathcal{T}, \forall \sigma \subseteq \tau,\quad \sigma \perp\!\!\!\perp (\mathrm{pa}_{\mathcal{D}}(\tau) \setminus \mathrm{pa}_G(\sigma)) | \mathrm{bd}_G(\sigma)\,[P]$.

The set of all LWF block-recursive G-Markovian P on $\mathbf{X}$ is denoted by $\mathcal{P}^b_{\mathrm{LWF}}(G; \mathbf{X})$.

A *complex* in G is an induced subgraph of the form $a \to c_1 — \cdots — c_n \leftarrow b$, $n \geq 1$. For $n = 1$, a complex reduces to an *immorality*. A complex is *moralized* by adding the line $a—b$. The *moralized* graph G^{m} is the UG obtained by moralizing all complexes and converting all arrows to lines.

DEFINITION 4: Let $G \equiv (V, E)$ be a chain graph. A probability P on $\mathbf{X}$ is said to be *LWF global G-Markovian* if $A \perp\!\!\!\perp B \mid S\,[P]$ whenever S separates A and B in $(G_{\mathrm{At}(A \dot{\cup} B \dot{\cup} S)})^m$. The set of all LWF global G-Markovian P on $\mathbf{X}$ is denoted by $\mathcal{P}^g_{\mathrm{LWF}}(G; \mathbf{X})$.

THEOREM 3: ([AMP] (1998a)). Let G be a chain graph. Then

$$\mathcal{P}^{\mathrm{b}}_{\mathrm{LWF}}(G; \mathbf{X}) = \mathcal{P}^{\mathrm{g}}_{\mathrm{LWF}}(G; \mathbf{X}) =: \mathcal{P}_{\mathrm{LWF}}(G; \mathbf{X}) \quad \forall \mathbf{X}$$

When applied to the CG in Figure 4, statement 1 of Definition 3 implies that $1 \perp\!\!\!\perp 2$, statement 2 is vacuous, while statement 3 yields the first two LWF Markov conditions in statement 1. To obtain instead the first two AMP Markov conditions in statement 2, we need only modify statement 3 by deleting the subset $\mathrm{nb}_G(\sigma)$ from the conditioning set $\mathrm{bd}_G(\sigma) \equiv \mathrm{pa}_G(\sigma) \dot{\cup} \mathrm{nb}_G(\sigma)$, as follows:

3.* $\forall \tau \in \mathcal{T}, \forall \sigma \subseteq \tau, \quad \sigma \perp\!\!\!\perp (\mathrm{pa}_{\mathcal{D}}(\tau) \setminus \mathrm{pa}_G(\sigma)) \mid pa_G(\sigma)\,[P]$

Statements 1, 2, and 3* constitute a block-recursive AMP for CGs which, unlike the LWF property, can be represented by a block-recursive linear system of normal random variates—see statement 3 and equation (8).

DEFINITION 5: Let $G \equiv (V, E)$ be a chain graph. A probability measure P on $\mathbf{X}$ is said to be *AMP block-recursive G-Markovian* if P satisfies statements 1, 2, and 3*. The set of all AMP block-recursive G-Markovian P on $\mathbf{X}$ is denoted by $\mathcal{P}^b_{\mathrm{AMP}}(G; \mathbf{X})$.

A *flag* in G is an induced subgraph of the form $a \to c—b$ or $a—c \leftarrow b$. A *triplex* is an ordered triple (a, c, b) that comprises either a flag or an immorality in G. This triplex is *augmented* by adding the line $a—b$ to the flag or immorality. A *2-biflag* in G is an induced subgraph of the form $a \to c—d \leftarrow b$. The 2-biflag is *augmented* by adding the lines $a—d$, $b—c$, *and* $a—b$. The *augmented* graph G^a is the UG obtained by augmenting all triplexes and 2-biflags and then converting all arrows to lines.

DEFINITION 6: Let $G \equiv (V, E)$ be a chain graph. A probability measure P on $\mathbf{X}$ is said to be *AMP global G-Markovian* if $A \perp\!\!\!\perp B \mid S\,[P]$ whenever S separates A and B in $(G_{\mathrm{An}(A\dot{\cup}B\dot{\cup}S)} \cup G^{\wedge}_{\mathrm{Co}(\mathrm{An}(A\dot{\cup}B\dot{\cup}S))})^a$. The set of all AMP global G-Markovian P on $\mathbf{X}$ is denoted by $\mathcal{P}^g_{\mathrm{AMP}}(G; \mathbf{X})$.

The AMP block-recursive and AMP global Markov properties are also equivalent.

THEOREM 4 ([AMP], 1998a): Let G be a chain graph. Then

$$\mathcal{P}^b_{\mathrm{AMP}}(G; \mathbf{X}) = \mathcal{P}^g_{\mathrm{AMP}}(G; \mathbf{X}) =: \mathcal{P}_{\mathrm{AMP}}(G; \mathbf{X}) \quad \forall X$$

For UGs and ADGs, the LWF and AMP Markov properties coincide. The simplest CG for which the LWF and AMP Markov properties differ is the flag $a \to c—b$. The LWF property for this graph is $a \perp\!\!\!\perp b \mid c$, while the AMP property is $a \perp\!\!\!\perp b$. The nonoccurrence of a flag is necessary and sufficient for these Markov properties to coincide.

THEOREM 5 ([AMP], 1998a): (i) If G has no flags, then

$$\mathcal{P}_{\text{LWF}}(G; \mathbf{X}) = \mathcal{P}_{\text{AMP}}(G; \mathbf{X}) \quad \forall \mathbf{X}$$

(ii) If G has at least one flag, then for every $\mathbf{X}$ such that $\mathbf{X}_v$ admits a non-degenerate probability measure for each $v \in V$,

$$\mathcal{P}_{\text{LWF}}(G; \mathbf{X}) \setminus \mathcal{P}_{\text{AMP}}(G; \mathbf{X}) \neq \emptyset$$
$$\mathcal{P}_{\text{AMP}}(G; \mathbf{X}) \setminus \mathcal{P}_{\text{LWF}}(G; \mathbf{X}) \neq \emptyset$$

The following questions, suggested by Theorem 5, are also investigated by [AMP] (1998a): for which CGs G does there exist G' such that $\mathcal{P}_{\text{LWF}}(G'; \mathbf{X}) = \mathcal{P}_{\text{AMP}}(G; \mathbf{X})$, and such that $\mathcal{P}_{\text{AMP}}(G'; \mathbf{X}) = \mathcal{P}_{\text{LWF}}(G; \mathbf{X})$?

4.3 The Pathwise Separation Criterion for AMP Chain Graphs

The standard computational method used to identify valid CIs in ADG models is based on a pathwise separation criterion, called *d-separation*, introduced by Pearl (1988), and then elegantly applied by Geiger and Pearl (1988) to establish the completeness of the global Markov property for ADGs (completeness guarantees that *d-separation* uncovers *all* conditional independences implied by the ADG model). Bouckaert and Studený (1995) and Studený and Bouckaert (1998) have generalized this to *c-separation*, a more complicated criterion for identifying valid CIs in LWF CG models, and then applied this to establish the completeness of the LWF global Markov property for CGs. Perlman et al. (1998) have obtained a new pathwise separation criterion for AMP CG models that is simpler than c-separation and more closely resembles the d-separation criterion, due again to the fact that triplexes involve only triples whereas complexes can be of arbitrary length. This new criterion can be used to establish the completeness of the AMP global Markov property for CGs by an argument similar to that of Geiger and Pearl (1988).

4.4 Markov Equivalence and Essential Graphs for AMP Chain Graph Models

Like ADGs (cf Section 3.1), two or more AMP CGs may determine the same Markov model, leading to computational inefficiency in model selection and to inappropriate specification of prior distributions in Bayesian model averaging. As for ADGs, the first step toward resolution of this problem must be to find a graphical criterion for the AMP Markov equivalence of CGs (see Theorem 6 below), complementing the results of [F] (1990) and [AMP] (1997a) characterizing LWF Markov equivalence.

Two CGs, $G_1 = (V, E_1)$ and $G_2 = (V, E_2)$, with the same vertex set V are called *AMP (or LWF) Markov equivalent* if $\mathcal{P}_{\text{AMP}}(G_1; \mathbf{X}) = \mathcal{P}_{\text{AMP}}(G_2; \mathbf{X})$ (or $\mathcal{P}_{\text{LWF}}(G_1; \mathbf{X}) = \mathcal{P}_{\text{LWF}}(G_2; \mathbf{X})$) for every product space $\mathbf{X}$ indexed by V. [F] (1990) and [AMP] (1997a) show that two CGs are LWF Markov equivalent iff they have the same skeleton and the same complexes. AMP Markov equivalence is characterized as follows:

THEOREM 6 ([AMP], 1998a): Two chain graphs $G_1 \equiv (V, E_1)$ and $G_2 \equiv (V, E_2)$ with the same vertex set V are AMP Markov equivalent iff they have the same skeleton and the same triplexes.

This condition for AMP Markov equivalence of CGs is closer to that for ADG Markov equivalence than is the condition for LWF Markov equivalence of CGs, in the sense that both immoralities and triplexes involve only three vertices, while complexes can involve arbitrarily many vertices. This suggests that determining AMP Markov equivalence requires lower computational complexity than does LWF Markov equivalence.

As was done for LWF Markov equivalence by [AMP] (1997a), this characterization of AMP Markov equivalence can be applied to determine when a given CG is AMP Markov equivalent to some UG, to some ADG, or to some decomposable UG, and, most importantly, to define, characterize, and construct the AMP *essential graph* G^*. We propose the following definition: G^* has the same skeleton as G, and an arrow occurs in G^* iff it occurs with the same orientation in at least one member G' of the AMP Markov equivalence class $[G]$ but with the opposite orientation in *no* $G'' \in [G]$.

When $G = D$ for some ADG D, then $G^* = D^*$, as defined in Section 3.1. By contrast, G_∞, the *largest CG* in the LWF Markov equivalence class $[G]_{\text{LWF}}$, introduced by [F] (1990) and used by Studený (1996, 1997) as a unique representative for $[G]_{\text{LWF}}$, need *not* reduce to D^* when $G = D$. The fourth ADG in Figure 2 provides an example: here $D^* = D$, but D_∞ replaces the essential arrow $c \to d$ by the line $c—d$. Thus, if essential arrows do indicate causal relations (Section 2.1), then the AMP essential graph G^* may represent more completely than G_∞ the set of causal relations determined by G.

Remarkably, *G^* may exhibit essential (= causal?) arrows that do not appear in G itself, nor in G_∞.* For example, if $G = a \to b—c$ then $G^* = a \to b \leftarrow c$, while $G_\infty = a—b—c$ (by contrast, if $G = a \to b \to c$ then $G^* = G_\infty = a—b—c$). More dramatically, *even if G (and therefore G_∞) is a fully undirected graph, G^* may possess essential (= causal?) arrows:*

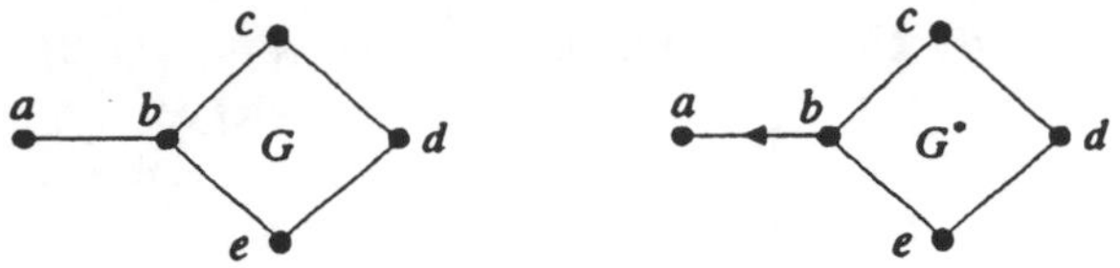

It is of interest to characterize those CGs in which this phenomenon occurs and to investigate its implications (if any) for causal inference.

Like the essential graph D^* for an ADG D, the AMP essential graph G^* will play a fundamental role for inference, model selection, and model averaging for AMP CGs. For these purposes, the results of [AMP] (1997b) and Madigan et al. (1996) can be extended to AMP CGs, in particular the characterization in Theorem 1 can be extended from essential graphs for ADGs to essential graphs for AMP CGs. This will lead to polynomial-time algorithms for constructing G^* from G and to irreducible Markov chain Monte Carlo algorithms for model searches over the space of AMP essential graphs (see also Section 4.5).

4.5 Bayesian and Likelihood Analysis of AMP and LWF Chain Graph Models

Methods for the implementation of Bayesian and likelihood analysis of LWF and AMP CG models for both discrete and continuous data are currently under study. Both exact and Monte Carlo Bayesian analysis will require the formulation of appropriate parameterizations and corresponding prior distributions, and no obvious analogy to the hyper-Dirichlet distributions of Dawid and Lauritzen (1993) exists. Recently, Roverato and Whittaker (1996) and Dellaportas et al. (1996) have proposed prior distributions for nondecomposable UG models, although it is not clear that these prior distributions would be amenable to expert elicitation. Their suggestions may nonetheless be useful in the CG context.

For both LWF and AMP CG models, model selection and Bayesian model averaging algorithms should be implemented and evaluated. Buntine (1995) has considered model selection for LWF models, while the Occam's Window algorithm of Madigan and Raftery (1994) and the Markov chain Monte Carlo model composition algorithm of Madigan and York (1995) may prove useful for both AMP and LWF models. The MCMC methods of Madigan et al. (1996) could be extended from ADG essential graphs D^* to AMP essential graphs G^*. Algorithms based on approximate Bayes factors may also prove feasible.

4.6. Equivalence of AMP Chain Graph Models to Block-Recursive Normal Linear Systems

If $P \in \mathcal{P}_{\mathrm{LWF}}(G; \mathbf{X})$ or $\mathcal{P}_{\mathrm{AMP}}(G; \mathbf{X})$ and P admits a probability density f wrt a σ-finite product measure on $\mathbf{X}$, then by statement 1 of Definitions 3 and 5, f admits the recursive factorization (Lauritzen et al., 1990)

$$f(x) = \prod (f(x_\tau | x_{\mathrm{pa}_\mathcal{D}(\tau)}) \mid \tau \in \mathcal{T}), \qquad x \in \mathbf{X} \tag{4}$$

This shows that the joint distribution of $X \equiv (X_v | v \in V)$ is determined by the family of conditional distributions $(X_\tau | X_{\mathrm{pa}_\mathcal{D}(\tau)} | \tau \in \mathcal{T})$.

Now assume in addition that f is a multivariate normal density. Then each of these conditional distributions has the form of a multivariate linear regression model

$$X_\tau | X_{\mathrm{pa}_\mathcal{D}(\tau)} \sim \mathcal{N}_\tau(\beta_\tau X_{\mathrm{pa}_\mathcal{D}(\tau)}, \Lambda_\tau) \tag{5}$$

where Λ_τ is the (nonsingular) $\tau \times \tau$ conditional covariance matrix of X_τ given $X_{\mathrm{pa}_\mathcal{D}(\tau)}$ and β_τ is the $\tau \times \mathrm{pa}_\mathcal{D}(\tau)$ matrix of regression coefficients. Statement 2 of Definitions 3 and 5 imposes the restriction that $\Lambda_\tau \in P(G_\tau)$, where $P(G_\tau)$ is the set of all Λ_τ that satisfy the *covariance selection* constraint (Dempster, 1972) determined by the UG G_τ

$$u, v \in \tau, \qquad u, v \text{ not adjacent in } G_\tau \;\Rightarrow\; (\Lambda_\tau^{-1})_{uv} = 0 \tag{6}$$

If $P \in \mathcal{P}_{\mathrm{AMP}}(G; \mathbf{X})$, statement 3* of Definition 5 imposes the restriction that $\beta_\tau \in \mathrm{B}_\tau(G)$, where $\mathrm{B}_\tau(G)$ is the set of all β_τ that satisfy

$$u \in \tau, \; v \in \mathrm{pa}_\mathcal{D}(\tau) \setminus \mathrm{pa}_G(u) \;\Rightarrow\; (\beta_\tau)_{uv} = 0 \tag{7}$$

Since this is a linear restriction, the conditional distribution in equation (5) retains the form of a multivariate linear regression model, a generalization of the *seemingly unrelated regression* (SUR) model (Zellner, 1962) with covariance selection constraints. The parameters β_τ and Λ_τ can be estimated by combining standard methods for SUR and covariance selection models.

By contrast, if $P \in \mathcal{P}_{\mathrm{LWF}}(G; \mathbf{X})$ then statement 3 of Definition 3 cannot be expressed in terms of zero restrictions on β_τ, but rather on $\Lambda_\tau^{-1}\beta_\tau$. This discussion shows that it is the AMP, rather than the LWF, Markov property for a CG G that is satisfied by the following block-recursive normal linear system naturally associated with G: G:

$$X_\tau = \beta_\tau X_{\mathrm{pa}_\mathcal{D}(\tau)} + \epsilon_\tau, \qquad \tau \in \mathcal{T} \tag{8}$$

Here, $\epsilon_\tau \sim \mathcal{N}_\tau(0, \Lambda_\tau)$ with $\Lambda_\tau \in \mathrm{P}(G_\tau)$, $\beta_\tau \in \mathrm{B}_\tau(G)$, and $(\epsilon_\tau | \tau \in \mathcal{T})$ are mutually independent. The model in equation (8) is a CW *concentration regression CG model* ([CW], 1996).

5. LINEAR STRUCTURAL EQUATION MODELS (SEM)

In a linear structural equation model (SEM), the variables are divided into two disjoint sets: substantive variables (**Y**) and error variables (ϵ). Associated with each substantive variable Y_i there is a unique error term ϵ_i. A linear SEM contains a set of linear equations, expressing each substantive variable Y_i as a linear function of the other substantive variables and of ϵ_i. In vector notation,

$$\mathbf{Y} = \Gamma \mathbf{Y} + \epsilon$$

where $\gamma_{ii} = 0$. In any given structural model, some off-diagonal entries in Γ may also be fixed at zero, depending on the form of the structural equations. If, under some rearrangement of the rows, Γ can be placed in lower triangular form, the system of equations is said to be *recursive*, otherwise it is said to be *nonrecursive*.

In addition, a SEM model specifies a multivariate normal distribution over the error terms: $\epsilon \sim \mathcal{N}(0, \Sigma)$. In any particular model, some off-diagonal (σ_{ij}) entries in Σ may be specified to be zero. If Σ is not diagonal, then the model is said to have *correlated errors*. If $(\mathbf{I} - \Gamma)$ is nonsingular, then the SEM determines a joint distribution over the substantive variables, given by the *reduced form* equations

$$\mathbf{Y} = (\mathbf{I} - \Gamma)^{-1} \epsilon$$

The free parameters of a SEM model fall into three categories: (i) the *linear coefficients* γ_{ij} such that γ_{ij} is not fixed at 0; (ii) the *error variances* $\sigma_{ii} \geq 0$; (iii) the *error covariances* σ_{ij} such that σ_{ij} is not fixed at 0. We restrict attention to the domain of the space of free parameters for which the model has a reduced form, and for which all variances and partial variances are finite and positive, and all partial correlations among the substantive variables are well defined.

In general the linear coefficients will *not* be regression coefficients, since it is possible for ϵ_i to be correlated with the other substantive variables occurring with nonzero coefficients in the equation for Y_i. However, in the relatively simple case of a recursive SEM with uncorrelated errors, the structural equations will be regression equations. It is also possible that two different sets of values of the free parameters may determine the same distribution over the substantive variables. In such a case the model is said to be *underidentified.* We do *not* restrict ourselves to identified models in what follows.

The following is an example of a nonrecursive SEM with correlated errors:

$$\begin{aligned} Y_1 &= \epsilon_1 \\ Y_2 &= \epsilon_2 \\ Y_3 &= \gamma_{31} Y_1 + \gamma_{34} Y_4 + \epsilon_3 \\ Y_4 &= \gamma_{42} Y_2 + \gamma_{43} Y_3 + \epsilon_4 \end{aligned} \qquad \Sigma = \begin{pmatrix} \sigma_{11} & \sigma_{12} & 0 & 0 \\ \sigma_{21} & \sigma_{22} & 0 & 0 \\ 0 & 0 & \sigma_{33} & 0 \\ 0 & 0 & 0 & \sigma_{44} \end{pmatrix} \tag{9}$$

5.1 Markov Properties of Path Diagrams for SEMs

There is a directed graph naturally associated with the set of structural equations in a SEM by the rule that there is an edge $b \to a$ in the graph if and only if $\gamma_{ab} \neq 0$, i.e., if the coefficient of Y_b in the structural equation for Y_a is not fixed at zero. Likewise, correlated errors have traditionally been represented by adding a bidirected edge $a \leftrightarrow b$ to the directed graph, if $\sigma_{ab} \neq 0$. The resulting graphical object containing directed and bidirected edges is known as a *path diagram*. Note that it is possible in a path diagram for there to be up to three edges between a given pair of variables, since it is possible for γ_{ab}, γ_{ba}, and σ_{ab} all to be free parameters in the model. Thus the path diagram corresponding to the SEM in equation (9) is

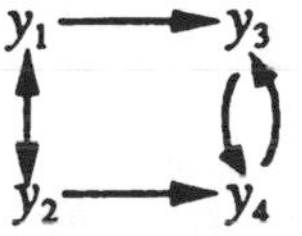

The path diagram for a *recursive* linear SEM with uncorrelated errors is an ADG. Pearl and Verma (1991) showed that in this case the SEM satisfies the local and global ADG Markov properties, and hence the d-separation criterion applied to the path diagram reveals the conditional independencies (CI) which hold for all values of the linear parameters. In general, however, path diagrams can differ from ADGs in two respects: directed cycles are allowed and bidirected edges may be present. Spirtes (1995) and Koster (1996) showed that nonrecursive SEMs with uncorrelated errors also satisfy the natural extension of the global Markov property (though not the local Markov property), and hence that d-separation may be applied to infer CIs entailed by the model. Spirtes et al. (1998) showed that (a natural extension of) d-separation can be applied to path diagrams with bidirected edges in order to infer the CIs entailed by SEMs with correlated errors.

These results show that as a calculus for calculating the CIs entailed by a SEM, d-separation is *sound*. It is also *complete*: if a pair of vertices a and b are d-connected conditional on some subset W of the other vertices in the path diagram, then there is some set of parameter values for which $Y_a \not\perp\!\!\!\perp Y_b \mid W$ (Geiger and Pearl, 1988; Spirtes et al., 1993; Meek, 1995b; Spirtes, 1995; Koster, 1996).

5.2. Markov Equivalence and Structural Underdetermination in SEMs

As with ADG and CG models (Sections 3.1, and 4.4), two different SEM path diagrams, embodying distinct structural hypotheses, may determine statistically equivalent models. This gives rise to one of the most serious criticisms of structural equation modeling: competing models that cannot be distinguished on the basis of the observational data, regardless of sample size. Thus we must examine the Markov equivalence of SEM models: when do two SEMs entail the same set of CIs, and what structural features are common to the path diagrams in a given Markov equivalence class? These questions directly address the issue of causal underdetermination, and their answers will facilitate the construction of efficient model specification algorithms that output a Markov equivalence class (or classes) of SEMs compatible with given data.

There is a further crucial complication concerning Markov-equivalent SEMs. It is often the case that only a subset of the variables occurring in the path diagram have actually been observed; those that have not been recorded are called *latent* variables. It is possible for two models, containing different sets of latent variables, to entail the same CIs on a set of observed variables that they share in common. Since an arbitrary number of latent variables may be postulated, it follows that there are an infinite number of different models in a Markov equivalence class of SEMs with latent variables. In spite of this, however, it is still possible to characterize such Markov equivalence classes in terms of invariant features.

Spirtes et al. (1998) show that any SEM model with correlated errors is Markov equivalent to a SEM without correlated errors but with extra latent variables (where the latent variables are either conditioned upon or marginalized). Thus, questions concerning the Markov equivalence classes of SEMs with latents *and* correlated errors can be reduced to questions about Markov equivalence classes of SEMs with latents but *without* correlated errors. For this reason, we shall restrict attention to SEM models whose path diagrams are directed graphs (DG), i.e., *without* bidirected edges, though possibly cyclic.

5.3 Characterizing Markov Equivalence of SEMs by Partial Ancestral Graphs

A directed edge $a \rightarrow b$ in a graph is usually defined as an ordered pair of vertices $\langle a, b \rangle$, and an undirected edge a—b as the pair $\{\langle a, b \rangle, \langle b, a \rangle\}$. An alternative way to define such edges is by pairs of vertices, each with associated endline symbols $-$ or $\succ$. Thus, under this scheme $a \rightarrow b$ would be defined as $\{\langle a, -\rangle, \langle b, \succ\rangle\}$, while a—b would be defined as $\{\langle a, -\rangle, \langle b, -\rangle\}$. We now extend this scheme to define a 3-*endline graph*, Ψ, as a graph containing three kinds of endline symbols: $-$, $\succ$, $\circ$, called *tails, heads, circles*, respectively:

$$\Psi \equiv (V, E), \qquad E \subseteq (V \times \{-, \succ, \circ\}) \times (V \times \{-, \succ, \circ\})$$

There are thus (up to symmetry) six distinct types of edge that can occur in a 3-endline graph:

o—o —o —— o—≻ —≻ ≺—≻

Richardson [R] (1996b) introduced the idea of a *partial ancestral graph* (PAG) Ψ for a directed graph (DG) $G \equiv (V, E)$, a (not necessarily unique) 3-*endline graph* that represents all invariant features common to the Markov equivalence class $[G]$ of SEMs whose path diagrams are DGs allowing latent variables (hence, allowing correlated errors). For $O \subseteq V$ ($O \equiv$ 'Observed'), let $\text{Cond}_G(O)$ denote the CIs among the variables O that are entailed by the global directed Markov property applied to G. Let Equiv(G, O) denote the class of DGs $G' \equiv (V', E')$ where $V' \supseteq O$ and such that the global directed Markov property applied to G' entails all and only the CIs in $\text{Cond}_G(O)$. We use "*" as a metasymbol indicating any of the endlines $-$, $\succ$, or $\circ$.

DEFINITION 7: A 3-endline graph Ψ is a *partial ancestral graph* (PAG) for the DG $G \equiv (V, E)$ and the observed set $O \subseteq V$ if Ψ has vertex set O and is such that:

1. there is an edge a*—*b in Ψ iff a and b are d-connected in G given all $W \subseteq O \backslash \{a, b\}$;
2. if there is an edge a—*b in Ψ, then $\forall G' \in$ Equiv(G, O), $a \ll_{G'} b$, i.e., a is an ancestor of b in G';
3. if there is an edge a≺—*b in Ψ, then $\forall G' \in$ Equiv(G, O), $a \not\ll_{G'} b$, i.e., a is not an ancestor of b in G';
4. if there is an edge ao—*b in Ψ then this asserts nothing about ancestor relationships in Equiv(G, O).

The preceding definitions are illustrated by the following three examples.

EXAMPLE 1: (a) A DG (in fact, ADG) G with $V = \{a, b, p\}$ and $O = \{a, b\}$, which entails no CIs over O; thus (b) $\text{Cond}_G(O) = \emptyset$; (c) Equiv$(G, O)$, the (infinite) class of DGs Markov equivalent to G over O; (d) a PAG Ψ for G over O (which says nothing about ancestor relations in Equiv(G, O)).

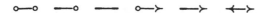

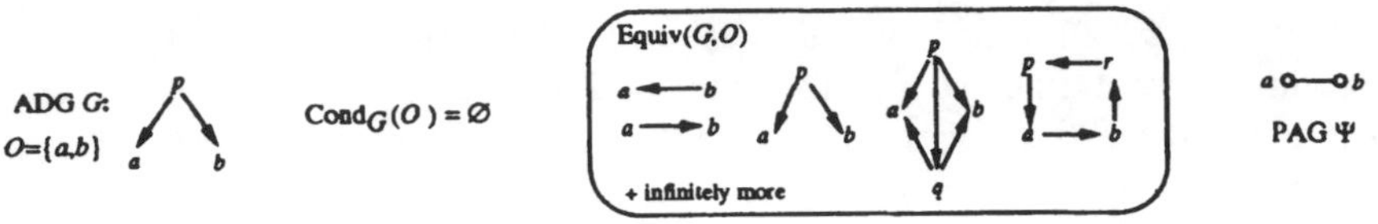

(a) (b) (c) (d)

EXAMPLE 2: (a) An ADG G with $V = \{a, b, c, d, q, r\}$ and $O = \{a, b, c, d\}$; (b) $\text{Cond}_G(O)$; (c) Equiv(G, O), the (infinite) class of DGs Markov equivalent to G over O; (d) a PAG Ψ for G over O. The heads at b indicate that b is not an ancestor of a or c in any DG $G' \in Equiv(G, O)$; likewise the heads at c indicate that c is not an ancestor of b or d in any DG $G' \in$ Equiv(G, O). (It follows that in every DG $G' \in$ Equiv(G, O), b and c have a common unmeasured ancestor.)

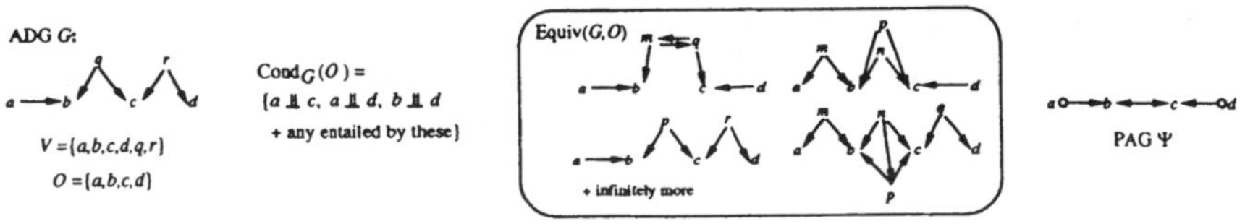

(a) (b) (c) (d)

EXAMPLE 3: (a) An ADG G with $V = O = \{a, b, c, d\}$; (b) $Cond_G(O)$; (c) $Equiv(G, O)$ the (infinite) class of DGs Markov equivalent to G over O; (d) a PAG Ψ for G over O. The heads at c indicate that c is not an ancestor of a or b in any DG $G' \in$ Equiv(G, O). The tail at c indicates that c is an ancestor of d in every DG $G' \in$ Equiv(G, O). (It follows that d is not an ancestor of c in any ADG $H' \in$ Equiv(G, O).)

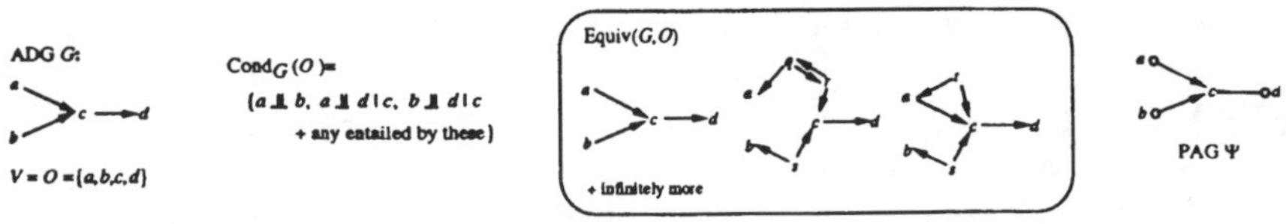

(a) (b) (c) (d)

5.4. Completeness of the Partial Ancestral Graph Representation of a SEM

Graphical representations display structural features of a model, possibly indicating a data generating process, and encode CIs entailed by the model. Ideally, a graphical representation of a *Markov equivalence class* of models should do the same, representing structural features common to all models in the class while encoding all CIs that hold in the class. These two goals are achieved for an ADG Markov equivalence class by the essential graph (Section 3.1); in fact by any *pattern* (Verma and Pearl, 1990), which has the same skeleton as the essential graph, but may include fewer directed edges. (Spirtes et al. (1993) developed the *PC algorithm* for generating patterns.)

By its definition, a PAG represents invariant features of its Markov equivalence class. However, are these invariant features sufficient to determine a *unique* Markov equivalence class? Is the PAG *complete* in this sense? In fact, the invariants given in the definition do not provide a sufficiently rich predicate language to distinguish among all Markov equivalence classes. With the addition of one extra symbol, however, representing the assertion *b is an ancestor of a or c in every graph in the Markov equivalence class*, the resulting PAG can distinguish between any two Markov equivalence classes of *recursive* SEMs, including latent variables and/or correlated errors (Spirtes and Verma, 1992). With the addition of a second additional symbol, the PAG can distinguish between any two Markov equivalence classes of *nonrecursive* SEMs *without* latent variables or selection bias ([R], 1996c; Richardson and Spirtes, 1998). It is conjectured that this set of symbols (the 3-endline graph plus two additional symbols) is sufficient for the full class of recursive or nonrecursive SEMs, including latent variables and/or correlated errors.

5.5. Construction of Partial Ancestral Graphs (PAGs) for SEMs

Since Markov equivalence classes of SEMs with latent variables may contain infinitely many DGs, and may grow superexponentially even without latent variables, it is infeasible to attempt to construct a PAG by checking each DG in turn. While no algorithm as yet exists for the full class of nonrecursive SEMs with latent variables, PAG construction algorithms exist for:

1. recursive SEMs with latents: the FCI algorithm (Spirtes et al., 1993, 1996);

2. nonrecursive SEMs without latents: the CCD algorithm, ([R], 1996b,c; Richardson and Spirtes, 1996).

These algorithms test a very small subset of the set of all d-separation relations, those that entail the results of all other such d-separation tests, then infer that certain invariant structural features hold in a given Markov equivalence class. (By definition, these tests will give the same result for every DG in the class.) Having established that an invariant feature holds, the relevant symbol is added to the PAG. (The CCD algorithm runs in polynomial-time on sparse graphs, while the FCI algorithm, although its computational complexity is worst-case exponential even on sparse graphs, can be used on models with up to 100 variables.)

The CCD and FCI algorithms are complete (with respect to their respective subclasses of SEMs): given as input two DGs that are not Markov equivalent, the algorithms will always output different PAGs. Thus the algorithms may be used as a decision procedure for checking the Markov equivalence of two SEMs (though there are faster, polynomial-time algorithms for determining Markov equivalence for these subclasses that are not based on PAGs (Spirtes and Verma, 1992; [R], 1996a; Spirtes and Richardson, 1997). Thus it is possible to read entailed CI relations from the PAG output by either of these algorithms.

We are currently developing a PAG construction algorithm that is complete in the above sense, for the full class of SEM models, including nonrecursive SEMs with latent variables and/or correlated errors.

5.6 The Essential Partial Ancestral Graph (PAG) for a SEM

The notion of completeness in Section 5.4 involves a 1–1 mapping between PAGs and Markov equivalence classes; the *pattern* of an ADG is complete in this sense. There is, however, a stronger sense of completeness. One may wish to construct the maximally informative PAG, in the sense that every invariant feature of the Markov equivalence class that can be expressed using the PAG formalism is represented. The essential graph (Sections 3.1 and 4.4) has this property for Markov equivalence classes of ADGs and CGs. By analogy, one can define an *essential PAG* and perhaps extend the FCI and CCD algorithms to output essential PAGs. (Spirtes conjectures that the PAG output by the FCI algorithm is essential.) As for ADG and CG models, essential PAGs should yield more accurate specification of priors, more efficient model selection procedures, and estimates of the number of Markov equivalence classes for various types of SEMs. The latter is a first step toward an expected case complexity analysis of Bayesian and constraint-based search algorithms.

Summary: Markov equivalence of SEMs (objectives still to be achieved are in italics)

Types of SEM	Recursive	Nonrecursive
Without latent variables or correlated errors	Characterization of Markov equivalence Construction of pattern (PC algorithm) Construction of essential graph (D^*)	Characterization of Markov equivalence Construction of PAG (CCD algorithm) *Construction of essential PAG*
With latent variables and/or correlated errors	Characterization of Markov equivalence Construction of PAG (FCI algorithm) *Construction of essential PAG*	*Characterization of Markov equivalence* *Construction of PAG* *Construction of essential PAG*

5.7 The Relation Between PAGs and CW Summary Graphs

Cox and Wermuth [CW] (1996) introduced a graphical representation, called a *summary graph*, designed to represent the CIs entailed by an ADG with vertex set $V = O\dot{\cup}S\dot{\cup}L$, where O = Observed, S = Selection and L = Latent). They consider all CIs of the form $A \perp\!\!\!\perp B \mid C\dot{\cup}S$ where $A\dot{\cup}B\dot{\cup}C \subseteq O$, i.e., those CIs entailed by the ADG over O conditioning on all the variables in S. The intuition here is that in an empirical setting, artifacts of the mechanism by which the units in the sample are selected may lead certain (unmeasured) variables to be conditioned (in the sample); such variables are called *selection variables*. See also (Cooper (1995)). One could then consider Markov equivalence classes of DGs with the same O, but different L and S. In fact, the PAG representation is complete for such Markov equivalence classes, and the same algorithm (FCI) can be used to construct such PAGs (Spirtes et al., 1996; Spirtes and Richardson, 1997). Hence the PAG achieves the objectives of the summary graph, yet also possesses a number of other desirable properties not shared by the summary graph: (i) a summary graph may contain two variables that are not adjacent but nonetheless are CI, while this cannot occur in a PAG; (ii) unlike PAGs, there is as yet no method for determining whether or not two summary graphs are Markov equivalent; (iii) the summary graph contains several different kinds of edge, but, unlike the endlines in a PAG, these have no simple interpretation as yet.

6. CONCLUDING REMARKS

This review of the origins, properties, and applications of GMMs and SEMs, although far from complete, hopefully conveys their pervasive occurrence in mathematical sciences and their highly valuable catalytic role in bringing together ideas from a broad range of scientific disciplines. We have surveyed only a portion of current research topics involving GMMs and SEMs. Additional challenges for new research include the development of diagnostic tools for model selection, sensitivity analysis for local misspecification of a GMM or SEM, computational methods for massive data sets and/or missing data, and the extension of present theory and methodology to graphical systems whose topology evolves with time.

ACKNOWLEDGMENTS

This research was supported in part by the US National Science Foundation and the Isaac Newton Institute for Mathematical Sciences. We are grateful to Michael Levitz and the referees for detailed and helpful comments.

REFERENCES

Almond, R. (1995). *Graphical Belief Modeling*. New York: Springer.

Andersen, H. H., Højbjerre, M., Sørensen, D., and Eriksen, P. S. (1995). *Linear and Graphical Models for the Multivariate Complex Normal Distribution*. New York: Springer.

Anderson, T. W. (1984). *An Introduction to Multivariate Statistical Analysis*, 2nd edn. New York: Wiley.

Andersson, S. A. (1975). Invariant normal models. *Ann. Stat.*, **3**, 132–154.

Andersson, S. A. (1982). Distributions of maximal invariant statistics using quotient measures. *Ann. Stat.*, **10**, 955–961.

Andersson, S. A. (1990). The lattice structure of orthogonal linear models and orthogonal variance component models. *Scand. J. Stat.*, **17**, 287–319.

Andersson, S. A. and Madsen, J. (1998). Symmetry and lattice conditional independence in a multivariate normal distribution. *Ann. Stat.*, **26**, 525-572.

Andersson, S. A. and Perlman, M. D. (1984). Two testing problems relating the real and complex multivariate normal distributions. *J. Mult. Anal.*, **14**, 21–51.

Andersson, S. A. and Perlman, M. D. (1988). Lattice models for conditional independence in a multivariate normal distribution. Technical Report, Institute of Mathematical Statistics, University of Copenhagen, and Department of Statistics, University of Washington.

Andersson, S. A. and Perlman, M. D. (1991). Lattice-ordered conditional independence models for missing data. *Stat. Probab. Lett.*, **12**, 465–486.

Andersson, S. A. and Perlman, M. D. (1993). Lattice models for conditional independence in a multivariate normal distribution. *Ann. Stat.*, **21**, 1318–1358.

Andersson, S. A. and Perlman, M. D. (1994). Normal linear models with lattice conditional independence restrictions. In: T. W. Anderson, K.-T. Fang, and I. Olkin, eds. *Proceedings of the International Symposium on Multivariate Analysis Applications*. IMS Lecture Notes-Monograph Series **24**, pp. 97–110.

Andersson, S. A. and Perlman, M. D. (1995a). Testing lattice conditional independence models. *J. Multivariate Anal.*, **53**, 18–38.

Andersson, S. A. and Perlman, M. D. (1995b). Unbiasedness of the likelihood ratio test for lattice conditional independence models. *J. Multivariate Anal.*, **53**, 1–17.

Andersson, S. A. and Perlman, M. D. (1998). Normal linear regression models with recursive graphical Markov structure. *J. Multivariate Anal.*, **66**, 133–187.

Andersson, S. A., Brøns, H, and Jensen, S. T. (1983). Distribution of eigenvalues in multivariate statistical analysis. *Ann. Stat.*, **11**, 392–415.

Andersson, S. A., Madigan, D., Perlman, M. D., and Triggs, C. M. (1995). On the relation between conditional independence models determined by finite distributive lattices and by directed acyclic graphs. *J. Stat. Plann. Inference*, **48**, 25–46.

Andersson, S. A., Madigan, D., and Perlman, M. D. (1996). An alternative Markov property for chain graphs. In: F. Jensen and E. Horvitz, eds. *Uncertainty in Artificial Intelligence: Proceeding of the Twelfth Conference*. San Francisco: Morgan Kaufmann, pp. 40–48.

Andersson, S. A., Madigan, D., and Perlman, M. D. (1997a). On the Markov equivalence of chain graphs, undirected graphs, and acyclic digraphs. *Scand. J. Stat.*, **24**, 81–102.

Andersson, S. A., Madigan, D., and Perlman, M. D. (1997b). A characterization of Markov equivalence classes for acyclic digraphs. *Ann. Stat.*, **25**, 505–541.

Andersson, S. A., Madigan, D., Perlman, M. D., and Triggs, C. M. (1997c). A graphical characterization of lattice conditional independence models. *Ann. Math. Artif. Intell.*, **21**, 27–50.

Andersson, S. A., Madigan, D., and Perlman, M. D. (1998a). Alternative Markov properties for chain graphs. Submitted.

Bertelsen, A. (1989). On non-null distributions connected with testing reality of a complex normal distribution. *Ann. Stat.*, **17**, 929–936.

Bollen, K. (1988). *Structural Equations with Latent Variables*. New York: Wiley.

Bouckaert, R. R. and Studený, M. (1995). Chain graphs: Semantics and expressiveness. In: Ch. Froidevaux and J. Kohlas, eds. *Symbolic and Qualitative Approaches to Reasoning and Uncertainty*, pp. 69–76. Berlin: Springer.

Buntine, W. L. (1994). Operations for learning with graphical models. *J. Artif. Intell. Res.*, **2**, 159–225.

Buntine, W. L. (1995). Chain graphs for learning. In: P. Besnard and S.Hanks, eds. *Uncertainty in Artificial Intelligence: Proceeding of the Eleventh Conference*, pp. 46–54. San Francisco: Morgan Kaufmann.

Castillo, E., Gutierrez, J.M., and Hadi, A.S. (1997). *Expert Systems and Probabilistic Network Models*. New York: Springer.

Chickering, D. M. (1995). A transformational characterization of equivalent Bayesian network structures. In: P. Besnard and S. Hanks, eds. *Uncertainty in Artificial Intelligence: Proceedings of the Eleventh Conference*, pp. 87–98. San Francisco: Morgan Kaufmann.

Chickering, D. M. (1996). Learning equivalence classes of Bayesian network structures. In: F. Jensen and E. Horvitz, eds. *Uncertainty in Artificial Intelligence: Proceeding of the Twelfth Conference*, pp. 150–157. San Francisco: Morgan Kaufmann.

Consonni, G. and Veronese, P. (1995). A Bayesian method for combining results from several binomial experiments. *J. Am. Stat. Assoc.*, **90**, 935–944.

Cooper, G. F. (1995). Causal discovery from data in the presence of selection bias. In *Fifth International Workshop on Artificial Intelligence and Statistics*, Fort Lauderdale, Florida, pp. 140–150.

Cooper, G. F. and Herskovits, E. (1992). A Bayesian method for the induction of probabilistic networks from data. *Mach. Learn.*, **9**, 309–347.

Cox, D. R. (1993). Causality and graphical models. *Bull. Int. Stat. Inst.*, Proceedings 49th Session, **1**, 363–372.

Cox, D. R. and Wermuth, N. (1993). Linear dependencies represented by chain graphs (with discussion). *Stat. Sci.*, **8**, 204–218; 247–277.

Cox, D. R. and Wermuth, N. (1996). *Multivariate Dependencies: Models, Analysis, and Interpretation.* London: Chapman and Hall.

Dawid, A. P. and Lauritzen, S. L. (1993). Hyper Markov laws in the statistical analysis of decomposable graphical models. *Ann. Stat.*, **21**, 1272–1317.

Dellaportas, P., Giudici, P., and Roberts, G. (1996). Bayesian inference for nondecomposable graphical Gaussian models. Submitted for publication.

Dempster, A. P. (1972). Covariance selection. *Biometrics*, **28**, 157–175.

Edwards, D. (1995). *Introduction to Graphical Modeling*. New York: Springer.

Eriksen, P. S. (1997). Tests in covariance selection models. *Scand. J. Stat.*, **3**, 275–284.

Frydenberg, M. (1990). The chain graph Markov property. *Scand. J. Stat.*, **17**, 333–353.

Geiger, D. and Heckerman, D. (1995). A characterization of the Dirichlet distribution with applications to learning Bayesian networks. In: P. Besnard and S. Hanks, eds. *Uncertainty in Artificial Intelligence: Proceedings of the Eleventh Conference*, pp. 196–207. San Francisco: Morgan Kaufmann.

Geiger, D. and Pearl, J. (1988). On the logic of influence diagrams. In: R. Shachter, T. Levitt, L. Kanal, J. Lemmer, eds. *Proceedings of the 4th Workshop on Uncertainty in Artificial Intelligence*, pp. 136–147. Amsterdam: North-Holland.

Geng, Z., Asano, Ch., Ichimura, M., Tao, F., Wan, K., and Kuroda, M. (1996). Partial imputation method in the EM algorithm. In: A. Prat, ed. *Compstat 96*, pp. 259–263. Physica-Verlag.

Geng, Z., Wan, K., Tao, F., and Guo, J.-H. (1997). Decomposition of mixed graphical models with missing data. *Contemporary Multivariate Anal. Appl.*, E.49–E.45.

Goldberger, A. S. (1991). *A Course in Econometrics*. Cambridge: Harvard University Press.

Golinolli, D., Madigan, D., and Consonni, G. (1999). Relaxing the local independence assumption for quantitative learning in acyclic directed graphical models. In: D. Heckerman and J. Whittaker, eds. *Proceedings of Uncertainty*, **99**, in press.

Heckerman, D. (1991). *Probabilistic Similarity Networks*. Cambridge: MIT Press.

Heckerman, D., Geiger, D., and Chickering, D. M. (1994). Learning Bayesian networks: The combination of knowledge and statistical data. In: B. Lopez de Mantaras and D. Poole, eds. *Uncertainty in Artificial Intelligence, Proceedings of the Tenth Conference*, pp. 293–301. San Francisco: Morgan Kaufmann.

Højsgaard, S. and Thiesson, B. (1995). BIFROST—Block recursive models induced from relevant knowledge, observations, and statistical techniques. *Comp. Stat. Data Anal.*, **19**, 155–175.

Hylleberg, B., Jensen, M., and Ørnbøl, E. (1993). Graphical symmetry models. M.Sc. Thesis, Aalborg University, Denmark.

Jensen, F. V. (1996). *An Introduction to Bayesian Networks*. London: University College London Press.

Jensen, S. T. (1988). Covariance hypotheses which are linear in both the covariance and the inverse covariance. *Ann. Stat.*, **16**, 302–322.

Kiiveri, H., Speed, T. P., and Carlin, J. B. (1984). Recursive causal models. *J. Aust. Math. Soc., Ser. A*, **36**, 30–52.

Koster, J. T. A. (1996). Markov properties of non-recursive causal models. *Ann. Stat.*, **24**, 2148–2177.

Lauritzen, S. L. (1989). Mixed graphical association models. *Scand. J. Stat.*, **16**, 273–306.

Lauritzen, S. L. (1996). *Graphical Models*. Oxford: Oxford University Press.

Lauritzen, S. L. and Spiegelhalter, D. J. (1988). Local computations with probabilities on graphical structures and their application to expert systems (with discussion.) *J. R. Stat. Soc., Ser. B*, **50**, 157–224.

Lauritzen, S. L. and Wermuth, N. (1989). Graphical models for association between variables, some of which are qualitative and some quantitative. *Ann. Stat.*, **17**, 31–57.

Lauritzen, S. L., Speed, T. P., and Vijayan, K. (1984). Decomposable graphs and hypergraphs. *J. Aust. Math. Soc., Series A*, **36**, 12–29.

Lauritzen, S. L., Dawid, A. P., Larsen, B. N., and Leimer, H.-G. (1990). Independence properties of directed Markov fields. *Networks*, **20**, 491–505.

Little, R. J. A. and Rubin D. B. (1987). *Statistical Analysis with Missing Data*. New York: Wiley.

Madigan, D. and Raftery, A. (1994). Model selection and accounting for model uncertainty in graphical models using Occam's window. *J. Am. Stat. Assoc.*, **89**, 1535–1546.

Madigan, D. and York, J. (1995). Bayesian graphical models for discrete data. *Int. Stat. Rev.*, **63**, 215–232.

Madigan, D. and York, J. (1997). Bayesian methods for estimating the size of a closed population. *Biometrika*, **84**, 19–31.

Madigan, D., Andersson, S. A., Perlman, M. D., and Volinsky, C. M. (1996). Bayesian model averaging and model selection for Markov equivalence classes of acyclic digraphs. *Commun. Stat. Theory Methods*, **25**, 2493–2519.

Madigan, D., Mosurski, K., and Almond, R. G. (1997). Explanation in belief networks. *J. Comput. Graphical Stat.*, **6**, 160–181.

Madsen, J. (1995). Invariant lattice conditional independence models. Preprint 12, Institute of Mathematical Statistics, University of Copenhagen.

Massam, H. (1994). An exact decomposition theorem and a unified view of some related distributions for a class of exponential transformation models on symmetric cones. *Ann. Stat.*, **22**, 369–394.

Massam, H and Neher, E. (1998). Estimation and testing for lattice conditional independence models on Euclidean Jordan algebras. *Ann. Stat.*, **26**, 1051–1082.

Meek, C. (1995a). Causal inference and causal explanation with background knowledge. In: P. Besnard, S. Hanks, eds. *Uncertainty in Artificial Intelligence: Proceedings of the Eleventh Conference*, pp. 403–410. San Mateo: Morgan Kaufmann.

Meek, C. (1995b). Strong completeness and faithfulness in Bayesian networks. In: P. Besnard, S. Hanks, eds. *Uncertainty in Artificial Intelligence: Proceedings of the Eleventh Conference*, pp. 411-418. San Mateo: Morgan Kaufmann.

Neapolitan, R. E. (1990). *Probabilistic Reasoning in Expert Systems.* New York: Wiley.

Oliver, R. E. and J. Q. Smith (eds.) (1990). *Influence Diagrams, Belief Nets, and Decision Analysis.* Wiley, London.

Pearl, J. (1988). *Probabilistic Reasoning in Intelligent Systems: Networks of Plausible Inference.* San Mateo: Morgan Kaufmann.

Pearl, J. and Verma, T. (1991). A theory of inferred causation. In: J. Allen, R. Fikes, E. Sandewall, eds. *Proceedings of the Second International Conference on Principles of Knowledge Representation*, pp. 441–452. San Francisco: Morgan Kaufmann.

Perlman, M. D. (1987). Group symmetry covariance models. *Stat. Sci.*, **2**, 421–425.

Perlman, M. D., Madigan, D., and Levits, M. (1998). A new pathwise separation for chain graphs. In preparation.

Perlman, M. D. and Wu, L. (1998a). Lattice conditional independence models for contingency tables with non-monotone missing data patterns. *J. Stat. Plan. Inference*), in press.

Perlman, M. D. and Wu, L. (1998b). Efficiency of lattice conditional independence models for multinormal data with non-monotone missing observations. Preliminary Report, Department of Statistics, University of Washington.

Perlman, M. D. and Wu, L. (1998c). Lattice conditional independence models for seemingly unrelated regressions. Preliminary Report, Department of Statistics, University of Washington.

Richardson, T. S. (1996a). A discovery algorithm for directed cyclic graphs. In: F. Jensen and E. Horvitz, eds. *Uncertainty in Artificial Intelligence: Proceedings of the Twelfth Conference*, pp. 454–461. San Francisco: Morgan Kaufmann.

Richardson, T. S. (1996b). A polynomial-time algorithm for deciding Markov equivalence of directed cyclic graphical models. In: F. Jensen and E. Horvitz, eds. *Uncertainty in Artificial Intelligence: Proceedings of the Twelfth Conference*, pp. 462–469. San Francisco: Morgan Kaufmann.

Richardson, T. (1996c). Feedback models: Interpretation and discovery. Ph.D. Thesis, Department of Philosophy, Carnegie–Mellon University.

Richardson, T. and Spirtes, P. (1998). Automated discovery of linear feedback models. To appear in: G. Cooper and C. Glymour, eds. Causation and Computation, Cambridge: MIT Press, in press.

Roverato, A. and Whittaker, J. (1998). The ISSERLIS MATRIX and its application to non-decomposable graphical gaussian models, *Biometrika*, **85**, 711–725.

Shachter, R. D. and Kenley, C. R. (1989). Gaussian influence diagrams. *Manage. Sci.*, **35**, 527–550.

Speed, T. P. and Kiiveri, H. (1986). Gaussian Markov distributions over finite graphs. *Ann. Stat.*, **14**, 138–150.

Spiegelhalter, D. J. and Lauritzen, S. L. (1990). Sequential updating of conditional probabilities on directed graphical structures. *Networks*, **20**, 579–605.

Spiegelhalter, D. J., Dawid, A. P., Lauritzen, S. L. and Cowell, R. G. (1993). Bayesian analysis in expert systems (with discussion). *Stat. Sci.*, **8**, 219–283.

Spirtes, P. (1995). Directed cyclic graphical representation of feedback models. In: P. Besnard, and S. Hanks, eds. *Uncertainty in Artificial Intelligence: Proceedings of the Eleventh Conference*, pp. 491–498. San Francisco: Morgan Kaufmann.

Spirtes, P. and Richardson, T. (1997). A polynomial-time algorithm for determining DAG equivalence in the presence of latent variables and selection bias. *Artif. Intell. Stat.*, 489–500.

Spirtes, P. and Verma, T. (1992). Equivalence of causal models with latent variables. Logic and Computation Technical Report CMU-PHIL 33, Carnegie–Mellon University.

Spirtes, P., Glymour, C., and Scheines, R. (1993). *Causation, Prediction, and Search*. New York: Springer.

Spirtes, P., Meek, C., and Richardson, T. (1995). Causal inference in the presence of latent variables and selection bias. In: P. Besnard and S.

Hanks, eds., *Uncertainty in Artificial Intelligence: Proceedings of the Eleventh Conference*, pp 403–410, San Francisco, 1995 Morgan Kaufmann.

Spirtes, P., Richardson, T., Meek, C., Scheines, R., and Glymour, C. (1998). Using path diagrams as a structural equation modeling tool. *Sociol. Methods Res.*, Vol 27, No. 2, November 1998, pp. 182–225.

Spirtes, P., Meek, C., and Richardson, T. (1998). An algorithm for making causal inferences in the presence of latent variables and selection bias. To appear in: G. Cooper and C. Glymour, eds. *Causation and Computation*, Cambridge: MIT Press, in press.

Studený, M. (1996). On separation criterion and recovery algorithm for chain graphs. In: F. Jensen and E. Horvitz, eds. *Uncertainty in Artificial Intelligence: Proceeding of the Twelfth Conference*, pp. 509–516. San Francisco: Morgan Kaufmann.

Studený, M. (1997). A recovery algorithm for chain graphs. *Internat. J. Approx. Reasoning*, **17**, 265–293.

Studený, M. and Bouckaert, R. R. (1998). On chain graph models for description of conditional independence structure. *Ann. Stat.*, **26**, 1434–1495.

Sung, H.-G. (1996). Hierarchical partition models: A MCMC implementation. Masters Thesis, Department of Statistics, University of Washington.

Verma, T. and Pearl, J. (1990). Equivalence and synthesis of causal models. In: M. Henrion, R. Shachter, L. Kanal, J. Lemmer, eds., *Uncertainty in Artificial Intelligence: Proceedings of the Sixth Conference*, pp. 220–227. San Francisco: Morgan Kaufmann.

Verma, T. and Pearl, J. (1992). An algorithm for deciding if a set of observed independencies has a causal explanation. In: D. Dubois, M. Wellman, B. D'Ambrosio and P. Smets, eds. *Proceedings of the Eighth Conference on Uncertainty in Artificial Intelligence*, pp. 323–330. San Francisco: Morgan Kaufmann.

Wermuth, N. (1980). Linear recursive equations, covariance selection, and path analysis. *J. Am. Stat. Assoc.*, **75**, 963–972.

Wermuth, N. (1992).On block-recursive linear regression equations (with discussion). *Revista Brasileira de Probabilidade e Estatistica*, **6**, 1–56.

Wermuth, N. and Lauritzen, S. L. (1983). Graphical and recursive models for contingency tables. *Biometrika*, **70**, 537–552.

Wermuth, N. and Lauritzen, S. L. (1990). On substantive research hypotheses, conditional independence graphs, and graphical chain models (with discussion). *J. R. Stat. Soc., Ser. B*, **52**, 21–72.

Whittaker, J. L. (1990). *Graphical Models in Applied Multivariate Statistics*. New York: Wiley.

Wright, S. (1921). Correlation and causation. *J. Agric. Res.*, **20**, 557–585.

York, J., Madigan, D., Heuch, I., and Lie, R. T. (1995). Estimating a proportion of birth defects by double sampling: A Bayesian approach incorporating covariates and model uncertainty. *Appl. Stat.*, **44**, 227–242.

Zellner, A. (1962). An efficient method of estimating seemingly unrelated regression equations and tests for aggregation bias. *J. Am. Stat. Assoc.*, **57**, 348–368.

9

Robust Regression with Censored and Truncated Data

CHUL-KI KIM Ewha Woman's University, Seoul, Korea

TZE LEUNG LAI Stanford University, Stanford, California

1. INTRODUCTION

Consider the linear regression model

$$y_j = \alpha + \beta^T x_j + \epsilon_j \quad (j = 1, 2, \ldots) \tag{1}$$

where the ϵ_j are i.i.d. random variables and the x_j are independent $p \times 1$ random vectors independent of $\{\epsilon_n\}$. Taking the location parameter α in equation (1) to be a minimizer of the function $E\rho(y_j - \beta^T x_j - a)$, Huber's (1973) M-estimators $\hat{\alpha}$, $\hat{\beta}$ of α, β based on $(x_1, y_1), \ldots, (x_n, y_n)$ are defined as a solution vector to the minimization problem

$$\sum_{j=1}^{n} \rho(y_j - a - b^T x_j)\left(= \int \rho(y - a)\mathrm{d}F_{n,b}(y)\right) = \min! \tag{2}$$

where $F_{n,b}$ is the empirical distribution constructed from $y_j(b) = y_j - b^T x_j$, $j = 1, \ldots, n$. In particular, when $\rho(u) = u^2$, $\hat{\alpha}$, and $\hat{\beta}$ reduce to the classical least-squares estimates, and they reduce to the maximum likelihood estimates of α and β when $\rho(u) = -\log h(u)$, where h is the density function of the ϵ_i. When ρ is differentiable, the M-estimators $\hat{\alpha}$ and $\hat{\beta}$ can also be defined as a solution of the system of estimating equations

$$\sum_{j=1}^{n} \rho'(y_j - a - b^T x_j) = 0, \quad \sum_{j=1}^{n} x_j \rho'(y_j - a - b^T x_j) = 0 \tag{3}$$

Choosing ρ suitably gives estimators that have desirable robustness properties. A well-known robust choice of ρ is Huber's score function which is continuous, with $\rho(0) = 0$ and

$$\rho'(u) = u \quad if \quad |u| \leq c \qquad \text{and} \qquad \rho'(u) = \pm c \qquad \text{if} \qquad |u| > c \tag{4}$$

where c represents some measure of the dispersion of F. Using equation (4) in equation (2) is tantamount to applying the method of least squares to "metrically Winsorized residuals", cf. Huber (1981, p. 180).

Another class of robust estimators of α and β is defined by the ranks of the residuals, cf. Jurečkova (1969, 1971). The idea is to first estimate β using linear rank statistics based on $y_j(b)$. Specifically, let $R_j(b)$ be the rank of $y_j(b)$ in the set $\{y_1(b), \ldots, y_n(b)\}$ and define the linear rank statistic

$$L(b) = \sum_{j=1}^{n} x_j a_n(R_j(b)) \tag{5}$$

where the scores $a_n(k)$ are generated from a score function $\phi : [0, 1] \to (-\infty, \infty)$ that satisfies $\int_0^1 \phi(u)du = 0$ and $\int_0^1 \phi^2(u)du < \infty$ via

$$a_n(k) = \phi(k/n), \quad \text{or} \quad a_n(k) = \phi(k/(n+1)), \quad \text{or} \quad a_n(k) = E\phi(U^n_{(k)}) \tag{6}$$

$U^n_{(1)} \leq \cdots \leq U^n_{(n)}$ being the order statistics of a sample of size n from the uniform distribution on [0,1]. The R-estimator $\hat{\beta}$ is defined by the estimating equation $L(b) = 0$. Since $L(b)$ is discontinuous and may not have a zero, $\hat{\beta}$ is usually defined as a zero-crossing of $L(b)$ in the case $p = 1$ and as a minimizer of $\|L(b)\|$ in the case $p > 1$, where $\|\cdot\|$ denotes some norm (e.g. Euclidean norm) of p-dimensional vectors. Once $\hat{\beta}$ is determined, the location parameter α can be estimated by using the Hodges–Lehmann or other rank estimators of location based on $\{y_1(\hat{\beta}), \ldots, y_n(\hat{\beta})\}$.

In many applications, the responses y_j in equation (1) are not completely observable due to constraints on the experimental design. For example, medical studies on chronic diseases and reliability studies of engineering systems are typically scheduled to last for a prespecified period of time. The response variables y_j in equation (1) for these studies represent failure times (or transformations thereof) whose relationship to the covariate vectors (explanatory variables) x_j is to be determined from the data via estimation of β. Typically only a fraction of the n items in a study fail before the prescheduled end of the study. Moreover, some of these items may be withdrawn from the study before they fail. Thus, in lieu of (x_j, y_j), one observes $(x_j, \tilde{y}_j, \delta_j)$, $j = 1, \ldots, n$, where $\tilde{y}_j = \min\{y_j, c_j\}$, $\delta_j = I(y_j \leq c_j)$ and the c_j represent censoring variables (cf. Schmee and Hahn, 1979; Kalbfleisch and Prentice, 1980; Lawless, 1982). Instead of censoring, the response variable may be subject to truncation so that (x_j, y_j) is observable only when y_j does not exceed some threshold y^*. For example, in astronomy, Hubble's law relates the magnitude (negative logarithm of luminosity) y_j of a distant celestial object to x_j, the logarithm of its velocity as measured by red shift, by equation (1) with $\alpha + \epsilon_j$ associated with the intrinsic luminosity that is independent of red shift. However, objects are only observable if they are sufficiently bright, say $y_j \leq y^*$ (cf. Nicoll and Segal, 1982; Bhattacharya et al., 1983). Truncated regression problems also arise in biostatistics (cf. Holgate, 1965; Kalbfleisch and Lawless, 1989; Gross and Lai, 1996a) and econometrics (cf. Tobin, 1958; Hausman and Wise, 1976, 1977; Goldberger, 1981; Amemiya, 1985). Multiplication of y_j and y^* by -1 converts right to left truncation.

A more general setting which incorporates both censoring and truncation is that of left-truncated and right-censored data. Here, in addition to right censoring of y_j by $-\infty < c_j \leq \infty$, we also assume left truncation in the sense that $(x_j, \tilde{y}_j, \delta_j)$ can be observed only when $\tilde{y}_j \geq t_j$. The data therefore consist of n observations

$$(x_i^o, \tilde{y}_i^o, \delta_i^o, t_i^o) \qquad \text{with} \qquad \tilde{y}_i^o \geq t_i^o, \qquad i = 1, \ldots, n \tag{7}$$

The special case $t_j \equiv -\infty$ corresponds to the "censored regression model," while the special case $c_j \equiv \infty$ corresponds to the "truncation regression model." Left-truncated data that are also right censored often occur in prospective studies of the natural history of a disease, where a sample of individuals who have experienced an initial event E_1 (such as being diagnosed as having the disease) enter the study at chronological time τ'. The study is terminated at time τ^*. For each individual, the survival time of interest is $y = \tau_2 - \tau_1$, where τ_2 is the chronological time of occurrence of a second event E_2 (such as death) and τ_1 is the chronological time of occurrence of event E_1. Because subjects cannot be recruited into the study unless $\tau_2 > \tau'$, y is truncated on the left by $\tau = \tau' - \tau_1$ (i.e., a subject is observed only if $y \geq \tau$), and then only the minimum $y \wedge c$ of y and the right-censoring variable $c = \tau^* - \tau_1$ is observed along with τ. Individuals in the study population who experienced E_2 as well as E_1 prior to the initiation of the study are not observable (cf. Keiding et al., 1989; Wang, 1991; Andersen et al., 1993; Gross and Lai, 1996a).

How can the M-estimators defined by equation (3) and R-estimators defined by minimizing the norm of equation (5) be extended to left-truncated and right-censored (l.t.r.c.) data? A relatively complete theory for M- and R-estimators of the parameters of equation (1) based on the l.t.r.c. data in equation (7) has emerged during the past decade. Section 2 gives a brief review of this theory and describes computational algorithms for the implementation of these regression methods and for statistical inference on the regression parameters. Section 3 considers ways to examine the residuals computed from l.t.r.c. data together with some simple regression diagnostics. Some concluding remarks are given in Section 4.

2. M-ESTIMATORS AND RANK REGRESSION WITH LTRC DATA

Throughout the sequel we shall use the following notation for the l.t.r.c. data in equation (7). Let $\tilde{y}_i^o(b) = \tilde{y}_i^o - b^T x_i^o$, $t_i^o(b) = t_i^o - b^T x_i^o$ and

$$N(b, u) = \sum_{i=1}^{n} I(t_i^o(b) \le u \le \tilde{y}_i^o(b)), \quad \Delta(b, u) = \sum_{i=1}^{n} I(\tilde{y}_i^o(b) = u, \delta_i^o = 1)$$

$$\hat{F}_b(u|v) = 1 - \prod_{i: v < \tilde{y}_i^o(b) \le u, \delta_i^o = 1} \{1 - \Delta(b, \tilde{y}_i^o(b))/N(b, \tilde{y}_i^o(b))\} \tag{8}$$

The notation $\hat{F}_b(u|v-)$ will be used to denote equation (8) in which "$v < \tilde{y}_i^o(b)$" is replaced by "$v \le \tilde{y}_i^o(b)$." The function $\hat{F}_b(u| -\infty)$ is the product-limit estimate of the common distribution function $F(u)$ of the $\epsilon_j + \alpha$ in equation (1).

2.1 Linear Rank Statistics and Associated Estimating Equations

The study of R-estimators of β in the censored regression model was initiated by Louis (1981) in the case of binary covariates $x_i = 0$ or 1 and the logrank score function $\phi(u) = 1 + \log(1 - u)$. Extensions to more general score functions and multivariate covariates were provided by Tsiatis (1990). For the truncated regression model, Bhattacharya et al. (1983) initiated the study of rank estimators of β for univariate x_i and the Wilcoxon score function $\phi(u) = 2u - 1$. To extend the linear rank statistics of equation (5) to the l.t.r.c. data of equation (7) and thereby to derive rank estimators of β from l.t.r.c. data, Lai and Ying (1991b, 1992) first introduced the following modification of the product-limit estimator:

$$\tilde{F}_b(u) = 1 - \prod_{i:\tilde{y}_i^o(b)<u,\delta_i^o=1} \{1 - p_n(N(b, \tilde{y}_i^o(b))/n)\, \Delta(b, \tilde{y}_i^o(b))/N(b, \tilde{y}_i^o(b))\} \tag{9}$$

where the factor $p_n(\cdot)$ is used to trim out the $\tilde{y}_i^o(b)$ with relatively small risk set sizes $N(b, \tilde{y}_i^o(b))$. Letting $\psi = \phi - \Phi$, in which ϕ is the score function in equation (6) and $\Phi(u) = (1-u)^{-1} \int_u^1 \phi(t)\mathrm{d}t$, $0 \le u < 1$, they defined

$$R(b) = \sum_{i=1}^{n} \delta_i^o \psi(\tilde{F}_b(\tilde{y}_i^o(b)))\, p_n\left(\frac{N(b, \tilde{y}_i^o(b))}{n}\right) \left\{x_i^o - \frac{\sum_{k=1}^{n} I(\tilde{y}_k^o(b) \ge \tilde{y}_i^o(b) \ge t_k^o(b))x_k^o}{\sum_{k=1}^{n} I(\tilde{y}_k^o(b) \ge \tilde{y}_i^o(b) \ge t_k^o(b))}\right\} \tag{10}$$

The R-estimator $\hat{\beta}$ of β is defined by the estimating equation $R(b) = 0$, or more precisely, as a zero-crossing of $R(b)$ in the case $p = 1$ and as a minimizer of $\|R(b)\|$ in the case $p > 1$, where $\|\cdot\|$ denotes some norm of p-dimensional vectors.

For technical reasons in the derivation of an asymptotic theory for such rank estimators, the trimming factor $p_n(w)$ in Lai and Ying (1991b) is taken to be of the form $g(n^\lambda(w - cn^{-\lambda}))$, $0 \le w \le 1$, for some $c > 0$, $0 < \lambda < 1$ and for some smooth function $g(t)$ that vanishes for $t \le 0$ and assumes the value 1 for $t \ge 1$. However, empirical studies have shown that the simple trimming $p_n(w) = I(w \ge r/n)$, or equivalently

$$p_n(N(b, \tilde{y}_i^o(b))/n) = I(N(b, \tilde{y}_i^o(b)) \ge r) \tag{11}$$

works well for r as small as 2 or 3 in moderately sized samples. Smoothing equation (11) into the form $p_n(w) = g(n^\lambda(w - cn^{-\lambda}))$ facilitates the proof of the asymptotic linearity of $R(b)$ under suitable regularity conditions. With probability 1,

$$R(b) - R(\beta) = nA(b - \beta) + o(\sqrt{n} \vee n\|b - \beta\|) \quad \text{as} \quad n \to \infty \quad \text{and} \quad b \to \beta \tag{12}$$

where A is a nonrandom matrix. Moreover, $n^{-1/2}R(\beta)$ has a limiting normal distribution with mean 0 and covariance matrix that can be consistently estimated by $\Lambda(\beta)$, where

$$\Lambda(b) = n^{-1}\sum_{i=1}^{n}\delta_i^o\psi^2(\widetilde{F}_b(\tilde{y}_i^o(b)))p_n^2\left(\frac{N(b,\tilde{y}_i^o(b))}{n}\right)$$
$$\times\left[\frac{\sum_{k=1}^{n} I(\tilde{y}_k^o(b)\geq\tilde{y}_i^o(b)\geq t_k^o(b))x_k^{o\otimes 2}}{\sum_{k=1}^{n} I(\tilde{y}_k^o(b)\geq\tilde{y}_i^o(b)\geq t_k^o(b))}\right.$$
$$\left.-\left\{\frac{\sum_{k=1}^{n} I(\tilde{y}_k^o(b)\geq\tilde{y}_i^o(b)\geq t_k^o(b))x_k^o}{\sum_{k=1}^{n} I(\tilde{y}_k^o(b)\geq\tilde{y}_i^o(b)\geq t_k^o(b))}\right\}^{\otimes 2}\right]$$

Here and later we use $a^{\otimes 2}$ to denote aa^T for a column vector a. Making use of this, Lai and Ying (1991b) showed that under certain regularity conditions $\sqrt{n}(\hat{\beta}-\beta)$ has a limiting normal distribution with mean 0 as $n\to\infty$. The covariance matrix of this limiting normal distribution involves A, which is difficult to estimate because one has to estimate the density function f of F and the derivative f'. Instead of using this normal distribution to construct confidence ellipsoids (for β) centered at $\hat{\beta}$, Wei et al. (1990) circumvented the difficulties in estimating A by considering $(1-\alpha)$-level confidence regions of the form

$$\{b : n^{-1}R(b)\Lambda^{-1}(\hat{\beta})R(b)\leq\chi_p^2(\alpha)\} \tag{13}$$

where $\chi_p^2(\alpha)$ is the $(1-\alpha)$-quantile of the chi-square distribution with p degrees of freedom. They also used a similar idea to construct confidence regions for subvectors of β.

2.2 M-Estimators Based on LTRC Data

For the special case $\rho(u)=u^2/2$ in equation (2) corresponding to least-squares estimates, Miller (1976) proposed the following extension of equation (2) to the censored regression model when $p=1$. Noting that the left-hand side of equation (2) can be expressed in the form $\int_{-\infty}^{\infty}\rho(z-a)\mathrm{d}F_{n,b}(z)$, where $F_{n,b}$ is the empirical distribution function of y_j-bx_j $(j=1,\ldots,n)$, he suggested replacing $F_{n,b}$ by the product-limit estimate $\hat{F}_{n,b}$ based on $(\tilde{y}_i-bx_i,\delta_i)_{1\leq i\leq n}$ and defined $(\hat{\alpha},\hat{\beta})$ as a solution of the minimization problem

$$\sum_{j=1}^{n} w_j(b)\rho(\tilde{y}_j-a-bx_j)\left(=\int_{-\infty}^{\infty}\rho(z-a)\mathrm{d}\hat{F}_{n,b}(z)\right)=\min!$$

where the weights $w_j(b)$ are given by the jumps of $\hat{F}_{n,b}$, and $\hat{F}_{n,b}(z)=\hat{F}_b(z|-\infty)$ is defined in equation (8). Lai and Ying (1994) showed that this approach typically does not yield consistent estimators of α and β.

Instead of equation (2), Buckley and James (1979) and Ritov (1990) worked with equation (3), and for censored data they proposed to replace the $\rho'(y_j - a - b^T x_j)$ in equation (3) by

$$E[\rho'(y_j - a - b^T x_j)|x_j, \tilde{y}_j, \delta_j]$$
$$= \delta_j \rho'(y_j - a - b^T x_j) + (1 - \delta_j) \frac{\int_{\tilde{y}_j - \beta^T x_j}^{\infty} \rho'(u - a - (b - \beta)^T x_j) \mathrm{d}F(u)}{1 - F(\tilde{y}_j - \beta^T x_j)} \tag{14}$$

and to replace the unknown F in equation (14) by $\hat{F}_{n,b}$ and the unknown β by b. A similar extension of the least-squares estimate to truncated (but uncensored) data was proposed by Tsui et al. (1988). Lai and Ying (1994) clarified the role of replacing F in equation (14) by $\hat{F}_{n,b}$ by using a generalization of the missing information principle of Orchard and Woodbury (1972), which led them to the following extension of equation (3) to l.t.r.c. data:

$$\sum_{i=1}^{n} \psi_i(a, b) = 0, \qquad \sum_{i=1}^{n} x_i^o \psi_i(a, b) = 0 \tag{15}$$

where

$$\psi_i(a, b) = \delta_i^o \rho'(\tilde{y}_i^o(b) - a) + (1 - \delta_i^o) \int_{u > \tilde{y}_i^o(b)} \rho'(u - a) \mathrm{d}\hat{F}_b(u|\tilde{y}_i^o(b))$$
$$- \int_{u \geq t_i^o(b)} \rho'(u - a) \mathrm{d}\hat{F}_b(u|t_i^o(b)-) \tag{16}$$

cf. equations (2.24) and (2.26) of Lai and Ying (1994), where it is also noted that the first equation in (2.24) there gives $\int_{-\infty}^{\infty} \psi(u - a) \mathrm{d}\hat{F}_b(u| - \infty) = 0$. Hence, for the censored regression model ($t_i \equiv -\infty$), the last term in equation (16) vanishes and the estimating equations (15) reduce to those of Buckley and James (1979) and Ritov (1990).

Since the product-limit estimate $\hat{F}_\beta(u|v)$ may be quite unstable when v is near $\min_{i \leq n} t_i^o(\beta)$ or near $\max_{i \leq n} \tilde{y}_i^o(\beta)$ (cf. Lai and Ying, 1991a), it is desirable to down-weight such v in equation (16). This is done in Section 4.1 of Lai and Ying (1994) by introducing smoothing kernels to dampen the instability due to small risk set sizes. Moreover, since $\hat{F}_b$ is a step function with jump discontinuities, equation (15) may not have solutions, in which case $(\hat{\alpha}, \hat{\beta})$ is defined as a minimizer of $|\sum_{i=1}^{n} \psi_i(a, b)| + \| \sum_{i=1}^{n} x_i^o \psi_i(a, b)\|$. Under certain regularity conditions, Lai and Ying (1994) showed that equation (15) with these modifications yields asymptotically normal estimates. They also analyzed the robustness properties of these M-estimators and derived formulas for influence functions based on l.t.r.c. data in this connection. They showed that the asymptotically optimal choice of ρ is given by $\rho' = f'/f$, where f is the density function of F. With this choice of ρ, one attains the asymptotic lower bound on minimax risks established in Lai and Ying (1992) for the semiparametric problem of estimating β from l.t.r.c. data in the presence of the infinite-dimensional nuisance parameter F. Earlier, Lai and Ying (1991b) showed that the R-estimator defined by equation (10) with $\psi \circ \tilde{F}_b$ replaced by $f'/f + f/(1-F)$ also attains this asymptotic minimax bound. Since f is unknown, however, the optimal score function is not available to form asymptotically efficient R- or M-estimators. An adaptive choice of score functions is provided by Lai and Ying (1991b, 1994) to construct asymptotically efficient estimators of β based on l.t.r.c. data.

2.3 Synthetic Data and Weighted M-Estimators

As an alternative to the Buckley and James (1979) method, Koul et al. (1981) proposed the following extension of least-squares estimates of α and β to the censored regression model. They assumed the censoring variables c_j to be i.i.d. with a common continuous survival function $K(y) = P\{c_j \geq y\}$ such that $\{c_j\}$ and $\{(x_j, y_j)\}$ are independent and $K(F_j^{-1}(1)) > 0$ for all j, where F_j is the distribution function of y_j. Noting that $y_{j.K}^* = \delta_j \tilde{y}_j / K(\tilde{y}_j)$ has the same conditional mean given x_j as y_j, they proposed to replace $K(y)$ by the product-limit estimate

$$\hat{K}(y) = \prod_{j:\tilde{y}_j < y} \{1 - \sum_{\nu=1}^{n} I(\tilde{y}_\nu = \tilde{y}_j, \delta_\nu = 0) / \sum_{\nu=1}^{n} I(\tilde{y}_\nu \geq \tilde{y}_j)\}$$

and to estimate α, β by

$$\begin{pmatrix} \hat{\alpha} \\ \hat{\beta} \end{pmatrix} = \left\{ \sum_{i=1}^{n} \begin{pmatrix} 1 \\ x_i \end{pmatrix} (1, x_i) \right\}^{-1} \sum_{i=1}^{n} \begin{pmatrix} 1 \\ x_i \end{pmatrix} y_{i.\hat{K}}^* \tag{17}$$

Leurgans (1986) called such $y_{j.\hat{K}}^*$ "synthetic data" and proposed to use an alternative choice of $y_{j.K}^*$ in equation (17):

$$y_{j.K}^* = \int_c^{\tilde{y}_j} (K(s))^{-1} \mathrm{d}s + c + \int_{-\infty}^{c} (1 - K(s))/K(s) \mathrm{d}s$$

which also satisfies $E(y_{j.K}^* | x_j) = E(y_j | x_j)$ for every choice of c. Replacing K by $\hat{K}$ and c by $\min_{1 \leq i \leq n} \tilde{y}_i$ yields Leurgans' synthetic data. Lai et al. (1995) showed that for censored data, a general class of synthetic data has the form

$$y_{j.K}^* = \delta_j \psi_K(\tilde{y}_j) + (1 - \delta_j) \Psi_K(\tilde{y}_j) \tag{18}$$

where $\psi_K(y) = y/K(y) + \int_{t<y}(h(t)/K(t))\mathrm{d}K(t)$, $\Psi_K(y) = h(y) + \int_{t\le y}(h(t)/K(t))\,\mathrm{d}K(t)$, and h is a continuous function on the real line such that $\int_{-\infty}^{t}(|h(s)|/K(s))\mathrm{d}K(s)$ is finite for all t. They also established under certain regularity conditions the asymptotic normality of $(\hat{\alpha}, \hat{\beta})$ defined by equations (17) and (18).

Extensions of this approach to l.t.r.c. data are discussed in Section 4 of Gross and Lai (1996a). Assuming the (c_j, t_j) to be i.i.d. bivariate vectors, a natural analogue of $K(y)$ is

$$G(y) = P\{c_j \ge y \ge t_j\}$$

However, there are certain technical difficulties caused by truncation, which can be circumvented by using a somewhat different approach introduced in Section 3 of their paper. Specifically, instead of trying to estimate α and β directly from synthetic data, they first considered consistent estimation of $E\rho(y_j - \alpha - \beta^T x_j)$ up to a multiplicative constant and then finding the minimizer $(\hat{\alpha}, \hat{\beta})$ of the estimate. Noting that the multiplicative constant does not affect the minimization problem, their weighted M-estimators $\hat{\alpha}$ and $\hat{\beta}$ are defined as the minimizer of $\sum_{i=1}^{n} \delta_i^o x_i^o \rho(\tilde{y}_i^o - a - b^T x_i^o)\hat{S}(\tilde{y}_i^o)/\#(\tilde{y}_i^o)$, where

$$\#(y) = \sum_{i=1}^{n} I(t_i^o \le y \le \tilde{y}_i^o), \qquad d(y) = \sum_{i=1}^{n} I(\tilde{y}_i^o = y, \delta_i^o = 1)$$

$$\hat{S}(y) = \prod_{i:\tilde{y}_i^o \le y, \delta_i^o = 1} \{1 - d(\tilde{y}_i^o) I(\#(\tilde{y}_i^o) \ge r)/\#(\tilde{y}_i^o)\} \tag{19}$$

in which $r \ge 2$ is some prescribed lower bound on the risk set size $\#(\cdot)$ to avoid instabilities in the product-limit estimator $\hat{S}(\cdot)$. For convex and twice differentiable ρ, Theorem 2 of Gross and Lai (1996a) establishes the asymptotic normality of $(\hat{\alpha}, \hat{\beta})$ when the (x_j, y_j) are i.i.d. and independent of the i.i.d. (c_j, t_j), such that

$$\inf\{t : S(t) < 1\} > \tau \qquad \text{and} \qquad G(\tau^*) > 0 \tag{20}$$

where $\tau = \inf\{t : G(t) > 0\}$, $\tau^* = \inf\{t > \tau : G(t)S(t) = 0\}$ and $S(t) = P\{y_j \ge t\}$ is the common survival distribution of the y_j.

As pointed out in Section 6 of Gross and Lai (1996a), when the (x_j, y_j) are i.i.d., the assumption of i.i.d. (c_j, t_j) can be removed by imposing a somewhat more complicated condition than equation (20) to accommodate for nonidentically distributed (c_j, t_j). Kim and Lai (1998) recently showed that when the (c_j, t_j) are i.i.d., the restrictive assumption of i.i.d. covariates x_j can be removed since $\hat{S}(y)/\#(y)$ in equation (19) in fact estimates $1/G(y)$ even when the y_j are not i.i.d. In the case of censored data (without truncation), Leurgans (1986) and Fygenson and Zhou (1994) proposed a stratification procedure to further extend the applicability of the synthetic data approach when the censoring variables are not i.i.d. The same idea can be used for weighted M-estimators based on l.t.r.c. data. Specifically, the data are divided into m strata (groups), so that within each stratum the (c_j, t_j) can be assumed to be i.i.d., at least approximately. From the n_k quadruples $(x_i^o, \tilde{y}_i^o, \delta_i^o, t_i^o)$, $i \in I_k$, of observed data within the kth stratum $(k = 1, \ldots, m, \sum_{k=1}^m n_k = n)$, define $\#_k(y)$, $d_k(y)$, $\hat{S}_k(y)$ by equation (19) but with i restricted to I_k. When ρ is differentiable, the weighted M-estimator $(\hat{\alpha}, \hat{\beta})$ of (α, β) is defined by the estimating equations

$$\sum_{k=1}^{m} \sum_{i:i \in I_k} \delta_i^o \rho'(\tilde{y}_i^o - a - b^T x_i^o) n_k \hat{S}_k(\tilde{y}_i^o) I(\#_k(\tilde{y}_i^o) \geq r)/\#_k(\tilde{y}_i^o) = 0 \tag{21}$$

$$\sum_{k=1}^{m} \sum_{i:i \in I_k} \delta_i^o x_i^o \rho'(\tilde{y}_i^o - a - b^T x_i^o) n_k \hat{S}_k(\tilde{y}_i^o) I(\#_k(\tilde{y}_i^o) \geq r)/\#_k(\tilde{y}_i^o) = 0 \tag{22}$$

For the special case $\rho(u) = u^2/2$, $\rho'(u) = u$ and therefore equations (21) and (22) reduce to linear equations with simple closed-form solutions.

2.4 Computation of R- and M-Estimators

Even for complete data, the computation of R-estimators of α and β involves linear programming in the case $p > 1$ and, as remarked by Huber (1981, p. 164), "appears to be formidable, however, unless p and n are quite small." For censored and/or truncated data in the where case $p > 1$, the R-estimators "require minimizing discrete objective functions with multiple local minima" and "conventional optimization algorithms cannot be used to solve such minimization problems," as noted by Lin and Geyer (1992), who used probabilistic algorithms (simulated annealing) to circumvent these difficulties.

For complete data, M-estimators of α and β involve much lower computational complexity than R-estimators, and efficient algorithms for computing them have been developed (cf. Huber 1977, 1981). These algorithms can readily be extended to the weighted M-estimators defined by equations (21) and (22). In particular, in the case when $\rho(u) = u^2/2$, equations (21) and (22) are the usual normal equations (with simple closed-form solutions) of weighted least-squares estimates of α and β based on $\{(x_i^o, \tilde{y}_i^o) : 1 \leq i \leq n\}$ with weights

$$w_i = \delta_i^o n_k \hat{S}_k(\tilde{y}_i^o) I(\#_k(\tilde{y}_i^o) \geq s)/\#_k(\tilde{y}_i^o) \qquad (i \in I_k) \tag{23}$$

The weights are introduced to adjust for the bias caused by censoring and truncation in the usual (unadjusted) least-squares estimates from $\{(x_i^o, \tilde{y}_i^o) : 1 \leq i \leq n\}$.

The M-estimators defined by the estimating equations (15), however, involve harder computational problems than those in the weighted M-estimators defined by equations (21) and (22). In the case where $\rho(u) = u^2/2$, Buckley and James (1979) and Miller and Halpern (1982) proposed to use the successive substitution algorithm to compute the estimate when the data are censored. Specifically, in this case, the estimating equations (15) can be written in the form

$$b = \left\{\sum_{i=1}^{n}(x_i - \bar{x})^{\otimes 2}\right\}^{-1}\sum_{i=1}^{n}(x_i - \bar{x})\{\delta_i\tilde{y}_i(b) + (1-\delta_i)\int_{\tilde{y}_i(b)}^{\infty} u\mathrm{d}\hat{F}_b(u|\tilde{y}_i(b))\} \tag{24}$$

$$a = n^{-1}\sum_{i=1}^{n}\left\{\delta_i\tilde{y}_i(b) + (1-\delta_i)\int_{\tilde{y}_i(b)}^{\infty} u\mathrm{d}\hat{F}_b(u|\tilde{y}_i(b))\right\} \tag{25}$$

where $\bar{x} = n^{-1}\sum_{i=1}^{n} x_i$. The successive substitution algorithm computes $\hat{\beta}$ via the iterative scheme

$$b_{k+1} = \left\{\sum_{i=1}^{n}(x_i - \bar{x})^{\otimes 2}\right\}^{-1}\sum_{i=1}^{n}(x_i - \bar{x})$$
$$\left\{\delta_i\tilde{y}_i(b_k) + (1-\delta_i)\int_{\tilde{y}_i(b_k)}^{\infty} u\mathrm{d}\hat{F}_{b_k}(u|\tilde{y}_i(b_k))\right\} \tag{26}$$

and then computes $\hat{\alpha}$ by equation (25) with $b = \hat{\beta}$. The algorithm in equation (26) is similar to the EM algorithm (cf. Dempster et al., 1977) in which the E-step corresponds to $E(y_i(\beta)|x_i, \tilde{y}_i, \delta_i) = \delta_i\tilde{y}_i(\beta) + (1-\delta_i)\int_{\tilde{y}_i(\beta)}^{\infty} u\mathrm{d}F(u|\tilde{y}_i(\beta))$ with $\mathrm{d}F(u|v) = \mathrm{d}F(u)/(1 - F(v))$ so that equation (26) corresponds to the M-step of the iterations.

Assuming the ϵ_i to be normally distributed with mean 0 and unknown variance σ^2 so that

$$\int_{y}^{\infty} u\mathrm{d}F(u|y) = \sigma\phi(y/\sigma)/(1 - \Phi(y/\sigma)) \tag{27}$$

in which ϕ and Φ denote the standard normal density and distribution functions, Schmee and Hahn (1979) used the iterative scheme in equation (26) but with $\int_{\tilde{y}_i(b_k)}^{\infty} u\mathrm{d}\hat{F}_{b_k}(u|\tilde{y}_i(b_k))$ replaced by the right-hand side of equation (27) with $y = \tilde{y}_i(b_k)$ and $\sigma = \hat{\sigma}_k$, where

$$\hat{\sigma}_{k+1}^2 = n^{-1}\sum_{i=1}^{n}\{e_i(b_k) - a_k\}^2, \qquad a_k = n^{-1}\sum_{i=1}^{n} e_i(b_k)$$
$$e_i(b_k) = \delta_i\tilde{y}_i(b_k) + (1-\delta_i)\hat{\sigma}_k\phi(\tilde{y}_i(b_k)/\hat{\sigma}_k)/[1 - \Phi(\tilde{y}_i(b_k)/\hat{\sigma}_k)]$$

Thus, for the Schmee–Hahn estimator in parametric (Gaussian) censored regression models, the M-step consists of updating the estimates of α, β, and σ, and the E-step corresponds to equation (27). Schmee and Hahn (1979) reported numerical studies showing excellent convergence properties of their algorithm, which is in fact an EM algorithm to evaluate the maximum likelihood estimates of α, β, and σ. Although the successive substitution algorithm in equation (26) for semiparametric censored regression models is similar to an EM algorithm, it does not maximize a likelihood, unlike the EM algorithm for which each step yields an increase in the (likelihood) criterion. Thus, the convergence theory of EM algorithms is not applicable to equation (26). In fact, Buckley and James (1979) and Miller and Halpern (1982) reported oscillatory behavior in equation (26) and considered estimating β by some kind of average of the range of oscillations to which the algorithm eventually settles. Lin and Geyer (1992) suggested that one should use the simulated annealing procedure to search for the minimizer of the difference between the left- and right-hand sides of equation (26) because of discontinuities and lack of monotonocity of this difference as a function of b.

Despite its potential nonconvergence, simulation studies reported by Buckley and James show good performance (as an estimator of β) of the iterative procedure in equation (26) initialized at the "naive least-squares estimate" ignoring the fact that the observed $\tilde{y}_i$ are subject to censoring. Similar successive substitution algorithms have also been used by Tsui et al. (1988) to compute their extension of the least-squares estimate to truncated (but uncensored) data. We now extend such algorithms to compute the more general M-estimators defined by equation (15) for l.t.r.c. data.

2.4.1 Initializing at a simple consistent estimator and a Gauss–Newton-type algorithm

Instead of initializing at the naive least-squares estimate, whose performance may be quite poor, as shown in the simulation study reported in Table 3 of Lai and Ying (1994), we propose to initialize equation (15) with the weighted least-squares estimates $\widetilde{\alpha}$ and $\widetilde{\beta}$ based on $\{(x_i^o, \tilde{y}_i^o) : i \leq i \leq n\}$ with weights w_i given by equation (23). Moreover, instead of down-weighting the extreme order statistics of the $\tilde{y}_i^o(b)$ and $t_i^o(b)$ via the elaborate smoothing scheme in Section 4.1 of Lai and Ying (1994), we can substantially reduce the computational complexity by simply trimming away the extreme order statistics of the $t_k^o(\widetilde{\beta})$ and $\tilde{y}_k^o(\widetilde{\beta})$, which do not involve b, so that we do not need the elaborate scheme to ensure trimming smoothly in b. Ordering the $\tilde{y}_k^o(\widetilde{\beta})$ as $\tilde{y}_{[1]}^o(\widetilde{\beta}) \geq \cdots \geq \tilde{y}_{[n]}^o(\widetilde{\beta})$ and the $t_k^o(\widetilde{\beta})$ as $t_{(1)}^o(\widetilde{\beta}) \leq \cdots \leq t_{(n)}^o(\widetilde{\beta})$, let

$$y_{[r]} = \tilde{y}^o_{[r]}(\widetilde{\beta}), \qquad t_{(r)} = t^o_{(r)}(\widetilde{\beta}) \tag{28}$$

Let X denote the $n \times (p+1)$ matrix whose ith row is $I(\tilde{y}^o_i(\widetilde{\beta}) \le y_{[r]}, t^o_i(\widetilde{\beta}) \ge t_{(r)})(1, x^{oT}_i)$. We now describe an iterative algorithm for computing the M-estimator defined by equation (15). Let $\theta = (\alpha, \beta^T)^T$ and let $\theta^{(k)} = (\alpha^{(k)}, \beta^{(k)T})^T$ denote the result after the kth iteration to compute the M-estimator of θ. The algorithm consists of the following steps.

1. For $k = 0$, set $\theta^{(0)} = (\widetilde{\alpha}, \widetilde{\beta}^T)^T$, which is the preliminary estimate given in Section 2.3 with $\rho'(u) = u$, and compute $y_{[r]}$, $t_{(r)}$ using $\widetilde{\alpha}$, $\widetilde{\beta}$ and equation (28). Also evaluate $X^T X$.
2. Compute $\tilde{y}^o_i(\beta^{(k)})$ and $t^o_i(\beta^{(k)})$ for $i = 1, \dots, n$.
3. Evaluate $\hat{F}_{\beta^{(k)}}(u|v)$ or $\hat{F}_{\beta^{(k)}}(u|v-)$ by equation (8) at $u \in \{\tilde{y}^o_i(\beta^{(k)}) : \delta^o_i = 1, i \le n\}$, $v \in \{\tilde{y}^o_i(\beta^{(k)}) : \delta^o_i = 0, i \le n\} \cup \{t^o_i(\beta^{(k)}) : \ i \le n$, with $u \ge v$.
4. Compute the $n \times 1$ vector $\Psi^{(k)}$ whose ith component is $\psi_i(\alpha^{(k)}, \beta^{(k)}) I(\tilde{y}^o_i(\widetilde{\beta}) \le y_{[r]}, t^o_i(\widetilde{\beta}) \ge t_{(r)})$ with $\psi_i(a, b)$ denoted by equation (16).
5. Solve the linear equation $X^T X z = X^T \Psi^{(k)}$ to find $z = z^{(k)}$.
6. Put $\theta^{(k+1)} = \theta^{(k)} + z^{(k)}$.
7. Increase counter from k to $k+1$ and go to step 2.

The algorithm is similar to that used to compute M-estimators for complete data, cf. Huber (1981, pp. 181–182). For complete data, the least-squares iterations of the type in steps 5 and 6 correspond to a Gauss–Newton procedure to solve equation (3), and each iteration of the procedure leads to a decrease in ρ if $0 \le \rho'' \le 1$, cf. Huber (1981, p.182). For l.t.r.c. data, the preceding algorithm uses at each iteration the reconstructed scores $\psi_i(\alpha^{(k)}, \beta^{(k)})$ instead of the scores $\rho'(y_i - \alpha^{(k)} - x_i^T \beta^{(k)})$ for the residuals from complete data. These iteratively reconstructed scores involve the estimates $\hat{F}_{\beta^{(k)}}$ in equation (8). Hence it is desirable to have a good starting value $\hat{\beta}^{(0)}$ to ensure the adequacy of $\hat{F}_{\hat{\beta}^{(0)}}$ as an estimate of F.

2.4.2 Termination criteria and supplementary searches

Let $\|u\|$ denote some norm of $(p+1) \times 1$ vectors u, e.g., the Euclidean norm or the maximum of the absolute values of the components of u. Define

$$C(\theta) = \left\| \sum_{i=1}^{n} \psi_i(\theta)(1, x^{oT}_i)^T \right\| \tag{29}$$

We propose to terminate the above iterative scheme if $C(\theta^{(k)}) < \eta$ or if $\|\theta^{(k+1)} - \theta^{(k)}\| < \eta'$ for some prescribed thresholds η and η', or at $k = K$ if such numerical convergence has not occurred within K iterations. Thus the procedure terminates if the $\theta^{(k)}$ change little or if the objective function becomes sufficiently small. In either case the estimate is given by the terminal value of these iterations. The thresholds η and η' depend on the sample size. Choosing $\eta = o(n^{1/2})$ and $\eta' = o(n^{-1/2})$ still preserves the asymptotic normality of the estimate.

As in the case of the successive substitution algorithm in equation (26), the above iterative scheme may not converge. Putting an upper bound K on these iterations prevents the procedure from taking too much time. If the procedure terminates with nonconvergence, one possibility is to estimate (α, β) by the consistent estimator $(\tilde{\alpha}, \tilde{\beta})$ that has been used to initialize the algorithm which has failed to converge to a more accurate estimate. This is similar to the approach in Section 4 of Tsui et al. (1984). Another possibility is to switch to a different kind of search for the minimizer of the objective function in equation (29).

One such search procedure is simulated annealing, which was advocated by Lin and Geyer (1992) in view of the discontinuities and lack of monotonocity of the objective function. The procedure consists of the following steps.

1. Set θ, c, and $\sigma_1^2, \ldots, \sigma_{p+1}^2$ to their initial values with $\theta^{(0)} = (\tilde{\alpha}, \tilde{\beta}^T)^T$.
2. Generate a $(p+1) \times 1$ vector of independent variates $\theta^\dagger$, its jth component being normally distributed with mean θ_j and variance σ_j^2.
3. Let $\Delta = C(\theta^\dagger) - C(\theta)$.
4. If $\Delta \leq 0$, set $\theta = \theta^\dagger$; otherwise reset $\theta = \theta^\dagger$ with probability $e^{-\Delta/c}$.
5. Decrease c and σ_j^2 $(j = 1, \ldots, p+1)$ and then go to step 2.

Lin and Geyer (1992) proposed to run this Markov chain Monte Carlo search for a fixed number H of steps with $H \approx 1000(p+1)$. They also proposed to choose the initial c_0 smaller than, but of the same order of magnitude as, $C(\theta^{(0)})$, and to adjust c in step 5 of the algorithm by multiplying it by ρ with $\rho^{H/2} \approx 0.0005$. For the standard deviations σ_j, they proposed to initialize at $\sigma_j^{(0)} = 1$ and to decrease them at a rate so that the final σ_j is about 0.0005.

The simulated annealing algorithm is computationally expensive and there is no guarantee that one ends up with a value close to the global minimum of the objective function after H steps. A much less expensive search procedure suited to the present setting is the simplex method of Nelder and Mead (1964). Let $d = p + 1$ be the dimension of θ, and let $P_0, \ldots, P_d$ denote the $d + 1$ points in $\mathbf{R}^d$ defining the current "simplex." Let $C_i = C(P_i)$, and use h (for "high") to denote the index for which $C_h = \max_{0 \leq i \leq d} C_i$, and l (for "low") to denote the index for which $C_l = \min_{0 \leq i \leq d} C_i$. Let $\bar{P}$ denote the centroid of the d points P_i with $i \neq h$. At each step of the iterative search P_h is replaced by a new point. Three operations are used to arrive at this new point—reflection, contraction, and expansion, and these are described below.

(i) Choose a positive number α to be the reflection coefficient. The "reflection" of P_h is defined by $P^* = (1+\alpha)\bar{P} - \alpha P_h$. If $C_l \le C(P^*) \le \max_{i \ne h} C(P_i)$, then P_h is replaced by P^* and we start anew with the new simplex.

(ii) If $C(P^*) < C_l$, i.e., if reflection has produced a new minimum, then we "expand" P^* to $P^{**} = \gamma P^* + (1-\gamma)\bar{P}$, where $\gamma > 1$ is the expansion coefficient. If $C(P^{**}) < C_l$, we replace P_h by P^{**} and start anew with the new simplex. If $C(P^{**}) \ge C_l$, then we have a failed expansion and replace P_h by P^* before restarting.

(iii) If $C(P^*) > \max_{i \ne h} C(P_i)$, we first define a new P_h to be either the old P_h or P^* whichever gives a lower value of $C(P)$ and then "contract" the P_h to $P^{**} = \beta P_h + (1-\beta)\bar{P}$, where $0 < \beta < 1$ is the contraction coefficient. If $C(P^{**}) \le C(P_h)$, we replace P_h by P^{**} and start anew with the new simplex. If $C(P^{**}) > C(P_h)$, then we have a failed contraction and replace all the P_is by $(P_i + P_l)/2$ to form the new simplex.

The simplex search can be initialized with $(P_0, \ldots, P_d)$ consisting of $\theta^{(0)}$ and the last d values obtained from the Gauss–Newton-type algorithm in Section 2.4. This terminates when $C(P_l) < \eta$ or when we have already carried out K^* such iterations. If the combined Gauss–Newton/simplex procedure fails to converge after $K + K^*$ steps, we can estimate θ by the value that gives the minimum of $C(\theta)$ over the θ values that have been obtained so far. Note that the minimum and minimizer can be recursively updated and therefore do not require extensive memory and sorting.

2.4.3 Some illustrative examples

The following numerical examples illustrate some of the issues discussed above. Example 1 considers an R-estimator, which is defined as a zero-crossing of equation (10) in the case of $p = 1$ and as a minimizer of the norm of equation (10) in the case of $p > 1$. In the case of $p = 1$, the function $R(b)$ may have multiple zero-crossings, while in the case of $p \geq 1$, $\|R(b)\|$ typically has multiple local minima. The preliminary estimate $\tilde{\beta}$ in Section 2.4 can be used to resolve which of these zero-crossings (or local minima of $\|R(b)\|$) should be chosen as the rank estimator. In practice, one can restrict the search for zero-crossings or minima within some neighborhood of $\tilde{\beta}$. For $p = 1$ or 2, a grid search within this region can be carried out without much difficulty. For larger p, we can use the simulated annealing algorithm in Lin and Geyer (1992) or the simplex method of Nelder and Mead (1964) to find the minimizer of $\|R(b)\|$ within this region. Examples 2 and 3 consider M-estimators.

EXAMPLE 1: Consider the simple regression model $y_j = \beta x_j + \epsilon_j$, in which the ϵ_j are i.i.d. $N(0, 1)$ and x_j are i.i.d. uniformly distributed on $[0, 2]$ and independent of the ϵ_j. The y_j are subject to left truncation by i.i.d. $N(0, 1)$ random variables t_j that are independent of the (x_j, ϵ_j), and to right censoring by $c_j = t_j + \max(e^{-t_j}, 0.5)u_j$, in which the u_j are i.i.d. uniformly distributed on $[0, 5]$ and independent of the (x_j, ϵ_j, t_j). A sample of 200 l.t.r.c. data $(x_i^o, \tilde{y}_i^o, \delta_i^o, t_i^o)$, $i = 1, \ldots, 200$, was generated from this model with $\beta = 1$ and the raw data are plotted in Figure 1(a). Taking $r = 2$, $m = 1$ in equations (11), (21), and (22) with $\rho'(u) = u$, the preliminary estimator was found to be $\widetilde{\alpha} = -0.0727$ and $\widetilde{\beta} = 0.962$. Figure 1(b) plots the function $R(b)$ with $\psi(u) = 1$ for $-25 \leq b \leq 25$. Although the function looks continuous because of the plotter and the relatively small jumps within the wide range of $R(b)$, $R(b)$ is actually a step function with a zero-crossing at $b = 0.952$ (where $R(b+) = 0.007$, $R(b-) = -0.008$). The function $R(b)$ approaches 0 as $|b| \to \infty$. In fact, for $b = -25$ or $b = 35$, $R(b)$ is equal to 0 to 6 decimal places. This illustrates the point noted by Lin and Geyer (1992) that $\|R(b)\|$ may have multiple local minima.

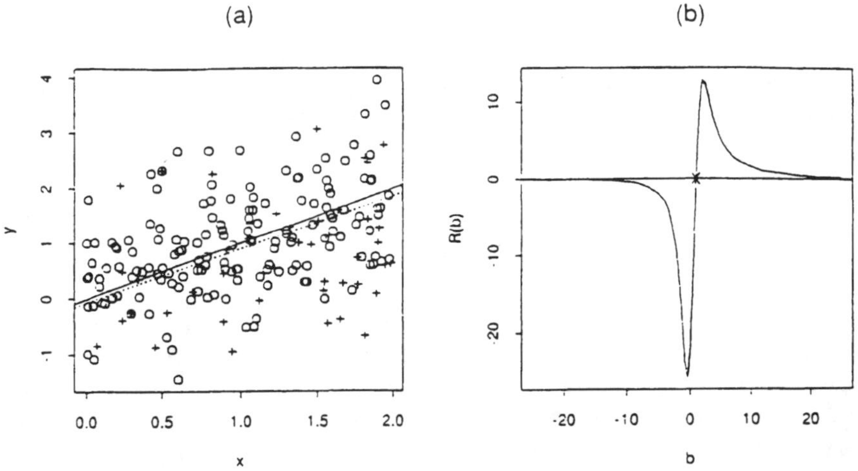

Figure 1 (a) Plot of raw data: o, uncensored data; +, censored data. The solid and the dotted lines represent the true and fitted (using the rank method) regression lines, respectively. (b) Plot of linear rank statistic $R(b)$ versus b

For complete data, after evaluating the R-estimator $\hat{\beta}$ as the zero-crossing (or minimizer of the norm) of equation (5), one can use the Hodges–Lehmann or other rank estimators of location based on the residuals $y_j - \hat{\beta}^T x_j$ to estimate the location parameter α in equation (1). Alternatively, one can also estimate α by the median, or some trimmed mean, or other linear combinations of the order statistics of the residuals $y_j - \hat{\beta}^T x_j$. For l.t.r.c. data, a convenient estimate of α after the R-estimator $\hat{\beta}$ has been computed is the trimmed mean

$$\hat{\alpha} = \left\{ \int u d\widetilde{F}_{\hat{\beta}}(u) \right\} / \widetilde{F}_{\hat{\beta}}(\max\{\tilde{y}_i^o(\hat{\beta}) : \delta_i^o = 1, N(\hat{\beta}, \tilde{y}_i^o(\hat{\beta})) \geq r\}) \tag{30}$$

Note that with $p_n(\cdot)$ given by equation (11), the modified product-limit estimator $\widetilde{F}_b(u)$ in equation (9) has no jumps at u with $N(b, u) < r$. For the simulated data in this example, the estimate of $\alpha(= 0)$ given by equation (30) is $\hat{\alpha} = -0.0686$.

EXAMPLE 2: Let $\rho'(u) = u$ and $p = 1$. Suppose the y_j are subject to both censoring and truncation (so that t_j and c_j are both finite). Then equation (16) reduces to

$$\begin{aligned} \psi_i(a, b) = \delta_i^o \tilde{y}_i^o(b) + (1 - \delta_i^o) \int_{u > \tilde{y}_i^o(b)} u d\hat{F}_b(u | \tilde{y}_i^o(b)) \\ - \int_{u \geq t_i^o(b)} u d\hat{F}_b(u | t_i^o(b)-) \end{aligned}$$

which does not depend on a. Thus $\psi_i(a, b) = \psi_i(0, b)$ in this case. Figure 2(a) plots the function $L(b) = \sum_{i=1}^n x_i^o \psi_i(0, b)$, in a neighborhood of the zero-crossing $b = 0.92835$, evaluated from a sample of 50 l.t.r.c. data $(x_i^o, \tilde{y}_i^o, \delta_i^o, t_i^o)$, $i = 1, \ldots, 50$, generated from the model in Example 1 (with $\beta = 1$). Rewriting the estimating equation $L(b) = 0$ as $L^*(b) = b$, where

$$L^*(b) = b + \left\{ \sum_{i=1}^n x_i^o \psi_i(0, b) \right\} \Big/ \sum_{i=1}^n x_i^{o2}$$

Figure 2(b) plots the function $L^*(b)$ in some neighborhood of $b = 0.92835$. The small rectangle centered on the diagonal line is similar to that in Figure 1 of Buckley and James (1979), in which the successive substitution algorithm in equation (26) oscillates between the upper left and lower right corners of the rectangle when it is initialized at either of these two points. The zero-crossing $b = 0.92835$ can easily be found by a bisection algorithm, an extension of which to higher dimensions is provided by the Nelder–Mead simplex method.

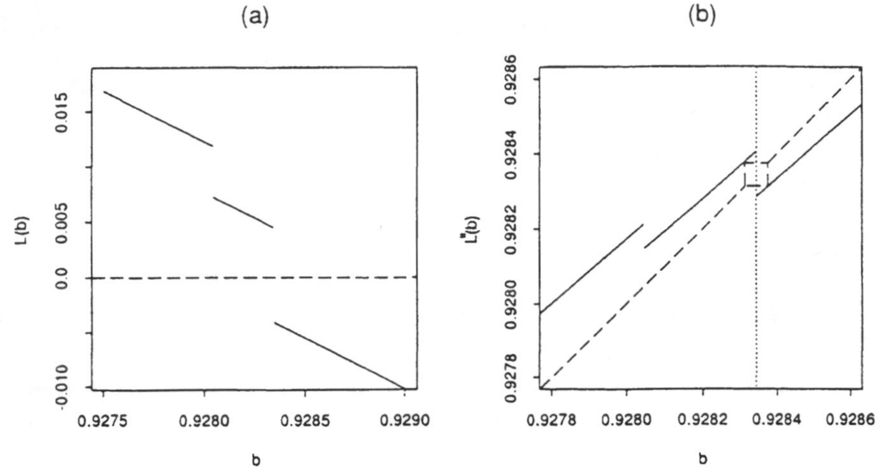

Figure 2 (a) Plot of $L(b)$ versus b. (b) Plot of $L^*(b)$ versus b

EXAMPLE 3: Consider the multiple regression model equation (1) in which the x_j are independent 3-dimensional vectors uniformly distributed in the cube $[-1, 1]^3$ and the ϵ_j are i.i.d. standard normal random variables. The censoring variables c_j are assumed to be i.i.d. normal with mean 4 and variance 1, and the truncation variables t_j are assumed to be i.i.d. normal with mean -4 and variance 1. This yields a 15% truncation rate and 22% censoring rate. It is assumed that $\alpha = 0$ is known and the unknown parameter vector is $\beta = (1, 3, 5)^T$. We first generated a sample of 50 l.t.r.c. data $(x_i^o, \tilde{y}_i^o, \delta_i^o, t_i^o)$, $i = 1, \ldots, 50$, from this model. Taking $m = 1$ and $r = 2$ in equations (21) and (22), the preliminary estimate was found to be $\widetilde{\beta}^T = (1.207, 2.810, 4.889)$. To compute the M-estimator defined by equations (15) and (16) with $\rho'(u) = u$, the Gauss–Newton-type algorithm in Section 2.4 converged after 10 iterations, yielding $\hat{\beta}^T = (1.046, 2.884, 5.011)$. The Nelder–Mead simplex algorithm also initialized at $\widetilde{\beta}$ was markedly slower, and converged after 207 iterations. We chose $\eta = \eta' = 10^{-5}$ in the convergence criterion described in Section 2.4. Setting $K = K^* = 1000$, the combined Gauss–Newton/simplex procedure described in Section 2.4 therefore converges for this data set without invoking the simplex component of the procedure. We next generated 100 samples of size $n = 50$ from this model and applied the same procedure to these simulated data. The procedure converged before entering the simplex phase for all these data sets, 90% of which yielded convergence within 20 iterations. The results of these 100 simulations are summarized as follows:

$$E(\hat{\beta}_1) = 1.0047, \qquad E(\hat{\beta}_2) = 3.0006, \qquad E(\hat{\beta}_3) = 4.9997$$
$$SD(\hat{\beta}_1) = 0.1167, \qquad S.D.(\hat{\beta}_2) = 0.0986, \qquad S.D.(\hat{\beta}_3) = 0.1503$$

2.5 Huber's Score Function and Concomitant Estimation of Scale

For complete data, a robust choice of $\psi = \rho'$ is Huber's score function in equation (4) which involves some scale parameter c. Accordingly estimating equations of robust M-estimators modify equation (3) as

$$\sum_{j=1}^{n} \psi\left(\frac{y_j - a - b^T x_j}{\sigma}\right)\begin{pmatrix} 1 \\ x_j \end{pmatrix} = 0 \tag{31}$$

in which σ is an unknown scale parameter to be estimated from the estimating equation

$$\sum_{j=1}^{n} \chi(\sigma^{-1}(y_j - a - b^T x_j)) = 0 \tag{32}$$

cf. Sections 7.7 and 7.9 of Huber (1981). In particular, the choice $\chi(u) = \text{sign}(|u| - 1)$ in equation (32) estimates σ by the median of absolute residuals, i.e., $\text{med}_{j \leq n}|y_j - a - b^T x_j|$. With this standardization, Huber's score function takes the form $\psi(x) = \min\{1, \max(-1, x)\}$.

Although in principle one can apply the "missing information principle" in Section 2 of Lai and Ying (1994) to extend equations (31) and (32) to l.t.r.c. data, this approach leads to simultaneous equations which are difficult to solve numerically. We can avoid this difficulty by using a separate scale estimate $\tilde{\sigma}$ based on $\hat{F}_{\tilde{\beta}}$, where $\tilde{\beta}$ is the preliminary estimate of β introduced in Section 2.4. Define

$$G_{\tilde{\beta}}(y) = \hat{F}_{\tilde{\beta}}(y|t_{(r)})/\hat{F}_{\tilde{\beta}}(y_{[r]}|t_{(r)}) \qquad \text{for} \qquad t_{(r)} \leq y \leq y_{[r]} \tag{33}$$

where $y_{[r]}$ and $t_{(r)}$ are defined in equation (28). Let $\tilde{\sigma}$ be the median absolute deviation of the $G_{\tilde{\beta}}$, where the median absolute deviation of the discrete distribution $G_{\tilde{\beta}}$ with atoms a_j is defined as the median of the discrete distribution that assigns mass $G_{\tilde{\beta}}(a_j)$ to $|a_j - \text{med}(G_{\tilde{\beta}})|$, with $\text{med}(G_{\tilde{\beta}})$ given by the median of the histogram of $G_{\tilde{\beta}}$. Replacing the σ in equation (31) by $\tilde{\sigma}$, we no longer have the problem of unknown σ in extending equation (31) to l.t.r.c. data via the estimating equations (15).

EXAMPLE 4: Consider the simple linear regression model $y_j = \beta x_j + \epsilon_j$ with $\beta = 1$, where the ϵ_j are i.i.d. random variables whose common distribution function F is contaminated normal of the form $F = 0.7\mathrm{N}(0, 1) + 0.3\mathrm{N}(0, 8^2)$. The x_j are independent, uniformly distributed on $[-1, 1]$, and independent of the ϵ_j. The y_j are subject to right censoring by i.i.d. $\mathrm{N}(6,5^2)$ random variables c_j that are independent of the (x_j, ϵ_j), and also to left truncation by $t_j = \min(c_j, u_j)$ in which the u_j are i.i.d. $\mathrm{N}(-6,5^2)$ and independent of (x_j, ϵ_j, t_j). This corresponds to 24% truncation rate and 28% censoring rate. A sample of 50 l.t.r.c. data $(x_i^o, \tilde{y}_i^o, \delta_i^o, t_i^o)$, $i = 1, \ldots, 50$, was generated from the model. The sample data are plotted in Figure 3(a). Since the (x_j, y_j) are i.i.d. and independent of the (t_j, c_j), no stratification is needed ($m = 1$) for the preliminary estimator in Section 2.3, which assumes unknown $\alpha(= 0)$. Taking $r = 2$ in equations (21) and (22) with $\rho'(u) = u$, the preliminary estimator was found to be $\widetilde{\alpha} = -1.45$, $\widetilde{\beta} = 3.06$. Initializing with this preliminary estimate, the scale estimate was found to be $\widetilde{\sigma} = 1.921$ and the Huber-type M-estimator defined above, with this scale estimate $\widetilde{\sigma}$ and $r = 2$, was found to be $(\hat{\alpha}^H, \hat{\beta}^H) = (0.38, 1.14)$ after 390 iterations by applying the algorithm in Section 2.4 with $\eta = \eta' = 10^{-5}$ and $K = K^* = 1000$.

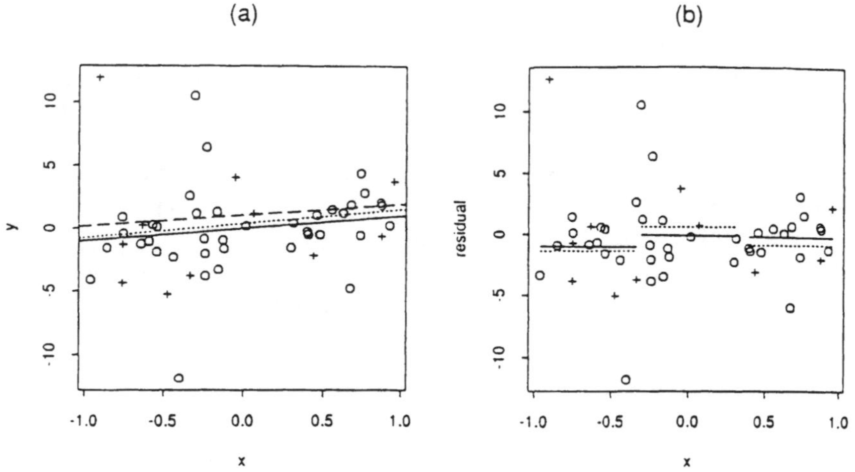

Figure 3 (a) Plot of l.t.r.c. data in Example 4: ○, uncensored data; +, censored data. The solid line corresponds to the true model, the dotted line to the Huber's M-estimate $(\hat{\alpha}^H, \hat{\beta}^H)$, and the broken line to $(\hat{\alpha}^{BJ}, \hat{\beta}^{BJ})$. (b) Plot of residuals based on Huber-type M-estimator versus x. The horizontal line segments represent the trimmed means of the residuals for three strata, with the solid lines corresponding to the M-estimate $(\hat{\alpha}^H, \hat{\beta}^H)$ and the dotted lines corresponding to $(\hat{\alpha}^{BJ}, \hat{\beta}^{BJ})$. ○, uncensored residual; +, censored residual

Using a similar algorithm to solve equations (15) and (16) with $\rho'(u) = u$ (the Buckley–James-type score function) gave $(\hat{\alpha}^{BJ}, \hat{\beta}^{BJ}) = (1.04, 0.82)$ after 338 iterations. Note that $\hat{\beta}^{H}$ is much closer to the true value $\beta = 1$ than $\hat{\beta}^{BJ}$. Moreover, both $\hat{\alpha}^{H}$ and $\hat{\alpha}^{BJ}$ underestimate $\alpha = 0$ since they are actually consistent estimates of certain trimmed means of the ϵ_j as shown in Section 4.3 of Lai and Ying (1994).

2.6 Bootstrap Methods for Standard Errors and Confidence Intervals

As mentioned in Section 2.1, the limiting covariance matrix of $\sqrt{n}(\hat{\beta} - \beta)$ of an R-estimator $\hat{\beta}$ is difficult to estimate, and this is also true for the M-estimators defined by equations (15) and (16), as shown by the results of Lai and Ying (1994), who suggest using a confidence region similar to equation (13) to circumvent the difficulty. The confidence region in equation (13), however, may have a strange shape and there may be numerical difficulties to finding the boundaries of the region, which need not be a connected set. Moreover, the coverage probability may differ considerably from the nominal value when the sample size is not large enough.

For weighted M-estimators defined by equations (21) and (22), the limiting covariance matrices have a much simpler representation, and Gross and Lai (1996a) showed how they can be estimated consistently so that one can use the asymptotic normality of the estimates to construct confidence ellipsoids for the regression parameters. When the (x_j, y_j, c_j, t_j) are i.i.d., Gross and Lai (1996a,b) showed that instead of relying on the normal approximation, which may be inadequate when the sample size is not large enough, one can use the so-called "simple" bootstrap method to estimate standard errors and construct confidence intervals. The method consists of drawing repeated random samples of size n, with replacement, from the n observed quadruples $(x_i^o, \tilde{y}_i^o, \delta_i^o, t_i^o)$ with $\tilde{y}_i^o \geq t_i^o$ $(i = 1, \ldots, n)$, and then using the bootstrap samples as one might do in the complete-data case to compute estimates of the standard deviation and the bias of an estimate, as well as to construct bootstrap confidence intervals for the parameters of interest. Making use of Edgeworth expansions for asymptotic U-statistics, Gross and Lai (1996b) have proved that the simple bootstrap approximations to the sampling distributions of the weighted M-estimators based on l.t.r.c. data are accurate to the order of $O_p(n^{-1})$.

When the covariates x_j are not i.i.d., the y_j can no longer be regarded as i.i.d. unless $\beta = 0$, since the ϵ_j are i.i.d. in the regression model in equation (1). In this case the simple bootstrap method fails even for complete data, for which one should instead resample from the residuals to form the bootstrap sample $y_j^* = \hat{\alpha} + \hat{\beta}^T x_j + \epsilon_j^*$, cf. Hall (1992). First consider the case where truncation is absent (i.e., $t_j \equiv -\infty$). Let $\hat{F}$ be the product-limit estimator of the common distribution function of the $\alpha + \epsilon_j$ based on $(\tilde{y}_i - \hat{\beta}^T x_i, \delta_i)_{1 \le i \le n}$ and let $\hat{G}$ be the product-limit estimator of the common distribution G of the c_j that are subject to censoring by the independent random variables y_j. Resampling $e_1^*, \ldots, e_n^*$ from $\hat{F}$ and $c_1^*, \ldots, c_n^*$ from $\hat{G}$ yields the bootstrap sample $(\tilde{y}_i^*, \delta_i^*)_{1 \le i \le n}$, where $\tilde{y}_i^* = \min\{\hat{\beta}^T x_i + e_i^*, c_i^*\}$. Kim and Lai (1998) have extended this idea to the case where the y_j are subject to both left truncation and right censoring. They also give an asymptotic justification of this bootstrap method and apply it to estimate standard errors of M-estimators and to construct confidence intervals for the regression parameters. Because of the computational efficiency of the combined Gauss–Newton/simplex algorithm in Section 2.4, the task of computing these bootstrap confidence intervals and standard errors is quite manageable.

3. EXAMINATION OF RESIDUALS AND REGRESSION DIAGNOSTICS

Let $(\hat{\alpha}, \hat{\beta})$ be an estimate of the parameter (α, β) of the assumed regression model in equation (1). For complete data, under equation (1) and suitable regularity conditions, the residuals $r_j = y_j - \hat{\alpha} - \hat{\beta}^T x_j$ provide approximations of the unobservable i.i.d. random variables ϵ_j with an $O_p(1/\sqrt{n})$ error. Therefore substantial deviations of these residuals from an i.i.d. pattern (e.g., conspicuous trends with respect to the components of the covariate vector x_j) suggest inadequacies and possible improvements of the assumed regression model. For the l.t.r.c. data in equation (7), we can likewise compute

$$r_i^o = \tilde{y}_i^o - \hat{\alpha} - \hat{\beta}^T x_i^o \; (= \tilde{y}_i^o(\hat{\beta}) - \hat{\alpha}), \; i = 1, \ldots, n \tag{34}$$

and regard them as "residuals" in the general sense of Cox and Snell (1968), namely, as approximations of the unobservable $\epsilon_i^o = \tilde{y}_i^o - \alpha - \beta^T x_i^o$, which comprise the observable subset of the complete sample of i.i.d. ϵ_j that are subject to left truncation by $t_j - \alpha - \beta^T x_j$ and right censoring by $c_j - \alpha - \beta^T x_j$. Substantial deviations of equation (34) from the observable part of an i.i.d. sample subject to truncation and censoring, therefore, again suggest inadequacies of the assumed regression model in equation (1), if we can quantify and detect these deviations.

As in the case of complete data, one can plot the residuals in equation (34), with different symbols to represent the uncensored ($\delta_i^o = 1$) and censored ($\delta_i^o = 0$) residuals, versus each component of the covariate vector x_i^o.

Although such plots are easy to do, direct examination of them is often not revealing because of the uncertainties in the censored residuals and because the corresponding ϵ_i^o need not be i.i.d. themselves due to truncation and censoring. We propose to (i) partition each covariate axis into subsets and thereby stratify the residuals, and (ii) compute from the l.t.r.c. residuals for each stratum the product-limit curve of the stratum. Specifically, noting that the kth stratum in the above stratification of the residuals can be described by a subset $\mathcal{X}_k$ of the covariate space, define $N_k(b, u)$, $\Delta_k(b, u)$, and $\hat{F}_{b,k}(u|v)$ as in equation (8) but with the sum or product over i restricted to $x_i^o \in \mathcal{X}_k$. Take $\nu \geq 2$, let R' be the smallest order statistic of $\tilde{y}_1^o(\hat{\beta}), \ldots, \tilde{y}_n^o(\hat{\beta})$ such that $\min_k N_k(\hat{\beta}, R') \geq \nu$, and let R'' be the largest order statistic of the $\tilde{y}_i^o(\hat{\beta})$ such that $\min_k N_k(\hat{\beta}, R'') \geq \nu$. Substantial differences of $\{\hat{F}_{\hat{\beta},k}(u|R') : R' \leq u \leq R''\}$ among the strata (indexed by k) suggest possible departures from the regression model in equation (1). Instead of plotting these estimated conditional distribution functions directly, one can plot their quantiles in the form of boxplots (or Q–Q plots in the case of $k = 2$, as in Figure 5(b) below). Alternatively one can plot the trimmed means

$$\mu_k(R', R'') = \left\{ \int_{R'}^{R''} u d\hat{F}_{\hat{\beta},k}(u|R') \right\} / \hat{F}_{\hat{\beta},k}(R''|R') - \hat{\alpha} \tag{35}$$

Gross and Lai (1995) showed that under equation (1) and certain regularity conditions, equation (35) is a consistent estimate of $E(\epsilon_j | c < \epsilon_j + \alpha < d)$ when ν is some fraction of n, where c and d are certain quantiles of the common distribution of the $\epsilon_j + \alpha$. We illustrate this idea in two examples, the first of which also shows how the influential-data diagnostics for linear regression in Belseley et al. (1980) can be extended to l.t.r.c. data.

EXAMPLE 4 (continued): Taking $\nu = 2$ to define R' and R'' for equation (35), we found $R' = -8.84$ and $R'' = 4.52$ for the residuals associated with $(\hat{\alpha}^H, \hat{\beta}^H)$ in Example 4, and $R' = -9.63$, $R'' = 5.78$ for those associated with $(\hat{\alpha}^{BJ}, \hat{\beta}^{BJ})$. Figure 3(b) plots the residuals associated with $(\hat{\alpha}^H, \hat{\beta}^H)$ and their trimmed means over three subintervals of the range of observed x-values. These trimmed means, represented by solid line segments, are close to 0 and show no trends of departure from the model. Also given for comparison are the trimmed means (dotted line segments) of the $(\hat{\alpha}^{BJ}, \hat{\beta}^{BJ})$-induced residuals. The dotted line segments are further from 0 than the solid line segments, but again show no conspicuous departure from the regression model.

The contaminated normal errors ϵ_j in this example are likely to yield influential observations, and the presence of censoring and truncation requires adjustments/reconstructions such as in equation (16), which may give certain observations even more influence than in the complete data case. For the present data set, what are the influential observations if we use $(\hat{\alpha}^{BJ}, \hat{\beta}^{BJ})$? Is Huber's score function robust enough against these influential observations? To address these questions, we can extend standard influential-data diagnostics for linear regression with complete data to l.t.r.c. data. The idea is to recompute, for each observation $(x_i^o, \tilde{y}_i^o, \delta_i^o, t_i^o)$, the M-estimator $(\hat{\alpha}_{(-i)}, \hat{\beta}_{(-i)})$ obtained by deleting this observation from the sample. Because of the computational efficiency of the combined Gauss–Newton/simplex algorithm in Section 2.4, the task of computing $(\hat{\alpha}_{(-i)}, \hat{\beta}_{(-i)})$, $i = 1, \ldots, n$, is quite manageable. Figure 4 plots the values $\hat{\alpha} - \hat{\alpha}_{(-i)}$ and

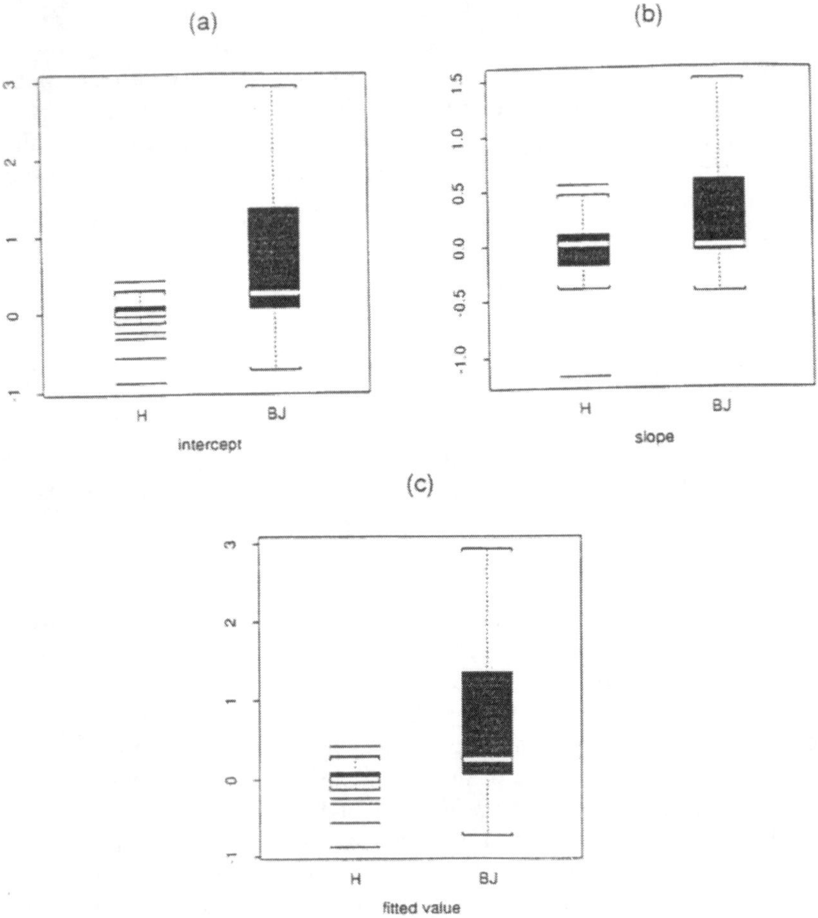

Figure 4 Boxplots of (a) $\hat{\alpha}^H - \hat{\alpha}^H_{(-i)}$ and $\hat{\alpha}^{BJ} - \hat{\alpha}^{BJ}_{(-i)}$, (b) $\hat{\beta}^H - \hat{\beta}^H_{(-i)}$ and $\hat{\beta}^{BJ} - \hat{\beta}^{BJ}_{(-i)}$, and (c) $\hat{y}_i^H - \hat{y}^H_{(-i)}$ and $\hat{y}_i^{BJ} - \hat{y}^{BJ}_{(-i)}$

$\hat{\beta} - \hat{\beta}_{(-i)}$ for the M-estimators using Huber's score function and the Buckley–James score function. In addition, it also plots the difference in the fitted values $\hat{y}_i - \hat{y}_{(-i)}$, where $\hat{y}_i = \hat{\alpha} + \hat{\beta}x_i^o$ and $\hat{y}_{(-i)} = \hat{\alpha}_{(-i)} + \hat{\beta}_{(-i)}x_i^o$.

EXAMPLE 5: This example shows how regression diagnostics can be used to reveal inadequacies in fitting equation (1) to real data. Table 1 of Kalbfleisch and Lawless (1989) reports the data from the Center for Disease Control (CDC) on 295 blood transfusion patients who were diagnosed with AIDS prior to July 1, 1986. It gives the month of infection (INF), with month 1 being January 1978, the age in years at the time of blood transfusion (AGE = age + 1), and the duration in months (DIAG) between infection with the human immunodeficiency virus by blood transfusion and the clinical manifestation of acquired immune deficiency syndrome (AIDS). Kalbfleisch and Lawless (1989) suggested using INF–0.5 as the time of infection and excluding two patients with INF exceeding 90 months from the study, leaving 293 patients with DIAG that exceeds 101–(INF–0.5), where 101 is the total number of months in the study.

Gross (1996) fitted the linear regression model in equation (1) to these data, taking $y = -\log(\text{DIAG})$ (with the modification that sets DIAG to be 0.5 if DIAG $= 0$), $x = -1$ if age ≤ 4, and $x = 1$ if age > 4. Since $y = -\log(\text{DIAG})$ is left truncated by $t = -\log(101.5 - \text{INF})$, Gross used a more robust version of the Tsui et al. (1988) method, which she denotes by TJW, and the rank regression method of Lai and Ying (1991b) to estimate the slope β. The rank regression method she used corresponds to equation (10) with the Wilcoxon-type score function $\psi(w) = 1 - w$. To get rid of ties and simplify the computation, she added independent random variables that are uniform on $(-0.5, 0.5)$ to all the DIAG values and used extensive search to find the TJW and rank estimators of β.

Without altering the data via this tie-breaking device, we recomputed the R- and the M-estimates. For the R-estimator, Gross found three zero-crossings of the function $R(b)$, evaluated from the altered data, at $b = -0.89$, -0.49, 0.92. She pointed out that since negative values of b do not make much sense, one should take 0.92 as the R-estimator. Using the original data instead, we evaluated $R(b)$ over a fine grid in $[0, 1.5]$. Figure 5(a) plots the function for $0.9 \leq b \leq 1.5$, showing zero-crossings at $b = 0.92$, 0.95, 1.25, 1.30, 1.40. Note that $R(b)$ is essentially zero for $b \geq 1.25$. Taking $m = 1$ and $r = 2$ in equations (21) and (22) with $\rho'(u) = u$, we found the preliminary estimator to be $\tilde{\alpha} = -3.58$ and $\tilde{\beta} = 0.53$ (which is in close agreement with the value $\tilde{\beta} = 0.59$ from the altered data reported by Gross). Initializing with this preliminary estimate and using $r = 2$, the algorithm in Section 2.4 converges to $(\hat{\alpha}^{BJ}, \hat{\beta}^{BJ}) = (-3.71, 0.93)$ after 10 iterations.

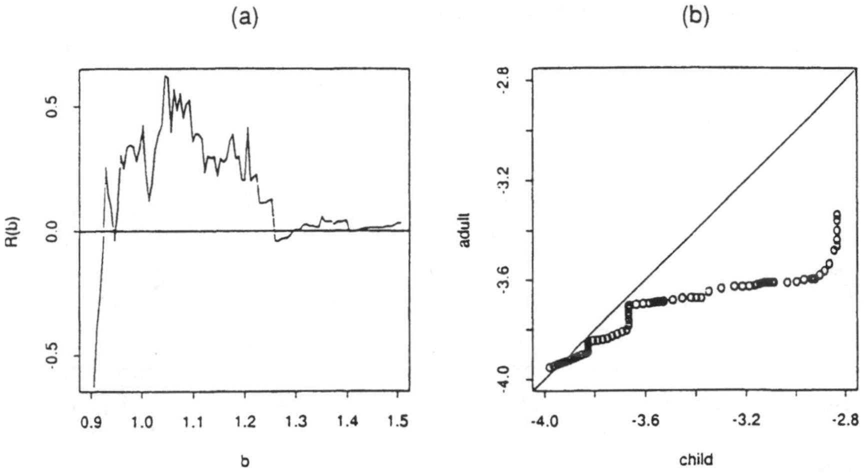

Figure 5 (a) Zero-crossings of rank estimator. (b) Q–Q plot of residuals of two age groups: child (age ≤ 4) and adult (age > 4)

Gross (1996) noted that the R-estimate $\hat{\beta} = 0.92$ appears to be too high in comparison with another estimate of around 0.55 obtained from the parametric modeling of Kalbfleisch and Lawless (1988), while her weighted least-squares estimate $\tilde{\beta} = 0.59$ (or $\tilde{\beta} = 0.53$ from the unaltered data) appears to be much closer. We use the method of stratified residuals described above to examine the adequacy of the assumed model. There are two natural strata, corresponding to $x = -1$ (or age ≤ 4) and $x = 1$ (or age > 4). Taking $R' = -3.95$ in equation (35), which corresponds to the choice $\nu = 3$ in the second paragraph of this section (where R' and R'' are defined), we computed the product-limit curves $\hat{F}_{\hat{\beta},x=-1}(u|R')$ and $\hat{F}_{\hat{\beta},x=1}(u|R')$. Figure 5(b), which gives the Q–Q plot of the quantiles of these distributions up to $u \leq R''$, shows marked departure from the diagonal line, showing that $\hat{F}_{\hat{\beta},x=-1}(\cdot|R')$ tends to be stochastically smaller than $\hat{F}_{\hat{\beta},x=1}(\cdot|R')$.

The assumption of i.i.d. ϵ_j in equation (1) does not seem to be valid. Since there are 33 patients with age ≤ 4 and 260 patients of age > 4, the sample sizes appear to be adequate for treating the two groups separately, as in Gross and Lai (1996), without making the questionable assumption in equation (1) that the two groups have the same distribution after adjusting for a mean difference between them.

In the case of censored data (without truncation), Smith and Zhang (1995) proposed to plot the "renovated residuals"

$$e_i = \delta_i(\tilde{y}_i(\hat{\beta}) - \hat{\alpha}) + (1 - \delta_i)\left\{\int_{u>\tilde{y}_i(\hat{\beta})} u\mathrm{d}\hat{F}_{\hat{\beta}}(u|\tilde{y}_i(\hat{\beta})) - \hat{\alpha}\right\} \quad (36)$$

versus each component of the covariate vector $x_i = (x_{i1}, \ldots, x_{ip})^T$. An example of such plots is shown in Figure 6(a), in which the vertical lines trace the renovations $\int_{u>\tilde{y}_i(\hat{\beta})} u\mathrm{d}\hat{F}_{\hat{\beta}}(u|\tilde{y}_i(\hat{\beta})) - \hat{\alpha}$ of the censored residuals $\tilde{y}_i(\hat{\beta}) - \hat{\alpha}$ (with $\delta_i = 1$). These vertical lines may vary considerably in length and may clutter the picture if the spacing between the censored residuals is small. A more revealing picture showing how the renovated residuals vary with each predictor variable u can be obtained by plotting a linear smoother of the form $\hat{g}(u) = \sum_{i=1}^{n} w_{ni}(u)e_i$ versus u. Let $u_1, \ldots, u_n$ denote the sample values of the kth predictor variable (i.e., $u_i = x_{ik}$). One such smoother is the locally weighted running-line smoother, which can be implemented by using *loess* in the S statistical computing language (Becker et al., 1988). Specifically, letting u_i, $i \in N(u)$, denote the J nearest neighbors of u, the weights $w_{ni}(u)$ for $i \in N(u)$ are those of weighted least-squares regression of e_i on u_i, $i \in N(u)$, with weights $w(|u - u_i|/\max_{j \in N(u)} |u_i - u_j|)$, where $w(z)$ is the tri-cube weight function defined on [0, 1] by $w(z) = (1 - z^3)^3$. For $i \notin N(u)$, $w_{ni}(u) = 0$.

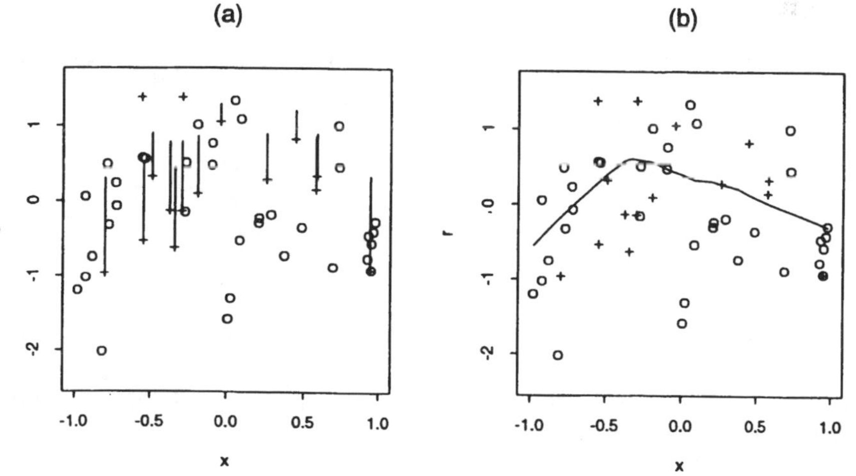

Figure 6 (a) Plot of renovated residuals. (b) Plot of residuals and locally weighted running-line smoother of renovated residuals. o, uncensored residual; +, censored residual

EXAMPLE 6: Consider the regression model $y_i = -x_i^2 + \epsilon_i$ ($i = 1, \ldots, n$), where the ϵ_i are i.i.d. standard normal random variables and the x_i are independent, uniformly distributed on $[-1, 1]$ and independent of the ϵ_i. The y_i are subject to right censoring by independent random variables $c_i = -x_i^2 + 1 + w_i$, where the w_i are i.i.d. standard normal random variables that are independent of $\{(x_n, \epsilon_n), n \geq 1\}$. A sample of size $n = 50$ was generated from this model, which has a censoring rate of 30%. Without assuming knowledge of the actual regression model, the Buckley–James method was used to fit a simple linear model $y_i = \beta_0 + \beta_1 x_i + \epsilon_i$, and the renovated residuals of the fit are plotted in Figure 6(a). The two largest residuals in the figure are censored and are therefore not prolonged under renovation. It is difficult to detect departures from the actual quadratic model in the plot of these renovated residuals computed from the fitted linear model. A more revealing plot showing a quadratic trend is given in Figure 6(b), where the solid curve represents the locally weighted running-line smoother of these renovated residuals, for which we took the default value $2n/3$ in *loess* as the smoothing parameter J (=number of nearest neighbors).

To explain the rationale behind our smoothing of the renovated residuals e_i, first replace them by $\hat{\epsilon}_i$ that are defined by replacing $\hat{\alpha}$ and $\hat{\beta}$ in the right hand side of equation (36) by the true parameter values α and β. Suppose that

$$H(s) = \lim_{n\to\infty} n^{-1} \sum_{i=1}^{n} P\{c_i - \beta^T x_i \geq s\} \qquad \text{exists for every } s \tag{37}$$

Let $\tau = \inf\{s : H(s) = 0\}$. Then it follows from Theorem 4 of Lai and Ying (1991a) that with probability 1,

$$\sup\left\{ |\hat{F}_\beta(u|v) - F(u|v)| : u \geq v, \sum_{i=1}^{n} I(\tilde{y}_i(\beta) \geq u) \geq n^\lambda \right\} \to 0$$

as $n \to \infty$, for every $0 < \lambda < 1$, where $F(u|v) = P\{\tilde{y}_i(\beta) \leq u | \tilde{y}_i(\beta) > v\}$ for $u \geq v$. Hence $\hat{\epsilon}_i$ can be approximated by

$$\eta_i = \delta_i(\tilde{y}_i(\beta) - \alpha) + (1 - \delta_i)\left\{ \int_{\tau > u > \tilde{y}_i(\beta)} u \mathrm{d}F(u|\tilde{y}_i(\beta)) - \alpha \right\}$$

Since $\epsilon_i = \tilde{y}_i(\beta) - \alpha$ has mean 0 and is independent of (x_i, c_i), integration by parts yields

$$E(\eta_i | x_i, c_i) = \int_{-\infty}^{\tau} (1 - F(u))\mathrm{d}u - \alpha = -\int_{\tau}^{\infty} (1 - F(u))\mathrm{d}u$$

which does not depend on x_i and c_i. Hence, the regression function of η_i on the covariate x_i is constant with respect to the covariate, and plots of linear smoothers of $\hat{\epsilon}_i$ ($\approx \eta_i$) versus components of the covariate vector should be trend-free if equation (1) indeed holds. Conspicuous trends in the plots of linear smoothers of the renovated residuals versus components of the covariate vector, therefore, suggest possible departures from the regression model in equation (1).

When the y_j are subject to both right censoring and left truncation, the renovated residuals can be defined as

$$e_i^o = \delta_i^o(\tilde{y}_i^o(\hat{\beta}) - \hat{\alpha}) + (1 - \delta_i^o)\left\{\int_{u > \tilde{y}_i^o(\hat{\beta})} u d\hat{F}_{\hat{\beta}}(u|\tilde{y}_i^o(\hat{\beta})) - \int_{u \geq t_i^o(\hat{\beta})} u d\hat{F}_{\hat{\beta}}(u|t_i^o(\hat{\beta})-) - \hat{\alpha}\right\} \quad (38)$$

and we can likewise apply a linear smoother of the form $\hat{g}(u) = \sum_{i=1}^n w_{ni}(u)e_i^o$ for each predictor variable u. Trends in the graph of $\hat{g}$ suggest possible departures from the regression model in equation (1).

4. CONCLUDING REMARKS

As pointed out in Section 1, the problem of robust regression with censored and truncated data has applications to various disciplines, including clinical medicine, epidemiology, reliability engineering, astronomy, and economics. Ideas and methods from different areas of statistics, including multivariate analysis, experimental design, sampling, robustness, survival analysis, counting processes and semiparametric theory, have contributed to its solution. In particular, truncation and censoring are themselves features of the experimental design. For example, in reliability experiments, life tests are conducted at accelerated stresses and the experimental data are used to estimate the parameters in a model on the relationship between time to failure and stress, such as the Arrhenius relationship or the inverse power law. However, at low stress conditions, which are of greatest practical interest, only a small fraction of the test units have failed by the prescheduled end of the experiment, thus resulting in censored failure times whose relationship to stress (the independent variable) has to be analyzed. Moreover, as discussed in Section 2.6, sampling techniques have proved effective, in the form of bootstrap resampling, to overcome the complexity of the sampling distributions of the M-estimators of regression parameters based on l.t.r.c. data. Another example of the use of sampling ideas is the simulated annealing algorithm proposed by Lin and Geyer (1992) to search for the minimizer of $\|R(b)\|$ in computing R-estimators from l.t.r.c. data.

This paper gives a unified exposition of recent work on R- and M-estimators of regression parameters based on l.t.r.c. data. It also introduces a relatively simple algorithm to compute M-estimators. Starting with a good preliminary estimate described in Section 2.4, the algorithm typically converges after a few Gauss–Newton-type iterations. A more extensive search involving the simplex method is used to supplement these iterations in the extraordinary situation when they have not converged before a prespecified upper bound on the number of iterations. This computational method makes M-estimators much more attractive than R-estimators, which have similar robustness properties but much higher computational complexity. It also makes the computationally intensive bootstrap methods in Section 2.6 and the "leave-one-out" regression diagnostics in Section 3 feasible for the M-estimators based on l.t.r.c. data. For multivariate covariates, graphical methods for displaying the censored and truncated residuals are given in Section 3 to assess the adequacy of the regression model.

REFERENCES

Amemiya, T. (1985). *Advanced Econometrics.* Cambridge, MA: Harvard University Press.

Andersen, P.K., Borgan, O., Gill, R.D., and Keiding, N. (1993). *Statistical Models Based on Counting Processes.* New York: Springer.

Becker, R.A., Chambers, J.M. and Wilks, A.R. (1988). *The New S Language: A Programming Environment for Data Analysis and Graphics.* Belmont, CA: Wadsworth.

Belseley, D.A., Kuh, E., and Welsch, R.E. (1980). *Regression Diagnostics.* New York: Wiley.

Bhattacharya, P.K., Chernoff, H., and Yang, S.S. (1983). Nonparametric estimation of the slope of a truncated regression. *Ann. Stat.*, **11**, 505–514.

Buckley, J. and James, I. (1979). Linear regression with censored data. *Biometrika*, **66**, 429–436.

Cox, D.R. and Snell, E.J. (1968). A general definition of residuals (with discussion). *J. R. Stat. Soc., Ser. B*, **30**, 248–275.

Dempster, A.P., Laird, N.M., and Rubin, D.B. (1977). Maximum likelihood from incomplete data via the EM algorithm (with discussion). *J. Roy. Stat. Soc., Ser. B*, **39**, 1–38.

Fygenson, M. and Zhou, M. (1994). On using stratification in the analysis of linear regression models with right censoring. *Ann. Stat.*, **22**, 747–762.

Goldberger, A.S. (1981). Linear regression after selection. *J. Econometrics*, **15**, 357–366.

Gross, S. (1996). Weighted estimation in linear regression for truncated survival data. *Scand. J. Stat.*, **23**, 179–193.

Gross, S. and Lai, T.L. (1996a). Nonparametric estimation and regression analysis with left truncated and right censored data. *J. Am. Stat. Assoc.*, **91**, 1166–1180.

Gross, S. and Lai, T.L. (1996b). Bootstrap methods for truncated and censored data. *Stat. Sinica*, **6**, 509–530.

Hall, P. (1992). *The Bootstrap and Edgeworth Expansions.* New York: Springer.

Hausman, J.A. and Wise, D.A. (1976). The evaluation of results from truncated samples: The New Jersey negative income tax experiment. *Ann. Econ. Soc. Meas.*, **5**, 421–445.

Hausman, J.A. and Wise, D.A. (1977). Social experimentation, truncated distributions and efficient estimation. *Econometrica*, **45**, 319–339.

Holgate, P. (1965). Fitting a straight line to data from a truncated population. *Biometrics*, **21**, 715–720.

Huber, P.J. (1973). Robust regression: Asymptotics, conjectures and Monte Carlo. *Ann. Stat.*, **1**, 799–821.

Huber, P.J. (1977). *Robust Statistical Procedures.* Philadelphia, PA: Society of Industrial and Applied Mathematics.

Huber, P.J. (1981). *Robust Statistics.* New York: Wiley.

Jurečkova, J. (1969). Asymptotic linearity of a rank statistic in regression parameter. *Ann. Math. Stat. 40*, 1889–1900.

Jurečkova, J. (1971). Nonparametric estimates of regression coefficients. *Ann. Math. Stat.*, **42**, 1328–1338.

Kalbfleisch, J.D. and Lawless, J.F. (1989). Inference based on retrospective ascertainment: An analysis of data on transfusion-related AIDS. *J. Am. Stat. Assoc.*, **84**, 360–372.

Kalbfleisch, J.D. and Prentice, R.L. (1980). *The Statistical Analysis of Failure Time Data.* New York: Wiley.

Kay, R. (1977). Proportional hazard regression models and the analysis of censored survival data. *Appl. Stat.*, **26**, 227–237.

Keiding, N., Holst, C., and Green, A. (1989). Retrospective estimation of diabetes incidence from information in a current prevalent population and historical mortality. *Am. J. Epidemiol.*, **130**, 588–600.

Kim, C.K. and Lai, T.L. (1998). Bootstrap methods for confidence intervals and standard errors in censored and truncated regression models. Technical Report, Department of Statistics, Stanford University.

Koul, H., Susarla, V., and Van Ryzin (1981). Regression analysis with randomly right-censored data. *Ann. Stat.*, **9**, 1276–1288.

Lai, T.L. and Ying, Z. (1991a). Estimating a distribution function with truncated and censored data. *Ann. Stat.*, **19**, 417–442.

Lai, T.L. and Ying, Z. (1991b). Rank regression methods for left-truncated and right-censored data. *Ann. Stat.*, **19**, 531–556.

Lai, T.L. and Ying, Z. (1992). Asymptotically efficient estimation in censored and truncated regression models. *Stat. Sinica*, **2**, 17–46.

Lai, T.L. and Ying, Z. (1994). A missing information principle and *M*-estimators in regression analysis with censored and truncated data. *Ann. Stat.*, **22**, 1222–1255.

Lai, T.L., Ying, Z., and Zheng., Z. (1995). Asymptotic normality of a class of adaptive statistics with applications to synthetic data methods for censored regression. *J. Multivariate Anal.*, **52**, 259–279.

Lawless, J.F. (1982). *Statistical Models and Methods for Lifetime Data.* New York: Wiley.

Leurgans, S. (1987). Linear models, random censoring and synthetic data. *Biometrika*, **74**, 301–309.

Lin, D.Y. and Geyer, C.J. (1992). Computational methods for semiparametric linear regression with censored data. *J. Comp. Graph. Stat.*, **1**, 77–90.

Louis, T.A. (1981). Nonparametric analysis of an accelerated failure time model. *Biometrika*, **68**, 381–390.

Miller, R.G. (1976). Least squares regression with censored data. *Biometrika*, **63**, 449–464.

Miller, R.G. and Halpern, J. (1982). Regression with censored data. *Biometrika*, **69**, 521–531.

Nelder, J.A. and Mead, R. (1964). A simplex method for function minimization. *Comput. J.*, **7**, 308–313.

Nicoll, J.F. and Segal, I.E. (1982). Spatial homogeneity and redshift distance law. *Proc. Natl. Acad. Sci. USA*, **79**, 3913–3917.

Orchard, T. and Woodbury, M.A. (1972). A missing information principle: Theory and applications. *Proceedings of the 6th Berkeley Symposium on Mathematics and Statistical Probability*, **1**, pp. 695–715.

Ritov, Y. (1990). Estimation in a linear regression model with censored data. *Ann. Stat.*, **18**, 303–328.

Schmee, J. and Hahn, G.J. (1979). A simple method for regression analysis with censored data. *Technometrics*, **21**, 417–432.

Smith, P.J. and Zhang, J. (1995). Renovated scatterplots for censored data. *Biometrika*, **82**, 447–452.

Tobin, J. (1958). Estimation of relationships for limited dependent variables. *Econometrica*, **26**, 24–36.

Tsiatis, A.A. (1990). Estimating regression parameters using linear rank tests for censored data. *Ann. Stat.*, **18**, 354–372.

Tsui, K.L., Jewell, N.P. and Wu, C.F.J. (1988). A nonparametric approach to the truncated regression problem. *J. Am. Stat. Assoc.*, **83**, 785–792.

Wang, M.C. (1991). Nonparametric estimation from cross-sectional survival data. *J. Am. Stat. Assoc.*, **86**, 130–143.

Wei, L.J., Ying, Z., and Lin, D.Y. (1990). Linear regression analysis of censored survival data based on rank tests. *Biometrika*, **77**, 845–851.

10

Multivariate Calibration

SAMUEL D. OMAN Hebrew University, Jerusalem, Israel

1. INTRODUCTION

In the calibration problem, two variables x and Y (possibly vector-valued) are related by

$$Y = g(x) + \varepsilon \tag{1}$$

for some function g and error term ε. Typically x is a more expensive measurement, while Y is cheaper or easier to obtain. A *calibration sample* $\{(x_i, Y_i), i = 1, \ldots, n\}$ of n independent observations from equation (1) is available, while at the *prediction step* only $Y = Y_0$ is observed. We wish to estimate the corresponding unknown x, denoted by ξ, which satisfies

$$Y_0 = g(\xi) + \varepsilon_0. \tag{2}$$

Here are a few examples.

EXAMPLE 1: In pregnancy monitoring, the week of pregnancy x is often estimated using ultrasound fetal bone measurements Y. In Oman and Wax

(1984, 1985), the two bone measurements F (femur length) and BPD (biparietal diameter) are quadratically related to week of pregnancy W, giving the model

$$\begin{bmatrix} F \\ BPD \end{bmatrix} = \begin{bmatrix} \beta_{10} & \beta_{11} & \beta_{12} \\ \beta_{20} & \beta_{21} & \beta_{22} \end{bmatrix} \begin{bmatrix} 1 \\ W \\ W^2 \end{bmatrix} + \begin{bmatrix} \varepsilon_1 \\ \varepsilon_2 \end{bmatrix}. \tag{3}$$

EXAMPLE 2: In agricultural science, Racine-Poon (1988) studied the degradation profile of an agrochemical using the model

$$Y = \frac{\theta_1}{1 + e^{\theta_2 + \theta_3 \log(x)}} + \varepsilon, \tag{4}$$

where x is the concentration of the chemical in a pot of soil and Y is the weight of a sensitive plant grown in the pot for a given period of time.

EXAMPLE 3: Naes (1985) discusses an application of near-infrared (NIR) spectrophotometry in which the concentration of fat (x) in a sample of fish is to be estimated from a vector Y of (logged) reflectances of light at $q = 9$ different wavelengths. The model is

$$\begin{bmatrix} Y_1 \\ \vdots \\ Y_q \end{bmatrix} = \begin{bmatrix} \beta_{10} & \beta_{11} \\ \vdots & \vdots \\ \beta_{q0} & \beta_{q1} \end{bmatrix} \begin{bmatrix} 1 \\ x \end{bmatrix} + \begin{bmatrix} \varepsilon_1 \\ \vdots \\ \varepsilon_q \end{bmatrix}. \tag{5}$$

In the simplest case, x and Y are scalar-valued and linearly related. If ε is normal, we then have the model

$$Y_i = \beta_0 + \beta_1 x_i + \varepsilon_i, \qquad i \le i \le n \tag{6}$$

$$Y_0 = \beta_0 + \beta_1 \xi + \varepsilon_0, \tag{7}$$

where $\varepsilon_i \sim N(0, \sigma^2)$ independently for $i = 0, 1, \ldots, n$. If $(\hat{\beta}_0, \hat{\beta}_1)$ and $s^2 = SSE/(n-2)$ denote the least-squares estimates using the calibration data, then the maximum likelihood estimate of ξ (assuming $\hat{\beta}_1 \neq 0$) is

$$\hat{\xi} = \frac{Y_0 - \hat{\beta}_0}{\hat{\beta}_1}. \tag{8}$$

A corresponding confidence region (Eisenhart, 1939; Fieller, 1954) is obtained by inverting the pivot

$$\frac{(Y_0 - \hat{\beta}_0 - \hat{\beta}_1 \xi)^2}{s^2 \left[1 + \frac{1}{n} + \frac{(\xi - \bar{x})^2}{s_{xx}} \right]} \sim F(1, n-2); \tag{9}$$

here, $\bar{x}$ is the mean of the x_i in the calibration sample and

$$s_{xx} = \sum_{i=1}^{n} (x_i - \bar{x})^2. \tag{10}$$

Graphically (Figure 1), we obtain $\hat{\xi}$ by drawing a horizontal line at Y_0 until it intersects the estimated regression line, and then projecting down to the x axis; the $(1-\alpha)$-level confidence limits ξ_l and ξ_u are similarly obtained from the intersection of the horizontal with the usual $(1-\alpha)$-level prediction envelope (for Y) about the regression line.

This simple example illustrates two nonstandard characteristics of the calibration problem.

1. $\hat{\xi}$, being the ratio of two normals, has no first moment and infinite mean-squared error (MSE) for ξ.
2. Fieller's region, although having exact $(1-\alpha)$-level coverage probability, need not be a finite interval. If the regression line in Figure 1 is too shallow, then the horizontal line may lie entirely within the curved envelope, giving the whole real line as the region. Even worse, the horizontal may intersect only one part (e.g., the upper) of the envel-

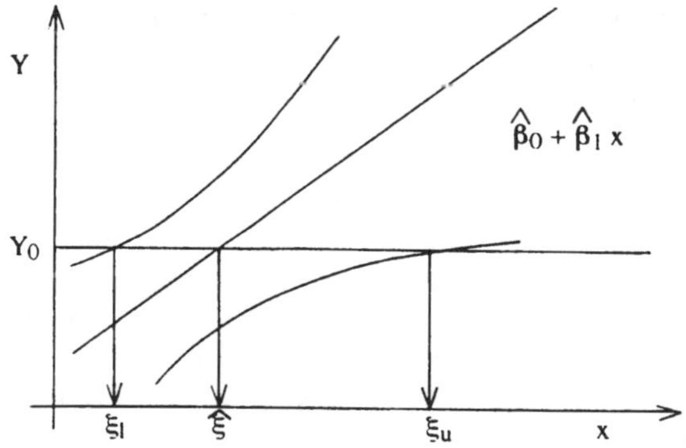

Figure 1 Simple linear calibration. Y_0 is a new observation and $\hat{\xi}$ is the corresponding estimate of ξ. Curved lines are the $(1-\alpha)$-level confidence envelope for predicting new Y, and (ξ_l, ξ_u) is the corresponding confidence interval for ξ.

ope, giving two semi-infinite intervals. A necessary and sufficient condition that these anomalies not occur is

$$\frac{\hat{\beta}_1^2}{s^2/s_{xx}} > F_\alpha(1, n-2), \tag{11}$$

i.e., that the slope be significant at the α level. When equation (11) is violated, the pathological regions are simply reminding us that we have no business trying to estimate ξ from Y_0 in equation (7) when $\beta_1 = 0$. We also remark that any procedure other than Fieller's is still bound to give arbitrarily long intervals for a certain percentage of the time. This is because Gleser and Hwang (1987) show that any confidence set for ξ with guaranteed $1-\alpha$ coverage probability must have infinite expected length.

Discussions and comprehensive bibliographies for the calibration problem in general may be found in Aitchison and Dunsmore (1975), Hunter and Lamboy (1981), Brown (1982, 1993), Martens and Naes (1989), Osborne (1991), and Sundberg (1997). Our intention here is to focus on the case where Y, and possibly x, are vector-valued. In the next section we assume that the x observations in the calibration sample have been randomly drawn from the same population from which the future ξ will arise. The problem thus essentially becomes one of prediction in multiple regression, ableit with some special characteristics. Section 3 treats point estimation in the alternative case of *controlled calibration*, in which the experimenter sets the x values in the calibration sample, and Section 4 treats interval estimation in the same context. Space limitations preclude treating important topics such as diagnostics, calibration with more variables than observations (as frequently occurs in NIR applications) and repeated-use intervals; see the above-mentioned articles for references.

2. RANDOM CALIBRATION

Suppose the observations (x_i, Y_i) at the calibration step comprise a random sample from a target population from which the future (ξ, Y_0) will come. For example, in NIR applications samples are often randomly taken from the production line of the product whose quality is to be monitored. In this case, even if equation (1) has a natural causal interpretation (e.g., in Example 3, the reflections at different wavelengths depend on the concentration of fat in the sample, and not the opposite), it makes statistical sense to model the x_i in terms of Y_i, say

$$x_i = h(Y_i) + \delta_i, \tag{12}$$

estimate h by regressing x_i on Y_i and then estimate ξ using $\hat{\xi} = \hat{h}(Y_0)$.

If h is linear, then we have a multiple regression problem; for the remainder of this section we reverse notation to agree with standard notation for the regression model. Thus $y_{n\times 1}$ denotes the vector of observations on the exact measurement (e.g., fat concentration), $X_{n\times q}$, comprising $n > q$ rows $x^t_{(i)}$, denotes the (full rank) matrix of observations on Y (*e.g.*, NIR readings), and we assume (ignoring the intercept) that

$$y \sim N(X\beta, \sigma^2 I).$$

Let $\hat{\beta}$ denote the maximum likelihood (OLS) estimator, so that

$$\hat{\beta} \sim N(\beta, \sigma S^{-1}) \tag{13}$$

where $S = X^t X$. The problem typically has the following special characteristics, especially in NIR applications:

1. There are a large number of explanatory variables. For example, each row $x^t_{(i)}$ of X may be a vector of NIR reflectances at hundreds of wavelengths (Martens and Naes, 1991).
2. The components of the spectrum tend to change together, resulting in a highly multicollinear or ill-conditioned regression problem. That is, if $\lambda_1 \geq \cdots \geq \lambda_q > 0$ denote the eigenvalues of S, then there is a large spread between them, and some may be quite close to 0.
3. The primary object is "black box" prediction (as opposed, for example, to inference about particular components β_i). Since the prediction is at point similar to the $x^t_{(i)}$ in X, an appropriate criterion for choosing an estimator β^* is its prediction MSE,

$$\begin{aligned} PMSE(\beta^*, \beta) &= \sum_{i-1}^{n} E(\beta^{*t} x_{(i)} - \beta^t x_{(i)})^2 \\ &= E(\beta^* - \beta)^t S(\beta^* - \beta). \end{aligned} \tag{14}$$

Note that $PMSE(\hat{\beta}, \beta) = q\sigma^2$, independently of the eigenvalues λ_i, so that multicollinearity *per se* poses no special problem when predicting with $\hat{\beta}$. We now briefly discuss four alternatives to $\hat{\beta}$ which have been proposed for this type of calibration problem.

2.1 Principal Components Regression

Let $v_1, \ldots, v_q$ denote orthonormal eigenvectors corresponding to $\lambda_1, \ldots, \lambda_q$, respectively, and let

$$z^{(i)} = Xv_i \tag{15}$$

denote the corresponding principal components. In principal components regression the q columns of X are transformed to $\{z^{(1)}, \ldots, z^{(q)}\}$, y is regressed on (usually the first) $k < q$ of the $z^{(i)}$, and the vector of regression coefficients is then transformed back to the original units, giving the q-vector $\hat{\beta}_{PC(k)}$. Thus essentially $q - k$ "variables" are eliminated, where each variable is a linear combination of the original variables which exhibits little variation in the x-space. Typically, the tuning parameter k is chosen using the data in y, e.g., by cross-validation. Leaving out variables decreases variance while increasing bias, and thus if the omitted components are in fact not too highly correlated with y, then $\hat{\beta}_{PC(k)}$ will have lower PMSE than $\hat{\beta}$. However, if important $z^{(i)}$ have been omitted, then the PMSE may increase substantially (Bock et al., 1973; Seber, 1977; Jennrich and Oman, 1986).

2.2 Ridge Regression

As typically used, ridge regression shrinks $\hat{\beta}$ towards the origin, giving the estimator

$$\hat{\beta}_{RR(k)} = (S + kI)^{-1} S\hat{\beta} \tag{16}$$

for a small "biasing parameter" $k > 0$ determined from the data. If some of the eigenvalues λ_i are very small, then even small values of k will result in dramatic shrinkage. Although Hoerl (1962) originally proposed ridge estimators for stabilizing numerical computations, a belief (based largely on simulations, and to an extent ignoring the effect of choosing k from the data) developed that $\hat{\beta}_{RR(k)}$ has desirable properties in terms of unweighted MSE, given by

$$MSE(\beta^*, \beta) = E\|\beta^* - \beta\|^2. \tag{17}$$

Note that $MSE(\hat{\beta}, \beta) = \sigma^2 \sum_{i=1}^{q} \lambda_i^{-1}$, which (as opposed to the PMSE of equation (14)) is very large for multicollinear data. The performance of equation (16) depends heavily, however, on the location of the true β. In particular, simulations and theoretical results (Newhouse and Oman, 1971; Thisted, 1976; Bingham and Larntz, 1977; Casella, 1980) show that, although ridge regression can dramatically reduce MSE for some β, $\hat{\beta}_{RR(k)}$ can perform much worse than OLS if β lies close to the space spanned by the eigenvector v_q (i.e., in the direction of least information). In fact, Casella (1980) essentially showed that a necessary and sufficient condition for a large class of ridge estimators (with data-dependent k) to be minimax with respect to equation (17) is that $\lambda_q^{-2} < \sum_{i=1}^{q-1} \lambda_i^{-1}$. This is often not satisfied with extremely multicollinear data, and as Casella points out, improving conditioning is to an extent incompatible with minimaxity. We also remark that the unweighted MSE criterion in equation (17) is more appropriate for parameter estimation or extrapolation than predicting in the region of the data.

A natural alternative justification for equation (16) is Bayesian. In fact, if a priori

$$\beta \sim N(\mu, \tau^2 D) \tag{18}$$

then combining with equation (13) immediately gives

$$E(\beta|\hat{\beta}) = (S + kD^{-1})^{-1}(S\hat{\beta} + kD^{-1}\mu) \tag{19}$$

where $k = \sigma^2/\tau^2$. Thus equation (16) is a Bayes estimator for the prior

$$\beta \sim N(0, \tau^2 I). \tag{20}$$

However, it seems difficult to justify equation (20) since the prior mean of 0 says that we believe there is no connection between y and X—clearly inappropriate for most calibration applications. Moreover, the diagonal prior covariance matrix in equation (20) means we have formed independent opinions of the components $\beta_i = \partial Y/\partial X_i$, even though the variables in X are very highly correlated in the sample and are expected to be so in the future as well. In particular, if we use a previous hypothetical experiment to elicit the prior on β (Raiffa and Schlaifer, 1961; Tiao and Zellner, 1964; Oman, 1985b), then the choice $D = I$ means that in the hypothetical experiment the explanatory variables were uncorrelated. It seems more natural to choose the observed S as the design matrix for the hypothetical experiment, giving $D = S^{-1}$ in equation (18).

2.3 Stein Estimation

Applying the original estimator of James and Stein (1961) gives (assuming $q \geq 3$)

$$\hat{\beta}_{ST} = (1 + k)^{-1}\hat{\beta}, \tag{21}$$

where k is chosen from the data in such a way to in fact guarantee that

$$PMSE(\hat{\beta}_{ST}, \beta) < PMSE(\hat{\beta}, \beta) \tag{22}$$

for all β. Unfortunately k decreases rapidly with the signal-to-noise ratio $\hat{\beta}'S\hat{\beta}/s^2$, so that in typical regression problems the improvement in equation (22) is minimal.

However, Sclove's (1968) variant of equation (21) can given substantial PMSE improvement. His estimator is

$$\hat{\beta}_L = \hat{\beta}_0 + \left\{ 1 - \frac{(q - \ell - 2)\tilde{\sigma}^2}{SSR - SSR_0} \right\}^+ (\hat{\beta} - \hat{\beta}_0), \tag{23}$$

where $\hat{\beta}_0$ is the maximum likelihood estimator of β under the hypothesis

$$H_0 : E(y) \in L$$

for a specified ℓ-dimenstional subspace L of span(X). SSR and SSR_0 are the regression sums of squares under the full model and H_0, $\tilde{\sigma}^2 = [SSE/(n - q + 2)]$ is almost the usual estimate s^2 of σ^2, and $\{\ \}^+$ denotes the positive part. For highly multicollinear data a natural choice of L is the space spanned by the first $\ell < q$ principal components, so that $\hat{\beta}_0 = \hat{\beta}_{PC(\ell)}$. James and Stein's (1961) results guarantee that the PMSE of $\hat{\beta}_L$ is less than that of $\hat{\beta}$ for *all* β, thus avoiding the overshrinking potential of $\hat{\beta}_{PC(k)}$; and the closer $E(y)$ is to L, the closer the PMSE gets to its minimum value of $(\ell + 2)\sigma^2$. Note from equation (23) that $\hat{\beta}_L$ is like a smoothed "pre-test" estimator, giving more weight to the submodel estimate $\hat{\beta}_0$ when SSR_0 is close to SSR. Also, $\hat{\beta}_L$ has an appealing empirical Bayes interpretation: shrinking towards $\hat{\beta}_0$ corresponds to a nonzero prior mean in equation (18), and comparing equations (23) and (19) shows that $D = S^{-1}$ in equation (18), corresponding to a hypothetical design matrix which reflects the observed correlations in the sample.

In principle, the number of components ℓ must be chosen without reference to the data vector y, in order to guarantee PMSE dominance. However, cross-validation evaluations on NIR calibration data (Oman, 1991; Oman et al., 1993) indicate that ℓ may be chosen from the data. Alternatively, one can use a multiple-shrinkage version of equation (23), with data-driven subspace choice, which is guaranteed to give lower PMSE than OLS (George, 1986; George and Oman, 1996).

2.4 Partial Least-Squares

Partial least squares (PLS) algorithms, originally developed in the context of general systems analysis (Jöreskog and Wold, 1982), are widely used in chemometrics (Frank, 1987). Statistical discussions of PLS and additional references may be found in Naes et al. (1986), Helland (1990), Stone and Brooks (1990), Naes and Helland (1993), Brown (1993), and Helland and Almøy (1994).

As with PCR, PLS constructs a sequence

$$w^{(i)} = Xu_i, \qquad i = 1, \ldots, m \leq q \tag{24}$$

of uncorrelated predictor variables and then regresses y on a subset of them. Unlike PCR, however, which chooses the $z^{(i)}$ in equation (15) to have maximum sample variance at each step, PLS chooses each $w^{(i)}$ to maximze the sample covariance with y. Observe that if the criterion were maximizing the sample *correlation* with y, then the first predictor variable would be the OLS predictor $w^{(1)} = X\hat{\beta}$. Moreover, the process would stop with $w^{(1)}$ since $w^{(2)}$ being uncorrelated with $w^{(1)}$ requires $u_2^t X^t X \hat{\beta} = 0$, and since $X^t X \hat{\beta} = X^t y$ this already implies zero correlation between $w_{(2)} = Xu_2$ and y. In fact, Stone and Brooks (1990) view PCR, PLS, and OLS as estimators in a continuum of methods which at the first stage choose the predictor variables in equation (24) to maximize

$$(u_i^t X^t y)^2 (u_i^t X^t X u_i)^{\alpha/(1-\alpha)-1}. \tag{25}$$

OLS corresponds to $\alpha = 0$, PLS to $\alpha = \frac{1}{2}$, and PCR to $\alpha = 1$. Although this unifying approach is elegant, it is not clear why equation (25) is a meaningful objective function. In particular, it is hard to justify why maximizing the sample covariance between $w^{(i)}$ and y is preferable to maximizing the sample correlation. A more appealing justification is the work of Naes and Martens (1985) and Helland (1990), who show that the space spanned by the $w^{(i)}$ is essentially that spanned by the "relevant components," i.e., those principal components $z^{(i)}$ having nonzero correlation with y.

Asymptotic and simulation comparisons of PLS and PCR have not indicated a clear superiority of one method over the other (Frank and Friedman, 1993; Helland and Almøy, 1994). Sundberg (1997) has a detailed illustration and comparison of ridge regression and PLS on a set of spectroscopic data for monitoring nitrate content in municpal waste water.

3. POINT ESTIMATES IN CONTROLLED CALIBRATION

We now return to the notation of Section 1. In equation (1), assume

$$\varepsilon \sim N(0, \Gamma) \tag{26}$$

and

$$g(x) = C_{q \times p} h(x) \tag{27}$$

where C and $\Gamma > 0$ are unknown and $h(x) = (h_1(x), \ldots, h_p(x))^t$ for known functions h_j. The calibration sample comprises n independent observations (x_i, Y_i) from equation (1), where the r-vectors x_i are considered fixed.

It is useful to distinguished two special cases.

1. *Linear calibration*. Here $h(x) = (1, x^t)^t$. Separating out the constant term, equation (2) becomes

   $$Y = a + B\xi \tag{28}$$

where $B_{q\times r}$ is assumed to be full rank, in order to ensure identifiability of ξ (this is analogous to assuming $\beta_1 \neq 0$ in equation (7)). Observe that equation (5) of Example 3 fits into this framework.

2. *Semilinear calibration.* In this case $h(x)$ is intrinsically nonlinear in x, e.g., $h(x) = (1, x, x^2)'$ as in equation (3) of Example 1. To ensure identifiability of ξ we assume equation (2) holds for all ξ in some region Ω, and that for all such ξ the matrix

$$Z(\xi) = \left(\frac{\partial g_i}{\partial \xi_j}(\xi)\right)_{q\times r}$$

is of full rank. Several points should be noted. First, equation (4) of Example 2 does not fit into the framework of equation (27) since the unknown parameters θ_i enter nonlinearly. Second, the distinction between cases 1 and 2 is not important for the calibration step, since in both cases estimating C and Γ is simply a problem in multivariate linear regression. At the prediction step, however, the distinction is important because in equation (28) the unknown parameter ξ appears linearly, while in the semilinear case it appears nonlinearly. Third, observe that equation (2) suggests a regression problem with Y_0 a vector of q correlated "observations" on a dependent variable. In the semilinear case, equation (2) gives a nonlinear regression problem with unknown parameter vector ξ, while equation (28) in the linear case suggests a linear regression with (by an unfortunate reversal of the usual notation) "design matrix" B, vector of "regression coefficients" ξ and vector of "observations" $Y_0 - a$. This analogy is very useful both in motivating different estimation techniques and in understanding their geometry.

We now discuss a number of estimators in the linear case, assuming first that $r = 1$. This is primarily for ease of exposition, and because results are more complete. Also, one can argue (Sundberg, 1982) that even when x is vector-valued, it may be preferable to treat its components separately. In Section 3.6 we discuss extensions to the semilinear and vector-valued cases.

Since the problem is invariant under translation, we may center the x_i and ξ by subtracting the sample mean, and similarly for Y_0. We then essentially have the following model:

$$\begin{aligned} Y_i &= \beta x_i + \varepsilon_i, \qquad 1 \le i \le n \\ Y_0 &= \beta\xi + \varepsilon_0 \end{aligned} \tag{29}$$

where $\varepsilon_0, \varepsilon_1, \ldots, \varepsilon_n$ are independent $N(0, \Gamma)$ and $\bar{x} = 0$. Sufficient statistics are Y_0,

$$\hat{\beta} = s_{xx}^{-1} \sum_{i=1}^{n} x_i Y_i \tag{30}$$

for s_{xx} defined in equation (10), and

$$\begin{aligned} S &= \sum_{i=1}^{n} (Y_i - \hat{\beta} x_i)(Y_i - \hat{\beta} x_i)^t \\ &= \sum_{i=1}^{n} Y_i Y_i^t - s_{xx} \hat{\beta} \hat{\beta}^t, \end{aligned} \tag{31}$$

and they are distributed independently as

$$\begin{aligned} \hat{\beta} &\sim N(\beta, s_{xx}^{-1} \Gamma), \\ S &\sim W_q(m, \Gamma) \end{aligned} \tag{32}$$

and

$$Y_0 \sim N(\beta\xi, \Gamma)$$

for $m = n - 1$. ($W_q(m, \Gamma)$ denotes the q-dimensional Wishart distribution with m degrees of freedom and shape parameter Γ.)

Before presenting the estimators, it is instructive to determine how much information on ξ is actually contained in the data. To do this, let

$$\Lambda = \Gamma^{-1}$$

denote the dispersion matrix and partition the full information matrix as

$$\mathcal{I} = \begin{bmatrix} \mathcal{I}_{\xi\xi} & \mathcal{I}_{\xi\beta} & \mathcal{I}_{\xi\Lambda} \\ & \mathcal{I}_{\beta\beta} & \mathcal{I}_{\beta\Lambda} \\ & & \mathcal{I}_{\Lambda\Lambda} \end{bmatrix}.$$

From equation (32) the log–likelihood is, except for an additive constant,

$$\begin{aligned} \ell(\xi, \beta, \Lambda) = -\tfrac{1}{2}\{ & s_{xx}(\beta - \hat{\beta})^t \Lambda (\beta - \hat{\beta}) + (\beta\xi - Y_0)^t \Lambda (\beta\xi - Y_0) \\ & + tr \Lambda S - (m+2) \log |\Lambda| \}. \end{aligned} \tag{33}$$

Differentiating equation (33), we see that $\mathcal{I}_{\xi\Lambda}$ and $\mathcal{I}_{\beta\Lambda}$ are expectations of linear functions of the centered vectors $\hat{\beta} - \beta$ and $Y_0 - \beta\xi$, and hence equal zero; it follows that $\mathcal{I}_{\xi|\beta,\Lambda} = \mathcal{I}_{\xi\xi} - \mathcal{I}_{\xi\beta}(\mathcal{I}_{\beta\beta})^{-1}\mathcal{I}_{\beta\xi}$. Straightforward computations, as in Kubokawa and Robert (1994), then give

$$\mathcal{I}_{\xi|\beta,\Lambda} = \beta^t \Gamma^{-1} \beta \left(\frac{s_{xx}}{s_{xx} + \xi^2} \right). \tag{34}$$

The first component in equation (34) reflects the ability of Y_0 to discriminate between different values of ξ when β and Γ are known. To see this, in Figure

1 replace the estimated regression line by the true line $\beta_0 + \beta_1 x$. Then large β_1^2/σ^2 means that typical deviations of Y_0 from $E(Y_0)$ will correspond to smaller deviations of $\hat{\xi}$ from ξ_0, either because the line is steeper or because the deviations $Y_0 - E(Y_0)$ are expected to be smaller. The second factor in equation (34) reflects the effect of not knowing β and Γ exactly. Defining the normalized squared deviation

$$\Delta = \frac{\xi^2}{s_{xx}/n}, \tag{35}$$

we can write the second factor as $1/(1 + \Delta/n)$. Thus inference is less accurate for more atypical ξ. For a large calibration sample with x_i which cover the expected range of values of ξ, Δ/n will be quite small even for atypical ξ. However, even as n becomes infinite, the information for ξ stays bounded owing to the uncertainty in the one observation Y_0.

3.1 The Classical Estimator

If β and Γ were known, then analogous to equation (8) we would try to "solve"

$$Y_0 = \beta\xi \tag{36}$$

for ξ. For $q > 1$ this generally cannot be done, so instead we minimize the weighted sum of squares

$$(Y_0 - \beta\xi)'\Gamma^{-1}(Y_0 - \beta\xi).$$

Replacing β and Γ by their estimates from the calibration step then gives the "classical" or "natural" estimator

$$\hat{\xi}_{clas} = (\hat{\beta}'S^{-1}\hat{\beta})^{-1}\hat{\beta}'S^{-1}Y_0 \tag{37}$$

(which reduces to equation (8) when $q = 1$). If Γ is known but β is not, then we replace S by Γ in equation (37).

As previously mentioned, when $q = 1$, $\hat{\xi}_{clas}$ has no first moment and infinite MSE. When $q > 1$, however, the denominator of $\hat{\xi}_{clas}$ is a quadratic form instead of a normal variable, and matters change. In particular, Lieftinck–Koeijers (1988) showed that $\hat{\xi}_{clas}$ has finite mean (when Γ is known) provided $q > 2$. Using the mixture representation for noncentral χ^2 variables (e.g., see Bock, 1975), she found relatively explicit expressions for the first two moments. Generalizing her results, Nishii and Krishnaiah (1988) showed that for unknown Γ, $\hat{\xi}_{clas}$ has finite mean if and only if $q \geq 2$ and finite MSE if and only if $q \geq 3$. They found relatively explicit expressions for these moments; thus

$$E(\hat{\xi}_{clas}) = \lambda E\left(\frac{1}{q+2K}\right) \cdot \xi \tag{38}$$

where

$$\lambda = \beta'\Gamma^{-1}\beta \cdot s_{xx} \tag{39}$$

and

$$K \sim \text{Poisson}(\lambda/2). \tag{40}$$

The MSE is a more complicated expression of the form $a_n(\lambda) + b_n(\lambda) \cdot \Delta$, where Δ is given by equation (35) and a_n, b_n involve expectations of rational functions of K defined as in equation (40). (We remark that (ii) of Theorem 5 in Nishii and Krishnaiah (1988) actually gives $E(\hat{\xi}^2)$ and not $E(\hat{\xi} - \xi)^2$ as stated.)

Since $\lambda E\left(\frac{1}{q+2K}\right) < 1$, equation (38) suggests that $\hat{\xi}_{clas}$ "underestimates" ξ. This is not too surprising if we view equation (36) as an inversion problem, and recall that $\hat{\beta}$ tends to "overestimate" β (Stein, 1956). Another way to see that $\hat{\xi}_{clas}$ is "too small" is by our previously mentioned regression analogy, in which equation (2) becomes

$$Y_0 - \beta\xi + \varepsilon_0$$

and $\hat{\beta}$ is the vector of "true values," β, observed with error. Thus $\hat{\xi}_{clas}$ in equation (37) being too small corresponds to the well-known phenomenon of attenuation in errors-in-variables models (Fuller, 1987). In fact, this analogy has been exploited by Thomas (1991) to provide alternative estimates to $\hat{\xi}_{clas}$; these will not be treated here.

3.2 The Inverse Estimator

In the context of equations (6) and (7), Krutchkoff (1967) proposed the inverse estimator $\hat{\xi}_{inv}$, obtained by regressing the x_i on Y_i as in Section 2 (even though the x_i are fixed) and then substituting the new Y_0 into the estimated regression equation. This proposal understandably generated some controversy; see Brown (1993) and Osborne (1991). We note only the following points:

1. Since $\hat{\xi}_{inv}$ is based on

$$f(x|Y) = \frac{f(Y|x)f(x)}{f(Y)} \tag{41}$$

   where $f(x)$ has been set by the experimenter, $\hat{\xi}_{inv}$ has no naturally associated confidence intervals.

2. If the primary objective is accurate prediction, MSE considerations (or alternative measures of accuracy) may still justify using $\hat{\xi}_{inv}$ instead of $\hat{\xi}_{clas}$.
3. Equation (41) suggests that $\hat{\xi}_{inv}$ may have a Bayesian interpretation; this is discussed in Section 3.4.

Generalizing to $q \geq 1$ gives the estimator

$$\hat{\xi}_{inv} = \hat{\gamma}^t Y_0 \tag{42}$$

where

$$\hat{\gamma} = \left[\sum_{i=1}^{n} Y_i Y_i^t\right]^{-1} \sum_{i=1}^{n} Y_i x_i.$$

From equations (30) and (31) we thus have

$$\hat{\xi}_{inv} = [(S + s_{xx}\hat{\beta}\hat{\beta}^t)^{-1} s_{xx}\hat{\beta}]^t Y_0, \tag{43}$$

and using the binomial inverse theorem (Press, 1972) and equation (37) then gives

$$\hat{\xi}_{inv} = \left(\frac{Q}{s_{xx}^{-1} + Q}\right)\hat{\xi}_{clas} \tag{44}$$

where

$$Q = \hat{\beta}^t S^{-1} \hat{\beta}. \tag{45}$$

Thus $\hat{\xi}_{inv}$ contracts $\hat{\xi}_{clas}$, which is already "too small," even further to the origin. Therefore if $\hat{\xi}_{inv}$ outperforms $\hat{\xi}_{clas}$, we would only expect this to occur when ξ is close to the origin. In fact, Sundberg (1985) compared approximations (which ignore the randomness in $\hat{\beta}$ and S) to the two MSEs and found that $\hat{\xi}_{inv}$ has lower (approximate) MSE than $\hat{\xi}_{clas}$ if and only if

$$\Delta \leq 2 + \frac{n}{\lambda} \tag{46}$$

where Δ and λ are defined by equations (35) and (39).

More precise comparisons of $\hat{\xi}_{inv}$ and $\hat{\xi}_{clas}$, accounting for the randomness in the calibration sample, have been made. Srivastava (1995), extending Kubokawa and Robert's (1994) results for $\Gamma = \sigma^2 I$ with σ^2 unknown, has shown that $\hat{\xi}_{inv}$ is admissible by exhibiting it as a Bayes estimate with respect to a proper prior. Oman and Srivastava (1996), generalizing earlier results of Oman (1985a) for $q = 1$, obtained exact formulas for the bias and MSE of $\hat{\xi}_{inv}$ in terms of moments of rational functions of the Poisson variable K in equation (40). Comparing their MSE formulas with those of Nishii and Krishnaiah (1988) for the MSE of $\hat{\xi}_{clas}$, they proved that $\hat{\xi}_{inv}$ is, in fact, guaranteed to have smaller MSE than $\hat{\xi}_{clas}$ for $|\xi|$ sufficiently small. Numerical cxamination of the MSEs indicates that the region of improvement of $\hat{\xi}_{inv}$ over $\hat{\xi}_{clas}$ is somewhat larger than in equation (46), and limited numerical calculations show that the actual decrease in MSE is not that great for large values of λ.

3.3 Maximum Likelihood Estimation

Let $(\hat{\xi}_{mle}, \hat{\beta}_{mle}, \hat{\Gamma}_{mle})$ denote a maximum likelihood estimate (MLE) using the full data set ($\{x_i, Y_i\}$ and Y_0). In equation (33) it is clear that if $q = 1$ then

$$(\beta\xi - Y_0)^t \Lambda(\beta\xi - Y_0) \tag{47}$$

attains a minimum of zero when $\xi = Y_0/\beta$, for any $\beta \neq 0$. Thus the full data MLEs of β and Γ are identical to those based on the calibration sample alone, $\hat{\beta}_{mle} = \hat{\beta}$, $\hat{\Gamma}_{mle} = \hat{\Gamma}$; and $\hat{\xi}_{mle}$ is the classical estimate $\hat{\xi}_{clas}$. If $q > 1$, however, the situation is more complicated as the minimum of equation (47) over ξ is not independent of β and Λ.

Brown and Sundberg (1987) developed the MLE using a profile likelihood approach. Specifically, let $L(\xi)$ denote the maximized value (up to a multiplicative constant) of the likelihood based on the full data, assuming that ξ is known. We then have a sample of $n+1$ observations $Y_i \sim N(\beta x_i, \Gamma)$ (where $x_0 = \xi$), so standard theory says that

$$L(\xi) = |\hat{\Gamma}_0|^{-(n+1)}$$

where $\hat{\Gamma}_0$ denotes the MLE of Γ using $\{(x_i, Y_i)\}$ and (ξ, Y_0). Expressing $\hat{\Gamma}_0$ and $\hat{\beta}_0$ as weighted expressions involving $\hat{\Gamma}$ and $\hat{\beta}$ then gives

$$|\hat{\Gamma}_0| \propto \frac{1 + \dfrac{\xi^2}{s_{xx}} + (Y_0 - \hat{\beta}\xi)^t S^{-1}(Y_0 - \hat{\beta}\xi)}{1 + \dfrac{\xi^2}{s_{xx}}} \tag{48}$$

Observe that we may partition

$$(Y_0 - \hat{\beta}\xi)^t S^{-1}(Y_0 - \hat{\beta}\xi) = \mathcal{R} + \mathcal{Q}(\xi - \hat{\xi})^2 \tag{49}$$

where

$$\mathcal{R} = (Y_0 - \hat{\beta}\hat{\xi})^t S^{-1}(Y_0 - \hat{\beta}\hat{\xi})$$

and Q is given in equation (45). $\mathcal{R}$ (normalized for degrees of freedom) may be used as a diagnostic statistic, since it indicates how different Y_0 is from its predicted value $\hat{\beta}\hat{\xi}$.

Substituting equation (49) into equation (48), it remains to maximize

$$\ell(\xi) = \left[\frac{1 + \frac{\xi^2}{s_{xx}} + \mathcal{R} + Q(\xi - \hat{\xi})^2}{1 + \frac{\xi^2}{s_{xx}}} \right]^{-1} \tag{50}$$

After further algebraic manipulations, Brown and Sundberg (1987) obtain an explicit expression for $\hat{\xi}_{mle}$ in terms of a root of a quadratic equation. They show that $\hat{\xi}_{mle}$ is unique, and expands $\hat{\xi}_{clas}$ in the sense that either $0 < \hat{\xi}_{clas} < \hat{\xi}_{mle}$ or $\hat{\xi}_{mle} < \hat{\xi}_{clas} < 0$. Although Brown and Sundberg view this property as undesirable from a Bayesian viewpoint, the comments at the end of Section 3.1 suggest that expansion might be useful from frequentist considerations. No MSE comparisons between $\hat{\xi}_{mle}$ and $\hat{\xi}_{clas}$ have been made; owing to the complicated form of $\hat{\xi}_{mle}$, such comparisons would most likely require simulations. Note that asymptotic properties of maximum likelihood estimators depend on increasing information on the parameter. Since the information on ξ stays bounded as $n \to \infty$ (see the discussion following equation (34)), the usual large-sample asymptotics are not valid. Brown and Sundberg (1987) assume instead that $\Gamma \to 0$, which in view of equation (34) guarantees increasing information and allows an asymptotic analysis.

3.4 Bayes Estimation

We describe results due to Brown and Sundberg (1987) and Brown (1993). Denoting $\mathcal{F} = (\beta, \Gamma)$ and $F = (\hat{\beta}, S)$, we wish to compute the posterior density

$$\begin{aligned} f(\xi|F, Y_0) &\propto f(F, Y_0|\xi)f(\xi) \\ &= f(Y_0|F, \xi)f(F|\xi)f(\xi). \end{aligned} \tag{51}$$

If we assume a priori that $\mathcal{F}$ and ξ are independent, then F (whose distribution is defined by $\mathcal{F}$) is independent of ξ as well. Thus $f(F|\xi)$ may be dropped from equation (51), giving

$$f(\xi|F, Y_0) \propto f(Y_0|F, \xi)f(\xi). \tag{52}$$

Using the independence of $\mathcal{F}$ and ξ, and the conditional independent of Y_0 and F given $(\mathcal{F}, \xi)$, we obtain

$$f(Y_0|F, \xi) \propto \int f(Y_0|\mathcal{F}, \xi)f(\mathcal{F}|F)\mathrm{d}\mathcal{F}. \tag{53}$$

Equation (53) lends itself nicely to an analysis using a natural conjugate family. First, recall (Press, 1972) the following results ("const" denotes a generic constant depending on scalar parameters [e.g., q and degrees of freedom] but not on matrix parameters):

1. If $A \sim W_q(k, \Omega)$ $(k \geq q)$ then A has density

$$f(A) = \text{const}\frac{|A|^{(k-q-1)/2}}{|\Omega|^{k/2}} e^{-(1/2)trA\Omega^{-1}}$$

2. $U \in \Re^q$ has a multivariate Student-t distribution with j degrees of freedom, scale matrix $C > 0$, and location parameter 0 if U has density

$$f(u) = \frac{\text{const}}{[j + u'C^{-1}u]^{(\frac{j+q}{2})}}. \tag{54}$$

$E(U) = 0$ if $j > 1$ and $\text{cov}(U) = \frac{j}{j-2}C$ if $j > 2$.

3. If $A \sim W_q(j + q - 1, \Omega)$ and $U \mid A \sim N(0, jA^{-1})$ then U has density as in equation (54) with $C = \Omega^{-1}$. (Observe that when $q = 1$, this simply says that if $Z \sim N(0, 1)$ and $a \sim \omega \cdot \chi^2(j)$ independently, then $\frac{Z}{(a/j)^{1/2}} \sim (1/\omega)t(j)$.)
4. If A is as in (3) and $Z|A \sim N(b, hA^{-1})$ for fixed b and $h > 0$, then Z has density

$$f(Z) = \frac{\text{const}h^{(j+q-1)/2}}{[h + (Z - b)'\Omega(Z - b)]^{(j+q)/2}}$$

and mean $E(Z) = b$ (this obtains from defining $U = (\frac{j}{h})^{1/2}(Z - b)$ in (3)).

Returning to equation (53), assume an invariant Jeffrey's prior for the dispersion matrix Λ,

$$f(\Lambda) \propto \frac{1}{|\Lambda|^{(q+1)/2}}$$

(Press, 1972, p. 75ff). From (1) and equation (32) we have

$$\begin{aligned}
f(\mathcal{F}|F) &\propto f(F|\mathcal{F})f(\mathcal{F}) \\
&\propto \frac{e^{-\frac{1}{2}s_{xx}(\hat{\beta}-\beta)'\Gamma^{-1}(\hat{\beta}-\beta)}}{|\Gamma|^{1/2}} \cdot \frac{e^{-\frac{1}{2}trS\Gamma^{-1}}}{|\Gamma|^{m/2}} \cdot \frac{1}{|\Lambda|^{\frac{q+1}{2}}} \\
&= \frac{e^{-\frac{1}{2}s_{xx}(\beta-\hat{\beta})'\Lambda(\beta-\hat{\beta})}}{|\Lambda^{-1}|^{1/2}} \cdot e^{-\frac{1}{2}tr\Lambda S}|\Lambda|^{(m-q-1)/2}
\end{aligned}$$

Comparing with (1) shows that

$$\Lambda|F \sim W_q(m, S^{-1}) \tag{55}$$

and

$$\beta|\Lambda, F \sim N(\hat{\beta}, (s_{xx}\Lambda)^{-1}). \tag{56}$$

Now write equation (53) as

$$f(Y_0|F, \xi) \propto \int\left[\int f(Y_0|\beta, \Lambda, \xi) f(\beta|\Lambda, F)\mathrm{d}\beta\right] f(\Lambda|F)\mathrm{d}\Lambda.$$

Since

$$Y_0|\beta, \Lambda, \xi \sim N(\beta\xi, \Lambda^{-1})$$

it follows from equation (56) that the inner integral above is the density of

$$Y_0|\Lambda, F, \xi \sim N(\hat{\beta}\xi, (1 + \frac{\xi^2}{s_{xx}})\Lambda^{-1}).$$

Together with equation (55) and (4), it follows that

$$f(Y_0|F, \xi) = \frac{\text{const}(1 + \frac{\xi^2}{s_{xx}})^{m/2}}{[1 + \frac{\xi^2}{s_{xx}} + (Y_0 - \hat{\beta}\xi)' S^{-1}(Y_0 - \hat{\beta}\xi)]^{(m+1)/2}}. \tag{57}$$

Substituting into equation (52) and comparing with equation (50), we see that ξ has posterior density

$$f(\xi|F, Y_0) \propto f(\xi) \cdot \frac{1}{(1 + \frac{\xi^2}{s_{xx}})^{1/2}} \cdot [\ell(\xi)]^{\frac{m+1}{2}} \tag{58}$$

where $\ell(\xi)$ is the function maximized by the MLE. If the vague prior $f(\xi) =$ constant is used, the $(1 + \xi^2/s_{xx})^{-1/2}$ term in equation (58) shrinks the MLE towards zero; i.e., the Bayes estimator satisfies

$$|\hat{\xi}_{Bayes}| < |\hat{\xi}_{mle}|.$$

If the prior for ξ assigns greater weights to values near the origin, then the shrinkage is even greater.

Since the inverse estimator is appropriate when (ξ, Y_0) and the (x_i, Y_i) constitute a random sample from the same joint distribution, it should be possible to exhibit $\hat{\xi}_{inv}$ as a Bayes estimate when the x_i are fixed, provided the prior for ξ is appropriately defined. This has been done by Brown (1982, 1993), extending the results of Hoadley (1970). If ξ has the prior

$$f(\xi) \propto \frac{1}{(1 + \frac{\xi^2}{s_{xx}})^{m/2}} \tag{59}$$

then we see from equations (50) and (58) that the Bayes estimator must minimize the denominator in equation (57), and from equation (49) this is equivalent to minimizing

$$\xi^2 + s_{xx}\hat{\beta}' S^{-1}\hat{\beta}(\xi - \hat{\xi})^2.$$

The minimum is attained at

$$\left(\frac{\hat{\beta}' S^{-1}\hat{\beta}}{s_{xx}^{-1} + \hat{\beta}' S^{-1}\hat{\beta}}\right)\hat{\xi}, \tag{60}$$

which is $\hat{\xi}_{inv}$ in view of equation (44). Brown (1993) also gives an interesting motivation for equation (59) by assuming that ξ and the x_i are independent samples from the same normal distribution, with a vague prior on its parameters.

Since our problem involves a parameter of interest (ξ) and nuisance parameters (β, Γ), the method of reference priors (Berger and Bernardo, 1992) may be appropriate. Using this approach, Kubokawa and Robert (1994) derived the prior distribution

$$f(\beta, \sigma^2, \xi) = \frac{1}{\sigma^{p+2}(1 + \xi^2)^{1/2}}$$

for the special case $\Gamma = \sigma^2 I$ with σ^2 unknown. The corresponding Bayes estimator avoids the problem of overshrinkage towards zero which the inverse estimator has.

3.5 Relation to the Control Problem

In the control problem (Zellner, 1971) an output is given by

$$\text{output} = h'\beta$$

where β is an unknown vector of parameters and $h = h(Z)$ is an input vector based on data Z, say

$$Z \sim N(\beta, I). \tag{61}$$

Our object is to choose the input so that the output is as close as possible to a given (nonzero) level. Taking the level (without loss of generality) to be one and using squared-error loss, we wish to choose h to minimize the risk

$$R_{\text{control}}(h, \beta) \equiv E(h'\beta - 1)^2. \tag{62}$$

A natural choice for h is the maximum likelihood estimator of $(\beta'\beta)^{-1}\beta$, namely

$$h(Z) = (Z'Z)^{-1}Z.$$

By studying more general estimators of the form

$$h(Z) = g(Z^t Z)Z$$

for different functions g, Berger et al. (1982), among others, obtained comprehensive results on admissibility as well as deriving estimators dominating the natural one.

To see the connection with calibration, observe from equations (37) and (44) that both the classical and inverse estimator are of the form

$$\xi^* = h^t \cdot S^{-1/2} Y_0 \tag{63}$$

where

$$h(\hat{\beta}, S) = g(\hat{\beta}^t S^{-1} \hat{\beta}) S^{-1/2} \hat{\beta}$$

for appropriate functions g. For an arbitrary estimator of this form, consider

$$\begin{aligned} E[(\xi^* - \xi)^2 | \hat{\beta}, S] &= E\{[h^t(Y_0 - \beta\xi) + \xi(h^t\beta - 1)]^2 | \hat{\beta}, S\} \\ &= h^t \Gamma h + \xi^2 (h^t \beta - 1)^2. \end{aligned}$$

It follows that

$$MSE(\xi^*, \xi) = Eh^t \Gamma h + \xi^2 R^*_{\text{control}}(h, \beta)$$

where R^*_{control} is defined as in equation (62), except that instead of equation (61) we now have the more complicated set-up $\hat{\beta} \sim N(\beta, s_{xx}^{-1}\Gamma)$ and $S \sim W_q(m, \Gamma)$ independently. Thus it is possible to compare estimators of the form of equation (63) by comparing their functions h and using domination results from control theory, provided the latter can be extended to the unknown variance case. This was done by Kubokawa and Robert (1994) (the extension to unknown variance is not trivial and requires some delicate arguments); in particular, they showed that $\hat{\xi}_{clas}$ is inadmissible.

3.6 Extensions

We now briefly discuss extensions of the preceding results to the general situation in equation (27). In this case, denote $h_i = h(x_i)$ and

$$H = \sum_{i=1}^{n} h_i h_i^t. \tag{64}$$

Then equations (30) and (31) become

$$\hat{C} = \sum_{i=1}^{n} Y_i h_i^t H^{-1} \tag{65}$$

and

$$S = \sum_{i=1}^{n} [Y_i - \hat{C}h_i][Y_i - \hat{C}h_i]'. \tag{66}$$

The classical estimator is just as easily defined in the linear case in equation (28) when $r > 1$, giving (analogous to equation (37))

$$\hat{\xi}_{clas} - \bar{x} = (\hat{B}'S^{-1}\hat{B})^{-1}\hat{B}'S^{-1}(Y_0 - \bar{Y}) \tag{67}$$

where $\hat{B}$ is computed from the linear regression of the Y_i on the (centered) x_i. Distributional results are much more complicated, however. Using the delta method, Fujikoshi and Nishii (1986) obtained asymptotic (in n) expansions for the mean and weighted (with respect to an arbitrary weight matrix) MSE of $\hat{\xi}_{clas}$. For example, analogous to equation (38), they obtain

$$E(\hat{\xi}_{clas}) = \xi - (q - r - 1)(B'\Gamma^{-1}B)^{-1}M^{-1}(\xi - \bar{x}) + o(n^{-1})$$

where

$$M_{r\times r} = \sum_{i=1}^{n} (x_i - \bar{x})(x_i - \bar{x})'. \tag{68}$$

Nishii and Krishnaiah (1988) proved that $\hat{\xi}_{clas}$ has finite mean if and only if $q \geq r + 1$, and finite MSE if and only if $q \geq r + 2$. Sundberg (1996) studied the effect of the uncertainty in S on the asymptotic covariance matrix of $\hat{\xi}_{clas}$, and also corrected an error in Fujikoshi and Nishii's (1986) results.

In the semilinear case (2) it is also simple to define the classical estimator: in view of equations (2), (26), and (27) we minimize

$$[Y_0 - \hat{g}(\xi)]'S^{-1}[Y_0 - \hat{g}(\xi)] \tag{69}$$

where $\hat{g}(\xi) = \hat{C}h(\xi)$. However, no exact distributional results are available for the resulting ξ_{clas}.

In the linear case, the natural extension of the inverse estimator in equation (42) to $r > 1$ is $\hat{\xi}_{inv} = \hat{G}'Y_0$, where $\hat{G}$ is computed from the multivariate regression of the x_i on the Y_i. It appears difficult, however, to extend the exact MSE results of Oman and Srivastava (1996) and thus compare $\hat{\xi}_{inv}$ and $\hat{\xi}_{clas}$ when $r > 1$. Sundberg (1985) compared approximate MSEs as described preceding equation (46) and found that the (approximate) MSE of $c'\hat{\xi}_{inv}$ for estimating $c'\xi$ is less than that of $c'\hat{\xi}_{clas}$, for all fixed vectors c, if and only if

$$\xi'\left[\frac{2}{n - r - 1}M + (B'\Gamma^{-1}B)^{-1}\right]\xi \leq 1$$

where M is defined in equation (68). Also, Brown (1982) obtained the analogue to equation (44) in this more general case as well:

$$\hat{\xi}_{inv} = [M^{-1} + \hat{B}^t S^{-1} \hat{B}]^{-1} (\hat{B}^t S^{-1} \hat{B})^t \hat{\xi}_{clas}. \tag{70}$$

In the semilinear case, the natural extension of the inverse estimator is to compute a nonlinear regression of the x_i on the Y_i and then, at the prediction step, substitute Y_0 into the resulting equation. There is no analogy with equation (70), in part because, for a given nonlinear g in equation (1), it is not clear how to define the corresponding nonlinear function when regressing x_i on Y_i. This is unfortunate, since equation (70) shows how close $\hat{\xi}_{inv}$ is to $\hat{\xi}_{clas}$, thus enabling us to use $\hat{\xi}_{inv}$ as a numerical approximation to the theoretically more justified $\hat{\xi}_{clas}$. This would be particularly useful for repeated predictions in the semilinear case, since it would obviate the need to solve a new "mini-nonlinear regression," as discussed following equation (28), for each new Y_0.

Brown and Sundberg (1987) used a profile likelihood approach in the more general context of equation (27) as well. In the linear case they were able to show that, under fairly general conditions, a unique MLE exists and "expands" $\hat{\xi}_{clas}$ in the sense discussed following equation (50). However, they were unable to obtain an explicit expression for the MLE.

Finally, the Bayesian analysis of Section 3.4 is essentially the same when $r > 1$ in the linear case in equation (28). A straightforward generalization of equation (58) obtains (formula (5.25) in Brown, 1993), and the analogue of equation (60) also holds when equation (59) is replaced by an appropriate multivariate t distribution (Theorem 5.2, Brown, 1993).

4. CONFIDENCE REGIONS IN CONTROLLED CALIBRATION

We now describe methods for computing confidence regions for ξ in the controlled calibration model. We shall exploit the analogy between equation (2) and regression, so we first briefly review confidence region methods for nonlinear regression.

4.1 Nonlinear Regression Confidence Regions

Consider the regression model

$$Y = f(\xi) + \varepsilon, \qquad \varepsilon \sim N(0, \sigma^2 I) \tag{71}$$

where $Y_{q\times 1}$ is a vector of observations on a dependent variable and f is a known (possibly nonlinear) function of $\xi_{r\times 1}$ satisfying certain smoothness requirements. Let

$$\mathcal{S} = \{f(\xi) : \xi \in \Omega\}$$

denote the expectation surface in $\Re^q$ and let

$$Z(\xi) = \left(\frac{\partial f_i}{\partial \xi_j}(\xi)\right)_{q \times r} \tag{72}$$

denote the matrix of first partial derivatives of f evaluated at ξ. Thus $Z(\xi_0)$ spans the tangent plane to $\mathcal{S}$ at the point $f(\xi_0)$. Let $\hat{\xi}$ denote a maximum likelihood estimate, so that $f(\hat{\xi})$ is a point on $\mathcal{S}$ closest to Y, and let $s^2 = \|Y - f(\hat{\xi})\|^2/(q - r)$ denote the usual estimate of σ^2.

The following three methods (which are identical and exact if f is linear) give confidence regions for ξ (see Donaldson and Schnabel, 1987, for a simulation evaluation of their coverage properties).

1. *Linearization* uses the asymptotic distribution

   $$\hat{\xi} \approx N(\xi,\ \sigma^2[Z'(\xi)Z(\xi)]^{-1})$$

   to give the approximate $1 - \alpha$ level confidence region

   $$\mathcal{C}_{\ell in} = \left\{\xi : \frac{(\xi - \hat{\xi})' Z'(\hat{\xi}) Z(\hat{\xi})(\xi - \hat{\xi})}{s^2} \le rF_\alpha(r, q - r)\right\}. \tag{73}$$

   $\mathcal{C}_{\ell in}$ is the most commonly used method, as it is relatively simple to compute and gives a well-behaved elliptical region. However, its coverage probability can be adversely affected if $\mathcal{S}$ is too curved.
2. If $\lambda(\xi)$ denotes the likelihood function, then a *likelihood-based* region is

   $$\mathcal{C}_{lik} = \left\{\xi : \frac{\lambda(\xi)}{\lambda(\hat{\xi})} \ge \text{ crit. value}\right\}$$
   $$= \{\xi : \|Y - f(\xi)\|^2 \le \|Y - f(\hat{\xi})\|^2 + rs^2 F_\alpha(r, q - r)\}.$$

   $\mathcal{C}_{lik}$ tends to give better coverage than $\mathcal{C}_{lin}$, but is more difficult to compute and may lead to strangely shaped regions. Figure 2(a) shows an example, for $q = 2$ and $f(\xi) = (\xi, \phi(\xi))'$, for nonlinear ϕ, in which $\mathcal{C}_{lik}$ is disconnected.
3. The *lack-of-fit* method is illustrated in Figure 2(b). For any ξ construct the tangent plane $\mathcal{M}_\xi$ to $\mathcal{S}$ at $f(\xi)$, and let

   $$P_\xi = Z(\xi)[Z'(\xi)Z(\xi)]^{-1}Z'(\xi) \tag{74}$$

   denote its orthogonal projection matrix. Then $\sigma^{-2}\|P_\xi[Y - f(\xi)]\|^2$ and $\sigma^{-2}\|(I - P_\xi)[Y - f(\xi)]\|^2$ are independent $\chi^2(r)$ and $\chi^2(q - r)$ variables, respectively, so that

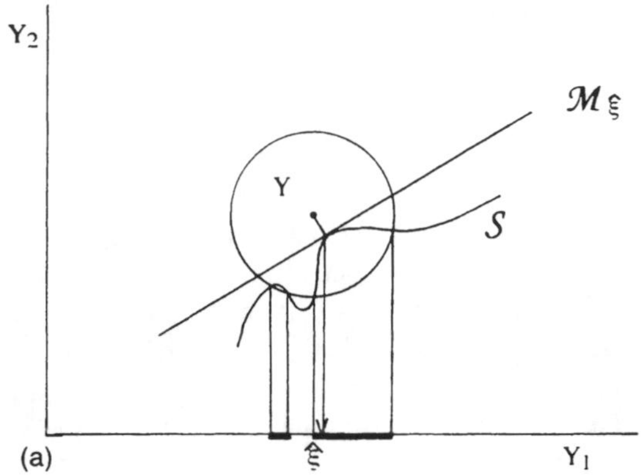

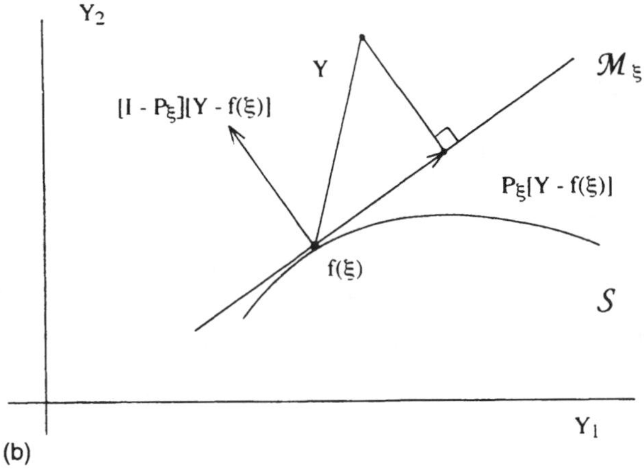

Figure 2 Confidence procedures in nonlinear regression. $\mathcal{S}$ is the expectation surface $= \{(\xi, \phi(\xi)\}$ for nonlinear function ϕ. $\mathcal{M}$ is the line tangential to the curve at the point indicated. (a) Likelihood-based region. (b) Lack-of-fit method.

$$\mathcal{C}_{lack} = \{\xi : \frac{[Y - f(\xi)]' P_{\xi}[Y - f(\xi)]}{[Y - f(\xi)]'(I - P_{\xi})[Y - f(\xi)]} \leq \frac{r}{q} F_{\alpha}(r, q - r)\} \tag{75}$$

is a $1 - \alpha$ level confidence region. Although $\mathcal{C}_{lack}$ has the advantage of being exact, it is quite difficult to compute, can give regions even more oddly shaped than $\mathcal{C}_{lik}$, and is little used in practice.

4.2 Calibration Confidence Regions

From equations (2), (26), and (27) we can write

$$Y_0 \doteq \hat{g}(\xi) + \varepsilon_0, \quad \varepsilon_0 \sim N(0, \Gamma) \tag{76}$$

where

$$\hat{g}(\xi) = \hat{C}h(\xi) \tag{77}$$

for $\hat{C}$ given by equation (65). Comparing with equation (71), it seems reasonable to define

$$\mathcal{S} = \{\hat{g}(\xi) : \xi \in \Omega\} \tag{78}$$

and try to adapt the methods of the preceding subsection. Observe that even when h is linear, so that equation (76) is similar to a linear regression model, the regions need not be exact since C and Γ are not known. We now discuss the various procedures which have been proposed to date.

Brown (1982) used the quantity

$$\begin{aligned} Y_0 - \hat{g}(\xi) &= Y_0 - \hat{C}h(\xi) \\ &\sim N_q(0, \ \{1 + \gamma(\xi)\}\Gamma), \end{aligned}$$

where

$$\gamma(\xi) = h'(\xi)H^{-1}h(\xi) \tag{79}$$

and H is defined in equation (64), to construct (analogous to Fieller's procedure in equation (9)) the region

$$\mathcal{C} = \{\xi : \frac{[Y_0 - \hat{C}h(\xi)]'S^{-1}[Y_0 - \hat{C}h(\xi)]}{1 + \gamma(\xi)} \leq \frac{q}{n - q - p} F_\alpha(q, n - q - p)\}. \tag{80}$$

This is an exact region and is relatively easy to compute; in fact, in the linear case Brown (1982) obtains conditions analogous to equation (11) guaranteeing that $\mathcal{C}$ be an ellipsoid. However, $\mathcal{C}$ has the disturbing property of giving smaller regions (implying greater "certainty" about ξ) for *atypical* Y_0. To see this, suppose for simplicity that the calibration sample is infinite and $\Gamma = I$. Then $\gamma(\xi) = 0$ in equation (79) and we compute $\mathcal{C}$ by first intersecting a sphere centered at Y_0 with $\mathcal{S}$ (given by equation (78)), and then finding all ξ such that $\hat{g}(\xi)$ is in the intersection. From Figure 2(a), it is clear that $\mathcal{C}$ gets smaller as Y_0 gets further from $\mathcal{S}$, and can even be empty. This paradoxical behavior, noted in Brown's (1982) discussion, in fact complicated the application of the procedure by Oman and Wax (1984, 1985). Observe that $\mathcal{C}$ compares with a $\chi^2(q)$ as opposed to a $\chi^2(r)$ distribution, and does not correspond to any of the methods discussed above for regression problems.

Fujikoshi and Nishii (1984) proposed a method similar in spirit to the linearization method (1), for the linear case in equation (28). Define the pivot (cf equation (73))

$$U_\xi = (\hat{\xi}_{clas} - \xi)^t \hat{B}^t \hat{\Gamma}^{-1} \hat{B}(\hat{\xi}_{clas} - \xi), \tag{81}$$

where the classical estimator $\hat{\xi}_{clas}$ is given by equation (67) and $\hat{\Gamma} = (n - r - 1)^{-1} S$ is the usual unbiased estimator of Γ. By conditioning arguments they obtain the following mixture representation for the distribution of U_ξ. Define $m = n - r - 1 \geq q$, $c = (m - q + r + 1)/[m(m - q)]$, $w = [(m - q + 1)n]/[m(n + 1)]$ and $B_0 = \Gamma^{-1/2} B$. If $\tilde{B}_{q \times r}$ is arbitrary, denote $\bar{B} = \Gamma^{-1/2} \tilde{B}$ and define for $r > 0$ the noncentrality parameters

$$\lambda_1(r, \tilde{B}) = \left(\frac{n}{n+1}\right)\left(1 + \frac{r}{m}\right)^{-1} \xi^t (\bar{B} - B_0)^t \bar{B}(\bar{B}^t \bar{B})^{-1} \bar{B}^t (\bar{B} - B_0)\xi$$

and

$$\lambda_2(\tilde{B}) = \left(\frac{n}{n+1}\right) \xi^t (\bar{B} - B_0)^t \{I - \bar{B}(\bar{B}^t \bar{B})^{-1} \bar{B}^t\} (\bar{B} - B_0)\xi.$$

(The distributions of λ_1 and λ_2 depend on the unknown ξ, B, and Γ, but we have supressed this in order to simplify notation.) Finally, let $V(k, \ell, \lambda)$ denote a noncentral $F(k, \ell, \lambda)$ variable and, by an abuse of notation, let $V(k, \ell)$ denote a central F variable. Then for $u > 0$,

$$P(U_\xi \leq u) = \int \left[\int P\left\{ rV\left(r, m - q + 1, \lambda_1\left(\frac{s}{c}, \tilde{B}\right)\right) \leq \frac{wu}{1 + \frac{s}{cm}} \right\} \right.$$

$$\left. \cdot P\left\{ V\left(q - r, n - q, \lambda_2\left(\frac{s}{c}, \tilde{B}\right)\right) \in \mathrm{d}s \right\} \right] P(\hat{B} \in \mathrm{d}\tilde{B}). \tag{82}$$

This formula is much too complicated for routine use. Assuming that $m \to \infty$ and

$$\frac{1}{m} \sum_{i=1}^{n} (x_i - \bar{x})(x_i - \bar{x})^t \to A > 0$$

Fujikoshi and Nishii expand the noncentral F variables in terms of noncentral χ^2 variables, use the mixture representation for noncentral χ^2 in terms of central χ^2 variables, and approximate the expectations with respect to the noncentrality parameters using the delta method. The result is an Edgeworth-like expansion of the form

$$P(U_\xi \leq u) = G_r(\tilde{u}) + g_r(\tilde{u})[N^{-1} a_1(\tilde{u}) + N^{-2} a_2(\tilde{u})] + O(N^{-3}) \tag{83}$$

where $\tilde{u} = nu/(n + 1)$, $N = n - r - q - 2$, G_r and g_r denote the CDF and pdf, respectively, of the $\chi^2(r)$ distribution, and the a_i are polynomials in $\tilde{u}$

with coefficients independent of N but involving the unknown parameters $\xi' A^{-1}\xi$ and $\xi'(AB'\Gamma^{-1}BA)^{-1}\xi$. Fujikoshi and Nishii also give a corresponding Cornish–Fisher-type expansion of the critical values of U_ξ.

Brown and Sundberg (1987) used the profile likelihood, analogous to method (2) above. This gives the region

$$\{\xi : \log \ell(\xi) - \log \ell(\hat{\xi}_{mle}) \geq \text{crit. value}\} \tag{84}$$

where, analogous to equation (50),

$$\ell(\xi) = \left\{ \frac{1 + \gamma(\xi) + \mathcal{R} + [h(\xi) - h(\hat{\xi})]'\hat{C}S^{-1}\hat{C}[h(\xi) - h(\hat{\xi})]}{1 + \gamma(\xi)} \right\}^{-1}$$

for

$$\mathcal{R} = [Y_0 - \hat{C}h(\hat{\xi})]'S^{-1}[Y_0 - \hat{C}h(\hat{\xi})]$$

and γ given by equation (79). This region is somewhat difficult to compute. Also, the χ^2 critical value in equation (84) must be justified asymptotically as $\Gamma \to 0$, as discussed at the end of Section 3.3. On the other hand, when $r = 1$ and h is linear, Brown and Sundberg (1987) showed that the size of the region given by equation (84) increases with $\mathcal{R}$. Thus, as opposed to equation (80), we obtain larger regions for atypical Y_0.

In the context of the general model in equation (27), Oman (1988) proposed a method similar in spirit to the lack-of-fit method (3). Analogous to equations (72) and (74), let

$$\hat{G}(\xi) = \left(\frac{\partial \hat{g}_i}{\partial \xi_j}(\xi)\right)_{q \times r} \tag{85}$$

where $\hat{g}$ is given by equation (77) and let $\hat{P}_\xi$ denote the orthogonal projection matrix onto the span of $\hat{G}(\xi)$. Then define the pivot

$$U_\xi = \{\hat{P}_\xi[Y_0 - \hat{g}(\xi)]\}'(\hat{P}_\xi\Gamma^*\hat{P}_\xi')^-\{\hat{P}_\xi[Y_0 - \hat{g}(\xi)]\} \tag{86}$$

where $\Gamma^* = (n - p - r + 1)^{-1}S$ and $(\ )^-$ denotes a generalized inverse. Analogous to the numerator in equation (75), U_ξ is based on $\hat{P}_\xi[Y_0 - \hat{g}(\xi)]$, the projection of $Y_0 - \hat{g}(\xi)$ onto the tangent plane to $\mathcal{S}$ at $\hat{g}(\xi)$; compare Figure 2(b) with Figure 1(b) in Oman (1988).

Oman shows that

$$P(U_\xi \leq x) = \int_0^\infty P\{rV(r, n - p - r + 1, \lambda) \leq x\}P\{\Lambda(\hat{C}) \in \mathrm{d}\lambda\} \tag{87}$$

where V is as in equation (82), $\hat{C}$ is defined in equation (65) and the noncentrality parameter Λ satisfies

$$\Lambda(\hat{C}) \prec_{st} \gamma(\xi) \cdot \chi^2(q). \tag{88}$$

Thus equation (87) gives

$$\begin{aligned} P\{rV(r, n-p-r+1) \leq x\} \\ \geq P(U_\xi \leq x) \\ \geq \int_0^\infty P\{rV(r, n-p-r+1, w\gamma(\xi)) \leq x\} P\{\chi^2(q) \in \mathrm{d}w\}. \end{aligned} \tag{89}$$

It is relatively easy to use the right-hand side of equation (89) to compute critical values giving a conservative region for ξ. The main disadvantage of equation (86) is that normalizing by the covariance after the projection, although allowing the inequality in equation (88), makes U_ξ not invariant under arbitrary linear transformations of Y.

For the linear model in equation (28), Mathew and Kasala (1994) proposed a method which is invariant and moreover exact, yet suffers from other drawbacks. To understand their approach, also similar in spirit to the lack-of-fit method (3), suppose that $a = 0$ in equation (28) and $H = I$ in equation (64). First define

$$\begin{aligned} v_\xi &= \frac{1}{(1 + \xi^t \xi)^{1/2}} (Y_0 - \hat{B}\xi) \\ &= [\hat{B} : Y_0] \cdot w_\xi \end{aligned}$$

where $w_\xi = (1 + \xi^t \xi)^{-1/2} (-\xi^t, 1)^t \in \Re^{r+1}$. If A_ξ is a $(r+1) \times r$ matrix whose columns are of length one and orthogonal to w_ξ, it follows that $V_\xi = [B : Y_0] A_\xi$ is independent of v_ξ. Then consider

$$T_\xi = v_\xi^t S^{-1} V_\xi (V_\xi^t S^{-1} V_\xi)^{-1} V_\xi^t S^{-1} v_\xi. \tag{90}$$

As opposed to the pivot in equation (86), T_ξ is invariant with respect to linear transformations of Y. Moreover, the independence of v_ξ and V_ξ makes the distribution of T_ξ more tractable than in equation (86), in which $\hat{P}_\xi$ involves $\hat{C}$ and is not independent of $Y_0 - \hat{g}(\xi)$. However, T_ξ still needs to be normalized to give a pivot. Mathew and Kasala (1994) therefore define

$$U_\xi = \frac{v_\xi^t S^{-1} V_\xi (V_\xi^t S^{-1} V_\xi)^{-1} V_\xi^t S^{-1} v_\xi}{1 + v_\xi^t W_\xi (W_\xi^t S W_\xi)^{-1} W_\xi^t v_\xi}, \tag{91}$$

where the columns of the $(q - r) \times r$ matrix W_ξ are of length one and orthogonal to V_ξ. By conditioning on v_ξ and using results for partitioned Wishart matrices, Mathew and Kasala are able to express the distribution of U_ξ in terms of a central F distribution, resulting in the exact confidence region

$$\{\xi : U_\xi \leq \text{crit. value}\}. \tag{92}$$

The main drawback to equation (92) is its extremely complicated shape (although Mathew and Kasala do give conditions gurantanteeing that the region be an interval when $r = 1$). Also, the intuition behind basing the pivot on equation (90) is not clear. Suppose that $q = 2$, $r = 1$, and the calibration sample is infinite, with $B = \beta$ and $S = I$ known. Then (Figure 3) it is natural to project $Y_0 - \beta\xi$ onto the span of β; but T_ξ uses the projection of $Y_0 - \beta\xi$ onto $\beta + Y_0\xi$, as shown.

Mathew and Zha (1996) proposed a method for the general model in equation (27) which combines the better aspects of the approaches of Fujikoshi and Nishii (1984) and Oman (1988). First consider the linear case in equation (28) where for simplicity $a = 0$, and let

$$U_\xi = \left(\frac{n-q-r+1}{r}\right)\frac{(Y_0 - \hat{B}\xi)^t S^{-1}\hat{B}(\hat{B}^t S^{-1}\hat{B})^{-1}\hat{B}^t S^{-1}(Y_0 - \hat{B}\xi)}{1+w} \tag{93}$$

where

$$w = Y_0^t[S^{-1} - S^{-1}\hat{B}(\hat{B}^t S^{-1}\hat{B})^{-1}\hat{B}^t S^{-1}]Y_0.$$

The quadratic form in the numerator of U_ξ equals

$$(\hat{\xi}_{clas} - \xi)^t \hat{B}^t S^{-1}\hat{B}(\hat{\xi}_{clas} - \xi),$$

as in equation (81). Fujikoshi and Nishii (1984) showed that, conditionally on $\hat{B}$ and w, U_ξ has a noncentral F distribution with noncentrality parameter

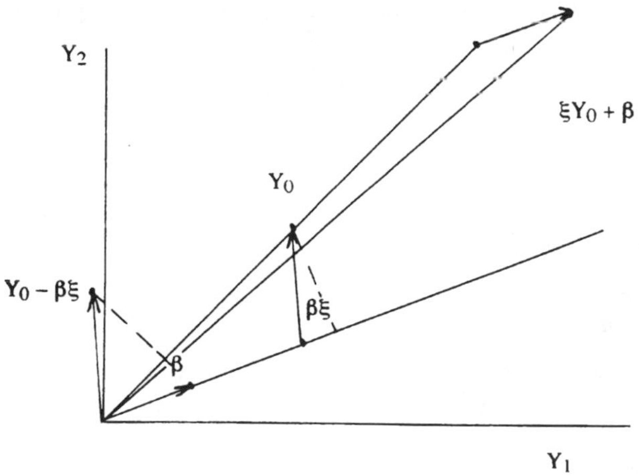

Figure 3 Geometry of Mathew and Kasala's (1994) method when $q = 2$.

$$\Lambda = \frac{(B\xi - \hat{B}\xi)'\Gamma^{-1}\hat{B}(\hat{B}'\Gamma^{-1}\hat{B})^{-1}\hat{B}'\Gamma^{-1}(B\xi - \hat{B}\xi)}{1 + w}$$

This is what leads to equation (82). Instead of using the exact distribution of Λ, Mathew and Zha (1996) bound the numerator of Λ from above by

$$(B\xi - \hat{B}\xi)'\Gamma^{-1}(B\xi - \hat{B}\xi)$$

Also, conditional on $\hat{B}$ w has a noncentral F distribution and hence is stochastically larger than a central F variable. Thus Λ can be stochastically bounded from above, and integrating out over the distribution of $\hat{B}$ gives the bounds

$$\begin{aligned} P\{V(r, n-q-r+1) \leq x\} &\geq P\{U_\xi \leq x\} \\ &\geq \int_0^\infty P\{V(r, n-q-r+1, w\gamma(\xi)) \leq x\} P\{\chi^2(q) \in \mathrm{d}w\}. \end{aligned} \tag{94}$$

As with equation (89), equation (94) can be used to obtain a conservative region. Mathew and Zha also obtain a tighter lower bound than in equation (94), but it involves a double integral and is less computationally feasible. For the more general semilinear case (2) this approach is still valid, except that in equation (93) the mean $\hat{B}\xi$ is replaced by $\hat{g}(\xi) = \hat{C}h(\xi)$, and $\hat{B}$ in the projection matrix is replaced by $\hat{G}(\xi)$ defined in equation (85).

The methods discussed above suffer from various drawbacks—strangely shaped regions, lack of invariance, inexact coverage probabilities, and computational complexity. No comprehensive comparisons have been made between the methods, and it is difficult to say which, if any, should be preferred.

REFERENCES

Aitchison, J., Dunsmore, I.R. *Statistical Prediction Analysis*. Cambridge: Cambridge University Press, 1975.

Berger, J.O., Bernardo, J.M. Ordered group reference priors with applications to the multinomial problem. *Biometrika*, **79**, 25–37, 1992.

Berger, J. O., Berliner, L.M., Zaman, A. General admissibility and inadmissibility results for estimation in a control problem. *Ann. Stat.*, **10**, 838–856, 1982.

Bingham, C., Larntz, K. Comment on Dempster et al. *J. Am. Stat. Assoc.*, **72**, 97–102, 1977.

Bock, M.E. Minimax estimators of the mean of a multivariate normal distribution. *Ann. Statist.*, **3**, 209–218, 1975.

Bock, M.E., Yancey, T.A., Judge, G.G. The statistical consequences of preliminary test estimators in regression. *J. Am. Stat. Assoc.*, **68**, 109–116, 1973.

Brown, P.J. Multivariate calibration (with discussion). *J. R. Stat. Soc., Ser. B*, **44**, 287–321, 1982.

Brown, P.J. *Measurement, Regression, and Calibration*. Oxford, Clarendon Press, 1993.

Brown, P.J., Sundberg, R. Confidence and conflict in multivariate calibration. *J. R. Stat. Soc., Ser. B*, **49**, 46–57, 1987.

Casella, G. Mimimax ridge regression estimation. *Ann. Stat.*, **8**, 1036–1056, 1980.

Donaldson, J.R., Schnabel, R.B. Computational experience with confidence regions and confidence intervals for nonlinear least squares. *Technometrics*, **29**, 67–82, 1987.

Eisenhart, C. The interpretation of certain regression methods and their use in biological and industrial research. *Ann. Math. Stat.*, **10**, 162–186, 1939.

Fieller, E.C. Some problems in interval estimation. *J. R. Stat. Soc., Ser. B*, **16**, 175–185, 1954.

Frank, I.E. Intermediate least squares regression method. *Chemometrics Intell. Lab. Syst.*, **1**, 1–38, 1987.

Frank, I.E., Friedman, J.H. A statistical view of some chemometrics regression tools (with discussion). *Technometrics*, **35**, 109–148, 1993.

Fujikoshi, Y., Nishii, R. On the distribution of a statistic in multivariate inverse regression analysis. *Hiroshima Math. J.*, **14**, 215–225, 1984.

Fujikoshi, Y., Nishii, R. Selection of variables in multivariate inverse regression problem. *Hiroshima Math. J.*, **16**, 269–277, 1986.

Fuller, W.A. *Measurement Error Models*. New York: Wiley, 1987.

George, E. Minimax multiple shrinkage estimation. *Ann. Stat.*, **14**, 188–205, 1986.

George, E.I., Oman, S.D. Multiple-shrinkage principal component regression. *Statistician*, **45**, 111–124, 1996.

Gleser, L.J., Hwang, J.T. The nonexistence of 100(1-α)% confidence sets of finite expected diameter in errors-in-variables and related models. *Ann. Stat.*, **15**, 1351–1362, 1987.

Helland, I.S. Partial least squares regression and statistical models. *Scand. J. Stat.*, **17**, 97–114, 1990.

Helland, I.S., Almøy, T. Comparison of prediction methods when only a few components are relevant. *J. Am. Stat. Assoc.*, **89**, 583–591, 1994.

Hoadley, B. A Bayesian look at inverse regression. *J. Am. Stat. Assoc.*, **65**, 356–369, 1970.

Hoerl, A.E. Application of ridge analysis to regression problems. *Chem. Eng. Prog.*, **60**, 54–59, 1962.

Hunter, W.G., Lamboy, W.F. A Bayesian analysis of the linear calibration problem (with discussion). *Technometrics*, **23**, 323–350, 1981.

James, W., Stein, C. Estimation with quadratic loss. Proceedings of the 4th Berkeley Symposium on Mathematical Statistics and Probability, Vol. 1. Berkeley, CA: University of California Press, pp. 361–379, 1961.

Jennrich, R.J., Oman, S.D. How much does Stein estimation help in multiple linear regression? *Technometrics*, **28**, 113-121, 1986.

Jöreskog, K.G., Wold, H. *Systems Under Indirect Observation. Causality—Structure—Prediction. Parts I and II.* Amsterdam: North-Holland, 1982.

Krutchkoff, R.G. Classical and inverse regression methods of calibration. *Technometrics*, **11**, 605–608, 1967.

Kubokawa, T., Robert, C.P. New perspectives on linear calibration. *J. Multivar. Anal.*, **51**, 178–200, 1994.

Lieftinck-Koeijers, C.A.J. Multivariate calibration: A generalization of the classical estimator. *J. Multivar. Anal.*, **25**, 31–44, 1988.

Martens, H., Naes, T. *Multivariate Calibration.* New York: Wiley, 1989.

Mathew, T., Kasala, S. An exact confidence region in multivariate calibration. *Ann. Statist* **22**, 94–105, 1994.

Mathew, T., Zha, W. Conservative confidence regions in multivariate calibration. *Ann. Stat.*, **24**, 707–725, 1996.

Naes, T. Multivariate calibration when the error covariance matrix is structured. *Technometrics*, **27**, 301–311, 1985.

Naes, T., Helland, I.S. Relevant components in regression. *Scand. J. Stat.*, **20**, 239–250, 1993.

Naes, T., Martens, H. Comparison of prediction methods for multicollinear data. *Commun. Stat. B, Simulation Comput.*, **14**, 545–576, 1985.

Naes, T., Irgens, C., Martens, H. Comparison of linear statistical methods for calibration of NIR instruments. *J. R. Stat. Soc. Ser. C*, **35**, 195–206

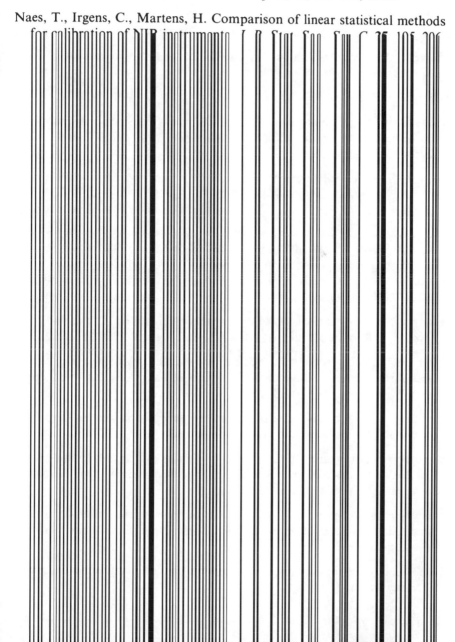

Press, S.J. *Applied Multivariate Analysis*. New York: Holt, Rinehart and Winston, 1972.

Racine-Poon, A. A Bayesian approach to nonlinear calibration problems. *J. Am. Stat. Assoc.*, **83**, 650–656, 1988.

Raiffa, H., Schlaifer, R. *Applied Statistical Decision Theory*. Boston, MA: Harvard University, Graduate School of Business Administration, 1961.

Sclove, S. Improved estimators for coefficients in linear regression. *J. Am. Stat. Assoc.*, **63**, 596–606, 1968.

Seber, G.A.F. *Linear Regression Analysis*. New York: Wiley, 1977.

Srivastava, M.S. Comparison of the inverse and classical estimators in multi-univariate linear calibration. *Commun. Stat. Theor. Meth.*, **24**, 2753–2767, 1995.

Stein, C. Inadmissibility of the usual estimator for the mean of a multivariate normal distribution. In: Neyman, J, ed. *Proceedings of the 3rd Berkeley Symposium*, Vol. 1. Berkeley and Los Angeles, CA: University of California Press, 1956, pp. 197–206.

Stone, M., Brooks, R.J. Continuum regression: Cross-validated sequentially constructed prediction embracing ordinary least squares, partial least squares and principal components regression. *J. R. Stat. Soc., Ser. B*, **52**, 237–269, 1990.

Sundberg, R. When does it pay to omit variables in multivariate calibration? Technical Report, Royal Institute of Technology, Stockholm, 1982.

Sundberg, R. When is the inverse regression estimator MSE-superior to the standard regression estimator in multivariate controlled calibration situations? *Stat. Probab. Lett.*, **3**, 75–79, 1985.

Sundberg, R. The precision of the estimated generalized least squares estimator in multivariate calibration. *Scand. J. Stat.*, **23**, 257–274, 1996.

Sundberg, R. Multivariate calibration—direct and indirect regression methodology. Report No. 200, Stockholm University Institute of Actuarial Mathematics and Mathematical Statistics, 1997.

Thisted, R.A. Ridge regression, minimax estimation and empirical Bayes methods. Technical Report 28, Stanford University Department of Biostatistics, 1976.

Thomas, E.V. Errors-in-variables estimation in multivariate calibration. *Technometrics*, **33**, 405–413, 1991.

Tiao, G.C., Zellner, A. Bayes' theorem and the use of prior knowledge in regression analysis. *Biometrika*, **51**, 219–230, 1964.

Zellner, A. *An Introduction to Bayesian Inference in Econometrics*. New York: Wiley, 1971.

11

Some Consequences of Random Effects in Multivariate Survival Models

YASUHIRO OMORI Tokyo Metropolitan University, Tokyo, Japan

RICHARD A. JOHNSON University of Wisconsin, Madison, Wisconsin

1. INTRODUCTION

Random effect models have been used for several decades in the context of linear models and the analysis of variance. They are introduced into the models as an additive term to model variation from subject to subject and, in most cases, their influence on the covariance structure of the response variables is well understood. It is also possible to perform many statistical analyses.

However, in the context of survival analysis, it has been relatively recently that random effect models have been widely used to model heterogeneity in a population (Clayton, 1978; Lancaster, 1979, 1985, 1990; Oakes, 1982, 1986a,b, 1989; Heckman and Singer, 1984; Clayton and Cuzick, 1985; Crowder, 1985, 1989; Hougaard, 1986a,b, 1987, 1989; Whitmore and Lee, 1991; Klein, 1992; Hougaard et al, 1992; Costigan and Klein, 1993; Yashin and Lachin, 1993; Lindeboom and Van Den Berg., 1994).

Before random effect survival models are placed in widespread use, it is necessary that the consequences and implications of the random effects be understood. The dependence structure created by the random effects in survival models is more complicated than that in linear models. The consequences can be striking. When considering survival times, the hazard rate may be modeled as $V\lambda_N(t)$, a positive random variable times a nominal hazard rate. Even in the univariate case, the p.d.f. (probability density function) and c.d.f. (cumulative distribution function) become nonlinear functions of the random effects.

In this paper we first derive several new results for multivariate survival models concerning the implications of random effects on the hazard rate, survival function, and various measures of association. We also suggest a few new parametric models and place them in the context of existing models which we give as examples.

Second, we exploit our approach to give a unified treatment that includes the existing results in a comprehensive presentation.

We begin by reviewing existing univariate results and then go on to consider multivariate random effect models. The common factor model, where a single random effect acts on several lifelengths, is considered first. Following this discussion in Section 3, we treat the more general model where each lifelength is influenced by different independent random effects. Multivariate survival times are assumed to be conditionally independent and have proportional hazard rates given the random effects.

The results pertain to the consequences on hazard rates, survival functions and various measures of dependence, including a multivariate extension of the odds ratio. The proofs, presented in Appendix 1, are almost all based on manipulations of conditional moments or cumulants. We have attempted to present a unified treatment by highlighting this common feature.

Before beginning the technical development, we review the key literature.

2. REVIEW OF THE LITERATURE

Most of the existing literature can be placed into four categories:

1. a univariate lifelength model with a univariate random effect;
2. a bivariate lifelength model with a univariate random effect;
3. a bivariate lifelength model with a bivariate random effect;
4. a general multivariate lifelength model induced by random effects.

Our literature survey will proceed in this order from the simplest case to the general fourth case.

For a univariate lifelength with a univariate random effect, Lancaster, 1979 found that the parameters of the Weibull regression model are affected by missing covariates (which is analogous to ignoring a random effect) in the analysis of unemployment data, and proved (Lancaster, 1985) that the maximum likelihood estimators are inconsistent when the existence of random effects is ignored (see Keiding et al., 1997 for a discussion of the effect of omitting covariates in survival analysis). Heckman and Singer (1984) showed the existence of a consistent nonparametric maximum likelihood estimator for parameters and distribution functions in a general class of proportional hazard models with a specified baseline hazard (but where the distribution of the random effect is unspecified). For the Weibull model, Honoré (1990) derived asymptotic normality for a class of estimators based on order statistics, and Ishwaran (1996) proposed an estimator which converges at a uniform rate under a finite-dimensional parameter space.

Further, it is known that the (nominal) hazard rates are underestimated and that survival times are stochastically larger than expected when the existence of random effects is ignored (Lee and Klein, 1988; Lancaster, 1990; Omori and Johnson, 1993).

For a bivariate lifelength with a univariate random effect, a random effect is used to derive the distribution of bivariate survival times that share common factors. Clayton (1978) introduced a multivariate survival model by assuming that multivariate survival times are conditionally independent given a random effect. This model has been studied with various random effect distributions including the gamma, inverse Gaussian and log–normal distributions, and the EM algorithm estimation methods have often been used (Oakes, 1982, 1986a,b, 1989; Clayton and Cuzick, 1985; Crowder, 1985, 1989; Hougaard, 1986a,b, 1987 1989; Lancaster, 1990; McGilchrist and Aisbett, 1991; Whitmore and Lee, 1991; Andersen et al., 1992; Hougaard et al., 1992 Guo and Rodoríguez, 1992; Klein, 1992; Nielsen et al., 1992; Klein et al., 1992; Costigan and Klein, 1993; Li et al., 1993; Wang et al., 1995). The asymptotic distribution of the semiparametric estimator of the cumulative hazard rate and variance of a gamma random effect was obtained by Murphy (1995).

For a bivariate lifelength with a bivariate random effect, Yashin and Lachine (1993), Lindeboom and Van Den Berg. (1994) and Yashin et al. (1995) considered a bivariate survival model with correlated random effects and found that the odds ratio and conditional hazard rates depend on the conditional expectation and the covariance of random effects.

Finally, the general form of a multivariate lifelength model induced by random effects was given by Marshall and Olkin (1988). Trivariate random effects are examined in Johnson and Omori (1998).

See Sinha and Dey (1997) for a review that includes a Bayesian approach.

3. UNIVARIATE AND BIVARIATE MODELS

3.1 Univariate Model with a Univariate Random Effect

Random effects can be introduced to model variation when observations arise from several heterogeneous subgroups. Such a model assumes that the hazard rate function from one group is parallel to that of other groups.

The first step is to accommodate all of the observations under a single model by taking account of covariate variables. This is often done by setting

$$\lambda_N(t) = \lambda_0(t)\exp\{\beta' x\}$$

where $\lambda_0(t)$ is a specified baseline hazard rate, x is a vector of fixed covariates, and the subscript N is attached to emphasize that this is a nominal hazard rate, i.e., without heterogeneity.

The most common way to introduce heterogeneity into the population via random effect, is to let the hazard rate for an individual be

$$V \times \lambda_N(t) = V \times \lambda_0(t)\exp\{\beta' x\}$$

where V is a random variable such that $P(V \geq 0) = 1$, $E(V) = 1$, and $P(V = 1) < 1$. We assume $E(V) = 1$ since we can always absorb a departure from 1 into the $\lambda_N(t)$. Conditional on $V = v$, an individual hazard rate is

$$\lambda_C(t|v) \equiv v\lambda_N(t)$$

and the conditional, nominal, and unconditional survival functions are

$$\begin{aligned} S_C(t|v) &= 1 - F_C(t|v) = \exp\{-v\Lambda_N(t)\} \\ S_N(t) &= 1 - F_N(t) = \exp\{-\Lambda_N(t)\} \\ S_U(t) &= 1 - F_U(t) = E[\exp\{-V\Lambda_N(t)\}] \end{aligned}$$

respectively, where $\Lambda_N(t) = \int_0^t \lambda_N(s)ds$. Then the unconditional hazard rate, $\lambda_U(t)$, is

$$\lambda_U(t) = \frac{E[V\exp\{-V\Lambda_N(t)]}{E[\exp\{-V\Lambda_N(t)]}$$

The popular distributions for a random effect, V, are the gamma, inverse Gaussian, and Hougaard's three-parameter family (including the positive stable family) of distributions.

3.1.1 Influence of random effects

When the existence of random effects is ignored, the nominal hazard rate will be underestimated and the survival times are stochastically larger than expected, as we show in the following theorem and corollary.

THEOREM 1 (Omori and Johnson, 1993): *Let T be a nonnegative random variable with probability density function $f_U(t)$. Suppose that $P(V \geq 0) = 1$, $E(V) = 1$, and $P(V = 1) < 1$. Then*

$$\lambda_U(t) - \lambda_N(t) = \lambda_N(t) \times [E(V|T \geq t) - 1],$$
$$\frac{\mathrm{d}}{\mathrm{d}t} E(V|T \geq t) = -\lambda_N(t) \times \mathrm{Var}(V|T \geq t)$$

Hence, $E(V|T \geq t)$ is nonincreasing in $t > 0$, with $E(V|T \geq 0) = E(V) = 1$.

Proof. See Appendix 1.

COROLLARY 1: (Omori and Johnson, 1993): *Let T be a nonnegative random variable with probability density function $f_U(t)$. Suppose that $P(V \geq 0) = 1$, $E(V) = 1$, and $P(V = 1) < 1$. Then the unconditional survival time is stochastically larger than the nominal survival time. That is,*

$$F_U(t) < F_N(t) \quad \text{for} \quad t > 0 \quad \text{and} \quad \Lambda_N(t) > 0$$

and $F_U(t) \leq F_N(t)$ otherwise.

Proof. See Appendix 1.

Since $\mathrm{d}E(V|T \geq t)/\mathrm{d}t < 0$, mixtures of distributions with a decreasing failure rate (DFR) family are always DFR, whereas some mixtures of distributions with an increasing failure rate (IFR) family might not be IFR, as in the following examples (also see Gurland and Sethuraman, 1995).

3.1.2 Examples

We assume that $E(V) = 1$ and $\mathrm{Var}(V) = c$.

1. *Gamma mixtures.* If V follows a gamma distribution, then

$$S_U(t) = (1 + c\Lambda_N(t))^{-1/c}, \qquad \lambda_U(t) = \frac{\lambda_N(t)}{1 + c\Lambda_N(t)}$$

2. *Inverse Gaussian mixtures.* If V follows an inverse Gaussian distribution, then

$$S_U(t) = \exp\left\{\frac{1}{c}\left(1 - \sqrt{1 + 2c\Lambda_N(t)}\right)\right\}, \qquad \lambda_U(t) = \frac{\lambda_N(t)}{\sqrt{1 + 2c\Lambda_N(t)}}$$

(see Whitmore and Lee (1991)).

3. *Three-parameter family mixtures.* Hougaard (1986a) gave the three-parameter family, $P(\alpha, \delta, \theta)$, whose Laplace transform is given by $E[e^{-sV}] = e^{-\{(\theta+s)^\alpha - \theta^\alpha\}\delta/\alpha}$. For $\alpha = 0$, this reduces to a gamma distribu-

tion. For $\alpha = 1/2$, this distribution reduces to an inverse Gaussian distribution. For $0 < \alpha < 1$, this family becomes an exponential family. If, further, $\alpha = \delta$ and $\theta = 0$, it is a positive stable distribution family (Hougaard, 1986b), for the case $\alpha = \delta$, see Crowder (1989). That is, many important special cases are included. If V follows the distribution $P(\alpha, \theta^{1-\alpha}, \theta)$ with $\theta = (1-\alpha)/c \geq 0$ so that $E(V) = 1$ and $\mathrm{Var}(V) = c$, then

$$S_U(t) = \exp\left[\frac{\theta}{\alpha}\left\{1 - \left(1 + \frac{\Lambda_N(t)}{\theta}\right)^{\alpha}\right\}\right] \qquad \lambda_U(t) = \lambda_N(t)\left(1 + \frac{\Lambda_N(t)}{\theta}\right)^{\alpha-1}$$

3.2 Bivariate Model with a Univariate Random Effect

Suppose that the j-th survival time, T_j, has baseline hazard $\lambda_{0,j}(t_j)$ and the nominal hazard rate is given by

$$\lambda_{N,j}(t_j) = \lambda_{0,j}(t_j)\exp\{\beta_j'(x_j)\} \qquad j = 1, 2$$

Then we incorporate the random effect, V, which is common to both survival times by setting hazard rates

$$V \times \lambda_{N,j}(t_j) = V \times \lambda_{0,j}(t_j)\exp\{\beta_j'(x_j)\}, \qquad j = 1, 2$$

where V is defined as in the previous subsection. The nominal and unconditional bivariate survival functions are given by

$$S_N(t_1, t_2) = \exp\left\{-\sum_{j=1}^{2} \Lambda_{N,j}(t_j)\right\},$$
$$S_U(t_1, t_2) = E\left[\exp\left\{-V\sum_{j=1}^{2} \Lambda_{N,j}(t_j)\right\}\right]$$

where $\Lambda_{N,j}(t) = \int_0^t \lambda_{N,j}(s)ds$ for $j = 1, 2$. Then unconditional hazard rates are

$$\lambda_{U,1}(t_1 | T_2 \geq t_2) = \frac{S_{U,1}}{S_U}, \qquad \lambda_{U,1}(t_1 | T_2 = t_2) = \frac{S_{U,12}}{S_{U,2}}$$
$$\lambda_U(t_1, t_2) = \frac{S_{U,12}}{S_U}$$

where $S_U = S_U(t_1, t_2)$ and

$$S_{U,i} = E\left[V \exp\left\{-V \sum_{j=1}^{2} \Lambda_{N,j}(t_j)\right\}\right],$$

$$S_{U,12} = E\left[V^2 \exp\left\{-V \sum_{j=1}^{2} \Lambda_{N,j}(t_j)\right\}\right]$$

for $i = 1, 2$.

This model imposes a dependence structure on the lifetimes which share a single common random effect, for example, lifetimes of twins that share a number of common factors.

3.2.1 Influences of random effects

The following corollary is obtained by Theorem 2 which we will prove in Section 3.3.

COROLLARY 2: *Suppose that* $P(V \geq 0) = 1$, $E(V) = 1$, *and* $P(V = 1) < 1$, *and let* $\{T \geq t\}$ *denote* $\{T_1 \geq t_1, T_2 \geq t_2\}$.

1. *The conditional hazard rate given* $T_2 \geq t_2$ *is*

$$\lambda_{U,1}(t_1 | T_2 \geq t_2) - \lambda_{N,1}(t_1) = \lambda_{N,1}(t_1)\{E(V|T \geq t) - 1\}$$

$$\frac{\partial}{\partial t_j} E(V|T \geq t) = -\lambda_{N,j}(t_j)\mathrm{Var}(V|T \geq t)$$

for $\lambda_{N,j}(t_j) > 0$, $j = 1, 2$. *Hence,* $E(V|T \geq t)$ *is nonincreasing in* $t_j > 0$ $(j = 1, 2)$, *with* $E(V|T \geq 0) = E(V) = 1$.

2. *The conditional hazard rate given* $T_2 = t_2$ *is*

$$\begin{aligned}\lambda_{U,1}(t_1 | T_2 = t_2) - \lambda_{N,1}(t_1) &= \lambda_{N,1}(t_1)(E(V_1 | T_1 \geq t_1, T_2 = t_2) - 1) \\ &= \{\lambda_{U,1}(t_1 | T_2 \geq t_2) - \lambda_{N,1}(t_1)\} \\ &\quad + \lambda_{N,1}(t_1)\frac{\mathrm{Var}(V|T \geq t)}{E(V|T \geq t)}\end{aligned}$$

for $\lambda_{N,2}(t_2) > 0$ *where*

$$\frac{\partial}{\partial t_j} E(V|T_1 \geq t_1, T_2 = t_2) = -\lambda_{N,j}(t_j)\mathrm{Var}(V|T_1 \geq t_1, T_2 = t_2)$$

for $\lambda_{N,j}(t_j) > 0$, $j = 1, 2$.

3. *The bivariate hazard rate is*

$$\lambda_U(t_1, t_2) - \lambda_{N,1}(t_1)\lambda_{N,2}(t_2) = \lambda_{N,1}(t_1)\lambda_{N,2}(t_2)\{E(V^2|T \geq t) - 1\}$$
$$\frac{\partial}{\partial t_j} E(V^2|T \geq t) = -\lambda_{N,j}(t_j)\mathrm{Cov}(V^2, V|T \geq t) < 0$$

for $\lambda_{N,j}(t_j) > 0, j = 1, 2.$

Proof. See Appendix 1.

The hazard rate $\lambda_{U,1}(t_1|T_2 \geq t_2)$ is decreasing in t_2 and the ratio $\lambda_{U,1}(t_1|T_2 \geq t_2)/\lambda_{N,1}(t_1)$ is also decreasing in t_1. As in the univariate model, the nominal hazard rate is underestimated. Similarly, the hazard rate $\lambda_{U,1}(t_1|T_2 = t_2)$ is decreasing in t_2, but the ratio $\lambda_{U,1}(t_1|T_2 \geq t_2)/\lambda_{N,1}(t_1)$ depends on the coefficient of variation of V given $T \geq t$ as well. However, since $E(V^2|T \geq t)$ is decreasing in t_1 and t_2 with $E(V^2|T \geq 0) = E(V^2) = 1 + \mathrm{Var}(V)$, the nominal bivariate hazard rate is overestimated for small t values and underestimated for large t values.

By setting $t_2 = 0$ in part 1 of Corollary 2, we obtain the result of Theorem 1. Now we give the definitions of some time-dependent measures of association and show the influences of random effects.

DEFINITION 1: Let $S = P(T_1 \geq t_1, T_2 \geq t_2)$, $S_1 = -\partial S/\partial t_1$, $S_2 = -\partial S/\partial t_2$, and $S_{12} = \partial^2 S/\partial t_1 \partial t_2$.

1. The odds ratio (Clayton, 1978), $\theta(t_1, t_2)$, for a bivariate survival time is defined by

$$\theta(t_1, t_2) \equiv \frac{SS_{12}}{S_1 S_2} = \frac{\lambda_U(t_1|T_2 = t_2)}{\lambda_U(t_1|T_2 \geq t_2)} = \frac{\lambda_U(t_2|T_1 = t_1)}{\lambda_U(t_2|T_1 \geq t_1)}$$

2. Densities of L-measure (Andersen et al., 1992; Gill, 1993), l_j, l_{12}, for univariate and bivariate survival times are defined by

$$l_j \equiv -\lambda_j, \qquad l_{12} \equiv \lambda_{12} - \lambda_1\lambda_2$$

where $\lambda_j = S_j/S$ for $j = 1, 2$ and $\lambda_{12} = S_{12}/S$. Univariate and bivariate hazard rates are expressed in terms of l_js and l_{12} as follows:

$$\lambda_j = -l_j, \qquad \lambda_{12} = l_{12} + l_1 l_2, \qquad j = 1, 2$$

The odds ratio, $\theta(t_1, t_2)$, is the ratio of hazard rates ($Pr(T_1 = t_1|T_1 \geq t_1, T_2 = t_2)$ and $Pr(T_1 = t_1|T_1 \geq t_1, T_2 \geq t_2)$). It shows how much the additional information, $\{T_2 = t_2\}$, changes the probability of $T_1 = t_1$ when $T \geq t$. It also measures a multiplicative interaction between two survival times in the bivariate hazard rate as follows:

$$\theta(t_1, t_2) = \frac{\lambda_{12}}{\lambda_1 \lambda_2}, \qquad \lambda_{12} = \lambda_1 \lambda_2 \theta(t_1, t_2)$$

where $\lambda_{12} = S_{12}/S$. An odds ratio $\theta(t_1, t_2) > 1$ implies a positive interaction between the two survival times, and $\theta(t_1, t_2) < 1$ implies a negative interaction. When two survival times are independent, $\theta(t_1, t_2) = 1$. Similarly, the density of L-measure, l_{12}, measures an additive interaction using their difference as in the calculation of a covariance. A positive value for l_{12} implies positive dependence, while a negative value implies negative dependence.

$$l_{12} = \lambda_{12} - \lambda_1 \lambda_2, \qquad \lambda_{12} = \lambda_1 \lambda_2 + l_{12}$$

COROLLARY 3: *For the common random effect model,*

1. *The odds ratio is*

$$\theta(t_1, t_2) = 1 + \frac{\mathrm{Var}(V|T \geq t)}{E^2(V|T \geq t)} - \frac{E(V|T_1 \geq t_1, T_2 = t_2)}{E(V|T_1 \geq t_1, T_2 \geq t_2)}$$

   when $\lambda_{N,j}(t_j) > 0$ for $j = 1, 2$.
2. *The density of the L-measure is*

$$l_{12} = \mathrm{Var}(V|T \geq t)\lambda_{N,1}(t_1)\lambda_{N,2}(t_2)$$

The odds ratio can be expressed as the ratio of conditional expectations of V. Both the odds ratio and the density of the L-measure quantify positive dependence in terms of the conditional expectation and variance of V given $T \geq t$. When the third conditional cumulant of V given $T \geq t$ is positive (negative) and $\lambda_{N,j}(t_j)$s are positive constants, l_{12} is decreasing (increasing) in both t_1 and t_2 (since $\partial \mathrm{Var}(V|T > t)/\partial t_j = -\lambda_{N,j}(t_j)E[\{V - E(V|T > t)\}^3|T > t]$).

3.1.2 Examples

We assume that $E(V) = 1$ and $\mathrm{Var}(V) = c$. For $S_U(t_1, t_2)$, see Section 4.5.

1. *Gamma mixtures.* If V follows a gamma distribution, then the hazard rate function, the conditional hazard rate of T_1 given $T_2 \geq t_2$, the odds ratio, and the density of L-measure are

$$\lambda_U(t_1, t_2) = (1+c)\prod_{j=1}^{2}\frac{\lambda_{N,j}(t_j)}{1 + c\sum_{i=1}^{2}\Lambda_{N,i}(t_i)}$$

$$\lambda_{U,1}(t_1|T_2 \geq t_2) = \frac{\lambda_{N,1}(t_1)}{1 + c\sum_{j=1}^{2}\Lambda_{N,j}(t_j)}$$

$$\theta(t_1, t_2) = 1 + c$$

$$l_{12} = c\prod_{j=1}^{2}\frac{\lambda_{N,j}(t_j)}{1 + c\sum_{i=1}^{2}\Lambda_{N,i}(t_i)}$$

The ratio $\lambda_U(t_1, t_2)/\lambda_{N,1}(t_1)\lambda_{N,2}(t_2)$ is greater than 1 for small values of t_1 and t_2, and is decreasing in both t_1 and t_2. On the other hand, $\lambda_{U,1}(t_1|T_2 \geq t_2)/\lambda_{N,1}(t_1)$ is less than 1 and is decreasing in t_1, t_2. Hence, when the $\lambda_{N,j}$s are positive constants, the additive interaction term, l_{12}, is also decreasing (and is a difference of decreasing terms). However, the multiplicative interaction term, $\theta(t_1, t_2)$, is always constant for all t_1, t_2. This suggests that $\theta(t_1, t_2)$ is a better measure of interaction for this model.

2. *Inverse Gaussian mixtures.* If V follows an inverse Gaussian distribution, then

$$\lambda_U(t_1, t_2) = \left[1 + \frac{c}{\sqrt{1 + 2c\sum_{i=1}^{2}\Lambda_{N,i}(t_i)}}\right] \times \prod_{j=1}^{2}\frac{\lambda_{N,j}(t_j)}{\sqrt{1 + 2c\sum_{i=1}^{2}\Lambda_{N,i}(t_i)}}$$

$$\lambda_{U,1}(t_1|T_2 \geq t_2) = \frac{\lambda_{N,1}(t_1)}{\sqrt{1 + 2c\sum_{j=1}^{2}\Lambda_{N,j}(t_j)}}$$

$$\theta(t_1, t_2) = 1 + \frac{c}{\sqrt{1 + 2c\sum_{j=1}^{2}\Lambda_{N,j}(t_j)}}$$

$$l_{12} = \frac{c}{\sqrt{1 + 2c\sum_{i=1}^{2}\Lambda_{N,i}(t_i)}}\prod_{j=1}^{2}\frac{\lambda_{N,j}(t_j)}{\sqrt{1 + 2c\sum_{i=1}^{2}\Lambda_{N,i}(t_i)}}$$

Overall, results are similar to those obtained in 1. (Gamma mixtures). The odds ratio is also decreasing in both t_1 and t_2 (but the decreases are more pronounced in l_{12} when the $\lambda_{N,j}$s are positive constants).

3. *Three-parameter family mixtures.* If V follows the distribution $P(\alpha, \theta^{1-\alpha}, \theta)$ with $\theta = (1-\alpha)/c \geq 0$ so that $E(V) = 1$ and $\text{Var}(V) = c$, then

$$\lambda_U(t_1, t_2) = \left[1 + \frac{c}{\left(1 + \sum_{j=1}^{2} \frac{\Lambda_{N,j}(t_j)}{\theta}\right)^{\alpha}}\right] \prod_{j=1}^{2} \frac{\lambda_{N,j}(t_j)}{\left(1 + \sum_{i=1}^{2} \frac{\Lambda_{N,i}(t_i)}{\theta}\right)^{1-\alpha}}$$

$$\lambda_{U,1}(t_1 | T_2 \geq t_2) = \frac{\lambda_{N,1}(t_1)}{\left(1 + \sum_{j=1}^{2} \frac{\Lambda_{N,j}(t_j)}{\theta}\right)^{1-\alpha}}$$

$$\theta(t_1, t_2) = 1 + \frac{c}{\left(1 + \sum_{j=1}^{2} \frac{\Lambda_{N,j}(t_j)}{\theta}\right)^{\alpha}}$$

$$l_{12} = \frac{c}{\left(1 + \sum_{j=1}^{2} \frac{\Lambda_{N,j}(t_j)}{\theta}\right)^{\alpha}} \prod_{j=1}^{2} \frac{\lambda_{N,j}(t_j)}{\left(1 + \sum_{i=1}^{2} \frac{\Lambda_{N,i}(t_i)}{\theta}\right)^{1-\alpha}}$$

3.3 Bivariate Model with a Bivariate Random Effect

As in the previous subsection, consider a nominal hazard rate given by

$$\lambda_{N,j}(t_j) = \lambda_{0,j}(t_j) \exp\{\beta_j' x_j\}, \qquad j = 1, 2$$

We then extend the bivariate model to have correlated random effects, (V_1, V_2), by setting hazard rates as follows:

$$V_j \times \lambda_{N,j}(t_j) = V_j \times \lambda_{0,j}(t_j) \exp\{\beta_j' x_j\}, \quad j = 1, 2$$

where V_1 and V_2 are possibly correlated. The nominal and unconditional bivariate survival functions are given by

$$S_N(t_1, t_2) = \exp\left\{-\sum_{j=1}^{2} \Lambda_{N,j}(t_j)\right\},$$

$$S_U(t_1, t_2) = E\left[\exp\left\{-\sum_{j=1}^{2} V_j \Lambda_{N,j}(t_j)\right\}\right]$$

where $\Lambda_{N,j}(t) = \int_0^t \lambda_{N,j}(s) ds$ for $j = 1, 2$. Then, (unconditional) hazard rates are

$$\lambda_{U,1}(t_1 | T_2 \geq t_2) = \frac{S_{U,1}}{S_U}, \qquad \lambda_{U,1}(t_1 | T_2 = t_2) = \frac{S_{U,12}}{S_{U,2}},$$

$$\lambda_U(t_1, t_2) = \frac{S_{U,12}}{S_U}$$

where $S_U = S_U(t_1, t_2)$ and

$$S_{U,i} = E\left[V_i \exp\left\{-\sum_{j=1}^{2} V_j \Lambda_{N,j}(t_j)\right\}\right],$$

$$S_{U,12} = E\left[V_1 V_2 \exp\left\{-\sum_{j=1}^{2} V_j \Lambda_{N,j}(t_j)\right\}\right]$$

for $i = 1, 2$.

3.3.1 Influences of random effects

Partial results were obtained by Lindeboom and Van Den Berg (1994) as indicated.

THEOREM 2 (Johnson and Omori, 1997): *Suppose that* $P(V_j \geq 0) = 1$, $E(V_j) = 1$ *and* $P(V_j = 1) < 1$ *for* $j = 1, 2$. *Then,*

1. *The conditional hazard rate given* $T_2 \geq t_2$ *is*

$$\lambda_{U,1}(t_1|T_2 \geq t_2) - \lambda_{N,1}(t_1) = \lambda_{N,1}(t_1)\{E(V_1|T_1 \geq t_1, T_2 \geq t_2) - 1\}$$

and

$$\frac{\partial}{\partial t_j} E(V_1|T_1 \geq t_1, T_2 \geq t_2) = -\lambda_{N,j}(t_j)\text{Cov}(V_1, V_j|T_1 \geq t_1, T_2 \geq t_2)$$

for $\lambda_{N,j}(t_j) > 0, j = 1, 2$ *(Lindeboom and Van Den Berg, 1994). If* V_1, V_2 *are associated, then* $E(V_1|T_1 \geq t_1, T_2 \geq t_2) \leq 1$.

2. *The conditional hazard rate given* $T_2 = t_2$ *is*

$$\begin{aligned}
&\lambda_{U,1}(t_1|T_2 = t_2) - \lambda_{N,1}(t_1) \\
&= \lambda_{N,1}(t_1)(E(V_1|T_1 \geq t_1, T_2 = t_2) - 1), \\
&= (\lambda_{U,1}(t_1|T_2 \geq t_2) - \lambda_{N,1}(t_1)) + \lambda_{N,1}(t_1)\frac{\text{Cov}(V_1, V_2|T_1 \geq t_1, T_2 \geq t_2)}{E(V_2|T_1 \geq t_1, T_2 \geq t_2)}
\end{aligned}$$

for $\lambda_{N,2}(t_2) > 0$ *where*

$$\frac{\partial}{\partial t_j} E(V_1|T_1 \geq t_1, T_2 = t_2) = -\lambda_{N,j}(t_j)\text{Cov}(V_1, V_j|T_1 \geq t_1, T_2 = t_2)$$

for $\lambda_{N,j}(t_j) > 0, j = 1, 2$.

3. *The bivariate hazard rate is*

$$\lambda_U(t_1, t_2) - \lambda_{N,1}(t_1)\lambda_{N,2}(t_2) = \lambda_{N,1}(t_1)\lambda_{N,2}(t_2) \times (E(V_1 V_2|T_1 \geq t_1, T_2 \geq t_2) - 1)$$

and

$$\frac{\partial}{\partial t_j} E(V_1 V_2 | T_1 \geq t_1, T_2 \geq t_2) = -\lambda_{N,j}(t_j)\text{Cov}(V_1 V_2, V_j | T_1 \geq t_1, T_2 \geq t_2)$$

for $\lambda_{N,j}(t_j) > 0,\ j = 1, 2$. *If* V_1, V_2 *are associated, then* $E(V_1 V_2 | T_1 \geq t_1, T_2 \geq t_2) \leq 1 + \text{Cov}(V_1, V_2)$ *and hence,*

$$\lambda_U(t_1, t_2) - \lambda_{N,1}(t_1)\lambda_{N,2}(t_2) \leq \lambda_{N,1}(t_1)\lambda_{N,2}(t_2)\text{Cov}(V_1, V_2).$$

Proof. See Appendix 1.

The ratio $\lambda_{U,1}(t_1 | T_2 \geq t_2)/\lambda_{N,1}(t_1)$, is decreasing in t_1, but whether the nominal hazard rate is underestimated depends on t_2 and $\text{Cov}(V_1, V_2 | T_1 \geq t_1, T_2 \geq t_2)$. For example, we expect $\text{Cov}(V_1, V_2 | T_1 \geq t_1, T_2 \geq t_2) > 0$ for ages at death of a son and his father. Then $\lambda_{U,1}(t_1 | T_2 \geq t_2)$ is decreasing in t_2. By ignoring the existence of positively correlated random effects (given $T_1 \geq t_1, T_2 \geq t_2$), we underestimate the nominal hazard rate. On the other hand, when $\text{Cov}(V_1, V_2 | T_1 \geq t_1, T_2 \geq t_2) < 0$, we would overestimate the nominal hazard rate for large values of t_2.

If, instead, we are given $T_2 = t_2$, the ratio of the unconditional and nominal hazard rates also depends on the fraction $\text{Cov}(V_1, V_2 | T_1 \geq t_1, T_2 \geq t_2)/E(V_2 | T_1 \geq t_1, T_2 \geq t_2)$. When $\text{Cov}(V_1, V_2 | T_1 \geq t_1, T_2 \geq t_2) > 0$, $\lambda_U(t_1 | T_2 = t_2)$ is greater than $\lambda_U(t_1 | T_2 \geq t_2)$. On the other hand, when $\text{Cov}(V_1, V_2 | T_1 \geq t_1, T_2 \geq t_2) < 0$, it is less than $\lambda_U(t_1 | T_2 \geq t_2)$.

Further, if V_1 and V_2 are associated, the difference between the unconditional and nominal bivariate hazard rate is less than or equal to the covariance of random effects multiplied by the nominal bivariate hazard rate.

Using the results above, we determine the influence of random effects on the following measures of dependency.

LEMMA 1:

1. *Odds ratio.*

$$\theta(t_1, t_2) = 1 + \frac{\text{Cov}(V_1, V_2 | T \geq t)}{E(V_1 | T \geq t)E(V_2 | T \geq t)} = \frac{E(V_1 | T_1 \geq t_1, T_2 = t_2)}{E(V_1 | T_1 \geq t_1, T_2 \geq t_2)}$$

when $\lambda_{N,j}(t_j) > 0$ *for* $j = 1, 2$.

2. *Density of L-measure.*

$$l_{12} = \text{Cov}(V_1, V_2 | T \geq t)\lambda_{N,1}(t_1)\lambda_{N,2}(t_2)$$

Both measures depend on the conditional covariance, $\mathrm{Cov}(V_1, V_2|T_1 \geq t_1, T_2 \geq t_2)$. By setting $V_1 = V_2 = V$ in Theorem 2 and Lemma 1, we obtain Corollaries 2 and 3, respectively. When $\lambda_{N,j}(t_j)$s are positive constants and $E[\{V_i - E(V_i|T > t)\}\{V_j - E(V_j|T > t)\}^2|T > t]$ $(i \neq j)$ are positive (negative), the l_{12} is decreasing (increasing) in t_j (since $\partial\mathrm{Cov}(V_1, V_2|T > t)/\partial t_j = -\lambda_{N,j}(t_j)E[\{V_i - E(V_i|T > t)\}\{V_j - E(V_j|T > t)\}^2$ $|T > t]$ $(i \neq j)$).

Finally we show a miscellaneous lemma for marginal and joint unconditional survival functions, which implies that positive (negative) quadrant dependence of (T_1, T_2) is induced by that of (V_1, V_2).

LEMMA 2: *If (V_1, V_2) is positively quadrant dependent, then*

1. *(T_1, T_2) is also positively quadrant dependent, i.e.,*

$$S_U(t_1, t_2) - S_U(t_1)S_U(t_2) \geq 0$$

2. *and further,*

$$S_U(t_1, t_2) - S_U(t_1)S_U(t_2) \leq \Lambda_{N,1}(t_1)\Lambda_{N,2}(t_2)\mathrm{Cov}(V_1, V_2)$$

If (V_1, V_2) is negatively quadrant dependent, then parts 1 and 2 hold with the inequality sign reversed and (T_1, T_2) is also negatively quadrant dependent.

Proof. See Appendix 1.

3.3.2 Examples

We assume $E(V_j) = 1$ and $\mathrm{Var}(V_j) = c_j$, $j = 1, 2$. For $S_U(t_1, t_2)$, see Section 4.5.

1. *Gamma mixtures*. Let $V_j = (Z_0 + Z_j)c_j$ where $Z_0 \sim Gamma(\alpha, 1)$, $Z_j \sim Gamma(c_j^{-1} - \alpha, 1)$ $(0 < \alpha < c_j^{-1})$ $j = 1, 2$, independently. Then V_1, V_2 follow a bivariate gamma distribution such that $E(V_j) = 1$, $\mathrm{Var}(V_j) = c_j$ for $j = 1, 2$ and $\mathrm{Cov}(V_1, V_2) = \alpha c_1 c_2$ $(Corr(V_1, V_2) = \alpha\sqrt{c_1 c_2})$. The conditional hazard rate of T_1 given $T_2 \geq t_2$, the odds ratio, and density of L-measure are

$$\lambda_{U,1}(t_1|T_2 \geq t_2) = \lambda_{N,1}(t_1)\left[\frac{1-\alpha c_1}{1+c_1\Lambda_{N,1}(t_1)} + \frac{\alpha c_1}{1+\sum_{j=1}^{2} c_j\Lambda_{N,j}(t_j)}\right]$$

$$\theta(t_1, t_2) = 1 + \alpha c_1 c_2 \prod_{j=1}^{2} \frac{1+c_j\Lambda_{N,j}(t_j)}{1+\sum_{i=1}^{2} c_i\Lambda_{N,i}(t_i) - \alpha c_1 c_2 \Lambda_{N,j}(t_j)}$$

$$l_{12} = \alpha c_1 c_2 \prod_{j=1}^{2} \frac{\lambda_{N,j}(t_j)}{1+\sum_{i=1}^{2} c_i\Lambda_{N,i}(t_i)}$$

Figure 1 shows $\lambda_{U,1}(t_1|T_2 \geq t_2)$, $\theta(t_1, t_2)$ and l_{12} when $\lambda_{N,j}(t_j) = 1$, $E(V_j) = 1$ and $\mathrm{Var}(V_j) = c_j = 0.5$ for $j = 1, 2$. Figure 1 (a1)–(a3) shows $\lambda_{U,1}(t_1|T_2 \geq t_2)$ for the values $\rho = Corr(V_1, V_2) = 0.0$ (hence V_1,V_2 are independent), 0.5 ($\alpha = 1$), and 1.0 ($\alpha = 2$). In all cases, $\lambda_{U,1}(t_1|T_2 \geq t_2)$ is decreasing in t_1. When V_1 and V_2 are uncorrelated, both nominal hazard rates are underestimated and this holds for all t_2. The amount of underestimation increases as the value of either t_2 or ρ increases.

Figure 1 (b1)–(b3) and (c1)–(c3) shows $\theta(t_1, t_2)$ and l_{12} for the values $\rho = 0.5, 0.8, 1.0$. When $\rho = 0.0$, $\theta(t_1, t_2) = 1.0$ and $l_{12} = 0$ for all t_1, t_2. As ρ increases, $\theta(t_1, t_2)$ increases overall and is equal to 1.5 for $\rho = 1.0$. On the other hand, the value of l_{12} at the origin increases. For the fixed value of ρ, they are mostly decreasing in t_1, t_2, except that $\theta(t_1, t_2)$ is increasing for small values of t_2 for a fixed large value of t_1.

2. *Inverse Gaussian mixtures.* Let $V_j = \beta_j Z_0 + Z_j$ where $Z_0 \sim IG(\mu, \lambda)$, $Z_j \sim IG(\mu_j, \lambda_j)$ with $\beta_j = c_j\lambda/\mu^2$, $\mu_j = 1 - (c_j\lambda/\mu)$, $\lambda_j = \mu_j^2/c_j$ for $j = 1, 2$. Then, the V_js follow a bivariate inverse Gaussian distribution with marginal distributions $IG(1, c_j^{-1})$ and $\mathrm{Cov}(V_1, V_2) = (\lambda/\mu)c_1c_2$ ($Corr(V_1, V_2) = (\lambda/\mu)\sqrt{c_1c_2}$). The conditional hazard rate of T_1 given $T_2 \geq t_2$, the odds ratio, and density of L-measure are

$$\lambda_{U,1}(t_1|T_2 \geq t_2) = \lambda_{N,1}(t_1)\left[\left(1-\frac{c_1\lambda}{\mu}\right)\frac{1}{\sqrt{1+2c_1\Lambda_{N,1}(t_1)}} + \frac{c_1\lambda}{\mu}\cdot\frac{1}{\sqrt{1+2\sum_{j=1}^{2} c_j\Lambda_{N,j}(t_j)}}\right]$$

$$\theta(t_1, t_2) = 1 + \frac{c_1c_2\lambda/\mu}{\sqrt{1+2\sum_{i=1}^{2} c_i\Lambda_{N,i}(t_i)}}$$

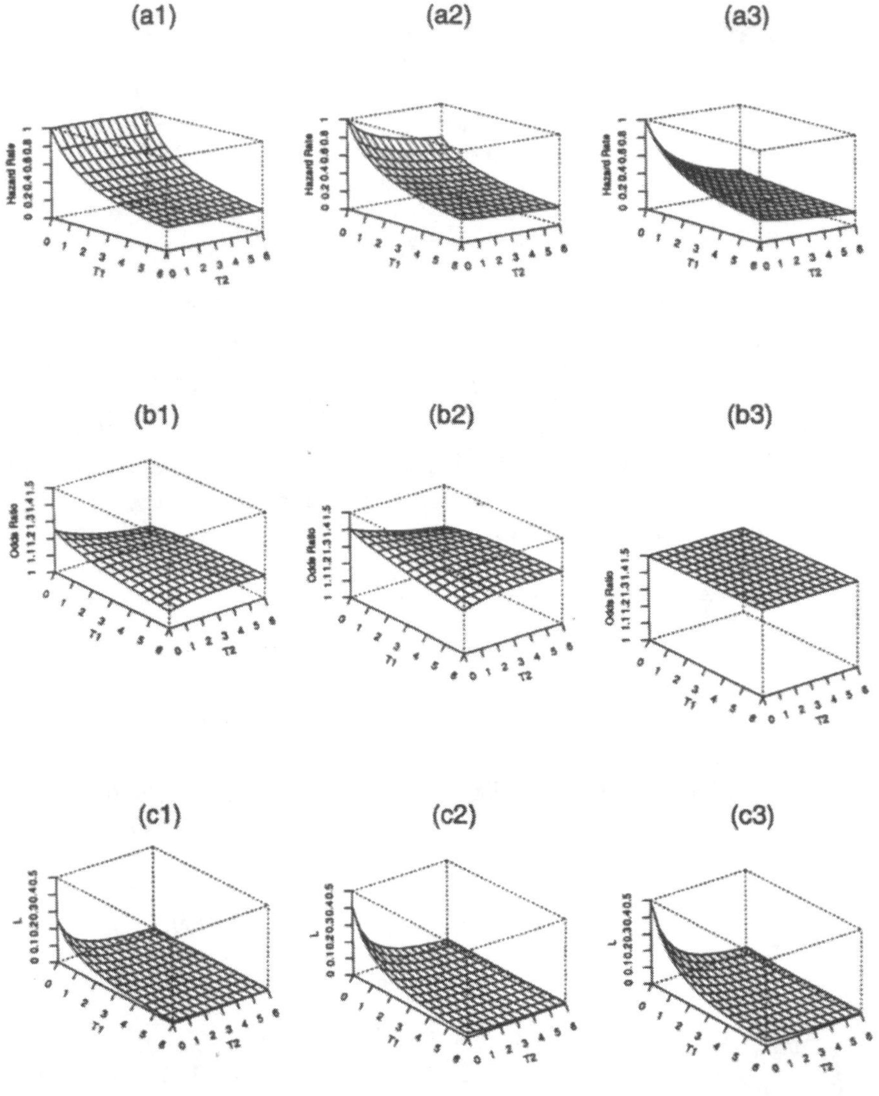

Figure 1 Hazard rate, odds ratio, and density of L-measure. Random effects: Gamma. $\lambda_{N,1}(t_1) = \lambda_{N,2}(t_2) = 1$. $\rho = Corr(V_1, V_2)$. $\lambda_{U,1}(t_1|T_2 \geq t_2)$:(a1)–(a3), $\theta(t_1, t_2)$:(b1)–(b3), l_{12} :(c1)–(c3). (a1) $\rho = 0.0$, (a2) $\rho = 0.5$, (a3) $\rho = 1.0$, (b1) $\rho = 0.5$, (b2) $\rho = 0.8$, (b3) $\rho = 1.0$, (c1) $\rho = 0.5$, (c2) $\rho = 0.8$, (c3) $\rho = 1.0$

$$\times \prod_{j=1}^{2} \frac{\sqrt{1+2c_j\Lambda_{N,j}(t_j)}}{\left\{\frac{c_j\lambda}{\mu}\sqrt{1+2c_j\Lambda_{N,j}(t_j)} + \left(1-\frac{c_j\lambda}{\mu}\right)\sqrt{1+2\sum_{i=1}^{2} c_i\Lambda_{N,i}(t_i)}\right\}}$$

$$l_{12} = \frac{c_1c_2\lambda/\mu}{\sqrt{1+2\sum_{i=1}^{2} c_i\Lambda_{N,i}(t_i)}} \prod_{j=1}^{2} \frac{\lambda_{N,j}(t_j)}{\sqrt{1+2\sum_{i=1}^{2} c_i\Lambda_{N,i}(t_i)}}$$

Figure 2 shows $\lambda_{U,1}(t_1|T_2 \geq t_2)$, $\theta(t_1, t_2)$ and l_{12} when $\lambda_{N,j}(t_j) = 1$, $E(V_j) = 1$ and $\mathrm{Var}(V_j) = c_j = 0.5$ for $j = 1, 2$. Figure 2 (a1)–(a3) shows $\lambda_{U,1}(t_1|T_2 \geq t_2)$ for the values $\rho = Corr(V_1, V_2) = 0.0$, $0.5(\lambda/\mu = 1)$, and 1.0 $(\lambda/\mu = 2)$. Overall, the plots are similar to those in Figure 1. The amount of underestimation increases as the value of either t_2 or ρ increases (but not as much as in Figure 1). Figure 2 (b1)–(b3) and (c1)–(c3) shows $\theta(t_1, t_2)$ and l_{12} for the values $\rho = 0.5, 0.8, 1.0$. As ρ increases, $\theta(t_1, t_2)$ increases overall (but not as much as in Figure 1). For the fixed value of ρ, it is generally decreasing in t_1, t_2. Otherwise, the plots look similar to those in Figure 1.

3. *Three parameter family mixtures*. Let $V_j = \beta_j Z_0 + Z_j$ for $j = 1, 2$ where $Z_0 \sim P(\alpha, \delta, \theta)$ and $Z_j \sim P(\alpha, \delta_j, \theta_j)$ for $j = 1, 2$ with $\beta_j = \theta/\theta_j > 0$, $\theta_j = (1-\alpha)/c_j \geq 0$, $\delta_j = (\theta_j - \delta\theta^\alpha)/\theta_j^\alpha > 0$. Then, the V_js follow the bivariate three-parameter family of distributions with marginal distributions $P(\alpha, \theta_j^{1-\alpha}, \theta_j)$ and $E(V_j) = 1$, $\mathrm{Var}(V_j) = c_j$, and $\mathrm{Cov}(V_1, V_2) = (1-\alpha)\delta\theta^\alpha/\theta_1\theta_2 = \delta\theta^\alpha c_1c_2/(1-\alpha)$ $(Corr(V_1, V_2) = \delta\theta^\alpha\sqrt{c_1c_2}/(1-\alpha))$. The conditional hazard rate of T_1 given $T_2 \geq t_2$, the odds ratio, and density of L-measure are

$$\lambda_{U,1}(t_1|T_2 \geq t_2) = \lambda_{N,1}(t_1)\left\{\left(1-\frac{\delta\theta^\alpha}{\theta_1}\right)\left(1+\frac{\Lambda_{N,1}(t_1)}{\theta_1}\right)^{\alpha-1} + \frac{\delta\theta^\alpha}{\theta_1}\left(1+\sum_{j=1}^{2}\frac{\Lambda_{N,j}(t_j)}{\theta_j}\right)^{\alpha-1}\right\}$$

$$\theta(t_1, t_2) = 1 + \frac{(1-\alpha)\delta\theta^\alpha/\theta_1\theta_2}{\left(1+\sum_{i=1}^{2}\frac{\Lambda_N(t_i)}{\theta_i}\right)^{\alpha}}$$

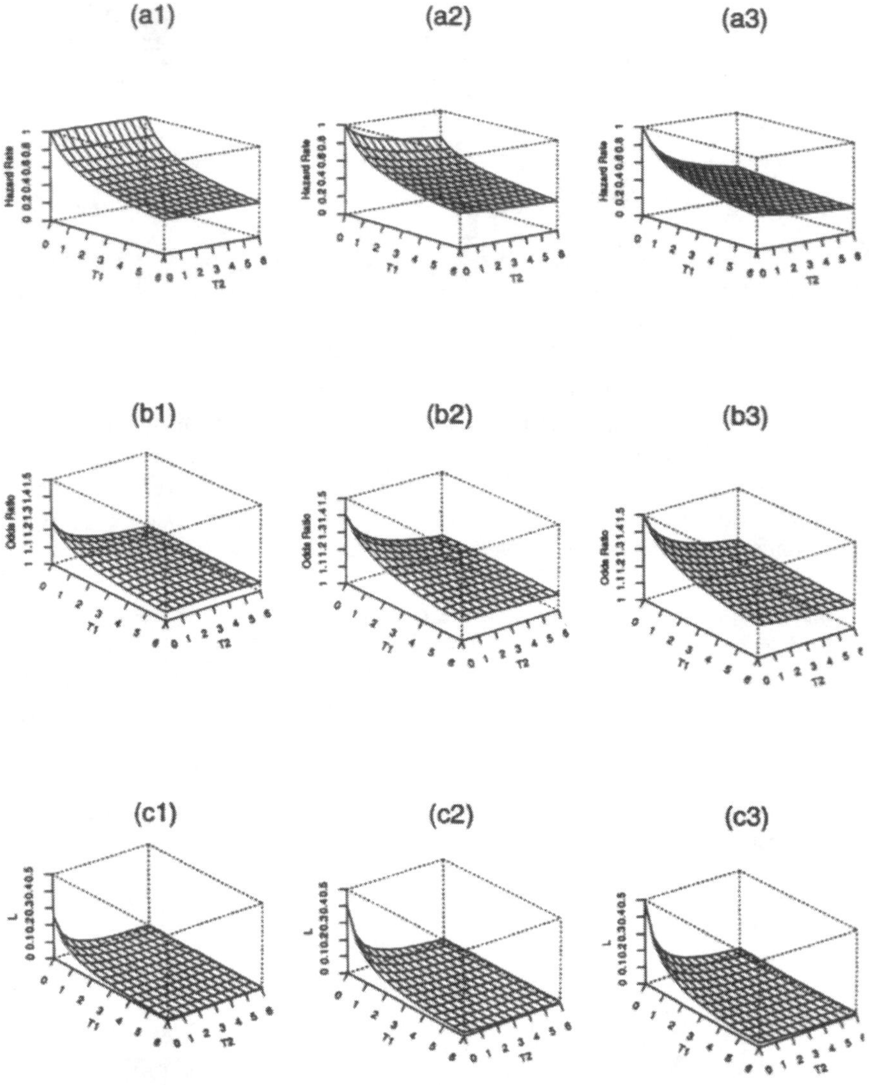

Figure 2 Hazard rate, odds ratio, and density of L-measure. Random effects: inverse Gaussian. $\lambda_{N,1}(t_1) = \lambda_{N,2}(t_2) = 1$. $\rho = Corr(V_1, V_2)$. $\lambda_{U,1}(t_1 | T_2 \geq t_2)$:(a1)–(a3), $\theta(t_1, t_2)$:(b1)–(b3), l_{12} :(c1)–(c3). (a1) $\rho = 0.0$, (a2) $\rho = 0.5$, (a3) $\rho = 1.0$, (b1) $\rho = 0.5$, (b2) $\rho = 0.8$, (b3) $\rho = 1.0$, (c1) $\rho = 0.5$, (c2) $\rho = 0.8$, (c3) $\rho = 1.0$

$$\times \prod_{j=1}^{2} \frac{\left(1+\frac{\Lambda_{N,j}(t_j)}{\theta_j}\right)^{1-\alpha}}{\frac{\delta\theta^\alpha}{\theta_j}\left(1+\frac{\Lambda_{N,j}(t_j)}{\theta_j}\right)^{1-\alpha}+\left(1-\frac{\delta\theta^\alpha}{\theta_j}\right)\left(1+\sum_{i=1}^{2}\frac{\Lambda_{N,i}(t_i)}{\theta_i}\right)^{1-\alpha}}$$

$$l_{12} = \frac{(1-\alpha)\delta\theta^\alpha/\theta_1\theta_2}{\left(1+\sum_{i=1}^{2}\frac{\Lambda_N(t_i)}{\theta_i}\right)^\alpha} \prod_{j=1}^{2} \frac{\lambda_{N,j}(t_j)}{\left(1+\sum_{i=1}^{2}\frac{\Lambda_N(t_i)}{\theta_i}\right)^{1-\alpha}}$$

4. *Lognormal mixtures*. Figure 3 shows $\lambda_{U,1}(t_1|T_2 \geq t_2)$ when $\lambda_{N,j}(t_j) = 1$ for $j = 1, 2$. The random effects, $(\log V_1, \log V_2)$, are assumed to follow a bivariate normal distribution with $E(V_j) = 1$ and $\mathrm{Var}(V_j) =$

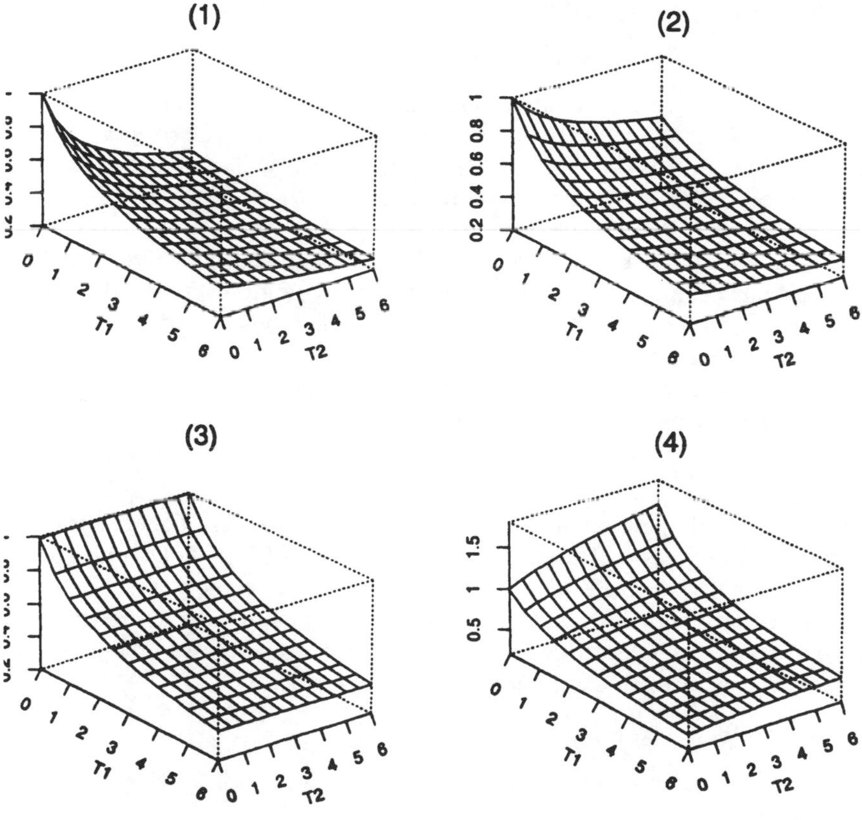

Figure 3 Hazard rate $\lambda_{U,1}(t_1|T_2 \geq t_2)$. Random effects: Lognormal. $\lambda_{N,1}(t_1) = \lambda_{N,2}(t_2) = 1$. $\rho = Corr(\log V_1, \log V_2)$. (1) $\rho = 1.0$, (2) $\rho = 0.5$, (3) $\rho = 0.0$, (4) $\rho = -0.5$

0.5 ($E(\log V_j) = -0.5\text{Var}(\log V_j)$, $\text{Var}(\log V_j) = \log(1 + \text{Var}(V_j)))$ for $j = 1, 2$ and the values $\rho = Corr(\log V_1, \log V_2) = 1.0, 0.5, 0, -0.5$. Overall, the plots are similar to those in Figure 1 and 2. When $\rho < 0$, the underestimation is smaller for small values of t_2, and there is even overestimation for small values of t_1 and large values of t_2.

Figures 4 and 5 show $\theta(t_1, t_2)$ and l_{12} for the values $\rho = 1.0, 0.5, -0.5, -1.0$. The plots look similar to those in Figure 2. When ρ (< 0) decreases, they decrease overall. Their patterns for $\rho < 0$ are similar to those for $\rho > 0$ but in the opposite direction.

4. MULTIVARIATE MODEL

4.1 General Random Effect Model

Since the 1960s, various multivariate failure-time distributions have been introduced. In the early stages of development, they were derived as direct generalizations of popular univariate distributions, such as an exponential distribution. Their marginal distributions reduce to the original univariate distribution which they generalize.

On the other hand, it has recently become popular to derive a multivariate failure–time distribution by introducing random effects that represent common unknown variables. This approach has the advantage that it is intuitive and easy to manipulate.

Marshall and Olkin (1988) gave the following general model for a multivariate distribution induced by random effects;

$$S_U(t_1, \ldots, t_J) = E_{V_1,\ldots,V_J}[K(S_{N,1}^{V_1}(t_1), \ldots, S_{N,J}^{V_J}(t_J))]$$

where $(V_1, \ldots, V_J)$ are random effects with $V_j > 0$, the $S_{N,j}^{V_j}(t_j)$s are survival functions, and K is a J-variate distribution function with all univariate marginals uniform on [0, 1]. This contains the wide class of distributions introduced by Genest and MacKay (1986a,b). As a natural generalization of the univariate model described in the previous subsection, we first consider the case where the j-th component has the hazard rate

$$\lambda_{N,j}(t) = \lambda_{0,j}(t)\exp\{\beta_j' x_j\}$$

where $\lambda_{0,j}(t)$ is a specified baseline hazard rate, and x_j is a vector of fixed covariates.

In order to introduce heterogeneity into the population via a random effect, we let the hazard rate for an individual be

$$V_j \times \lambda_{N,j}(t) = V_j \times \lambda_{0,j}(t)\exp\{\beta_j' x_j\}$$

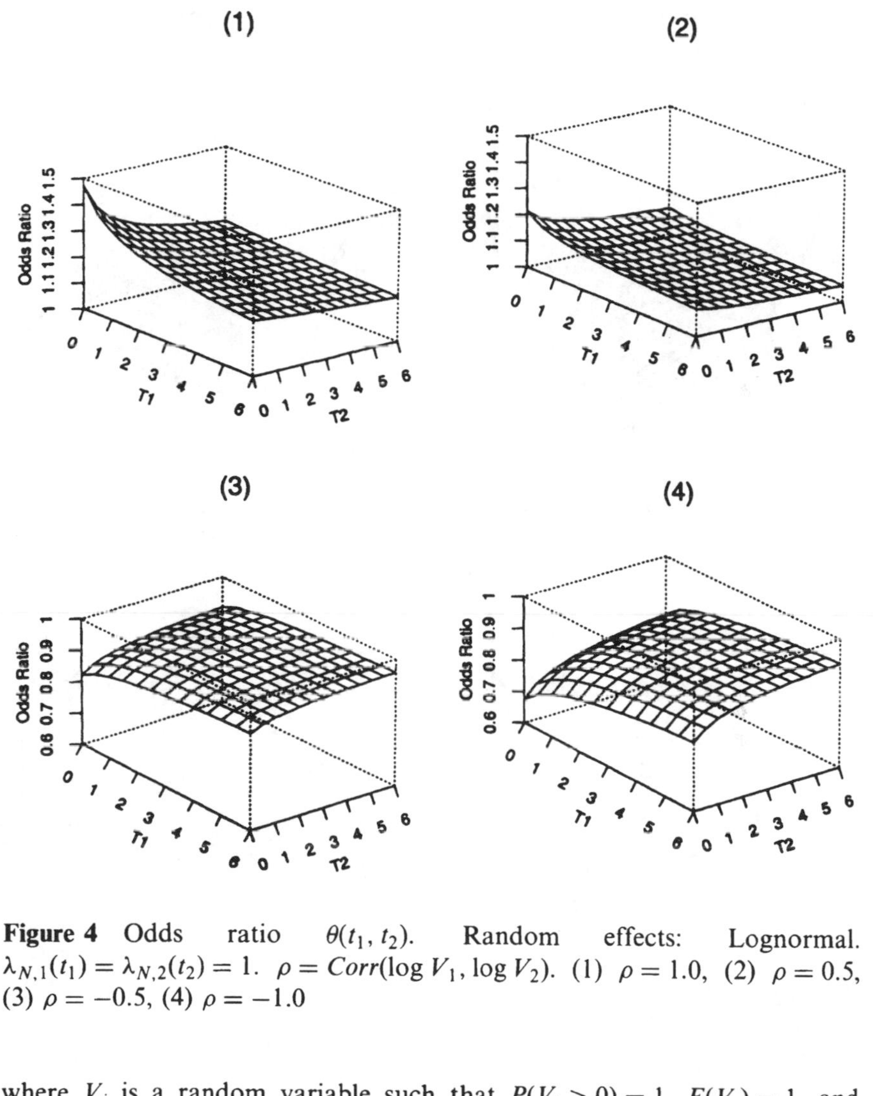

Figure 4 Odds ratio $\theta(t_1, t_2)$. Random effects: Lognormal. $\lambda_{N,1}(t_1) = \lambda_{N,2}(t_2) = 1$. $\rho = Corr(\log V_1, \log V_2)$. (1) $\rho = 1.0$, (2) $\rho = 0.5$, (3) $\rho = -0.5$, (4) $\rho = -1.0$

where V_j is a random variable such that $P(V_j \geq 0) = 1$, $E(V_j) = 1$, and $P(V_j = 1) < 1$.

Suppose that, conditional on $V_j = v_j$, the individual hazard rates

$$\lambda_{C,j}(t|v_j) \equiv v_j \lambda_{N,j}(t) \qquad \textit{for } j = 1, 2, \ldots, J$$

are independent. Then the conditional, nominal, and unconditional marginal survival functions are

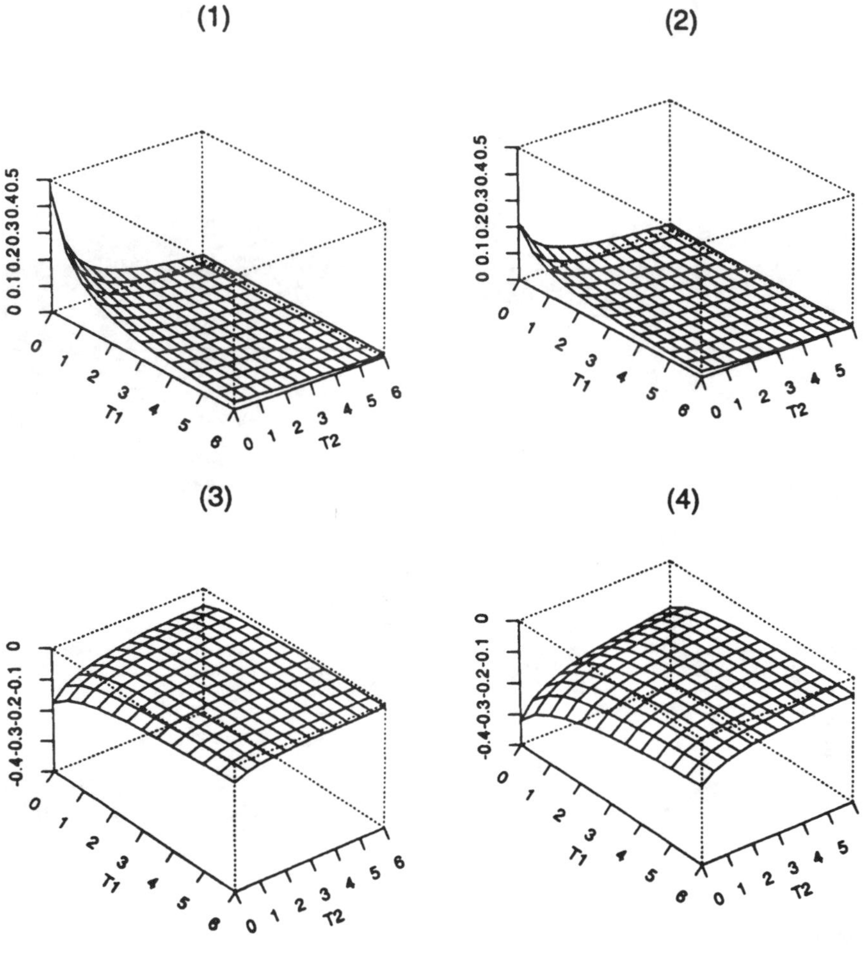

Figure 5 Density of L-measure. Random effects: Lognormal. $\lambda_{N,1}(t_1) = \lambda_{N,2}(t_2) = 1$. $\rho = Corr(\log V_1, \log V_2)$. (1) $\rho = 1.0$, (2) $\rho = 0.5$, (3) $\rho = -0.5$, (4) $\rho = -1$.

$$S_{C,j}(t|v_j) = 1 - F_{C,j}(t|v_j) = \exp\{-v_j \Lambda_N(t)\}$$
$$S_{N,j}(t) = 1 - F_{N,j}(t) = \exp\{-\Lambda_{N,j}(t)\}$$
$$S_{U,j}(t) = 1 - F_{U,j}(t) = E_{V_j}[\exp\{-V_j \Lambda_{N,j}(t)\}]$$

where $\Lambda_{N,j}(t) = \int_0^t \lambda_{N,j}(s) \mathrm{d}s$, respectively. The conditional, nominal, and unconditional joint survival functions are

$$S_C(t|v) = \exp\left\{-\sum_{j=1}^{J} v_j \Lambda_{N,j}(t_j)\right\}$$

$$S_N(t) = \exp\left\{-\sum_{j=1}^{J} \Lambda_{N,j}(t_j)\right\}$$

$$S_U(t) = E_V\left[\exp\left\{-\sum_{j=1}^{J} V_j \Lambda_{N,j}(t_j)\right\}\right]$$

This model imposes a dependence structure on the lifetimes which have correlated factors, such as ages at death of a mother, her son, and daughter. When $V_j = V$ for $j = 1, \ldots, J$, the multivariate survival times share a single common random effect. This may be an appropriate model for association among lifetimes, such as lifetimes of twins, that share a number of common factors. This extension is discussed in detail in Costigan and Klein (1993). In the following examples, we consider a multivariate random effect ($V_1, \ldots, V_J$) that follows a multivariate distribution with $E(V_j) = 1$ for $j = 1, \ldots, J$.

4.2 Inconsistency of Maximum Likelihood Estimators

We establish, in a Weibull model, that ignoring random effects and finding maximum likelihood estimators based on the nominal likelihood can lead to inconsistent estimators as in Lancaster (1985). For the Weibull model with random effects, we have

$$\Lambda_{N,j}(t_j) = V_j t_j^{\alpha_j} \exp\{\beta_j' x_j\}, \quad j = 1, \ldots, J$$

and the nominal log likelihood (where $V_j \equiv 1$ for all j) for n independent observations is given by

$$\log L = \sum_{j=1}^{J} \left\{ n \log \alpha_j + (\alpha_j - 1) \sum_{i=1}^{n} \log t_{ij} + \sum_{i=1}^{n} \beta_j' x_{ij} - \sum_{i=1}^{n} t_{ij}^{\alpha_j} \exp\{\beta_j' x_{ij}\} \right\}$$

When we ignore the random effects, the maximum likelihood estimators will converge to the maximizers of $E(\log L)$ (since $\log L/n$ converges to $E(\log L)/n$ as $n \to \infty$). We can find them by taking expected values of $\log L$, and its calculation reduces to the univariate case as in Lancaster (1985), and hence the maximum likelihood estimators can be shown to be inconsistent under mild conditions.

4.3 Measures of Multivariate Dependence

To investigate the influence of random effects on multivariate dependence, we consider multivariate hazard rate functions, densities of L-measures, and odds ratios. Let

$$S = P(T_1 \geq t_1, T_2 \geq t_2, \ldots, T_J \geq t_J), \quad S_{i_1 i_2 \cdots i_m} = (-1)^m \frac{\partial^m S}{\partial t_{i_1} \partial t_{i_2} \cdots \partial t_{i_m}}$$

where $\{i_1, i_2, \ldots, i_{J-n}, i_1', \ldots, i_n'\}$ is a permutation of $\{1, 2, \ldots, J\}$.

1. *Multivariate hazard rate functions.* We denote multivariate hazard rate functions as

$$\lambda_{i_1 \cdots i_m | i_1' \cdots i_n'} = \frac{S_{i_1 \cdots i_m i_1' \cdots i_n'}}{S_{i_1' \cdots i_n'}}$$

   and $\lambda_{i_1 \cdots i_m} = \lambda_{i_1 \cdots i_m | \phi}$ for $\{i_1', \ldots, i_n'\} = \phi$.
2. *Density of L-measures* (Andersen et al., 1992; Gill, 1993) . A density of L-measure, $l_{1 \cdots m}$, for m-variate survival times is defined by

$$l_1 = -\lambda_1, \qquad l_{1 \cdots m} = \frac{\partial^{m-1} l_1}{\partial t_2 \cdots \partial t_m}, \qquad m \geq 2$$

   Hence, if $l_{i_1 \cdots i_m}$ exists, a density of L-measure for m-variate survival times can be obtained as

$$l_{i_1 \cdots i_m} = \frac{\partial^m \log S}{\partial t_{i_1} \cdots \partial t_{i_m}}$$

3. *Odds ratio.* Finally, we define an odds ratio for multivariate survival times.

DEFINITION 2: *Denote*

$$\theta_{i_1 \cdots i_m | i_1' \cdots i_n'} = \theta(t_{i_1}, \ldots, t_{i_m} | T_{i_{m+1}} \geq t_{i_{m+1}}, \ldots, T_{i_{J-n}} \geq t_{i_{J-n}}, \\ T_{i_1'} = t_{i_1'}, \ldots, T_{i_n'} = t_{i_n'})$$

and $\theta_{i_1 \cdots i_m} = \theta_{i_1 \cdots i_m | \phi}$. Then we define an odds ratio for m-variate survival times recursively as

$$\theta_{i_1 | i_1' \cdots i_n'} \equiv \frac{S_{i_1 i_1' \cdots i_n'}}{S_{i_1' \cdots i_n'}}, \qquad \theta_{i_1 \cdots i_{m+1} | i_1' \cdots i_n'} \equiv \frac{\theta_{i_1 \cdots i_m | i_{m+1} i_1' \cdots i_n'}}{\theta_{i_1 \cdots i_m | i_1' \cdots i_n'}}$$

Examples. For trivariate survival times, we obtain the following relationships among above measures.

$$l_{123} = -\lambda_{123} + \lambda_{12}\lambda_3 + \lambda_{23}\lambda_1 + \lambda_{31}\lambda_2 - 2\lambda_1\lambda_2\lambda_3)$$
$$\lambda_{123} = -(l_{123} + l_{12}l_3 + l_{23}l_1 + l_{31}l_2 + l_1l_2l_3) = \theta_1\theta_2\theta_3\theta_{12}\theta_{23}\theta_{31}\theta_{123}$$

and

$$\theta_j = \frac{S_j}{S} = \lambda_j, \qquad \theta_{ij} = \frac{SS_{ij}}{S_iS_j} = \frac{\lambda_{ij}}{\lambda_i\lambda_j}, \qquad \theta_{123} = \frac{S_1S_2S_3S_{123}}{SS_{12}S_{23}S_{31}} = \frac{\lambda_1\lambda_2\lambda_3\lambda_{123}}{\lambda_{12}\lambda_{23}\lambda_{31}}$$

The trivariate odds ratio, θ_{123}, is the ratio of bivariate odds ratios, $\theta_{12|3}$ and θ_{12}, which shows how much the information $\{T_3 = t_3\}$ changes the ratio θ_{12} when $T \geq t$. It measures a multiplicative third-order interaction among three survival times in the trivariate hazard rate as above. The trivariate hazard rate, λ_{123}, is a product of three marginal $(\theta_1, \theta_2, \theta_3)$ and three second-order interaction $(\theta_{12}, \theta_{23}, \theta_{31})$, and one third-order interaction (θ_{123}) terms. We see that $\theta_{123} > 1$ implies positive interaction among three survival times, and $\theta_{123} < 1$ implies negative interaction. When three survival times are independent, $\theta_{123} = 1$. Higher-order odds ratios can be interpreted similarly. In general, the m-variate odds ratio measures the m-th order multiplicative interaction among the m survival times in m-variate hazard rate. The condition $\theta_{12\cdots m} > 1$ implies positive interaction among the m survival times, and $\theta_{12\cdots m} < 1$ implies negative interaction. When m survival times are independent, $\theta_{12\cdots m} = 1$.

Similarly, the density of L-measures, $l_{12\cdots m}$, measures the m-th order additive interaction among the m survival times, but using differences of hazard rates as in the calculation of the cumulants. Note that the expression for $(-1)^m l_{12\cdots m}$ in terms of the λs corresponds to that of the cumulants in terms of the moments of random variables. The trivariate hazard rate, λ_{123}, is a sum of the product of three marginal $(-l_1l_2l_3)$ and the second-order interactions (multiplied by the marginal term), $(-l_{12}l_3, -l_{23}l_1, -l_{31}l_2)$, and a third-order interaction $(-l_{123})$ term. While the odds ratio decomposes λ into the product of lower order ratios, the density of L-measure decomposes it into the sum of terms expressed by lower order ls.

In the following theorems, we generalize the above relationship among hazard rates functions, densities of L-measures and odds ratios.

THEOREM 3: *An odds ratio for m-variate survival times* $(T_{i_1}, \ldots, T_{i_m})$ *can be expressed as*

$$\theta_{i_l\cdots i_m|i'_l\cdots i'_n} = \left[S_{i'_1\cdots i'_n} \prod_{j=1}^{m} \left\{ \prod_{k_1<\cdots<k_j} S_{i'_1\cdots i'_n k_1\cdots k_j} \right\}^{(-1)^j} \right]^{(-1)^m}$$

$$= \left[\prod_{j=1}^{m} \left\{ \prod_{k_1<\cdots<k_j} \lambda_{k_1\cdots k_j|i'_1\cdots i'_n} \right\}^{(-1)^j} \right]^{(-1)^m}$$

where $\{k_1, \ldots, k_j\} \subseteq \{i_1, \ldots, i_m\} \subseteq \{1, \ldots, J\}$ for $j = 1, \ldots, m$.

Proof. See Appendix 1.

THEOREM 4: *Hazard rate functions for m-variate survival times* $(T_{i_1}, \ldots, T_{i_m})$ *can be expressed in terms of odds ratios.*

$$\lambda_{i_l\cdots i_m|i'_l\cdots i'_n} = \theta_{i_1\cdots i_m|i'_1\cdots i'_n} \prod_{j=1}^{m-1} \left(\prod_{k_1<\cdots<k_j} \theta_{k_1\cdots k_j|i'_1\cdots i'_n} \right)$$

where $\{k_1, \ldots, k_j\} \subseteq \{i_1, \ldots, i_m\} \subseteq \{1, \ldots, J\}$ *for* $j = 1, \ldots, m$.

Proof. See Appendix 1.

THEOREM 5: *Hazard rate functions for m-variate survival times* $(T_{i_1}, \ldots, T_{i_m})$ *can be expressed in terms of densities of L-measures.*

$$\lambda_{i_1\cdots i_m} = (-1)^m \sum_{\pi} \prod_{j=1}^{m} l_{\pi_j}$$

where a summation is taken over all combinations of possible $\{\pi_1, \ldots, \pi_m\}$ *such that*

$$\bigcup_{j=1}^{m} \pi_j = \{i_1, \ldots, i_m\} \subseteq \{1, \ldots, J\}$$

$$\pi_i \bigcap \pi_j = \phi \qquad \textit{for } i \neq j,$$

with $l_\phi \equiv 1$.

Proof. See Appendix 1.

4.4 Influence of Random Effects

We generalize the results of Section 3 in the following theorem. The unconditional hazard rate λ, odds ratio θ, and density of L-measure l are functions of nominal hazard rates λ_N, conditional moments and cumulants of V_j's given $T \geq t$.

THEOREM 6: *Let $\{T \geq t\}$ denote $\{T_1 \geq t_1, T_2 \geq t_2, \ldots, T_J \geq t_J\}$. For the multivariate random effects model,*

1. *The multivariate hazard rate function for $(T_{i_1}, \ldots, T_{i_m})$ is given by*

$$\lambda_{i_1 \cdots i_m} = E(V_{i_1} \cdots V_{i_m} | T \geq t) \times \prod_{j=1}^{m} \lambda_{N,i_j}(t_{i_j})$$

2. *The odds ratio is given by*

$$\theta_{i_1 i_2 \cdots i_m} = \left[\prod_{j=1}^{m} \left\{ \prod_{k_1 < \cdots < k_j} E(V_{k_1} \cdots V_{k_j} | T \geq t) \right\}^{(-1)^j} \right]^{(-1)^m}, \qquad m \geq 2$$

3. *If $l_{i_1 \cdots i_m}$ exists, then*

$$l_{i_1 \cdots i_m} = (-1)^m \kappa_{\delta_1, \ldots, \delta_J}(V_1, \ldots, V_J | T \geq t) \times \prod_{j=1}^{m} \lambda_{N,i_j}(t_{i_j})$$

where $\kappa_{\delta_1, \ldots, \delta_J}(V_1, \ldots, V_J | T \geq t)$ is a cumulant of $(V_1, \ldots, V_J)$ given $T \geq t$ and

$$\delta_j = \begin{cases} 1 & \text{when } j \in \{i_1, \ldots, i_m\} \\ 0 & \text{otherwise} \end{cases}$$

where $\{k_1, k_2, \ldots, k_j\} \subseteq \{i_1, i_2, \ldots, i_m\} \subseteq \{1, 2, \ldots, J\}$, for $j = 1, \ldots, J$.

Proof. See Appendix 1.

Note that the dependence structure is described explicitly by the conditional cumulants for $(-1)^m l_{i_1 \cdots i_m}$. When $V_j = V$ for all j, we obtain the following corollary.

COROLLARY 4:

1. *The multivariate hazard rate function for* $(T_{i_1}, \ldots, T_{i_m})$ *is given by*

$$\lambda_{i_1 \cdots i_m} = E(V^m | T \geq t) \times \prod_{j=1}^{m} \lambda_{N,i_j}(t_{i_j})$$

2. *The multivariate odds ratio is given by*

$$\theta_{i_1 \cdots i_m} = \left[\prod_{j=1}^{m} \left\{ \prod_{k=1}^{j} E(V^k | T \geq t)^{C_k^m} \right\}^{(-1)^j} \right]^{(-1)^m}$$

for $m \geq 2$.

3. *If* $l_{i_1 \cdots i_m}$ *exists, then*

$$l_{i_1 \cdots i_m} = (-1)^m \kappa_m(V | T \geq t) \times \prod_{j=1}^{m} \lambda_{N,i_j}(t_{i_j})$$

where $\kappa_m(V|T \geq t)$ *is the m-th cumulant of* V *given* $T \geq t$.

Finally, we show that unconditional survival function is greater than or equal to the nominal survival function.

LEMMA 3:

$$S_U(t_1, t_2, \ldots, t_J) \geq \prod_{j=1}^{J} S_{N,j}(t_j)$$

Proof. See Appendix 1.

4.5 Examples

4.5.1 Common random effect

Let $E(V) = 1$ and $\mathrm{Var}(V) = c$. Denote $t = (t_1, t_2, \ldots, t_J)$.

1. *Gamma mixtures.* If V follows a gamma distribution, then

$$S_U(t) = \left(1 + c \sum_{j=1}^{J} \Lambda_{N,j}(t_j) \right)^{-1/c}$$

and the density of its L-measure is

$$l_{12\cdots J} = (-1)^J \frac{c^{J-1}(J-1)!}{\left(1 + c\sum_{j=1}^J \Lambda_{N,j}(t_j)\right)^J} \times \prod_{j=1}^J \lambda_{N,j}(t_j)$$

2. *Inverse Gaussian mixtures* (see Whitmore and Lee, 1991). If V follows inverse Gaussian distribution, then

$$S_U(t) = \exp\left\{\frac{1}{c}\left(1 - \sqrt{1 + 2c\sum_{j=1}^J \Lambda_{N,j}(t_j)}\right)\right\}$$

and the density of its L-measure is

$$l_{12\cdots J} = \frac{(-c)^{J-1}\prod_{j=1}^J(2j-3)}{\left(1 + 2c\sum_{j=1}^J \Lambda_{N,j}(t_j)\right)^{J-1/2}} \times \prod_{j=1}^J \lambda_{N,j}(t_j)$$

3. *Three-parameter family mixtures*. If V follows a $P(\alpha, \theta^{1-\alpha}, \theta)$ distribution, with $\theta = (1-\alpha)/c \geq 0$, then

$$S_U(t) = \exp\left[\frac{\theta}{\alpha}\left\{1 - \left(1 + \sum_{j=1}^J \frac{\Lambda_{N,j}(t_j)}{\theta}\right)^\alpha\right\}\right]$$

and the density of its L-measure is

$$l_{12\cdots J} = \frac{\prod_{j=1}^J (j-1-\alpha)}{\alpha(-\theta)^{J-1}\left(1 + \theta^{-1}\sum_{j=1}^J \Lambda_{N,j}(t_j)\right)^{J-\alpha}} \times \prod_{j=1}^J \lambda_{N,j}(t_j)$$

4.5.2 Multivariate random effect

In the following examples, we consider a multivariate random effect $(V_1, \ldots, V_J)$ that follows a multivariate distribution with $E(V_j) = 1$ for $j = 1, \ldots, J$.

1. *Gamma mixtures*. Let Z_o, Z_j be independently distributed as standard gamma distributions, Gamma $(\alpha, 1)$, Gamma $(\alpha_j, 1)$ where $\alpha_j = c_j^{-1} - \alpha$ for $j = 1, \ldots, J$. Then the $V_j = (Z_0 + Z_j)c_j$, $j = 1, \ldots, J$, have a multivariate gamma distribution with $E(V_j) = 1$ and $\mathrm{Var}(V_j) = c_j$ for $j = 1, \ldots, J$. Its Laplace transform, $M(s)$, is given by

$$M(s) = \left(1 + \sum_{j=1}^{J} c_j s_j\right)^{-\alpha} \prod_{j=1}^{J} (1 + c_j s_j)^{\alpha - 1/c_j}$$

Hence, the unconditional survival function is

$$S_U(t) = \left(1 + \sum_{j=1}^{J} c_j \Lambda_{N,j}(t_j)\right)^{-\alpha} \prod_{j=1}^{J} (1 + c_j \Lambda_{N,j}(t_j))^{\alpha - 1/c_j}$$

and the density of its L-measure is

$$l_{12\cdots J} = \frac{(-1)^J \alpha (J-1)!}{\left(1 + c \sum_{j=1}^{J} \Lambda_{N,j}(t_j)\right)^J} \times \prod_{j=1}^{J} c_j \lambda_{N,j}(t_j)$$

2. *Inverse Gaussian mixtures.* Let $V_j = \beta_j Z_0 + Z_j$ for $j = 1, \ldots, J$ where $Z_0 \sim IG(\mu, \lambda)$ and $Z_j \sim IG(\mu_j, \lambda_j)$ for $j = 1, \ldots, J$ with $\beta_j = c_j \lambda / \mu^2$, $\mu_j = 1 - (c_j \lambda / \mu)$, $\lambda_j = \mu_j^2 / c_j$. Then, the V_j's follow a multivariate inverse Gaussian distribution with marginal $IG(1, c_j^{-1})$. Its Laplace transform is given by

$$M(s) = \exp\left[\frac{\lambda}{\mu}\left(1 - \sqrt{1 + 2\sum_{j=1}^{J} c_j s_j}\right) + \sum_{j=1}^{J}\left(\frac{1}{c_j} - \frac{\lambda}{\mu}\right)\left(1 - \sqrt{1 + 2c_j s_j}\right)\right]$$

Hence, the unconditional survival function is

$$S_U(t) = \exp\left[\frac{\lambda}{\mu}\left(1 - \sqrt{1 + 2\sum_{j=1}^{J} c_j \Lambda_{N,j}(t_j)}\right) + \sum_{j=1}^{J}\left(\frac{1}{c_j} - \frac{\lambda}{\mu}\right)\left(1 - \sqrt{1 + 2c_j \Lambda_{N,j}(t_j)}\right)\right]$$

and the density of its L-measure is

$$l_{12\cdots J} = \frac{(-1)^{J-1} (\lambda/\mu) \prod_{j=1}^{J} (2j-3)}{\left(1 + 2\sum_{j=1}^{J} c_j \Lambda_{N,j}(t_j)\right)^{J-1/2}} \times \prod_{j=1}^{J} c_j \lambda_{N,j}(t_j)$$

3. *Three-parameter family mixtures.* Let $V_j = \beta_j Z_0 + Z_j$ for $j = 1, \ldots, J$ where $Z_0 \sim P(\alpha, \delta, \theta)$ and $Z_j \sim P(\alpha, \delta_j, \theta_j)$ for $j = 1, \ldots, J$ with $\beta_j = \theta/\theta_j > 0$, $\theta_j = (1-\alpha)/c_j \geq 0$, $\delta_j = (\theta_j - \delta\theta^\alpha)/\theta_j^\alpha > 0$. Then, the V_j's follow a multivariate three parameter family of distributions with marginals $P(\alpha, \theta_j^{1-\alpha}, \theta_j)$ and $E(V_j) = 1$, $\mathrm{Var}(V_j) = c_j$. Its Laplace transform is given by

$$M(s) = \exp\left[\frac{\delta\theta^\alpha}{\alpha}\left\{1 - \left(1 + \sum_{j=1}^{J} \frac{s_j}{\theta_j}\right)^\alpha\right\} + \sum_{j=1}^{J}\left(\frac{\theta_j}{\alpha} - \frac{\delta\theta^\alpha}{\alpha}\right)\left\{1 - \left(1 + \frac{s_j}{\theta_j}\right)^\alpha\right\}\right]$$

Hence, the unconditional survival function is

$$S_U(t) = \exp\left[\frac{\delta\theta^\alpha}{\alpha}\left\{1 - \left(1 + \sum_{j=1}^{J} \frac{\Lambda_{N,j}(t_j)}{\theta_j}\right)^\alpha\right\} + \sum_{j=1}^{J}\left(\frac{\theta_j}{\alpha} - \frac{\delta\theta^\alpha}{\alpha}\right)\left\{1 - \left(1 + \frac{\Lambda_{N,j}(t_j)}{\theta_j}\right)^\alpha\right\}\right]$$

and the density of its L-measure is

$$l_{12\cdots J} = \frac{(-1)^{J-1}\delta\theta^\alpha \prod_{j=1}^{J}(j-1-\alpha)}{\alpha\left(1 + \sum_{j=1}^{J} \Lambda_{N,j}(t_j)/\theta_j\right)^{J-\alpha}} \times \prod_{j=1}^{J} \frac{\lambda_{N,j}(t_j)}{\theta_j}$$

APPENDIX 1

Proof of Theorem 1. Let $I(T \geq t)$ be the indicator function of the event $T \geq t$. Then,

$$E[V^k I(T \geq t)] = E[V^k E\{I(T \geq t) | V\}] = E\left[V^k e^{-V\Lambda_N(t)}\right]$$

for $k = 0, 1$. Hence,

$$\lambda_U(t) = \lambda_N(t)\frac{E\left[Ve^{-V\Lambda_N(t)}\right]}{E\left[e^{-V\Lambda_N(t)}\right]} = \lambda_N(t)\frac{E[VI(T \geq t)]}{E[I(T \geq t)]} = \lambda_N(t)E(V | T \geq t)$$

For all t where $\Lambda_N(t) > 0$, $\exp(-v\Lambda_N(t))$, $v\exp(-v\Lambda_N(t))$, and $v^2\exp(-v\Lambda_N(t))$ are uniformly bounded. By the dominated convergence theorem,

$$\frac{\partial}{\partial t}E(V|T \geq t) = -\lambda_N(t)\left\{\frac{E\left[V^2e^{-V\Lambda_N(t)}\right]}{E\left[e^{-V\Lambda_N(t)}\right]} - \frac{E^2\left[Ve^{-V\Lambda_N(t)}\right]}{E^2\left[e^{-V\Lambda_N(t)}\right]}\right\}$$

$$= -\lambda_N(t)\text{Var}(V|T \geq t)$$

and the result follows. □

Proof of Corollary 1. By Theorem 1, the result follows. □

Proof of Theorem 2.

1. Let $I(T_1 \geq t_1, T_2 \geq t_2)$ be the indicator function of the event $T_1 \geq t_1, T_2 \geq t_2$. Then,

$$E[V_1^k I(T_1 \geq t_1, T_2 \geq t_2)] = E[V_1^k E\{I(T_1 \geq t_1, T_2 \geq t_2)|V_1 = v_1, V_2 = v_2\}]$$

$$= E\left[V_1^k e^{-\sum_{j=1}^2 V_j\Lambda_{N,j}(t_j)}\right]$$

for $k = 0, 1$. Hence,

$$\lambda_{U,1}(t_1|T_2 \geq t_2) = \lambda_{N,1}(t_1)\frac{E\left[V_1e^{-\sum_{j=1}^2 V_j\Lambda_{N,j}(t_j)}\right]}{E\left[e^{-\sum_{j=1}^2 V_j\Lambda_{N,j}(t_j)}\right]}$$

$$= \lambda_{N,1}(t_1)\frac{E[V_1I(T_1 \geq t_1, T_2 \geq t_2)]}{E[I(T_1 \geq t_1, T_2 \geq t_2)]}$$

$$= \lambda_{N,1}(t_1)E(V_1|T_1 \geq t_1, T_2 \geq t_2)$$

For all t_j where $\Lambda_{N,j}(t_j) > 0$, the functions $\exp(-v_j\Lambda_{N,j}(t_j))$, $v_j^2\exp(-v_j\Lambda_{N,j}(t_j))$, and $v_j^2\exp(-v_j\Lambda_{N,j}(t_j))$ are uniformly bounded for $j = 1, 2$. By the dominated convergence theorem,

$$\frac{\partial}{\partial t_j} E(V_1 | T_1 \geq t_1, T_2 \geq t_2)$$

$$= -\lambda_{N,j}(t_j) \left\{ \frac{E\left[V_1 V_j e^{-\sum_{j=1}^{2} V_j \Lambda_{N,j}(t_j)}\right]}{E\left[e^{-\sum_{j=1}^{2} V_j \Lambda_{N,j}(t_j)}\right]} - \frac{E\left[V_1 e^{-\sum_{j=1}^{2} V_j \Lambda_{N,j}(t_j)}\right]}{E\left[e^{-\sum_{j=1}^{2} V_j \Lambda_{N,j}(t_j)}\right]} \frac{E\left[V_j e^{-\sum_{j=1}^{2} V_j \Lambda_{N,j}(t_j)}\right]}{E\left[e^{-\sum_{j=1}^{2} V_j \Lambda_{N,j}(t_j)}\right]} \right\}$$

$$= -\lambda_{N,j}(t_j) \mathrm{Cov}(V_1, V_j | T_1 \geq t_1, T_2 \geq t_2)$$

If V_1 and V_2 are associated,

$$E\left\{V_1\left(-e^{-\sum_{j=1}^{2} V_j \Lambda_{N,j}(t_j)}\right)\right\} \geq E(V_1) E\left(-e^{-\sum_{j=1}^{2} V_j \Lambda_{N,j}(t_j)}\right)$$

and hence,

$$E(V_1 | T_1 \geq t_1, T_2 \geq t_2) = \frac{E\left(V_1 e^{-\sum_{j-1}^{2} V_j \Lambda_{N,j}(t_j)}\right)}{E\left(e^{-\sum_{j=1}^{2} V_j \Lambda_{N,j}(t_j)}\right)} \leq 1$$

2. Similarly,

$$\begin{aligned} E[V_1^k I(T_1 \geq t_1, T_2 = t_2)] &= E[V_1^k E\{I(T_1 \geq t_1, T_2 = t_2) \\ &\quad | V_1 = v_1, V_2 = v_2\}] \\ &= \lambda_{N,2}(t_2) E\left[V_1^k V_2 e^{-\sum_{j=1}^{2} V_j \Lambda_{N,j}(t_j)}\right] \end{aligned}$$

for $k = 0, 1, 2$. Hence,

$$\begin{aligned} \lambda_{U,1}(t_1 | T_2 = t_2) &= \lambda_{N,1}(t_1) \frac{E\left[V_1 V_2 e^{-\sum_{j=1}^{2} V_j \Lambda_{N,j}(t_j)}\right]}{E\left[V_2 e^{-\sum_{j=1}^{2} V_j \Lambda_{N,j}(t_j)}\right]} \\ &= \lambda_{N,1}(t_1) \frac{E[V_1 I(T_1 \geq t_1, T_2 = t_2)]}{E[I(T_1 \geq t_1, T_2 = t_2)]} \\ &= \lambda_{N,1}(t_1) E(V_1 | T_1 \geq t_1, T_2 = t_2) \end{aligned}$$

for $\lambda_{N,2}(t_2) > 0$. Also,

$$\lambda_{U,1}(t_1|T_2 = t_2) - \lambda_{N,1}(t_1)$$

$$= \lambda_{N,1}(t_1)\frac{E\left[V_1 V_2 e^{-V\sum_{j=1}^{2}\Lambda_{N,j}(t_j)}\right]}{E\left[V_2 e^{-V\sum_{j=1}^{2}\Lambda_{N,j}(t_j)}\right]} - \lambda_{N,1}(t_1)$$

$$= \lambda_{N,1}(t_1)\frac{E(V_1 V_2|T_1 \geq t_1, T_2 \geq t_2)}{E(V_2|T_1 \geq t_1, T_2 \geq t_2)} - \lambda_{N,1}(t_1)$$

$$= \lambda_{N,1}(t_1)(E(V_1|T_1 \geq t_1, T_2 \geq t_2) - 1)$$

$$+ \lambda_{N,1}(t_1)\frac{\text{Cov}(V_1, V_2|T_1 \geq t_1, T_2 \geq t_2)}{E(V_2|T_1 \geq t_1, T_2 \geq t_2)}$$

$$= (\lambda_{U,1}(t_1|T_2 \geq t_2) - \lambda_{N,1}(t_1))$$

$$+ \lambda_{N,1}(t_1)\frac{\text{Cov}(V_1, V_2|T_1 \geq t_1, T_2 \geq t_2)}{E(V_2|T_1 \geq t_1, T_2 \geq t_2)}$$

For all t_j where $\Lambda_{N,j}(t_j) > 0$, the functions $\exp(-v_j\Lambda_{N,j}(t_j))$, $v_j\exp(-v_j\Lambda_{N,j}(t_j))$, and $v_j^2\exp(-v_j\Lambda_{N,j}(t_j))$ are uniformly bounded for $j = 1, 2$. By the dominated convergence theorem,

$$\frac{\partial}{\partial t_j}E(V_1|T_1 \geq t_1, T_2 = t_2)$$

$$= -\lambda_{N,j}(t_j)\left\{\frac{E\left[V_j V_1 V_2 e^{-\sum_{j=1}^{2}V_j\Lambda_{N,j}(t_j)}\right]}{E\left[V_2 e^{-\sum_{j=1}^{2}V_j\Lambda_{N,j}(t_j)}\right]}\right.$$

$$\left. - \frac{E\left[V_1 V_2 e^{-\sum_{j=1}^{2}V_j\Lambda_{N,j}(t_j)}\right]}{E\left[V_2 e^{-\sum_{j=1}^{2}V_j\Lambda_{N,j}(t_j)}\right]}\,\frac{E\left[V_j V_2 e^{-\sum_{j=1}^{2}V_j\Lambda_{N,j}(t_j)}\right]}{E\left[V_2 e^{-\sum_{j=1}^{2}V_j\Lambda_{N,j}(t_j)}\right]}\right\}$$

$$= -\lambda_{N,j}(t_j)\text{Cov}(V_1, V_j|T_1 \geq t_1, T_2 = t_2)$$

3. Similarly,

$$\lambda_{U,1}(t_1, t_2) = \lambda_{N,1}(t_1)\lambda_{N,2}(t_2)\frac{E\left[V_1 V_2 e^{-\sum_{j=1}^{2} V_j \Lambda_{N,j}(t_j)}\right]}{E\left[e^{-\sum_{j=1}^{2} V_j \Lambda_{N,j}(t_j)}\right]}$$

$$= \lambda_{N,1}(t_1)\lambda_{N,2}(t_2)\frac{E[V_1 V_2 I(T_1 \geq t_1, T_2 \geq t_2)]}{E[I(T_1 \geq t_1, T_2 \geq t_2)]}$$

$$= \lambda_{N,1}(t_1)\lambda_{N,2}(t_2)E(V_1 V_2 | T_1 \geq t_1, T_2 \geq t_2)$$

Also, for all t_j where $\Lambda_{N,j}(t_j) > 0$, the functions $\exp(-v_j\Lambda_{N,j}(t_j))$, $v_j \exp(-v_j\Lambda_{N,j}(t_j))$, and $v_j^2 \exp(-v_j\Lambda_{N,j}(t_j))$ are uniformly bounded for $j = 1, 2$. By the dominated convergence theorem,

$$\frac{\partial}{\partial t_j} E(V_1 V_2 | T_1 \geq t_1, T_2 \geq t_2)$$

$$= -\lambda_{N,j}(t_j)\left\{\frac{E\left[V_j V_1 V_2 e^{-\sum_{j=1}^{2} V_j \Lambda_{N,j}(t_j)}\right]}{E\left[e^{-\sum_{j=1}^{2} V_j \Lambda_{N,j}(t_j)}\right]} - \frac{E\left[V_1 V_2 e^{-\sum_{j=1}^{2} V_j \Lambda_{N,j}(t_j)}\right]}{E\left[e^{-\sum_{j=1}^{2} V_j \Lambda_{N,j}(t_j)}\right]} \frac{E\left[V_j e^{-\sum_{j=1}^{2} V_j \Lambda_{N,j}(t_j)}\right]}{E\left[e^{-\sum_{j=1}^{2} V_j \Lambda_{N,j}(t_j)}\right]}\right\}$$

$$= -\lambda_{N,j}(t_j)\mathrm{Cov}(V_1 V_2, V_j | T_1 \geq t_1, T_2 \geq t_2)$$

If V_1 and V_2 are associated,

$$E\left\{V_1 V_2\left(-e^{-\sum_{j=1}^{2} V_j \Lambda_{N,j}(t_j)}\right)\right\} \geq E(V_1 V_2)E\left(-e^{-\sum_{j=1}^{2} V_j \Lambda_{N,j}(t_j)}\right)$$

and hence,

$$E(V_1 V_2 | T_1 \geq t_1, T_2 \geq t_2) = \frac{E\left[V_1 V_2 e^{-\sum_{j=1}^{2} V_j \Lambda_{N,j}(t_j)}\right]}{E\left[e^{-\sum_{j=1}^{2} V_j \Lambda_{N,j}(t_j)}\right]} \leq E(V_1 V_2)$$

$$= 1 + \mathrm{Cov}(V_1, V_2)$$

□

Proof of Lemma 1.

1. By parts 1 and 2 of Theorem 2,

$$\lambda_{U,1}(t_1|T_2 \geq t_2) = \lambda_{N,1}(t_1)E(V_1|T_1 \geq t_1, T_2 \geq t_2),$$
$$\lambda_{U,1}(t_1|T_2 = t_2) - \lambda_{U,1}(t_1|T_2 \geq t_2) = \lambda_{N,1}(t_1) \times \frac{\mathrm{Cov}(V_1, V_2|T_1 \geq t_1, T_2 \geq t_2)}{E(V_2|T_1 \geq t_1, T_2 \geq t_2)}$$

when $\lambda_{N,j}(t_j) > 0$ for $j = 1, 2$ and the result follows.

2. Since

$$\lambda_j = -\frac{\lambda_{N,j}(t_j)E[V_j e^{-\sum_{j=1}^{2} V_j \Lambda_{N,j}(t_j)}]}{E[e^{-\sum_{j=1}^{2} V_j \Lambda_{N,j}(t_j)}]} = -\lambda_{N,j}(t_j)E(V_j|T_1 \geq t_1, T_2 \geq t_2)$$

for $j = 1, 2$ and by part 3 of Theorem 2,

$$\lambda_{12} = \lambda_{N,1}(t_1)\lambda_{N,2}(t_2)E(V_1, V_2|T_1 \geq t_1, T_2 \geq t_2)$$

and the result follows. □

Proof of Lemma 2.

1. Suppose that (V_1, V_2) is positively quadrant dependent. For $j = 1, 2$, the $r(v_j) = -\exp\{-\Lambda_{N,j}(t_j)v_j\}$ are nondecreasing functions so, by Lemma 1 of Lehmann (1966), $(r(V_1), r(V_2))$ is positively quadrant dependent. Hence, by Lemma 3 of Lehmann (1966),

$$S_U(t_1, t_2) = E\big[\exp\{-\Lambda_{N,1}(t_1)V_1\}\exp\{-\Lambda_{N,2}(t_2)V_2\}\big]$$
$$\geq E\big[\exp\{-\Lambda_{N,1}(t_1)V_1\}\big]E\big[\exp\{-\Lambda_{N,2}(t_2)V_2\}\big] = S_U(t_1)S_U(t_2)$$

When (V_1, V_2) is negatively quadrant dependent, then (V_1, V_2^*) $(V_2^* = -V_2)$ is positively quadrant dependent. Since $r(V_1) = -\exp\{-\Lambda_{N,1}(t_1)V_1\}$ and $s(V_2^*) = \exp\{\Lambda_{N,j}(t_j)V_2^*\}$ are nondecreasing functions, we have

$$\begin{aligned}
-S_U(t_1, t_2) &= -E\big[\exp\{-\Lambda_{N,1}(t_1)V_1\}\exp\{\Lambda_{N,2}(t_2)(-V_2)\}\big] \\
&= E(r(V_1)s(V_2^*)) \\
&\geq E(r(V_1))E(s(V_2^*)) \\
&= -E\big[\exp\{-\Lambda_{N,1}(t_1)V_1\}\big]E\big[\exp\{-\Lambda_{N,2}(t_2)V_2\}\big] \\
&= -S_U(t_1)S_U(t_2)
\end{aligned}$$

and the result follows.

2. Suppose (V_1, V_2) and (V_1', V_2') are independently distributed with the same distribution. Then,

$$S_U(t_1, t_2) - S_U(t_1)S_U(t_2)$$

$$= E\big[\exp\{-\Lambda_{N,1}(t_1)V_1\}\exp\{-\Lambda_{N,2}(t_2)V_2\}\big]$$

$$- E\big[\exp\{-\Lambda_{N,1}(t_1)V_1\}\big]E\big[\exp\{-\Lambda_{N,2}(t_2)V_2\}\big]$$

$$= \tfrac{1}{2}E\big(\exp\{-\Lambda_{N,1}(t_1)V_1\} - \exp\{-\Lambda_{N,1}(t_1)V_1'\}\big)\big(\exp\{-\Lambda_{N,2}(t_2)V_2\}$$

$$- \exp\{-\Lambda_{N,2}(t_2)V_2'\}\big)$$

$$= \Lambda_{N,1}(t_1)\Lambda_{N,2}(t_2)$$

$$\times \tfrac{1}{2}E\int_{-\infty}^{\infty}\int_{-\infty}^{\infty}\exp\left\{-\sum_{j=1}^{2}\Lambda_{N,j}(t_j)x_j\right\}[I(x_1, V_1) - I(x_1, V_1')]$$

$$[I(x_2, V_2) - I(x_2, V_2')]\mathrm{d}x_1\mathrm{d}x_2$$

$$= \Lambda_{N,1}(t_1)\Lambda_{N,2}(t_2)$$

$$\times \int_{-\infty}^{\infty}\int_{-\infty}^{\infty}\exp\left\{-\sum_{j=1}^{2}\Lambda_{N,j}(t_j)x_j\right\}[P(V_1 > x_1, V_2 > x_2)$$

$$- P(V_1 > x_1)Pr(V_2 > x_2)]\mathrm{d}x_1\mathrm{d}x_2$$

where $I(x_j, V_j) = 1$ if $x_j \leq V_j$ and 0 otherwise, for $j = 1, 2$. If (V_1, V_2) is positively (negatively) quadrant dependent, then

$$\int_{-\infty}^{\infty}\int_{-\infty}^{\infty}\exp\left\{-\sum_{j=1}^{2}\Lambda_{N,j}(t_j)x_j\right\}[P(V_1 > x_1, V_2 > x_2)$$

$$- P(V_1 > x_1)P(V_2 > x_2)]\mathrm{d}x_1\mathrm{d}x_2$$

$$\leq (\geq) \int_{-\infty}^{\infty}\int_{-\infty}^{\infty}[P(V_1 > x_1, V_2 > x_2) - P(V_1 > x_1)P(V_2 > x_2)]\mathrm{d}x_1\mathrm{d}x_2$$

$$= \mathrm{Cov}(V_1, V_2)$$

and the result follows. □

Proof the Theorem 3. Without loss of generality, it suffices to show that

$$\theta_{1\cdots m|i'_1\cdots i'_n} = \left[S_{i'_1\cdots i'_n} \prod_{j=1}^{m} \left\{ \prod_{1\le k_1<\cdots<k_j\le m} S_{i'_1\cdots i'_n k_1\cdots k_j} \right\}^{(-1)^j} \right]^{(-1)^m}$$
$$= \left[\prod_{j=1}^{m} \left\{ \prod_{1\le k_1<\cdots<k_j\le m} \lambda_{k_1\cdots k_j|i'_1\cdots i'_n} \right\}^{(-1)^j} \right]^{(-1)^m}$$

for all $\{i'_1, \ldots, i'_n\} \subseteq \{m+1, \ldots, J\}$. We establish the first equality by induction. For $m = 1$,

$$\theta_{1|i'_1\cdots i'_n} = \frac{S_{i'_1\cdots i'_n 1}}{S_{i'_1\cdots i'_n}}$$

Suppose that for $m = m_0$, and all $\{i'_1, \ldots, i'_n\} \subseteq \{m_0+1, \ldots, J\}$,

$$\theta_{1\cdots m_0|i'_1\cdots i'_n} = \left[S_{i'_1\cdots i'_n} \prod_{j=1}^{m_0} \left\{ \prod_{1\le k_1<\cdots<k_j\le m_0} S_{i'_1\cdots i'_n k_1\cdots k_j} \right\}^{(-1)^j} \right]^{(-1)^{m_0}}$$

Then for $m = m_0+1$, and all $\{i'_1, \ldots, i'_{n-1}\} \subseteq \{m_0+2, \ldots, J\}$,

$$\theta_{1\cdots m_0(m_0+1)|i'_1\cdots i'_{n-1}} = \frac{\theta_{1\cdots m_0|i'_1\cdots i'_{n-1}(m_0+1)}}{\theta_{1\cdots m_0|i'_1\cdots i'_{n-1}}}$$
$$= \left[S_{i'_1\cdots i'_{n-1}(m_0+1)} \prod_{j=1}^{m_0} \left\{ \prod_{1\le k_1<\cdots<k_j\le m_0} S_{i'_1\cdots i'_{n-1}(m_0+1)k_1\cdots k_j} \right\}^{(-1)^j} \right]^{(-1)^{m_0}}$$
$$\times \left[S_{i'_1\cdots i'_{n-1}} \prod_{j=1}^{m_0} \left\{ \prod_{1\le k_1<\cdots<k_j\le m_0} S_{i'_1\cdots i'_{n-1}k_1\cdots k_j} \right\}^{(-1)^j} \right]^{(-1)^{m_0+1}}$$
$$= \Bigg[S_{i'_1\cdots i'_{n-1}} S^{-1}_{i'_1\cdots i'_{n-1}(m_0+1)} \prod_{j=1}^{m_0}$$

$$\left\{\prod_{1\le k_1<\cdots<k_j\le m_0} S_{i'_1\cdots i'_{n-1}k_1\cdots k_j} S^{-1}_{i'_1\cdots i'_{n-1}k_1\cdots k_j(m_0+1)}\right\}^{(-1)^j}\Bigg]^{(-1)^{m_0+1}}$$

$$= \left[S_{i'_1\cdots i'_{n-1}} \prod_{j=1}^{m_0+1}\left\{\prod_{1\le k_1<\cdots<k_j\le m_0+1} S_{i'_1\cdots i'_{n-1}k_1\cdots k_j}\right\}^{(-1)^j}\right]^{(-1)^{m_0+1}}$$

and the first equality follows. For the second equality, using $S_{i'_1\cdots i'_n k_1\cdots k_j} = S_{i'_1\cdots i'_n} \times \lambda_{k_1\cdots k_j|i'_1\cdots i'_n}$, we have

$$\theta_{12\cdots m|i'_1\cdots i'_n} = \left[S_{i'_1\cdots i'_n}\prod_{j=1}^{m}\left\{\prod_{1\le k_1<\cdots<k_j\le m} S_{i'_1\cdots i'_n}\times\lambda_{k_1\cdots k_j|i'_1\cdots i'_n}\right\}^{(-1)^j}\right]^{(-1)^m}$$

$$= \left[\prod_{j=1}^{m}\left\{\prod_{1\le k_1<\cdots<k_j\le m}\lambda_{k_1\cdots k_j|i'_1\cdots i'_n}\right\}^{(-1)^j} S_{i'_1\cdots i'_n}^{\sum_{j=0}^{m}(-1)^j C_j^m}\right]^{(-1)^m}$$

$$= \left[\prod_{j=1}^{m}\left\{\prod_{1\le k_1<\cdots<k_j\le m}\lambda_{k_1\cdots k_j|i'_1\cdots i'_n}\right\}^{(-1)^j} S_{i'_1\cdots i'_n}^{(1+(-1))^m}\right]^{(-1)^m}$$

and the result follows. □

Proof of Theorem 4. Without loss of generality, it suffices to show that

$$\lambda_{12\cdots m|i'_1\cdots i'_n} = \theta_{12\cdots m|i'_1\cdots i'_n}\prod_{j=1}^{m-1}\left[\prod_{1\le k_1<\cdots<k_j\le m}\theta_{k_1\cdots k_j|i'_1\cdots i'_n}\right]$$

for all $\{i'_1,\ldots,i'_n\}\subseteq\{m+1,\ldots,n\}$ where $n\le J-m$. The proof is by induction. For $m=1$,

$$\lambda_{1|i'_1\cdots i'_n} = \frac{S_{i'_1\cdots i'_n 1}}{S_{i'_1\cdots i'_n}} = \theta_{1|i'_1\cdots i'_n}$$

for all $\{i'_1,\ldots,i'_n\}\subseteq\{2,\ldots,n\}$ where $n\le J-1$. Suppose that, for $m=m_0$,

$$\lambda_{12\cdots m_0|i'_1\cdots i'_n} = \theta_{12\cdots m_0|i'_1\cdots i'_n}\prod_{j=1}^{m_0-1}\left[\prod_{1\le k_1<\cdots<k_j\le m_0}\theta_{k_1\cdots k_j|i'_1\cdots i'_n}\right]$$

for all $\{i'_1, \ldots, i'_n\} \subseteq \{m_0+1, \ldots, n\}$, $(n \leq J - m_0)$. Then,

$$\frac{S_{i'_1 \cdots i'_{n-1}(m_0+1)1\cdots m_0}}{S_{i'_1\cdots i'_{n-1}(m_0+1)}} = \lambda_{12\ldots m_0|i'_1\ldots i'_{n-1}(m_0+1)}$$

$$= \theta_{12\cdots m_0|i'_1\cdots i'_{n-1}(m_0+1)} \prod_{j=1}^{m_0-1} \left[\prod_{1\leq k_1<\cdots<k_j\leq m_0} \theta_{k_1\cdots k_j|i'_1\cdots i'_{n-1}(m_0+1)} \right]$$

$$= \theta_{12\cdots m_0|i'_1\cdots i'_{n-1}(m_0+1)} \prod_{j=1}^{m_0-1} \left[\prod_{1\leq k_1<\cdots<k_j\leq m_0} \theta_{k_1\cdots k_j|i'_1\cdots i'_{n-1}} \theta_{k_1\cdots k_j(m_0+1)|i'_1\cdots i'_{n-1}} \right]$$

for all $\{i'_1, \ldots, i'_{n-1}\} \subseteq \{m_0+2, \ldots, n\}$, where $n \leq J - m_0 + 1$.

$$\lambda_{12\cdots m_0(m_0+1)|i'_1\cdots i'_{n-1}} = \frac{S_{i'_1\cdots i'_{n-1}12\cdots m_0(m_0+1)}}{S_{i'_1\cdots i'_{n-1}}}$$

$$= \frac{S_{i'_1\cdots i'_{n-1}(m_0+1)}}{S_{i'_1\cdots i'_{n-1}}} \times \frac{S_{i'_1\cdots i'_{n-1}12\cdots m_0(m_0+1)}}{S_{i'_1\cdots i'_{n-1}(m_0+1)}}$$

$$= \theta_{m_0+1|i'_1\cdots i'_{n-1}} \times \theta_{12\cdots m_0|i'_1\cdots i'_{n-1}(m_0+1)} \times \prod_{j=1}^{m_0-1} \left[\prod_{1\leq k_1<\cdots<k_j\leq m_0} \theta_{k_1\cdots k_j|i'_1\cdots i'_{n-1}} \theta_{k_1\cdots k_j(m_0+1)|i'_1\cdots i'_{n-1}} \right]$$

$$= \theta_{m_0+1|i'_1\cdots i'_{n-1}} \times \theta_{12\cdots m_0(m_0+1)|i'_1\cdots i'_{n-1}} \times \theta_{12\cdots m_0|i'_1\cdots i'_{n-1}} \times \prod_{j=1}^{m_0-1} \left[\prod_{1\leq k_1<\cdots<k_j\leq m_0} \theta_{k_1\cdots k_j|i'_1\cdots i'_{n-1}} \theta_{k_1\cdots k_j(m_0+1)|i'_1\cdots i'_{n-1}} \right]$$

$$= \theta_{12\cdots m_0(m_0+1)|i'_1\cdots i'_{n-1}} \prod_{j=1}^{m_0} \left[\prod_{1\leq k_1<\cdots<k_j\leq m_0+1} \theta_{k_1\cdots k_j|i'_1\cdots i'_{n-1}} \right]$$

for all $\{i'_1, \ldots, i'_{n-1}\} \subseteq \{m_0+2, \ldots, n\}$, where $n \leq J - m_0 + 1$ and the result follows. □

Proof of Theorem 5. Without loss of generality, it suffices to show that

$$\lambda_{1\cdots m} = (-1)^m \sum_{\pi} \prod_{j=1}^{m} l_{\pi_j}$$

where the summation is taken over all possible combinations of $\{\pi_1, \pi_2, \ldots, \pi_m\}$ such that

$$\bigcup_{j=1}^{m} \pi_j = \{1, 2, \ldots, m\}, \qquad \pi_i \bigcap \pi_j = \phi, \qquad \textit{for } i \neq j$$

The proof is by induction. For $m = 1$,

$$\lambda_1 = \frac{S_1}{S} = -l_1$$

Suppose that, for $m = m_0$,

$$\lambda_{1\cdots m_0} = (-1)^{m_0} \sum_{\pi} \prod_{j=1}^{m_0} l_{\pi_j}$$

where the summation is taken over all possible combinations of $\{\pi_1, \pi_2, \ldots, \pi_{m_0}\}$ such that

$$\bigcup_{j=1}^{m_0} \pi_j = \{1, 2, \ldots, m_0\}, \qquad \pi_i \bigcap \pi_j = \phi, \qquad \textit{for } i \neq j$$

Since

$$\frac{d\lambda_{1\cdots m_0}}{dt_{m_{0+1}}} = -\lambda_{1\cdots m_{0+1}} + \lambda_{1\cdots m_0}\lambda_{m_{0+1}}$$

Then,

$$\begin{aligned}
\lambda_{1\cdots m_{0+1}} &= -\lambda_{1\cdots m_0} l_{m_{0+1}} - \frac{d\lambda_{1\cdots m_0}}{dt_{m_0+1}} \\
&= (-1)^{m_0+1} \sum_{\pi} l_{m_{0+1}} \prod_{j=1}^{m_0} l_{\pi_j} + (-1)^{m_0+1} \sum_{\pi'} \prod_{j=1}^{m_0} l_{\pi'_j} \\
&= (-1)^{m_0+1} \sum_{\pi^*} \prod_{j=1}^{m_0+1} l_{\pi^*_j}
\end{aligned}$$

where

$$\bigcup_{j=1}^{m_0} \pi_j' = \{1, 2, \ldots, m_0 + 1\}, \qquad \pi_i' \bigcap \pi_j' = m_0 + 1, \qquad \textit{for } i \neq j$$

$$\bigcup_{j=1}^{m_0+1} \pi_j^* = \{1, 2, \ldots, m_0 + 1\}, \qquad \pi_i^* \bigcap \pi_j^* = \phi, \qquad \textit{for } i \neq j$$

and the summation is taken over all possible combinations of $\{\pi_1^*, \pi_2^*, \ldots, \pi_{m_0+1}^*\}$ and the result follows. □

Proof of Theorem 6.

1. Let $\{T \geq t\}$ denote $\{T_1 \geq t_1, T_2 \geq t_2, \ldots, T_J \geq t_J\}$.

$$\begin{aligned}
\lambda_{i_1 i_2 \cdots i_m} &= \frac{S_{i_1 i_2 \cdots i_m}}{S} \\
&= \prod_{j=1}^{m} \lambda_{N,i_j}(t_{i_j}) \times \frac{E\left[V_{i_1} \cdots V_{i_m} e^{-\sum_{j=1}^{J} V_j \Lambda_{N,j}(t_j)}\right]}{E\left[e^{-\sum_{j=1}^{J} V_j \Lambda_{N,j}(t_j)}\right]} \\
&= \prod_{j=1}^{m} \lambda_{N,i_j}(t_{i_j}) \times E(V_{i_1} \cdots V_{i_m} | T \geq t)
\end{aligned}$$

2. By part 1 and Theorem 3,

$$\begin{aligned}
\theta_{i_1 i_2 \cdots i_m} &= \left[\prod_{j=1}^{m} \left\{ \prod_{k_1 < \cdots < k_j} E(V_{k_1} \cdots V_{k_j} | T \geq t) \lambda_{N,k_1}(t_{k_1}) \cdots \lambda_{N,k_j}(t_{k_j}) \right\}^{(-1)^j} \right]^{(-1)^m} \\
&= \left[\prod_{j=1}^{m} \left\{ \prod_{k_1 < \cdots < k_j} E(V_{k_1} \cdots V_{k_j} | T \geq t) \right\}^{(-1)^j} \right]^{(-1)^m} \\
&\quad \times \left[\prod_{j=1}^{m} \left\{ \prod_{k_1 < \cdots < k_j} \lambda_{N,k_1}(t_{k_1}) \cdots \lambda_{N,k_j}(t_{k_j}) \right\}^{(-1)^j} \right]^{(-1)^m}
\end{aligned}$$

On the other hand,

$$\prod_{j=1}^{m}\left\{\prod_{k_1<\cdots<k_j}\lambda_{N,k_1}(t_{k_1})\cdots\lambda_{N,k_j}(t_{k_j})\right\}^{(-1)^j}$$

$$=\prod_{j=1}^{m}\{\lambda_{N,k_1}(t_{k_1})\cdots\lambda_{N,k_m}(t_{k_m})\}^{C_{j-1}^{m-1}(-1)^j}$$

$$=\{\lambda_{N,k_1}(t_{k_1})\cdots\lambda_{N,k_m}(t_{k_m})\}^{\sum_{j=1}^{m}C_{j-1}^{m-1}(-1)^j}$$

$$=\{\lambda_{N,k_1}(t_{k_1})\cdots\lambda_{N,k_m}(t_{k_m})\}^{(-1)(1+(-1))^{m-1}}$$

and the result follows. □

3. Let $I(T \geq t)$ be the indicator function of the events $\{T \geq t\}$. Then

$$E\left[e^{\sum_{j=1}^{J}s_jV_j}I(T\geq t)\right]=E\left[e^{\sum_{j=1}^{J}s_jV_j}E\{I(T\geq t)|V_1=v_1,\ldots,V_J=v_J\}\right]$$
$$=E\left[e^{\sum_{j=1}^{J}(s_j-\Lambda_{N,j}(t_j))V_j}\right]$$

and

$$E[I(T\geq t)]=E[E\{I(T\geq t)|V_1=v_1,\ldots,V_J=v_J\}]$$
$$=E\left[e^{-\sum_{j=1}^{J}V_j\Lambda_{N,j}(t_j)}\right]$$

Since $E\left[e^{\sum_{j=1}^{J}s_jV_j}\middle|T\geq t\right]=E\left[e^{\sum_{j=1}^{J}(s_j-\Lambda_{N,j}(t_j))V_j}\right]/E\left[e^{-\sum_{j=1}^{J}V_j\Lambda_{N,j}(t_j)}\right]$

$$\frac{\partial^m}{\partial t_{i_1}\cdots\partial t_{i_m}}\log E\left[e^{-\sum_{j=1}^{J}V_j\Lambda_{N,j}(t_j)}\right]=(-1)^m\frac{\partial^m}{\partial s_{i_1}\cdots\partial s_{i_m}}\log E\left[e^{\sum_{j=1}^{J}s_jV_j}\middle|T\geq t\right]\Big|_{s_1=\cdots=s_J=0}\times\prod_{j=1}^{m}\lambda_{N,i_j}(t_{i_j})$$
$$=(-1)^m\kappa_{\delta_1,\ldots,\delta_J}(V_1,\ldots,V_J|T\geq t)\times\prod_{j=1}^{m}\lambda_{N,i_j}(t_{i_j})$$

and the result follows. □

Proof of Lemma 3. For $v_j > 0$ for $j = 1, 2, \ldots, J$, $g(v_1, \ldots, v_J) = \prod_{j=1}^{J} \exp\{-\Lambda(t_j)v_j\}$ is a convex function on a convex subset of J dimensional Euclidean space R^J. The result follows by Jensen's inequality. □

REFERENCES

Andersen, P. K., Borgan, Ø., Gill, R. D., and Keiding, N. (1992), *Statistical Models Based on Counting Processes*. New York: Springer.

Clayton, D. G. (1978), A model for association in bivariate life tables and its application in epidemiological studies of familial tendency in chronic disease incidence, *Biometrika*, 65, 141–151.

Clayton, D. G. and Cuzick, J. (1985), Multivariate generalizations of the proportional hazards model (with discussion), *J. R. Stat. Soc., Ser. A*, 148, 82–117.

Costigan, T. M. and Klein, J. P. (1993), Multivariate survival analysis based on frailty models. In: A. P. Basu, ed. *Advances in Reliability*, Amsterdam: North Holland, pp. 43–58.

Crowder, M. (1985), A distributional model for repeated failure time measurements, *J. R. Stat. Soc., Ser. B*, 47, 447–452.

Crowder, M. (1989), Multivariate distribution with Weibull connections, *J. R. Stat. Soc., Ser. B*, 51, 93–107.

Genest, C. and MacKay, R. J. (1986a), Couples Archimédiennes et Familles de Lois Bidmensionells Dont les Marges Sont Données, *Can. J. Stat.*, 14, 145–159.

Genest, C. and MacKay, R. J. (1986b), The joy of copulas: bivariate distributions with uniform marginals, *Am. Stat.*, 40, 280–283.

Gill, R. D. (1993), Multivariate survival analysis, *Theory Probab. Appl.*, 37, 18–31.

Guo, G. and Rodoríguez, G. (1992), Estimating a proportional hazards model for clustered data using EM algorithm with an application to child survival in Guatemala, *J. Am. Stat. Assoc.*, 87, 969–976.

Gurland, J. and Sethuraman, J. (1995), How pooling failure data may reverse increasing failure rates, *J. Am. Stat. Assoc.*, 90, 1416–1423.

Heckman, J. and Singer, B. (1984), A method for minimizing the impact of distributional assumptions in econometric models for duration data, *Econometrica*, 52, 271–320.

Honoré, B. E. (1990), Simple estimation of a duration model with unobserved heterogeneity, *Econometrica*, 58, 453–473.

Hougaard, P. (1986a), Survival models for heterogeneous populations derived from stable distributions, *Biometrika*, 73, 387–396.

Hougaard, P. (1986b), A class of multivariate failure time distributions, *Biometrika*, 73, 671–678.

Hougaard, P. (1987), Modelling multivariate survival, *Scand. J. Stat.*, 14, 291–304.

Hougaard, P. (1989), Fitting multivariate failure time distribution, *IEEE Trans. Reliab.*, 38, 444–448.

Hougaard, P., Holm, N., and Harvard, B. (1992), Measuring the similarities between the lifetimes of adult Danish twins born between 1881 and 1930, *J. Am. Stat. Assoc.*, 87, 17–24.

Ishwaran, H. (1996), Uniform rates of estimation in the semiparametric Weibull mixture model, *Ann. Stat.*, 24, 1572–1585.

Johnson, R. A. and Omori, Y. (1998), Influence of random effects on bivariate and trivariate survival models, *J. Nonparametric Stat.*, in press.

Keiding, N., Andersen, K. and Klein, J. P. (1997), The role of frailty models and accelerated failure time models in describing heterogeneity due to omitted variables, *Stat. Med.*, 16, 215–224.

Klein, J. P. (1992), Semiparametric estimation of random effects using the Cox model based on the EM algorithm, *Biometrics*, 48, 795–806.

Klein, J. P., Moeschberger, M. L., Li, Y. H., and Wang, S. T. (1992), Estimating random effects in the Framingham heart study. In: J. P. Klein and P. K. Goel (eds.) *Survival Analysis*. Dordrecht: Kluwer, pp. 99–118.

Lancaster, T. (1979), Econometric methods for the duration of unemployment, *Econometrica*, 47, 939–956.

Lancaster, T. (1985), Generalised residuals and heterogeneous duration models with applications to the Weibull model, *J. Econometrics*, 28, 155–169.

Lancaster, T. (1990), *The Econometric Analysis of Transition Data.* Cambridge: Cambridge University Press.

Lee, S., and Klein, J. P. (1988), Bivariate models with a random environmental factor, *IAPQR Trans.*, 13(2), 1–18.

Lehmann, E. L. (1966), Some concepts of dependence, *Ann. Math. Stat.*, 37, 1137–1153.

Li, Y. H., Klein, J. P., and Moeschberger, M. L. (1993), Semiparametric estimation of covariate effects using the inverse Gaussian frailty model, Technical Report, Department of Statistics, Ohio State University.

Lindeboom, M. and Van Den Berg, G. J. (1994), Heterogeneity in models for bivariate survival: The importance of the mixing distribution, *J. R. Stat. Soc., Ser. B*, 56, 49–60.

Marshall, A. W. and Olkin, I. (1988), Families of multivariate distributions, *J. Am. Stat. Assoc.*, 83, 834–841.

McGilchrist, C. A. and Aisbett, C. W. (1991), Regression with frailty in survival analysis, *Biometrics*, 47, 461–466.

Murphy, S. A. (1995), Asymptotic theory of the frailty model, *Ann. Stat.*, 23, 182–198.

Nielsen, G. G., Gill, R. D., Andersen, P. K., and Sørensen, T. I. A. (1992), A counting process approach to maximum likelihood estimation in frailty models, *Scand. J. Stat.*, 19, 25–44.

Oakes, D. (1982), A model for association in bivariate survival data, *J. R. Stat. Soc., Ser. B.*, 44, 414–422.

Oakes, D. (1986a), Semiparametric inference in a model for association in bivariate survival data, *Biometrika*, 73, 353–361.

Oakes, D. (1986b), A model for bivariate survival data. In: R. L. Prentice and S. H. Moolgavkar (eds), *Modern Statistical Methods in Chronic Disease Epidemiology*, New York: John Wiley and Sons, pp. 151–166.

Oakes, D. (1989), Bivariate survival models induced by frailties, *J. Am. Stat. Assoc.*, 84, 487–493.

Omori, Y. and Johnson, R. A. (1993), The influence of random effects on the unconditional hazard rate and survival functions, *Biometrika*, 80, 910–914.

Sinha, D. and Dey, D. K. (1997), Semiparametric Bayesian estimation of survival data, *J. Am. Stat. Assoc.*, 92, 1195–1212.

Wang, S. T., Klein, J. P., and Moeschberger, M. L. (1995), Semiparametric estimation of covariate effects using the positive stable frailty model, *Appl. Stochastic Models Data Anal.*, 11, 121–133.

Whitmore, G. A. and Lee, M. L. T. (1991), A multivariate survival distribution generated by an inverse Gaussian mixtures of exponentials, *Technometrics*, 33, 39–50.

Yashin, A. I. and Lachine, I. A. (1993), Survival of related individuals: An extension of some fundamental results of heterogeneity analysis, unpublished manuscript.

Yashin, A. I., Vaupel, J. W., and Lachine, I. A. (1995), Correlated individual frailty: An advantageous approach to survival analysis of bivariate data, *Math. Popul. Stud.*, 5, 145–159.

12

A Unified Methodology for Constructing Multivariate Autoregressive Models

SERGIO G. KOREISHA University of Oregon, Eugene, Oregon

TARMO PUKKILA Ministry of Social Affairs and Health, Helsinki, Finland

1. INTRODUCTION

The use and popularity of multivariate autoregressive models, also known as vector autoregressive models (VAR), to study and forecast macroeconomic phenomena have grown immensely in the last two decades. Sims (1980) in summarizing the limitations of large-scale structural models, particularly the imposition of "incredible identification restrictions," initiated a new stream of economic research based on vector autoregressions, which from an econometric viewpoint can be interpreted as a fairly unrestricted system of reduced form equations from an unknown structural system of simultaneous equations. Many empirical studies using VAR models have been published, including those of Litterman and Weiss (1985) (money, real interest rates, and output), Reagan and Sheehan (1985) (manufacturer's inventories and back orders), and Funke (1990) (industrial production in OECD countries). Doan et al. (1984), Litterman (1986), and Lupoletti and Webb (1986), among others, have strongly advocated the merits of unrestricted (VAR) as well as Bayesian VAR models (models for which prior values are assigned to the weights of lagged variables) (see Rose, 1989, for a more detailed explanation). To support such applications, much theoretical work has been conducted in the areas of model identification (e.g., Hannan and Quinn, 1979; Quinn, 1980; Tiao and Box, 1981; Lutkepohl, 1985; Tiao and Tsay, 1989; Koreisha and Pukkila, 1993) and estimation (e.g., Zellner, 1962; Hillmer and Tiao, 1979; Litterman, 1981; Lysne and Tjostheim, 1987; Reinsel and Ahn, 1992).

A vector autoregressive process of order p (VAR(p)) for a system of K variables $\boldsymbol{Y}_t = (Y_{1t}, Y_{2t}, \ldots, Y_{Kt})^T$ with N observations is defined as

$$\boldsymbol{Y}_t = \boldsymbol{\phi}_1 \boldsymbol{Y}_{t-1} + \boldsymbol{\phi}_2 \boldsymbol{Y}_{t-2} + \cdots + \boldsymbol{\phi}_p \boldsymbol{Y}_{t-p} + \boldsymbol{a}_t \tag{1}$$

where

$$\boldsymbol{\phi}_m = \begin{bmatrix} \phi_{m,1,1} & \phi_{m,1,2} & \cdots & \phi_{m,1,K} \\ \phi_{m,2,1} & \phi_{m,2,2} & \cdots & \phi_{m,2,K} \\ \vdots & \vdots & \vdots & \vdots \\ \phi_{m,K,1} & \phi_{i,K,2} & \cdots & \phi_{m,K,K} \end{bmatrix}, \qquad m = 1, 2, \ldots, p \tag{2}$$

and $\boldsymbol{a}_t^T = (a_{1t}, a_{2t}, \ldots, a_{Kt})$ is a vector of random shocks which are independently, identically, and normally distributed with mean zero and covariance $\Sigma_{\boldsymbol{a}} = E[\boldsymbol{a}_t \boldsymbol{a}_t^T]$ for all t, i.e.,

$$\Sigma_{\boldsymbol{a}} = \begin{bmatrix} \sigma_{11} & \sigma_{12} & \cdots & \sigma_{1K} \\ \sigma_{21} & \sigma_{22} & \cdots & \sigma_{2K} \\ \vdots & \vdots & \ddots & \vdots \\ \sigma_{K1} & \sigma_{K2} & \cdots & \sigma_{KK} \end{bmatrix} \tag{3}$$

The system of equations (1) can also be defined in terms of backshift operators as follows:

$$\underset{(K \times K)}{\boldsymbol{\phi}(\boldsymbol{B})} \ \underset{(K \times 1)}{\boldsymbol{Y}_t} = \underset{(K \times 1)}{\boldsymbol{a}_t} \tag{4}$$

where $\boldsymbol{\phi}(\boldsymbol{B}) = \boldsymbol{I} - \boldsymbol{\phi}_1 B - \boldsymbol{\phi}_2 B^2 - \cdots - \boldsymbol{\phi}_p B^p$, and $\boldsymbol{B}$ is defined as $B^j Y_{i,t} = Y_{i,t-j}$. The process is stationary if the roots of the determinantal equation $|\boldsymbol{\phi}(\boldsymbol{B})| = 0$ are outside the unit circle.

The selection of the order of vector autoregressive processes is an important but difficult step in the construction of models used to explain economic phenomena and to generate forecasts. Several order-determination criteria possessing different statistical properties such as consistency and efficiency have been proposed to facilitate the model building process. Lutkepohl (1985) contrasted the performance of as many as 12 different identification approaches, and concluded that for small samples (40–200 observations), the penalty–function–type order-determination criteria proposed by

Schwarz (1978), Rissanen (1978), and Hannan and Quinn (1979) could more accurately estimate the order of VAR processes than any of the other methods. Koreisha and Pukkila (1993) proposed a new identification procedure based on a multivariate white noise test that for small samples outperformed other penalty function type of criteria. This new approach will form the basis of our unified methodology for constructing multivariate autoregressive models, and will be discussed more fully in Section 2.

Estimation of VAR structures can also be problematic, particularly for the sample sizes generally available to analysts because the number of unknown parameters increases quadratically with the number of component series in the model. To facilitate the estimation of such models various "restrictive" Bayesian-constraint-type approaches (see Funke, 1990) have been proposed which often (but not always) require lagged lengths for each variable and parametrization of each equation to be identical. Constraining parameters to zero is not a simple matter, frequently requiring the use of likelihood functions (which can consume large amounts of CPU time), or, as in the case of Bayesian approaches, *single* equation mixed estimation techniques (see Theil, 1971, p. 347), which is a way to combine sample information with prior judgments using Aitken's estimators. In this article we present an estimation method that can cope with models whose parametrizations are either restricted a priori or found to be insignificant according to some criterion. This new approach, as we shall show in Section 3, is based on a generalized least-squares procedures applied to a "parsimoniously" rearranged component series design matrix.

In Section 4, using simulated as well as economic data, we demonstrate the small sample performance of our unified approach for identifying and estimating VAR models. Finally, in Section 5, we offer some concluding remarks.

2. IDENTIFICATION OF VAR(p) PROCESSES

In the past 15–20 years several order determination criteria have been proposed based on the minimization of functions of the form

$$\delta(p) = N \log|\hat{\Sigma}_p| + pg(N), \qquad p = 0, 1, 2, \ldots, p^* \tag{5}$$

where $\hat{\Sigma}_p$ is an approximation of the residual covariance matrix associated with the fitted VAR(p) structure, $g(N)$ is a prescribed nonnegative penalty function, and p^* is an a priori determined upper autoregressive order limit. Note that as the number of parameters increases, $N \log|\hat{\Sigma}_p|$ will have a tendency to decrease, while $pg(N)$ will always increase. If $g(N)$ is set equal to $2K^2$, then equation (5) becomes the AIC(p) criterion (Akaike, 1974). This criterion is known to overestimate p asymptotically with positive probability. Selecting $g(N)$ to be $K^2 \log N$ defines the BIC(p) criterion (Rissanen, 1978; Schwarz, 1978). The BIC criterion estimates the VAR(p) order consistently provided that $p^* \geq p$. The Hannan and Quinn (1979) criterion HQ(p) is obtained when $g(N) = c \log \log N$. In generalizing the criterion for multivariate autoregressions, Quinn (1980) has shown that when c is set at $2 + \varepsilon, \varepsilon > 0$, the criterion will be strongly consistent. However, he points out that this is a "borderline criterion of strong consistency" because it depends on the "law of the iterated logarithm," generally identifying an order greater than or equal to the one selected by the AIC as long as the sample size N is less than $\exp(\exp K^2)$. Thus he suggests relating c to K^2. The optimal values for p are obtained by successively testing the residuals of all VAR(p) models up to p^*.

Another altogether different way of using these criteria to select the order of the process is to fit increasing order VAR(p) structures to the data systematically, and then to verify that the resulting residuals behave like white-noise processes. Since a vector white-noise process can be viewed as a VAR(0) process, and conversely a non-white-noise process can be approximated as a VAR(k) process, $k > 0$, Pukkila and Krishnaiah (1988) proposed using consistent VAR order determination criteria on the residual series of fitted models as a test for white noise. Thus, if one were to apply this multivariate white-noise test to the residuals of the VAR(p) fits to the data, then the order of the process governing the time series could be ascertained at the first opportunity, indicating that the residuals of the fitted VAR(p) structure follow a VAR(0) process. In other words, by establishing the order of the VAR(k) process governing the residuals of the VAR(p) structure fitted to the observed series $Y_1, Y_2, \ldots, Y_N$, namely $\hat{a}_t(p)$, one can determine whether or not the autoregressive fit to the data is adequate. If it is determined that $k = 0$ from the minimization of

$$\delta(p) = N \log |\hat{\Sigma}_{k,p}| + pg(N), \qquad k = 0, 1, 2, \ldots, k^* \tag{6}$$

where $\hat{\Sigma}_{k,p}$ is an estimate of the residual covariance matrix associated with the VAR(k)th fit of the series $\hat{a}_t(p)$ and k^* is an a priori determined autoregressive upper bound, then the residuals are white noise and the order p is identified. If $k > 0$, the white-noise hypothesis can be rejected and the order of the VAR structure fitted to the data changed. However, to demonstrate that the residuals $\hat{a}_t(p)$ are white noise, all VAR(k) fits up to k^* must be checked. We shall refer to this approach as the residual white-noise autoregressive (RWNAR) order determination criterion; when appropriate, a suffix denoting the particular penalty function used will be appended to RWNAR, e.g., RWNAR-BIC. (For a discussion on how this approach can be modified to identify the order of univariate ARMA(p, q) processes, see Pukkila et al., 1990.) In this study we let $k^* = p^*$, and set their upper limit at $\sqrt{N}$.

3. ESTIMATION OF VECTOR AUTOREGRESSIONS

Suppose the VAR(p) system of equations (1) are "restacked" such that the dependent variables are grouped together and aligned with their corresponding regressors for each of the $N-p$ time periods. That is, let $0_{pK} = (0, 0, \ldots, 0)$ be a pK-dimensional zero vector, and let $x_t = (Y_{t-1}^T, Y_{t-2}^T, \ldots, Y_{t-p}^T,)$; furthermore, let

$$\boldsymbol{Z}_t = \begin{bmatrix} x_t^T & 0_{pK}^T & \ldots & 0_{pK}^T \\ 0_{pK}^T & x_t^T & \ldots & 0_{pK}^T \\ \vdots & \vdots & \ddots & \vdots \\ 0_{pK}^T & 0_{pK}^T & \ldots & x_t^T \end{bmatrix}, \qquad t = p+1, \ldots, N \tag{7}$$

be a $(K \times pK^2)$ matrix, and define the $(K \times pK)$ matrix $\boldsymbol{\phi}$ as $[\boldsymbol{\phi}_1 \boldsymbol{\phi}_2 \ldots \boldsymbol{\phi}_p]$. Finally, let $\boldsymbol{\beta}$ be a pK^2 column vector containing the elements of $\boldsymbol{\phi}$ stacked on top of each other by row, i.e.,

$$\begin{aligned} \boldsymbol{\beta}^T = (&\phi_{1,1,1}\,\phi_{1,1,2}\cdots\phi_{1,1,K}\ \cdots\ \phi_{p,1,1}\,\phi_{p,1,2}\cdots\phi_{p,1,K}\ \cdots \\ &\phi_{1,2,1}\phi_{1,2,2}\cdots\phi_{1,2,K}\ \cdots\ \phi_{p,2,1}\,\phi_{p,2,2}\cdots\phi_{p,2,K}\ \cdots \\ &\phi_{1,K,1}\,\phi_{1,K,2}\cdots\phi_{1,K,K}\ \cdots\ \phi_{p,K,1}\,\phi_{p,K,2}\cdots\phi_{p,K,K}) \end{aligned} \tag{8}$$

With this new set of notations the VAR system in equation (1) can be rewritten as

$$\boldsymbol{Y}_t = \boldsymbol{Z}_t\boldsymbol{\beta} + \boldsymbol{a}_t \tag{9}$$

or equivalently, in matrix form, as

$$\mathbf{Y} = \mathbf{Z}\boldsymbol{\beta} + \mathbf{a} \tag{10}$$

where

$$\mathbf{Y} = \begin{bmatrix} \boldsymbol{Y}_{p+1} \\ \boldsymbol{Y}_{p+1} \\ \vdots \\ \boldsymbol{Y}_N \end{bmatrix}, \qquad \mathbf{Z} = \begin{bmatrix} \boldsymbol{Z}_{p+1} \\ \boldsymbol{Z}_{p+2} \\ \vdots \\ \boldsymbol{Z}_N \end{bmatrix}, \qquad \mathbf{a} = \begin{bmatrix} \boldsymbol{a}_{p+1} \\ \boldsymbol{a}_{p+2} \\ \vdots \\ \boldsymbol{a}_N \end{bmatrix} \tag{11}$$

Thus, for given values of Σ_a, the generalized least-squares estimate for $\boldsymbol{\beta}$

$$\boldsymbol{\beta} = (\mathbf{Z}^T\mathbf{V}^{-1}\mathbf{Z})^T\mathbf{Z}^T\mathbf{V}^{-1}\mathbf{Y} \tag{12}$$

and its variance–covariance is given by

$$\text{Var}(\boldsymbol{\beta}) = (\mathbf{Z}^T\mathbf{V}^{-1}\mathbf{Z})^T \tag{13}$$

where

$$\mathbf{V} = \begin{bmatrix} \Sigma_a & 0 & \dots & 0 \\ 0 & \Sigma_a & \dots & 0 \\ \vdots & \vdots & \dots & \vdots \\ 0 & 0 & \dots & \Sigma_a \end{bmatrix} \tag{14}$$

Moreover, because the structure of the covariance matrix $\mathbf{V}$ is so simple, its inverse can easily be computed using only the inverse of Σ_a which has dimensions $(K \times K)$. Initial estimates for Σ_a are obtained from least-squares residuals. Note also that new estimates for Σ_a can be generated from the GLS estimates of the AR coefficients. Consequently, a new $\mathbf{V}^{-1}$ matrix can be constructed which in turn can yield new estimates for the nonzero AR coefficients. The algorithm can be allowed to iterate until some sort of convergence criterion is reached. (In this study we will present only the results associated with one iteration of the procedure. This is because, as we shall demonstrate in the next section, the estimates obtained from just one iteration are quite good, and do not merit additional computational time.)

Now if any of the coefficients ϕ_{mij} are zero, then the corresponding columns associated with these coefficients can be deleted from the design matrix of explanatory variables to facilitate the estimation of the system of equations. As an example, consider the following VAR(2) system with three variables:

$$\begin{bmatrix} y_{1,t} \\ y_{2,t} \\ y_{3,t} \end{bmatrix} = \begin{bmatrix} \phi_{111} & \phi_{112} & \phi_{113} \\ \phi_{121} & \phi_{122} & \phi_{123} \\ \phi_{131} & \phi_{132} & \phi_{133} \end{bmatrix} \begin{bmatrix} y_{1,t-1} \\ y_{2,t-1} \\ y_{3,t-1} \end{bmatrix} + \begin{bmatrix} \phi_{211} & \phi_{212} & \phi_{213} \\ \phi_{221} & \phi_{222} & \phi_{223} \\ \phi_{231} & \phi_{232} & \phi_{233} \end{bmatrix} \begin{bmatrix} y_{1,t-2} \\ y_{2,t-2} \\ y_{3,t-2} \end{bmatrix} + \begin{bmatrix} a_{1,t} \\ a_{2,t} \\ a_{3,t} \end{bmatrix} \tag{15}$$

Using our stacking scheme, which groups together the dependent variables for each time period along with their corresponding regressor lagged variables, this system can be expressed as

$$
\begin{bmatrix} Y_{1,3} \\ Y_{2,3} \\ Y_{3,3} \\ Y_{1,4} \\ Y_{2,4} \\ Y_{3,4} \\ \vdots \\ Y_{1,N} \\ Y_{2,N} \\ Y_{3,N} \end{bmatrix}
=
\begin{bmatrix}
Y_{1,2} & Y_{2,2} & Y_{3,2} & Y_{1,1} & Y_{2,1} & Y_{3,1} & 0 & 0 & 0 & 0 & 0 & 0 & 0 & 0 & 0 & 0 & 0 & 0 \\
0 & 0 & 0 & 0 & 0 & 0 & Y_{1,2} & Y_{2,2} & Y_{3,2} & Y_{1,1} & Y_{2,1} & Y_{3,1} & 0 & 0 & 0 & 0 & 0 & 0 \\
0 & 0 & 0 & 0 & 0 & 0 & 0 & 0 & 0 & 0 & 0 & 0 & Y_{1,2} & Y_{2,2} & Y_{3,2} & Y_{1,1} & Y_{2,1} & Y_{3,1} \\
Y_{1,3} & Y_{2,3} & Y_{3,3} & Y_{1,2} & Y_{2,2} & Y_{3,2} & 0 & 0 & 0 & 0 & 0 & 0 & 0 & 0 & 0 & 0 & 0 & 0 \\
0 & 0 & 0 & 0 & 0 & 0 & Y_{1,3} & Y_{2,3} & Y_{3,3} & Y_{1,2} & Y_{2,2} & Y_{3,2} & 0 & 0 & 0 & 0 & 0 & 0 \\
0 & 0 & 0 & 0 & 0 & 0 & 0 & 0 & 0 & 0 & 0 & 0 & Y_{1,3} & Y_{2,3} & Y_{3,3} & Y_{1,2} & Y_{2,2} & Y_{3,2} \\
 & \vdots & & & & & & \vdots & & & & & & \vdots & & & & \\
Y_{1,N-1} & Y_{2,N-1} & Y_{3,N-1} & Y_{1,N-2} & Y_{2,N-2} & Y_{3,N-2} & 0 & 0 & 0 & 0 & 0 & 0 & 0 & 0 & 0 & 0 & 0 & 0 \\
0 & 0 & 0 & 0 & 0 & 0 & Y_{1,N-1} & Y_{2,N-1} & Y_{3,N-1} & Y_{1,N-2} & Y_{2,N-2} & Y_{3,N-2} & 0 & 0 & 0 & 0 & 0 & 0 \\
0 & 0 & 0 & 0 & 0 & 0 & 0 & 0 & 0 & 0 & 0 & 0 & Y_{1,N-1} & Y_{2,N-1} & Y_{3,N-1} & Y_{1,N-2} & Y_{2,N-2} & Y_{3,N-2}
\end{bmatrix}
\begin{bmatrix} \phi_{111} \\ \phi_{112} \\ \phi_{113} \\ \phi_{211} \\ \phi_{212} \\ \phi_{213} \\ \phi_{121} \\ \phi_{122} \\ \phi_{123} \\ \phi_{221} \\ \phi_{222} \\ \phi_{223} \\ \phi_{131} \\ \phi_{132} \\ \phi_{133} \\ \phi_{231} \\ \phi_{232} \\ \phi_{233} \end{bmatrix}
+
\begin{bmatrix} a_{1,3} \\ a_{2,3} \\ a_{3,3} \\ a_{1,4} \\ a_{2,4} \\ a_{3,4} \\ \vdots \\ a_{1,N} \\ a_{2,N} \\ a_{3,N} \end{bmatrix}
$$

If we assume, for instance, that only $\phi 111$, $\phi 223$, and $\phi 231$ are nonzero, then the design matrix of explanatory variables reduces to just,

$$\mathbf{z} = \begin{bmatrix} y_{1,2} & 0 & 0 \\ 0 & y_{3,1} & 0 \\ 0 & 0 & y_{1,1} \\ y_{1,3} & 0 & 0 \\ 0 & y_{3,2} & 0 \\ 0 & 0 & y_{1,2} \\ \vdots & \vdots & \vdots \\ y_{1,N-1} & 0 & 0 \\ 0 & y_{3,N-2} & 0 \\ 0 & 0 & y_{1,N-2} \end{bmatrix} \tag{16}$$

To determine if it is possible to delete some of the parameters from the VAR structure identified as having order p without affecting the whiteness of the residuals, we developed the following strategy based on t-ratios calculated from the fully parametrized VAR(p) model. First, we calculated the t-ratios of all parameters and ranked them in descending order. We then estimated a VAR(p) structure with just one nonzero element in the pth polynomial matrix, i.e., the one with the highest t-ratio in absolute value in the *polynomial matrix of order p*, to see if the residuals from this parsimonious model were white noise. If the VAR(p) structure with just one parameter passed the RWNAR test, then the identification procedure stopped; if not, a second parameter was incorporated into the model based on the highest value of all the remaining t-ratios. The two-parameter VAR(p) was then estimated using the GLS approach (as was the single parameter model) to determine if the residuals were white noise. (The covariance matrix $\mathbf{V}$ was estimated from the fully parametrized VAR(p) structure.) Additional parameters continued to be introduced into the polynomial structure (in order of magnitude of remaining t-ratios) until residuals were demonstrated to be white noise. Note, therefore, that this method of introducing parameters into the model can lead to incorporation of parameters whose t-ratios may be lower than those typically used in significance tests. No special cut-off values for the t-ratios were used. Residual randomness was determined solely on the basis of RWNAR tests.

We recognize that there are many other approaches that can be used to identify "significant" elements in the parameter polynomial matrices. One possibility would be to systematically delete parameters based on the lowest t-ratios. Such a backward-type approach, however, would be computationally more intensive than the method we selected. We chose this method because of its simplicity and ease of programmability. Moreover, it cleverly circumvents problems associated with the estimation of models with a large number of parameters. No doubt there may be other approaches which may yield even better results.

4. SIMULATION AND EMPIRICAL RESULTS

The small sample performance of the RWNAR identification method has been documented by Koreisha and Pukkila (1993), using simulated data for a variety of model structures and sample sizes. They have found that the RWNAR approach can be used effectively to identify the order of VAR models. Moreover, it was shown that the method's performance is superior to traditional order determination criteria such as AIC, BIC, and HQ.

To demonstrate the small sample performance of the new generalized least-squares estimation method, we simulated thousands of realizations for several VAR model structures which had only a fraction of nonzero elements in their polynomial matrices. The procedure we used to generate the data and construct models containing varying number of component series and observations is identical to the one described in Koreisha and Pukkila (1993, p. 50).

Table 1 contains the mean number of parameter estimates along with sample standard errors for several VAR(p) processes with a varying number of component series K and parameter polynomial configurations for sample sizes ranging from 50 to 200 observations. Unless otherwise stated, for all simulated structures the elements of the covariance matrix Σ_a were defined as $\sigma_{ij} = 0.5^{|i-j|}$. These particular parametrizations were chosen in order to encompass a wide range of parameter values, different feedback mechanisms, and a wide spectrum of numbers of nonzero elements in the AR polynomial matrices.

As can be seen, the mean value of the parameter estimates obtained by our new procedure closely matches the simulated values regardless of parametrization, sample size, or number of component series. Also, as the sample size increases, the variation in the parameter estimates decreases, as would be expected. We also simulated some VAR(1) structures containing three component series that had roots that were nearly on the unit circle. Table 2 presents the percentage of the structures (based on 200 replications) which contained at least one nonstationary parameter estimate in the polynomial matrices of the model. Notice that even for small sample sizes the procedure very seldom generates nonstationary values. When $N = 50$, only 0.5% of the cases were identified as nonstationary models when $\phi_{1ii} = 0.995$, and just 5.5% when $\phi_{1ii} = 0.999$. Increasing the sample size to 200 observations does not significantly reduce the percentage of nonstationary

Table 1. Mean value of parameter estimates and their sample standard errors based on 200 replications

K=3

N = 50 Observations			N = 100 Observations			N = 200 Observations		
VAR(1) Diagonal Model								
$\phi_{1ij} = (0.5)(-1)^i,\ j=i;\ \phi_{1ij} = 0.0,\ j \neq i$								
-0.5072	0.0000	0.0000	-0.5039	0.0000	0.0000	-0.5063	0.0000	0.0000
(0.0969)	0.0000	0.0000	(0.0689)	0.0000	0.0000	(0.0453)	0.0000	0.0000
0.0000	0.4625	0.0000	0.0000	0.4798	0.0000	0.0000	0.4838	0.0000
0.0000	(0.1036)	0.0000	0.0000	(0.0606)	0.0000	0.0000	(0.0454)	0.0000
0.0000	0.0000	-0.5006	0.0000	0.0000	-0.4992	0.0000	0.0000	-0.5016
0.0000	0.0000	(0.0909)	0.0000	0.0000	(0.0643)	0.0000	0.0000	(0.0425)
VAR(1) Lower Triangular Model								
$\phi_{1ij} = (0.7)^{(i-j)+1}(-1)^{i-j},\ j<i;\ \phi_{1ij} = 0,\ j>i$								
0.6136	0.0000	0.0000	0.6506	0.0000	0.0000	0.6534	0.0000	0.0000
(0.1175)	0.0000	0.0000	(0.0688)	0.0000	0.0000	(0.0535)	0.0000	0.0000
-0.5291	0.6329	0.0000	-0.5158	0.6574	0.0000	-0.5206	0.7088	0.0000
(0.1215)	(0.0803)	0.0000	(0.0800)	(0.0525)	0.0000	(0.0554)	(0.0353)	0.0000
0.3314	-0.5424	0.6601	0.3364	-0.5279	0.6765	0.3944	-0.4641	0.6515
(0.1317)	(0.0993)	0.0612)	(0.0790)	(0.0668)	0.0370)	(0.0519)	(0.0420)	(0.0298)
VAR(2) Triangular Model								
$\phi_{1ij} = (0.8)^{(j-i)+1}(-1)^{j-i},\ j>i;\ \phi_{1ij} = 0,\ j<i$ $\phi_{1ij} = (-0.7)^{(j-i)+1}(-1)^{j-i},\ j>i;\ \phi_{1ij} = 0,\ j<i$								
0.7958	-0.6574	0.5330	0.8029	-0.6381	0.5315	0.8036	-0.6333	0.5298
(0.0473)	(0.1097)	(0.1262)	(0.0307)	(0.0729)	(0.0870)	(0.0197)	(0.0496)	(0.0668)
0.0000	0.7837	-0.6271	0.0000	0.7903	-0.6156	0.0000	0.7942	-0.6120
0.0000	(0.0681)	(0.1104)	0.0000	(0.0419)	(0.0738)	0.0000	(0.0295)	(0.0542)
0.0000	0.0000	0.7959	0.0000	0.0000	0.8065	0.0000	0.0000	0.8113
0.0000	0.0000	(0.0927)	0.0000	0.0000	(0.0612)	0.0000	0.0000	(0.0379)
-0.6961	0.4938	-0.3643	-0.7037	0.4918	-0.3558	-0.7046	0.4879	-0.3533
(0.0444)	(0.1066)	(0.1504)	(0.0273)	(0.0744)	(0.1028)	(0.0193)	(0.0509)	(0.0724)
0.0000	-0.6984	0.4667	0.0000	-0.7005	0.4665	0.0000	-0.7039	0.4642
0.0000	(0.0546)	(0.1235)	0.0000	(0.0356)	(0.0913)	0.0000	(0.0261)	(0.0641)
0.0000	0.0000	-0.7020	0.0000	0.0000	-0.7074	0.0000	0.0000	-0.7129
0.0000	0.0000	(0.0933)	0.0000	0.0000	(0.0708)	0.0000	0.0000	(0.0493)

VAR(1) Diagonal Model

$\phi_{1ij} = (0.5)(-1)^{i},\ j=i;\ \phi_{1ij} = 0.0,\ j\neq i$

N=100 Observations					N=200 Observations				
-0.4951	0.0000	0.0000	0.0000	0.0000	-0.4950	0.0000	0.0000	0.0000	0.0000
(0.0828)	0.0000	0.0000	0.0000	0.0000	(0.0586)	0.0000	0.0000	0.0000	0.0000
0.0000	0.4880	0.0000	0.0000	0.0000	0.0000	0.4947	0.0000	0.0000	0.0000
0.0000	(0.0786)	0.0000	0.0000	0.0000	0.0000	(0.0582)	0.0000	0.0000	0.0000
0.0000	0.0000	-0.4971	0.0000	0.0000	0.0000	0.0000	-0.4967	0.0000	0.0000
0.0000	0.0000	(0.0768)	0.0000	0.0000	0.0000	0.0000	(0.0499)	0.0000	0.0000
0.0000	0.0000	0.0000	0.4886	0.0000	0.0000	0.0000	0.0000	0.4949	0.0000
0.0000	0.0000	0.0000	(0.0784)	0.0000	0.0000	0.0000	0.0000	(0.0589)	0.0000
0.0000	0.0000	0.0000	0.0000	-0.5005	0.0000	0.0000	0.0000	0.0000	-0.5027
0.0000	0.0000	0.0000	0.0000	(0.0757)	0.0000	0.0000	0.0000	0.0000	(0.0499)

VAR(1) Lower Triangular Model

$\phi_{1ij} = (0.7)^{(i-j)+1}(-1)^{i-j},\ j<i;\ \phi_{1ij} = 0,\ j>i$

0.6612	0.0000	0.0000	0.0000	0.0000	0.6815	0.0000	0.0000	0.0000	0.0000
(0.0838)	0.0000	0.0000	0.0000	0.0000	(0.0592)	0.0000	0.0000	0.0000	0.0000
-0.5000	0.6730	0.0000	0.0000	0.0000	-0.4950	0.6857	0.0000	0.0000	0.0000
(0.0715)	(0.0517)	0.0000	0.0000	0.0000	(0.0509)	(0.0342)	0.0000	0.0000	0.0000
0.3491	-0.5287	0.6864	0.0000	0.0000	0.3435	-0.5059	0.6963	0.0000	0.0000
(0.0850)	(0.0745)	(0.0434)	0.0000	0.0000	(0.0475)	(0.0463)	(0.0300)	0.0000	0.0000
-0.2431	0.3215	-0.5113	0.6938	0.0000	-0.2493	0.3286	-0.5016	0.6987	0.0000
(0.0874)	(0.0653)	(0.0655	(0.0314)	0.0000	(0.0563)	(0.0359)	(0.0502)	(0.0237)	0.0000
0.1724	-0.2658	0.3238	-0.5035	0.6962	0.1644	-0.2641	0.3298	-0.4953	0.6984
(0.0920)	(0.0739)	(0.0620	(0.0564)	(0.0239	(0.0621)	(0.0460)	(0.0348)	(0.0367)	(0.0157)

VAR(2) Triangular Model

$\phi_{1ij} = (0.8)^{(j-i)+1}(-1)^{j-i},\ j>i;\ \phi_{1ij} = 0,\ j<i$

$\phi_{2ij} = (-0.7)^{(j-i)+1}(-1)^{j-i},\ j>i;\ \phi_{2ij} = 0,\ j<i$

0.7956	-0.6659	0.4961	-0.3753	0.3123	0.8000	-0.6490	0.4996	-0.3793	0.3220
(0.0164)	(0.0499)	(0.0797)	(0.1044)	(0.1106)	(0.0106)	(0.0347)	(0.0495)	(0.0728)	(0.0757)
0.0000	0.7958	-0.6588	0.5131	-0.3956	0.0000	0.7980	-0.6481	0.5173	-0.4031
0.0000	(0.0204)	(0.0600)	(0.0845)	(0.0989)	0.0000	(0.0120)	(0.0364)	(0.0561)	(0.0635)
0.0000	0.0000	0.7948	-0.6422	0.4984	0.0000	0.0000	0.8000	-0.6348	0.4957
0.0000	0.0000	(0.0342)	(0.0730)	(0.0920)	0.0000	0.0000	(0.0195)	(0.0486)	(0.0623)
0.0000	0.0000	0.0000	0.8017	-0.6638	0.0000	0.0000	0.0000	0.8064	-0.6571
0.0000	0.0000	0.0000	(0.0438)	(0.0828)	0.0000	0.0000	0.0000	(0.0354)	(0.0535)
0.0000	0.0000	0.0000	0.0000	0.7813	0.0000	0.0000	0.0000	0.0000	0.7858
0.0000	0.0000	0.0000	0.0000	(0.0690)	0.0000	0.0000	0.0000	0.0000	(0.0482)
-0.6955	0.4950	-0.3763	0.2085	-0.1492	-0.6987	0.4937	-0.3653	0.1921	-0.1484
(0.0104)	(0.0497)	(0.0792)	(0.0936)	(0.1260)	(0.0073)	(0.0305)	(0.0507)	(0.0516)	(0.0905)
0.0000	-0.6959	0.4876	-0.3873	0.2109	0.0000	-0.6996	0.4799	-0.3838	0.2149
0.0000	(0.0205)	(0.0640)	(0.0807)	(0.1119)	0.0000	(0.0119)	(0.0343)	(0.0466)	(0.0647)
0.0000	0.0000	-0.7022	0.4613	-0.3654	0.0000	0.0000	-0.7048	0.4634	-0.3618
0.0000	0.0000	(0.0318)	(0.0733)	(0.0945)	0.0000	0.0000	(0.0189)	(0.0474)	(0.0688)
0.0000	0.0000	0.0000	-0.7068	0.4770	0.0000	0.0000	0.0000	-0.7105	0.4776
0.0000	0.0000	0.0000	(0.0400)	(0.0844)	0.0000	0.0000	0.0000	(0.0282)	(0.0639)
0.0000	0.0000	0.0000	0.0000	-0.6974	0.0000	0.0000	0.0000	0.0000	-0.7026
0.0000	0.0000	0.0000	0.0000	(0.0756)	0.0000	0.0000	0.0000	0.0000	(0.0568)

Table 1. (*continued*)

N=200 Observations

$\phi_{1ij} = (0.5)(-1)^i,\ j=i;\ \phi_{1ij} = 0.0,\ j \neq i$

-0.4914	0.0000	0.0000	0.0000	0.0000	0.0000	0.0000
(0.0506)	0.0000	0.0000	0.0000	0.0000	0.0000	0.0000
0.0000	0.4793	0.0000	0.0000	0.0000	0.0000	0.0000
0.0000	(0.0515)	0.0000	0.0000	0.0000	0.0000	0.0000
0.0000	0.0000	-0.5023	0.0000	0.0000	0.0000	0.0000
0.0000	0.0000	(0.0442)	0.0000	0.0000	0.0000	0.0000
0.0000	0.0000	0.0000	0.4800	0.0000	0.0000	0.0000
0.0000	0.0000	0.0000	(0.0482)	0.0000	0.0000	0.0000
0.0000	0.0000	0.0000	0.0000	-0.5022	0.0000	0.0000
0.0000	0.0000	0.0000	0.0000	(0.0450)	0.0000	0.0000
0.0000	0.0000	0.0000	0.0000	0.0000	0.4810	0.0000
0.0000	0.0000	0.0000	0.0000	0.0000	(0.0493)	0.0000
0.0000	0.0000	0.0000	0.0000	0.0000	0.0000	-0.5117
0.0000	0.0000	0.0000	0.0000	0.0000	0.0000	(0.0519)

$\phi_{1ij} = (0.7)^{(i-j)+1}(-1)^{i-j},\ j<i;\ \phi_{1ij} = 0,\ j>i$

0.6743	0.0000	0.0000	0.0000	0.0000	0.0000	0.0000
(0.0498)	0.0000	0.0000	0.0000	0.0000	0.0000	0.0000
-0.5189	0.6796	0.0000	0.0000	0.0000	0.0000	0.0000
(0.0497)	(0.0376)	0.0000	0.0000	0.0000	0.0000	0.0000
0.3375	-0.5256	0.6892	0.0000	0.0000	0.0000	0.0000
(0.0599)	(0.0409)	(0.0293)	0.0000	0.0000	0.0000	0.0000
-0.2328	0.3241	-0.5175	0.6922	0.0000	0.0000	0.0000
(0.0458)	(0.0410)	(0.0415)	(0.0198)	0.0000	0.0000	0.0000
0.1833	-0.2334	0.3224	-0.5074	0.6976	0.0000	0.0000
(0.0506)	(0.0392)	(0.0357)	(0.0337)	(0.0122)	0.0000	0.0000
-0.1153	0.1880	-0.2388	0.3263	-0.5033	0.6969	0.0000
(0.0650)	(0.0465)	(0.0384)	(0.0343)	(0.0292)	(0.0094)	0.0000
0.0823	-0.1097	0.1814	-0.2366	0.3279	-0.5016	0.6979
(0.0479)	(0.0539)	(0.0434)	(0.0393)	(0.0299)	(0.0236)	(0.0066)

Table 2 Percentage of structures containing at least one nonstationary parameter estimate in VAR(1) diagonal structures with three component series (based on 200 replications)

ϕ_1	$N = 50$	$N = 100$	$N = 200$	$N = 500$
0.90	0.0	0.0	0.0	0.0
0.95	0.5	0.0	0.0	0.0
0.99	5.0	2.5	0.5	0.0
0.999	5.5	5.5	5.0	1.0

estimates when $\phi_{1ii} = 0.999$; when $N = 500$, however, only 1% of the structures contained nonstationary elements

An in-depth study is now underway to show how effective the new methodology is in identifying the nonzero elements of the autoregressive polynomial matrices. Table 3 contains some preliminary results showing the frequency distribution of the identified nonzero elements in the polynomial matrices of a VAR(1) process with three component series with a wide range of parameter values based on sample sizes of 50, 100, 200, and 500 observations. It also contains similar results for a VAR(2) process with five component series based on 150 observations. As can be seen, the GLS approach in conjunction with the RWNAR order determination criterion appears to perform extremely well. As one would expect, correct identification of the location of the parameters improves as the magnitude of the parameters increases and as the sample size increases. When sample sizes are small, parameters having values near $|0.3|$ or less appear to be difficult to identify. (Note, however, how the mean number of identified parameters increases as the sample size increases.) The loss in predictive efficiency due to using a model without the low-valued parameters can be quite small (Bhansali, 1991; Koreisha and Pukkila, 1994).

Finally, we also applied our new methodology on the 3-monthly flour price series studied by Tiao and Tsay (1985). The data are logarithms of indices of monthly flour prices (August 1972 to November 1980) in three American cities (Buffalo, Minneapolis, and Kansas City) that have been normalized into canonical variates which, according to Tiao and Tsay, facilities the "finding of simple structures hidden in vector processes." Table 4 not only contains the unrestricted fully parametrized VAR (2) models identified by Tiao and Tsay and by our procedures, but also Tiao and Tsay's final restricted model (formulated on the basis of the estimated standard errors using a 5% significance level), and also the one identified by our approach when it was permitted to identify the order as well as the

Table 3 Frequency distribution of identified nonzero elements in the AR polynomial matrices (based on 100 replications)

Simulated structure							
	0.75	−0.5	0.25		1.0	0.5	0.25
$\phi_1 =$	0	0.5	−0.25	$\Sigma =$	0.5	1.0	0.5
	0	0	0.25		0.25	0.5	1.0

$N = 50$

Identified polynomial structure			Means of parameter estimates and their corresponding standard errors based only on identified element positions		
100	84	33	0.6895	−0.4648	0.1549
			(0.1138)	(0.2695)	(0.2528
2	62	22	−0.0004	0.2827	−0.1087
			(0.0332)	(0.2471)	(0.2147)
3	3	12	−0.0021	0.0055	0.0564
			(0.0386)	(0.0542)	(0.1563)
Identified order: VAR(1) = 100%			Mean number of estimated parameters = 3.21		

$N = 100$

Identified polynomial structure			Means of parameter estimates and their corresponding standard errors based only on identified element positions		
100	97	52	0.7276	−0.4884	0.2096
			(0.0564)	(0.1693)	(0.2093)
0	92	36	0	0.3986	−0.1642
				(0.1610)	(0.2252)
5	2	31	−0.0043	0.0061	0.1142
			(0.0319)	(0.0425)	(0.1749)
Identified order: VAR(1) = 100%			Mean number of estimated parameters = 4.15		

$N = 200$					
Identified polynomial structure			Means of parameter estimates and their corresponding standard errors based only on identified element positions		
100	100	67	0.7368	−0.4833	0.2321
			(0.0357)	(0.1011)	(0.1762)
0	100	49	0	0.4495	−0.1921
				(0.0858)	(0.2016)
4	0	60	−0.006	0	0.1933
			(0.0298)		(0.1645)
Identified order: VAR(1) = 100%			Mean number of estimated parameters = 4.80		

$N = 500$					
Identified polynomial structure			Means of parameter estimates and their corresponding standard errors based only on identified element positions		
100	100	81	0.7474	−0.4973	0.2789
			(0.0233)	(0.0664)	(0.1476)
0	100	39	0	0.4491	−0.1356
				(0.0662)	(0.1725)
0	0	87	0	0	0.2611
					(0.1240)
Identified order: VAR(1) = 100%			Mean number of estimated parameters = 5.07		

Table 3 (*continued*)

$N = 150$

Simulated structure[a]					Identified polynomial structure				
ϕ_1					ϕ_1				
0.8	0.0	0.0	0.0	0.0	100	0	0	0	0
−0.64	0.8	0.0	0.0	0.0	100	100	0	0	0
0.512	−0.64	0.8	0.0	0.0	90	100	100	0	0
−0.4096	0.512	−0.64	0.8	0.0	92	99	100	100	0
0.32768	−0.4096	0.512	−0.64	0.8	43	92	100	100	100
ϕ_2					ϕ_2				
−0.7	0.0	0.0	0.0	0.0	100	0	0	0	0
0.49	−0.7	0.0	0.0	0.0	97	100	0	0	0
−0.343	0.49	−0.7	0.0	0.0	28	98	100	0	1
0.2401	−0.343	0.49	−0.7	0.0	6	59	100	100	0
−0.16807	0.2401	−0.343	0.49	−0.7	0	15	83	100	100

Identified order: VAR(2) = 100%

Mean number of estimated parameters = 25.03

Means of parameter estimates based only on identified element positions					Standard deviations of parameter estimates based only on identified element positions				
ϕ_1					ϕ_1				
0.7785	0.0000	0.0000	0.0000	0.0000	0.0654	0.0000	0.0000	0.0000	0.0000
−0.6917	0.8096	0.0000	0.0000	0.0000	0.0941	0.0487	0.0000	0.0000	0.0000
0.3642	−0.5868	0.8110	0.0000	0.0000	0.1788	0.0967	0.0427	0.0000	0.0000
−0.4360	0.3732	−0.5950	0.8314	0.0000	0.1919	0.1017	0.0924	0.0240	0.0000
0.1851	−0.3438	0.4116	−0.6511	0.8132	0.2217	0.1300	0.1158	0.0511	0.0184
ϕ_2					ϕ_2				
−0.6771	0.0000	0.0000	0.0000	0.0000	0.0977	0.0000	0.0000	0.0000	0.0000
0.5759	−0.6848	0.0000	0.0000	0.0000	0.1520	0.0521	0.0000	0.0000	0.0000
−0.1447	0.5107	0.7143	0.0000	0.0000	0.2382	0.1286	0.0313	0.0000	0.0004
0.0323	−0.2344	0.5797	−0.7006	0.0000	0.1287	0.2062	0.0712	0.0194	0.0000
0.0000	0.0704	−0.3014	0.5772	−0.6894	0.0000	0.1699	0.1462	0.0721	0.0113

[a]The elements of the covariance matrix Σ_a were defined as $\sigma_{ij} = 0.5^{|i-j|}$.

Table 4 Models identified for the flour price data using different procedures (dependent variables are Buffalo, Minneapolis, and Kansas City indices, respectively)

Unrestricted VAR(2) model

Tiao and Tsay (1985)						GLS and RWNAR					
ϕ_1			ϕ_2			ϕ_1			ϕ_2		
0.88	0.01	0.01	−0.03	0.02	−0.00	0.88	0.00	0.01	−0.03	0.02	−0.01
(0.11)	(0.04)	(0.06)	(0.10)	(0.04)	(0.06)	(0.11)	(0.04)	(0.06)	(0.10	(0.04)	(0.06)
0.25	0.97	−0.22	−0.59	0.03	0.28	0.25	0.97	−0.23	−0.59	0.02	0.30
(0.33)	(0.11)	(0.20)	(0.33)	(0.12)	(0.20)	(0.33)	(0.11)	(0.20)	(0.34)	(0.12)	(0.20)
−0.59	−0.22	0.58	0.78	0.18	0.36	−0.59	−0.21	0.59	0.77	0.18	0.34
(0.18)	(0.06)	(0.11)	(0.18)	(0.06)	(0.11)	(0.18)	(0.06)	(0.11)	(0.18)	(0.06)	(0.10)

Restricted VAR(2) model

Tiao and Tsay (1985)						GLS and RWNAR					
ϕ_1			ϕ_2			ϕ_1			ϕ_2		
1.00[a]	0.00	0.00	0.00	0.00	0.00	0.89	0.00	0.00	0.00	0.00	0.00
(0.01)	(0.00)	(0.00)	(0.00)	(0.00)	(0.00)	(0.05)	(0.00)	(0.00)	(0.00)	(0.00)	(0.00)
0.00	1.00[b]	0.00	0.00	0.00	0.00	0.00	0.97	0.00	0.00	0.00	0.00
(0.00)	(0.00)	(0.00)	(0.00)	(0.00)	(0.00)	(0.00)	(0.03)	(0.00)	(0.00)	(0.00)	(0.00)
−0.57	−0.22	0.52	0.61	0.19	0.43	−0.45	−0.04	0.65	0.55	0.00	0.26
(0.14)	(0.05)	(0.09)	(0.14)	(0.05)	(0.09)	(0.14)	(0.02)	(0.08)	(0.14)	(0.00)	(0.00)

$\Sigma_a \times 10^3$			$\Sigma_a \times 10^3$		
0.122			0.115		
0.026	1.276		0.016	1.248	
−0.068	−0.373	0.340	−0.057	−0.332	0.339

[a]Rounded up from 0.998.
[b]Rounded down from 1.001.

nonzero elements of the AR polynominal matrices completely *automatically*. As can be seen, both approaches yield very similar parameterizations, but the RWNAR criterion seems to identify a slightly more parsimonious structure in that it only gives the net effect that Minneapolis prices have on the Kansas City flour index ϕ_{132}, whereas the "manual" approach based on t-ratios spreads the effects over the two lag periods. Note that these effects are comparatively small in relation to the other effects. The elements of the residual covariance matrix obtained from the completely automatic approach are slightly smaller than those generated by Tiao and Tsay. (The values have been multiplied by 1000 in order to conform with the estimates presented in Table 4.)

5. CONCLUDING REMARKS

Our simulation studies show that for moderate size samples the new generalized least-squares method can accurately and efficiently estimate VAR(p) processes with parameters constrained to zero. Moreover, when combined with the RWNAR order identification approach, preliminary studies suggest that it can also accurately identify the nonzero elements of the polynomial matrices of such processes. The method's estimation and identification performance improve as the sample size increases, particularly when the number of component series is large. The approach seldom generates nonstationary estimates even for parametrizations with roots near the unit circle.

REFERENCES

Akaike, H. (1974). A new look at the statistical model identification, *IEEE Transactions on Automatic Control* 19, 716-723.

Bhansali, R. J. (1991). Consistent recursive estimation of the order of an autogressive moving average process. *Int. Stat. Rev.*, **59**, 81–96.

Doan, T., Litterman, R., and Sims, C. (1984). Forecasting and conditional projection using prior distributions. *Econometric Rev.*, **3**, 1–100.

Funke, M. (1990). Assessing the forecasting accuracy of monthly vector autoregressive models. The case of five OECD countries. *Int. J. Forecasting*, **6**, 363–378.

Hannan, E. J. and Quinn, B.G. (1979). The determination of the order of an autoregression. *J. R. Stat. Soc. B*, **41**, 190–195.

Hillmer, S. and Tiao, G. (1979). Likelihood function of stationary multiple autoregressive moving average models. *J. Am. Stat. Assoc.*, **74**, 652–660.

Koreisha, S. and Pukkila, T. (1993). Determining the order of a vector autoregression when the number of component series is large. *J. Time Ser. Anal.*, **14**, 47–69.

Koreisha, S. and Pukkila, T. (1994). Identification of seasonal autoregressive models. *J. Time Ser. Anal.*, **16**, 267–290.

Litterman, R. (1981). A Bayesian procedure for forecasting with vector autoregressions. Federal Reserve Bank Working Paper.

Litterman, R. (1986). Forecasting with Bayesian autoregressions-Five years experience. *Journal of Business and Economic Statistics*, **4**, 25–38.

Litterman, R. and Weiss, L. (1985). Money, real interest rates, and output: A reinterpretation of postwar US data. *Econometrica*, **53**, 129–156.

Lupoletti, W. and Webb, R. (1986). Defining and improving the accuracy of macroeconomic forecasts: Contributions from a VAR model. *J. Bus.*, **59**, 263–285.

Lutkepohl, H. (1985). Comparison of criteria for estimating the order of a vector autoregressive process. *J. Time Ser. Anal.*, **6**, 35–62.

Lysne, D. and Tjostheim, D. (1987). Loss of spectral peaks in autoregressive spectral estimation. *Biometrika*, **74**, 200–206.

Pukkila, T., Koreisha, S. and Kallinen, A. (1990). The identification of ARMA models. *Biometrika* **77**, 537–548.

Pukkila, T. and Krishnaiah, P. (1988). On the use of autoregressive order determination criteria in multivariate white noise tests. *IEEE Transaction on Acoustics, Speech, and Signal Processing*, **36**, 1396–1403

Quinn, B. (1980). Order determination for a multivariate autoregression. *J. R. Stat. Soc. B*, **42**, 182–185.

Reagan, P. and Sheehan, D. (1985). The stylized facts about the behavior of manufacturers' inventories and back orders over the business cycle. *J. Monetary Econ.*, **15**, 21–46.

Reinsel, G. and Ahn, K. (1992). Vector autoregressions with unit roots and reduced rank structure: estimation, likelihood ratio test, and forecasting. *J. Time Ser. Anal.*, **13**, 353–375.

Rissanen, J. (1978). Modeling by shortest data description. *Automatica*, **14**, 465–471.

Rose, E. (1989). Forecasting with vector autoregressive models. *Proceedings of the Economic and Business Statistics Section of the American Statistical Association*, 104–109.

Schwarz, G. (1978). Estimating the dimension of a model. *Annals of Statistics* **6**, 461–464.

Sims, C. (1980). Macroeconomics and reality. *Econometrica,* **48**, 1–48.

Theil, H. (1971). *Principles of Econometrics.* Wiley, New York.

Tiao, G. and Box G. (1981). Modeling multiple time series with applications. *J. Am. Stat. Assoc.*, **75**, 802–816.

Tiao, G. and Tsay, R. (1985). A canonical correlation approach to modelling multivariate time series. *Proceedings of the Economic & Business Statistics Section of the American Statistical Association*, pp.112–120.

Tiao, G. and Tsay, R. (1989). Model specification in multivariate time series with discussion. *J. R. Stat. Soc. B*, **51**, 157–213.

Zellner, A. (1962). An efficient method of estimating seemingly unrelated regressions and tests of aggregation bias. *J. Am. Stat. Assoc.*, **57**, 348–368.

13

Statistical Model Evaluation and Information Criteria

SADANORI KONISHI Kyushu University, Fukuoka, Japan

1. INTRODUCTION

The problem of evaluating the goodness of statistical models is fundamental and of importance in various fields of statistics, natural sciences, neural networks, engineering, economics, etc. Akaike's (1973, 1974) information criterion, known as AIC, provides a useful tool for constructing statistical models, and a number of successful applications of AIC in statistical data analysis have been reported (see, e.g., Bozdogan, 1994; Kitagawa and Gersch, 1996).

AIC is a criterion for evaluating the models estimated by the maximum likelihood method, and it can be derived mathematically under the assumption that a specified parametric family of densities contains the true distribution generating the data. With the development of various modeling techniques, the construction of criteria which enable us to evaluate various types of statistical models has been required. Several attempts have been made to construct information criteria, relaxing the assumptions imposed on AIC, and the criteria proposed have been examined in both theoretical and practical aspects (Konishi and Kitagawa, 1996; Ishiguro et al., 1997).

The main aim of the present paper is to give a systematic account of some recent developments in model evaluation criteria from an information-theoretic point of view. Section 2 presents a unified information-theoretic approach to statistical model evaluation problems. We intend to provide a basic expository account of the fundamental principles behind information criteria. The use of the criteria is illustrated by examples. We also discuss the application of the bootstrap methods in model evaluation problems. In Section 3, the information criteria proposed are applied to the evaluation of the various types of models based on robust, maximum penalized likelihood. In Section 4 we give the derivation of information criteria and investigate their asymptotic properties with theoretical and numerical improvements.

2. STATISTICAL MODELING AND INFORMATION CRITERIA

2.1 Statistical Models

Suppose that we have n independent observations whose underlying probability distribution is to be estimated. We consider a parametric family of probability distributions $\{f(x|\boldsymbol{\theta}); \boldsymbol{\theta} \in \Theta \subset R^p\}$ as a candidate model. The probability distribution is then estimated by finding a suitable estimate, $\hat{\boldsymbol{\theta}}$, of unknown parameters from the data and replacing $\boldsymbol{\theta}$ in $f(x|\boldsymbol{\theta})$ by the sample estimate $\hat{\boldsymbol{\theta}}$. We call this probability distribution $f(x|\hat{\boldsymbol{\theta}})$ a statistical model.

Another example of statistical models is the regression model. Consider the familiar Gaussian linear regression model of the form $y = \boldsymbol{x}'\boldsymbol{\beta} + \varepsilon$, where $\boldsymbol{x}$ is a p-dimensional vector of known explanatory variables, $\boldsymbol{\beta}$ is a p-dimensional vector of unknown parameters and the error is assumed to have a normal distribution with mean zero and variance σ^2. The parametric model may be represented as

$$f(y|\boldsymbol{\theta}) = \frac{1}{\sigma}\phi\left(\frac{y - \boldsymbol{x}'\boldsymbol{\beta}}{\sigma}\right); \qquad \boldsymbol{\theta} = (\boldsymbol{\beta}', \sigma)' \in \Theta \subset R^{p+1} \tag{1}$$

where $\phi(x)$ is the density of the standard normal distribution. This specified model may be estimated by various types of procedures which include maximum likelihood, penalized likelihood, robust procedures, etc. We then have the statistical model

$$f(y|\hat{\boldsymbol{\theta}}) = \frac{1}{\hat{\sigma}}\phi\left(\frac{y - \boldsymbol{x}'\hat{\boldsymbol{\beta}}}{\hat{\sigma}}\right); \qquad \hat{\boldsymbol{\theta}} = (\hat{\boldsymbol{\beta}}', \hat{\sigma})' \tag{2}$$

There are a variety of candidate models which may be fitted to the observed data. The problem is how to choose the best-fitting model from among these competing models. We investigate this model evaluation or model selection problem, putting emphasis on the following considerations:

1. In practical situations it seems to be difficult to obtain precise information on distributional form from a finite number of observations. Hence one usually selects an approximating parametric family to the true distribution. This requires the assumption that a specified parametric family of probability distributions may or may not contain the true distribution generating the data.
2. Statistical models are constructed to draw information from the observed data in various ways. It is desired to construct criteria which enable us to evaluate statistical models estimated by various types of procedures, including the methods of maximum likelihood.
3. A statistical model is used for making inferences on the future observation which might be obtained on the same random structure in the future. Hence, we consider from the predictive point of view that the true density or probability function $g(z)$ for a future observation z is estimated by $f(z|\hat{\boldsymbol{\theta}})$.

In the next subsection we present a general framework for constructing information criteria which may deal with a variety of statistical models.

2.2 Kullback–Leibler Measure and Model Evaluation

Suppose that $\mathbf{X}_n = \{X_1, \ldots, X_n\}$ is a random sample from an unknown distribution $G(x)$ having density (or probability) function $g(x)$. In general, the X_α may take values in an arbitrary sample space, which is a subset of the real line or the d-dimensional Euclidean space R^d. On the basis of the information contained in the observations, we choose a parametric model which consists of a family of probability distributions $\{f(x|\boldsymbol{\theta});\ \boldsymbol{\theta} \in \Theta\}$, where $\boldsymbol{\theta} = (\theta_1, \ldots, \theta_p)'$ is the p-dimensional vector of unknown parameters, and Θ is an open subset of R^p. This specified family of probability distributions may or may not contain the true density $g(x)$, but it is expected that its deviation from the parametric model is not large.

The adopted parametric model is estimated by replacing the unknown parameter vector $\boldsymbol{\theta}$ by some estimate $\hat{\boldsymbol{\theta}}$, for which maximum likelihood, penalized likelihood, or robust procedures may be used for parameter estimation. Then a future observation z from the true density g is predicted by using the statistical model $f(z|\hat{\boldsymbol{\theta}})$.

After fitting a parametric model to the data, one would like to assess the closeness of $f(z|\hat{\boldsymbol{\theta}})$ to the true density $g(z)$ generating the data. The Kullback–Leibler information (Kullback–Leibler, 1951)

$$I\{g(z); f(z|\hat{\boldsymbol{\theta}})\} = E_G\left[\log \frac{g(Z)}{f(Z|\hat{\boldsymbol{\theta}})}\right]$$

is used as an overall measure of the divergence of $f(z|\hat{\boldsymbol{\theta}})$ from $g(z)$, conditional on the observed data $\boldsymbol{X}_n$. For a distribution with continuous density, it can be expressed as

$$I\{g(z); f(z|\hat{\boldsymbol{\theta}})\} = \int g(z) \log g(z) \mathrm{d}z - \int g(z) \log f(z|\hat{\boldsymbol{\theta}}) \mathrm{d}z \tag{3}$$

When we consider the problem of choosing a statistical model from among different models, the first term in the right-hand side of equation (3) does not depend on the model, and only the second term

$$\eta(G; \hat{\boldsymbol{\theta}}) = \int g(z) \log f(z|\hat{\boldsymbol{\theta}}) \mathrm{d}z = \int \log f(z|\hat{\boldsymbol{\theta}}) \mathrm{d}G(z) \tag{4}$$

is relevant. The $\eta(G; \hat{\boldsymbol{\theta}})$, called the expected log likelihood, is conditional on the observed data $\boldsymbol{X}_n$ and depends on the true unknown distribution G. The aim is to estimate the expected log likelihood on the basis of the observations. An obvious and easily computed estimate of $\eta(G; \hat{\boldsymbol{\theta}})$ is the (average) log likelihood

$$\eta(\hat{G}; \hat{\boldsymbol{\theta}}) = \int \log f(z|\hat{\boldsymbol{\theta}}) \mathrm{d}\hat{G}(z) = \frac{1}{n} \sum_{\alpha=1}^{n} \log f(X_\alpha|\hat{\boldsymbol{\theta}}) \tag{5}$$

obtained by replacing the unknown distribution G in $\eta(G; \hat{\boldsymbol{\theta}})$ by the empirical distribution, $\hat{G}$, putting mass $1/n$ on each observation X_α. Usually the log likelihood provides an optimistic assessment (overestimation) of the expected log likelihood $\eta(G; \hat{\boldsymbol{\theta}})$, because the same data are used both to estimate the parameters of the model and to evaluate $\eta(G; \hat{\boldsymbol{\theta}})$. We therefore consider the bias correction of the log likelihood $\eta(\hat{G}; \hat{\boldsymbol{\theta}})$.

The bias of the log likelihood in estimating the expected log likelihood is given by

$$b(G) = E_G\left[\frac{1}{n} \sum_{\alpha=1}^{n} \log f(X_\alpha|\hat{\boldsymbol{\theta}}) - \int \log f(z|\hat{\boldsymbol{\theta}}) \mathrm{d}G(z)\right] \tag{6}$$

where expectation is taken over the true distribution G, that is, $\prod_{\alpha=1}^{n} dG(x_\alpha)$. If the bias $b(G)$ can be estimated by appropriate procedures, then the bias corrected log likelihood is given by

$$\frac{1}{n}\sum_{\alpha=1}^{n} \log f(X_\alpha|\hat{\boldsymbol{\theta}}) - \hat{b}(G)$$

which is usually used in the form

$$\begin{aligned} IC(\boldsymbol{X}_n; \hat{\boldsymbol{\theta}}) &= -2n\left\{\frac{1}{n}\sum_{\alpha=1}^{n} \log f(X_\alpha|\hat{\boldsymbol{\theta}}) - \hat{b}(G)\right\} \\ &= -2\sum_{\alpha=1}^{n} \log f(X_\alpha|\hat{\boldsymbol{\theta}}) + 2n\hat{b}(G) \end{aligned} \tag{7}$$

The estimated bias $\hat{b}(G)$ is generally given as an asymptotic bias and an approximation to $b(G)$. We choose a statistical model for which the value of the information criterion is minimized over a set of competing models.

2.3 Information Criteria

According to the assumptions made on model estimation and the relationship between the model and true distribution, the bias $\hat{b}(G)$ in equation (7) takes a different form, and consequently we obtain the information criteria proposed previously.

(1) Akaike's (1973, 1974) information criterion. Akaike's information criterion, known as AIC, is a criterion for evaluating statistical models estimated by maximum likelihood methods, and is given by

$$\mathrm{AIC} = -2\sum_{\alpha=1}^{n} \log f(X_\alpha|\hat{\boldsymbol{\theta}}_{ML}) + 2p \tag{8}$$

where $\hat{\boldsymbol{\theta}}_{ML}$ is the maximum likelihood estimate, and p is the number of estimated parameters within the model.

As will be shown in Section 4, the bias $b(G)$ in equation (6) is asymptotically given by p/n, and it can be derived under the assumptions that:

the parametric model is estimated by the method of maximum likelihood; the specified parametric family of probability distributions contains the true density, that is, $g(z) = f(z|\boldsymbol{\theta}_0)$ (i.e., $G = F_{\boldsymbol{\theta}_0}$) for some $\boldsymbol{\theta}_0$ in Θ.

Akaike (1973, 1974) indicated that the bias of the log likelihood $\eta(\hat{G}; \hat{\boldsymbol{\theta}}_{ML})$ can be approximated by p/n if the specified parametric model is not far from the true density $g(x)$. AIC has some attractive properties in practical applications. The bias correction term does not require any analytical derivation, and it can be applied in an automatic way in various situations. Also, the bias approximated by the number of estimated parameters is constant and has no variability.

(2) Takeuchi's (1976) information criterion. Because model classes used can be incorrect, it is useful to relax the second assumption above. Without assuming that the true distribution belongs to the specified parametric family of probability distributions, the bias of the log likelihood in equation (6) is asymptotically given by

$$\begin{aligned} b(G) &= E_G\left[\frac{1}{n}\sum_{\alpha=1}^{n}\log f(X_\alpha|\hat{\boldsymbol{\theta}}_{ML}) - \int \log f(z|\hat{\boldsymbol{\theta}}_{ML})\mathrm{d}G(z)\right] \\ &= \frac{1}{n}\mathrm{tr}\{J(G)^{-1}I(G)\} + O(n^{-2}) \end{aligned} \tag{9}$$

where $J(G)$ and $I(G)$ are $p \times p$ matrices defined by

$$\begin{aligned} J(G) &= -E_G\left[\frac{\partial^2 \log f(Z|\boldsymbol{\theta})}{\partial\boldsymbol{\theta}\partial\boldsymbol{\theta}'}\right] \quad \text{and} \\ I(G) &= E_G\left[\frac{\partial \log f(Z|\boldsymbol{\theta})}{\partial\boldsymbol{\theta}}\frac{\partial \log f(Z|\boldsymbol{\theta})}{\partial\boldsymbol{\theta}'}\right] \end{aligned} \tag{10}$$

Hence, by correcting the asymptotic bias of the log likelihood, we have

$$TIC(\boldsymbol{X}_n; \hat{\boldsymbol{\theta}}_{ML}) = -2\sum_{\alpha=1}^{n}\log f(X_\alpha|\hat{\boldsymbol{\theta}}_{ML}) + 2\mathrm{tr}\{\hat{J}(G)^{-1}\hat{I}(G)\} \tag{11}$$

where $\hat{J}$ and $\hat{I}$ are consistent estimates of J and I, respectively. This criterion was originally introduced by Takeuchi (1976) and also by Stone (1977), and was later discussed extensively by Shibata (1989).

It is now assumed that the true distribution belongs to the parametric family of densities. In this special situation, we have the well-known identity

$$\int \frac{\partial \log f(z|\boldsymbol{\theta})}{\partial\theta_i}\frac{\partial \log f(z|\boldsymbol{\theta})}{\partial\theta_j} f(z|\boldsymbol{\theta})\mathrm{d}z = -\int \frac{\partial^2 \log f(z|\boldsymbol{\theta})}{\partial\theta_i\partial\theta_j} f(z|\boldsymbol{\theta})\mathrm{d}z$$

for $i, j = 1, \ldots, p$, so $I(F_{\boldsymbol{\theta}_0}) = J(F_{\boldsymbol{\theta}_0})$, called the Fisher information matrix. Hence the asymptotic decomposition equation (9) of the bias $b(G)$ can be reduced to

$$b(G) = \frac{1}{n}\mathrm{tr}\{J(F_{\boldsymbol{\theta}_0})^{-1}I(F_{\boldsymbol{\theta}_0})\} + O(n^{-2}) = \frac{1}{n}p + O(n^{-2})$$

and we then have AIC given by equation (8).

(3) Generalized information criterion (Konishi and Kitagawa, 1996). To construct an information-theoretic criterion which enables us to evaluate various types of statistical models, we have to remove both of the assumptions imposed on AIC. Konishi and Kitagawa (1996) considered this problem by employing a functional estimator $\hat{\boldsymbol{\theta}} = \boldsymbol{T}(\hat{G})$ with Fisher consistency, where $\boldsymbol{T}(\cdot) = (T_1(\cdot), \ldots, T_p(\cdot))' \in R^p$ is a functional on the space of distribution functions, and obtained the generalized information criterion in the form

$$GIC(\boldsymbol{X}_n; \hat{\boldsymbol{\theta}}) = -2\sum_{\alpha=1}^{n} \log f(X_\alpha \mid \hat{\boldsymbol{\theta}}) + \frac{2}{n}\sum_{\alpha=1}^{n} \mathrm{tr}\left\{ \boldsymbol{T}^{(1)}(X_\alpha; \hat{G}) \left.\frac{\partial \log f(X_\alpha \mid \boldsymbol{\theta})}{\partial \boldsymbol{\theta}'}\right|_{\hat{\boldsymbol{\theta}}} \right\} \tag{12}$$

Here $\boldsymbol{T}^{(1)}(X_\alpha; \hat{G}) = (T_1^{(1)}(X_\alpha; \hat{G}), \ldots, T_p^{(1)}(X_\alpha; \hat{G}))'$ and $T_i^{(1)}(X_\alpha; \hat{G})$ is the empirical influence function defined by

$$T_i^{(1)}(X_\alpha; \hat{G}) = \lim_{\varepsilon \to 0} \frac{T_i((1-\varepsilon)\hat{G} + \varepsilon\delta_\alpha) - T_i(\hat{G})}{\varepsilon}$$

with δ_α being a point mass at X_α.

We observe that the bias $b(G)$ in equation (6) is approximated as a function of the empirical influence function of the estimator and the score function of the parametric model. Simple examples of functional estimators and their influence functions are given in the next subsection.

The maximum likelihood estimator, $\hat{\boldsymbol{\theta}}_{ML}$, is given as the solution of the likelihood equation

$$\sum_{\alpha=1}^{n} \left.\frac{\partial \log f(X_\alpha | \boldsymbol{\theta})}{\partial \boldsymbol{\theta}}\right|_{\hat{\boldsymbol{\theta}}_{ML}} = \boldsymbol{0}$$

The solution $\hat{\boldsymbol{\theta}}_{ML}$ can be written as $\hat{\boldsymbol{\theta}}_{ML} = \boldsymbol{T}_{ML}(\hat{G})$, where $\boldsymbol{T}_{ML}$ is the p-dimensional functional implicitly defined by

$$\int \left.\frac{\partial \log f(z|\boldsymbol{\theta})}{\partial \boldsymbol{\theta}}\right|_{\boldsymbol{T}_{ML}(G)} \mathrm{d}G(z) = \boldsymbol{0} \tag{13}$$

Replacing G in equation (13) by $G_\varepsilon = (1-\varepsilon)G + \varepsilon\,\delta_\alpha$ and differentiating with respect to ε yields the p-dimensional influence function of the maximum likelihood estimator $\hat{\boldsymbol{\theta}}_{ML} = \boldsymbol{T}_{ML}(\hat{G})$ in the form

$$\boldsymbol{T}_{ML}^{(1)}(z; G) = J(G)^{-1} \left.\frac{\partial \log f(z|\boldsymbol{\theta})}{\partial \boldsymbol{\theta}}\right|_{\boldsymbol{T}_{ML}(G)}$$

where $J(G)$ is defined in equation (10). Taking $\boldsymbol{T}_{ML}^{(1)}(z; G)$ in equation (12), we have the asymptotic bias $\mathrm{tr}J(G)^{-1}I(G)/n$ for the log likelihood of the model $f(z|\hat{\boldsymbol{\theta}}_{ML})$. If $J(G)$ and $I(G)$ are estimated by $J(\hat{G})$ and $I(\hat{G})$, respectively, then we have

$$GIC(\boldsymbol{X}_n; \hat{\boldsymbol{\theta}}_{ML}) = TIC(\boldsymbol{X}_n; \hat{\boldsymbol{\theta}}_{ML}) \tag{14}$$

The detailed derivation and asymptotic properties of the information-theoretic criteria AIC, TIC, and GIC will be given in Section 4.

(4) Bootstrap bias-corrected log likelihood. Advances in the performance of computers in recent years enable us to construct complicated models for analyzing data with complex structure, and consequently we need to solve complex inferential problems including estimation and evaluation. For example, finite mixtures of multivariate distributions are widely used as models in various fields of statistics and the EM algorithm (Dempster et al., 1977) is used for their estimation. When applying the information criteria to the problem of estimating the number of components, it seems to be difficult to derive asymptotic biases analytically. In such cases the bootstrap methods, introduced by Efron (1979), provide a useful tool for solving problems numerically.

Conditional on the observed data $\boldsymbol{X}_n$, let $\boldsymbol{X}_n^* = \{X_1^*, \ldots, X_n^*\}$ be the bootstrap sample generated according to the empirical distribution $\hat{G}$ of $\boldsymbol{X}_n$. The basic idea behind the bootstrap methods is to consider $\hat{G}$ to be the true distribution, so the model constructed is $f(z|\hat{\boldsymbol{\theta}}^*)$ where $\hat{\boldsymbol{\theta}}^* = \hat{\boldsymbol{\theta}}(\boldsymbol{X}_n^*)$. Hence the bootstrap analogue of the expected log likelihood of $f(z|\hat{\boldsymbol{\theta}})$ defined by equation (4) is

$$\eta(\hat{G}; \hat{\boldsymbol{\theta}}^*) = \int \log f(z|\hat{\boldsymbol{\theta}}^*)\mathrm{d}\hat{G}(z) = \frac{1}{n}\sum_{\alpha=1}^{n} \log f(X_\alpha|\hat{\boldsymbol{\theta}}^*)$$

By noting that the same bootstrap sample $\boldsymbol{X}_n^*$ is used to estimate the expected log likelihood, the log likelihood of $f(z|\hat{\boldsymbol{\theta}}^*)$ is constructed as

$$\eta(\hat{G}^*; \hat{\boldsymbol{\theta}}^*) = \int \log f(z|\hat{\boldsymbol{\theta}}^*)\mathrm{d}\hat{G}^*(z) = \frac{1}{n}\sum_{\alpha=1}^{n} \log f(X_\alpha^*|\hat{\boldsymbol{\theta}}^*)$$

where $\hat{G}^*$ is the empirical distribution function of the bootstrap sample $\boldsymbol{X}_n^*$. Then the bootstrap bias estimate of the log likelihood is given by

$$\hat{b}_B(\hat{G}) = \mathrm{E}_{\hat{G}}[\eta(\hat{G}^*; \hat{\boldsymbol{\theta}}^*) - \eta(\hat{G}; \hat{\boldsymbol{\theta}}^*)] \tag{15}$$

the bootstrap estimate of $b(G) = \mathrm{E}_G[\eta(\hat{G}; \hat{\boldsymbol{\theta}}) - \eta(G; \hat{\boldsymbol{\theta}})]$ in equation (6).

The bias $\hat{b}_B(\hat{G})$ is approximated by averaging $\eta(\hat{G}^*; \hat{\boldsymbol{\theta}}^*) - \eta(\hat{G}; \hat{\boldsymbol{\theta}}^*)$ over a large number of repeated bootstrap samples. By subtracting the bootstrap bias estimate from the log likelihood, we have

$$\mathrm{EIC}(\boldsymbol{X}_n; \hat{\boldsymbol{\theta}}) = -2\sum_{\alpha=1}^{n} \log f(X_\alpha|\hat{\boldsymbol{\theta}}) + 2n\hat{b}_B(\hat{G})$$

The use of the bootstrap in model evaluation problems was introduced by Wong (1983) and Efron (1986). Ishiguro et al. (1997) examined the performance of the bootstrap bias corrected information criterion called EIC through numerical examples and the analysis of real data (see also Shibata, 1997). Konishi and Kitagawa (1996) proposed a method of reducing the variance of bootstrap bias estimate caused by simulation.

The predictive density function in the Bayesian framework is defined by

$$h(z|\boldsymbol{X}_n) = \int f(z|\boldsymbol{\theta})\pi(\boldsymbol{\theta}|\boldsymbol{X}_n)\mathrm{d}\boldsymbol{\theta}$$

where $\pi(\boldsymbol{\theta}|\boldsymbol{X}_n)$ is the posterior density function for $\boldsymbol{\theta}$ based on a prior $\pi(\boldsymbol{\theta})$ and the data $\boldsymbol{X}_n$. If the Bayesian predictive density is approximated by $f(z|\hat{\boldsymbol{\theta}})$ with an error of order $O_p(n^{-1})$, that is, $h(z|\boldsymbol{X}_n) = f(z|\hat{\boldsymbol{\theta}}) + O_p(n^{-1})$, then the same argument as given in this subsection may be applied to the problem of evaluating the goodness of the predictive density $h(z|\boldsymbol{X}_n)$ (see Konishi and Kitagawa, 1996, p. 883).

2.4 Examples

EXAMPLE 1: Suppose that n independent observations $\{x_1, \ldots, x_n\}$ are generated from the true distribution $G(x)$ having the density function $g(x)$. Consider, as a candidate model, a parametric family of normal densities

$$f(x|\boldsymbol{\theta}) = \frac{1}{\sigma}\phi\left(\frac{x-\mu}{\sigma}\right) = \frac{1}{\sqrt{2\pi\sigma^2}}\exp\left\{-\frac{(x-\mu)^2}{2\sigma^2}\right\}, \qquad \boldsymbol{\theta} = (\mu, \sigma^2)' \in \Theta$$

If the parametric model is correctly specified, the family $\{f(x|\boldsymbol{\theta});\ \boldsymbol{\theta} \in \Theta\}$ contains the true density as an element $g(x) = \sigma_0^{-1}\phi((x-\mu_0)/\sigma_0)$ for some $\boldsymbol{\theta}_0 = (\mu_0, \sigma_0^2)' \in \Theta$. The statistical model estimated by the method of maximum likelihood is

$$f(x|\hat{\boldsymbol{\theta}}) = \frac{1}{s}\phi\left(\frac{x-\bar{x}_n}{s}\right) = \frac{1}{\sqrt{2\pi s^2}}\exp\left\{-\frac{(x-\bar{x}_n)^2}{2s^2}\right\}, \qquad \hat{\boldsymbol{\theta}} = (\bar{x}_n, s^2)' \tag{16}$$

where $\bar{x}_n$ and s^2 are the sample mean and variance, respectively. Then for the log likelihood

$$\eta(\hat{G};\hat{\boldsymbol{\theta}}) = \frac{1}{n}\sum_{\alpha=1}^{n}\log f(x_\alpha|\hat{\boldsymbol{\theta}}) = -\frac{1}{2}(c+\log s^2) \tag{17}$$

with $c = \log(2\pi)+1$, we have the information criteria AIC, TIC, and GIC introduced in the last subsection as follows.
(i) AIC. Since the two parameters (μ, σ^2) are estimated, we have, under the assumption that the true density is $g(x) = \sigma_0^{-1}\phi((x-\mu_0)/\sigma_0)$,

$$\mathrm{AIC} = n(c+\log s^2) + 2\times 2$$

(ii) TIC. For the criterion TIC defined by equation (11), $I(G)$ and $J(G)$ in the bias of the log likelihood can be calculated as

$$I(G) = E_G\left[\frac{\partial \log f(z|\boldsymbol{\theta})}{\partial \boldsymbol{\theta}}\frac{\partial \log f(z|\boldsymbol{\theta})}{\partial \boldsymbol{\theta}'}\right]$$

$$= \begin{bmatrix} \dfrac{1}{\sigma^2(G)} & \dfrac{\mu_3(G)}{2\sigma^6(G)} \\ \dfrac{\mu_3(G)}{2\sigma^6(G)} & -\dfrac{1}{4\sigma^4(G)}\left(1-\dfrac{\mu_4(G)}{\sigma^4(G)}\right) \end{bmatrix}$$

$$J(G) = -E_G\left[\frac{\partial^2 \log f(z|\boldsymbol{\theta})}{\partial\boldsymbol{\theta}\partial\boldsymbol{\theta}'}\right] = \begin{bmatrix} \dfrac{1}{\sigma^2(G)} & 0 \\ 0 & \dfrac{1}{2\sigma^4(G)} \end{bmatrix}$$

where $\sigma^2(G)$, $\mu_3(G)$, and $\mu_4(G)$ are the variance, and the third and fourth central moments, respectively, of the true distribution G, not the parametric model $f(x|\boldsymbol{\theta})$. Hence the asymptotic bias of the log likelihood is

$$\mathrm{tr}J(G)^{-1}I(G) = \frac{1}{2} + \frac{\mu_4(G)}{2\sigma^4(G)}$$

Estimating $\mu_4(G)$ and $\sigma^4(G)$ by

$$\mu_4(\hat{G}) = \frac{1}{n}\sum_{\alpha=1}^{n}(x_\alpha - \bar{x}_n)^4 \quad \text{and} \quad \sigma^4(\hat{G}) = s^4$$

we have from equation (11) that

$$\mathrm{TIC} = n(c+\log s^2) + 2\left\{\frac{1}{2} + \frac{1}{2ns^4}\sum_{\alpha=1}^{n}(x_\alpha - \bar{x}_n)^4\right\}$$

In a particular situation where the parametric model contains the true density, that is, $g(x) = \sigma_0^{-1}\phi((x-\mu_0)/\sigma_0)$ for some $\boldsymbol{\theta}_0 = (\mu_0, \sigma_0^2)' \in \Theta$, the fourth central moment μ_4 equals $3\sigma_0^4$, and hence we have

$$\frac{1}{2} + \frac{\mu_4}{2\sigma_0^4} = 2$$

the asymptotic bias for AIC.

(iii) GIC. In GIC the asymptotic bias of the log likelihood of the statistical model $f(x|\hat{\boldsymbol{\theta}})$ was derived in the context of functional statistics. The sample mean $\bar{x}_n$ can be written as $\bar{x}_n = T_\mu(\hat{G})$ for the functional given by

$$T_\mu(G) = \int x\mathrm{d}G(x) \tag{18}$$

It can easily be seen that the influence function of the sample mean at the distribution G is

$$T_\mu^{(1)}(z; G) = \lim_{\varepsilon \to 0} \frac{T_\mu((1-\varepsilon)G + \varepsilon\delta_z) - T_\mu(G)}{\varepsilon} = z - T_\mu(G)$$

with δ_α being a point mass at z. The functional defined sample variance $s^2 = T_\sigma(\hat{G})$ and its influence function are, respectively,

$$T_\sigma(G) = \frac{1}{2}\int\int (x-y)^2 \mathrm{d}G(x)\mathrm{d}G(y) \quad \text{and}$$
$$T_\sigma^{(1)}(z; G) = (z - T_\mu(G))^2 - T_\sigma(G)$$

Then $\boldsymbol{T}_{ML}(G)$ satisfying equation (13) is given by $\boldsymbol{T}_{ML}(G) = (T_\mu(G), T_\sigma(G))'$, and we have

$$\left.\frac{\partial \log f(z|\boldsymbol{\theta})}{\partial\mu}\right|_{\boldsymbol{T}_{ML}(G)} = \frac{z - T_\mu(G)}{T_\sigma(G)},$$
$$\left.\frac{\partial \log f(z|\boldsymbol{\theta})}{\partial\sigma^2}\right|_{\boldsymbol{T}_{ML}(G)} = -\frac{1}{2T_\sigma(G)} + \frac{(z - T_\mu(G))^2}{2T_\sigma^2(G)}$$

Evaluating at $\hat{\boldsymbol{\theta}}_{ML} = \boldsymbol{T}_{ML}(\hat{G})$, the expression defining GIC's bias correction in equation (12) is given by

$$\sum_{\alpha=1}^{n} \mathrm{tr}\left\{ \boldsymbol{T}^{(1)}(x_\alpha; \hat{G}) \frac{\partial \log f(x_\alpha \mid \boldsymbol{\theta})}{\partial \boldsymbol{\theta}'}\bigg|_{\hat{\boldsymbol{\theta}}} \right\}$$

$$= \sum_{\alpha=1}^{n} T_\mu^{(1)}(x_\alpha; \hat{G}) \frac{\partial \log f(x_\alpha|\boldsymbol{\theta})}{\partial \mu}\bigg|_{\boldsymbol{T}_{ML}(\hat{G})}$$

$$+ \sum_{\alpha=1}^{n} T_\sigma^{(1)}(x_\alpha; \hat{G}) \frac{\partial \log f(x_\alpha|\boldsymbol{\theta})}{\partial \sigma^2}\bigg|_{\boldsymbol{T}_{ML}(\hat{G})}$$

$$= n\left\{ \frac{1}{2} + \frac{1}{2ns^4} \sum_{\alpha=1}^{n} (x_\alpha - \bar{x}_n)^4 \right\} \tag{19}$$

Hence, it follows from equation (12) that we have

$$\mathrm{GIC} = n(c + \log s^2) + 2\left\{ \frac{1}{2} + \frac{1}{2ns^4} \sum_{\alpha=1}^{n} (x_\alpha - \bar{x}_n)^4 \right\}$$

with c given in equation (17), and consequently the result as in equation (14).

EXAMPLE 2: *Linear model.* Suppose that we have n independent observations $\{y_1, \ldots, y_n\}$, and that the distribution of y_i for the i-th response is $g(y_i)$. Consider the Gaussian linear regression model

$$y_i = \boldsymbol{x}_i'\boldsymbol{\beta} + \varepsilon_i, \qquad \varepsilon_i \sim n(0, \sigma^2) \quad (i = 1, \ldots, n) \tag{20}$$

where $\boldsymbol{x}_i$ is a known $p \times 1$ vector of covariates and $\boldsymbol{\beta}$ is a $p \times 1$ vector of unknown parameters. The maximum likelihood estimates of the parameters $\boldsymbol{\theta} = (\boldsymbol{\beta}', \sigma^2)' \in \Theta \subset R^{p+1}$ are given by

$$\hat{\boldsymbol{\beta}} = (X'X)^{-1}X'\boldsymbol{y} \qquad \text{and} \qquad \hat{\sigma}^2 = \frac{1}{n}(\boldsymbol{y} - X\hat{\boldsymbol{\beta}})'(\boldsymbol{y} - X\hat{\boldsymbol{\beta}}) \tag{21}$$

where $\boldsymbol{y} = (y_1, \ldots, y_n)'$ and $X = (\boldsymbol{x}_1, \ldots, \boldsymbol{x}_n)'$. Then the statistical model estimated by the maximum likelihood method is

$$f(z|\hat{\boldsymbol{\theta}}) = \left(2\pi\hat{\sigma}^2\right)^{-1/2} \exp\left\{ -\frac{1}{2\hat{\sigma}^2}(z - \boldsymbol{x}'\hat{\boldsymbol{\beta}})^2 \right\}$$

where $\hat{\boldsymbol{\theta}} = (\hat{\boldsymbol{\beta}}', \hat{\sigma}^2)'$.

The first and second derivatives of $\ell_f(\boldsymbol{\theta}) = \log f(z|\boldsymbol{\theta})$ are given by

$$\frac{\partial \ell_f(\boldsymbol{\theta})}{\partial \boldsymbol{\beta}} = \frac{1}{\sigma^2}\boldsymbol{x}(z - \boldsymbol{x}'\boldsymbol{\beta}), \qquad \frac{\partial \ell_f(\boldsymbol{\theta})}{\partial \sigma^2} = -\frac{1}{2\sigma^2}\left\{ 1 - \frac{1}{\sigma^2}(z - \boldsymbol{x}'\boldsymbol{\beta})^2 \right\}$$

$$\frac{\partial^2 \ell_f(\boldsymbol{\theta})}{\partial \boldsymbol{\beta} \partial \boldsymbol{\beta}'} = -\frac{1}{\sigma^2}\boldsymbol{x}\boldsymbol{x}', \qquad \frac{\partial^2 \ell_f(\boldsymbol{\theta})}{\partial \sigma^2 \partial \sigma^2} = \frac{1}{2\sigma^4}\left\{1 - \frac{2}{\sigma^2}(z - \boldsymbol{x}'\boldsymbol{\beta})^2\right\}$$

$$\frac{\partial^2 \ell_f(\boldsymbol{\theta})}{\partial \boldsymbol{\beta} \partial \sigma^2} = -\frac{1}{\sigma^4}\boldsymbol{x}(z - \boldsymbol{x}'\boldsymbol{\beta})$$

We then have

$$n\hat{J}(G) = -\sum_{\alpha=1}^{n} \left.\frac{\partial^2 \log f(y_\alpha|\boldsymbol{\theta})}{\partial \boldsymbol{\theta} \partial \boldsymbol{\theta}'}\right|_{\hat{\boldsymbol{\theta}}} = \begin{bmatrix} \frac{1}{\hat{\sigma}^2}X'X & 0 \\ 0 & \frac{n}{2\hat{\sigma}^4} \end{bmatrix}$$

$$n\hat{I}(G) = \sum_{\alpha=1}^{n} \left.\frac{\partial \log f(y_\alpha|\boldsymbol{\theta})}{\partial \boldsymbol{\theta}} \frac{\partial \log f(y_\alpha|\boldsymbol{\theta})}{\partial \boldsymbol{\theta}'}\right|_{\hat{\boldsymbol{\theta}}}$$
$$= \begin{bmatrix} \frac{1}{\hat{\sigma}^4}X'D^2X & \frac{1}{2\hat{\sigma}^6}\sum_{\alpha}^{n}\boldsymbol{x}_\alpha e_\alpha^3 \\ \frac{1}{2\hat{\sigma}^6}\sum_{\alpha}^{n}\boldsymbol{x}'_\alpha e_\alpha^3 & \frac{1}{4\hat{\sigma}^4}\left\{\frac{1}{\hat{\sigma}^4}\sum_{\alpha}^{n} e_\alpha^4 - n\right\} \end{bmatrix}$$

where $e_\alpha = y_\alpha - \boldsymbol{x}'_\alpha\hat{\boldsymbol{\beta}}$ and $D^2 = \mathrm{diag}[e_1^2, \ldots, e_n^2]$. Hence it follows from equation (11) that

$$\mathrm{TIC} = n\{\log(2\pi) + 1\} + n\log\hat{\sigma}^2 + 2\left[\frac{1}{\hat{\sigma}^2}\mathrm{tr}\{X(X'X)^{-1}X'D^2\} + \frac{1}{2n}\left(\frac{1}{\hat{\sigma}^4}\sum_{\alpha=1}^{n} e_\alpha^4 - n\right)\right]$$

EXAMPLE 3: *Numerical comparison.* Suppose that the true density $g(x)$ and the parametric model $f(x|\boldsymbol{\theta})$ are, respectively,

$$g(x) = (1-\varepsilon)\frac{1}{\sigma_{01}}\phi\left(\frac{x-\mu_{01}}{\sigma_{01}}\right) + \varepsilon\frac{1}{\sigma_{02}}\phi\left(\frac{x-\mu_{02}}{\sigma_{02}}\right), \quad 0 \le \varepsilon \le 1 \tag{22}$$
$$f(x|\boldsymbol{\theta}) = \frac{1}{\sigma}\phi\left(\frac{x-\mu}{\sigma}\right), \qquad \boldsymbol{\theta} = (\mu, \sigma^2)'$$

where $\phi(x)$ denotes the density function of a standard normal distribution. The statistical model is constructed based on n independent observations from the mixture distribution $g(x)$, and is given by equation (16).

Under this situation, the expected log likelihood in equation (4) can be written as

$$\int g(z)\log f(z|\bar{x}_n, s^2)dz = -\frac{1}{2}\log(2\pi) - \frac{1}{2}\log s^2 - \frac{1}{2s^2}\int (z-\bar{x}_n)^2 g(z)dz$$
$$= -\frac{1}{2}\log(2\pi) - \frac{1}{2}\log s^2$$
$$-\frac{1}{2s^2}[(1-\varepsilon)\{\sigma_{01}^2 + (\mu_{01}-\bar{x}_n)^2\}$$
$$+\varepsilon\{\sigma_{02}^2 + (\mu_{02}-\bar{x}_n)^2\}]$$

From equation (9) and the formula for $trJ(G)^{-1}I(G)$ derived in Example 1 (ii), the bias of the log likelihood in estimating this expected log likelihood is approximated by

$$b(G) = E_G\left[\frac{1}{n}\sum_{i=1}^{n}\log f(X_i|\overline{X}, S^2) - \int g(z)\log f(z|\overline{X}, S^2)dz\right]$$
$$= E_G\left[-\frac{1}{2} + \frac{1}{2S^2}\{(1-\varepsilon)(\sigma_{01}^2 + (\mu_{01}-\overline{X}_n)^2)\right.$$
$$\left.+\varepsilon(\sigma_{02}^2 + (\mu_{02}-\overline{X}_n)^2)\}\right]$$
$$\approx \frac{1}{n}\left\{\frac{1}{2} + \frac{\mu_4(G)}{2\sigma^4(G)}\right\}$$

where $\sigma^2(G)$ and $\mu_4(G)$ are the variance and fourth central moment of the mixture distribution $g(x)$, respectively. Hence, we have the bias estimate

$$nb(\hat{G}) \approx \frac{1}{2} + \frac{1}{2ns^4}\sum_{\alpha=1}^{n}(x_\alpha - \bar{x}_n)^4 \tag{23}$$

A Monte Carlo simulation was performed to examine the accuracy of the asymptotic biases. Repeated random samples were generated from a mixture of normal distributions in equation (22) for different combinations of parameters, in which we took (i) $(\mu_{01}, \mu_{02}, \sigma_{01}, \sigma_{02}) = (0, 0, 1, 3)$ in Figure 1, and (ii) $(\mu_{01}, \mu_{02}, \sigma_{01}, \sigma_{02}) = (0, 5, 1, 1)$ in Figure 2.

Figures 1 and 2 plot the true bias $nb(G)$ and the asymptotic bias estimate given by equation (23) with standard errors for various values of mixing proportion ε. The quantities are evaluated by a Monte Carlo simulation with 100,000 repetitions.

It can be seen from the figures that the log likelihood of a fitted model has a significant bias as an estimate of the expected log likelihood, and that the bias is considerably larger than 2, the approximation by AIC, if the values of

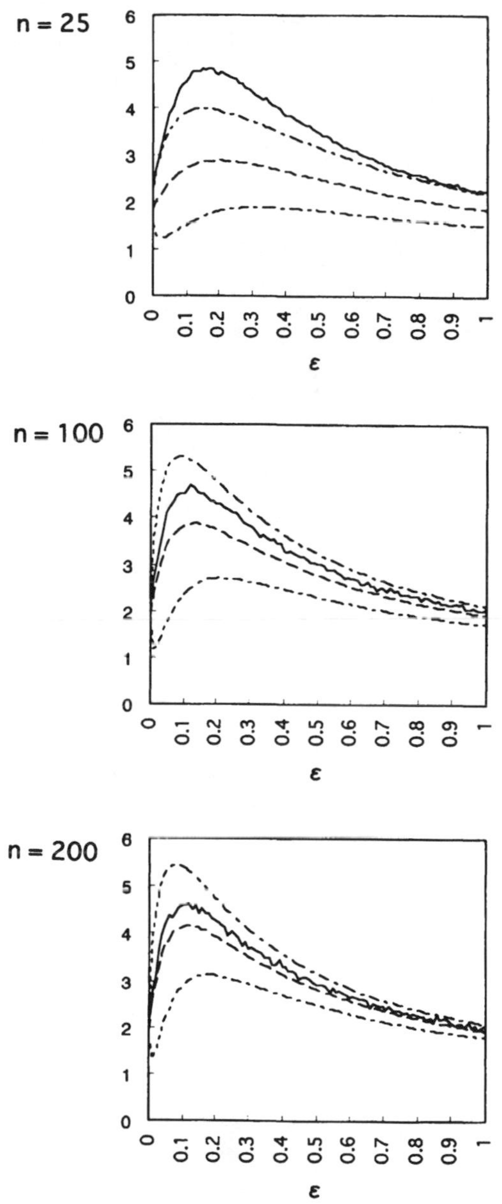

Figure 1 Comparison of the true bias $n \times b(G)$ (——) and the estimated asymptotic bias (– – –) with standard errors (- - - - -) for the sample sizes $n = 25$, 100, and 200

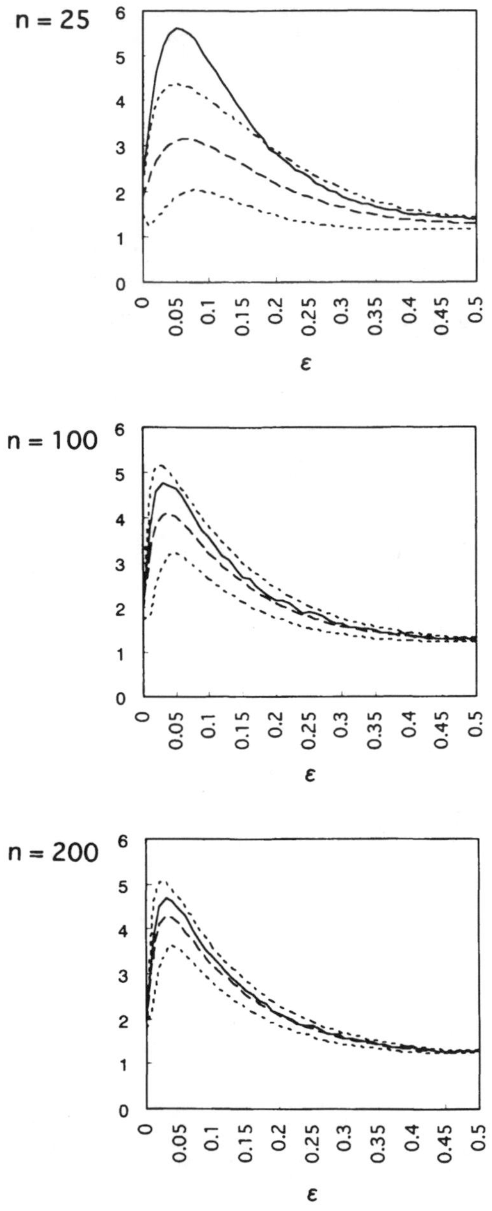

Figure 2 Comparison of the true bias $n \times b(G)$ (——) and the estimated asymptotic bias (– – –) with standard errors (- - - - -) for the sample sizes $n = 25$, 100, and 200

the mixing proportion ε are around 0.05–0.1. In the case where $\varepsilon = 0$ or 1, the true distribution $g(x)$ belongs to the specified parametric model and the bias is well approximated by the number of estimated parameters. We also see that for larger sample sizes, the true bias and the estimated asymptotic bias in equation (23) coincide well. On the other hand, for smaller sample sizes such as $n = 25$, the estimated asymptotic bias still underestimates the true bias.

3. APPLICATIONS OF GENERALIZED INFORMATION CRITERION

In this section the generalized information criterion GIC given by equation (12) in Section 2 is applied to the evaluation of statistical models estimated by robust and maximum penalized likelihood procedures.

3.1 Statistical Models Estimated by Robust Procedures

We recall that the maximum likelihood estimator, $\hat{\boldsymbol{\theta}}_{ML}$, defined as the solution of the likelihood equation

$$\sum_{\alpha=1}^{n} \left. \frac{\partial \log f(X_\alpha|\boldsymbol{\theta})}{\partial \theta_i} \right|_{\hat{\boldsymbol{\theta}}_{ML}} = 0, \qquad (i = 1, \ldots, p) \tag{24}$$

can be written as $\hat{\boldsymbol{\theta}}_{ML} = \boldsymbol{T}_{ML}(\hat{G})$, where $\boldsymbol{T}_{ML}$ is the p-dimensional functional implicitly defined by

$$\int \left. \frac{\partial \log f(z|\boldsymbol{\theta})}{\partial \theta_i} \right|_{\boldsymbol{T}_{ML}(G)} \mathrm{d}G(z) = 0, \qquad (i = 1, \ldots, p) \tag{25}$$

Huber (1964) generalized the maximum likelihood estimator to the M-estimator, $\hat{\boldsymbol{\theta}}_M$, defined as the solution of the implicit equations

$$\sum_{\alpha=1}^{n} \psi_i(X_\alpha, \hat{\boldsymbol{\theta}}_M) = 0, \qquad (i = 1, \ldots, p) \tag{26}$$

with ψ_is being some function on $\mathcal{X} \times \Theta(\Theta \subset R^p)$. Let $\boldsymbol{\psi} = (\psi_1, \ldots, \psi_p)'$ be the p-dimensional $\boldsymbol{\psi}$-function. Then the M-estimator $\hat{\boldsymbol{\theta}}_M$ can be expressed as $\hat{\boldsymbol{\theta}}_M = \boldsymbol{T}_M(\hat{G})$ for the p-dimensional functional $\boldsymbol{T}_M$ given by

$$\int \psi_i(z, \boldsymbol{T}_M(G))\mathrm{d}G(z) = 0, \qquad (i = 1, \ldots, p) \tag{27}$$

corresponding to the functional $\boldsymbol{T}_{ML}$ in equation (25) for the maximum likelihood estimator. Equations (24) and (26) can be obtained by replacing

G in equations (25) and (27), respectively, by the empirical distribution function $\hat{G}$.

It is known (see, e.g., Hampel et al., 1986, p. 230) that the influence function of the M-estimator is

$$\boldsymbol{T}_M^{(1)}(z; G) = M(\boldsymbol{\psi}, G)^{-1}\boldsymbol{\psi}(z, \boldsymbol{T}_M(G)) \tag{28}$$

where $M(\boldsymbol{\psi}, G)$ is the $p \times p$ nonsingular matrix given by

$$M(\boldsymbol{\psi}, G) = -\int \left.\frac{\partial\boldsymbol{\psi}(z, \boldsymbol{\theta})'}{\partial\boldsymbol{\theta}}\right|_{\boldsymbol{T}_M(G)} \mathrm{d}G(z) \tag{29}$$

Substituting the influence function equation (28) into GIC given by equation (12), we have an information criterion for the evaluation of the statistical model $f(z|\hat{\boldsymbol{\theta}}_M)$ estimated by robust procedures in the form

$$-2\sum_{\alpha=1}^{n} \log f(X_\alpha|\hat{\boldsymbol{\theta}}_M) + \frac{2}{n}\sum_{\alpha=1}^{n} \operatorname{tr}\left\{ M(\boldsymbol{\psi}, \hat{G})^{-1}\boldsymbol{\psi}(X_\alpha, \hat{\boldsymbol{\theta}}_M)\left.\frac{\partial \log f(X_\alpha|\boldsymbol{\theta})}{\partial\boldsymbol{\theta}'}\right|_{\hat{\boldsymbol{\theta}}_M}\right\} \tag{30}$$

where

$$M(\boldsymbol{\psi}, \hat{G}) = -\frac{1}{n}\sum_{\alpha=1}^{n} \left.\frac{\partial\boldsymbol{\psi}(X_\alpha, \boldsymbol{\theta})'}{\partial\boldsymbol{\theta}}\right|_{\hat{\boldsymbol{\theta}}_M}$$

We now consider the situations that the parametric family of densities $\{f(x|\boldsymbol{\theta}); \boldsymbol{\theta} \in \Theta \subset R^p\}$ contains the true density and that the functional $\boldsymbol{T}_M$ is Fisher consistent, so that $\boldsymbol{T}_M(F_{\boldsymbol{\theta}}) = \boldsymbol{\theta}$ for all $\boldsymbol{\theta} \in \Theta$, where $F_{\boldsymbol{\theta}}(x)$ is the distribution function of $f(x|\boldsymbol{\theta})$. Then the matrix M in equation (29) can be rewritten as

$$M(\boldsymbol{\psi}, F_{\boldsymbol{\theta}}) = \int \boldsymbol{\psi}(z, \boldsymbol{\theta})\frac{\partial \log f(z \mid \boldsymbol{\theta})}{\partial\boldsymbol{\theta}'}\mathrm{d}F_{\boldsymbol{\theta}}(z) \tag{31}$$

(see Hampel et al., 1986, p. 231). By replacing $M(\boldsymbol{\psi}, \hat{G})$ in equation (30) by $M(\boldsymbol{\psi}, \hat{F}_{\boldsymbol{\theta}})$, we see that the information criterion in equation (30) can be reduced to

$$-2\sum_{\alpha=1}^{n} \log f(X_\alpha|\hat{\boldsymbol{\theta}}_M) + 2p \tag{32}$$

where p is the number of estimated parameters within the model $f(x|\hat{\boldsymbol{\theta}}_M)$. We observe that AIC may be used for evaluating statistical models estimated by robust procedures.

The maximum likelihood estimator is an M-estimator, corresponding to

$$\psi(x, \boldsymbol{T}_{ML}(G)) = \left.\frac{\partial \log f(x|\boldsymbol{\theta})}{\partial \boldsymbol{\theta}}\right|_{\boldsymbol{T}_{ML}(G)}$$

Hence TIC given by equation (11) can also be obtained by taking $\psi(x, \boldsymbol{T}_{ML}(G))$ in equation (30) as the ψ-function, since $M(\psi, \hat{G}) = J(\hat{G})$ for $J(G)$ defined in equation (10).

EXAMPLE 4: Consider the parametric model $F_{\boldsymbol{\theta}}(x) = \Phi((x-\mu)/\sigma)$, where Φ is the standard normal distribution function. It is assumed that the parametric family of distributions $\{F_{\boldsymbol{\theta}}(x); \boldsymbol{\theta} \in \Theta \subset R^2\}$ contains the true distribution generating the data $\boldsymbol{X}_n = \{X_1, \ldots, X_n\}$. The location and scale parameters are estimated by the median, $\hat{\mu}_m$, and the median absolute deviation, $\hat{\sigma}_m$, respectively, given by

$$\hat{\mu}_m = \text{med}_i\{X_i\} \qquad \text{and} \qquad \hat{\sigma}_m = \frac{1}{c}\text{med}_i\{|X_i - \text{med}_j(X_j)|\}$$

where $c = \Phi^{-1}(0.75)$ is chosen to make $\hat{\sigma}_m$ Fisher consistent for Φ. The M-estimators $(\hat{\mu}_m, \hat{\sigma}_m)$ are defined by the ψ-function

$$\psi(z; \mu, \sigma) = \left(\text{sign}(z-\mu), \qquad c^{-1}\text{sign}(|z-\mu| - c\sigma)\right)'$$

and their influence functions are

$$(T_\mu^{(1)}(z; F_{\boldsymbol{\theta}}),\ T_\sigma^{(1)}(z; F_{\boldsymbol{\theta}})) = \left(\frac{\text{sign}(z-\mu)}{2\phi(0)}, \quad \frac{\text{sign}(|z-\mu| - c\sigma)}{4c\phi(c)}\right)$$

where ϕ is the standard normal density function (see Huber, 1981, p. 137).

Then the bias approximation term in equation (12) for the (average) log likelihood

$$\frac{1}{n}\sum_{\alpha=1}^{n} \log\left\{\frac{1}{\hat{\sigma}_m}\phi\left(\frac{X_\alpha - \hat{\mu}_m}{\hat{\sigma}_m}\right)\right\}$$

is, writing $y = (z-\mu)/\sigma$,

$$\frac{1}{n}\left\{\int \frac{\text{sign}(y)}{2\phi(0)} y d\Phi(y) + \int \frac{\text{sign}(|y| - c)}{4c\phi(c)}(y^2 - 1)d\Phi(y)\right\} = \frac{1}{n}2$$

which yields the result in the criterion (32).

3.2 Maximum Penalized Likelihood Estimation

Konishi and Kitagawa (1996) introduced a criterion for statistical models constructed by maximum penalized likelihood estimation (Good and Gaskins, 1971) in a general situation. We consider the criterion through the special but important case of the logistic regression model and discuss the choice of a smoothing parameter through a model evaluation problem.

Suppose that we have n observations $\{(Y_\alpha, \boldsymbol{x}_\alpha); \alpha = 1, \ldots, n\}$, where Y_α are independent random variables coded as either 0 or 1, and $\boldsymbol{x}_\alpha$ is a vector of p covariates. The logistic regression model assumes that

$$\Pr(Y_\alpha = 1|\boldsymbol{x}_\alpha) = \pi(\boldsymbol{x}_\alpha) \qquad \text{and} \qquad \Pr(Y_\alpha = 0|\boldsymbol{x}_\alpha) = 1 - \pi(\boldsymbol{x}_\alpha)$$

where

$$\pi(\boldsymbol{x}_\alpha) = \frac{\exp(\boldsymbol{x}'_\alpha \boldsymbol{\beta})}{1 + \exp(\boldsymbol{x}'_\alpha \boldsymbol{\beta})}$$

with $\boldsymbol{\beta}$ a p-dimensional parameter vector. Under this model, the log likelihood function for y_α in terms of $\boldsymbol{\beta}$ is

$$\begin{aligned}\ell(\boldsymbol{\beta}) &= \sum_{\alpha=1}^{n} \left[y_\alpha \log \pi(\boldsymbol{x}_\alpha) + (1 - y_\alpha) \log\{1 - \pi(\boldsymbol{x}_\alpha)\} \right] \\ &= -\sum_{\alpha=1}^{n} \left[\log\{1 + \exp(\boldsymbol{x}'_\alpha \boldsymbol{\beta})\} - y_\alpha \boldsymbol{x}'_\alpha \boldsymbol{\beta} \right] \end{aligned} \tag{33}$$

We consider the maximum penalized likelihood estimator, $\hat{\boldsymbol{\beta}}_\lambda$, of the unknown parameter $\boldsymbol{\beta}$ given as the value which maximizes the penalized likelihood

$$\ell_\lambda(\boldsymbol{\beta}) = \ell(\boldsymbol{\beta}) - \frac{n}{2} \lambda \boldsymbol{\beta}' A \boldsymbol{\beta} \tag{34}$$

where λ is a smoothing parameter and A is some fixed $p \times p$ nonnegative-definite matrix. If $\lambda = 0$, the solution $\hat{\boldsymbol{\beta}}_0$ becomes the usual maximum likelihood estimate, while if $\lambda \to +\infty$, the estimates tend to 0.

The first and second derivatives of $\ell_\lambda(\boldsymbol{\beta})$ are

$$\frac{\partial \ell_\lambda(\boldsymbol{\beta})}{\partial \boldsymbol{\beta}} = \sum_{\alpha=1}^{n} [\{y_\alpha - \pi(\boldsymbol{x}_\alpha)\} \boldsymbol{x}_\alpha - \lambda A \boldsymbol{\beta}]$$

$$\frac{\partial^2 \ell_\lambda(\boldsymbol{\beta})}{\partial \boldsymbol{\beta} \partial \boldsymbol{\beta}'} = -\sum_{\alpha=1}^{n} \pi(\boldsymbol{x}_\alpha)\{1 - \pi(\boldsymbol{x}_\alpha)\} \boldsymbol{x}_\alpha \boldsymbol{x}'_\alpha - n\lambda A$$

Hence, put

$$\psi(y_\alpha, \hat{\boldsymbol{\beta}}_\lambda) = \{y_\alpha - \hat{\pi}(\boldsymbol{x}_\alpha)\} \boldsymbol{x}_\alpha - \lambda A \hat{\boldsymbol{\beta}}_\lambda$$

for the ψ-function in equation (30), where

$$\hat{\pi}(\boldsymbol{x}_\alpha) = \frac{\exp(\boldsymbol{x}'_\alpha \hat{\boldsymbol{\beta}}_\lambda)}{1 + \exp(\boldsymbol{x}'_\alpha \hat{\boldsymbol{\beta}}_\lambda)}$$

Then we have a criterion for evaluating the model $f(y|\hat{\boldsymbol{\beta}}_\lambda)$ with maximum penalized likelihood estimate in

$$IC(\lambda) = -2\sum_{\alpha=1}^{n}\left[y_\alpha \log \hat{\pi}(\boldsymbol{x}_\alpha) + (1 - y_\alpha)\log\{1 - \hat{\pi}(\boldsymbol{x}_\alpha)\}\right] + 2b_\lambda \tag{35}$$

where

$$b_\lambda = \frac{1}{n}\sum_{\alpha=1}^{n} \mathrm{tr}\left[M(\psi, \hat{G})^{-1}\psi(y_\alpha, \hat{\boldsymbol{\beta}}_\lambda)\{y_\alpha - \hat{\pi}(\boldsymbol{x}_\alpha)\}\boldsymbol{x}'_\alpha\right]$$

with

$$M(\psi, \hat{G}) = \frac{1}{n}\sum_{\alpha=1}^{n} \hat{\pi}(\boldsymbol{x}_\alpha)\{1 - \hat{\pi}(\boldsymbol{x}_\alpha)\}\boldsymbol{x}_\alpha\boldsymbol{x}'_\alpha + \lambda A$$

We choose the value of λ that minimizes the criterion $IC(\lambda)$.

An interesting application of the penalized likelihood for the logistic regression model was given by le Cessie and van Houwelingen (1992), in which they analyzed the DNA histogram data and investigated the choice of a smoothing parameter based on prediction errors estimated by cross-validation.

4. THEORETICAL DEVELOPMENT AND ASYMPTOTIC PROPERTIES

In Section 2, we gave the fundamental concept of constructing model evaluation criteria based on the Kullback–Leibler (1951) measure of discriminatory information between two probability distributions. The ideas for the construction of AIC given in Akaike (1973) have been reconsidered by Takeuchi (1976), Findley (1985), Bozdogan (1987), Shibata (1983, 1989), Kitagawa and Gersch (1996), Konishi and Kitagawa (1996), Fujikoshi and Satoh (1997), etc., including the applications of information-theoretic criteria for model selection in various fields of statistics. Bozdogan (1988, 1990) proposed an information-theoretic measure of complexity called ICOMP which takes into account lack of fit, lack of parsimony, and the profusion of complexity.

The main aim of this section is to outline a derivation of AIC, TIC, and GIC which is a little different from those in previous papers, except for that of Konishi and Kitagawa (1996). The relationship between the criteria is given from a theoretical point of view. We also examine the asymptotic properties of information criteria in the estimation of the expected log likelihood of a statistical model.

4.1 Derivation of AIC, TIC, and GIC

Let $f(z|\hat{\boldsymbol{\theta}}_{ML})$ be a statistical model fitted to the observed data by maximum likelihood. We derive the bias of the log likelihood of $f(z|\hat{\boldsymbol{\theta}}_{ML})$ in estimating the expected log likelihood $E_G[\log f(Z|\hat{\boldsymbol{\theta}}_{ML})]$ under the assumption that the parametric family does not contain the true density $g(z)$ underlying the observations.

Suppose that there exists $\boldsymbol{\theta}_0$ giving both a global and a local maximum of the expected log likelihood $E_G[\log f(Z|\boldsymbol{\theta})]$ for $f(z|\boldsymbol{\theta})$. Under some regularity conditions, the maximum likelihood estimate $\hat{\boldsymbol{\theta}}_{ML}$ converges to $\boldsymbol{\theta}_0$ as n tends to infinity. It can be seen that in the functional estimation situation, $\hat{\boldsymbol{\theta}}_{ML}$ converges to $\boldsymbol{\theta}_0 = \boldsymbol{T}_{ML}(G)$ for the p-dimensional functional $\boldsymbol{T}_{ML}$ given as the solution of the implicit equation (13).

Let $\ell(\boldsymbol{\theta}) = \sum_{\alpha=1}^{n} \log f(X_\alpha|\boldsymbol{\theta})$ and

$$J(G) = (j_{rs}) = -\frac{1}{n} E_G\left[\frac{\partial^2 \ell(\boldsymbol{\theta})}{\partial\boldsymbol{\theta}\partial\boldsymbol{\theta}'}\right]_{\boldsymbol{\theta}=\boldsymbol{\theta}_0} \qquad \text{and} \qquad J^{-1}(G) = (j^{rs})$$

Then under suitable regularity conditions the r-th element of $\hat{\boldsymbol{\theta}}_{ML} = (\hat{\theta}_1, \ldots, \hat{\theta}_p)'$ can be expanded in a Taylor's series around $\hat{\boldsymbol{\theta}}_{ML} = \boldsymbol{\theta}_0 = (\theta_{01}, \ldots, \theta_{0p})'$ in the form

$$\hat{\theta}_r = \theta_{0r} + \frac{1}{n} a_{r,n}^{(1)} + \frac{1}{n^2} a_{r,n}^{(2)} + o_p(n^{-1}) \tag{36}$$

where

$$a_{r,n}^{(1)} = \sum_{k=1}^{p} j^{rk} \ell_k(\boldsymbol{\theta}_0)$$

$$a_{r,n}^{(2)} = \frac{1}{2n} \sum_{s,t,u,v,w}^{p} j^{ru} j^{sv} j^{tw} E_G[\ell_{uvw}(\boldsymbol{\theta}_0)] \ell_s(\boldsymbol{\theta}_0) \ell_t(\boldsymbol{\theta}_0)$$
$$+ \sum_{u,v,w}^{p} j^{ru} j^{vw} \{\ell_{uv}(\boldsymbol{\theta}_0) + n j_{uv}\} \ell_w(\boldsymbol{\theta}_0)$$

Here, $\ell_k(\boldsymbol{\theta}_0)$, $\ell_{uv}(\boldsymbol{\theta}_0)$, and $\ell_{uvw}(\boldsymbol{\theta}_0)$ are partial derivatives of $\ell(\boldsymbol{\theta})$ evaluated at $\boldsymbol{\theta} = \boldsymbol{\theta}_0$, i.e.,

$$\ell_{i_1 \ldots i_r}(\boldsymbol{\theta}_0) = \sum_{\alpha=1}^{n} \left. \frac{\partial^r \log f(X_\alpha|\boldsymbol{\theta})}{\partial\theta_{i_1} \ldots \theta_{i_r}} \right|_{\boldsymbol{\theta}_0}$$

By expanding $\log f(z|\hat{\boldsymbol{\theta}}_{ML})$ in a Taylor series around $\hat{\boldsymbol{\theta}}_{ML} = \boldsymbol{\theta}_0$ and substituting equation (36) in the resulting expansion, we have stochastic expansions for the expected log likelihood $\eta(G; \hat{\boldsymbol{\theta}}_{ML})$ and the log likelihood $\eta(\hat{G}; \hat{\boldsymbol{\theta}}_{ML})$ as follows:

$$\begin{aligned}\eta(G; \hat{\boldsymbol{\theta}}_{ML}) &= \int g(z) \log f(z|\hat{\boldsymbol{\theta}}_{ML}) \mathrm{d}z \\ &= \int g(z) \log f(z \mid \boldsymbol{\theta}_0) \mathrm{d}z + \frac{1}{n} \sum_{r=1}^{p} a_{r,n}^{(1)} \int g(z) \left.\frac{\partial \log f(z \mid \boldsymbol{\theta})}{\partial \theta_r}\right|_{\boldsymbol{\theta}_0} \mathrm{d}z \\ &+ \frac{1}{n^2} \left\{ \sum_{r=1}^{p} a_{r,n}^{(2)} \int g(z) \left.\frac{\partial \log f(z|\boldsymbol{\theta})}{\partial \theta_r}\right|_{\boldsymbol{\theta}_0} \mathrm{d}z \right. \\ &\left. + \frac{1}{2} \sum_{r=1}^{p} \sum_{s=1}^{p} a_{r,n}^{(1)} a_{s,n}^{(1)} \int g(z) \left.\frac{\partial^2 \log f(z \mid \boldsymbol{\theta})}{\partial \theta_r \partial \theta_s}\right|_{\boldsymbol{\theta}_0} \mathrm{d}z \right\} + o_p(n^{-1}) \\ \eta(\hat{G}; \hat{\boldsymbol{\theta}}_{ML}) &= \frac{1}{n} \sum_{\alpha=1}^{n} \log f(X_\alpha|\hat{\boldsymbol{\theta}}_{ML}) \\ &= \frac{1}{n} \sum_{\alpha=1}^{n} \log f(X_\alpha|\boldsymbol{\theta}_0) + \frac{1}{n^2} \sum_{r=1}^{p} a_{r,n}^{(1)} \ell_r(\boldsymbol{\theta}_0) \\ &+ \frac{1}{n^3} \sum_{r=1}^{p} a_{r,n}^{(2)} \ell_r(\boldsymbol{\theta}_0) + \frac{1}{2n^3} \sum_{r=1}^{p} \sum_{s=1}^{p} a_{r,n}^{(1)} a_{s,n}^{(1)} \ell_{rs}(\boldsymbol{\theta}_0) + o_p(n^{-1})\end{aligned}$$

Noting that

$$E_G[\ell_r(\boldsymbol{\theta}_0)] = 0 \qquad \text{and} \qquad E_G[\ell_r(\boldsymbol{\theta}_0)\ell_s(\boldsymbol{\theta}_0)] = n i_{rs}$$

and taking expectations term by term yield

$$\begin{aligned}&E_G\left[\int g(z) \log f(z|\hat{\boldsymbol{\theta}}_{ML}) \mathrm{d}z\right] \\ &= \int g(z) \log f(z|\boldsymbol{\theta}_0) \mathrm{d}z + \frac{1}{2n} \sum_{r,s,t,u}^{p} j^{rt} j^{su} (-j_{rs}) i_{tu} + o(n^{-1}) \\ &= E_G[\log f(Z|\boldsymbol{\theta}_0)] - \frac{1}{2n} \sum_{r=1}^{p} \sum_{t=1}^{p} j^{rt} i_{tr} + o(n^{-1}) \\ &= E_G[\log f(Z|\boldsymbol{\theta}_0)] - \frac{1}{2n} \mathrm{tr}\{J(G)^{-1} I(G)\} + o(n^{-1}) \qquad (37)\end{aligned}$$

and

$$
\begin{aligned}
E_G&\left[\frac{1}{n}\sum_{\alpha=1}^{n}\log f(X_\alpha|\hat{\boldsymbol{\theta}}_{ML})\right]\\
&=\int g(z)\log f(z|\boldsymbol{\theta}_0)\mathrm{d}z+\frac{1}{n}\left\{\sum_{r=1}^{p}\sum_{s=1}^{p}j^{rs}i_{rs}+\frac{1}{2}\sum_{r,s,t,u}^{p}j^{rt}j^{su}(-j_{rs})i_{tu}\right\}+o(n^{-1})\\
&=E_G[\log f(Z|\boldsymbol{\theta}_0)]+\frac{1}{2n}\sum_{r=1}^{p}\sum_{t=1}^{p}j^{rt}i_{tr}+o(n^{-1})\\
&=E_G[\log f(Z|\boldsymbol{\theta}_0)]+\frac{1}{2n}\mathrm{tr}\{J(G)^{-1}I(G)\}+o(n^{-1})
\end{aligned}
\tag{38}
$$

where $I(G)$ and $J(G)$ are given by equation (10). This shows immediately that the asymptotic bias of the log likelihood is

$$
\begin{aligned}
b(G)&=E_G\left[\frac{1}{n}\sum_{\alpha=1}^{n}\log f(X_\alpha|\hat{\boldsymbol{\theta}}_{ML})-\int g(z)\log f(z|\hat{\boldsymbol{\theta}}_{ML})\mathrm{d}z\right]\\
&=\frac{1}{n}\mathrm{tr}\{J(G)^{-1}I(G)\}+o(n^{-1})
\end{aligned}
\tag{39}
$$

which yields TIC given by equation (11).

We recall that AIC was derived under the assumptions that model estimation is by maximum likelihood and the parametric model contains the true distribution. In order to remove both of the assumptions imposed to AIC, Konishi and Kitagawa (1996) used a functional approach to model estimation and evaluation problems, which will be introduced briefly.

The essential methods in the derivation of TIC are the use of a stochastic expansion of a maximum likelihood estimator and its properties

$$
\sum_{\alpha=1}^{n}\left.\frac{\partial\log f(X_\alpha|\boldsymbol{\theta})}{\partial\boldsymbol{\theta}}\right|_{\hat{\boldsymbol{\theta}}_{ML}}=\mathbf{0},\qquad \int g(z)\left.\frac{\partial\log f(z|\boldsymbol{\theta})}{\partial\boldsymbol{\theta}}\right|_{\boldsymbol{\theta}_0}\mathrm{d}z=\mathbf{0}
$$

To derive GIC, we use instead a functional Taylor series expansion of an estimator $\hat{\boldsymbol{\theta}}=(\hat{\theta}_1,\ldots,\hat{\theta}_p)'$, for which there exists a p-dimensional functional $\boldsymbol{T}$ such that $\hat{\boldsymbol{\theta}}=\boldsymbol{T}(\hat{G})=(T_1(\hat{G}),\ldots,T_p(\hat{G}))'$.

Suppose that T_r is the second-order compact differentiable at G for $r=1,\ldots,p$ (see, for example, Reeds, 1976; Fernholz, 1983). Then the functional Taylor series expansion for $\hat{\theta}_r=T_r(\hat{G})$ is, up to order n^{-1},

$$
\hat{\theta}_r=T_r(G)+\frac{1}{n}c_{r,n}^{(1)}+\frac{1}{n^2}c_{r,n}^{(2)}+o_p(n^{-1})
\tag{40}
$$

where

$$c_{r,n}^{(1)} = \sum_{\alpha=1}^{n} T_r^{(1)}(X_\alpha; G) \qquad \text{and} \qquad c_{r,n}^{(2)} = \frac{1}{2}\sum_{\alpha=1}^{n}\sum_{\beta=1}^{n} T_r^{(2)}(X_\alpha, X_\beta; G)$$

Here $T_r^{(m)}$ is defined as a symmetric function such that for an arbitrary distribution H on R^d

$$\frac{\mathrm{d}^m}{\mathrm{d}\varepsilon^m} T_r((1-\varepsilon)G + \varepsilon H) = \int\cdots\int T_r^{(m)}(x_1, \ldots, x_m; G)\prod_{j=1}^{m} \mathrm{d}\{H(x_j) - G(x_j)\}$$

at $\varepsilon = 0$, and

$$\int T_r^{(m)}(x_1, \ldots, x_m; G)\mathrm{d}G(x_j) = 0 \qquad \text{for} \qquad 1 \le j \le m$$

In particular, it is known that $T_r^{(1)}(X_\alpha; G)$ is an influence function defined by

$$T_r^{(1)}(X_\alpha; G) = \lim_{\varepsilon\to 0}\frac{T_r((1-\varepsilon)G + \varepsilon\delta_\alpha) - T_r(G)}{\varepsilon}$$

where δ_α is the point mass at X_α. For theoretical work on the functional Taylor series expansion, we refer to von Mises (1947), Withers (1983), and Konishi (1991).

Replacing $a_{r,n}^{(i)}$ in equation (36) by $c_{r,n}^{(i)}$ in equation (40) and using an approach similar to that developed for $f(z|\hat{\boldsymbol{\theta}}_{ML})$ give the bias of the log likelihood of the estimated density $f(z|\hat{\boldsymbol{\theta}})$ with functional estimator in

$$\begin{aligned} b(G) &= E_G\left[\frac{1}{n}\sum_{\alpha=1}^{n}\log f(X_\alpha|\hat{\boldsymbol{\theta}}) - \int g(z)\log f(z|\hat{\boldsymbol{\theta}})\mathrm{d}z\right] \\ &= \frac{1}{n}\mathrm{tr}\left\{\int \boldsymbol{T}^{(1)}(z; G)\frac{\partial \log f(z|\boldsymbol{\theta})}{\partial \boldsymbol{\theta}'}\bigg|_{\boldsymbol{T}(G)}\mathrm{d}G(z)\right\} + o\left(\frac{1}{n}\right) \end{aligned}$$

By subtracting the asymptotic bias estimate $b(\hat{G})$ from the log likelihood, we have the generalized information criterion GIC given by equation (12).

4.2 Asymptotic Accuracy

The information criteria given in Section 2 were constructed by correcting the asymptotic biases of log likelihoods of statistical models. We discuss the asymptotic accuracy of an information criterion in the estimation of the expected log likelihood.

EXAMPLE 5: As an illustration, let us first consider the Gaussian linear regression model $\boldsymbol{y} = X\boldsymbol{\beta} + \boldsymbol{\epsilon}$, where $\boldsymbol{y}$ is an n-dimensional observation vector of response, X is a known $n \times p$ matrix of covariates, $\boldsymbol{\beta}$ is a p-dimensional vector of parameters, and $\boldsymbol{\epsilon}$ is an n-dimensional random error vector whose elements are independently, normally distributed with zero mean and variance σ^2. The maximum likelihood estimates of the parameters $\boldsymbol{\theta} = (\boldsymbol{\beta}', \sigma^2)' \in \Theta \subset R^{p+1}$ are given by

$$\hat{\boldsymbol{\beta}} = (X'X)^{-1}X'y \qquad \text{and} \qquad \hat{\sigma}^2 = \frac{1}{n}(y - X\hat{\boldsymbol{\beta}})'(y - X\hat{\boldsymbol{\beta}})$$

Then the statistical model is for a future observation vector z

$$f(z|\hat{\boldsymbol{\theta}}) = \left(2\pi\hat{\sigma}^2\right)^{-n/2}\exp\left\{-\frac{1}{2\hat{\sigma}^2}(z - X\hat{\boldsymbol{\beta}})'(z - X\hat{\boldsymbol{\beta}})\right\} \tag{41}$$

Suppose that the true model $g(z)$ is included in the specified model, and hence $g(z)$ is distributed according to an n-dimensional normal distribution with mean vector $X\boldsymbol{\beta}_0$ and covariance matrix $\sigma_0^2 I_n$ for some $\boldsymbol{\beta}_0 = (\boldsymbol{\beta}_0', \sigma_0^2)' \in \Theta$. The log likelihood and expected log likelihood of the statistical model in equation (41) are, respectively,

$$\log f(y|\hat{\boldsymbol{\theta}}) = -\frac{n}{2}\{\log(2\pi\hat{\sigma}^2) + 1\},$$

$$\int g(z)\log f(z|\hat{\boldsymbol{\theta}})dz = -\frac{n}{2}\left\{\log(2\pi\hat{\sigma}^2) + \frac{\sigma_0^2}{\hat{\sigma}^2} + \frac{1}{n\hat{\sigma}^2}(X\boldsymbol{\beta}_0 - X\hat{\boldsymbol{\beta}})'(X\boldsymbol{\beta}_0 - X\hat{\boldsymbol{\beta}})\right\} \tag{42}$$

Then the bias of the log likelihood can be calculated exactly as

$$E_G\left[\log f(y|\hat{\boldsymbol{\theta}}) - \int g(z)\log f(z|\hat{\boldsymbol{\theta}})dz\right] = \frac{n(p+1)}{n-p-2} \tag{43}$$

Hence, the information criterion is

$$-2\log f(y|\hat{\boldsymbol{\theta}}) + 2\frac{n(p+1)}{n-p-2} = n\log(2\pi\hat{\sigma}^2) + \frac{n(n+p)}{n-p-2} \tag{44}$$

which was originally obtained by Sugiura (1978). This implies that the criterion in equation (44) in the normal regression model achieves bias reduction exactly under the assumption that the true distribution belongs to the specified parametric model.

We recall that in the normal linear regression model, AIC is given by

$$\text{AIC} = -2\log f(y|\hat{\boldsymbol{\theta}}) + 2(p+1)$$

It follows from equation (43) that the bias can be expanded as a power series in terms of order $1/n$ in

$$\frac{n(p+1)}{n-p-2} = (p+1)\left\{1 + \frac{1}{n}(p+2) + \frac{1}{n^2}(p+2)^2 + \cdots\right\}$$

Hence, we see that AIC is a criterion obtained by correcting the asymptotic bias of the log likelihood. By taking the higher order bias correction terms into account, we can obtain more refined criterion.

Hurvich and Tsai (1989, 1991, 1993) have investigated the asymptotic properties of AIC and demonstrated the effectiveness of bias reduction in regression and autoregressive time series models in both theoretical and numerical aspects.

We now consider the asymptotic accuracy of information-theoretic criteria in a general setting. The expected log likelihood $\eta(G; \hat{\boldsymbol{\theta}}) = E_G[\log f(Z|\hat{\boldsymbol{\theta}})]$ of $f(z|\hat{\boldsymbol{\theta}})$ in equation (4) is conditional on the observed data $\boldsymbol{X}_n$, and also depends on the unknown distribution G generating the data.

Suppose that its expectation over the sampling distribution G of $\boldsymbol{X}_n$ admits an expansion of the form

$$E_G[\eta(G; \hat{\boldsymbol{\theta}})] = \int g(z) \log f(z|\boldsymbol{T}(G))dz + \frac{1}{n}\eta_1(G) + \frac{1}{n^2}\eta_2(G) + O(n^{-3}) \quad (45)$$

The aim is to estimate the above quantity based on the observed data $\boldsymbol{X}_n$. Ideally we would like to obtain an estimator, $\hat{\eta}(\hat{G}; \hat{\boldsymbol{\theta}})$, which satisfies the condition that

$$E_G[\hat{\eta}(\hat{G}; \hat{\boldsymbol{\theta}}) - \eta(G; \hat{\boldsymbol{\theta}})] = O(n^{-j})$$

for larger values of j. This formula is used to assess the effect of an estimator in estimating the expected log likelihood $\eta(G; \hat{\boldsymbol{\theta}})$.

An obvious estimator was the log likelihood $\eta(\hat{G}; \hat{\boldsymbol{\theta}}) - E_{\hat{G}}[f(Z|\hat{\boldsymbol{\theta}})]$ of $f(z|\hat{\boldsymbol{\theta}})$, obtained by replacing the unknown distribution G by the empirical distribution $\hat{G}$ in $\eta(G; \hat{\boldsymbol{\theta}})$. Under suitable regularity conditions, usually moment conditions for $\log f(z|\boldsymbol{\theta})$ and smoothness conditions for $\hat{\boldsymbol{\theta}}$, the expectation of $\eta(\hat{G}; \hat{\boldsymbol{\theta}})$ admits a valid expansion of the following form:

$$E_G[\eta(\hat{G}; \hat{\boldsymbol{\theta}})] = \int g(z) \log f(z|\boldsymbol{T}(G))dz + \frac{1}{n}L_1(G) + \frac{1}{n^2}L_2(G) + O(n^{-3}) \quad (46)$$

Then, the asymptotic expansions in equations (45) and (46) differ in term of order n^{-1}, that is,

$$E_G[\eta(\hat{G}; \hat{\boldsymbol{\theta}}) - \eta(G; \hat{\boldsymbol{\theta}})] = \frac{1}{n}\{L_1(G) - \eta_1(G)\} + O(n^{-2})$$

This implies that the log likelihood is a consistent estimator of $E_G[\log f(Z|\boldsymbol{T}(G))]$, and that the optimism of the log likelihood $\eta(\hat{G};\hat{\boldsymbol{\theta}})$ declines as n increases, but is usually of practical concern for finite sample sizes.

The asymptotic bias of the log likelihood given by $\{L_1(G)-\eta_1(G)\}/n$ $(=b_1(G)/n)$ may be estimated by $b_1(\hat{G})/n=\{L_1(\hat{G})-\eta_1(\hat{G})\}/n$, and a bias-corrected version of the log likelihood is

$$\eta_{bc}(\hat{G};\hat{\boldsymbol{\theta}})=\eta(\hat{G};\hat{\boldsymbol{\theta}})-b_1(\hat{G})/n \tag{47}$$

Noting that the difference between $E_G[b_1(\hat{G})]$ and $b_1(G)$ is usually of order n^{-1}, that is, $E_G[b_1(\hat{G})]=b_1(G)+O(n^{-1})$, we have

$$E_G[\eta_{bc}(\hat{G};\hat{\boldsymbol{\theta}})-\eta(G;\hat{\boldsymbol{\theta}})]=O(n^{-2})$$

Hence, the bias-corrected log likelihood $\eta_{bc}(\hat{G};\hat{\boldsymbol{\theta}})$ is second-order correct or accurate for $\eta(G;\hat{\boldsymbol{\theta}})$ in the sense that the expectation of two quantities are in agreement up to and including the term of order n^{-1}.

It can readily be seen that the $-(2n)^{-1}$ times generalized information criterion GIC given by equation (12) is second-order correct for the corresponding expected log likelihood. In contrast, the log likelihood itself is only first-order correct.

If the parametric model is correctly specified, and the maximum likelihood estimate is used to estimate the underlying density, then the asymptotic bias of the log likelihood is given by the number of estimated parameters, and we have AIC $=-2n\{\eta(\hat{F};\hat{\boldsymbol{\theta}}_{ML})-p/n\}$. In such a case, the bias-corrected version of the log likelihood is given by

$$\eta_{ML}(\hat{F};\hat{\boldsymbol{\theta}}_{ML})=\eta(\hat{F};\hat{\boldsymbol{\theta}}_{ML})-\frac{1}{n}p-\frac{1}{n^2}\{L_2(\hat{F})-\eta_2(\hat{F})\}$$

It can readily be checked that $E_F[\eta_{ML}(\hat{F};\hat{\boldsymbol{\theta}}_{ML})-\eta(F;\hat{\boldsymbol{\theta}}_{ML})]=O(n^{-3})$, which implies that $\eta_{ML}(\hat{F};\hat{\boldsymbol{\theta}}_{ML})$ is third-order correct for $\eta(F;\hat{\boldsymbol{\theta}}_{ML})$.

In practice, we need to derive the second-order bias-corrected term $L_2(\hat{F})-\eta_2(\hat{F})$ analytically for each estimator. However, this seems to be of no practical use. In such cases, the bootstrap methods may be applied to estimate the bias of the log likelihood, and the same asymptotic order as for $\eta_{ML}(\hat{F};\hat{\boldsymbol{\theta}}_{ML})$ can be achieved by bootstrapping $\eta(\hat{F};\hat{\boldsymbol{\theta}}_{ML})-p/n$ or equivalently $\eta(\hat{F};\hat{\boldsymbol{\theta}}_{ML})$.

ACKNOWLEDGMENTS

I would like to thank two referees for helpful suggestions and comments which improved the quality of this chapter considerably.

REFERENCES

Akaike, H. (1973). Information theory and an extension of the maximum likelihood principle. In: Petrov, B.N. and Csaki, F., eds. *2nd International Symposium on Information Theory*. Budapest: Akademiai Kiado, pp. 267–281. (Reproduced in S. Kotz and N.L. Johnson, eds. *Breakthroughs in Statistics*, Vol. 1. New York: Springer, 1992, pp. 610–624.)

Akaike, H. (1974). A new look at the statistical model identification. *IEEE Trans. Autom. Control*, **AC-19**: 716–723.

Bozdogan, H. (1987). Model selection and Akaike's information criterion (AIC): The general theory and its analytical extensions. *Psychometrika*, **52**: 345–370.

Bozdogan, H. (1988). ICOMP: A new model-selection criterion. In: Hans H. Bock, ed. *Classification and Related Methods Data Analysis*. Amsterdam: Elsevier, pp. 599–608.

Bozdogan, H. (1990). On the information-based measure of covariance complexity and its application to the evaluation of multivariate linear models. *Commun. Stat. Theory and Methods*, **19**: 221–278.

Bozdogan, H. (ed.) (1994). *Proceedings of the First US/Japan Conference on the Frontiers of Statistical Modeling: An Informational Approach*. Dordrecht: Kluwer.

le Cessie, S. and van Houwelingen, J. C. (1992). Ridge estimators in logistic regression. *Appl. Stat.*, **41**: 191–201.

Dempster, A. P., Laird, N. M., and Rubin, D. B. (1977). Maximum likelihood from incomplete data via the EM algorithm (with discussion). *J. R. Stat. Soc.*, **B39**: 1–38.

Efron, B. (1979). Bootstrap methods: Another look at the jackknife. *Ann. Stat.*, **7**: 1–26.

Efron, B. (1986). How biased is the apparent error rate of a prediction rule? *J. Am. Stat. Assoc.*, **81**: 461–470.

Fernholz, L. T. (1983). *von Mises Calculus for Statistical Functionals*. Lecture Notes in Statistics 19, New York: Springer.

Findley, D. F. (1985). On the unbiasedness property of the AIC for linear stochastic time series models. *J. Time Series Anal.*, **6**: 229–252.

Fujikoshi, Y. and Satoh, K. (1997). Modified AIC and C_p in multivariate linear regression. *Biometrika*, **84**, 707–716.

Good, I. J. and Gaskins, R. A. (1971). Nonparametric roughness penalties for probability densities. *Biometrika*, **58**: 255–277.

Hampel, F. R., Ronchetti, E. M., Rousseeuw, P.J. and Stahel, W. A. (1986). *Robust Statistics. The Approach Based on Influence Functions*. New York: Wiley.

Huber, P. J. (1964). Robust estimation of a location parameter. *Ann. Math. Stat.*, **35**: 73–101.

Huber, P. J. (1981). *Robust Statistics*. New York: Wiley.

Hurvich, C. M. and Tsai, C.-L. (1989). Regression and time series model selection in small samples. *Biometrika*, **76**: 297–307.

Hurvich, C. M. and Tsai, C.-L. (1991). Bias of the corrected AIC criterion for underfitted regression and time series models. *Biometrika*, **78**: 499 - 509.

Hurvich, C. M. and Tsai, C.-L. (1993). A corrected Akaike information criterion for vector autoregressive model selection. *J. Time Series Anal.*, **14**: 271–279.

Ishiguro, M., Sakamoto, Y., and Kitagawa, G. (1997). Bootstrapping log likelihood and EIC, an extension of AIC. *Ann. Inst. Stat. Math.*, **49**: 411–434.

Kitagawa, G. and Gersch, W. (1996). *Smoothness Priors Analysis of Time Series*. New York: Springer.

Konishi, S. (1991). Normalizing transformations and bootstrap confidence intervals. *Ann. Stat.*, **19**: 2209–2225.

Konishi, S. and Kitagawa, G. (1996). Generalised information criteria in model selection. *Biometrika*, **83**: 875–890.

Kullback, S. and Leibler, R. A. (1951). On information and sufficiency. *Ann. Math. Stat.*, **22**: 79–86.

Reeds, J. (1976). On the definition of von Mises functionals. Ph.D. Dissertation, Harvard University.

Shibata, R. (1983). A theoretical view of the use of the AIC. In: O. D. Anderson ed. *Time Series Analysis: Theory and Practice, IV*. Amsterdam: North-Holland, pp. 237–244.

Shibata, R. (1989). Statistical aspects of model selection. In: J. C. Willems ed. *From Data to Model.* New York: Springer, pp. 215–240.

Shibata, R. (1997). Bootstrap estimate of Kullback–Leibler information for model selection. *Stat. Sinica*, **7**: 375–394.

Stone, M. (1977). An asymptotic equivalence of choice of model by cross-validation and Akaike's criterion. *J. R. Stat. Soc., B*, **39**: 44-47.

Sugiura, N. (1978). Further analysis of the data by Akaike's information criterion and the finite corrections. *Commun. Stat. Theory Methods*, **A7**: 13–26.

Takeuchi, K. (1976). Distribution of information statistics and criteria for adequacy of models. *Math. Sci.*, **153**: 12–18 (in Japanese).

von Mises, R. (1947). On the asymptotic distribution of differentiable statistical functions. *Ann. Math. Stat.*, **18**: 309–348.

Withers, C. S. (1983). Expansions for the distribution and quantiles of a regular functional of the empirical distribution with applications to nonparametric confidence intervals. *Ann. Stat.*, **11**: 577–587.

Wong, W. (1983). A note on the modified likelihood for density estimation. *J. Am. Stat. Assoc.*, **78**: 461–463.

14

Multivariate Rank Tests

JOHN I. MARDEN University of Illinois at Urbana–Champaign, Champaign, Illinois

1. INTRODUCTION

Rank-based nonparametric procedures have served statisticians well, providing easy-to-use, intuitive, efficient, and robust alternatives to the common normal-theory procedures. The original rank procedures were univariate. Multivariate procedures, in which each variable is ranked separately, immediately affords practitioners many of the benefits of the univariate procedures in more complicated situations. In recent years, there has been development of other methods of multivariate ranking that use the variables together rather than separately. This paper shows how to use particular definitions of multivariate sign and rank to generate a multitude of multivariate procedures that not only mimic the popular univariate ones, but go beyond to treat genuinely multivariate problems.

For a number x, denote the sign function by S, that is, $S(x) = -1, 0, 1$, as $x < 0, x = 0, x > 0$, respectively. Given a sample $x_1, x_2, \ldots, x_n$, the rank of observation x_i among those in the sample is k if x_i is the kth smallest. The ranks are functions of the signs of the differences between the observations, i.e.,

$$RANK(x_i) = \frac{1}{2}\left(\sum_{j=1}^{n} S(x_i - x_j) + n + 1\right) \tag{1}$$

(With this definition, if there are ties in the observations, then the ranks will be midranks.) A popular and natural multivariate generalization of the sign of x is the unit vector in the direction of x, which we will also denote S:

$$S(x) = \begin{cases} \dfrac{x}{\|x\|} & \text{if } x \neq 0 \\ 0 & \text{if } x = 0 \end{cases} \tag{2}$$

A multivariate generalization of rank is given by equation (1) using the multivariate definition of sign. We find it more convenient to work with just the summation, and hence will define multivariate rank by

$$R(x_i; F_n) = \frac{1}{n}\sum_{j=1}^{n} S(x_i - x_j) \tag{3}$$

The "F_n" is the empirical distribution based on the sample of multivariate x_is. For a given distribution function F, the rank of x relative to F is given by

$$R(x; F) = E_F(S(x - X)), \qquad X \sim F \tag{4}$$

Small (1990) alluded to this definition of multivariate rank when he noted that the gradient vector relative to x of $E_F(\|x - X\|)$ is a multivariate analog of the univariate rank (which he called quantile). This gradient is $R(x; F)$. The definition in equation (3) is given explicitly in Möttönen and Oja (1995), Chaudhuri (1996), and Koltchinskii (1997). Other notions of multivariate sign and rank, in addition to the obvious coordinatewise one, can be found in Hettmansperger et al. (1992).

We cover many multivariate tests based on these multivariate signs and ranks. Most of the procedures described in this paper are fairly easy to understand and implement. One downside is that although the procedures are invariant under rotations and scalar multiplication, they are not invariant under the general linear group. Other multivariate generalizations also lead to multivariate tests. Coordinatewise sign and rank vectors, where one looks at the usual signs or ranks within each variable, are very easy to use. Puri and Sen (1971) is the major reference for such procedures. They are invariant under coordinatewise monotone transformations, but not under rotations. Bickel (1964, 1965) notes that these tests can have poor power when the variables are highly correlated.

There is a large body of work on affine-invariant multivariate sign and rank procedures. Especially interesting is Randles' (1989) notion of "interdirections," which can be used to adaptively estimate the angles between vectors. Closely related is the use of data-dependent axes in, e.g., Chaudhuri and Sengupta (1993) and Chakraborty and Chaudhuri (1997). See also Blumen (1958), Dietz (1982), Brown and Hettmansperger (1989), and Oja and Nyblom (1989) for sign tests, Peters and Randles (1990) and Jan and Randles (1994) for signed-rank tests, and Brown (1983), Brown and Hettmansperger (1987), Peters and Randles (1991), Hettmansperger and Oja (1994), and Hettmansperger et al. (1994, 1997) for one-, two-, and multi-sample tests. Friedman and Rafsky (1979, 1983) use graph-theoretic notions, such as minimal spanning trees, to develop multivariate generalizations of two-sample tests (including runs tests and the Kolmogorov–Smirnov test), and measures of association. Other approaches involve counting the numbers of points in various half-spaces. For these, see Hodges (1955), Liu (1990, 1992), Liu and Singh (1993), and the survey paper by Liu et al. (1997). Tukey (1975) and Barnett (1976) contain many additional notions of multivariate ordering.

Some authors have taken a somewhat more parametric approach that gives additional flexibility in the linear models context. Assume a location-family model on the data, where the location parameters satisfy a linear model. Then tests of linear hypotheses on the coefficient parameters are carried out by estimating the parameters under the null hypothesis using an L_1-type objective function, and applying appropriate multivariate sign or rank tests to the resulting residuals. These procedures can handle the testing problems in Sections 2–6, and more (see Hettmansperger and Oja, 1994; Naranjo and Hettmansperger, 1994; Hössjer and Croux, 1995; Möttönen and Oja, 1995).

Many of the results described in subsequent sections are already in the literature. Möttönen and Oja (1995) give a nice review that covers much of the material we present in the one-, two-, and many-sample situations below, as well as corresponding estimation problems. Koltchinskii (1997) proposes alternative multivariate Kolmogorov–Smirnov tests. Some results seem to be new, e.g., the multivariate runs test in Section 2.3, McWilliams (1990) test for symmetry in one dimension extended to several dimensions, a class of generalizations of the Wilcoxon signed rank test, and a general class of tests for the group invariant models found in Andersson (1975). In fact, the last case is still in the formative stages.

In the following, we narrow our focus to tests based on the signs, equation (2), and ranks, equation (3), and do not assume any linear models on location parameters. Presumably, most of what we present here that is new can also be generalized to other definitions of multivariate signs and ranks, in particular to the affine-invariant definitions. A clear drawback to not being fully affine invariant is that the relative scaling of the individual variables affects the statistic. In practice, it is reasonable to first scale the individual variables (e.g., by their median absolute deviation from the median) before finding the signs and ranks, in much the same way one often scales the covariance matrix before using principle components. This preliminary scaling should ameliorate some of the worst drawbacks, without increasing the difficulty in calculation.

The main hope is that people will be encouraged to use these techniques to continue to find new and useful multivariate procedures that are easy to implement and yield robust alternatives to multivariate normal methods.

2. ONE-SAMPLE TESTS

2.1 Sign Tests

We assume we have a sample $X_1, X_2, \ldots, X_n$ of independent and identically distributed (iid) p-dimensional observations from distribution F, and we wish to test the null hypothesis that the spatial median of F is 0, i.e.,

$$H_0 : \mu = 0 \tag{5}$$

where μ is the population *spatial median*, i.e., it satisfies

$$R(\mu; F) \equiv E_F(S(\mu - X)) = 0 \tag{6}$$

It is possible that there is no solution to equation (6). An alternative definition is the μ that minimizes $E_F\|X - \mu\|$ (equivalently, $E_F(\|X - \mu\| - \|X\|)$, in case $E_F\|X\|$ does not exist). Unless the points are collinear, this equation has a unique solution, and the solution satisfies equation (6) if anything does. Small (1990) gives a very interesting overview of multivariate medians, including this spatial median.

Many other multivariate medians have been proposed, many of which are affine equivariant; see, for example, Tukey (1975), Seheult et al. (1976), Green (1981), Donoho (1982), Oja (1983), Liu (1990, 1992), and Liu and Singh (1993).

In the univariate situation, the *sign test* is based on the sum of the signs of the observations. If the population median is 0, then the sum should be near 0. The multidimensional analog does the same, basing the test on

$$\overline{S} \equiv \frac{1}{n}\sum_{i=1}^{n} S(X_i) \tag{7}$$

The covariance of this statistic depends on F, but it is easily estimated under the null hypothesis by the sample dispersion matrix of the signs about 0. An alternative estimator first adjusts the data so that the sample spatial median is 0. That is, let $\widehat{\mu}_n$ be the solution to

$$R(\widehat{\mu}_n; F_n) = \frac{1}{n}\sum_{i=1}^{n} S(X_i - \widehat{\mu}_n) = 0 \tag{8}$$

(or the minimizer of $\sum \|X_i - \mu\|$ over μ). Then the estimate of $\mathrm{Cov}(S(X_i))$ is

$$\hat{\Sigma}_{Sn} = \frac{1}{n-1}\sum_{i=1}^{n} S(X_i - \widehat{\mu}_n) S(X_i - \widehat{\mu}_n)' \tag{9}$$

The test statistic is $T^2 \equiv n\bar{S}'\hat{\Sigma}_{Sn}^{-1}\bar{S}$, which is asymptotically χ_p^2 under the null hypothesis. See Brown (1983) for the bivariate case, in which case the test is called an "angle test," and Möttönen and Oja (1995) for the general case.

2.2 Signed Rank Tests

The univariate *Wilcoxon signed rank test* is an alternative test for situations in which the distribution is assumed to be symmetric about μ. The idea is to take account of the relative sizes of the positive and negative observations, not just count them. Two equivalent forms of the statistic are

$$T_1 = \binom{n+1}{2}^{-1} \sum\sum_{1 \le i \le j \le n} S(x_i + x_j) \quad \text{and} \quad T_2 = \frac{1}{n}\sum_{i=1}^{n} RANK(\|x_i\|) S(x_i) \tag{10}$$

where $RANK(||x_i||)$ is the usual rank of the magnitude of observation i among the magnitudes of all the observations in the sample. A third form, which is not exactly the same, averages the ranks of the observations relative to the original sample combined with its negatives, $x_1, \ldots, x_n$; $-x_1, \ldots, -x_n$:

$$T_3 = \frac{1}{n}\sum_{i=1}^{n} R(x_i; F_n^{(\pm)}) = \frac{1}{2n^2}\sum_{i=1}^{n}\sum_{j=1}^{n} (S(x_i - x_j) + S(x_i + x_j)) \tag{11}$$

where $F_n^{(\pm)}$ is the empirical distribution function for the augmented sample. Multivariate extensions of these three statistics are immediate. Chaudhuri (1992) suggests T_1, Möttönen and Oja (1995) and Möttönen et al. (1997) propose both T_2 and T_3, while Hössjer and Croux (1995) present T_2. Hössjer and Croux also allow score functions for the $RANK(\|x_i\|)$s.

The symmetry in one dimension can be extended in a number of ways to multiple dimensions. Let us restrict to F in the null hypothesis, i.e., with spatial median 0. Then the simplest generalization to the symmetry above requires that X and $-X$ have the same distribution. The most restrictive generalization requires that the distribution of X be spherically symmetric. The appropriate test statistic will thus depend on how much symmetry one is willing to suppose. We describe symmetries by subgroups $\mathcal{G}$ of $\mathcal{O}(p)$, the group of p–dimensional orthogonal matrices. The minimal nontrivial group we consider is the simple sign change, $\mathcal{G}_{\pm}(p) = \{I_p, -I_p\}$, where I_p is the p-dimensional identity matrix. Assume that for given $\mathcal{G}$, $\mathcal{G}_{\pm}(p) \subseteq \mathcal{G} \subseteq \mathcal{O}(p)$, F is $\mathcal{G}$ invariant in the sense that X and gX have the same distribution for each $g \in \mathcal{G}$. Then the corresponding ranks we use are defined relative to the sample combined with all the gX_is, that is,

$$\{gX_i \mid 1 \leq i \leq n, g \in \mathcal{G}\} \tag{12}$$

Letting $\mathcal{G}F_n$ denote the empirical distribution function for this augmented sample, we have

$$R(x; \mathcal{G}F_n) = \frac{1}{n}\sum_{i=1}^{n}\int_{\mathcal{G}} S(x - gx_i)\nu(\mathrm{d}g) \tag{13}$$

where ν is uniform measure over the group $\mathcal{G}$. Often, $\mathcal{G}$ will be a finite group, so that the integral is actually the average over the gs. Note that in equation (11), $F_n^{(\pm)} = \mathcal{G}_{\pm}(p)F_n$. In addition to $\mathcal{G}_{\pm}(p)$ and the full orthogonal group, plausible choices for $\mathcal{G}$ include the group that allows different sign changes for each variable, $\mathcal{G}_{\pm}^{*}(p) \equiv \{diag(\epsilon_1, \ldots, \epsilon_p) | (\epsilon_1, \ldots, \epsilon_p) \in \{-1, +1\}^p\}$ (see Chaudhuri, 1992), and the groups which consist of the elements in $\mathcal{G}_{\pm}(p)$ or $\mathcal{G}_{\pm}^{*}(p)$ plus each of those elements multiplied by each of the $p \times p$ permutation matrices. See Section 7 for more use of group symmetries.

Once the group has been chosen, the test statistic is based on the sample average of the signed ranks:

$$\bar{R} = \frac{1}{n}\sum_{i=1}^{n} R(x_i; \mathcal{G}F_n) \tag{14}$$

The covariance of the statistic can be estimated by the sample covariance matrix based on the signed ranks,

$$\widehat{\Sigma}_{SRn} = \frac{1}{n-1}\sum_{i=1}^{n}(R(x_i; \mathcal{G}F_n) - \bar{R})(R(x_i; \mathcal{G}F_n) - \bar{R})' \tag{15}$$

However, the symmetry assumed in F transfers to the covariance, so that a symmetrized version of the estimate may be used:

$$\hat{\Sigma}_{\mathcal{G}n} = \int_{\mathcal{G}} g\hat{\Sigma}_{SRn}g'\nu(\mathrm{d}g) \tag{16}$$

If the group is $\mathcal{G}_{\pm}(p)$, then this step has no effect. Using $\mathcal{G}_{\pm}^{*}(p)$, this step involves setting the covariances to 0 and leaving the variances alone. Adding the permutation matrices sets the variances equal to the average variance, and the covariances equal to the average covariance. The test statistic is $n\bar{R}'\hat{\Sigma}_{\mathcal{G}n}^{-1}\bar{R}$, which is asymptotically χ_p^2 under the null hypothesis.

Turn to the statistic T_2. Hössjer and Croux (1995) propose it for multiple dimensions when the distribution F is spherically symmetric, so that under the null hypothesis, X and gX have the same distribution for any $g \in \mathcal{O}(p)$. When X has a spherically symmetric distribution, then $S(X)$ and $||X||$ are independent, and $S(X)$ is uniformly distributed on the unit sphere. Thus $\mathrm{Cov}(S(X)) = (1/p)I_p$, and

$$\mathrm{Cov}(T_2) = \frac{1}{pn^2}\sum_{i=1}^{n} E(RANK(\|x_i\|)^2)I_p \tag{17}$$

If the distribution of $||X||$ is continuous, then the sum of squares of their ranks is $n(n+1)(2n+1)/6$. If the distribution is not continuous, then the expected value of the sum of squares can be estimated by the sum of squares of observed midranks. For the continuous case, the test statistic is then $6np\|T_2\|^2/((n+1)(2n+1))$, which is asymptotically χ_p^2 under the null hypothesis that F is spherically symmetric about 0.

There is a genuine improvement in efficiency when going from the sign tests to the signed rank tests, at least for distributions close to the normal. The different versions of the signed rank test corresponding to different choices of the group $\mathcal{G}$ will yield the same asymptotic Pitman efficiency, if the distribution F satisfies the requisite symmetry. We conjecture that the small sample properties are improved if one uses the largest group possible.

2.3 Runs Test

A popular test for the independence of the elements in a series is the runs test. Assume the sample index i represents time or some other serial arrangement of the observations. In the univariate case, the observed values are changed to signs, and the *runs* of positive and negative signs are noted. A run is a consecutive set of observations that have the same sign. If there is positive serial correlation, then there will tend to be fewer and longer runs. If there is negative serial correlation, there will tend to be many small runs. Tests have been developed based on such measures as the number of runs and the length of the longest run. We will look at an extension of the test based on the number of runs, the Wald–Wolfowitz test.

The number of runs among the x_is is one plus the number of sign changes in the sequence, where a sign change occurs at point i if $x_i x_{i+1} < 0$. Thus

$$(\text{No. of runs}) = \frac{1}{2}(n+1-V_n), \text{ where } V_n = \sum_{i=1}^{n-1} S(x_i)'S(x_{i+1}) \tag{18}$$

For this univariate case, the prime is superfluous. However, if we move to multidimensional observations, we have an immediate extension of the runs test. This multidimensional statistic V_n will tend to be large in magnitude if either consecutive observations tend to be pointing in the same direction, or they tend to be pointing in opposite directions, suggesting types of positive and negative multivariate serial correlation. Assuming that the spatial median of the observations is 0, under the null hypothesis of independence of the observations, we have that

$$E(V_n) = 0 \quad \text{and} \quad \text{Var}(V_n) = (n-1)tr(\text{Cov}(S(X_i))) \tag{19}$$

It can be shown that under these assumptions,

$$W_n \equiv \frac{V_n^2}{n \cdot tr(\widehat{\Sigma}_{Sn})} \longrightarrow \chi_1^2 \tag{20}$$

The assumption that the spatial median is 0 is generally too strong, but it is easy to modify the test for cases that the median is not 0 by subtracting the sample spatial median $\widehat{\mu}_n$ from each observation.

2.4 McWilliams Test for Symmetry

McWilliams (1990) exploited the runs test to develop a test of symmetry about 0 in one dimension that improves upon many popular tests for common alternatives. The basic idea is to note that in one dimension, when F is symmetric about 0, then $\|X\|$ and $S(X)$ are independent. McWilliams' test is the usual runs test, except that the $S(x_i)$s are taken in order of the ranks of the $\|x_i\|$s. For the p-dimensional case, this procedure can be used to test the null hypothesis of whether F has a spherically symmetric distribution about 0. Now we let

$$V_n^0 = \sum_{i=1}^{n-1} S(x_{o(i)})' S(x_{o(i+1)}) \tag{21}$$

where $o(i) = j$ if $RANK(\|x_j\|) = i$. (If there are ties, break them randomly.) Under the null hypothesis, equation (19) again holds, but now we know that $\text{Cov}(S(X)) = (1/p)I_p$. Thus from equation (19), the McWilliams test for spherical symmetry is based on

$$W_n^0 \equiv \frac{1}{n}\left(V_n^0\right)^2 \longrightarrow \chi_1^2 \tag{22}$$

under the null.

3. TWO- AND MULTI-SAMPLE TESTS: BASICS

As in the one-sample situations, there are both sign tests and rank tests for testing the equality of two or more populations. The sign tests are analogs of the Mood median test in one dimension. The data consist of K independent groups of observations, group k consisting of $n^{(k)}$ iid observations with distribution function $F^{(k)}$:

$$\left\{X_i^{(k)} | k = 1, \ldots, K, i = 1, \ldots, n^{(k)}\right\} \tag{23}$$

The general null hypothesis for this section is

$$H_0 : F^{(1)} = \cdots = F^{(K)} \tag{24}$$

Set $n = n^{(1)} + \cdots + n^{(K)}$. Tests are developed by first finding the signs or ranks of the combined sample, then comparing these quantities between the groups. The next two subsections describe the basic sign and rank statistics we will base the tests upon. Sections 4 and 5 will look at specific tests. The asymptotics will assume that for each k,

$$\frac{n^{(k)}}{n} \longrightarrow \lambda^{(k)}, \qquad \text{for } 0 < \lambda^{(k)} < 1 \tag{25}$$

3.1 Sign Statistics

Let $\widehat{\mu}_n$ be the spatial median of the combined sample in equation (23), and let $S_i^{(k)} = S(x_i^{(k)} - \widehat{\mu}_n)$for each i and k. The sign tests for equation (24) are based on linear combinations of the $S_i^{(k)}$s, or more simply on the $\overline{S}^{(k)}$s, where

$$\overline{S}^{(k)} = \frac{1}{n^{(k)}} \sum_{i=1}^{n^{(k)}} S_i^{(k)}$$

Let

$$\mathbb{S} = (S_1^{(1)}, \ldots, S_{n^{(1)}}^{(1)}, S_1^{(2)}, \ldots, S_{n^{(2)}}^{(2)}, \ldots, S_1^{(K)}, \ldots, S_{n^{(K)}}^{(K)}) \tag{26}$$

be the matrix of signs. Under the null hypothesis, the distribution of $\mathbb{S}$ is invariant under permutation of the sign vectors. Let $\Sigma_{Sn} = \mathrm{Cov}(S_i^{(k)})$ under the null. By Theorem 2.1.2 of Chaudhuri (1996), the sum of the elements is bounded by the number of observations that equal $\widehat{\mu}_n$, and hence if F is continuous,

$$\mathrm{Cov}\left(\sum_{k=1}^{K} \sum_{i=1}^{n_k} S_i^{(k)}\right) = O(1) \tag{27}$$

The $O(1)$ is a bound on the trace of the matrix as $n \to \infty$.Expanding the covariance, we have that

$$n\Sigma_{Sn} + n(n-1)\text{Cov}\left(S_i^{(k)}, S_j^{(l)}\right) = O(1) \tag{28}$$

Hence for $(i, k) \neq (j, l)$,

$$\text{Cov}\left(S_i^{(k)}, S_j^{(l)}\right) = -\frac{1}{n-1}\Sigma_{Sn} + O\left(\frac{1}{n^2}\right) \tag{29}$$

Thus we can write

$$\text{Cov}(\mathbb{S}) = \frac{n}{n-1}\Sigma_{Sn} \otimes H_n + O\left(\frac{1}{n^2}\right) \tag{30}$$

where H_n is the $n \times n$ centering matrix, $H_n = I_n - (1/n)J_n$, and J_n is the $n \times n$ matrix of all ones. For an $a \times b$ matrix W, Cov(W) is defined to be the $ab \times ab$ matrix Cov(vec(W)), where vec(W) is the $ab \times 1$ vector obtained by stacking the columns of W, that is,

$$\text{vec}(W) = (w_{11}, \ldots, w_{a1}, w_{12}, \ldots, w_{a2}, \ldots, w_{1b}, \ldots, w_{ab})' \tag{31}$$

We are using the left direct, or Kronecker, products of matrices. If Z is $c \times d$, then $W \otimes Z$ is the $ac \times bd$ matrix given by

$$W \otimes Z = \begin{pmatrix} z_{11}W & z_{12}W & \ldots & z_{1d}W \\ z_{21}W & z_{22}W & \ldots & z_{2d}W \\ \vdots & \vdots & \ddots & \vdots \\ z_{c1}W & z_{c2}W & \ldots & z_{cd}W \end{pmatrix} \tag{32}$$

(see Graybill, 1969). We use a pooled estimate of Σ_{Sn},

$$\widehat{\Sigma}_{Sn} = \frac{1}{n-K}\sum_{k=1}^{K}\sum_{j=1}^{n^{(k)}} S(x_j^{(k)} - \widehat{\mu}^{(k)})S(x_j^{(k)} - \widehat{\mu}^{(k)})' \tag{33}$$

where $\widehat{\mu}^{(k)}$ is the spatial median for group k.

3.2 Rank Statistics

Now we have the ranks defined for the combined sample, so that for each k and i,

$$R_i^{(k)} = \frac{1}{n}\sum_{l=1}^{K}\sum_{j=1}^{n^{(l)}} S(x_i^{(k)} - x_j^{(l)}) \tag{34}$$

Let

$$\mathbb{R} = (R_1^{(1)}, \ldots, R_{n^{(1)}}^{(1)}, R_1^{(2)}, \ldots, R_{n^{(2)}}^{(2)}, \ldots, R_1^{(K)}, \ldots, R_{n^{(K)}}^{(K)}) \tag{35}$$

These ranks sum up to exactly 0, so as for the signs, if we define $\Sigma_{Rn} = \mathrm{Cov}(R_i^{(k)})$ under the null hypothesis, we have that

$$\mathrm{Cov}(\mathbb{R}) = \frac{n}{n-1} \Sigma_{Rn} \otimes H_n \tag{36}$$

The pooled estimate of Σ_{Rn} is given by

$$\widehat{\Sigma}_{Rn} = \frac{1}{n-K} \sum_{k=1}^{K} \sum_{j=1}^{n^{(k)}} R(x_j^{(k)}; \widehat{F}_n^{(k)}) R(x_j^{(k)}; \widehat{F}_n^{(k)})' \tag{37}$$

where $\widehat{F}_n^{(k)}$ is the empirical distribution function for the observations in group k. Thus the inner summation is the sum of squares of the ranks defined within group k.

4. TWO-SAMPLE TESTS

In this subsection, we take $K = 2$. The test based on the signs is the multivariate analog of the *Mood median test*, and that based on the ranks is an analog of the *Mann–Whitney/Wilcoxon two-sample test*. The median test for bivariate samples is the two-sample angle test in Brown (1983). Möttönen and Oja (1995) present the general two-sample median test as well as the rank test.

4.1 The Median Test

The median test is based on the mean of the signs in equation (26) that correspond to the first group, $\overline{S}^{(1)}$, or, equivalently, on the difference in means, $T_{Sn} = \overline{S}^{(1)} - \overline{S}^{(2)}$. Under the null hypothesis, the mean is 0 (or possibly $O(1/n)$), and from equation (30),

$$\mathrm{Cov}(T_{Sn}) = \frac{n}{n-1} \left(\frac{1}{n^{(1)}} + \frac{1}{n^{(2)}} \right) \Sigma_{Sn} + O\left(\frac{1}{n^3} \right) \tag{38}$$

It can then be shown that under the null hypothesis,

$$\frac{n^{(1)} n^{(2)}}{n} T'_{Sn} (\widehat{\Sigma}_{Sn})^{-1} T_{Sn} \longrightarrow \chi_p^2 \tag{39}$$

4.2 Mann–Whitney/Wilcoxon Test

This test parallels the median test but with ranks in place of the signs. The statistic, the difference of the means of the ranks for the two groups, has a couple of familiar forms:

$$T_{Rn} = \overline{R}^{(1)} - \overline{R}^{(2)} = \frac{n^{(1)} + n^{(2)}}{n^{(2)}} \overline{R}^{(1)} = \frac{1}{n^{(1)} n^{(2)}} \sum_{i=1}^{n^{(1)}} \sum_{j=1}^{n^{(2)}} S\left(x_i^{(1)} - x_j^{(2)}\right) \quad (40)$$

We then have that under the null hypothesis,

$$\frac{n^{(1)} n^{(2)}}{n} T'_{Rn} (\widehat{\Sigma}_{Rn})^{-1} T_{Rn} \longrightarrow \chi_p^2 \quad (41)$$

4.3 Kolmogorov–Smirnov Tests

The hypothesis in equation (5) and the hypothesis of spherical symmetry in Section 2.4 are used to test whether a particular aspect of the distribution F is true. One may also wish to test whether a specific distribution F_0 holds, i.e.,

$$H_0 : F = F_0 \quad (42)$$

In the univariate case, a popular statistic is the one-sample Kolmogorov–Smirnov statistic, which is based on $KS_{F_0} = \sup_x |F_n(x) - F_0(x)|$. A multivariate analog of the test replaces the univariate Fs with their multivariate versions. Koltchinskii (1997) proposes an alternative. Note that in the univariate case, from equation (4), $2F_0(x) - 1 = R(x; F_0)$ and $2F_n(x) - 1 = R(x; F_n)$. Hence, the statistic $KS_R = \sup_x |R(x; F_n) - R(x; F_0)|$ yields another multivariate generalization. I know of no comparisons of the two tests, but KS_R will be invariant under rotations and the same scale change for all variables, while KS_{F_0} is invariant under different monotonic transformations for each variable. (Both are location-invariant.) Koltchinskii also suggests the statistic $\int |F_n(x) - F_0(x)|^2 \mathrm{d}F(x)$.

Similarly, an analog of the two-sample Kolmogorov–Smirnov is based on $KS_R^{(2)} = \sup_x |R(x; F_n^{(1)}) - R(x; F_n^{(2)})|$.

5. ANALYSIS OF VARIANCE

The generic test for the hypothesis in equation (24) in the normal case is the F-test. The Kruskal–Wallis test is the rank analog for univariate data, wherein one calculates the F-statistic using the ranks instead of the original data. Such a test is known as a *rank transform test*, that is, one uses the normal-theory statistic on the ranks. There is quite a bit of discussion in the literature about when rank transform tests are valid and when not; see Conover and Iman (1981) and Fligner (1981) for some insight, Thompson (1991) for a negative example, and Akritas and Arnold (1994) for a positive approach. The next four subsections consider testing main effects in one-way layouts. Sections 5.5 and 5.6 consider testing main effects and interactions in two-way layouts.

5.1 Median and Kruskal–Wallis Tests

The statistics for the multivariate tests are analogs of the Mood median and Kruskal–Wallis tests:

$$W_{Sn} = \sum_{k=1}^{K} n^{(k)} \overline{S}^{(k)\prime} \widehat{\Sigma}_{Sn}^{-1} \overline{S}^{(k)} \qquad \text{and} \qquad W_{Rn} = \sum_{k=1}^{K} n^{(k)} \overline{R}^{(k)\prime} \widehat{\Sigma}_{Rn}^{-1} \overline{R}^{(k)} \tag{43}$$

Under the null hypothesis, both of these statistics are asymptotically $\chi^2_{(K-1)p}$. This result for the ranks follows by noting that

$$\mathrm{Cov}\left(\sqrt{n^{(1)}}\overline{R}^{(1)}, \ldots, \sqrt{n^{(K)}}\overline{R}^{(K)}\right) = \frac{n}{n-1} \Sigma_{Rn} \otimes (I_K - \lambda_n \lambda_n') \tag{44}$$

where $\lambda_n = \left(\sqrt{n^{(1)}/n}, \ldots, \sqrt{n^{(K)}/n}\right)'$; hence the matrix $I_K - \lambda_n \lambda_n'$ is idempotent of rank $K - 1$. That for the signs follows similarly. See Brown (1983) for the bivariate sign test, and Choi and Marden (1997) for the rank test.

5.2 Contrasts

In normal-based analysis of variance, one often looks at contrasts of the group means. (A contrast is a linear combination whose coefficients sum to 0.) With a set of orthogonal contrasts, one can decompose the overall statistic into independent (or at least asymptotically independent) statistics. Crouse (1967) proposes the same idea for rank tests, but the contrasts tested are contrasts on the expected ranks, not the means of the raw variables. Thus these contrasts may be difficult to interpret. The fully nonparametric models of Akritas and Arnold (1994) can be used to justify such contrasts, where the models assume that the contrasts of the distribution functions are 0.

Let C be an $n \times q$ matrix whose columns are orthonormal contrasts, that is, they are orthogonal, have length 1, and their elements sum to 0.Then the statistics we consider are $\mathbb{S}C$ for the median tests, and $\mathbb{R}C$ for the rank tests. Under the null hypothesis, they have mean 0 (or $O(1/n)$), and

$$\text{Cov}(\mathbb{S}C) = \frac{n}{n-1}\Sigma_{Sn} \otimes I_q + O\left(\frac{1}{n^2}\right) \quad \text{and} \quad \text{Cov}(\mathbb{R}C) = \frac{n}{n-1}\Sigma_{Rn} \otimes I_q \tag{45}$$

Letting c_r be the rth column of C, the test statistics based on the contrasts are

$$\sum_{r=1}^{q}(\mathbb{S}c_r)'\widehat{\Sigma}_{Sn}^{-1}(\mathbb{S}c_r) \quad \text{and} \quad \sum_{r=1}^{q}(\mathbb{R}c_r)'\widehat{\Sigma}_{Rn}^{-1}(\mathbb{R}c_r) \tag{46}$$

where the individual summands within each statistic are asymptotically independent χ_p^2, and the overall statistics themselves are asymptotically χ_{qp}^2, under the null.

5.3 Jonckheere–Terpstra Tests: Qualitative Contrasts

The Jonckheere–Terpstra test of equation (24) in the univariate case is developed to be sensitive to alternatives in which there is a trend (either increasing or decreasing) in the groups, that is, the $F^{(k)}$s are stochastically increasing or decreasing in k. We term this type of comparison a *qualitative* contrast because there is no particular functional form to the differences, e.g., it is not assumed that the mean of group i equals $a + b \cdot i$ for some constants a and b. The Jonckheere–Terpstra statistic is

$$JT \equiv \sum\sum_{1 \le k < l \le K} \sum_{i=1}^{n^{(k)}} \sum_{j=1}^{n^{(l)}} S\left(x_i^{(k)} - x_j^{(l)}\right) \tag{47}$$

In the multivariate case, this statistic sums all the unit vectors pointing from an observation in a later group l towards an observation in an earlier group k (i.e., $k < l$), and hence would tend to be powerful if there is a trend in the groups in some particular direction. Note that for each value of k, the triple summation over (l, i, j) is the same as the Wilcoxon/Mann–Whitney statistic (equation (40), without the constant) for testing the group k versus the "metagroup" consisting of the combined groups $k + 1$ through K, which is the sum of the ranks of the observations in group k relative to the observations in groups k through K. An alternative expression for JT is thus

$$JT = \sum_{k=1}^{K-1} n^{(k+)} \sum_{i=1}^{n^{(k)}} R\left(x_i^{(k)}; F_n^{(k+)}\right) \tag{48}$$

where $F_n^{(k+)}$ is the empirical distribution function for the combined samples $k, \ldots, K$, and $n^{(k+)} = n^{(k)} + \cdots + n^{(K)}$. Choi and Marden (1997) show that under the null hypothesis,

$$\frac{1}{\gamma_n}(JT)'\widehat{\Sigma}_{Rn}^{-1}(JT) \longrightarrow \chi_p^2 \tag{49}$$

where

$$\gamma_n = 2\left(\binom{n}{3} - \sum_{k=1}^{K}\binom{n^{(k)}}{3}\right) \tag{50}$$

There are many other possible qualitative contrasts defined by combining the groups into metagroups. For example, if there are four groups, one could compare the combined first two groups to the combined last two groups. See Marden and Muyot (1995, 1996) for the univariate case, and Choi and Marden (1997) for the multivariate case. These papers also explain a qualitative notion of orthogonality for qualitative contrasts. If one has a set of qualitatively orthogonal contrasts, then under the null hypothesis their Jonckheere–Terpstra statistics are asymptotically independent.

5.4 Two-Way Analysis of Variance: Main Effects

Friedman's test is a univariate rank test designed to test the main effect of the row factor in a two-way layout, where the column factor is generally thought of as a blocking factor. The idea is to rank the observations within each column, then compare the average of these ranks for each row. An alternative approach is to form a χ^2 test statistic for testing row effect within each column, and then sum the statistics. The Friedman approach is best when the row effects are similar for each column, and the second approach is best when the row effects could be quite different in the different columns. We will call these two cases "low interaction" and "general," respectively.

The two approaches are quite easy to extend to the multivariate case, whether using signs or ranks. Suppose there are K rows and L columns, and that $F^{(kl)}$ is the distribution function for the data from the $(kl)^{th}$ cell. The null hypothesis conforming to no row effect is

$$H_{Row} : F^{(kl)} = F^{(k'l)} \qquad \text{for all } l \text{ and } k \neq k' \tag{51}$$

Denote the observations in cell (kl) by $X_i^{(kl)}, i = 1, \ldots, n^{(kl)}$. We will consider the two cases separately.

5.4.1 General case

Within each column, use any of the above test statistics, e.g., equations (43), (46), or (49). For the sign (median) tests, a separate sample spatial median is used for each column. Similarly, for the rank tests, the rank for any observation is calculated relative to the observations within its column. Because the observations from different columns are independent, the individual test statistics are independent. Thus by summing the statistics, one obtains an overall statistic that is asymptotically χ^2 under the null hypothesis, where the degrees of freedom are the sum of those for the individual statistics.

5.4.2 Low interaction case

The statistic in the general case is sensitive to any kind of row effects. When the row effects are approximately the same for the different columns, then one may be able to increase the power of the test by combining the effects before finding the quadratic forms. This is the thinking behind Friedman's test. The null hypothesis assumes no row effects but allows for certain column effects:

$$H_0^{(Friedman)} : F^{(kl)}(x) = F\left(a^{(l)}x + b^{(l)}\right) \tag{52}$$

where $a^{(l)} > 0$ and $b^{(l)}$ is a vector. We will present just the balanced case, that is,

$$n^{(kl)} = n^{(k)}, \qquad \text{for all } k, l \tag{53}$$

For the sign test, let $S_i^{(k|l)} = S(X_i^{(kl)} - \widehat{\eta}^{(l)})$, where $\widehat{\eta}^{(l)}$ is the sample spatial median among the observations in column l. For the rank test, let $R_i^{(k|l)} = R\left(X_i^{(kl)}; F_n^{(l)}\right)$, where $F_n^{(l)}$ is the distribution function for the observations in column l. Let $\overline{S}^{(k)}$ $\left(\overline{R}^{(k)}\right)$ be the averages of the $S_i^{(k|l)}\left(R_i^{(k|l)}\right)$ for row k, averaging over observations in all columns. Then the analogs of the Kruskal–Wallis statistics in equation (43) are

$$W_{Sn} = \sum_{k=1}^{K} Ln^{(k)}\overline{S}^{(k)\prime}\widehat{\Sigma}_{Sn}^{-1}\overline{S}^{(k)} \quad \text{and} \quad W_{Rn} = \sum_{k=1}^{K} Ln^{(k)}\overline{R}^{(k)\prime}\widehat{\Sigma}_{Rn}^{-1}\overline{R}^{(k)} \tag{54}$$

respectively, where the covariance matrices are pooled over rows and columns. These statistics are asymptotically χ_p^2 under the null hypothesis if the limiting covariance matrices are nonsingular.

5.5 Two-Way Analysis of Variance: Interaction

For notational simplicity, we will restrict attention to the 2×2 case, that is, $K = L = 2$, but everything we say can be extended to the general case. There does not appear to be a unique way to define lack of row by column interaction. In the usual normal-theory case, or in any location-family situation, interaction is defined additively, so that no interaction means that $\mu^{(11)} - \mu^{(21)} = \mu^{(12)} - \mu^{(22)}$, where $\mu^{(kl)}$ is the location parameter (e.g., the mean in the normal case) for the observations in cell (kl). For general distributions F, it is not obvious how to define explicitly the notion that "F_{11} is to F_{21} as F_{12} is to F_{22}." For example, one could take any functional $\rho(F)$ and define no interaction to mean

$$\rho(F^{(11)}) - \rho(F^{(21)}) = \rho(F^{(12)}) - \rho(F^{(22)}). \tag{55}$$

There can still be quite a bit of "interaction" not captured by the function ρ. Akritas and Arnold (1994) use the identity function, that is, the no-interaction hypothesis is

$$H_{NI} : F^{(11)} - F^{(21)} = F^{(12)} - F^{(22)} \tag{56}$$

This definition also makes sense in the multivariate case, and in particular implies equation (55) for any linear functional ρ, including the expected value of any function of the random variable.

Another definition of no interaction extends the univariate definition in Marden and Muyot (1995, 1996). That definition is motivated by the one in Patel and Hoel (1973), where no interaction means

$$P\left(X^{(11)} > X^{(21)}\right) = P\left(X^{(12)} > X^{(22)}\right) \tag{57}$$

That is, the chance that an observation from row 1 exceeds that from row 2 is the same for each column. In a location-family model, this definition is equivalent to the usual definition, but also makes sense outside of location-families. The multivariate analog of equation (57) is

$$E\left(S\left(X_i^{(11)} - X_j^{(21)}\right)\right) = E\left(S\left(X_i^{(12)} - X_j^{(22)}\right)\right) \tag{58}$$

which means that the average direction pointing from an observation in row 1 to one in row 2 is the same for each column. Again, this definition is equivalent to the usual one in the location family case.

The formal definition from Marden and Muyot is a bit more complicated, but it implies equation (57). Our multivariate analog follows. For positive integers s and t, consider the set of independent random variables $X_1^{(11)}, \ldots, X_s^{(11)}, X_1^{(21)}, \ldots, X_t^{(21)}, X_1^{(12)}, \ldots, X_s^{(12)}$, and $X_1^{(22)}, \ldots, X_t^{(22)}$, where the $X_i^{(kl)} \sim F^{(kl)}$. It is said that "rows are concordant within columns" if the two sets of vectors $\mathbb{R}^{(1)}$ and $\mathbb{R}^{(2)}$ have the same distribution, where

$$\begin{aligned} \mathbb{R}^{(j)} = \Big\{ & R\left(X_1^{(1j)}; F_{t+s}^{(j)}\right), \ldots, R\left(X_s^{(1j)}; F_{t+s}^{(j)}\right), \\ & R\left(X_1^{(2j)}; F_{t+s}^{(j)}\right), \ldots, R\left(X_t^{(2j)}; F_{t+s}^{(j)}\right)\Big\}, j = 1, 2 \end{aligned} \tag{59}$$

and $F_{t+s}^{(j)}$ is the empirical distribution function for the observations in the combined two cells of column j. The idea is that the set of ranks for row 1 relative to rows 1 and 2 have the same distribution for both columns. The hypothesis of no interaction is then

$$H_{NI}: \text{ Rows are concordant within columns} \tag{60}$$

which implies equation (58).

The two hypotheses in equations (56) and (60) are distinctly different, and have different test statistics associated with them. The first is amazingly invariant. For any function h, if we let $Y_i^{(kl)} = h\left(X_i^{(kl)}\right)$ for each i, k, l, then equation (56) also holds for the distribution functions of the $Y_i^{(kl)}$s. The second is invariant under rotations, location shift, and scalar multiplication; it has a little extra invariance, however, because one can apply different shifts and scalar multiples to the observations in different columns. By the same token, equation (60) is not symmetric in rows and columns, that is, one can have rows concordant within columns without having columns concordant within rows. Both hypotheses are easy to test using rank statistics.

The test for the hypothesis in equation (56) is based on

$$T_n^{(I)} = \overline{R}^{(11)} - \overline{R}^{(21)} - \overline{R}^{(12)} + \overline{R}^{(22)} \tag{61}$$

where $\overline{R}^{(kl)}$ is the average of the ranks in cell (kl),where the ranks are relative to the entire sample. When equation (56) holds, the expected value of $T_n^{(I)}$ can be shown to be zero. We conjecture that the covariance of $T_n^{(I)}$ can be estimated, and an asymptotic χ_p^2 test statistic can be constructed as in the univariate case treated in Akritas and Arnold (1994).

The concordant hypothesis in equation (60) is based on the statistic

$$T_n^{(C)} = \frac{n^{(11)} + n^{(21)}}{n^{(21)}} \overline{R}^{(11)} - \frac{n^{(12)} + n^{(22)}}{n^{(22)}} \overline{R}^{(12)} \tag{62}$$

where $\overline{R}^{(1l)}$ is the average rank for observations in cell $(1l)$, and the ranks are calculated relative to the observations in row l. That is,

$$\frac{n^{(1l)} + n^{(2l)}}{n^{(2l)}} \overline{R}^{(1l)} = \frac{1}{n^{(1l)}} \frac{n^{(1l)} + n^{(2l)}}{n^{(2l)}} \sum_{i=1}^{n^{(1l)}} R^{(l)}\left(X_i^{(1l)}\right) \tag{63}$$

$$= \frac{1}{n^{(1l)} n^{(2l)}} \sum_{i=1}^{n^{(1l)}} \sum_{j=1}^{n^{(2l)}} S\left(X_i^{(1l)} - X_j^{(2l)}\right) \tag{64}$$

$$\text{where } R^{(l)}(X_i^{(1l)}) \equiv R\left(X_i^{(1l)}; F^{(l)}_{n^{(1l)} + n^{(2l)}}\right) \tag{65}$$

Then by equation (58), the expected value under the null hypothesis of $T_n^{(C)}$ in equation (62) is 0. Note that because the ranks are relative to the separate columns, the statistics $\overline{R}^{(11)}$ and $\overline{R}^{(12)}$ are independent, so that

$$\operatorname{Cov}\left(T_n^{(C)}\right) = \left(\frac{n^{(11)} + n^{(21)}}{n^{(21)}}\right)^2 \operatorname{Cov}\left(\overline{R}^{(11)}\right) + \left(\frac{n^{(12)} + n^{(22)}}{n^{(22)}}\right)^2 \operatorname{Cov}\left(\overline{R}^{(12)}\right) \tag{66}$$

Even within columns, the observations are not iid because the distributions in the two rows are not the same, so we cannot directly use equation (36) to find the covariances of the average ranks. Instead we have that

$$\begin{aligned} \operatorname{Cov}\left(\overline{R}^{(1l)}\right) = \frac{1}{n^{(1l)}} \Big(& \operatorname{Cov}\left(R^{(l)}\left(X_i^{(1l)}\right)\right) \\ & + \left(n^{(1l)} - 1\right) \operatorname{Cov}\left(R^{(l)}\left(X_i^{(1l)}\right), R^{(l)}\left(X_j^{(1l)}\right)\right)\Big) \end{aligned} \tag{67}$$

where $i \neq j$. The two covariance matrices can be estimated using their sample analogs,

$$\widehat{\operatorname{Cov}}\left(R^{(l)}\left(X_i^{(1l)}\right)\right) = \frac{1}{n^{(1l)} - 1} \sum_{i=1}^{n^{(1l)}} \left(R^{(l)}\left(X_i^{(1l)}\right) - \overline{R}^{(l)}\right)\left(R^{(l)}\left(X_i^{(1l)}\right) - \overline{R}^{(l)}\right)' \tag{68}$$

and

$$\begin{aligned} \widehat{\operatorname{Cov}}\left(R^{(l)}\left(X_i^{(1l)}\right), R^{(l)}\left(X_j^{(1l)}\right)\right) = & \frac{1}{n^{(1l)}\left(n^{(1l)} - 1\right)} \sum_{i=1}^{n^{(1l)}} \sum_{j=1, j \neq i}^{n^{(1l)}} \left(R^{(l)}\left(X_i^{(1l)}\right) - \overline{R}^{(l)}\right) \\ & \times \left(R^{(l)}\left(X_j^{(1l)}\right) - \overline{R}^{(l)}\right)' \end{aligned}$$

The test statistic is then the quadratic form

$$W_n^{(C)} = T_n^{(C)\prime} \widehat{\operatorname{Cov}}\left(T_n^{(C)}\right)^{-1} T_n^{(C)} \tag{69}$$

which is asymptotically χ_p^2 under the null hypothesis as long as the limiting covariance in equation (67) is nonsingular.

5.6 Higher-Way Layouts

Testing main and interaction effects in higher-way analysis of variance layouts is also possible. Akritas and Arnold's (1994) models consider arbitrary sets of linear constraints on the distribution functions, the interaction constraint in equation (56) being a special case. The statistics then mirror the usual normal-theory statistics, but with ranks replacing the original observations. The concordance definition in equation (60) can also be extended to models in higher-way layouts, but they do not cover all possibilities. For example, there is no provision for testing the hypothesis of no three-way interaction in the presence of all two-way interactions.

6. MULTIPLE/MULTIVARIATE REGRESSION AND INDEPENDENCE

In this section we have two sets of variables, the Xs and the Ys. It may be a regression context, with the Xs the independent variables and the Ys the dependent ones, or the two sets may be considered symmetrically. The data consist of n iid pairs,

$$(X_1, Y_1), \ldots, (X_n, Y_n) \tag{70}$$

where the X_is are $p \times 1$, and the Y_is are $q \times 1$. The two models considered are the *regression model*, where the X_is are fixed and the distribution function of the Y_is depends on the X_is:

$$Y_i \mid X_i = x_i \sim F(\cdot; x_i) \tag{71}$$

and the *correlation model*, where the pair (X_i, Y_i) has a joint distribution J. The null hypotheses corresponding to the two models are that the Y_is are independent of the X_is, i.e., for the regression model,

$$H_R : F(\cdot; x_i) = F \text{ for } i = 1, \ldots, n \text{ for some } F \tag{72}$$

and for the correlation model,

$$H_C : J(x, y) = G(x)F(y) \tag{73}$$

where G and F are the distribution functions for X and Y, respectively. We note that the hypothesis in equation (73) is that for the canonical correlations problem in the multivariate normal case.

Two popular test statistics for these hypotheses in the univariate case are Kendall's τ and Spearman's ρ. The multivariate versions are

$$T_n = \binom{n}{2}^{-1} \sum\sum_{1 \le i < j \le n} S(Y_i - Y_j)S(X_i - X_j)', \tag{74}$$

and

$$S_n = \frac{1}{n}\sum_{i=1}^{n} R(Y_i; F_n)R(X_i; G_n)' \tag{75}$$

respectively, where F_n and G_n are the empirical marginal distribution functions for the Y_is and X_is, respectively. Under either the hypothesis in equation (72) or (73), both statistics have expected value 0. Choi and Marden

(1998) treat the case that $p = 1$ and $q > 1$, and Choi(1997) looks at the general case, for Kendall's τ.

Define $\mathcal{S}$ to be the $p \times \binom{n}{2}$ matrix defined by

$$\mathcal{S}^{(Y)} = (\mathcal{S}_{11}, \mathcal{S}_{12}, \ldots, \mathcal{S}_{1n}, \mathcal{S}_{23}, \ldots, \mathcal{S}_{2n}, \ldots, \mathcal{S}_{n-1,n}) \tag{76}$$

where $S_{ij} = S(Y_i - Y_j)$. Define $\mathcal{S}^{(X)}$ similarly, with the X_is instead of the Y_is. Choi and Marden (1998) show that

$$\text{Cov}(\mathcal{S}^{(Y)}) = (\Lambda^{(Y)} - 2\Sigma_R^{(Y)}) \otimes I_{\binom{n}{2}} + \Sigma_R^{(Y)} \otimes \Gamma_n \Gamma_n' \tag{77}$$

where $\Sigma_R^{(Y)} = \text{Cov}(R(Y; F))$, $\Lambda^{(Y)} = \text{Cov}(S(Y_1 - Y_2))$,and Γ_n is the $\binom{n}{2} \times n$ matrix whose rows are indexed in concert with the columns of $\mathcal{S}^{(Y)}$, i.e., $(12), (13), \ldots, (n-1, n)$, and $(\Gamma_n)_{(ij)k} = +1$ if $i = k$, -1 if $j = k$, and 0 otherwise. From equation (77), because from equation (74) we have that

$$T_n = \binom{n}{2}^{-1} \mathcal{S}^{(Y)} \mathcal{S}^{(X)\prime} \tag{78}$$

$$\text{Cov}(T_n \mid \mathbb{X}) = \binom{n}{2}^{-2} \left((\Lambda^{(Y)} - 2\Sigma_R^{(Y)}) \otimes \mathcal{S}^{(X)\prime} \mathcal{S}^{(X)} + \Sigma_R^{(Y)} \otimes n^3 \widehat{\Sigma}_{Rn}^{(X)} \right) \tag{79}$$

because

$$\mathbb{R}^{(X)} = \frac{1}{n} \mathcal{S}^{(X)} \Gamma_n. \tag{80}$$

(Here, $\mathbb{X} = (X_1, \ldots, X_n)$.) For the regression model, we assume that

$$\lim_{n \to \infty} \widehat{\Sigma}_{Rn}^{(X)} = \Sigma_R^{(X)} \tag{81}$$

for some nonsingular matrix $\Sigma_R^{(X)}$. Under the correlation model and null equation (73), we find the covariance in equation (79) by taking the expected value over $\mathbb{X}$:

$$\text{Cov}(T_n) = \binom{n}{2}^{-2} \left((\Lambda^{(Y)} - 2\Sigma_R^{(Y)}) \otimes nE(S_{ij}^{(X)} S_{ij}^{(X)\prime}) + \Sigma_R^{(Y)} \otimes n^3 \Sigma_{Rn}^{(X)} \right) \tag{82}$$

Then under either equation (72) or (73), it can be shown that

$$4\widehat{\Sigma}_R^{(Y)} \otimes \widehat{\Sigma}_{Rn}^{(X)} \tag{83}$$

is a consistent estimate of the limit of $n \cdot \text{Cov}(T_n \mid \mathbb{X})$ in the regression model, and of $n \cdot \text{Cov}(T_n)$ in the correlation model. The test statistic is then

$$
\begin{aligned}
W_n^{(Kendall)} &= \frac{n}{4} vec(T_n)' \left(\widehat{\Sigma}_R^{(Y)} \otimes \widehat{\Sigma}_{Rn}^{(X)} \right)^{-1} vec(T_n) \\
&= \frac{n}{4} tr \left(\widehat{\Sigma}_R^{(Y)} \right)^{-1} T_n \left(\widehat{\Sigma}_R^{(X)} \right)^{-1} T_n'
\end{aligned} \tag{84}
$$

and it has an asymptotic χ^2_{pq} distribution under the null.

The covariance for the Spearman statistic in equation (75) can be obtained using equation (36), so that

$$
\mathrm{Cov}(S_n \mid \mathbb{X}) = \left(\frac{n}{n-1} \right)^2 \Sigma_{Rn}^{(Y)} \otimes (n-1) \widehat{\Sigma}_{Rn}^{(X)} \tag{85}
$$

Then we can proceed as for Kendall's statistic to obtain

$$
W_n^{(Spearman)} = n \cdot tr \left(\widehat{\Sigma}_R^{(Y)} \right)^{-1} S_n \left(\widehat{\Sigma}_R^{(X)} \right)^{-1} S_n' \tag{86}
$$

which is also asymptotically χ^2_{pq} under the null.

We note that if the dependent variables are one-dimensional ($q = 1$), then Kendall's τ statistic in equation (74) is the Jonckheere–Terpstra statistic in equation (47), where the groups are defined by the unique values of the x_is.

7. GROUP SYMMETRY MODELS

7.1 Introduction

One area where there has been little work on rank procedures is that of structural hypotheses for covariance matrices. Andersson (1975) lays out the theory for a wide class of models based on group symmetries. These models are given for the normal model as follows. Suppose $X \sim N_M(0, \Sigma)$, and $\mathcal{G}$ is a subgroup of $\mathcal{O}(M)$, the $M \times M$ group of orthogonal matrices. Then the *invariant normal* model given by $\mathcal{G}$ is

$$
\{N_M(0, \Sigma) \mid \Sigma \quad \text{such that} \quad \Sigma = g\Sigma g' \quad \text{for all} \quad g \in \mathcal{G}\} \tag{87}
$$

These models include the general iid model, intraclass correlation models (where all variances are equal, and all covariances are equal), spherical symmetry and the other common symmetries mentioned in Section 2.2, models for equality of covariance matrices, independence of blocks of variables, complex normal structure, quaternion normal structure, and others. Andersson(1975) analyzes all possible hypothesis tests in which both hypotheses are invariant normal models, with the null model being contained in the alternative model. He shows that any such testing problem can be decomposed into a series of smaller testing problems, each one being one of a set of ten types of problems.

The work depends strongly on the normality assumption. We would like to create rank analogs of the testing procedures. Note that the model in equation (87) implies that gX and X have the same distribution for any $g \in \mathcal{G}$. This fact motivates our more general model, or models. For $\mathcal{G}$ as above and $p \times 1$ vector γ, let $\mathbf{F}(\mathcal{G}, \gamma)$ consist of all distribution functions F such that

$$\begin{aligned} &X \sim F \Longrightarrow g(X-\gamma) \text{ and } (X-\gamma) \\ &\text{have the same distribution for any } g \in \mathcal{G} \end{aligned} \tag{88}$$

The γ is there because we do not necessarily wish to demand that the location parameter satisfy the symmetry condition. Our general model is that

$$X_1, \dots, X_n \text{ are iid } F \text{ for } F \in \mathbf{F}(\mathcal{G}, \gamma) \tag{89}$$

The key fact to note is that if equation (89) holds, then the distribution of the signs and the distribution of the ranks are also $\mathcal{G}$ invariant.

LEMMA 1: Suppose equation (89) holds. Then for each $g \in \mathcal{G}$,

$$\begin{aligned} &(S(gX_1 - \widehat{\eta}_n^{(g)}), \dots, S(gX_n - \widehat{\eta}_n^{(g)})) \text{ has the same distribution as} \\ &(S(X_1 - \widehat{\eta}_n), \dots, S(X_n - \widehat{\eta}_n)) \end{aligned} \tag{90}$$

where $\widehat{\eta}_n$ is the spatial median of $\mathbb{X}$, and $\widehat{\eta}_n^{(g)}$ is the spatial median of $g\mathbb{X}$, and

$$\begin{aligned} &(R(gX_1; F_n^{(g)}), \dots, R(gX_n; F_n^{(g)})) \text{ has the same distribution as} \\ &(R(X_1; F_n), \dots, R(X_n; F_n)) \end{aligned} \tag{91}$$

where $F_n^{(g)}$ is the empirical distribution function for $gX_1, \dots, gX_n$. □

The lemma can be proved by noting that for any orthogonal matrix, $S(gz) = gS(z)$, hence in equation (90), $\widehat{\eta}_n^{(g)} = g\widehat{\eta}_n$, and so $S(gX_i - \widehat{\eta}_n^{(g)}) = gS(X_i - \widehat{\eta}_n)$. We also have that $R(gX_i; F_n^{(g)}) = gR(X_i; F_n)$. The lemma shows that the covariance matrices of the signs and ranks must be $\mathcal{G}$–invariant, that is,

$$\Sigma_{Sn} = g\Sigma_{Sn}g' \qquad \text{and} \qquad \Sigma_{Rn} = g\Sigma_{Rn}g' \tag{92}$$

Thus these covariance matrices have the same structure as those in the invariant normal model in equation (87).

The rest of this section will concentrate on tests based on the ranks. Corresponding tests based on the signs can be developed similarly. The general hypothesis testing problem we will consider tests

$$H_0 : \mathcal{G} = \mathcal{G}_0 \quad \text{versus} \quad H_A : \mathcal{G} = \mathcal{G}_A \tag{93}$$

based on equation (89) for $\mathcal{G}_A \subset \mathcal{G}_0 \subset O(p)$. For example, one may wish to test for spherical symmetry given the intraclass correlation structure, as in Section 7.4. In the normal situation, the maximum likelihood estimator of the covariance matrix is the usual sample covariance matrix averaged over the group $\mathcal{G}$. We use this fact to motivate our tests. Let $\widehat{\Sigma}_{Rn}$ be the sample covariance matrix in equation (37), where there is only one group ($K = 1$). Then the estimate of Σ_{Rn} under the model in equation (89) is

$$\widehat{\Sigma}_n(\mathcal{G}) = \int_{\mathcal{G}} g\hat{\Sigma}_{Rn}g'\nu(\mathrm{d}g) \tag{94}$$

as in equation (16). The statistics for testing equation (93) are based on linear functions of the difference in the covariance estimates for the two hypotheses:

$$t^{(l)} = tr\left(B^{(l)}\left(\widehat{\Sigma}_n(\mathcal{G}_A) - \widehat{\Sigma}_n(\mathcal{G}_0)\right)\right) \tag{95}$$

for fixed $p \times p$ matrices $B^{(1)}, \ldots, B^{(L)}$. These matrices are chosen to capture interesting differences between the hypotheses. At this point, they need to be chosen in an ad hoc manner. Under the null hypothesis, the expected value of $\hat{\Sigma}_{Rn}$ is invariant under both hypotheses, i.e.,

$$E_0(\hat{\Sigma}_{Rn}) = gE_0(\hat{\Sigma}_{Rn})g' \qquad \text{for all } g \in \mathcal{G}_0 \tag{96}$$

Because $t^{(j)}$ is a linear function of the sample covariance, $E_0(t^{(j)}) = 0$. The overall test statistic is then

$$W_n = T_n'\widehat{\mathrm{Cov}}(T_n)T_n \tag{97}$$

where $T = (t^{(1)}, \ldots, t^{(L)})$.

The covariance matrix for T_n can be estimated from the estimate of the covariance of $\hat{\Sigma}_{Rn}$. For $p \times p$ symmetric matrix Λ, let

$$vec^*(\Lambda)' = (\lambda_{11}, \ldots, \lambda_{p1}, \lambda_{22}, \ldots, \lambda_{p2}, \ldots, \lambda_{pp}) \tag{98}$$

the $p(p+1)/2$–dimensional vector containing the elements on the diagonal and above. Let

$$V_i = vec^*(R(X_i; F_n)R(X_i; F_n)') \tag{99}$$

Then $vec^*(\hat{\Sigma}_{Rn})$ is the sample average of the V_is, and hence an estimate of its covariance is given by

$$n \cdot \widehat{\mathrm{Cov}}(vec^*(\hat{\Sigma}_{Rn})) = \frac{1}{n-1}\sum_{i=1}^{n}(V_i - \overline{V})(V_i - \overline{V})' \tag{100}$$

It can be shown that the statistic in equation (97) is asymptotically χ^2_L under the null hypothesis if $\mathrm{Cov}(T_n)$ has a nonsingular limit.

We note that there are many famous procedures for testing spherical symmetry. See Mardia (1972) for an overview. Chaudhuri (1996) studied the quantile function associated with the ranks in equation (3), which is basically the inverse function of the rank function. (The γth quantile for $\|\gamma\| < 1$ is the point η that minimizes $E(\|X - \eta\| - (X - \eta)'\gamma)$.) Among many other results, he suggests an interesting approach based on a multivariate generalization of the interquartile range. Koltchinskii (1997) proposes another test based on a measure of asymmetry.

The next subsection presents a simple example.

7.2 Example: Testing Intraclass Correlation Structure

In the normal problem, we wish to test whether all the variances are equal to σ^2, and all the covariances are equal to $\rho\sigma^2$. This model can arise in a repeated-measures context, where there are p measurements of the same quantity on each individual, and the distribution of the measurements conditional on the individual are assumed iid. We will take the alternative to be the general one. Translating into groups, we wish to test equation (93) where $\mathcal{G}_0$ is the group of $p \times p$ permutation matrices, and $\mathcal{G}_A = \{I_p\}$ (hence places no restriction on the distribution). The estimate in equation (94) for the null hypothesis has diagonal elements equal to the average of the diagonals, $\widehat{\sigma^2} = tr(\hat{\Sigma}_{Rn})/p$, and off-diagonal elements equal to the average of the off-diagonals,

$$\widehat{(\sigma^2\rho)} = \binom{p}{2}^{-1} \sum\sum_{1<i<j<p} (\hat{\Sigma}_{Rn})_{ij} \tag{101}$$

A natural set of $t^{(l)}$s in equation (95) consists of

$$(\hat{\Sigma}_{Rn})_{ii} - \widehat{\sigma^2} \qquad \text{for } i = 1, \ldots, p-1; \qquad \text{and} \qquad (\hat{\Sigma}_{Rn})_{ij} - \widehat{(\sigma^2\rho)}$$
$$\text{for } 1 \le i < j \le p-1 \tag{102}$$

7.3 Nested Hypotheses

A key feature of Andersson's (1975) work is that one can decompose testing problems into smaller, independent ones. The main fact used is that when the null hypothesis holds, the likelihood ratio statistic is independent of the maximum likelihood estimator calculated under the null. We have a similar but weaker result, namely that the linear test statistics in equation (95) are uncorrelated with $\widehat{\Sigma}_n(\mathcal{G}_0)$ under the null. That is, for any $p \times p$ matrices B and C, under the null hypothesis that $\mathcal{G} = \mathcal{G}_0$,

$$\mathrm{Cov}_0\Big(tr\Big(B\Big(\widehat{\Sigma}_{Rn} - \widehat{\Sigma}_n(\mathcal{G}_0)\Big)\Big), tr\Big(C\widehat{\Sigma}_n(\mathcal{G}_0)\Big)\Big) = 0 \tag{103}$$

To see this fact, note that under the null hypothesis, $\mathbb{X}$ and $g\mathbb{X}$ have the same distribution for each $g \in \mathcal{G}$, and that $\widehat{\Sigma}_{Rn}(g\mathbb{X}) = g\widehat{\Sigma}_{Rn}(\mathbb{X})g'$ and $\widehat{\Sigma}_n(\mathcal{G}_0)(g\mathbb{X}) = \widehat{\Sigma}_n(\mathcal{G}_0)(\mathbb{X})$. Because the covariance is linear, we have that

$$\begin{aligned}
&\mathrm{Cov}_0\Big(tr\Big(B\Big(\widehat{\Sigma}_{Rn} - \widehat{\Sigma}_n(\mathcal{G}_0)\Big)\Big), tr\Big(C\widehat{\Sigma}_n(\mathcal{G}_0)\Big)\Big)\\
&= \int_{\mathcal{G}_0} \mathrm{Cov}_0\Big(tr\Big(B\Big(g\widehat{\Sigma}_{Rn}g' - \widehat{\Sigma}_n(\mathcal{G}_0)\Big)\Big), tr\Big(C\widehat{\Sigma}_n(\mathcal{G}_0)\Big)\Big)\nu(dg)\\
&= \mathrm{Cov}_0\left(tr\left(B(\int_{\mathcal{G}_0} g\widehat{\Sigma}_{Rn}g'\nu(dg) - \widehat{\Sigma}_n(\mathcal{G}_0))\right), tr\Big(C\widehat{\Sigma}_n(\mathcal{G}_0)\Big)\right)\\
&= 0
\end{aligned}$$

by definition of $\widehat{\Sigma}_n(\mathcal{G}_0)$ in equation (94).

Now suppose we wish to test equation (93) , but we have an intermediate group $\mathcal{G}_1, \mathcal{G}_A \subset \mathcal{G}_1 \subset \mathcal{G}_0$. Consider the two hypothesis testing problems:

$$\begin{aligned}
H_0^{(1)} : \mathcal{G} = \mathcal{G}_1 \quad &\text{versus} \quad H_A^{(1)} : \mathcal{G} = \mathcal{G}_A;\\
\text{and } H_0^{(2)} : \mathcal{G} = G_0 \quad &\text{versus} \quad H_A^{(2)} : \mathcal{G} = \mathcal{G}_1
\end{aligned} \tag{104}$$

Let $T_n^{(1)}$ and $T_n^{(2)}$ be statistics as in equation (97) for the two problems, respectively. Then if either null hypothesis is true, $T_n^{(1)}$ and $T_n^{(2)}$ are uncorrelated, and hence asymptotically independent if asymptotic normality holds. We can then decompose the overall testing problem in equation (93) into two smaller ones whose test statistics are asymptotically independent under the null.

7.4 Example: Testing Spherical Symmetry and Intraclass Correlation Structure

Consider equation (93) where $\mathcal{G}_0 = \mathcal{O}(p)$ and $\mathcal{G}_A = \{I_p\}$, so that we are testing the null hypothesis of spherical symmetry versus the general hypothesis. An intermediate hypothesisis the intraclass correlation structure. (Any group hypothesis will be intermediate, in fact.) Thus we take $\mathcal{G}_1$ to be the set of permutation matrices. In this case, $\widehat{\Sigma}_n(\mathcal{G}_0) = \widehat{\sigma^2} I_p$, and $\widehat{\Sigma}_n(\mathcal{G}_1)$ has diagonals $\widehat{\sigma^2}$ and off-diagonals $(\widehat{\sigma^2\rho})$ as in equation (101). To test $H_0^{(1)}$ in equation (104), we just use the single linear statistic $(\widehat{\sigma^2\rho})$ because we are testing spherical symmetry when we know that the variances are equal and the covariances are equal. Testing the second hypothesis in equation (104) is what we did in Section 7.2. Under the null, the two test statistics are asymptotically independent.

8. CONCLUSIONS

We have not said much about the operating characteristics of the procedures. The references given provide many simulations and asymptotic power comparisons. Generally, the procedures appear to have good Type I errors. We will mention a few particular studies to give the flavor of the power results. We are not trying to be at all exhaustive.

Möttönen et al. (1997) compare the asymptotic efficiencies relative to the Hotelling test under the multivariate t distribution of the tests based on the sign statistic in equation (7) and the signed-rank statistic (T_1 in equation (10)). In the univariate ($p = 1$) case, the signed-rank test has efficiency relative to Hotelling of 0.955 at the normal, which rises as the degrees of freedom v in the t declines, until at $v = 3$ the relative efficiency is 1.9. As the dimension p increases, the relative efficiency of the signed-rank test increases slowly, so that at the normal, it is 0.989, while at $v = 3$, it is 2.093. The sign test reveals a similar pattern, but when $p = 1$ it is worse than the signed-rank test (efficiencies of 0.637 and 1.621 at the normal and $v = 3$, respectively), and as p increases, it becomes relatively better, so that for $p = 10$, it has efficiency of 0.951 at the normal, and 2.422 at $v = 3$.

Choi and Marden (1997) consider the two-sample case, finding the asymptotic efficiency relative to the Hotelling test for the Mann–Whitney/Wilcoxon rank test based on equation (40), the coordinatewise Mann-Whitney/Wilcoxon test, and the (affine-invariant) median test of Hettmansperger and Oja (1994). In the spherical normal case, when $p = 1$, the rank and coordinatewise test are the same, with efficiency 0.955, and the median test has efficiency 0.637 (cf. the previous paragraph). As p increases to 100, the efficiencies of the rank and median tests increase to 0.999 and 0.995, respectively, while the coordinatewise test remains at 0.955. Chaudhuri (1992) noticed a similar effect when estimating the mean of the spherical normal, that is, that the efficiency (relative to the mean) of the coordinatewise median was the same for all dimensions, while the efficiency of the spatial median increased to 1 as the dimension increased.

For nonnormal distributions, Choi and Marden used a spherically symmetric distribution found in Randles (1989), in which the parameter controlled the heaviness of the tails. The median test performed better than the rank and coordinate test for heavy-tailed distributions, while the reverse held for slightly heavy and light-tailed distributions. In almost all cases when $p > 1$, the rank test was better than the coordinatewise test. As the dimension increased, all efficiencies became closer to 1.

That paper also contained some simulations, including simulations using contaminated distributions. Generally speaking, the rank and coordinatewise tests outperformed the median test. The rank test tended to be a little better than the coordiatewise test for low levels of contamination, while the coordinatewise test was a little better for heavy contamination.

The power characteristics generally parallel those for their univariate analogs. Thus the sign and median tests tend to have poorer power than the rank-based procedures near the normal, but perform relatively better as the distributions have heavier tails. Interestingly, as the dimension increases, the sign and rank procedures tend to improve relative to the normal-theory procedures when the normal assumption is reasonable. Thus, if anything, the multivariate procedures are even better than the univariate ones, which have proven very useful.

There is much work to be done evaluating procedures mentioned in this paper, as well as developing new procedures. Interesting problems include finding rank-based tests for conditional independence models and other structural models such as arise in factor analysis and time series analysis.

ACKNOWLEDGMENTS

Research supported in part by National Science Foundation Grants NSF DMS 93-04244 and NSF DMS 95-04525. The author thanks the referees for their helpful comments.

REFERENCES

Akritas, M. G., and Arnold, S. F. Fully nonparametric hypotheses for factorial designs. I. Multivariate repeated measures designs. *J. Am. Stat. Assoc.*, 89:336–343, 1994.

Andersson, S. Invariant normal models. *Ann. Stat.*, 3:132–154, 1975.

Barnett, V. The ordering of multivariate data (with discussions). *J. R. Stat. Soc. B*, 139:318–354, 1976.

Bickel, P. J. On some alternative estimates for shift in the p-variate one sample problem. *Ann. Math. Stat.*, 35:1079–1090, 1964.

Bickel, P. J. On some asymptotically nonparametric competitors of Hotelling's T^2. *Ann. Math. Stat.*, 36:160–173, 1965.

Blumen, I. A new bivariate sign test. *J. Am. Stat. Assoc.*, 53:448–456, 1958.

Brown, B. M. Statistical uses of the spatial median. *J. R. Stat. Soc. B*, 45:25–30, 1983.

Brown, B. M., and Hettmansperger, T. P. Affine invariant methods in the bivariate location model. *J. R. Stat. Soc. B*, 49:301–310, 1987.

Brown, B. M., and Hettmansperger, T. P. An affine invariant bivariate version of the sign test. *J. R. Stat. Soc. B*, 51:117–125, 1989.

Chakraborty, B. and Chaudhauri, P. On multivariate rank regression. Technical Report, Indian Statistical Institute, Calcutta, 1997.

Chaudhuri, P. Multivariate location estimation using extension of R-estimates through U-statistics type approach. *Ann. Stat.*, 20:897–916, 1992.

Chaudhuri, P. On a geometric notion of quantiles for multivariate data. *J. Am. Stat. Assoc.*, 91:862–872, 1996.

Chaudhuri, P. and Sengupta, D. Sign tests in multidimension: Inference based on the geometry of the point cloud. *J. Am. Stat. Assoc.*, 88: 1363–1370, 1993.

Choi, K. M. Multivariate rank regression. Technical Report, Hong Ik University, South Korea, 1997.

Choi, K. M., and Marden, J. I. An approach to multivariate rank tests in multivariate analysis of variance. *J. Am. Stat. Assoc.*, 92:1581–1590, 1997.

Choi, K. M., and Marden, J. I. A multivariate version of Kendall's tau. *Nonparametric Stat.*, 9:261–293, 1998.

Conover, W. J., and Iman, R. L. Rank transformations as a bridge between parametric and nonparametric statistics. *Am. Stat.*, 35:124–129, 1981.

Crouse, C. F. A class of distribution-free analysis of variance tests. *S. Afr. Stat. J.*, 1:75–80, 1967.

Dietz, E. J. Bivariate nonparametric tests for the one-sample location problem. *J. Am. Stat. Assoc.*, 77:163–169, 1982.

Donoho, D. L. Breakdown properties of multivariate location estimators. Technical Report, Department of Statistics, Harvard University, 1982.

Fligner, M. A. Comments on "Rank transformations as a bridge between parametric and nonparametric statistics". *Am. Stat.*, 35:131–132, 1981.

Friedman, J. H., and Rafsky, L. C. Multivariate generalizations of the Wald–Wolfowitz and Smirnov two-sample tests. *Ann. Stat.*, 7:697–717, 1979.

Friedman, J. H., and Rafsky, L. C. Graph-theoretic measures of multivariate association and prediction. *Ann. Stat.*, 11:377–391, 1983.

Graybill, F. A. *Introduction to Matrices with Applications in Statistics.* Belmont: Wadsorth, 1969.

Green, P. J. Peeling bivariate data. In: V. Barnett, ed., *Interpreting Multivariate Data.* New York: Wiley, 1981, pp. 3–20.

Hettmansperger, T. P., and Oja, H. Affine invariant multivariate multisample sign tests. *J. R. Stat. Soc. B*, 56:235–249, 1994.

Hettmansperger, T. P., Nyblom, J., and Oja, H. On multivariate notions of sign and rank. In: Y. Dodge. ed. *L_1-Statistical Analysis and Related Methods*. Amsterdam: North Holland, 1992, pp. 267–278.

Hettmansperger, T. P., Nyblom, J., and Oja, H. Affine invariant multivariate one-sample sign tests. *J. R. Stat. Soc. B*, 56:221–234, 1994.

Hettmansperger, T. P., Nyblom, J. and Oja, H. Affine-invariant multivariate one-sample signed-rank tests. *J. Am. Stat. Assoc.*, 92:1591–1600, 1997.

Hodges, J. L. A bivariate sign test. *Ann. Math. Stat.*, 26:523–527, 1955.

Hössjer, O., and Croux, C. Generalizing univariate signed rank statistics for testing and estimating a multivariate location parameter. *Nonparametric Stat.*, 4:293–308, 1995.

Jan, S.-L., and Randles, R. H. A multivariate signed sum test for the one-sample location problem. *Nonparametric Stat.*, 4:49–63, 1994.

Koltchinskii, V. M-estimation, convexity, and quantiles. *Am. Stat.*, 25:435–477, 1997.

Liu, R. Y. On a notion of data depth based on random simplices. *Ann. Stat.*, 18:405–414, 1990.

Liu, R. Y. Data depth and multivarate rank tests. In: Y. Dodge, Ed. *L_1-Statistical Analysis and Related Measures.* Amsterdam: North-Holland, 1992, pp. 279–294.

Liu, R. Y., and Singh, K. A quality index based on data depth and multivariate rank tests. *J. Am. Stat. Assoc.*, 88:252–260, 1993.

Liu, R. Y., Parelius, J. M., and Singh, K. Multivariate analysis by data depth: Descriptive statistics, graphics and inference. Technical Report, Rutgers University, 1997.

Marden, J. I., and Muyot, M. E. T. Rank tests for main and interaction effects in analysis of variance. *J. Am. Stat. Assoc.*, 90:138–1398, 1995.

Marden, J. I., and Muyot, M. E. T. Correction: Rank tests for main and interaction effects in analysis of variance. *J. Am. Stat. Assoc.*, 91: 916, 1996.

Mardia, K. V. *The Statistics of Directional Data.* New York: Academic Press, 1972.

McWilliams, T. P. A distribution-free test for symmetry based on a runs test. *J. Am. Stat. Assoc.*, 85:1130–1133, 1990.

Möttönen, J. and Oja, H. Multivariate spatial sign and rank methods. *Nonparametric Stat.* 5:201--213, 1995.

Möttönen, J., Oja, H. and Tienari, J. On the efficiency of multivariate spatial sign and rank tests. *Ann. Stat.*, 25:542–552, 1997.

Naranjo, J. D., and Hettmansperger, T. P. Bounded influence rank regression. *J. R. Stat. Soc. B*, 56:209–220, 1994.

Oja, H. Descriptive statistics for multivariate distribution. *Stat. Probab. Lett.*, 47:372–377, 1983.

Oja, H. and Nyblom, J. Bivariate sign tests. *J. Am. Stat. Assoc.*, 84:249–259, 1989.

Patel, K. M., and Hoel, D. G. A nonparametric test for interaction in factorial experiments. *J. Am. Stat. Assoc.*, 68:615–620, 1973.

Peters, D. and Randles, R. H. A multivariate signed-rank test for the one-sampled location problem. *J. Am. Stat. Assoc.*, 85:552–557, 1990.

Peters, D., and Randles, R. H. A bivariate signed rank test for the two-sample location problem. *J. R. Stat. Soc. B*, 53:493–504, 1991.

Puri, M. L., and Sen, P. K. *Nonparametric Methods in Multivariate Analysis.* New York: Wiley, 1971.

Randles, R. H. A distribution-free multivariate sign test based on interdirections. *J. Am. Stat. Assoc.*, 84:1045–1050, 1989.

Seheult, A. H., Diggle, P. J., and Evans, D. A. Discussion of paper by V. Barnett. *J. R. Stat. Soc. A*, 139:351–352, 1976.

Small, C. G. A survey of multidimensional medians. *Int. Stat. Rev.*, 58:263–277, 1990.

Thompson, G. L. A note on the rank transform for interactions. *Biometrika*, 78:697–701, 1991.

Tukey, J. W. Mathematics and the picturing data. In: R. D. James, ed. *Proceedings of the International Congress of Mathematicians, Vancouver 1974*, Vol. 2. Montreal: Canadian Mathematical Congress, 1975, pp. 523–531.

15

Asymptotic Expansions of the Distributions of Some Test Statistics for Elliptical Populations

TAKESI HAYAKAWA Hitotsubashi University, Tokyo, Japan

1. INTRODUCTION

Let a sample based on $N = n + 1$ observations come from an m-dimensional normal population with mean μ and covariance matrix Σ, $N_m(\mu, \Sigma)$, and let S be a Wishart matrix with n degrees of freedom, $S \sim W_m(\Sigma, n)$. The spectral decompositions of S and Σ are expressed as

$$S = HLH' \qquad \text{and} \qquad \Sigma = \Gamma\Lambda\Gamma'$$

where $L = diag(l_1, l_2, \ldots, l_m), l_1 > l_2 > \cdots > l_m$ and $\Lambda = diag(\lambda_1, \lambda_2, \ldots, \lambda_m)$, $\lambda_1 > \lambda_2 > \cdots > \lambda_m$. H is an orthogonal matrix whose first element of each column vector is positive and Γ is also an orthogonal matrix.

The exact probability density function of L was studied by Fisher (1939), Girshick (1938), Hsu (1939), and Roy (1939) when Σ is a scalar matrix, and James (1960) gave it for a general case of Σ by use of a ${}_0F_0$-type hypergeometric function in terms of zonal polynomials. More details may be found in James (1964). The distributions of the maximum latent root l_1 and the mimium latent root l_m were obtained by Sugiyama (1967), Muirhead and Chikuse (1975), and Khatri (1972).

It is well known that the joint distribution of latent vectors $H = [h_1, h_2, \ldots, h_m]$ is uniformly distributed on the orthogonal group (James, 1954; Anderson, 1958).

There are several approaches to finding the asymptotic expansions of the distribution function and the probability density function of latent roots.

Anderson (1965) considered the asymptotic expansion of ${}_0F_0$-type hypergeometric functions by use of Hsu's theorem (1948) and obtained the asymptotic expansion of the joint probability density function of latent roots of a Wishart matrix when Σ has simple root λ_is. James (1969) also considered the case for $\Lambda = diag(\lambda_1, \ldots, \lambda_k, \lambda, \ldots, \lambda)$.

The partial differential equation method developed by Muirhead (1970a) is a useful technique to treat the asymptotic expansions of the distribution functions of the normalized latent roots $\sqrt{\frac{n}{2}}\left(\frac{l_i}{\lambda_i} - 1\right), i = 1, 2, \ldots, m$. Muirhead (1970b) and Muirhead and Chikuse (1975) handled the case when λ_is are simple, and Constantine and Muirhead (1976) and Chikuse (1976) considered the case of multiple roots. The reader will find more details in Muirhead (1978). The other method is to use the inversion formula of the characteristic function. Lawley (1956) gave the asymptotic expansion of a latent root by elements of a Wishart matrix (perturbation method). Sugiura (1973b) considered a Taylor expansion and showed that it agreed with Lawley's result. Fujikoshi (1977) and Konishi (1977) obtained the same results as mentioned above. Lawley (1953) considered asymptotic variances and covariances of latent roots for the case of some small roots of a covariance matrix equal. Anderson (1951) also considered asymptotic normality of latent roots under a similar situation.

Nagao (1978) considered an asymptotic behavior for the case of a noncentral Wishart matrix whose noncentrality parameter matrix is of the order of degrees of freedom n.

The limiting distribution of the latent vectors $h_1, \ldots, h_m$ of the sample covariance matrix corresponding to the latent roots $l_1 > l_2 > \cdots > l_m$ is studied in many publications (see, for example, Girshick, 1939; Anderson, 1963; Sugiura, 1976).

2. ASYMPTOTIC DISTRIBUTION OF A SAMPLE MEAN AND A SAMPLE COVARIANCE MATRIX UNDER THE ELLIPITCAL POPULATION

Let x be an m-dimensional random vector having a probability density function of the form

$$c_m|V|^{-1/2}g((x-\mu)'V^{-1}(x-\mu)) \tag{1}$$

for some positive function g and a positive definite matrix V. x is known as an elliptical random vector with parameters μ and V, and the distribution law is denoted as $E_m(\mu, V)$. The characteristic function $\phi(x)$ is assumed to be expressed by some function ψ as

$$\phi(t) = E[\exp(it'x)] = \exp(it'\mu)\psi(t'Vt) \tag{2}$$

assuming the existence of

$$E[x] = \mu, \qquad \mathrm{Cov}(x) = \Sigma = -2\psi'(0)V = \alpha V$$

The kurtosis of $x_j, j = 1, 2, \ldots, m$ are all the same as

$$\kappa = \{\psi''(0) - [\psi'(0)]^2\}/[\psi'(0)]^2 \tag{3}$$

A detail discussion may be seen in Kelker (1970).

The limiting distribution problem was first studied by Muirhead and Waternaux (1980). They considered limiting distributions of a likelihood ratio criterion for testing the hypothesis that k largest canonical correlation coefficients are zero, and of some test criteria for the covariance structures. Muirhead (1980) gave a good review paper on the effects of the elliptical population for some statistics.

To handle the asymptotic theory of a statistic based on a sample covariance matrix, the asymptotic expansion of the probability density function (pdf) of a sample covariance matrix plays a fundamental role. Hayakawa and Puri (1985) considered it for $\mu = 0$, and Iwashita (1997) and Wakaki (1994) extended it for $\mu \neq 0$ as follows.

THEOREM 1 (Iwashita, 1997; Wakaki, 1994): *Let $x_1, \ldots, x_N$ be the random sample from $E_m(\mu, V)$. Let $z = \sqrt{N}\,\Sigma^{-1/2}(\bar{x} - \mu)$ and $Z = \sqrt{n}\,\Sigma^{-1/2}\left(\frac{S}{n} - \Sigma\right)\Sigma^{-1/2}$, where $\bar{x} = \sum_{\alpha=1}^{N} x_\alpha/N$ and $S = \sum_{\alpha=1}^{N}(x_\alpha - \bar{x})(x_\alpha - \bar{x})'$, $n = N - 1$, then the asymptotic expansion of the joint pdf of z and $Z = (z_{ij})$ is given by*

$$f(z, Z) = (2\pi)^{-\frac{1}{4}m(m+3)}|\Sigma_1|^{-\frac{1}{2}}(1+\kappa)^{-\frac{1}{4}m(m-1)}$$
$$\times \exp\left(-\frac{1}{2}z'z\right)\exp\left[-\frac{1}{2}z_1'\Sigma_1^{-1}z_1\right]\exp\left[-\frac{1}{2(1+\kappa)}z_2'z_2\right]$$

$$\times \left[1 + \frac{1}{\sqrt{N}} \left\{ \left(a_0 - \frac{1}{2} + (u + mv) \right) trZ - a_1 (trZ)^3 \right. \right.$$
$$- a_2 trZtrZ^2 - a_3 trZ^3$$
$$\left. \left. + \left(\frac{1}{2} - u \right) z'Zz - vz'ztrZ \right\} + o\left(\frac{1}{\sqrt{N}} \right) \right]$$

where

$$\begin{aligned}
z_1' &= (z_{11}, z_{22}, \ldots, z_{mm}), \qquad z_2' = (z_{12}, z_{13}, \ldots, z_{m-1m}) \\
\Sigma_1^{-1} &= uI_m + vll', \qquad l' = (1, 1, \ldots, 1) \\
u &= \frac{1}{2(1+\kappa)}, \qquad v = \frac{1}{m}\left\{ \frac{1}{2 + (m+2)\kappa} - \frac{1}{2(1+\kappa)} \right\} \\
l_1 &= \frac{4}{3} \frac{\psi^{(3)}(0)}{\alpha^3}, \qquad l_2 = \frac{2\psi''(0)}{\alpha^2}, \qquad l_3 = -\frac{1}{3} \\
a_0 &= (u + mv)[3(m+2)(m+4)(u+v)l_1 \\
&\quad + \{(m+2)^2 u + 3m(m+2)v\} l_2 + 3m(u + mv) l_3] \\
a_1 &= \{u^3 + 3(m+4)u^2 v + 3(m+2)(m+4)uv^2 + m(m+2)(m+4)v^3\} l_1 \\
&\quad + \{u^3 + (3m+4)u^2 v + 3m(m+2)uv^2 + m^2(m+2)v^3\} l_2 \\
&\quad + (u + mv)^3 l_3 \\
a_2 &= 6\{u^3 + (m+4)u^2 v\} l_1 + 2u^2(u + mv) l_2 \\
a_3 &= 8u^3 l_1
\end{aligned}$$

The marginal pdf of z and Z are expanded as follows:

$$\begin{aligned}
f(Z) &= (2\pi)^{-\frac{1}{4}m(m+1)} |\Sigma_1|^{-\frac{1}{2}} (1+\kappa)^{-\frac{1}{4}m(m-1)} \\
&\quad \times \exp\left[-\frac{1}{2} z_1' \Sigma_1^{-1} z_1 \right] \exp\left[-\frac{1}{2(1+\kappa)} z_2' z_2 \right] \\
&\quad \times \left[1 + \frac{1}{\sqrt{N}} \{ a_0 trZ - a_1 (trZ)^3 - a_2 trZ^2 trZ \right. \\
&\quad \left. - a_3 trZ^3 \} + o\left(\frac{1}{\sqrt{N}} \right) \right]
\end{aligned} \tag{5}$$

and

$$f(z) = (2\pi)^{-\frac{1}{2}m} \exp\left(-\frac{1}{2} z'z\right)$$
$$\times \left[1 + \frac{\kappa}{8N}\{m(m+2) - 2(m+2)z'z + (z'z)^2\} + o\left(\frac{1}{N}\right)\right] \tag{6}$$

Equation (5) agrees with the one of Hayakawa and Puri (1985) up to the order $1/\sqrt{n}$. This implies that the results due to Hayakawa and Puri (1985) up to the order $1/\sqrt{n}$ can be extended to the case $\mu \neq 0$.

Recently the expression in equation (4) was extended up to the order $1/N$ by Wakaki, (1997).

It is sometimes useful to use the exponential transformation due to Nagao (1973c):

$$\frac{S}{n} = \Sigma^{1/2} \exp\left(\frac{1}{\sqrt{n}} W\right) \Sigma^{1/2} \tag{7}$$

that is,

$$\frac{Z}{\sqrt{n}} + I = \exp\left(\frac{W}{\sqrt{n}}\right)$$

The asymptotic expansion of the joint pdf of (z, W) is given as follows:

$$f(z, W) = (2\pi)^{-\frac{1}{4}m(m+3)} |\Sigma_1|^{-\frac{1}{2}} (1+\kappa)^{-\frac{1}{4}m(m+1)}$$
$$\times \exp\left(-\frac{1}{2} z'z\right) \exp\left(-\frac{1}{2} w_1' \Sigma_1^{-1} w_1\right) \exp\left(-\frac{1}{2(1+\kappa)} w_2' w_2\right)$$
$$\times \left[1 + \frac{1}{\sqrt{N}}\left\{-\frac{1}{2} w_1^{(2)'} \Sigma_1^{-1} w_1 - \frac{w_2^{(2)'} w_2}{2(1+\kappa)}\right.\right.$$
$$+ \left(a_0 + \frac{m}{2} + (u + mw)\right) trW$$
$$- a_1 (trW)^3 - a_2 trW trW^2 - a_3 trW^3$$
$$\left.\left. + \left(\frac{1}{2} - u\right) z'Wz - vz'ztrW\right\} + o\left(\frac{1}{\sqrt{N}}\right)\right] \tag{8}$$

where

$$W^2 = \left(w_{ij}^{(2)}\right)$$

and

$$w_1' = (w_{11}, \ldots, w_{mm}), \qquad w_2' = (w_{12}, \ldots, w_{(m-1)m})$$
$$w_1^{(2)'} = \left(w_{11}^{(2)}, \ldots, w_{mm}^{(2)}\right), \qquad w_2^{(2)'} = \left(w_{12}^{(2)}, \ldots, w_{(m-1)m}^{(2)}\right)$$

3. THE ROBUSTNESS OF TESTS

Let a sample come from $E_m(\mu, V)$, and let nS_n and Σ be a sample covariance matrix and a population covariance matrix, respectively. Let $h(\Omega)$ be a q-dimensional vector whose elements are functions on the set of $m \times m$ positive definite symmetric matrices possessing continuous first derivatives in the neighborhood $\Omega = \Sigma$. If h satisfies $h(\Omega) = h(\alpha\Omega)$ for all $\alpha > 0$, then we say that h satisfies the "condition h." Let

$$H_d(\Omega) = \frac{1}{2}\left\{\frac{\partial h(\Omega)}{\partial\, vec(\Omega)}\right\}(I + J_m) \tag{9}$$

where $vec(\Omega)$ is an $m^2 \times 1$ vector by stacking the columns of Ω, and $H_d(\Omega)$ is a $q \times m^2$ matrix. $J_m = \sum_{i=1}^m J_{ii} \otimes J_{ii}$ and J_{ij} is an $m \times m$ matrix with one in the (i, j) position and zero elsewhere.

Consider the hypothesis $H_0 : h(\Sigma) = 0$, where h satisfies rank $(H_d(\Sigma)) = q$ for all Σ. Here we define for symmetric positive definite matrices A and Ω:

$$f_n(A, \Omega) = |\Omega|^{-n/2} etr\left(-\frac{n}{2}\,\Omega^{-1}A\right)$$

Let

$$f_{n,h}(A) = \sup_{h(\Omega)=0} f_n(A, \Omega)$$
$$L_{n,h}(A) = f_{n,h}(A)/f_n(A, A) \tag{10}$$

If nS_n represents a Wishart matrix with covariance matrix $\Omega = \Sigma$, then $L_{n,h}(S_n)$ is the likelihood ratio statistic for testing the hypothesis H_0. Tyler showed the following fundamental results.

THEOREM 2 (Tyler, 1983a): *If the condition h holds and a sample comes from* $E_m(\mu, V)$, *then*

1. *under* H_0, $-2\log\{L_{n,h}(S_n)\}/(1+\kappa)$ *converges to* χ_q^2 *in law, where* χ_q^2 *is a central chi-square random variable with q degrees of freedom;*
2. *under the sequence of alternative* K_n,

   $$K_n : \Sigma_n = \Sigma + B/\sqrt{n} \qquad \text{with} \qquad h(\Sigma) = 0$$

$-2\log\{L_{n,h}(S_n)\}/(1+\kappa)$ converges to $\chi_q^2(\delta_h(\Sigma, B)/(1+\kappa))$, where $\chi_q^2(\delta_h(\Sigma, B)/(1+\kappa))$ is a noncentral chi-square random variable with q degrees of freedom and noncentrality parameter

$$\delta_h(\Sigma, B) = \{vec(B)\}'\{H_d(\Sigma)\}'\{C_h(\Sigma)\}^{-1}\{H_d(\Sigma)\}\{vec(B)\}$$

and

$$C_h(\Sigma) = 2\{H_d(\Sigma)\}(\Sigma \otimes \Sigma)\{H_d(\Sigma)\}'$$

THEOREM 3 (Tyler, 1983a): *If the condition h does not hold and a sample comes from $E_m(\mu, V)$, then under H_0 $-2\{\log L_{n,h}(S_n)\}$ converges in law to $(1+\kappa)\chi_{q-1}^2 + \{(1+\kappa) + \kappa\delta_h(\Sigma, \Sigma)\}\chi_1^2$. The necessary and sufficient condition for noncentrality parameter $\delta_h(\Sigma, \Sigma) = 0$ is $h(\Omega) = h(\alpha\Omega)$ for some neighborhood of Σ,*

The results of Muirhead and Waternaux (1980) are closely related to Theorem 3. Tyler (1983a) suggested examples satisfying the condition h as follows:

1. the sphericity for a covariance matrix;
2. the ratio of principal component roots;
3. the principal component for vectors;
4. correlations, canonical vectors and correlations, and multiple correlation coefficients;
5. multivariate linear regression coefficients;
6. the ratio of marginal variances.

Tyler (1983b,c) considered some test problems concerning the structure of a covariance matrix and principal component vectors. However, it seemed to be difficult to handle an asymptotic expansion of a power function of a statistic raised from the multivariate normal theory under elliptical populations when we used the approach mentioned above.

4. ASYMPTOTIC EXPANSIONS UNDER ELLIPTICAL POPULATIONS

In this section we consider the asymptotic behaviors of latent roots of a Wishart matrix, and of likelihood ratio criteria and related statistics for several hypotheses concerning a covariance matrix, latent roots, and latent vectors of it.

4.1 Latent Roots of a Sample Covariance Matrix

Fujikoshi (1980) and Waternaux (1976) studied the asymptotic expansion of the pdf of the latent root l_i of $nS_n = \sum_{\alpha=1}^{N}(x_\alpha - \bar{x})(x_\alpha - \bar{x})'$ for a nonnormal population. Here we consider this problem for an elliptical population.

THEOREM 4 (Hayakawa, 1987): *Let $f(u)$ be a one to one and twice differentiable function in the neighborhood of the population latent root λ_i. The asymptotic expansion of the distribution of a normalized statistic of $f(l_i)$ is given as*

$$\begin{aligned} &P[\sqrt{n}\{f(l_i) - f(\lambda_i) - c/n\}/\lambda_i f'(\lambda_i)(6b_2 - 1)^{1/2} \leq x] \\ &= \Phi(x) - \frac{1}{\sqrt{n(6b_2 - 1)}}\left[2b_2 \sum_{j \neq i}^{m} \frac{\lambda_j}{\lambda_i - \lambda_j} - \frac{1}{6(6b_2 - 1)}(-90b_1 - 18b_2 + 2)\right. \\ &\quad - \frac{c}{\lambda_i f'(\lambda_i)} + x^2\left\{\frac{1}{6(6b_2 - 1)}(-90b_1 - 18b_2 + 2)\right. \\ &\quad \left.\left. + \frac{6b_2 - 1}{2}\lambda_i \frac{f''(\lambda_i)}{f'(\lambda_i)}\right\}\right]\phi(x) + o\left(\frac{1}{n}\right) \end{aligned} \tag{11}$$

where b_1 and b_2 correspond to l_1 and l_2 in Theorem 1.

The transformation which makes the coefficient of x^2 in equation (11) vanish is given by

$$f(x) = \begin{cases} x^d/d : d \neq 0 \\ \log x : d = 0 \end{cases}$$

where

$$d = \frac{90b_1 + 108b_2^2 - 18b_2 + 1}{3(6b_2 - 1)^2}$$

The correction term c which makes the term of order $1/\sqrt{n}$ vanish is chosen as

$$c = \lambda_i^d\left[2b_2 \sum_{j \neq i} \frac{\lambda_j}{\lambda_i - \lambda_j} + \frac{45b_1 + 9b_2 - 1}{3(6b_2 - 1)}\right]$$

The variance stabilization transformation given by $\lambda_i f'(\lambda_i) = 1$ is $f(\lambda_i) = \log \lambda_i$.

For the normal population we have $d = 1/3$ and $c = \lambda_i^{1/3}\left[\sum_{j \neq i} \lambda_j/(\lambda_i - \lambda_j) - 2/3\right]$, which agrees with the result by Konishi (1981).

4.2 Testing the Hypothesis that a Covariance Matrix is Equal to a Given Matrix

Let $x_1, \ldots, x_N$ be a random sample from $N_m(\mu, \Sigma)$. To test the hypothesis

$$H_1 : \Sigma = \Sigma_0 \qquad \text{(a given matrix)} \qquad \text{against} \qquad K_1 : \Sigma_1 \neq \Sigma_0$$

the modified likelihood ratio criterion is given by

$$L_1 = \left(\frac{e}{n}\right)^{\frac{1}{2}mn} |S\Sigma_0^{-1}|^{\frac{n}{2}} etr\left(-\frac{1}{2}\Sigma_0^{-1}S\right) \tag{12}$$

where $S = \sum_{\alpha=1}^{N}(x_\alpha - \bar{x})(x_\alpha - \bar{x})'$, $\bar{x} = \sum_{\alpha=1}^{N} x_\alpha/N$, $n = N - 1$. The properties of L_1 have been extensively studied. The asymptotic expansion of the distribution function of L_1 under the hypothesis was given by Korin (1968), Sugiura (1969), and Davis (1971). The exact percentile points are tabulated by Nagarsenker and Pillai (1973a). The asymptotic expansion of a power function of a normalized likelihood ratio criterion under a fixed alternative was given by Sugiura (1973a). The leading term is a standard normal distribution. However, for a parameter near the null hypothesis, the expression does not give a good approximation. So the local alternative hypotheses $K_n : \Sigma = \Sigma_0 + \Theta/n^{\gamma/2}$, $\gamma > 0$, are considered. Sugiura (1973a) gave the asymptotic expansion of the likelihood ratio criterion for the case $\gamma = 1$, in which the leading term was a noncentral chi-square distribution. It may be worth noting that for $\gamma > 1$, the leading term of the power function becomes a central chi-square distribution with same degrees of freedom, and for $1 > \gamma > 0$, the normalized likelihood ratio criterion converges to a standard normal one.

When a sample comes from $E_m(\mu, V)$, one of the measures of robustness of the behavior of test statistics would be a power function of it. Here we would like to consider the asymptotic expansion of a power function of it. When we use equation (4) under H_1, the term of order $1/\sqrt{n}$ of the asymptotic expansion of the power function vanishes. It is worth having the second term to understand the effects of the ellipticity of the population. So we consider this problem under the squence of alternatives $K_{1n} : \Sigma = \Sigma_0 + \Theta/\sqrt{n}$, $\Theta = (\theta_{ij}) = \Theta'$. Without loss of generality, we assume $\Sigma_0 = I$.

By use of the exponential transformation (Nagao, 1973c)

$$\frac{S}{n} = \left(I + \frac{1}{\sqrt{n}}\,\Theta\right)^{1/2} \exp\left(\frac{1}{\sqrt{n}}\,W\right)\left(I + \frac{1}{\sqrt{n}}\,\Theta\right)^{1/2}$$

the likelihood ratio criterion L_1 is expanded as

$$-2\log L_1 = \frac{1}{2}\,tr(W+\Theta)^2 + \frac{1}{\sqrt{n}}\left\{\frac{1}{6}\,trW^2 + \frac{1}{2}\,trW^2\Theta - \frac{1}{3}\,tr\Theta^3\right\} + o_p\left(\frac{1}{\sqrt{n}}\right) \tag{13}$$

It should be noted that the leading term is expressed as the weighted sum of independent noncentral chi-square random variables

$$\frac{1}{2u}\,\chi^2_{\frac{1}{2}m(m+1)}\left(\frac{\theta_2'\theta_2}{1+\kappa}\right) + \frac{1}{2u}\,\chi^2_{m-1}\left(u\sum_{i=1}^{m}(\theta_{ii}-\bar{\theta})^2\right) + \frac{1}{2(u+mv)}\,\chi^2_1((u+mv)m\bar{\theta}^2) \tag{14}$$

where $\bar{\theta} = tr\Theta/m$, $\theta_2'\theta_2 = \sum_{i<j}\theta_{ij}^2$.

THEOREM 5 (Hayakawa, 1986): *When a sample comes from* $E_m(\mu, V)$, *the power function of* $-2\log L_1$ *under* $K_{1n} : \Sigma = I + \Theta/\sqrt{n}$ *is given as*

$$P\{-2\log L_1 \ge x|K_{1n}\} = \bar{F}^{(1)}_{0,0}(x) + \frac{1}{\sqrt{n}}\sum_{\alpha=0}^{3}\sum_{\beta=0}^{3} d^{(1)}_{\alpha,\beta}\bar{F}^{(1)}_{\alpha,\beta}(x) + o\left(\frac{1}{\sqrt{n}}\right) \tag{15}$$

where

$$d^{(1)}_{3,0} = -\left(\frac{1}{3}\,u + a_3\right)\left(s_3 - \frac{3}{m}\,s_1s_2 + \frac{2}{m^2}\,s_1^3\right)$$

$$d^{(1)}_{2,1} = -\frac{1}{2}(2u + mv + 2a_2m + 6a_3)\left(\frac{s_1s_2}{m} - \frac{s_1^3}{m^2}\right), \qquad d^{(1)}_{1,2} = 0$$

$$d^{(1)}_{0,3} = -\left\{\frac{1}{3}(u+mv) + a_1m^2 + a_2m + a_3\right\}\frac{s_1^3}{m^2}$$

$$d^{(1)}_{2,0} = \left(\frac{4}{3}u + 3a_3\right)s_3 - \left(\frac{5}{2}u - \frac{1}{2}mv - a_2 m + 6a_3\right)\frac{s_1 s_2}{m} + \left(\frac{7}{6}u - \frac{1}{2}mv - a_2 m + 3a_3\right)\frac{s_1^3}{m^2}$$

$$d^{(1)}_{1,1} = \left(\frac{5}{2}u + mv + 2a_2 m + 6a_3\right)\left(\frac{s_1 s_2}{m} - \frac{s_1^3}{m^2}\right) - \frac{1}{4u}(m-1)(m+2)(2u + mv + 2a_2 m + 6a_3)\frac{s_1}{m}$$

$$d^{(1)}_{0,2} = \left\{\frac{4}{3}(u + mv) + 3(a_1 m^2 + a_2 m + a_3)\right\}\frac{s_1^3}{m^2} - \left\{1 + \frac{3}{u + mv}(a_1 m^2 + a_2 m + a_3)\right\}\frac{s_1}{m}$$

$$d^{(1)}_{1,0} = -(2u + 3a_3)s_3 + (2u - mv - 2a_2 m + 3a_3)\frac{s_1 s_2}{m} + (mv + 2a_2 m)\frac{s_1^3}{m^2} + \frac{1}{4u}(m-1)(m+2)(3u + mv + 2a_2 m + 6a_3)\frac{s_1}{m}$$

$$d^{(1)}_{0,1} = -(2u + mv + a_2 m + 3a_3)\frac{s_1 s_2}{m} - (mv + 3a_1 m^2 + 2a_2 m)\frac{s_1^3}{m^2} + \left\{\frac{m^2 + m + 6}{4} + a_0 m + \frac{1}{u + mv}(a_1 m^2 + a_2 m + a_3)\right\}\frac{s_1}{m}$$

$$d^{(1)}_{0,0} = (u + a_3)s_3 + (v + a_2)s_1 s_2 + a_1 s_1^3 - \left(\frac{m+1}{2} + a_0\right)s_1$$

$$s_\alpha = tr\Theta^\alpha, \qquad \alpha = 1, 2, 3$$

It should be noted that $\sum_{\alpha=0}^{3}\sum_{\beta=0}^{3} d^{(1)}_{\alpha,\beta} = 0.$

$$\bar{F}^{(1)}_{\alpha,\beta}(x) = 1 - \sum_{k=0}^{\infty} e_k P\{\chi^2_{f+2\alpha+2\beta+2k} \le x/w\}$$

$$e_0 = \exp\left(-\frac{1}{2}(us_2 + vs_1^2)\right)(2w)^{\frac{f}{2}+\alpha+\beta} u^{\frac{1}{4}(m-1)(m+2)+\alpha}(u + mv)^{\frac{1}{2}+\beta}$$

$$e_k = \frac{1}{2k}\sum_{j=0}^{k-1} G_{k-j} e_j, \qquad k \ge 1$$

$$
\begin{aligned}
G_k = &\left\{\frac{1}{4}(m-1)(m+2)+2\alpha\right\}(1-2uw)^k \\
&+(1+2\beta)\{1-2(u+mv)w\}^k \\
&+kw\Bigg[2(u+mv)^2\frac{s_1^2}{m}\{1-2(u+mv)w\}^{k-1} \\
&+2u^2\left(s_2-\frac{s_1^2}{m}\right)(1-2uw)^{k-1}\Bigg]
\end{aligned}
$$

where $f = \frac{1}{2}m(m+1)$. w is a suitably chosen constant for a rapid convergence.

4.3 Testing the Hypothesis of Sphericity of a Covariance Matrix

To test the hypothesis $H_2 : \Sigma = \sigma^2 I_m$ against $K_2 : \Sigma \neq \sigma^2 I_m$, the modified likelihood ratio criterion L_2 for the normal population is given by

$$L_2 = |S|^{n/2}/\{trS/m\}^{mn/2} \tag{16}$$

It is well known that L_2 is an unbiased test (Gleser, 1966; Sugiura and Nagao, 1968). The asymptotic expansion of the distribution function of L_2 under the null hypothesis was given by Anderson (1958), and the tables of the percentile points were given by Nagarsenker and Pillai (1973b).

Sugiura (1969) gave the asymptotic expansion of the power function of L_2 under a fixed alternative, and Nagao (1973a) gave it under the local alternatives $K_{2n} : \Sigma = \sigma^2 I_m + \Theta/\sqrt{n}$. When a sample comes from $E_m(\mu, V)$, the asymptotic expansion of the power function of $-2\log L_2$ under the sequence of local alternatives was given by Hayakawa (1986), and was expressed by use of the representation of the distribution of a quadratic form. In Hayakawa (1986), the sequence of local alternatives should be read as $\tilde{K}_{2n} : \Sigma = \sigma^2(I + \Theta/\sqrt{n})$.

By noting the fact that the hypothesis H_2 enjoys Tyler's condition h, a limiting distribution of $-2\log L_2/(1+\kappa)$ is a noncentral chi-square one.

THEOREM 6: *When a sample comes from $E_m(\mu, V)$, the asymptotic expansion of a power function of $-2\log L_2/(1+\kappa)$ under $\tilde{K}_{2n} : \Sigma = \sigma^2(I + \Theta/\sqrt{n})$ is expressed as*

$$P\{-2\log L_2/(1+\kappa) \geq x|\tilde{K}_{2n}\} = \bar{P}_f(\delta_2^2) + \frac{1}{\sqrt{n}}\sum_{\alpha=0}^{3} d_\alpha^{(2)}\bar{P}_{f+2\alpha}(\delta_2^2) + o\left(\frac{1}{\sqrt{n}}\right) \tag{17}$$

where

$$d_3^{(2)} = -\left(\frac{u}{3}+a_3\right)\left(s_3 - \frac{3s_1s_2}{m} + \frac{2s_1^3}{m^2}\right)$$

$$d_2^{(2)} = \left(\frac{4}{3}u + a_3\right)s_3 - (4u+9a_3)\,\frac{s_2s_2}{m} + \left(\frac{8}{3}u + 6a_3\right)\frac{s_1^3}{m^2}$$

$$d_1^{(2)} = -(2u+3a_3)s_3 + (5u+9a_3)\frac{s_1s_2}{m} - (3u+6a_3)\frac{s_1^3}{m^2}$$

$$d_0^{(2)} = (u+a_3)s_3 - (2u+3a_3)\frac{s_1s_2}{m} + (u+2a_3)\frac{s_1^3}{m^2}$$

$$f = \frac{1}{2}(m-1)(m+2), \qquad \delta_2^2 = \frac{u}{2}\left(s_2 - \frac{1}{m}s_1^2\right), \quad s_\alpha = tr\Theta^\alpha, \quad \alpha = 1.2.3$$

$$\bar{P}_f(\delta_2^2) = P\{\chi_f^2(\delta_2^2) \geq x\}$$

Nagao (1993) obtained the same representation with slightly different notations for the case $K_{2n} : \Sigma = \sigma^2 I + \Theta/\sqrt{n}$ and $s_1 = tr\Theta = 0$. Nagao and Srivastava (1992) handled the asymptotic distribution problem of some test criteria for a covariance matrix including a sphericity structure under a more general set up.

4.4 Testing the Independence of Sets of Variates

Let x be distributed as $E_m(\mu, V)$ and let μ be partitioned into q subvectors with $m_1, m_2, \ldots, m_q$ components. The covariance matrix Σ is partitioned similarly, i.e.,

$$\Sigma = \begin{bmatrix} \Sigma_{11} & \Sigma_{12} & \cdots & \Sigma_{1q} \\ \Sigma_{21} & \Sigma_{22} & \cdots & \Sigma_{2q} \\ \vdots & \vdots & & \vdots \\ \Sigma_{q1} & \Sigma_{q2} & \cdots & \Sigma_{qq} \end{bmatrix}$$

To test the hypothesis

$$H_3 : \Sigma_{\alpha\beta} = 0 \qquad (\alpha \neq \beta)$$

against

$$K_3 : \Sigma_{\alpha\beta} \neq 0 \qquad \text{for at least one pair } \alpha, \beta(\alpha \neq \beta)$$

the modified likelihood ratio criterion based on an $N = n + 1$ normal sample is given as

$$L_3 = \left\{ |S| / \prod_{\alpha=1}^{q} |S_{\alpha\alpha}| \right\}^{n/2} \tag{18}$$

where the $S_{\alpha\alpha}$s are submatrices of S partitioned in the same manner as the covariance matrix. The asymptotic expansion of the distribution of L_3 under H_3 was obtained by Box (1949). The asymptotic expansions of a power function of it under the fixed alternative and local alternatives $K_{3n} : \Sigma = \Sigma_D + \Theta/\sqrt{n}$, where $\Sigma_D = diag(\Sigma_{11}, \ldots, \Sigma_{qq})$ and $\Theta = (\Theta_{\alpha\beta})$ with $\Theta_{\alpha\alpha} = O$ and $\Theta'_{\alpha\beta} = \Theta_{\beta\alpha}, \alpha \neq \beta$, were obtained by Nagao (1972) and Nagao (1973b), respectively. As the hypothesis H_3 enjoys Tyler's condition h, the limiting distribution of $-2\log L_3/(1+\kappa)$ is a noncentral chi-square one.

THOEREM 7 (Hayakawa, 1986): *When a sample comes from $E_m(\mu, V)$, the asymptotic expansion of the power function of it under K_{3n} is given as*

$$P\{-2\log L_3/(1+\kappa) \geq x | K_{3n}\} = \bar{P}_f(\delta_3^2) + \frac{s_3}{\sqrt{n}} \sum_{\alpha=0}^{3} d_\alpha^{(3)} \bar{P}_{f+2\alpha}(\delta_3^2) + o\left(\frac{1}{\sqrt{n}}\right) \tag{19}$$

where

$$d_3^{(3)} = -\left(\frac{1}{3}u + a_3\right), \qquad d_2^{(3)} = \left(\frac{4}{3}u + 3a_3\right), \qquad d_1^{(3)} = -(2u + 3a_3)$$

$$d_0^{(3)} = u + a_3, \qquad f = \frac{1}{2}(m^2 - \sum_{\alpha=1}^{q} m_\alpha^2), \qquad \delta_3^2 = \frac{u}{2} s_2 \qquad s_\alpha tr\Theta^\alpha, \ \alpha = 2, 3$$

Nagao (1993) also obtained this with slightly different notations.

4.5 Testing the Hypothesis that the Latent Roots of a Covariance Matrix are Equal to Given Values

Let x be an m-dimensional normal random vector with mean μ and covariance matrix Σ. Let Λ be a diagonal matrix $diag(\lambda_1, \ldots, \lambda_m)$, where λ_is are the latent roots of Σ with $\lambda_1 > \cdots > \lambda_m$ and we assume that the multiplicity of each latent root is one. Let Γ be an $m \times m$ orthogonal matrix whose i-th column vector corresponds to the i-th latent root λ_i.

To test the hypothesis

$$H_4 : \Lambda = \Lambda_0 \qquad \text{(a given diagonal matrix)} \qquad \text{against} \qquad K_4 : \Lambda \neq \Lambda_0$$

the modified likelihood ratio criterion L_4 is given by

$$L_4 = \left(\frac{e}{n}\right)^{\frac{1}{2}mn} \left(\prod_{i=1}^{m} \frac{l_i}{\lambda_{i0}}\right)^{\frac{n}{2}} \exp\left(-\frac{1}{2}\sum_{i=1}^{m} \frac{l_i}{\lambda_{i0}}\right) \tag{20}$$

where $l_1 > l_2 > \cdots > l_m$ are the latent roots of S.

THEOREM 8 (Hayakawa and Puri, 1985): *When a sample comes from $N_m(\mu, \Sigma)$, the asymptotic expansions of the distribution of $-2\log L_4$ under H_4 and $K_{4n} : \Lambda = \Lambda_0 + \Theta/\sqrt{n}$, $\Theta = diag(\theta_1, \theta_2, \ldots, \theta_m)$ are given as*

$$P\{-2\log L_4 \le x | H_4\}$$
$$= P\{\chi_m^2 \le x\} + \frac{g}{n}[P\{\chi_{m+2}^2 \le x\} - P\{\chi_m^2 \le x\}] + o\left(\frac{1}{n}\right) \tag{21}$$

where

$$g = \frac{1}{12}(3m^2 - m) + \frac{1}{2}\, trA^2 + \frac{1}{4}\sum_{j=1}^{m}\left(\sum_{i \ne j} a_{ij}\right)^2$$

$$A = (a_{ij}), \qquad a_{ij} = \lambda_{ij0}\lambda_{j0}, \qquad \lambda_{ij0} = (\lambda_{i0} - \lambda_{j0})^{-1}$$

$$P\{-2\log L_4 \ge x | K_{4n}\}$$
$$= \bar{P}_m(\delta_4^2) + \frac{1}{\sqrt{n}}\sum_{\alpha=0}^{2} b_\alpha^{(4)} \bar{P}_{m+2\alpha}(\delta_4^2) + o\left(\frac{1}{\sqrt{n}}\right) \tag{22}$$

where

$$b_2^{(4)} = \frac{1}{6}\sum_{i=1}^{m} \theta_i^3$$

$$b_1^{(4)} = \frac{1}{2}\sum_{i=1}^{m} \theta_i \sum_{j \ne i} \lambda_{ij0}\lambda_{j0} - \frac{1}{2}\sum_{i=1}^{m} \theta_i^3$$

$$b_0^{(4)} = -\frac{1}{2}\sum_{i=1}^{m} \theta_i \sum_{j \ne i} \lambda_{ij0}\lambda_{j0} + \frac{1}{3}\sum_{i=1}^{m} \theta_i^3$$

$$\delta_4^2 = \frac{1}{4}\sum_{i=1}^{m} \theta_i^2$$

THEOREM 9 (Hayakawa and Puri, 1985): *When a sample comes from $E_m(\mu, V)$, the asymptotic expansion of the power function of $-2\log L_4$ under K_{4n} is given as*

$$P\{-2\log L_4 \geq x|K_{4n}\}$$
$$= \bar{F}^{(4)}_{0,0}(x) + \frac{1}{\sqrt{n}}\sum_{\alpha=0}^{3}\sum_{\beta=0}^{3} d^{(4)}_{\alpha,\beta}\bar{F}^{(4)}_{\alpha,\beta}(x) + o\left(\frac{1}{\sqrt{n}}\right) \tag{23}$$

where

$$\bar{F}^{(4)}_{\alpha,\beta}(x) = 1 - \sum_{k=0}^{\infty} e_k P\{\chi^2_{m+2\alpha+2\beta+2k} \leq x/w\}$$
$$e_0 = \exp\left\{-\frac{1}{2}\left(us_2 + (u+mv)m\bar{\theta}^2\right)\right\}(2w)^{\frac{1}{2}m+\alpha+\beta}u^{\frac{1}{2}(m-1)+\alpha}(u+mv)^{\frac{1}{2}+\beta}$$
$$e_k = \frac{1}{2k}\sum_{j=0}^{k-1} G_{k-j}e_j, \qquad (k \geq 1)$$
$$G_k = (m-1+2\alpha)(1-2uw)^k + (1+2\beta)\{1-2(u+mv)w\}^k$$
$$+ kw\left[2(u+mv)^2 m\bar{\theta}^2\{1-2(u+mv)w\}^{k-1} + 2u^2 s_2(1-2uw)^{k-1}\right]$$
$$d^{(4)}_{3,0} = -\left(\frac{1}{3}u + a_3\right)\bar{s}_3, \qquad d^{(4)}_{2,1} = -(u + a_2 m + 3a_3)\bar{\theta}\bar{s}_2$$
$$d^{(4)}_{1,2} = 0, \qquad d^{(4)}_{0,3} = -\left\{\frac{1}{3}(u+mv) + (a_1 m^2 + a_2 m + a_3)\right\}m\bar{\theta}^3$$

$$d^{(4)}_{2,0} = \left(\frac{4}{3}u + 3a_3\right)\bar{s}_3 + (u + a_2 m + 3a_3)\bar{\theta}\bar{s}_2$$
$$d^{(4)}_{1,1} = (3u + 2a_2 m + 6a_3)\bar{\theta}\bar{s}_2 - \frac{1}{2}(m-1)\frac{1}{u}(2u + mv + 2a_2 m + 6a_3)\bar{\theta}$$
$$d^{(4)}_{0,2} = \left\{\frac{4}{3}(u+mv) + 3(a_1 m^2 + a_2 m + a_3)\right\}m\bar{\theta}^3$$
$$- \left\{1 + \frac{3}{u+mv}(a_1 m^2 + a_2 m + a_3)\right\}\bar{\theta}$$
$$d^{(4)}_{1,0} = -(2u + 3a_3)s_3 - (3u + 2a_2 m + 6a_3)\bar{\theta}\bar{s}_2$$
$$+ \frac{1}{2}(m-1)\left\{\frac{m}{2} + \frac{1}{u}(3u + mv + 2a_2 m + 6a_3)\right\}\bar{\theta}$$
$$+ \frac{1}{2}\sum_{i=1}^{m}\theta_i\left(\sum_{j\neq i}\lambda_{ij0}\lambda_{j0}\right)$$

$$d_{0,1}^{(4)} = -(3u + mv + a_2 m + 3a_3)\bar{\theta}\bar{s}_2$$
$$- (2u + 2mv + 3a_1 m^2 + 3a_2 m + 3a_3) m\bar{\theta}^3$$
$$+ \left\{ \frac{3}{u + mv} (a_1 m^2 + a_2 m + a_3) + \frac{1}{2}(m^2 + 3) + a_0 m \right.$$
$$\left. - \frac{1}{2} m(m-1)(1+\kappa)(u + mv + 2a_2 m + 6a_3) \right\} \bar{\theta}$$
$$d_{0,0}^{(4)} = (u + a_3)\bar{s}_3 + (3u + mv + a_2 m + 3a_3)\bar{\theta}\bar{s}_2$$
$$+ (u + mv + a_1 m^2 + a_2 m + a_3) m\bar{\theta}^3$$
$$+ \left\{ \frac{1}{2} m(m-1)(1+\kappa)(mv + 2a_2 m + 6a_3) - \frac{1}{2} m(m+1) - a_0 m \right\} \bar{\theta}$$
$$- \frac{1}{2} \sum_{i=1}^{m} \theta_i \left(\sum_{j \neq i} \lambda_{ij0} \lambda_{j0} \right)$$
$$d_{\alpha,\beta}^{(4)} = 0 \quad \text{for} \quad \alpha + \beta \geq 4, \qquad \bar{s}_k = \sum_{i=1}^{k} (\theta_i - \bar{\theta})^k, \qquad \bar{\theta} = tr\Theta/m$$

w is a suitably chosen constant for a rapid convergence.

4.6 Testing the Hypothesis that the Latent Vectors of a Covariance Matrix are Equal to Given Vectors

Under the same condition as in Section 4.5, the modified likelihood ratio criterion L_5 for testing the hypothesis $H_5 : \Gamma_1 = [\gamma_1, \ldots, \gamma_k] = [\gamma_{10}, \ldots, \gamma_{k0}] = \Gamma_{10}$, where Γ_{10} is a specified $m \times k$ matrix whose columns are orthonormal vectors, is given by

$$L_5 = \left[|S| \Bigg/ \left\{ \prod_{i=1}^{k} (\Gamma_{10}' S \Gamma_{10})_{ii} |\Gamma_2' S \Gamma_2| \right\} \right]^{\frac{n}{2}} \tag{24}$$

where $(A)_{ii}$ denotes the i-th diagonal element of the matrix and an $m \times (m-k)$ matrix Γ_2 is such that $\Gamma = [\Gamma_{10}, \Gamma_2]$ is orthogonal. L_5 was obtained by Mallows (1961) and Gupta (1967). Here we consider the sequence of local alternatives

$$K_{5n} : \Gamma_1 = \Gamma_{10} \exp(\Theta_{11}/\sqrt{n})$$

where Θ_{11} is a $k \times k$ skew symmetric matrix.

THEOREM 10 (Hayakawa and Puri, 1985): *When a sample comes from $N_m(\mu, \Sigma)$, the asymptotic expansion of the power function of L_5 under K_{5n} is given as*

$$P\{-2\log L_5 \geq x | K_{5n}\}$$
$$= \bar{P}_f(\xi^2) + \frac{m_3}{2\sqrt{n}}\left[2\bar{P}_{f+4}(\xi^2) - 3\bar{P}_{f+2}(\xi^2) + \bar{P}_f(\xi^2)\right] + o\left(\frac{1}{\sqrt{n}}\right) \tag{25}$$

where $f = \frac{1}{2}k(k-1) + k(m-k)$, $\quad m_3 = \sum_{i,j,l}^{k}\left(\frac{\lambda_j}{\lambda_i}\right)\theta_{ij}\theta_{jl}\theta_{li}$

$$\Xi_{11} = (\xi_{ij}), \qquad \xi_{ij} = \begin{cases} \left\{\left(\frac{\lambda_j}{\lambda_i}\right)^{1/2} - \left(\frac{\lambda_i}{\lambda_j}\right)^{1/2}\right\}\theta_{ij}, & (1 \leq i \neq j \leq k) \\ 0 \quad , & \text{otherwise} \end{cases}$$

$$\xi^2 = \frac{1}{2(1+\kappa)}\, tr\Xi_{110}^2$$

THEOREM 11 (Hayakawa and Puri, 1985): *When a sample comes from $E_m(\mu, V)$, the asymptotic expansion of the power function of L_5 under K_{5n} is given by*

$$P\{-2\log L_5/(1+\kappa) \geq x | K_{5n}\}$$
$$= \bar{P}_f(\xi^2) - \frac{m_3}{\sqrt{n}}\sum_{\alpha=0}^{3} d_\alpha^{(5)}\bar{P}_{f+2\alpha}(\xi^2) + o\left(\frac{1}{\sqrt{n}}\right) \tag{26}$$

where

$$d_3^{(5)} = \frac{1}{1+\kappa} + \frac{6l_1}{(1+\kappa)^3}, \qquad d_2^{(5)} = -\frac{4}{1+\kappa} - \frac{18l_1}{(1+\kappa)^3}$$
$$d_1^{(5)} = \frac{9}{2(1+\kappa)} + \frac{18l_1}{(1+\kappa)^3}, \qquad d_0^{(5)} = -\frac{3}{2(1+\kappa)} - \frac{6l_1}{(1+\kappa)^3}$$

4.7 Testing the Hypothesis that the First Latent Vector Corresponding to the Largest Latent Root of a Covariance Matrix is Equal to a Given Vector

Under the same condition as Section 4.5, the modified likelihood ratio criterion L_5 for testing the hypothesis $H_5' : \gamma_1 = \gamma_{10}$ against $K_5' : \gamma_1 \neq \gamma_{10}$ is reduced to

$$L_5' = [|S|/\gamma_{10}' S \gamma_{10} |\Gamma_2' S \Gamma_2|]^{n/2} \tag{27}$$

Anderson's test statistic T for H_5' is given as

$$T = n\{\gamma_{10}' S \gamma_{10}/l_1 - 2 + l_1 \gamma_{10}' S^{-1} \gamma_{10}\} \tag{28}$$

which corresponds to the Wald statistic (Anderson, 1963).

THEOREM 12 (Hayakawa, 1978): *When a sample comes from $N_m(\mu, \Sigma)$, the asymptotic expansions of the distributions of L_5' and T are given under H_5' as*

$$P\{-2\log L_5' \le x|H_5'\} = P_f + \frac{1}{n}\frac{m^2-1}{4}\{P_{f+2} - P_f\} + o\left(\frac{1}{n}\right) \tag{29}$$

$$P\{T \le x|H_5'\} - P_f + \frac{1}{n}\sum_{\alpha=0}^{2} b_\alpha^{(5)} P_{f+2\alpha} + o\left(\frac{1}{n}\right) \tag{30}$$

where

$$f = m-1, \qquad P_f = Pr\{\chi_f^2 \le x\}$$

$$b_2^{(5)} = \frac{1}{4}(m^2-1) + D, \qquad b_1^{(5)} = -D, \qquad b_0^{(5)} = -\frac{1}{4}(m^2-1)$$

$$D = \sum_{i=2}^{m} a_{1i}^2 + \frac{1}{2}\left(\sum_{i=2}^{m} a_{1i}\right)^2 + \frac{m+1}{2}\sum_{i=2}^{m} a_{1i} \, (> 0)$$

$$a_{1i} = \lambda_i/(\lambda_1 - \lambda_i), \qquad i = 2, 3, \ldots, m$$

The Cornish–Fisher-type generalized expansion due to Hill and Davis (1968) gives the upper α percentile points x_L of $-2\log L_5'$ and x_T of T.

$$x_L = \eta + \frac{1}{n}\frac{m+1}{2}\eta + o\left(\frac{1}{n}\right) \tag{31}$$

$$x_T = \eta + \frac{1}{n}\left[\frac{1}{(m^2-1)}\left\{\frac{1}{4}(m^2-1) + D\right\}\eta^2 + \frac{1}{2}(m+1)\eta\right] + o\left(\frac{1}{n}\right) \tag{32}$$

where η is the upper α percentile point of a chi-square random variable with $f = m-1$ degrees of freedom. Noting that D is positive, we have

$$x_T > x_L + o\left(\frac{1}{n}\right) \tag{33}$$

To compare the powers of $-2\log L_5'$ and T under the elliptical population as well as the normal population, we use the following sequence of alternatives:

$$K_{5n}' : \Gamma = [\gamma_{10}, \Gamma_2]\exp(\Theta/\sqrt{n}), \qquad \Theta' = -\Theta$$

It is of interest to note that the expansions of $-2\log L_5'$ and T are the same up to the order $1/\sqrt{n}$ under K_{5n}', as follows:

$$
\begin{aligned}
-2\log L_5' &= T \\
&= \sum_{i=2}^{m}\left\{w_{i1} + \left(\sqrt{\frac{\lambda_1}{\lambda_i}} - \sqrt{\frac{\lambda_i}{\lambda_1}}\right)\theta_{i1}\right\}^2 \\
&\quad + \frac{1}{\sqrt{n}}\left[-\sum_{i=2}^{m}\left(\sqrt{\frac{\lambda_1}{\lambda_i}} + \sqrt{\frac{\lambda_i}{\lambda_1}}\right)\theta_{i1}w_{i1}^{(2)}\right. \\
&\quad + 2w_{11}\sum_{i=2}^{m}\left(\sqrt{\frac{\lambda_1}{\lambda_i}} + \sqrt{\frac{\lambda_i}{\lambda_1}}\right)\theta_{i1}w_{i1} \\
&\quad - \sum_{i=2}^{m}\left(\sqrt{\frac{\lambda_1}{\lambda_i}} - \sqrt{\frac{\lambda_i}{\lambda_1}}\right)\theta_{i1}^{(2)}w_{i1} + w_{11}\sum_{i=2}^{m}\left(\frac{\lambda_1}{\lambda_i} - \frac{\lambda_i}{\lambda_1}\right)\theta_{i1}^2 \\
&\quad \left. - \sum_{i=2}^{m}\left(\frac{\lambda_1}{\lambda_i} + \frac{\lambda_i}{\lambda_1}\right)\theta_{i1}^{(2)}\theta_{i1} + \sum_{i=2}^{m}\sum_{j=2}^{m}\left(\frac{\sqrt{\lambda_i\lambda_j}}{\lambda_1} - \frac{\lambda_1}{\sqrt{\lambda_i\lambda_j}}\right)w_{ij}\theta_{i1}\theta_{j1}\right] \\
&\quad + o_p\left(\frac{1}{\sqrt{n}}\right)
\end{aligned}
\tag{34}
$$

where w_{ij}s are elements of W in equation (7), and $\theta_{i1}^{(2)}$ is the $(i, 1)$-th element of Θ^2.

THEOREM 13: *Under the sequence of alternatives K_{5n}', the likelihood ratio criterion is more powerful than T up to the order $1/\sqrt{n}$.*

THEOREM 14: *When a sample comes from $E_m(\mu, V)$, the asymptotic expansion of the power function of L_5' and T is expressed under K_{5n}' as follows:*

$$
\begin{aligned}
&P\{T/(1+\kappa) \geq x | K_{5n}'\} \\
&= \bar{P}_f(\delta_5^2) - \frac{u}{\sqrt{n}}\sum_{i=2}^{m}\left(\frac{\lambda_1}{\lambda_i} + \frac{\lambda_i}{\lambda_1}\right)\theta_{i1}\theta_{i1}^{(2)}[\bar{P}_{f+2}(\delta_5^2) - \bar{P}_f(\delta_5^2)] + o\left(\frac{1}{\sqrt{n}}\right)
\end{aligned}
\tag{35}
$$

where

$$
f = m - 1 \quad \text{and} \quad \delta_5^2 = \frac{1}{1+\kappa}\sum_{i=2}^{m}\left(\sqrt{\frac{\lambda_i}{\lambda_1}} - \sqrt{\frac{\lambda_1}{\lambda_i}}\right)^2\theta_{i1}^2
$$

4.8 Testing the Hypothesis that the Latent Roots and Latent Vectors of a Covariance Matrix are Equal to the Given Values and Vectors

Under the same situation as Section 4.5, to test hypothesis

$$H_6 : \Lambda_1 = \Lambda_{10}, \qquad \Gamma_1 = \Gamma_{10} \qquad \text{against} \qquad K_6 : H_6 \text{ is not true}$$

the modified likelihood ratio criterion is given by

$$L_6 = \left[|S|/|\Lambda_{10}||\Gamma_2' S \Gamma_2|\right]^{\frac{n}{2}} etr\left(-\frac{1}{2}\Lambda_{10}^{-1}\Gamma_{10}' S \Gamma_{10}\right)\left(\frac{e}{n}\right)^{\frac{1}{2}nk} \tag{36}$$

as in Mallows (1961) and Gupta (1967). Nagao (1970) gave the asymptotic expansions of the distribution of L_6 under the null hypothesis, fixed alternative, and the sequence of certain types of alternatives.

In this paper we consider the following alternatives:

$$K_{6n} : \Lambda_1 = \Lambda_{10} + \frac{1}{\sqrt{n}}\Lambda_{10}^{1/2}\Delta_1\Lambda_{10}^{1/2}, \qquad \Gamma_1 = \Gamma_{10}\exp(\Theta_{11}/\sqrt{n})$$

$$\Delta_1 = diag(\delta_1, \ldots, \delta_k)$$

The limiting distribution of $-2\log L_6$ under the elliptical population is the one of

$$\frac{1}{2u}\chi_{k-1}^2\left(u\sum_{i=1}^{k}(\delta_i - \bar{\delta})^2\right) + \frac{u+(m-k)v}{2u(u+mv)}\chi_1^2\left(\frac{u(u+mv)}{u+(m-k)v}k\bar{\delta}^2\right)$$
$$+\frac{1}{2u}\chi_{\frac{1}{2}k(k-1)}^2\left(\frac{1}{1+\kappa}\sum_{i<j}\xi_{ij}^2\right) + \frac{1}{2u}\chi_{k(m-k)}^2 \tag{37}$$

where

$$\xi_{ij0} = \left(\sqrt{\frac{\lambda_{j0}}{\lambda_{i0}}} - \sqrt{\frac{\lambda_{i0}}{\lambda_{j0}}}\right)\theta_{ij}, \qquad 1 \le i \ne j \le k \qquad \text{and} \qquad \bar{\delta} = \frac{1}{k}\sum_{i=1}^{k}\delta_i$$

Laborious calculation gives the asymptotic expansion of the power function of L_6 under the elliptical population as

THEOREM 15: *When a sample comes from $E_m(\mu, V)$, the power function of L_6 under a sequence of alternative K_{6n} is given as*

$$P\{-2\log L_6 \ge x|K_{6n}\}$$
$$= \bar{F}_{0,0,0} + \frac{1}{\sqrt{n}}\sum_{\alpha=0}^{3}\sum_{\beta=0}^{3}\sum_{\gamma=0}^{3}d_{\alpha,\beta,\gamma}\bar{F}_{\alpha,\beta,\gamma}(x) + o\left(\frac{1}{\sqrt{n}}\right) \tag{38}$$

where

$$d_{0,3,3} = -\frac{(m-k)^3k^4v^6\bar{\delta}^3}{[u(u+mv)]^3}Q_1 + \frac{1}{6}\frac{(m-k)^3k^3v^6\bar{\delta}^3}{[u(u+mv)]^2[u+(m-k)v]}$$

$$d_{0,3,2} = -\frac{3(m-k)^2k^3v^4\bar{\delta}^3}{[u(u+mv)]^2}Q_1 + \bar{\delta}^3\left\{\frac{1}{2}\frac{(m-k)^2k^2v^4}{u(u+mv)[u+(m-k)v]} - \frac{1}{6}\frac{(m-k)^3k^3v^6}{[(u+mv)]^2[u+(m-k)v]}\right\}$$

$$d_{0,2,3} = \frac{3(m-k)^3k^4v^6\bar{\delta}^3}{[u(u+mv)]^3}Q_1 + \frac{(m-k)^3k^4v^5\bar{\delta}^3}{[u(u+mv)]^2[u+(m-k)v]}R_1 - \frac{1}{2}\frac{(m-k)^3k^3v^6\bar{\delta}^3}{[u(u+mv)]^2[u+(m-k)v]}$$

$$d_{0,3,1} = -\frac{3(m-k)k^2v^2\bar{\delta}^3}{u(u+mv)}Q_1 + \bar{\delta}^3\left\{-\frac{1}{2}\frac{(m-k)^2k^2v^4}{u(u+mv)[u+(m-k)v]} + \frac{1}{2}\frac{(m-k)kv^2}{u+(m-k)v}\right\}$$

$$d_{0,2,2} = \left\{\frac{3(m-k)^2k^2v^4\bar{\delta}}{[u(u+mv)]^2(u+kv)} + \frac{9(m-k)^2k^3v^4\bar{\delta}^3}{[u(u+mv)]^2}\right\}Q_1 + \left\{\frac{2(m-k)^2k^3v^3\bar{\delta}^3}{u(u+mv)[u+(m-k)v]} - \frac{(m-k)^3k^4v^5\bar{\delta}^3}{[u(u+mv)]^2[u+(m-k)v]}\right\}R_1 + \frac{1}{2}\frac{(m-k)^2k^2v^4\bar{\delta}}{[u(u+mv)]^2} + \bar{\delta}^3\left\{\frac{1}{2}\frac{(m-k)^2k^2(k-3)v^4}{u(u+mv)[u+(m-k)v]} + \frac{1}{2}\frac{(m-k)^3k^3v^6}{[u(u+mv)]^2[u+(m-k)v]}\right\}$$

$$d_{0,1,3} = -\frac{3(m-k)^3k^4v^6\bar{\delta}^3}{[u(u+mv)]^2}Q_1 - \frac{2(m-k)^3k^4v^5\bar{\delta}^3}{[u(u+mv)]^2[u+(m-k)v]}R_1 - \frac{(m-k)^2k^4v^4\bar{\delta}^3}{u(u+mv)[u+(m-k)v]^2}R_2 + \frac{1}{2}\frac{(m-k)^3k^3v^6\bar{\delta}^3}{[u(u+mv)]^2[u+(m-k)v]}$$

$$d_{2,1,1} = -\frac{(m-k)kv^2\bar{\delta}}{u(u+mv)}(s_2 + tr\Xi_{110}^2)S_1 - \frac{3}{2}\frac{(m-k)kv^2\bar{\delta}}{u+mv}(s_2 + tr\Xi_{110}^2) + \frac{1}{2}\frac{(m-k)kv^2\bar{\delta}}{u+(m-k)v}(s_2 + tr\Xi_{110}^2)$$

$$d_{3,0,0} = -\left(a_3 + \frac{u}{3}\right)\{s_3 + 3tr\Delta_{10}\Xi_{110}^2 - 3\bar{\delta}tr\Xi_{110}^2 + 6tr\Lambda_{10}^{-1}\Theta_{11}\Lambda_{10}\Theta_{11}^2\}$$

$$d_{0,3,0} = -k\bar{\delta}^3 Q_1 + \bar{\delta}^3\left\{\frac{1}{6}(u+kv)k - \frac{1}{2}\frac{(m-k)kv^2}{u+(m-k)v}\right\}$$

$$d_{0,0,3} = \frac{(m-k)^3k^4v^6\bar{\delta}^3}{[u(u+mv)]^3}Q_1 + \frac{(m-k)k^3v^3\bar{\delta}^3}{[u+(m-k)v]^3}Q_2$$
$$+\frac{(m-k)^3k^4v^5\bar{\delta}^3}{[u(u+mv)]^2[u+(m-k)v]}R_1 + \frac{(m-k)^2k^4v^4\bar{\delta}^3}{u(u+mv)[u+(m-k)v]^2}R_2$$
$$-\frac{1}{6}\frac{(m-k)^3k^3v^6\bar{\delta}^3}{[u(u+mv)]^2[u+(m-k)v]}$$

$$d_{2,1,0} = -\bar{\delta}(s^2 + tr\Xi_{110}^2)S_1 - \left\{u\bar{\delta} + \frac{1}{2}\frac{(m-k)kv^2\bar{\delta}}{u+(m-k)v}\right\}(s_2 + tr\Xi_{110}^2)$$

$$d_{2,0,1} = \frac{(m-k)kv^2\bar{\delta}}{u(u+mv)}(s_2 + tr\Xi_{110}^2)S_1 + \left\{\frac{3}{2}\frac{(m-k)kv^2}{u+mv} - \frac{1}{2}\frac{(m-k)kv^2}{u+(m-k)v}\right\}$$
$$\times\bar{\delta}(s_2 + tr\Xi_{110}^2) + \left(a_2 + \frac{v}{2}\right)\frac{(m-k)kv}{u+(m-k)v}\bar{\delta}(s_2 + tr\Xi_{110}^2)$$

$$d_{0,2,1} = \left\{\frac{9(m-k)k^2v^2\bar{\delta}^3}{u(u+mv)} - \frac{6(m-k)kv^2\bar{\delta}}{u(u+mv)(u+kv)}\right\}Q_1$$
$$+\left\{\frac{(m-k)k^2v\bar{\delta}^3}{u+(m-k)v} - \frac{2(m-k)^2k^3v^3\bar{\delta}^3}{u(u+mv)[u+(m-k)v]}\right\}R_1 + \bar{\delta}\frac{(m-k)kv^2}{u(u+mv)}$$
$$+\bar{\delta}^3\left\{\frac{3}{2}\frac{(m-k)^2k^2v^4}{u(u+mv)[u+(m-k)v]} + \frac{(m-k)k^2v^2}{u+(m-k)v} - \frac{3}{2}\frac{(m-k)kv^2}{u+(m-k)v}\right.$$
$$\left.-\frac{1}{2}\frac{(m-k)^2k^3v^4}{u(u+mv)[u+(m-k)v]}\right\}$$

$$d_{0,1,2} = -\left\{\frac{3(m-k)^2k^2v^4\bar{\delta}}{[u(u+mv)](u+kv)} + \frac{9(m-k)^2k^3v^4\bar{\delta}^3}{[u(u+mv)]^2}\right\}Q_1$$
$$+\left\{\frac{2(m-k)^3k^4v^5\bar{\delta}^3}{[u(u+mv)]^2[u+(m-k)v]} - \frac{4(m-k)^2k^3v^3\bar{\delta}^3}{u(u+mv)[u+(m-k)v]}\right.$$
$$\left.+\frac{2(m-k)^2k^2v^3\bar{\delta}}{[u(u+mv)]^2} + \frac{(m-k)^2k^2v^3\bar{\delta}}{u(u+mv)(u+kv)[u+(m-k)v]}\right\}R_1$$

$$+\left\{\frac{2(m-k)^2k^4v^4\bar{\delta}^3}{u(u+mv)[u+(m-k)v]^2}-\frac{(m-k)^2k^3v^4\bar{\delta}}{[u(u+mv)]^2[u+(m-k)v]}\right.$$
$$\left.-\frac{(m-k)k^3v^2\bar{\delta}^3}{[u+(m-k)v]^2}\right\}R_2-\bar{\delta}\,\frac{(m-k)^2k^2v^4}{[u(u+mv)]^2}$$
$$+\bar{\delta}^3\left\{\frac{3}{2}\,\frac{(m-k)^2k^2v^4}{u(u+mv)[u+(m-k)v]}-\frac{1}{2}\,\frac{(m-k)^3k^3v^6}{[u(u+mv)]^2[u+(m-k)v]}\right.$$
$$\left.-\frac{(m-k)^2k^3v^4}{u(u+mv)[u+(m-k)v]}\right\}$$

$$d_{1,1,1}=\left\{\frac{2(m-k)kv^2\bar{\delta}}{u(u+mv)}(s_2+tr\Xi_{110}^2)-\frac{1}{2}\,\frac{(m-k)(mk+k-2)kv^2\bar{\delta}}{u^2(u+mv)}\right\}S_1$$
$$-\frac{1}{2}\,\frac{(m-k)(mk+k-2)kv^2\bar{\delta}}{u(u+mv)}-\frac{1}{2}\left(a_2+\frac{v}{2}\right)\frac{(m-k)^2k^3v^2\bar{\delta}}{u^2(u+mv)}$$
$$+\frac{3(m-k)kv^2\bar{\delta}}{u+mv}\,(s_2+tr\Xi_{110}^2)$$

$$d_{2,0,0}=\bar{\delta}(s_2+tr\Xi_{110}^2)S_1+\left\{\frac{3}{2}u+\frac{1}{2}\,\frac{(m-k)kv^2}{u+(m-k)v}\right\}\bar{\delta}(s_2+tr\Xi_{110}^2)$$
$$-\left(a_2+\frac{v}{2}\right)\frac{(m-k)kv}{u+(m-k)v}\,\bar{\delta}(s_2+tr\Xi_{110}^2)$$
$$+3\left(a_3+\frac{4}{9}u\right)(s_3+3tr\Delta_{10}\Xi_{110}^2-3\bar{\delta}tr\Xi_{110}^2+6tr\Lambda_{10}^{-1}\Theta_{11}\Lambda_{10}\Theta_{11}^2)$$

$$d_{0,2,0}=\left(3k\bar{\delta}^3-\frac{3\bar{\delta}}{u+kv}\right)Q_1-\frac{(m-k)k^2v\bar{\delta}^3}{u+(m-k)v}\,R_1+\frac{\bar{\delta}}{2}$$
$$+\bar{\delta}^3\left\{-\frac{1}{6}(u+kv)k+\frac{3}{2}\,\frac{(m-k)kv^2}{u+(m-k)v}-\frac{(m-k)k^2v^2}{u+(m-k)v}\right\}$$

$$d_{0,0,2}=\frac{3(m-k)^2k^3v^4\bar{\delta}^3}{[u(u+mv)]^2}\,Q_1+\left\{\frac{3(m-k)k^2v^3\bar{\delta}}{u(u+mv)[u+(m-k)v]}\right.$$
$$\left.-\frac{3(m-k)k^3v^3\bar{\delta}^3}{[u+(m-k)v]^3}\right\}Q_2$$
$$+\left\{-\frac{(m-k)^3k^4v^5\bar{\delta}^3}{[u(u+mv)]^2[u+(m-k)v]}+\frac{2(m-k)^2k^3v^3\bar{\delta}^3}{u(u+mv)[u+(m-k)v]}\right.$$

$$-\frac{2(m-k)^2k^2v^3\bar{\delta}}{[u(u+mv)]^2}\Bigg\}R_1+\Bigg\{-\frac{2(m-k)^2k^4v^4\bar{\delta}^3}{u(u+mv)[u+(m-k)v]^2}$$

$$+\frac{(m-k)^2k^3v^4\bar{\delta}}{[u(u+mv)]^2[u+(m-k)v]}+\frac{(m-k)k^3v^2\bar{\delta}^3}{[u+(m-k)v]^2}$$

$$-\frac{2(m-k)k^2v^2\bar{\delta}}{u(u+mv)[u+(m-k)v]}\Bigg\}R_2+\bar{\delta}\frac{1}{2}\frac{(m-k)^2k^2v^4}{[u(u+mv)]^2}$$

$$+\bar{\delta}^3\Bigg\{-\frac{1}{2}\frac{(m-k)^2k^2v^4}{u(u+mv)[u+(m-k)v]}+\frac{1}{6}\frac{(m-k)^3k^3v^6}{[u(u+mv)]^2[u+(m-k)v]}$$

$$+\frac{1}{2}\frac{(m-k)^2k^3v^4}{u(u+mv)[u+(m-k)v]}\Bigg\}$$

$$d_{1,1,0}=\Bigg\{2\bar{\delta}(s_2+tr\Xi_{110}^2)-\frac{1}{2}(mk+k-2)\frac{\bar{\delta}}{u}\Bigg\}S_1+\frac{5}{2}u\bar{\delta}(s_2+tr\Xi_{110}^2)$$

$$-\frac{1}{2}(mk+k-2)\bar{\delta}-\left(a_2+\frac{v}{2}\right)\frac{1}{2}\frac{(m-k)k^2}{u}\bar{\delta}$$

$$d_{0,1,1}=\Bigg\{-\frac{9(m-k)k^2v^2\bar{\delta}^3}{u(u+mv)}+\frac{6(m-k)kv^2\bar{\delta}}{u(u+mv)(u+kv)}\Bigg\}Q_1$$

$$+\Bigg\{\frac{4(m-k)^2k^3v^3\bar{\delta}^3}{u(u+mv)[u+(m-k)v]}-\frac{2(m-k)k^2v\bar{\delta}^3}{u+(m-k)v}+\frac{2(m-k)kv\bar{\delta}}{u(u+mv)}$$

$$-\frac{(m-k)^2k^2v^3\bar{\delta}}{u(u+mv)(u+kv)[u+(m-k)v]}+\frac{(m-k)kv\bar{\delta}}{u(u+mv)[u+(m-k)v]}\Bigg\}R_1$$

$$+\Bigg\{\frac{2(m-k)k^3v^2\bar{\delta}^3}{[u+(m-k)v]^2}-\frac{(m-k)^2k^4v^4\bar{\delta}^3}{u(u+mv)[u+(m-k)v]^2}$$

$$-\frac{2(m-k)k^2v^2\bar{\delta}}{u(u+mv)[u+(m-k)v]}\Bigg\}R_2-\frac{(m-k)kv^2\bar{\delta}}{u(u+mv)}(s_2+tr\Xi_{110}^2)S_1$$

$$+\left(a_0+\frac{m+1}{2}\right)\frac{(m-k)k^2v^2\bar{\delta}}{u(u+mv)}$$

$$-\left(a_2+\frac{v}{2}\right)\frac{(m-k)(m-k-1)(m-k+2)k^2v^2\bar{\delta}}{2u^2(u+mv)}$$

$$
- \left\{ \frac{3}{2} \frac{(m-k)kv^2}{u+mv} + \frac{1}{2} \frac{(m-k)kv^2}{u+(m-k)v} \right\} \bar{\delta}(s_2 + tr\Xi_{110}^2)
$$

$$
- \bar{\delta} \left\{ \frac{3}{2} \frac{(m-k)kv^2}{u(u+mv)} + \frac{1}{4} \frac{(m-k)(mk+k-2)kv^2}{u(u+mv)} \right\}
$$

$$
+ \bar{\delta}^3 \left\{ -\frac{3}{2} \frac{(m-k)^2k^2v^4}{u(u+mv)[u+(m-k)v]} + \frac{3}{2} \frac{(m-k)kv^2}{u+(m-k)v} \right.
$$

$$
\left. + \frac{(m-k)^2k^3v^4}{u(u+mv)[u+(m-k)v]} - \frac{2(m-k)k^2v^2}{u+(m-k)v} \right\}
$$

$$
d_{1,0,1} = \left\{ \frac{1}{2} \frac{(m-k)(mk+k-2)kv^2\bar{\delta}}{u^2(u+mv)} - \frac{2(m-k)kv^2}{u(u+mv)} \bar{\delta}(s_2 + tr\Xi_{110}^2) \right\} S_1
$$

$$
+ \frac{1}{2} \frac{(m-k)k^2v\bar{\delta}}{u[u+(m-k)v]} S_2
$$

$$
+ \left(a_2 + \frac{v}{2}\right) \left\{ \frac{1}{2} \frac{(m-k)^2k^3v^2\bar{\delta}}{u^2(u+mv)} + \frac{1}{2} \frac{(m-k)(mk+k-2)kv\bar{\delta}}{u[u+(m-k)v]} \right\}
$$

$$
- \left(a_2 + \frac{v}{2}\right) \frac{2(m-k)kv\bar{\delta}}{u+(m-k)v} (s_2 + tr\Xi_{110}^2)
$$

$$
- 3 \frac{(m-k)kv^2}{u+mv} \bar{\delta}(s_2 + tr\xi_{110}^2)
$$

$$
+ \bar{\delta} \left\{ \frac{1}{2} \frac{(m-k)(mk+k-2)kv^2}{u(u+mv)} + \frac{3}{4} \frac{(m-k)k^2v}{u+(m-k)v} \right\}
$$

$$
d_{1,0,0} = \left\{ -2\bar{\delta}(S_2 + tr\Xi_{110}^2) + \frac{1}{2}(mk+k-2) \frac{\bar{\delta}}{u} \right\} S_1
$$

$$
- \frac{1}{2} \frac{(m-k)k^2v\bar{\delta}}{u[u+(m-k)v]} S_2
$$

$$
+ \left(a_2 + \frac{v}{2}\right) \left\{ \frac{1}{2} \frac{(m-k)k^2\bar{\delta}}{u} - \frac{(m-k)(mk+k-2)kv\bar{\delta}}{2u[u+(m-k)v]} \right\}
$$

$$
+ 2\left(a_2 + \frac{v}{2}\right) \frac{(m-k)kv}{u+(m-k)v} \bar{\delta}(s_2 + tr\Xi_{110}^2)
$$

$$
- 4u\bar{\delta}(s_2 + tr\Xi_{110}^2)
$$

$$+\bar{\delta}\left\{\frac{3}{4}(mk+k-2)-\frac{3}{4}\frac{(m-k)k^2v}{u+(m-k)v}\right\}$$

$$-3\left(a_3+\frac{2}{3}u\right)\{s_3+3tr\Delta_{10}\Xi_{110}^2-3\bar{\delta}tr\Xi_{110}^2+6tr\Lambda_{10}^{-1}\Theta_{11}\Lambda_{10}\Theta_{11}^2\}$$

$$+u\{3tr\Delta_{10}\Theta_{11}^2+3tr\Lambda_{10}^{-1}\Theta_{11}\Lambda_{10}\Theta_{11}^2-2tr\Delta_{10}\Lambda_{10}\Theta_{11}\Lambda_{10}^{-1}\Theta_{11}$$

$$-tr\Delta_{10}\Lambda_{10}^{-1}\Theta_{11}\Lambda_{10}\Theta_{11}-\bar{\delta}(tr\Theta_{11}^2-tr\Lambda_{10}^{-1}\Theta_{11}\Lambda_{10}\Theta_{11})\}$$

$$d_{0,1,0}=-\left\{3k\bar{\delta}^3-\frac{3\bar{\delta}}{u+kv}\right\}Q_1+\left\{\frac{2(m-k)k^2v\bar{\delta}^3}{u+(m-k)v}\right.$$

$$\left.-\frac{(m-k)kv\bar{\delta}}{u(u+kv)[u+(m-k)v]}\right\}R_1$$

$$-\left\{\frac{(m-k)k^3v^2\bar{\delta}^3}{[u+(m-k)v]^2}+\frac{k\bar{\delta}}{u+(m-k)v}\right\}R_2-\bar{\delta}(s_2+tr\Xi_{110}^2)S_1$$

$$+\left(a_0+\frac{m+1}{2}\right)k\bar{\delta}-\left(a_2+\frac{v}{2}\right)\frac{1}{2u}(m-k-1)(m-k+2)k\bar{\delta}$$

$$-\frac{1}{2}(u+kv)\bar{\delta}s_2-\frac{(m-k)kv^2\bar{\delta}}{u+(m-k)v}(tr\Theta_{11}^2-tr\Lambda_{10}^{-1}\Theta_{11}\Lambda_{10}\Theta_{11})$$

$$+\bar{\delta}(s_2+tr\Xi_{110}^2)\left\{-\frac{3}{2}u+\frac{1}{2}\frac{(m-k)kv^2}{u+(m-k)v}\right\}$$

$$-\bar{\delta}\left\{\frac{1}{2}+\frac{1}{4}(mk+k-2)\right\}$$

$$-\bar{\delta}^3\left\{\frac{1}{2}(u+kv)k+\frac{3}{2}\frac{(m-k)kv^2}{u+(m-k)v}-\frac{2(m-k)k^2v^2}{u+(m-k)v}\right\}$$

$$d_{0,0,1}=\frac{3(m-k)k^2v^2\bar{\delta}^3}{u(u+mv)}Q_1$$

$$+\left\{\frac{3(m-k)k^3v^3\bar{\delta}^3}{[u+(m-k)v]^3}+\frac{3kv\bar{\delta}}{[u+(m-k)v]^2}\right.$$

$$\left.-\frac{3(m-k)k^2v^3\bar{\delta}}{u(u+mv)[u+(m-k)v]}\right\}Q_2$$

$$+\left\{-\frac{2(m-k)^2k^3v^3\bar{\delta}^3}{u(u+mv)[u+(m-k)v]}+\frac{(m-k)k^2v\bar{\delta}^3}{u+(m-k)v}-\frac{2(m-k)kv\bar{\delta}}{u(u+mv)}\right\}R_1$$

$$+\left\{\frac{(m-k)^2k^4v^4\bar{\delta}^3}{u(u+mv)[u+(m-k)v]^2}\right.$$

$$\left.-\frac{2(m-k)k^3v^2\bar{\delta}^3}{[u+(m-k)v]^2}+\frac{4(m-k)k^2v^2\bar{\delta}}{u(u+mv)[u+(m-k)v]}\right\}R_2$$

$$+\frac{(m-k)kv^2\bar{\delta}}{u(u+mv)}(s_2+tr\Xi_{110}^2)S_1+\frac{1}{2}\frac{(m-k-1)(m-k+2)kv\bar{\delta}}{u(u+mv)}S_2$$

$$-\left(a_0+\frac{m+1}{2}\right)\left\{\frac{(m-k)k^2v^2\bar{\delta}}{u(u+mv)}+\frac{(m-k)kv\bar{\delta}}{u+(m-k)v}\right\}$$

$$+\left(a_2+\frac{v}{2}\right)\frac{(m-k)(m-k-1)(m-k+2)k^2v^2\bar{\delta}}{2u^2(u+mv)}$$

$$+\left(a_2+\frac{v}{2}\right)\frac{(m-k)kv}{u+(m-k)v}\bar{\delta}(s_2+tr\Xi_{110}^2)$$

$$+\left\{\frac{3}{2}\frac{(m-k)kv^2}{u+mv}+\frac{1}{2}\frac{(m-k)kv^2}{u+(m-k)v}\right\}\bar{\delta}(s_2+tr\Xi_{110}^2)$$

$$-\frac{(m-k)kv^2\bar{\delta}}{u+(m-k)v}\left\{tr\Theta_{11}^2-tr\Lambda_{10}^{-1}\Theta_{11}\Lambda_{10}\Theta_{11}\right\}$$

$$+\bar{\delta}\left\{\frac{1}{2}\frac{(m-k)kv^2}{u(u+mv)}+\frac{1}{4}\frac{(m-k)(mk+k-2)kv^2\bar{\delta}}{u(u+mv)}\right.$$

$$\left.+\frac{3}{4}\frac{(m-k-1)(m-k+2)kv\bar{\delta}}{u+(m-k)v}\right\}$$

$$+\bar{\delta}^3\left\{\frac{1}{2}\frac{(m-k)^2k^2v^4}{u(u+mv)[u+(m-k)v]}-\frac{1}{2}\frac{(m-k)^2k^3v^4}{u(u+mv)[u+(m-k)v]}\right.$$

$$\left.-\frac{1}{2}\frac{(m-k)kv^2}{u+(m-k)v}+\frac{(m-k)k^2v^2}{u+(m-k)v}\right\}$$

$$d_{0,0,0} = k\bar{\delta}^3 Q_1 - \left\{ \frac{(m-k)k^3 v^3 \bar{\delta}^3}{[u+(m-k)v]^3} + \frac{3kv\bar{\delta}}{[u+(m-k)v]^2} \right\} Q_2$$
$$- \frac{(m-k)k^2 v \bar{\delta}^3}{u+(m-k)v} R_1 + \left\{ \frac{(m-k)k^3 v^2 \bar{\delta}^3}{[u+(m-k)v]^2} + \frac{k\bar{\delta}}{u+(m-k)v} \right\} R_2$$
$$+ \bar{\delta}(S_2 + tr\Xi_{110}^2) S_1 - \frac{1}{2} \frac{(m-k-1)(m-k+2)kv\bar{\delta}}{u[u+(m-k)v]} S_2$$
$$+ \left(a_0 + \frac{m+1}{2}\right) \left\{ \frac{(m-k)kv\bar{\delta}}{u+(m-k)v} - k\bar{\delta} \right\}$$
$$+ \left(a_0 + \frac{v}{2}\right) \frac{1}{2u} (m-k-1)(m-k+2)k\bar{\delta}$$
$$- \left(a_2 + \frac{v}{2}\right) \frac{(m-k)kv}{u+(m-k)v} \bar{\delta}(S_2 + tr\Xi_{110}^2) + \frac{1}{2}(u+kv)\bar{\delta}s_2$$
$$+ (a_3 + u)(s_3 + 3tr\Delta_{10}\Xi_{110}^2 - 3\bar{\delta} tr\Xi_{110}^2 + 6tr\Lambda_{10}^{-1}\Theta_{11}\Lambda_{10}\Theta_{11}^2)$$
$$- u\{3tr\Delta_{10}\Theta_{11}^2 + 3tr\Lambda_{10}^{-1}\Theta_{11}\Lambda_{10}\Theta_{11}^2 - 2tr\Delta_{10}\Lambda_{10}\Theta_{11}\Lambda_{10}^{-1}\Theta_{11}$$
$$- tr\Delta_{10}\Lambda_{10}^{-1}\Theta_{11}\Lambda_{10}\Theta_{11} - \bar{\delta}(tr\Theta_{11}^2 - tr\Lambda_{10}\Theta_{11}\Lambda^{-1}\Theta_{11})\}$$
$$+ \frac{(m-k)kv^2\bar{\delta}}{u+(m-k)v} (tr\Theta_{11}^2 - tr\Lambda_{10}\Theta_{11}\Lambda_{10}^{-1}\Theta_{11})$$
$$+ \left\{ \frac{5}{2}u - \frac{1}{2} \frac{(m-k)kv^2}{u+(m-k)v} \right\} \bar{\delta}(s_2 + tr\Xi_{110}^2)$$
$$- \frac{3}{4} \frac{(m-k-1)(m-k+2)kv}{u+(m-k)v} \bar{\delta}$$
$$+ \bar{\delta}^3 \left\{ \frac{1}{2}(u+kv)k + \frac{1}{2} \frac{(m-k)kv^2}{u+(m-k)v} - \frac{(m-k)k^2v^2}{u+(m-k)v} \right\}$$

$f = \frac{1}{2}(k+1) + k(m-k)$, $\Xi_{110} = (\xi_{ij0})$,
$\xi' = (\xi_{120}, \xi_{130}, \ldots, \xi_{k-1,k0})$

$$Q_1 = a_1 k^2 + \left(a_2 + \frac{v}{2}\right)k + a_3 + \frac{u}{2}$$
$$Q_2 = a_1(m-k)^2 + \left(a_2 + \frac{v}{2}\right)(m-k) + a_3 + \frac{u}{2}$$
$$R_1 = 3a_1 k + a_2 + \frac{v}{2}, \qquad R_2 = 3a_1(m-k) + a_2 + \frac{v}{2}$$
$$S_1 = \left(a_2 + \frac{v}{2}\right)k + 3a_3, \qquad S_2 = \left(a_2 + \frac{v}{2}\right)(m-k) + 3a_3$$

$$\bar{F}_{\alpha,\beta,\gamma}(x) = 1 - \sum_{l=0}^{\infty} e_l P\{\chi^2_{f+2\alpha+2\beta+2\gamma+2l} \le x/w\}$$

$$e_0 = \exp\left[-\frac{1}{2}\left\{us_2 + 2u\xi'\xi + \frac{u(u+mv)}{u+(m-k)v}k\bar{\delta}^2\right\}\right]$$
$$\times (2w)^{\frac{1}{2}f+\alpha+\beta+\gamma} u^{\frac{1}{2}(f-1)+\alpha}(u+kv)^{\beta}\left\{\frac{u(u+mv)}{u+(m-k)v}\right\}^{\frac{1}{2}+\gamma}$$

$$e_l = \frac{1}{2l}\sum_{j=0}^{l-1} G_{l-j}e_j \qquad (l \ge 1)$$

$$G_l = (1-2uw)^{l-1}\{f-1+2\alpha\} + \{1-2(u+kv)w\}^{l-1}(2\beta)$$
$$+ \left\{1 - \frac{2u(u+mv)}{u+(m-k)v}\right\}^{l-1}(1+2\gamma)$$
$$+ lw\left[2u^2(s_2+2\xi'\xi)(1-2uw)^{l-1}\right.$$
$$\left. + 2\left\{\frac{u(u+mv)}{u+(m-k)v}\right\}^2 k\bar{\delta}^2\left\{1-\frac{2u(u+mv)}{u+(m-k)v}\right\}^{l-1}\right]$$

and w is an arbitrary constant which is chosen so that series may converge rapidly. For $k = m$, equation (38) is reduced to equation (4) in Hayakawa (1986). In the case of a normal population ($\kappa = 0$), the power function is expressed simply as

$$P\{-2\log L_6 \ge x | K_{6n}, \text{ Normal population}\}$$
$$= \bar{P}_f(\tau^2) + \frac{1}{\sqrt{n}}\sum_{\alpha=0}^{2} d_\alpha \bar{P}_{f+\alpha}(\tau^2) + o\left(\frac{1}{\sqrt{n}}\right) \tag{39}$$

where

$$d_2 = \tfrac{1}{6}(s_3 + 3\bar{\delta}s_2) + \tfrac{1}{2}\left[2tr\Delta_{10}\Theta_{11}^2 - tr\Delta_{10}\Lambda_{10}\Theta_{11}\Lambda_{10}\Theta_{11}\right.$$
$$\left. - tr\Delta_{10}\Lambda_{10}^{-1}\Theta_{11}\Lambda_{10}\Theta_{11} + 2tr\Lambda_{10}^{-1}\Theta_{11}\Lambda_{10}\Theta_{11}^2\right]$$
$$d_1 = -\tfrac{1}{2}(s_3 + 3\bar{\delta}s_2) + \tfrac{1}{2}\left[-3tr\Delta_{10}\Theta_{11}^2 + tr\Delta_{10}\Lambda_{10}\Theta_{11}\Lambda_{10}\Theta_{11}\right.$$
$$\left. + 2tr\Delta_{10}\Lambda_{10}^{-1}\Theta_{11}\Lambda_{10}\Theta_{11} - 3tr\Lambda_{10}^{-1}\Theta_{11}\Lambda_{10}\Theta_{11}^2\right]$$
$$d_0 = \tfrac{1}{3}(s_3 + 3\bar{\delta}s_2) + \tfrac{1}{2}\left[tr\Delta_{10}\Theta_{11}^2 - tr\Delta_{10}\Lambda_{10}^{-1}\Theta_{11}\Lambda_{10}\Theta_{11}\right.$$
$$\left. - tr\Lambda_{10}^{-1}\Theta_{11}\Lambda_{10}\Theta_{11}^2\right]$$

and $\bar{P}_f(\tau^2) = P\{\chi_f^2(\tau^2) \geq x\}$ and $\chi_f^2(\tau^2)$ is a noncentral chi-square random variable with $f = \frac{1}{2}k(k+1) + k(m-k)$ degrees of freedom and a noncentrality parameter $\tau^2 = \frac{1}{4}(s_2 + k\bar{\delta}^2 + 2\xi'\xi)$.

REFERENCES

Anderson, GA. An asymptotic expansion for the distribution of the latent roots of the estimated covariance matrix. *Ann. Math. Stat.*, 36:1153–1173, 1965.

Anderson, TW. The asymptotic distribution of certain characteristic roots and vectors. *Proceedings of the Second Berkeley Symposium on Mathematical Statistics and Probability*, Berkeley, 1951, pp. 103–130.

Anderson, TW. *An Introduction to Multivariate Statistical Analysis.* New York: Wiley, 1958.

Anderson, TW. Asymptotic theory for principal component analysis. *Ann. Math. Stat.*, 34:122–148, 1963.

Box, GEP. A generalized distribution theory for a class of likelihood criteria. *Biometrika*, 36:317–346, 1949.

Chikuse, Y. Asymptotic distributions of the latent roots of covariance matrix with multiple population roots. *J. Multivar. Anal.*, 6:237–249, 1976.

Constantine, AG, Muirhead, RJ. Asymptotic expansions for distributions of latent roots in multivariate analysis. *J. Multivar. Anal.*, 6:369–391, 1976.

Davis, AW. Percentile approximations for a class of likelihood ratio criteria. *Biometrika*, 58:349–356, 1971.

Fisher, RA. The sampling distribution of some statistics obtained from nonlinear equations. *Ann. Eugenics*, 9:238–249, 1939.

Fujikoshi, Y. Asymptotic expansions of the distributions of the latent roots of the Wishart matrix with multiple population roots. *Ann. Inst. Stat. Math.*, 29(Part A):379–387, 1977.

Fujikoshi, Y. Asymptotic expansions for the distributions of the sample roots under nonnormality. *Biometrika*, 67:45–51, 1980.

Girshick, MA. On the sampling theory of roots of determinantal equations. *Ann. Math. Stat.*, 10:203–224, 1939.

Gleser, LJ. A note on the sphericity test. *Ann. Math. Stat.*, 37:464–467, 1966.

Gupta, RP. Latent roots and vectors of a Wishart matrix. *Ann. Inst. Stat. Math.*, 19:157–165, 1967.

Hayakawa, T. The asymptotic expansion of the distribution of Anderson's statistic for testing a latent vector of a covariance matrix. *Ann. Inst. Stat. Math.*, 30 (Part A): 51–55, 1978.

Hayakawa, T. On testing hypothesis of covariance matrices under an elliptical population. *J. Stat. Plann. Inference*, 13:193–202, 1986.

Hayakawa, T. Normalizing and variance stabilizing transformation of multivariate statistics under an elliptical population. *Ann. Inst. Stat. Math.*, 39(Part A):299–306, 1987.

Hayakawa, T, Puri, ML. Asymptotic distributions of likelihood ratio criteria for testing latent roots and latent vectors of a covariance matrix under an elliptical population. *Biometrika*, 72:331–338, 1985.

Hill, GW, Davis, AW. Generalized asymptotic expansions of Cornish–Fisher type. *Ann. Math. Stat.*, 39:1264–1273, 1968.

Hsu, LC. A theorem on the asymptotic behavior of a multiple integral. *Duke Math. J.*, 15:623–632, 1948.

Hsu, PL. On the distribution of the roots of certain determinantal equations. *Ann. Eugenics*, 9:250–258, 1939.

Iwashita, T. Asymptotic null and nonnull distribution of Hotelling's T^2-statistic under the elliptical distribution. *J. Stat. Plann. Inference*, 61: 85–104, 1997.

James, AT. Normal multivariate analysis and the orthogonal group. *Ann. Math. Stat.*, 25:40–75, 1954.

James, AT. The distribution of the latent roots of the covariance matrix. *Ann. Math. Stat.*, 31:151–158, 1960.

James, AT. Distributions of matrix variates and latent roots derived from normal samples. *Ann. Math. Stat.*, 35:475–501, 1964.

James, AT. Test of equality of the latent roots of the covariance matrix. In: Krishnaiah, PR, ed. *Multivariate Analysis II*. London: Academic Press, 1969, pp. 205–218.

Kelker, D. Distribution theory of spherical distributions and a location-scale parameter generalization. *Sankhyā A*, 32:419–430, 1970.

Khatri, CG. On the exact finite series distribution of the smallest or the largest root of matrices in three situations. *J. Multivar. Anal.*, 2:201–207, 1972.

Konishi, S. Asymptotic expansion for the distribution of a function of latent roots of the covariance matrix. *Ann. Inst. Stat. Math.*, 29:389–396, 1977.

Konishi, S. Normalizing transformations of some statistics in multivariate analysis. *Biometrika*, 68:647–651, 1981.

Korin, BP. On the distribution of a statistic used for testing a covariance matrix. *Biometrika*, 55:171–178, 1968.

Lawley, DN. A modified method of estimation in factor analysis and some large sample results. Uppsala Symposium on Psychological Factor Analysis, March 17–19, 1953, Uppsala, pp. 33–42.

Lawley, DN. Test of significance for the latent roots of covariance and correlation matrices. *Biometrika*, 43:128–136, 1956.

Mallows, CL. Latent vectors of random symmetric matrices. *Biometrika*, 48:133–149, 1961.

Muirhead, RJ. Systems of partial differential equations for hypergeometric functions of matrix argument. *Ann. Math. Stat.*, 41:991–1001, 1970a.

Muirhead, RJ. Asymptotic distributions of some multivariate tests. *Ann. Math. Stat.*, 41:1002–1010, 1970b.

Muirhead, RJ. Latent roots and matrix variates: A review of some asymptotic results. *Ann. Stat.*, 6:5–33, 1978.

Muirhead, RJ. The effects of elliptical distributions on some standard procedures involving correlation coefficients: A review. In: Gupta RP, ed. *Multivariate Statistical Analysis*. Amsterdam: North Holland, 1980, pp. 143–159.

Muirhead, RJ, Chikuse, Y. Asymptotic expansions for the joint and marginal distributions of the latent roots of a covariance matrix. *Ann. Stat.*, 3:1011–1017, 1975.

Muirhead, RJ, Waternaux, CM. Asymptotic distributions in canonical correlation analysis and other multivariate procedures for nonnormal populations. *Biometrika*, 67:31–43, 1980.

Nagao, H. Asymptotic expansions of some test criteria for homogeneity of variances and covariance matrices from normal populations. *J. Sci. Hiroshima Univ. Ser. A-I*, 34:153–247, 1970.

Nagao, H. Non-null distributions of the likelihood ratio criteria for independence and equality of mean vectors and covariance matrices. *Ann. Inst. Stat. Math.*, 24:67–79, 1972.

Nagao, H. Asymptotic expansions of the distributions of Bartlett's test and sphericity test under the local alternatives. *Ann. Inst. Stat. Math.*, 25:407–422, 1973a.

Nagao, H. Nonnull distributions of two test criteria for independence under local alternatives. *J. Multivar. Anal.*, 3:435–444, 1973b.

Nagao, H. On some test criteria for covariance matrix. *Ann. Stat.*, 1:700–709, 1973c.

Nagao, H. An asymptotic expansion for the distribution of a function of latent roots of the noncentral Wishart matrix, when $\Omega = O(n)$. *Ann. Inst. Stat. Math.*, 30(Part A):377–383, 1978.

Nagao, H. Asymptotic expansions of some test criteria for sphericity test and test of independence under local alternatives from an elliptical distribution. *Math. Japonica*, 38:165–170, 1993.

Nagao, H, Srivastava, MS. On the distribution of some test criteria for a covariance matrix under local alternatives and bootstrap approximations. *J. Multivar. Anal.*, 43:331–350, 1992.

Nagarsenker, BN, Pillai, KCS. Distribution of the likelihood ratio criterion for testing a hypothesis specifying a covariance matrix. *Biometrika*, 60:359–364, 1973a.

Nagarsenker, BN, Pillai, KCS. The distribution of the sphericity test criterion. *J. Multivar. Anal.*, 3:226–235, 1973b.

Roy, SN. *p*-statistics, or some generalization in analysis of variance appropriate to multivariate problems. *Sankhyā*, 61:15–34, 1939.

Sugiura, N. Asymptotic expansions of the distributions of the likehood ratio criteria for covariance matrix. *Ann. Math. Stat.*, 40:2051–2063, 1969.

Sugiura, N. Asymptotic non-null distributions of the likelihood ratio criteria for covariance matrix under local alternatives. *Ann. Stat.*, 1:718–728, 1973a.

Sugiura, N. Derivatives of the characteristic roots of a symmetric or Hermitian matrix with two applications in multivariate analysis. *Commun. Stat.*, 1:393–417, 1973b.

Sugiura, N. Asymptotic expansions of the distributions of the latent roots and latent vectors of the Wishart and multivariate F matrices. *J. Multivar. Anal.*, 6:500–525, 1976.

Sugiura, N., Nagao, H. Unbiasedness of some test criteria for the equality of one or two covariance matrices. *Ann. Math. Stat.*, 39:1686–1692, 1968.

Sugiyama, T. On thc distribution of the largest root of the covariance matrix. *Ann. Math. Stat.*, 38:1148–1151, 1967.

Tyler, DE. Robustness and efficiency properties of scatter matrices. *Biometrika*, 70:411–420, 1983a.

Tyler, DE. Radial estimates and the test for sphericity. *Biometrika*, 69:429–436, 1983b.

Tyler, DE. A class of asymptotic tests for principal components vectors. *Ann. Stat.*, 11:1243–1250, 1983c.

Wakaki, H. Discriminant analysis under elliptical populations. *Hiroshima Math. J.*, 24:257–298, 1994.

Waternaux, CM. Asymptotic distribution of the sample roots for a non-normal population. *Biometrika*, 63:639–645, 1976.

16

A Review of Variance Estimators with Extensions to Multivariate Nonparametric Regression Models

HOLGER DETTE, AXEL MUNK, and THORSTEN WAGNER Ruhr-University at Bochum, Bochum, Germany

1. INTRODUCTION

Consider the multivariate regression model where the experimenter observes n outcomes $Y_i = (Y_{i,1}, \ldots, Y_{i,d})' \in \mathbb{R}^d$ at design points $t_i \in T$ given by

$$Y_i := Y(t_i) = g(t_i) + \epsilon_i \qquad (i = 1, \ldots, n) \tag{1}$$

Here, $T \subseteq \mathbb{R}$ is the design space, $g : T \to \mathbb{R}^d$ denotes an unknown regression function, and $\epsilon_i := (\epsilon_{i,1}, \ldots, \epsilon_{i,d})'$, $i = 1, \ldots, n$, is an i.i.d. sequence of centered d-dimensional random vectors with existing second moments. Chemical experiments can serve as the simplest example for these models (see, e.g., Allen (1983) or Stewart et al. (1992) for some parameter multiresponse models). In this article we consider the problem of estimating the covariance structure Σ of the response variable,

$$\Sigma := \text{Cov}[\epsilon_1] \in \mathbb{R}^{d \times d}$$

which will be assumed to be independent of the specific design point $t \in T \subseteq \mathbb{R}$. The design space T will be assumed, without loss of generality, to be the unit interval [0, 1].

Although the main work in nonparametric regression considers the problem of estimating the regression curve g itself, methods for estimating the variance have been discussed by numerous authors in recent years. To our knowledge, Breiman and Meisel (1976) were the first who considered the problem of estimating the variance in nonparametric regression as a topic in its own right. Their motivation was twofold. On the one hand, if the variability of the data is known, this can be used to judge the goodness-of-fit of a specified regression function. On the otherhand, such knowledge is useful for selecting a subset of the independent variables which best determine the dependent variable in a setup with high dimensional design space T. The variance is also required for the plug-in estimation of the bandwidth in nonparametric regression (see, e.g., Rice, 1984; Müller and Stadtmüller, 1987a) and as a measure for the accuracy of a prespecified linear regression model to be validated (see, e.g., Härdle and Marron, 1990; Eubank and Hart, 1992; Dette and Munk, 1998). For further applications of variance estimators such as in quality control, in immunoassay, or in calibration of variance estimation see Carroll and Ruppert (1988).

The purpose of this paper is to construct estimates for the covariance matrix of the error distribution in the regression model in equation (1). Because in multivariate models the computational effort for calculations of such estimators increases significantly, we propose a computationally "simple" method of estimating the covariance matrix which can easily be obtained from the approach for univariate data, i.e., $d = 1$. In this case, several estimators of the variance $\sigma^2 = V[\epsilon_1]$ have been suggested during the last two decades. These estimators may be roughly classified into three different types.

The first class consists of kernel-type estimators which can even be applied when the variance is a function of the explanatory variable t, $V[Y_i] = \sigma^2(t_i)$ say. Such estimators were considered by Müller and Stadtmüller (1987b, 1993), Hall and Carroll (1989), and Neumann (1994) among others and can be used, of course, in the homoscedastic case $\sigma^2(t) \equiv \sigma^2$. A further kernel-type estimator was suggested by Hall and Marron (1990) for the specific situation of homoscedastic errors and shown to be asymptotically first- and second-order optimal. The method requires additional estimation of the bandwidth, which can be done by least-squares cross-validation (Silverman, 1984a). However, for multivariate data the application of kernel methods is computationally extensive and the performance of these methods may become rather poor.

The second class of variance estimators is based on spline smoothing methods, as suggested by Wahba (1978, 1983), Silverman (1985), and Buckley et al. (1988). Conceptually these estimators are related to the kernel-type methods (see Silverman, 1984b). A detailed comparison of different spline smoothing methods can be found in Carter and Eagleson (1992), whereas Carter et al. (1992) compared estimators of the variance which use the splines introduced by Reinsch (1967) and Speckman (1985). In particular, Buckley et al. (1988) made an interesting contribution to the understanding of this technique by constructing a minimax estimator for finite sample sizes. For the practical merits of these estimators the reader should consult Eagleson (1989), who gave data-driven guidelines on how to choose the smoothing parameter. The performance of these estimators in the univariate case is well investigated and also depends on a smoothing parameter, which has to be determined from the data. Therefore the application of these methods in multivariate regression models also becomes computationally extensive.

The third class of estimators are difference-type estimators which can conveniently be described in a quadratic form by

$$\hat{\sigma}_D^2 = Y'DY/tr(D) \tag{2}$$

where D is a matrix which does not depend on the data. Rice (1984), Gasser et al. (1986), and Hall et al. (1990) suggested estimators based on this approach for the univariate case. For the applied statistician, these estimators are particularly appealing because they do not require the additional specification of a smoothing parameter (bandwidth) or kernel. In many cases the performance of these estimators is comparable to the second-order optimal method of Hall and Marron (1990) (see the discussion in Section 2). Moreover, a generalization to the multivariate case is attractive from the computational point of view. On the other hand, the necessity of choosing a smoothing parameter is a practical drawback to the use of kernel-type and spline smoothing estimators of the variance. In particular, the generalization of these methods to the multivariate nonparametric regression model becomes computationally extremely extensive. For this reason we focus our attention in the following discussion mainly on the class of estimators of the form in equation (2).

In this chapter we propose a simple generalization of difference-type estimators for the estimation of the covariance matrix Σ in the multivariate nonparametric regression in equation (1). Because every difference-type estimator of the variance in the univariate setup produces a corresponding estimator of the covariance matrix in the multivariate model, we first provide a comparison of the different methods in the univariate case. An understanding of the procedures in this situation is crucial for the discussion of the multivariate case.

In the next section we investigate the finite sample behavior of various estimators for the variance in univariate nonparametric regression. We compare the mean-squared error (MSE) for small to moderate sample sizes

($n \leq 200$) and give data-driven guidelines for the choice of an appropriate estimator from a practical point of view. We also refer in this context to the work of Buckley and Eagleson (1989), who proposed a graphical procedure which indicates whether the bias of a particular estimator is small. Our simulation study shows that in many situations the difference-based class of estimators suggested by Hall et al. (1990) represents a good compromise between an estimator with minimal MSE and an estimator which allows a simple computation. The weights for the terms in the differences are characterized by asymptotically minimizing the MSE within the class of difference-type estimators

$$\hat{\sigma}^2_{D,r} := \frac{1}{n-r} \sum_{k=m_1+1}^{n-m_2} \left(\sum_{j=-m_1}^{m_2} d_j Y_{j+k} \right)^2 \tag{3}$$

where $m_1, m_2 \geq 0$; $r = m_1 + m_2$ denotes the order of $\hat{\sigma}^2_{D,r}$ and $\{d_j\}_{j=-m_1,\ldots,m_2}$ is a difference sequence of real numbers such that

$$\sum_{j=-m_1}^{m_2} d_j = 0, \qquad \sum_{j=-m_1}^{m_2} d_j^2 = 1, \qquad d_{-m_1}, d_{m_2} \neq 0 \tag{4}$$

This method depends on a parameter $r = 1, 2, \ldots$ which corresponds to the number of terms included in the calculation of the residuals. An increasing value of r decreases the asymptotic MSE (up to terms of order $o(n^{-1})$). However, although a large r will decrease the MSE asymptotically, we found in our simulation study that for realistic sample sizes ($n \leq 200$) $r \leq 5$ will always lead to a better finite sample performance. Data-driven guidelines in order to select the most appropriate r are given in Section 2.

As a rule of thumb, the finite sample performance of Hall et al.'s (1990) class of estimators only becomes inefficient when the response function g is highly oscillating. In this case, difference-type estimators which reflect a high local curvature of g are more appropriate and are proposed in Section 2. These methods correspond to a class of polynomial fitting estimators which were mentioned by Buckley and Eagleson (1989, p. 205), and Hall et al. (1990). Although these estimators are asymptotically inefficient, we still find that in concrete applications they may become a reasonable choice because in a high frequency case they improve substantially on the estimators of Hall et al. (1990) and in some case also on the second-order optimal estimtor of Hall and Marron (1990) with asymptotically optimal bandwidth.

In Section 3 we generalize the difference-type estimators in order to estimate the covariance matrix in the multivariate nonparametric regression model in equation (1). We present a representation of the asymptotic covariance matrix of these estimators and asymptotic normality is established. An "optimal" difference scheme (similar to that in Hall et al., 1990) can be determined under the assumption of normally distributed errors. In this case, these "optimal" weights minimize the generalized variance of the estimators for the diagonal elements of the covariance matrix Σ. However, for an arbitrary error distribution the "optimal" weights do depend on the elements of Σ and cannot be used in practice. Finally, some practical guidelines are given for the choice of a difference-type estimator of the covariance matrix in multivariate nonparametric regression models.

2. ESTIMATING THE INTRINSIC VARIABILITY IN UNIVARIATE NONPARAMETRIC REGRESSION—A REVIEW AND COMPARISON

In this section we are interested in estimating $\sigma^2 = V[\epsilon_1]$ in the univariate nonparametric regression model in equation (1). It turns out that an understanding of the univariate case is crucial for a comparison of the estimators in the multivariate case. On the one hand, most of the univariate variance estimators can be generalized for estimating the covariance matrix of multivariate responses. On the other hand, most of the qualitative differences of the estimators in the univariate case carry over to the multivariate case.

One of the first variance estimators was proposed by Rice (1984), who suggested adding up the squared differences between the responses at two successive design points, i.e.,

$$\hat{\sigma}_R^2 = \frac{1}{2(n-1)} \sum_{i=2}^{n} (Y_i - Y_{i-1})^2 \tag{5}$$

Following Rice (1984), Gasser et al. (1986) calculated the squared difference between the responses at the design points t_i, $i = 2, \ldots, n-1$, and a straight line through the responses at their left and right neighbor design points, i.e.,

$$\hat{\sigma}_G^2 = \frac{1}{n-2} \sum_{i=2}^{n-1} c_i^2 [(t_{i+1} - t_i) Y_{i-1} + (t_i - t_{i-1}) Y_{i+1} - (t_{i+1} - t_{i-1}) Y_i]^2 \tag{6}$$

where

$$c_i^2 = [(t_{i+1} - t_i)^2 + (t_i - t_{i-1})^2 + (t_{i+1} - t_{i-1})^2]^{-1}, \qquad i = 2, \ldots, n-1$$

In the case of an equidistant design $t_i = i/n, i = 1, \dots, n$, this simplifies to

$$\hat{\sigma}_G^2 = \frac{2}{3(n-2)} \sum_{i=3}^{n} \left(\frac{1}{2} Y_{i-2} - Y_{i-1} + \frac{1}{2} Y_i \right)^2 \tag{7}$$

Note that both variance estimators have the form

$$\hat{\sigma}_D^2 = Y'DY/tr(D) \tag{8}$$

where $Y = (Y_1, \dots, Y_n)'$ denotes the vector of observations and D is a symmetric $n \times n$ nonnegative-definite matrix, $D \neq 0$.

Generalizing this idea, Buckley et al. (1988) considered the class of estimators $\hat{\sigma}^2$ of σ^2 which are

(A1) quadratic in the data, i.e.,

$$\hat{\sigma}^2 = Y'AY \text{ for some matrix } A \in \mathbb{R}^{n \times n}$$

(A2) always nonnegative, i.e.,

$$\hat{\sigma}^2 \geq 0 \text{ almost surely}$$

(A3) unbiased for σ^2 if the regression function $g(.)$ is a straight line, i.e.,

$$\exists a, b, \in \mathbb{R} : g(t) = a + bt, \ \forall t \in \mathbb{R} \Rightarrow E\hat{\sigma}^2 = \sigma^2$$

In the following, we assume further that $E[\epsilon^4] < \infty$ and that the regression function g is Hölder-continuous of order $\gamma > 1/4$, i.e.,

(A4) $|g(x) - g(y)| \leq c|x - y|^\gamma, \qquad x, y \in [0, 1], \qquad \gamma > 1/4$

Buckley et al. (1988) (in the normal case) and Ullah and Zinde-Walsh (1992) for nonnormal errors gave an explicit representation of the matrix D^* of $\hat{\sigma}_{D^*}^2$ which minimizes

$$\max_{g'\Omega g \leq C} MSE[\hat{\sigma}_D^2] \tag{9}$$

within the class of estimators of the form (A1)–(A3). The maximum in equation (9) is calculated with respect to all functions g satisfying $g'\Omega g \leq C$ where g denotes the mean vector $g = (g(t_1), \dots, g(t_n))'$ and Ω is a particular member of the class of matrices such that (A1)–(A3) hold.

However, the practical merits of this result are rather limited because even under the assumption of normally distributed errors, the minimizer of equation (9) does still depend on the signal-to-noise ratio nC/σ^2, which in general is unknown. In the case of a nonsymmetric error distribution, $\hat{\sigma}_{D^*}^2$ depends in addition on the standardized third and fourth moment

$$\gamma_i := E[(\epsilon/\sigma)^i], \qquad i = 3, 4$$

For these reasons we restrict our considerations in the following simulation study to the class of estimators which can be represented by a quadratic form such that the matrix depends only on the design and the sample size. In view of Rice's (1984) and Gasser et al.'s (1986) variance estimators, Hall et al. (1990) developed such an estimator for the variance which is optimal in the class of estimators described by equations (3) and (4) and will be explained next. These estimators allow a representation

$$\hat{\sigma}^2_{D,r} = \frac{1}{n-r} Y'DY \tag{10}$$

where $D = \tilde{D}'\tilde{D} \geq 0$ and

$$\tilde{D} = \begin{pmatrix} d_{-m_1} & \dots & d_{m_2} & 0 & \dots & 0 \\ & \ddots & & \ddots & & \\ & & \ddots & & \ddots & \\ 0 & \dots & 0 & d_{-m_1} & \dots & d_{m_2} \end{pmatrix} \tag{11}$$

so that the entries in each diagonal of the matrix D are identical except for a finite number of elements in the upper left and lower right part. If $D = \tilde{D}'\tilde{D}$ with $\tilde{D}$ given by equation (11) and the condition in equation (4) holds, it follows that the quadratic form estimator $\hat{\sigma}^2_D$ (defined in equation (8)) and $\hat{\sigma}^2_{D,r}$ (defined in equations (3) and (10)) are asymptotically equivalent. Throughout this paper we will call these variance estimators "difference-type estimators." We assume, without loss of generality, that $m_1 = 0$, and the condition in equation (4) determines the first-order differences ($r = 1$) uniquely as

$$(d_0, d_1) = \left(\frac{1}{\sqrt{2}}, -\frac{1}{\sqrt{2}}\right)$$

which yields the estimator in equation (5) suggested by Rice (1984).

In order to explain the approach of Hall et al. (1990) for a general order r, we observe that the mean-squared error of a difference-type estimator $\hat{\sigma}^2_{D,r}$ defined in equation (3) is given by (see Hall et al., 1990, formula (2.1))

$$MSE[\hat{\sigma}^2_{D,r}] = n^{-1}\sigma^4 \left(\kappa + 2 \sum_{|k| \leq r} \left(\sum_{j=\max(0,\sigma-k)}^{\min(r,r-k)} d_j d_{j+k} \right)^2 \right) + o(n^{-1}) \tag{12}$$

where $\kappa = \gamma_4 - 3$ denotes the kurtosis of ϵ/σ. In order to establish equation (12), one needs a regularity assumption on the design points, namely

(A5) $\max_{1 \leq i \leq n-1} |t_{i+1} - t_i| = O(n^{-1})$

Observe that this condition is fulfilled when the design is asymptotically generated by a Sacks–Ylvisaker density (see Sacks and Ylvisaker, 1970), i.e.,

$$\max_{1\le i\le n}\left|\int_0^{t_i} h(t)\mathrm{d}t - \frac{i}{n}\right| = o(n^{-1}), \qquad i = 1, \dots n$$

where h denotes a positive density.

Minimizing the MSE of $\hat{\sigma}^2_{D,r}$ in equation (12) now becomes tantamount to minimizing

$$\delta = \sum_{k\neq 0}\left(\sum_j d_j d_{j+k}\right)^2 \tag{13}$$

subject to the constraint in equation (4). The minimizing difference sequences for fixed $r = 1, \dots, 10$ were calculated numerically by Hall et al. (1990) and can be found in Table 1 for $r = 1, \dots, 5$. Thoughout this paper we denote by $\hat{\sigma}^2_{ow,r}$ the difference type estimator in equation (3) of order r obtained by the "optimal" weights minimizing the asymptotic MSE in equation (12).

Observe that in practice there is no need to evaluate the sequence of optimal weights $\{d_j\}$ for a given order r because the estimator $\hat{\sigma}^2_{ow,r}$ is asymptotically first-order equivalent (the difference in MSE is $o(n^{-1})$) to the estimator

$$S_r^2 = \frac{2}{2n - r - 1} Y'DY \tag{14}$$

Table 1. Optimal weights for difference-type variance estimators of order r

r	d_0	d_1	d_2	d_3	d_4	d_5
1	0.7071	−0.7071				
2	0.8090	−0.5	−0.3090			
3	0.1942	0.2809	0.3832	−0.8582		
4	0.2798	−0.0142	0.6909	−0.4858	−0.4617	
5	0.9064	−0.2600	−0.2167	−0.1774	−0.1420	−0.1103

where the matrix D is of the form

$$
D = \begin{pmatrix}
\frac{1}{2} & \frac{-1}{2r} & \cdots & \frac{-1}{2r} & & & & & & 0 \\
\frac{-1}{2r} & \frac{1}{2}+\frac{1}{2r} & \frac{-1}{2r} & \cdots & \frac{-1}{2r} & & & & & \\
\vdots & \ddots & \ddots & \ddots & \ddots & \ddots & & & & \\
\frac{-1}{2r} & \cdots & \frac{-1}{2r} & \frac{1}{2r}+\frac{r-1}{2r} & \frac{-1}{2r} & \cdots & \frac{-1}{2r} & & & \\
\frac{-1}{2r} & \cdots & \cdots & \frac{-1}{2r} & 1 & \frac{-1}{2r} & \cdots & \frac{-1}{2r} & & \\
 & \ddots & \ddots & \ddots & \ddots & \ddots & \ddots & \ddots & \ddots & \\
 & & \frac{-1}{2r} & \cdots & \cdots & \frac{-1}{2r} & 1 & \frac{-1}{2r} & \cdots & \frac{-1}{2r} \\
 & & & \ddots & \ddots & \ddots & \ddots & \ddots & \ddots & \vdots \\
 & 0 & & & \frac{-1}{2r} & \cdots & \cdots & \frac{-1}{2r} & \frac{1}{2}+\frac{1}{2r} & \frac{-1}{2r} \\
 & & & & & \frac{-1}{2r} & \cdots & \cdots & \frac{-1}{2r} & \frac{1}{2}
\end{pmatrix} \tag{15}
$$

(all other entries in this matrix are 0). This follows from equation (11), and the fact that the optimal weights minimizing equation (12) must satisfy

$$\sum_{i=0}^{r} d_i^2 = 1 \quad \text{and} \quad \sum_{i=\max(0,-k)}^{\min(r,r-k)} d_i d_{i+k} = -\frac{1}{2r} \quad (k = -r, \ldots, -1, 1, \ldots, r)$$

as is shown in Hall et al. (1990), Appendix 2, pp. 526/527. Hence, this estimator becomes computationally extremely simple. Hall et al. (1990) further proved (see the theorem in Appendix 1 of their paper) that

$$MSE[\hat{\sigma}^2_{ow,r}] = n^{-1}\sigma^4(\kappa + 2 + r^{-1}) + o(n^{-1}) \tag{16}$$

which shows that with increasing order r, the asymptotic MSE decreases.

It is interesting to note that the class of estimators considered in Buckley et al. (1988) is neither contained in Hall et al.'s (1990) class of difference scheme estimators nor vice versa. On the one hand, an arbitrary positive definite matrix D associated to $\hat{\sigma}^2_D$ cannot be decomposed such that $D = \tilde{D}'\tilde{D}$ where $\tilde{D}$ is a matrix of the form of equation (11). On the other hand, the condition in equation (4) does not guarantee that $\hat{\sigma}^2_{ow,r}$ is unbiased whenever g is an arbitrary straight line. Only when $g \equiv c$ does this property hold in general. For example, Gasser et al.'s (1986) estimator in equation (6) does not belong to Hall et al.'s (1990) class but to Buckley et al.'s (1988) class, while Rice's (1984) estimator in equation (5) belongs only to Hall et al.'s (1990) class.

Another class of variance estimators considered in this connection is the class of difference-type estimators in equation (3) of order r with weights

$$d_j = \binom{2r}{r}^{-1/2} \binom{r}{j} (-1)^j, \qquad 0 \leq j \leq r \tag{17}$$

Note that the weights in equation (17) are commonly employed for numerical differentiation. A different motivation for these weights can be obtained as follows. For a design and $(r+1)$ successive pairs of observations $(t_i, Y_i), \ldots, (t_{i+r}, Y_{i+r})$, we interpolate r of the points by a polynomial of degree $r-1$, $P_{r,i}(.)$ say, where one point, (t_{i^*}, Y_{i^*}) say, is left out. Then we use the vertical distance between this polynomial and the point (t_{i^*}, Y_{i^*}) to define the local residuals as

$$\delta_{i^*} = Y_{i^*} - P_{r,i}(t_{i^*})$$

From the Lagrange representation of the polynomial $P_{r,i}$,

$$P_{r,i}(t) = \sum_{\substack{j=0 \\ i+j \neq i^*}}^{r} Y_{i+j} \prod_{\substack{k=0 \\ k \neq j,(i^*-i)}}^{r} \frac{t - t_{i+k}}{t_{i+j} - t_{i+k}} \tag{18}$$

one can easily recognize that the local residuals can be rewritten as

$$\delta_{i^*} = \sum_{k-0}^{r} \delta_{k,i^*,i} Y_{i+k}$$

with an appropriate choice of $\delta_{0,i^*,i}, \ldots, \delta_{r,i^*,i}$. Now define

$$\bar{\delta}_{k,i} = \delta_{k,i^*,i} \left(\sum_{j=0}^{r} \delta^2_{j,i^*,i} \right)^{-1/2}$$

and the difference-based estimator of order r by

$$\hat{\sigma}_{p,r} = \frac{1}{n-r} \sum_{i=1}^{n-r} \left(\sum_{k=0}^{r} \bar{\delta}_{k,i} Y_{i+k} \right)^2 \tag{19}$$

Note that the definition of $\bar{\delta}_{k,i}$ does not depend on the particular pair (t_{i^*}, Y_{i^*}) which is left out among $(t_i, Y_i), \ldots, (t_{i+r}, Y_{i+r})$ in the interpolating polynomial. This is easily seen from the Lagrange representation in equation (18) by changing i^* to i° and observing that

$$\delta_{l,i^*,i} = \delta_{l,i^\circ,i}\left(- \prod_{\substack{k=0 \\ i+k \neq i^*, i^\circ}}^{r} \frac{t_{i^\circ} - t_{i+k}}{t_{i^*} - t_{i+k}} \right), \qquad l = 0, \ldots, r$$

Observe that this class of variance estimators contains Rice's (1984) estimator ($r = 1$) as well as Gasser et al.'s (1986) estimator ($r = 2$) defined in equations (5) and (6), respectively.

If the design is equidistant, i.e., $t_i = i/n$, $i = 1, \ldots, n$, we obtain the weights in equation (17). More precisely, we put $i = i^*$ and calculate

$$\delta_{j,i,i} = -\prod_{\substack{k=1 \\ k \neq j}} \frac{i/n - (i+k)/n}{(i+j)/n - (i+k)/n} = -\prod_{k=1}^{j-1} \frac{-k}{j-k} \times \prod_{k=j+1}^{r} \frac{k}{k-j} = (-1)^j \binom{r}{j}$$

for $j = 0, \ldots, r$ and $i = 1, \ldots, n - r$. For the determination of the weights $\bar{\delta}_{k,i}$, we note that

$$\sum_{k=0}^{r} \delta_{k,i,i}^2 = \sum_{k=0}^{r} \binom{r}{k}^2 = \binom{2r}{r}$$

Therefore, for an equidistant design the local residuals are given by

$$\hat{\epsilon}_{i^*} = \sum_{k=0}^{r} \bar{\delta}_{k,i} Y_{i+k} = \sum_{k=0}^{r} (-1)^k \binom{2r}{r}^{-1/2} \binom{r}{k} Y_{i+k}$$

and from equation (19) we obtain

$$\hat{\sigma}_{p,r}^2 = \frac{1}{n-r} \sum_{i=1}^{n-r} \hat{\epsilon}_{i^*}^2$$

which is the difference-type estimator in equation (3) with the weights in equation (17). Note that $\hat{\sigma}_{p,r}^2$ is the squared Euclidean distance from the vector of observations to the local interpolation polynomials, scaled by $(n-r)^{-1}$. Throughout this article we will call $\hat{\sigma}_{p,r}^2$ the polynomial fitting or interpolating polynomial estimator of order r. It can be shown that the estimator $\hat{\sigma}_{p,r}^2$ in equation (19) is unbiased if the regression function g is a polynomial of degree $r - 1$. The price which has to be paid for this nice property is a relatively large variance if the order of differences increases.

Hall et al. (1990) argued forcefully that the weights in equation (17) perform rather badly for increasing r because the asymptotic MSE of the corresponding difference-type estimator is given by

$$MSE[\hat{\sigma}_{p,r}^2] \sim n^{-1}\sigma^4 \left(\kappa + 2\frac{\binom{4r}{2r}}{\binom{2r}{r}^2} \right) \tag{20}$$

However, we will now see that this method may become useful if the unknown regression function g is strongly oscillating. In this case the bias of the difference-type estimators is more important than the variance, especially for small sample size.

Because Hall and Marron's (1990) kernel-type estimator is asymptotically first- and second-order optimal, we will also include this approach in the comparison of different variance estimators. It will be used to see how the difference-type estimators perform compared with the (asymptotically) best variance estimator. The kernel method of Hall and Marron (1990) can also be regarded as a quadratic form, although conceptually it is a little bit different because the scheme $\{d_j\}$ does depend on the observations and is generated by a kernel. More precisely, consider a kernel function, say K, and a bandwidth h. Then the kernel estimator of the regression function g at the point t_i is defined by

$$\hat{g}(t_i) = \sum_{j=1}^{n} w_{ij} Y_j$$

where

$$w_{ij} = K\{(t_i - t_j)/h\} / \sum_{l=1}^{n} K\{(t_i - t_l)/h\}$$

We therefore define residuals $\hat{\epsilon}_i = Y_i - \sum_{j=1}^{n} w_{ij} Y_j$ and an estimator of the variance by

$$\hat{\tau} := \nu^{-1} S$$

where $\nu = n - 2\sum_{i=1}^{n} w_{ii} + \sum_{i,j=1}^{n} w_{ij}^2$ denotes a normalizing constant and $S = \sum_{i=1}^{n} \hat{\epsilon}_i^2$ is the sum of the squared residuals. For more details regarding kernel-type estimators for nonparametric regression we refer to Härdle (1990). In our simulation, the kernel K was chosen to be $K(x) = 3/4\ (1 - x^2)$ on $[-1, 1]$ and 0 otherwise, which corresponds to the optimal kernel of order two for estimating the mean response function g (see Epanechnikov, 1969).

We investigated the estimator with its optimal theoretical bandwidth h_0, described in Hall and Marron (1990), and in addition one with a bandwidth h_{cv} determined by a least-squares cross-validation. This was computed by the Fortran nag-routine e04fdf as the minimum of the function

$$cv(h) = \sum_{i=1}^{n} \left(Y_i - \hat{g}_{h,n}^{(i)}(t_i) \right)^2 \tag{21}$$

where $\hat{g}_{h,n}^{(i)}$ denotes the "leave one out" kernel estimator for g drawn from the sample without the pair (t_i, Y_i) with the same kernel K. See Silverman (1984a) for more details regarding cross-validation.

In order to get some insight into the finite sample behavior of these estimators, we considered the following setup in our simulation study.

(E1) Hall et al.'s (1990) difference-type estimators defined in equation (3) with the optimal weights of Table 1 for different orders $r = 1, \dots 5$. Note that these estimators are denoted by $\hat{\sigma}^2_{ow,r}$ and include Rice's (1984) estimator in equation (5) for $r = 1$.

(E2) The polynomial fitting estimator in equation (19), and more precisely, the difference type estimators defined in equation (3) with weights

$$d_j = \binom{2r}{r}^{-1/2} \binom{r}{j} (-1)^j, \qquad 0 \leq j \leq r$$

where $r = 1, \dots, 5$. These estimators are denoted by $\hat{\sigma}^2_{p,r}$ and also include Rice's (1984) estimator in equation (5) for $r = 1$ as well as Gasser et al.'s (1986) estimator in equation (6) for $r = 2$ (if the design is equidistant).

(E3) The kernel estimator of Hall and Marron (1990) with asymptotically optimal bandwidth and with a bandwidth determined by least-squares cross-validation (see equation (21)). The optimal bandwidth can be obtained from Hall and Marron (1990, p. 417), but is unknown in general. Because this estimator achieves the second-order optimal rate

$$MSE[\hat{\tau}^2] = n^{-1}\sigma^4(\kappa + 2) + o(n^{-1})$$

it serves as a good reference point for the MSE in general. It is worthwhile mentioning that the optimal rate is also achieved for Buckley et al.'s (1988) minimax estimator.

We performed each simulation study with 5000 replications for the classes of estimators (E1)–(E3), except for the estimator of Hall and Marron (1990) with bandwidth determined by cross-validation, where 1000 replications were used. We considered the following five mean functions:

$$\begin{aligned} g_1(x) &= 2 - 5x + 5\exp(-25(x-5)^2) \\ g_2(x) &= 5\exp(-5x) \\ g_3(x) &= 5\sin(2\pi x) \\ g_4(x) &= 5\sin(6\pi x) \\ g_5(x) &= 20x^3 - 40x^2 + 25x - 3 \end{aligned} \tag{22}$$

where g_1, g_2, g_3 were drawn from Gasser et al. (1986, p. 629). All functions are displayed in Figure 1. This shows that g_1 is nearly a straight line, whereas g_2 serves as an example for an exponentially decreasing function and g_5 for a polynomial; g_4 was chosen as an example for a high frequency function, whereas g_3 is a low frequency function.

Sample sizes were chosen as $n = 25, 50, 100, 200$ and an equidistant design in [0, 1] was always assumed. Some representative results are displayed in Figures 2–10 which show the MSE of various estimators as a function of the variance. Further simulation results are available from the third author.

The legend is read as follows. The polynomial fitting estimators (E2) are denoted by $\hat{\sigma}^2_{p,1}, \ldots, \hat{\sigma}^2_{p,4}$, where the second index denotes the order r. For example, $\hat{\sigma}^2_{p,1}$ gives Rice's (1984) estimator and $\hat{\sigma}^2_{p,2}$ corresponds to Gasser et al.'s (1986) variance estimator. The estimators (E1) with optimal weighting scheme are denoted by $\hat{\sigma}^2_{ow,1}, \ldots, \hat{\sigma}^2_{ow,5}$, where the second index also denotes the order r (note that $\hat{\sigma}^2_{p,1} = \hat{\sigma}^2_{ow,1}$). Finally, Hall and Marron's (1990) estimator is denoted by $\hat{\sigma}^2_{hm}$ and by $\hat{\sigma}^2_{cv}$, reflecting the choice of the (asymptotically) optimal bandwidth (hm) and the bandwidth obtained by cross-validation (cv). The estimator $\hat{\sigma}^2_{hm}$ is displayed in Figure 2–8 (boldface line) because it is used to investigate the behavior of the proposed estimates with respect to the (asymptotically) best variance estimator (with respect to the MSE criterion).

The error distribution was always assumed as normal with mean 0 and variance σ^2, which is the label on the horizontal axis in each figure. The variance σ^2 was chosen as 0.01, 0.05, 0.1, 0.3, 0.5, 1.0, 1.5, 2.0, ...,5.0. The random generator was the standard Fortran nag-routine g05ddf. The vertical axis displays the MSE of the different estimators. The conclusions from our simulation study are summarized below.

- It is indicated by equation (16) that a large-order r in the class of Hall et al.'s (1990) optimal difference variance estimators yields a smaller asymptotic MSE. However, for the finite samples the performance of these estimators for various orders is different from the asymptotic behavior. When the sample size is rather small, Rice's (1984) estimator ($r = 1$) is superior in class (E1), independent of the mean function. This effect is substantial for small values of the variance σ^2 (Figures 2 and 3). In some cases ($n \leq 50$, σ^2 small) Rice's (1984) estimator even improves the method of Hall and Marron (1990) (see, for example, Figure 2). If $n \geq 50$, a large order becomes reasonable, where for $n \geq 100$ the highest-order estimator ($r = 5$) is always the best as long as the mean function g is not too strongly oscillating and σ^2 is not too small (Figure 4). It is important that in these cases the best difference-type estimator yields nearly the same results as the estimator of Hall and Marron (1990) with asymptotically optimal bandwidth (which is

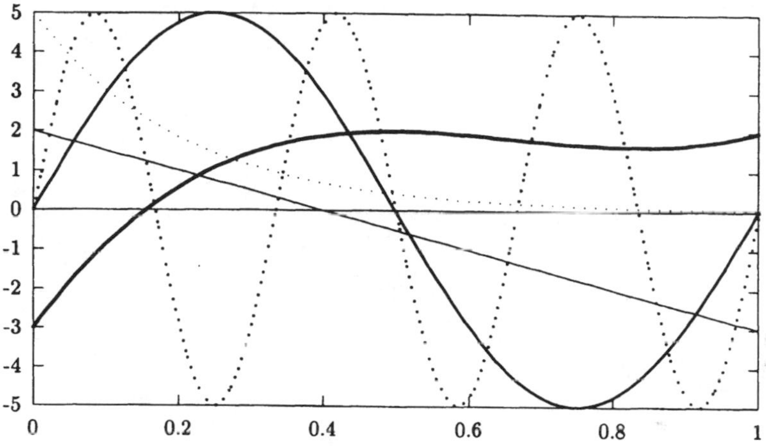

Figure 1. The regression functions of equation (22) used in the simulation study. $g_1(x)$ ——; $g_2(x)$ · · · ·; $g_3(x)$ ——; $g_4(x)$ · · · ·; $g_5(x)$ ——

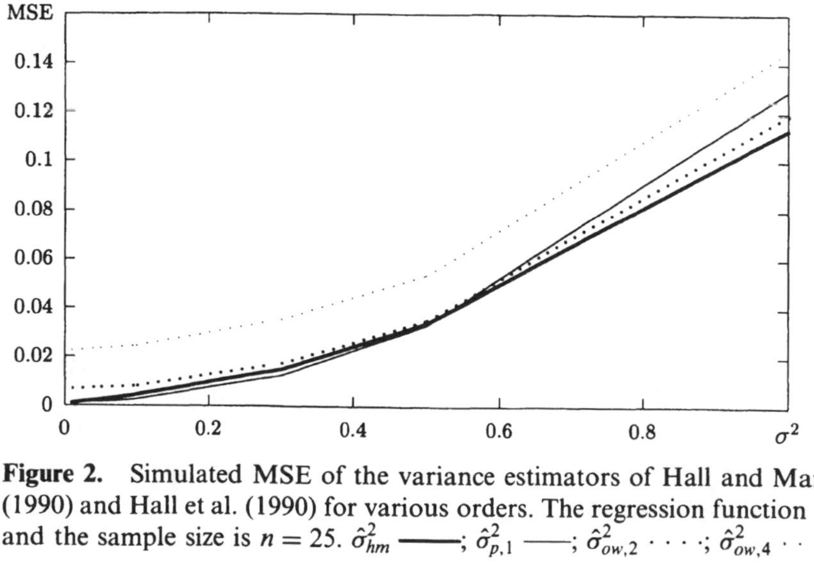

Figure 2. Simulated MSE of the variance estimators of Hall and Marron (1990) and Hall et al. (1990) for various orders. The regression function is g_2 and the sample size is $n = 25$. $\hat{\sigma}^2_{hm}$ ——; $\hat{\sigma}^2_{p,1}$ ——; $\hat{\sigma}^2_{ow,2}$ · · · ·; $\hat{\sigma}^2_{ow,4}$ · · · ·.

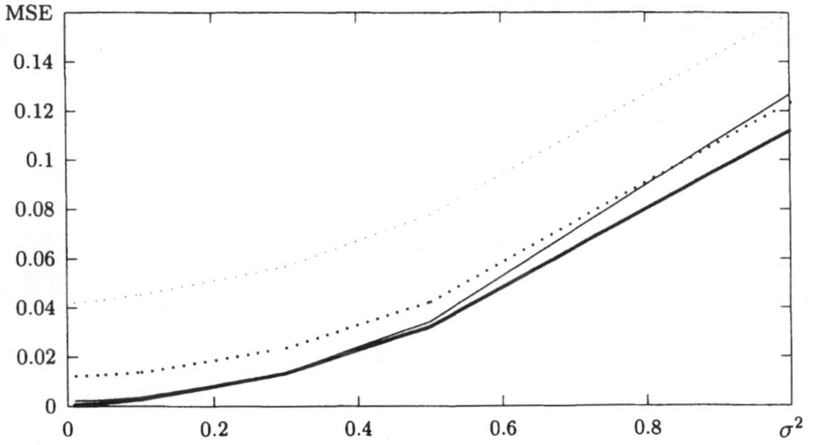

Figure 3. Simulated MSE of the variance estimators of Hall and Marron (1990) and Hall et al. (1990) for various orders. The regression function is g_5 and the sample size is $n = 25$. $\hat{\sigma}^2_{hm}$ ——; $\hat{\sigma}^2_{p,1}$ ——; $\hat{\sigma}^2_{ow,2}$ · · · ·; $\hat{\sigma}^2_{ow,4}$ · · · ·.

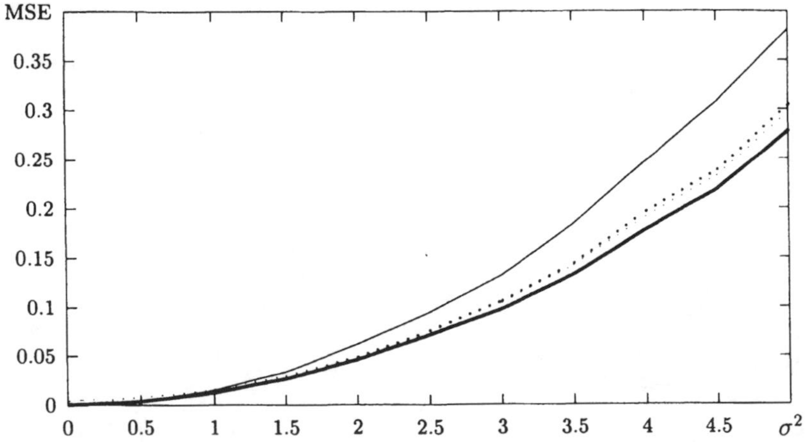

Figure 4. Simulated MSE of the variance estimators of Hall and Marron (1990) and Hall et al. (1990) for various orders. The regression function is g_3 and the sample size is $n = 200$. $\hat{\sigma}^2_{hm}$ ——; $\hat{\sigma}^2_{p,1}$ ——; $\hat{\sigma}^2_{ow,3}$ · · · ·; $\hat{\sigma}^2_{ow,5}$ · · · ·.

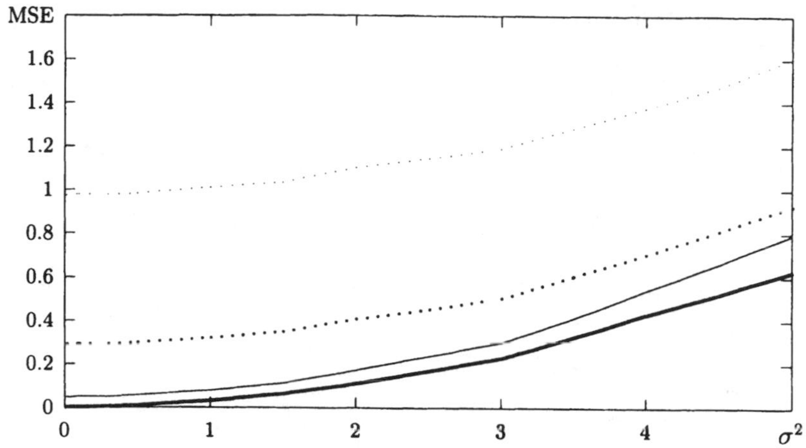

Figure 5. Simulated MSE of the variance estimators of Hall and Marron (1990) and Hall et al. (1990) for various orders. The regression function is g_4 and the sample size is $n = 100$. $\hat{\sigma}^2_{hm}$ ——; $\hat{\sigma}^2_{p,1}$ ——; $\hat{\sigma}^2_{ow,2}$ · · · ·; $\hat{\sigma}^2_{ow,3}$ · · · ·.

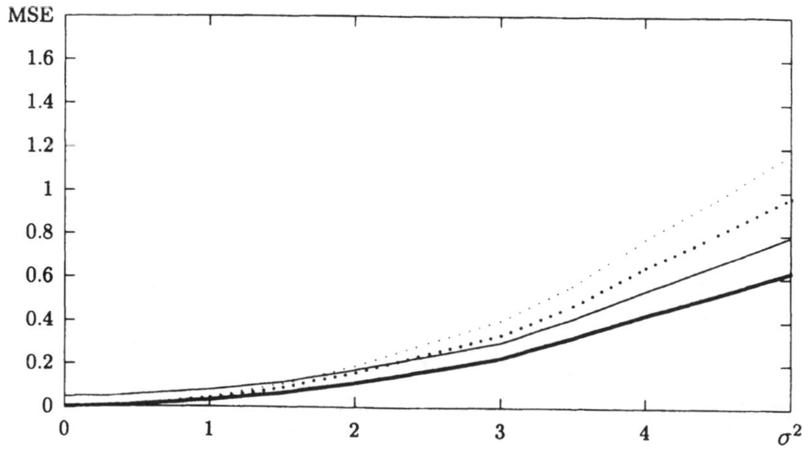

Figure 6. Simulated MSE of the variance estimator of Hall and Marron (1990) and of the polynomial fitting estimators for various orders. The regression function is g_4 and the sample size is $n = 100$. $\hat{\sigma}^2_{hm}$ ——; $\hat{\sigma}^2_{p,1}$ ——; $\hat{\sigma}^2_{p,2}$ · · · ·; $\hat{\sigma}^2_{p,3}$ · · · ·.

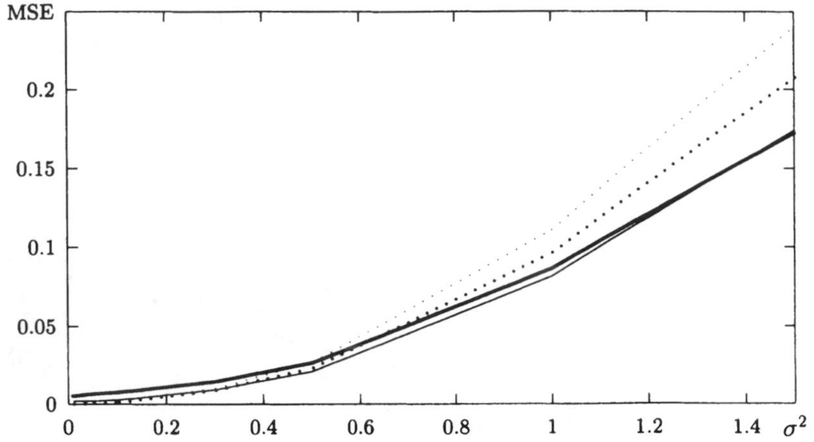

Figure 7. Simulated MSE of the variance estimators of Hall and Marron (1990) and of the polynomial fitting estimators for various orders. The regression function is g_4 and the sample size is $n = 50$. $\hat{\sigma}^2_{hm}$ ——; $\hat{\sigma}^2_{p,2}$ —— $\hat{\sigma}^2_{p,3}$ · · · ·; $\hat{\sigma}^2_{p,4}$ · · · ·.

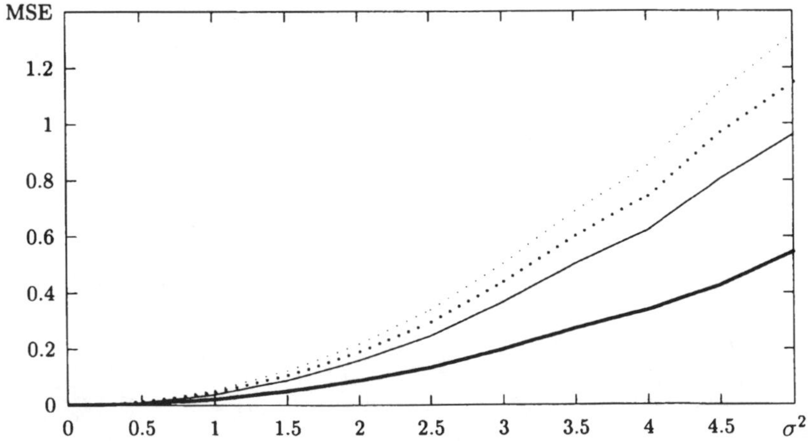

Figure 8. Simulated MSE of the variance estimators of Hall and Marron (1990) and of the polynomial fitting estimators for various orders. The regression function is g_5 and the sample size is $n = 100$. $\hat{\sigma}^2_{hm}$ ——; $\hat{\sigma}^2_{p,2}$ ——; $\hat{\sigma}^2_{p,3}$ · · · ·; $\hat{\sigma}^2_{p,4}$ · · · ·.

unknown in practice). In the high-frequency case, Rice's (1984) estimator is still superior for large sample size, and the difference from Hall and Marron's (1990) estimator with asymptotically optimal bandwidth as well as from the polynomial fitting estimators of class (E2) is significant (Figures 5 and 6). The difference from the last-named estimators in class (E2) becomes more obvious if the sample size or the value of the variance σ^2 decreases.

- The polynomial fitting estimators with order $r \geq 2$ perform very well in the high-frequency case and nearly achieve the MSE of the optimal estimator (E3), especially for small values of σ^2 (see Figure 6). They also yield a much smaller MSE then the best estimator of class (E1), in contrast to the asymptotic theory (see Figures 5 and 6 and note that the MSE axes in these figures are of the same scale). This is to be expected since the polynomial fitting estimators represent the most flexible way to deal with a high local curvature. Note that they even improve on Hall and Marron's (1990) estimator with asymptotically optimal bandwidth for small values of σ^2. Here "small" is relative to the sample size n: for a smaller sample size the polynomial fitting estimators improve on Hall and Marron's (1990) estimator even for larger values of σ^2 provided that the regression function g is oscillating (Figure 7).
- The variance σ^2 does not essentially affect the order of the estimators in class (E1). However, if σ^2 increases, the polynomial fitting estimators in class (E2) with increasing order r perform rather poorly. In nearly all cases investigated in our study, a small order estimator ($r = 2$) is more appropriate (Figure 8).
- The polynomial fitting estimators for $r \geq 2$ are very robust in the sense that the MSE is nearly independent of the regression function g. This explains why for all five regression functions simulated in our study the plots of the MSE of the polynomial fitting estimators are nearly the same as in Figure 8. In contrast, the asymptotically optimal difference-type estimators of Hall et al. (1990) are more sensitive with respect to the regression function g, especially for small sample sizes (Figure 9).
- Hall and Marron's (1990) estimator with least-squares cross-validation bandwidth nearly achieves the estimator with the optimal theoretical bandwidth (Figure 10) except for the strong oscillating regression curve g_4. In this case, cross-validation leads to completely inefficient results, in contrast to the case where the (asymptotically) optimal bandwidth is known. On the other hand, if the regression curve is almost a straight line, cross-validation results in an estimator with small MSE, whereas surprisingly the theoretical optimal bandwidth does not yield satisfactory results.

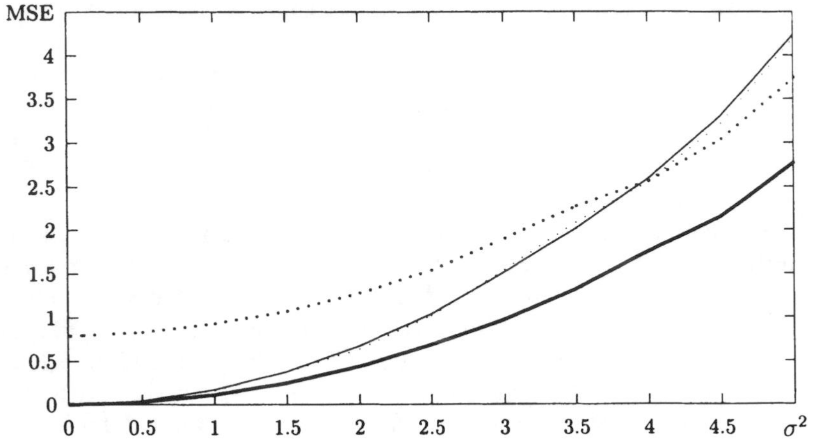

Figure 9. Simulated MSE of the variance estimator of Hall et al. (1990) and of the polynomial fitting estimator, each for order $r = 2$. The figure illustrates the impact of the function g on the performance of the optimal difference and the polynomial fitting estimator. The regression functions are g_1, g_3 and the sample size is $n = 25$. $(\hat{\sigma}^2_{ow,2}, g_1)$ ——; $(\hat{\sigma}^2_{ow,2}, g_3)$ · · · ·; $(\hat{\sigma}^2_{p,2}, g_1)$ ——; $(\hat{\sigma}^2_{p,2}, g_3)$ · · · ·

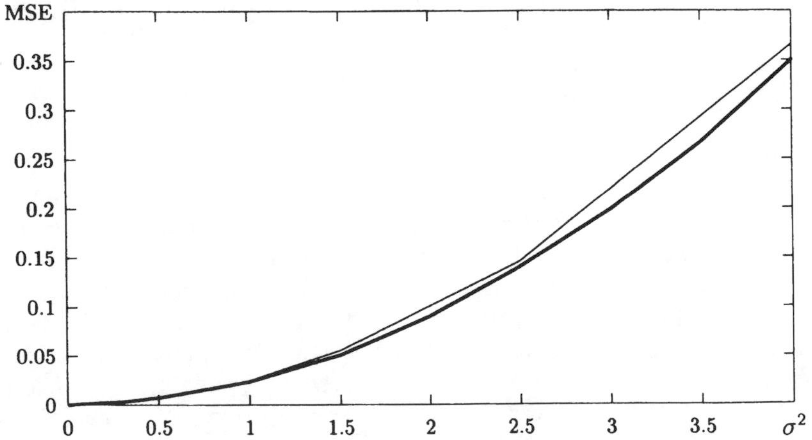

Figure 10. Simulated MSE of the variance estimators of Hall and Marron (1990) with asymptotically optimal bandwidth and bandwidth determined by least-squares cross-validation. The regression function is g_2 and the sample size is $n = 100$. $\hat{\sigma}^2_{hm}$ ——; $\hat{\sigma}^2_{cv}$ ——

Based on these observations, the following guidelines for the choice of the variance estimator are given. If there is no information available concerning the magnitude of the oscillation of g (e.g., by pilot estimation) Gasser et al.'s (1986) estimator is a good compromise between an efficient estimator in the case of a smooth regression function g and an estimator with acceptable MSE in the high-frequency case. Moreover, this estimator is very easy to compute.

If a high-oscillating regression function g can be excluded, or one is willing to tolerate a large contribution from the bias to the MSE in this case, Hall et al.'s (1990) estimator as well as Hall and Marron's (1990) estimator with cross-validated bandwidth are reasonable. The last-named method is in general slightly more efficient, but the computational effort is larger, in particular considering the representation in equation (15) for the asymptotically equivalent estimator S_r^2 in equation (14). If one decides for Hall et al.'s (1990) estimtor, a choice of $3 \leq r \leq 5$ is appropriate when the sample size $n \geq 100$, $\sigma^2 \geq 0.5$, and the regression function is not too strongly oscillating. In other cases, $r = 2$ should not be exceeded.

If the data are extremly noisy and one has to assume that this is caused by the oscillation of the regression function g rather than by a large variance, then the polynomial fitting estimators become necessary.

3. MULTIVARIATE REGRESSION

We will now address the problem of estimating the covariance matrix Σ in the multivariate regression model defined in the introduction. To this end, let $g_1, \ldots, g_d$ denote the components of the response function g which is assumed to be Hölder-continuous of order $\gamma > 1/4$ in each coordinate,

(A6) $|g_j(x) - g_j(y)| \leq c|x - y|^\gamma, \qquad \forall x, y \in [0, 1], j = 1, \ldots, d$

In principle, all proposed estimators in Section 2 can be generalized for the estimation of the covariance matrix Σ in the multivariate regression in equation (1). This can easily be seen by observing that all estimators in the univariate regression are based on pseudo-residuals $\hat{\epsilon}_1, \ldots, \hat{\epsilon}_{n-r}$. In the multivariate case we obtain residuals for each regression function g_j, say $\hat{\epsilon}_{1,j}, \ldots, \hat{\epsilon}_{n-r,j}$, $j = 1, \ldots, d$, and the covariance between the jth and the lth component of the response can be consistently estimated by

$$\frac{1}{n-r} \sum_{i=1}^{n-r} \hat{\epsilon}_{i,j} \hat{\epsilon}_{i,l}$$

For pseudo-residuals based on kernel-type estimators or spline smoothing methods, this approach requires the estimation of (at least) d smoothing parameters and the computational effort increases significantly. This is a serious drawback from a practical point of view. The simulation study of Section 2 for the univariate case shows that for finite samples an appropriate difference-type estimator always yields an MSE which is comparable with the MSE obtained by the first- and second-order optimal kernel-type estimator of Hall and Marron (1990). For these reasons we restrict ourselves in the following discussion to the multivariate generalization of the difference-type estimators proposed in Section 2. If $D = \tilde{D}'\tilde{D}$ is the matrix corresponding to a difference-type estimator (see Section 2, equations (3), (4), and (11)), then an estimator of the covariance matrix $\Sigma = (\rho_{ij})_{i,j=1,\dots,d}$ can now be conveniently written as

$$\hat{\Sigma} = (\hat{\rho}_{ij})_{i,j=1,\dots,d} = Y'DY/tr(D) \tag{23}$$

where $Y = (Y_1, \dots, Y_n)' \in \mathbb{R}^{n \times d}$ and $Y_i = (Y_{i,1}, \dots, Y_{i,d})'$ is the response at the point t_i, $i = 1, \dots, n$. For example, if Rice's (1984) estimator is used, from equation (5) we obtain

$$\tilde{D} = \frac{1}{\sqrt{2}} \begin{pmatrix} 1 & -1 & 0 & \dots & 0 \\ 0 & 1 & -1 & \ddots & \vdots \\ & & \ddots & & \\ 0 & \dots & 0 & 1 & -1 \end{pmatrix} \in \mathbb{R}^{n-1 \times n}$$

and for the quadratic matrix in equation (23)

$$D = \tilde{D}'\tilde{D} = \frac{1}{2} \begin{pmatrix} 1 & -1 & 0 & \dots & \dots & 0 \\ -1 & 2 & -1 & 0 & \dots & 0 \\ 0 & \ddots & \ddots & \ddots & \ddots & \vdots \\ \vdots & \ddots & \ddots & \ddots & \ddots & 0 \\ 0 & \dots & 0 & -1 & 2 & -1 \\ 0 & \dots & \dots & 0 & -1 & 1 \end{pmatrix} \in \mathbb{R}^{n \times n}$$

This gives as estimator for the covariance between the jth and lth component of the response

$$\hat{\rho}_{jl} = \frac{1}{2(n-1)} \sum_{i=2}^{n} (Y_{i,j} - Y_{i-1,j})(Y_{i,l} - Y_{i-1,l}), \qquad 1 \le j, l \le d$$

Other difference-type estimators for the covariance matrix are obtained in exactly the same way from the various methods described in Section 2 for the univariate case. The following gives the asymptotic distribution of the class of estimators $\hat{\Sigma}$ obtained by this approach.

THEOREM 1: *Assume that conditions (A5) and (A6) are satisfied and that the vector* ϵ_i *in the nonparametric regression model in equation (1) has moments up to order four. Consider a difference-type estimator of the covariance matrix* Σ *defined in equation (23) with matrix* $D = \tilde{D}'\tilde{D}$, *where* $\tilde{D}$ *is given by equations (11) and (4). The covariance between the components* $\hat{\rho}_{ij}$ *and* $\hat{\rho}_{kl}$ $(\leq i, j, k, l \leq d)$ *is given by*

$$\mathrm{Cov}(\hat{\rho}_{ij}, \hat{\rho}_{kl}) = \frac{\rho_{ik}\rho_{jl} + \rho_{il}\rho_{jk}}{(tr(D))^2}\, tr(D^2)$$
$$+ \frac{\delta^{(i,j,k,l)} - \rho_{ik}\rho_{jl} - \rho_{il}\rho_{jk} - \rho_{ij}\rho_{kl}}{(tr(D))^2} \sum_{s=1}^{n} D_{ss}^2 + o(n^{-1}) \tag{24}$$

where $D_{11}, \ldots, D_{nn}$ *denote the diagonal elements of the matrix D in equation (23), and for* $\epsilon_s = (\epsilon_{s,1}, \ldots, \epsilon_{s,d})'$, $s = 1, \ldots, n$,

$$\delta^{(i,j,k,l)} := E[\epsilon_{1,i}\epsilon_{1,j}\epsilon_{1,k}\epsilon_{1,l}]$$

Moreover, $tr(D^2) = \alpha n + O(1)$ *with*

$$\alpha = 1 + 2\sum_{k=1}^{r}\left(\sum_{j} d_j d_{j+k}\right)^2 \tag{25}$$

and the vector of dimension $d(d+1)/2$

$$\sqrt{n}\big[(\hat{\rho}_{1,1}, \hat{\rho}_{1,2}, \ldots, \hat{\rho}_{d-1,d}, \hat{\rho}_{d,d})' - (\rho_{1,1}, \rho_{1,2} \ldots, \rho_{d-1,d}, \rho_{d,d})'\big]$$

is asymptotically normally distributed with mean 0 *and covariance matrix with elements given by* $(1 \leq i \leq j \leq d, 1 \leq k \leq j \leq d)$

$$\lim_{n\to\infty} n\,\mathrm{Cov}(\hat{\rho}_{ij}, \hat{\rho}_{kl}) = \alpha(\rho_{ik}\rho_{jl} + \rho_{il}\rho_{jk}) + \left(\delta^{(i,j,k,l)} - \rho_{ik}\rho_{jl} - \rho_{il}\rho_{jk} - \rho_{ij}\rho_{kl}\right)$$

Proof. Without loss of generality, assume that the matrix in the quadratic form is given by $D = \tilde{D}'\tilde{D}$ as in equation (11), with $m_1 = 0$ and $m_2 = r$ where the sequence of weights $\{d_i\}$ satisfies the conditions stated in equation (4). Let

$$\tilde{\epsilon}_{\mu,\nu} = \sum_{s=0}^{r} d_s Y_{s+\mu,\nu}, \qquad \nu = 1, \ldots, d, \qquad \mu = 1, \ldots, n-r$$
$$\tilde{\epsilon}_{\nu} = (\tilde{\epsilon}_{1,\nu}, \ldots, \tilde{\epsilon}_{n-r,\nu})', \qquad \nu = 1, \ldots, d$$
$$\epsilon_{(\nu)} = (\epsilon_{1,\nu}, \ldots, \epsilon_{n,\nu})', \qquad \nu = 1, \ldots, d$$

so that

$$\tilde{\epsilon}_{\nu} = \tilde{D}\epsilon_{(\nu)} + B_{\nu}$$

where B_ν is the $(n-r)$-dimensional vector representing the bias. A simple calculation and equation (4) show that the jth component of B_ν is given by

$$\sum_{s=0}^{r} d_s g_\nu(t_{j+s}) = \sum_{s=0}^{r-1}\left(\sum_{k=0}^{s} d_k\right)[g_\nu(t_{j+s}) - g_\nu(t_{j+s+1})]$$

and by the Hölder continuity of $g_\nu (\nu = 1, \ldots, d)$ all components of B_ν are of order $O(n^{-\gamma})$. This leads to

$$\begin{aligned}\hat{\rho}_{ij} &= \frac{1}{tr(D)}(\epsilon'_{(i)}\tilde{D}' + B'_i)(\tilde{D}\epsilon_{(j)} + B_j) \\ &= \frac{1}{tr(D)}(\epsilon'_{(i)}D\epsilon_{(j)} + \epsilon_{(i)}\tilde{D}'B_j + B'_i\tilde{D}\epsilon_{(j)}) + o(n^{-1/2}) \qquad (26)\end{aligned}$$

and gives for the expected value

$$E[\hat{\rho}_{ij}] = \frac{1}{tr(D)}\sum_{s=1}^{n} D_{ss}\rho_{ij} + o(n^{-1/2}) = \rho_{ij} + o(n^{-1/2}) \qquad (27)$$

Here and throughout this proof, the elements of the matrix D are denoted by D_{ij} $(i, j = 1, \ldots, n)$ and we used the independence of $\epsilon_{r,i}$ and $\epsilon_{s,j}$ if $r \neq s$ $(r, s = 1, \ldots, n; i, j = 1, \ldots, d)$. For the expected value of $(\hat{\rho}_{ij}\hat{\rho}_{kl})$ we find that

$$E[\hat{\rho}_{ij}\hat{\rho}_{kl}] = \frac{1}{(tr(D))^2}E\left[(\epsilon'_{(i)}D\epsilon_{(j)})(\epsilon'_{(k)}D\epsilon_{(l)})\right] + O(n^{-\gamma-1}) \qquad (28)$$

where we used the assumption $E[\epsilon_{1,\nu}^4] < \infty (\nu = 1, \ldots, d)$, the fact that the components of B_ν are of order $O(n^{-\gamma})$, and the definition of $D = \tilde{D}'\tilde{D}$ by equation (11). Some straightforward calculations show that the term on the right-hand side of equation (28) can be rewritten as

$$\begin{aligned}E\left[(\epsilon'_{(i)}D\epsilon_{(j)})(\epsilon'_{(k)}D\epsilon_{(l)})\right] &= \delta^{(i,j,k,l)}\sum_{s=1}^{n} D_{ss}^2 + \sum_{s\neq m} D_{ss}D_{mm}\rho_{ij}\rho_{kl} \\ &\quad + \sum_{s\neq m} D_{sm}^2(\rho_{ik}\rho_{jl} + \rho_{il}\rho_{jk})\end{aligned}$$

and from equations (27) and (28) the first assertion of the theorem will follow easily.

For the second part of the theorem we note that by the condition in equation (4) the diagonal elements of the matrix D are equal to 1 except for a finite number of terms which do not depend on the sample size n, so that $tr(D) = n + O(1)$ and $\sum_{s=1}^{n} D_{ss}^2 = n + O(1)$. Furthermore, a straightforward calculation shows $tr(D^2) = \alpha n + O(1)$ where α is given by equation (25). This implies the existence of

$$\lim_{n\to\infty} n\,\mathrm{Cov}(\hat{\rho}_{ij}, \hat{\rho}_{kl}) = \alpha(\rho_{ik}\rho_{jl} + \rho_{il}\rho_{jk}) + \left(\delta^{(i,j,k,l)} - \rho_{ik}\rho_{jl} - \rho_{il}\rho_{jk} - \rho_{ij}\rho_{kl}\right)$$

Therefore, asymptotic normality follows from the Cramér–Wold device and the central limit theorem for r-dependent sequences (see Orey, 1958). □

Examples where the constant in $tr(D^2) = \alpha n + O(1)$ is simple are the estimators of Hall et al. (1990) and the polynomial fitting estimators introduced in Section 2. More precisely, for the asymptotically optimal difference-type estimators of class (E1) we obtain

$$tr(D) = n + O(1), \qquad tr(D^2) \sim n(1 + (2r)^{-1})$$

and for the polynomial estimators of class (E2)

$$tr(D) = n + O(1), \qquad tr(D^2) \sim n \frac{\binom{4r}{2r}}{\binom{2r}{r}^2}$$

where r denotes the order of the estimator in classes (E1) and (E2), respectively.

COROLLARY 1: *In the case of a normally distributed error, the second term in equation (24) vanishes and the covariance between $\hat{\rho}_{ij}$ and $\hat{\rho}_{kl}$ simplifies to*

$$\mathrm{Cov}(\hat{\rho}_{ij}, \hat{\rho}_{kl}) = \frac{tr(D^2)}{(tr(D))^2}(\rho_{ik}\rho_{jl} + \rho_{il}\rho_{jk}) + o(n^{-1})$$

Proof. We have to show that

$$\delta^{(i,j,k,l)} = E[\epsilon_{1,i}\epsilon_{1,j}\epsilon_{1,k}\epsilon_{1,l}] = \rho_{ik}\rho_{jl} + \rho_{il}\rho_{jk} + \rho_{ij}\rho_{kl}, \qquad 1 \le i, j, k, l \le d$$

when the error is normal. Here we will only consider the case where i, j, k, l are pairwise different. The other cases are treated in exactly the same way and omitted for the sake of brevity. To this end, consider a normally distributed random variable

$$X = \begin{pmatrix} X_1 \\ X_2 \\ X_3 \\ X_4 \end{pmatrix} \sim \mathcal{N}_4(\mathbf{0}, \Sigma)$$

with covariance matrix $\Sigma = (\rho_{ij})_{i,j=1}^4$. For $u = (u_1, u_2, u_3, u_4)'$ the characteristic function of X is known as

$$\varphi_X(u) = E[e^{iu'X}] = \exp\left(-\frac{u'\Sigma u}{2}\right)$$

which gives

$$\begin{aligned}
E[X_1 X_2 X_3 X_4] &= E\left[\frac{\partial^4}{\partial u_1 \partial u_2 \partial u_3 \partial u_4} e^{iu'X}\Big|_{u=0}\right] \\
&= \frac{\partial^4}{\partial u_1 \partial u_2 \partial u_3 \partial u_4} E\left[e^{iu'X}\right]\Big|_{u=0} \\
&= \frac{\partial^4}{\partial u_1 \partial u_2 \partial u_3 \partial u_4} \exp\left(-\frac{1}{2}\sum_{i,j=1}^{4} \rho_{ij} u_i u_j\right)\Big|_{u=0}
\end{aligned}$$

The result now follows after some simple calculations. □

A reasonable criterion for the choice of an optimal difference-type estimator for the covariance matrix is the minimization of the generalized variance of the estimator of the diagonal elements of Σ, i.e.,

$$\hat{Q} = diag(\hat{\Sigma}) = (\hat{\rho}_{11}, \dots, \hat{\rho}_{dd})'$$

By Corollary 3.2 and the fact that $tr(D) = n + O(1)$ for all difference-type estimators, one can observe that for normally distributed errors

$$\text{Cov}(\hat{\rho}_{ii}, \hat{\rho}_{ll}) = 2\,\frac{tr(D^2)}{n^2}\,\rho_{il}^2 = o(n^{-1}), \qquad i, l = 1, \dots, d$$

and so it follows that the generalized variance $det(\text{Cov}(\hat{Q}))$ can easily be minimized by minimizing $tr(D^2)$. Hence the minimization problem remains the same as in the univariate case, and therefore Hall et al.'s (1990) weighting scheme and the equivalent quadratic form of equation (15) remain optimal in the multivariate case when the generalized variance of $\hat{Q}$ is chosen as a measure of accuracy for different difference-type estimators for the covariance matrix Σ. The same result is true when the determinant of the MSE matrix of the estimator $\hat{Q}$ is used as an optimality criterion. However, surprisingly we note that by equation (24) this optimality property does not hold for general nonnormal error distributions. In this case an optimal choice of the weights in the difference-type estimator depends sensitively on the magnitude of the elements in the (unknown) covariance matrix Σ.

Based on these observations and the results of the simulation study in Section 2, we give the following guidelines for the choice of an estimator of the covariance matrix in a multivariate regression.

- Because of the computational simplicity we recommend difference-type estimators of the form of equation (23) where the matrix $D =$

$\tilde{D}'\tilde{D}$ is obtained from a difference-type estimator for the variance in the univariate nonparametric regression. An asymptotically equivalent estimator is given by

$$\tilde{\Sigma} = \frac{1}{n-r} Y'DY \tag{29}$$

(see the condition in equation (4) and the discussion in Section 2). In this case Theorem 1 remains true, where in equation (24) $tr(D)$ has to be replaced by $n - r$.

- For the discrimination between different estimators of the form of equations (23) or (29), the following rules can be applied.
  —If the error distribution can be approximated by a normal law, then the discussion for the univariate case can be transferred to the present multivariate case. In particular, for weakly oscillating regression functions the asymptotically optimal difference-type estimators of Hall et al. (1990) should be applied, where lower-order differences should be applied for small sample sizes and higher-order estimators should be used for large sample sizes. For strongly oscillating regression functions, polynomial fitting estimators should be applied. In particular, Gasser et al.'s (1986) estimator seems to be useful in this situation because it yields a reasonable MSE for a broad class of regression functions.
  —If the assumption of a normally distributed error cannot be justified, an optimal choice of the difference-type estimator for the covariance matrix Σ depends on the elements of Σ and cannot be determined in practice. Based on our experience and additional simulation results, in this case we recommend the application of equations (23) or (29) with the matrix D corresponding to the asymptotically optimal difference-type estimator of Hall et al. (1990). At least this estimator of the covariance matrix Σ minimizes the individual (asymptotic) variances of the diagonal elements of the covariance matrix Σ. The simulation study of Section 2 shows that the asymptotic comparison can be transferred to a finite sample case provided that the regression functions are not too strongly oscillating. If $n \leq 50$, a lower order r should be used in the estimator of Hall et al. (1990), whereas for larger samples higher orders become reasonable. In the high-frequency case, these estimators should be replaced by polynomial fitting methods as described in Section 2.

ACKNOWLEDGMENT

We would like to thank G.K. Eagleson for some helpful comments.

REFERENCES

Allen, D.M. Parameter estimation for nonlinear models with emphasis on compartmental models. *Biometrics*, 39: 629–637, 1983.

Breiman, L., Meisel, W. S. General estimates of the intrinsic variability of data in nonlinear regression models. *J. Am. Stat. Assoc.*, 71: 301–307, 1976.

Buckley, M.J., Eagleson, G.K. A graphical method for estimating the residual variance in nonparametric regression. *Biometrika*, 76: 203–210, 1989.

Buckley, M.J., Eagleson, G.K., Silverman, B.W. The estimation of residual variance in nonparametric regression. *Biometrika*, 75: 189–199, 1988.

Carroll, R.J., Ruppert, D. *Transforming and Weighting in Regression*. London: Chapman & Hall, 1988.

Carter, C.K., Eagleson, G.K. A comparison of variance estimators in nonparametric regression. *J. R. Stat. Soc. B*, 54: 773–780, 1992.

Carter, C.K., Eagleson, G.K., Silverman, B.W. A comparison of the Reinsch and Speckman splines. *Biometrika*, 79: 81–91, 1992.

Dette, H., Munk, A. Validation of linear regression models. *Ann. Stat.*, 26: 778–800, 1998.

Eagleson, G.K. Curve estimation–whatever happened to the variance? *Proceedings of the 47th Session of the International Statistical Institute*, pp. 535–551, Paris, 1989.

Epanechnikov, V.A. Nonparametric estimation of a multidimensional probability density. *Theor. Probab. Appl.*, 14: 153–158, 1969.

Eubank, R.L., Hart, J.D. Testing goodness of fit in regression via order selection criteria. *Ann. Stat.*, 20: 1412–1425, 1992.

Gasser, T., Sroka, L., Jennen-Steinmetz, C. Residual variance and residual pattern in nonlinear regression. *Biometrika*, 73: 625–633, 1986.

Hall, P., Carroll, R.J. Variance function estimation in regression: The effect of estimating the mean. *J. R. Stat. Soc. B*, 51: 3–14, 1989.

Hall, P., Marron, J.S. On variance estimation in nonparametric regression. *Biometrika*, 77: 415–419, 1990.

Hall, P., Kay, J.W., Titterington, D.M. Asymptotically optimal difference-based estimation of variance in nonparametric regression. *Biometrika*, 77: 521–528, 1990.

Härdle, W. *Applied Nonparametric Regression*. Cambridge: Cambridge University Press, 1990.

Härdle, W., Marron, J.S. Semiparametric comparison of regression curves. *Ann. Stat.*, 18: 63–89, 1990.

Müller, H.G., Stadtmüller, U. Variable bandwidth kernel estimators of regression curves. *Ann. Stat.*, 15: 182–201, 1987a.

Müller, H.G., Stadtmüller, U. Estimation of heteroscedasticity in regression analysis. *Ann. Stat.*, 15: 610–635, 1987b.

Müller, H.G., Stadtmüller, U. On variance function estimation with quadratic forms. *J. Stat. Plann. Inference*, 35: 213–231, 1993.

Neumann, M.H. Fully data-driven nonparametric variance estimators. *Statistics*, 25: 189–212, 1994.

Orey, S. A central limit theorem for *m*-dependent random variables. *Duke Math. J.*, 52: 543–546, 1958.

Reinsch, C. Smoothing by spline functions. *Numer. Math.*, 10: 177–183, 1967.

Rice, J. Bandwidth choice for nonparametric kernel regression. *Ann. Stat.*, 12: 1215–1230, 1984.

Sacks, J., Ylvisaker, D. Designs for regression problems with correlated errors. III. *Ann. Math. Stat.*, 41: 2057–2074, 1970.

Silverman, B.W. A fast and efficient cross-validation method for smoothing parameter choice in spline regression. *J. Am. Stat. Assoc.*, 79: 584–589, 1984a.

Silverman, B.W. Spline smoothing: The equivalent variable kernel method. *Ann. Stat.*, 12: 898–916, 1984b.

Silverman, B.W. Some aspects of the spline smoothing approach to non-parametric regression curve fitting. *J. R. Stat. Soc. B*, 47: 1–52, 1985.

Speckman, P. Spline smoothing and optimal rates of convergence in non-parametric regression models. *Ann. Stat.*, 13: 970–983, 1985.

Stewart, W.E., Caracotsios, M., Sørenson, Y.P. Parametric estimation from multiresponse data. *AIChE J.*, 38: 641–650, 1992.

Ullah, A., Zinde-Walsh, V. On the estimation of residual variance in nonparametric regression. *Nonparametric Stat.*, 1: 263–265, 1992.

Wahba, G. Improper priors, spline smoothing and the problems of guarding against model errors in regression. *J. R. Stat. Soc. B*, 40: 364–372, 1978.

Wahba, G. Bayesian confidence intervals for the cross-validated smoothing spline. *J. R. Stat. Soc. B*, 45: 133–150, 1983.

17

On Affine Invariant Sign and Rank Tests in One- and Two-Sample Multivariate Problems

BIMAN CHAKRABORTY National University of Singapore, Republic of Singapore, and Indian Statistical Institute, Calcutta, India

PROBAL CHAUDHURI Indian Statistical Institute, Calcutta, India

1. INTRODUCTION

The simplicity and widespread popularity of univariate sign and rank tests for one-sample and two-sample location problems have motivated numerous statisticians to explore several possibilities for their multivariate generalizations. In the 1950s and the 1960s, Bennett (1962), Bickel (1965), Hedges (1955), Blumen (1958) and Chatterjee (1966) developed some multivariate sign- and rank-based methods. More recent attempts in that direction have been made by Brown and Hettmansperger (1987, 1989), Brown et al. (1992), Liu (1992), Liu and Singh (1993), Oja and Nyblom (1989), Randles (1989), Hettmanspergen et al. (1994), Hettmansperger et al. (1996a,b), and others. Readers are referred to Hettmansperger et al. (1992) and Chaudhuri and Sengupta (1993) for some recent detailed reviews. The popularity of univariate sign- and rank-based methods has its roots in their distribution-free nature and their wide applicability for solving a number of practical problems for which more traditional techniques cannot be used as they often require the assumption of normality of the data, which may be hard to justify in practice. It is a well-known fact that the sample median is the estimate of location associated with the univariate sign test, and in the same way the Hodges–Lehmann (1963) estimate is associated with Wilcoxon's signed rank test. So it is reasonable to expect that a multivariate version of the median will be associated with a multivariate sign test, and a multivariate analog of the Hodges–Lehmann (HL) estimate will be associated with a multivariate rank test. Among different versions of multivariate median and HL-type estimates of location, the vector of coordinatewise median and the vector of coordinatewise HL-estimates are perhaps the simplest. In one sample multivariate location problem, the tests that are associated with it are the coordinatewise sign test and the coordinatewise rank test. Both tests have been studied extensively by Bickel (1965), Puri and Sen (1971), etc.

One serious drawback of coordinatewise sign as well as rank tests is that they are not invariant under arbitrary affine transformations of the data. In addition to being an undesirable geometric feature, this lack of invariance is known to have some negative impact on the statistical performance of these tests, especially when the real-valued components of the multivariate data are substantially correlated. This is also a handicap for both coordinatewise median and HL-estimates, as none of them is equivariant under affine transformation of the data. This issue was first raised by Bickel (1964, 1965), and subsequently investigated by Brown and Hettmansperger (1987, 1989).

Recently, Chakraborty and Chaudhuri (1996, 1998) have proposed a general technique for constructing affine equivariant versions of multivariate median based on a data-based transformation–retransformation strategy, which provides a resolution for this problem of lack of affine equivariance and loss in statistical efficiency in the case of coordinatewise median. The main idea behind the transformation–retransformation technique originated from the concept of "data driven coordinate systems" introduced by Chaudhuri and Sengupta (1993) as an effective general tool for constructing affine invariant versions of multivariate sign tests. In this chapter, we intend to explore affine invariant versions of sign tests and rank tests in one- and two-sample location problems and their related estimates of location, which are constructed by applying the transformation–retransformation technique to the coordinatewise sign and rank tests and coordinatewise HL-estimates. We have been motivated to use a transformation and retransformation approach primarily by its appealing geometric interpretation and computational simplicity, as well as the elegant mathematical theory for the statistical properties of the resulting tests and estimates. Our proposed procedures are quite easy to implement to analyze data in practice. In Section 2, we propose the transformation–retransformation technique for constructing affine invariant sign and rank tests in one- and two-sample problems, and then we proceed to generalize the technique to obtain affine equivariant versions of multivariate HL-estimates. In the same section, we will see a very encouraging common feature of all of our tests and location estimates—they inherit the same impressive efficiency properties of coordinatewise rank tests and coordinatewise HL-estimates in spherically symmetric multivariate normal models and extend them to more general elliptically symmetric situations. In Section 3, we will discuss the statistical performance of the proposed tests and location estimates under various elliptically symmetric distributions with the help of a simulation study. In the same section, we will indicate how one can estimate the *P*-value when our proposed tests are applied to real data sets by simulating the permutation distributions of the proposed test statistics under the null hypothesis.

2. THE TRANSFORMATION–RETRANSFORMATION METHODOLOGY

In the univariate set-up, sign tests and rank tests in one-sample and two-sample location problems are invariant under affine transformations of the observations. Thus it is a natural requirement that the analogs of them in multidimensions should be affine invariant in nature. To construct multivariate tests that are affine invariant, we begin by observing a simple geometric fact about any given affine transformation of a set of multivariate observations. We can identify an affine transformation $\mathbf{A}\boldsymbol{X} + \boldsymbol{b}$ as something that rewrites an observation using a new coordinate system determined by $\mathbf{A}$ and $\boldsymbol{b}$. The new origin is located at $-\mathbf{A}^{-1}\boldsymbol{b}$, and depending on whether $\mathbf{A}$ is an orthogonal matrix or not, this new coordinate system may or may not be an orthonormal system. The basic idea of data based transformation–retransformation is to form an appropriate "data-driven coordinate system" (see also Chaudhuri and Sengupta, 1993) and to transform all data points in that coordinate system first. Then one computes a test statistic or a location estimate based on those transformed observations. Finally, the location estimate is retransformed back to the original coordinate system (see also Chakraborty and Chaudhuri, 1996, 1998). We can appreciate the invariance property of the transformed observations better if we identify a nonsingular transformation as some kind of a motion on the data cloud and "the data-driven coordinate system" associated with the transformation matrix as a reference frame for an observer. As the data-centric reference frame moves with the motion of the data cloud, the data points will appear to be stationary (i.e., invariant) to the observer. However, it is not necessary for this "data-driven coordinate system" to be an orthonormal system. We will discuss in detail in the following subsections how this transformation–retransformation approach is used in constructing sign and rank tests in one- and two-sample location problems and their related estimates.

2.1 One-Sample Location Problem

Suppose that we have n data points $\boldsymbol{X}_1, \dots, \boldsymbol{X}_n$ in $\mathbb{R}^d$, and assume that $n > d+1$. From now on, all vectors in this paper will be assumed to be column vectors unless specified otherwise, and the superscript T will be used to denote transpose of vectors and matrices. Let us begin by introducing some notation. Define

$$S_n = \{\alpha | \alpha \subseteq \{1, 2, \dots, n\} \text{ and } \#\{i : i \in \alpha\} = d+1\},$$

which is the collection of all subsets of size $d+1$ of $\{1, 2, \dots, n\}$. For a fixed subset $\alpha = \{i_0, i_1, \dots, i_d\}$, consider the data points $\boldsymbol{X}_{i_0}, \boldsymbol{X}_{i_1}, \dots, \boldsymbol{X}_{i_d}$, which will form a "data-driven coordinate system" as described before, and the $d \times d$ matrix $\mathbf{X}(\alpha)$ containing the columns $\boldsymbol{X}_{i_1} - \boldsymbol{X}_{i_0}, \dots, \boldsymbol{X}_{i_d} - \boldsymbol{X}_{i_0}$ can be taken as the transformation matrix for transforming the data points $\boldsymbol{X}_j$ such that $1 \le j \le n, j \notin \alpha$ in order to express them in terms of the new coordinate system as $\boldsymbol{Y}_j^{(\alpha)} = \{\mathbf{X}(\alpha)\}^{-1}\boldsymbol{X}_j$. If the observations $\boldsymbol{X}_i$s are generated from a common distribution which is absolutely continuous w.r.t the Lebesgue measure on $\mathbb{R}^d$, the transformation matrix $\mathbf{X}(\alpha)$ must be an invertible matrix with probability one. Now it is easy to verify that the transformed observations $\boldsymbol{Y}_j^{(\alpha)}$s with $1 \le j \le n$ and $i \notin \alpha$ are invariant under the group of nonsingular transformations and actually they form a maximal invariant with respect to the group of nonsingular linear transformations on $\mathbb{R}^d$ (cf. Chakraborty and Chaudhuri, 1996). One can now define the following multivariate sign and signed rank test statistics:

$$T_n^{(\alpha)} = \sum_{i \notin \alpha} Sign\left(\boldsymbol{Y}_i^{(\alpha)}\right) \tag{1}$$

and

$$R_n^{(\alpha)} = \frac{1}{n} \sum_{i \notin \alpha} \sum_{\substack{j \notin \alpha \\ j \ne i}} Sign\left(\boldsymbol{Y}_i^{(\alpha)} + \boldsymbol{Y}_j^{(\alpha)}\right), \tag{2}$$

where $Sign(\boldsymbol{X}_i) = (Sign(X_{i1}), \dots, Sign(X_{id}))^{\mathrm{T}}$ with $\boldsymbol{X}_i = (X_{i1}, \dots, X_{id})^{\mathrm{T}}$. In view of the construction, it is clear that $T_n^{(\alpha)}$ and $R_n^{(\alpha)}$ are affine invariant. Note that they are nothing but coordinatewise sign test statistic and coordinatewise Wilcoxon's signed rank statistic, respectively, based on the transformed observations $\boldsymbol{Y}_j^{(\alpha)}$s. A question that naturally arises at this stage is how to choose the "data-driven coordinate system," or equivalently the data-based transformation matrix $\mathbf{X}(\alpha)$. An answer to this question is provided in the following discussion.

Let us consider the elliptically symmetric density $h(\boldsymbol{x}) = \{\det(\Sigma)\}^{-1/2} f(\boldsymbol{x}^{\mathrm{T}}\Sigma^{-1}\boldsymbol{x})$, where Σ is a $d \times d$ positive definite matrix, and $f(\boldsymbol{x}^{\mathrm{T}}\boldsymbol{x})$ is a spherically symmetric density around the origin in $\mathbb{R}^d$. $\boldsymbol{X}_1, \ldots, \boldsymbol{X}_n$ are assumed to be i.i.d. observations with common elliptically symmetric density $h(\boldsymbol{x} - \boldsymbol{\theta})$, where $\boldsymbol{\theta} \in \mathbb{R}^d$ is the location of elliptic symmetry for the data. Suppose that we have two competing hypotheses $H_0 : \boldsymbol{\theta} = \mathbf{0}$ and $H_A : \boldsymbol{\theta} \neq \mathbf{0}$. We now state a result that summarizes the main features of the affine invariant sign test statistic $T_n^{(\alpha)}$.

THEOREM 1: *Under the null hypothesis $H_0 : \boldsymbol{\theta} = \mathbf{0}$, the conditional distribution of $n^{-1/2}T_n^{(\alpha)}$ given the $\boldsymbol{X}_i$s with $i \in \alpha$ does not depend on f, and it converges to d-variate normal with zero mean and a variance covariance matrix $\Psi_1\{\Sigma^{-1/2}\mathbf{X}(\alpha)\}$ as $n \to \infty$, where Ψ_1 depends only on $\Sigma^{-1/2}\mathbf{X}(\alpha)$. When $\log f$ is twice differentiable almost everywhere (w.r.t. Lebesgue measure) on $\mathbb{R}^d$ and satisfies the Cramér-type regularity conditions, the alternatives $H_A^{(n)} : \boldsymbol{\theta} = n^{-1/2}\boldsymbol{\delta}$ such that $\boldsymbol{\delta} \in \mathbb{R}^d$ and $\boldsymbol{\delta} \neq \mathbf{0}$ will form a contiguous sequence, and the conditional limiting distribution of $n^{-1/2}T_n^{(\alpha)}$ under that sequence of alternatives is normal with the same variance covariance matrix Ψ_1 and a mean vector $\Lambda_1\{f, \Sigma^{-1/2}\boldsymbol{\delta}, \Sigma^{-1/2}\mathbf{X}(\alpha)\}$ that depends on f, $\Sigma^{-1/2}\boldsymbol{\delta}$, and $\Sigma^{-1/2}\mathbf{X}(\alpha)$. Also, the limiting conditional power of the test under such a sequence of contiguous alternatives increases monotonically with the noncentrality parameter*

$$\begin{aligned}&\Delta_1\{f, \Sigma^{-1/2}\boldsymbol{\delta}, \Sigma^{-1/2}\mathbf{X}(\alpha)\}\\ &= [\Lambda_1\{f, \Sigma^{-1/2}\boldsymbol{\delta}, \Sigma^{-1/2}\mathbf{X}(\alpha)\}]^{\mathrm{T}}[\Psi_1\{\Sigma^{-1/2}\mathbf{X}(\alpha)\}]^{-1}\Lambda_1\{f, \Sigma^{-1/2}\delta, \Sigma\mathbf{X}(\alpha)\}.\end{aligned} \tag{3}$$

Here Δ_1 is such that for any f, $\boldsymbol{\delta}$, Σ and any invertible matrix $\mathbf{A}$, we have $\Delta_1\{f, \Sigma^{-1/2}\boldsymbol{\delta}, \Sigma^{-1/2}\mathbf{A}\} = \beta(f)\boldsymbol{\delta}^{\mathrm{T}}\Sigma^{-1}\boldsymbol{\delta}$, where β is a scalar depending only on f whenever $\mathbf{A}^{\mathrm{T}}\Sigma^{-1}\mathbf{A}$ is a diagonal matrix. Further, for any invertible matrix $\mathbf{B}$ we will have

$$\inf_{\delta:\delta^T\Sigma^{-1}\delta=c} \Delta_1\{f, \Sigma^{-1/2}\,\boldsymbol{\delta}, \Sigma^{-1/2}\mathbf{B}\} \leq c\beta(f).$$

The main implication of the above theorem is that whatever f may be, it is possible to simulate the conditional finite sample null distribution of $T_n^{(\alpha)}$ after obtaining an appropriate estimate of Σ in small-sample situations, and one can use normal approximation when the sample size is large. It is interesting to note that in order to maximize the minimum power of the test one needs to choose $\mathbf{X}(\alpha)$ in such a way that $\{\mathbf{X}(\alpha)\}^{\mathrm{T}}\Sigma^{-1}\mathbf{X}(\alpha)$ becomes as close as possible to a diagonal matrix (especially for alternatives close to the null). We now make an observation, which will become more transparent when we present the proof of Theorem 1 in Appendix 1.

OBSERVATION 1: The Pitman efficiency of the test based on $T_n^{(\alpha)}$ with the above choice of $\mathbf{X}(\alpha)$ would be close to that of a coordinatewise sign test when the coordinate variables are uncorrelated (i.e., Σ is diagonal). Further, it will outperform the coordinatewise sign test in elliptically symmetric models in a minimax sense, i.e., if the parameter $\boldsymbol{\theta}$ is nonzero and oriented in the worst possible direction, giving rise to the minimum power of the tests, our invariant test will have more power than the noninvariant sign test. Also, this version of the affine invariant sign test will be more (or less) efficient than the usual Hotelling's T^2-test if the tail of the density f is "heavy" (or "light").

Note that the transformation–retransformation strategy leads to an affine invariant version of the multivariate sign test which is "distribution free" in nature in the sense that the conditional null distribution of the statistic under elliptically symmetric models does not depend on the unknown density f. Next we state a result on the behavior of the affine invariant rank test statistic $R_n^{(\alpha)}$ defined in equation (2).

THEOREM 2: *As $n \longrightarrow \infty$, under the null hypothesis $H_0 : \boldsymbol{\theta} = \mathbf{0}$, the conditional distribution of $n^{-1/2}R_n^{(\alpha)}$ given the $\mathbf{X}_i$s with $i \in \alpha$ converges to a d-variate normal distribution with zero mean and a variance covariance matrix $\Psi_2\{f, \Sigma^{-1/2}\mathbf{X}(\alpha)\}$ that depends on f and $\Sigma^{-1/2}\mathbf{X}(\alpha)$. When* $\log f$ *is twice differentiable almost everywhere (w.r.t. Lebesgue measure) on $\mathbb{R}^d$ and satisfies Cramér-type regularity conditions, the conditional limiting distribution of $n^{-1/2}R_n^{(\alpha)}$ under the sequence of contiguous alternatives $H_A^{(n)} : \boldsymbol{\theta} = n^{-1/2}\boldsymbol{\delta}$ such that $\boldsymbol{\delta} \in \mathbb{R}^d$ and $\boldsymbol{\delta} \neq \mathbf{0}$ is normal with the same variance covariance matrix Ψ_2 and a mean vector $\Lambda_2\{f, \Sigma^{-1/2}\boldsymbol{\delta}, \Sigma^{-1/2}\mathbf{X}(\alpha)\}$ that depends on f, $\Sigma^{-1/2}\boldsymbol{\delta}$, and $\Sigma^{-1/2}\mathbf{X}(\alpha)$. Also, the limiting conditional power of the test under such a sequence of alternatives depends monotonically on the noncentrality parameter*

$$
\begin{aligned}
&\Delta_2\{f, \Sigma^{-1/2}\boldsymbol{\delta}, \Sigma^{-1/2}\mathbf{X}(\alpha)\} \\
&= [\Lambda_2\{f, \Sigma^{-1/2}\boldsymbol{\delta}, \Sigma^{-1/2}\mathbf{X}(\alpha)\}]^{\mathrm{T}}[\Psi_2\{f, \Sigma^{-1/2}\mathbf{X}(\alpha)\}]^{-1} \\
&\quad \Lambda_2\{f, \Sigma^{-1/2}\boldsymbol{\delta}, \Sigma^{-1/2}\mathbf{X}(\alpha)\}.
\end{aligned} \tag{4}
$$

Here Δ_2 is such that for any f, $\boldsymbol{\delta}$, Σ and any invertible matrix $\mathbf{A}$, we have $\Delta_2\{f, \Sigma^{-1/2}\boldsymbol{\delta}, \Sigma^{-1/2}\mathbf{A}\} = \gamma(f)\boldsymbol{\delta}^{\mathrm{T}}\Sigma^{-1}\boldsymbol{\delta}$, where γ is a scalar depending only on *f, whenever* $\mathbf{A}^T\Sigma^{-1}\mathbf{A}$ *is a diagonal matrix. Further, normality of the density f implies that for any invertible matrix* $\mathbf{B}$, *we have*

$$\inf_{\delta:\delta^T\Sigma^{-1}\delta=c} \Delta_2\{f, \Sigma^{-1/2}\boldsymbol{\delta}, \Sigma^{-1/2}\mathbf{B}\} \leq c\gamma(f).$$

From Theorem 2, it is clear that when the underlying distribution is normal, the optimal choice for $\mathbf{X}(\alpha)$ is such that $\{\mathbf{X}(\alpha)\}^T\Sigma^{-1}\mathbf{X}(\alpha)$ is as close as possible to a diagonal matrix in order to maximize the minimum power of the test. We now make the following observation.

OBSERVATION 2: The asymptotic Pitman efficiency of the test based on $R_n^{(\alpha)}$ can be made close to that of the coordinatewise Wicoxon's signed rank test when Σ is diagonal by selecting $\mathbf{X}(\alpha)$ as described above, and it will outperform the coordinatewise signed rank test in a minimax sense when Σ is not a diagonal matrix and the data follow normal distribution.

Whether the underlying distribution is normal or not, it is easy to simulate the null permutation distribution of $R_n^{(\alpha)}$, and based on that one can carry out the permutation rank test. Furthermore, one can use normal approximation to the null distribution of $R_n^{(\alpha)}$ for large values of n. In Section 3, we will compare the finite sample performance of our affine invariant test statistics $T_n^{(\alpha)}$ and $R_n^{(\alpha)}$ with some other standard one-sample tests for multivariate location.

2.2 Two-Sample Location Problem

Let $\boldsymbol{X}_1, \ldots, \boldsymbol{X}_m$ and $\boldsymbol{Y}_1, \ldots, \boldsymbol{Y}_n$ be two independent samples from two d-dimensional distributions. For the purpose of comparing these two samples, we will develop an affine invariant multivariate analog of Wilcoxon's two-sample rank-sum test based on the transformation and retransformation approach. Let us define $\boldsymbol{X}_{m+1} = \boldsymbol{Y}_1, \ldots, \boldsymbol{X}_{m+n} = \boldsymbol{Y}_n$ and the transformation matrix $\mathbf{X}(\alpha)$ whose columns are $\boldsymbol{X}_{i_1} - \boldsymbol{X}_{i_1^*}, \ldots, \boldsymbol{X}_{i_d} - \boldsymbol{X}_{i_d^*}$ where $\alpha = \{j_1, j_2, i_1, \ldots, i_d\} \in S_{m+n}$ with $1 \leq j_1 \leq m$, $m+1 \leq j_2 \leq m+n$ and $i_k^* = j_1$ or j_2 depending on whether $i_k \in \{1, \ldots, m\}$ or $i_k \in \{m+1, \ldots, m+n\}$, respectively, for all $1 \leq k \leq d$. As before, define transformed observations $\mathbf{Z}_j^{(\alpha)} = \{\mathbf{X}(\alpha)\}^{-1}\boldsymbol{X}_j$, $1 \leq j \leq m+n, j \notin \alpha$. One can now compute the coordinatewise Wilcoxon's rank-sum statistic based on $\mathbf{Z}_j^{(\alpha)}$s, which is equivalent to the statistic

$$W_{m,n}^{(\alpha)} = \frac{1}{m+n}\sum_{\substack{i=1 \\ i\notin\alpha}}^{m}\sum_{\substack{j=1 \\ m+j\notin\alpha}}^{n} Sign\left(\mathbf{Z}_i^{(\alpha)} - \mathbf{Z}_{m+j}^{(\alpha)}\right). \tag{5}$$

The following result describes the main features of the test statistic $W_{m,n}^{(\alpha)}$. Let $\boldsymbol{X}_i$s be i.i.d. with density $h(\boldsymbol{x}-\boldsymbol{\theta})$, $\boldsymbol{Y}_i$s be i.i.d. with density $h(\boldsymbol{x}-\boldsymbol{\theta}-\boldsymbol{\delta})$, where h is as before, and we want to test $H_0 : \boldsymbol{\delta} = 0$ against $H_A : \boldsymbol{\delta} \neq 0$.

THEOREM 3: *Under the null hypothesis $H_0 : \boldsymbol{\delta} = 0$, the conditional distribution of $(m+n)^{-1/2} W_{m,n}^{(\alpha)}$ given the $\boldsymbol{X}_i$s with $i \in \alpha$ converges to a normal distribution with zero mean and variance covariance matrix $\Psi_3\{f, \Sigma^{-1/2}\mathbf{X}(\alpha), \lambda\}$ as $m, n \longrightarrow \infty$ such that $m/(m+n) \rightarrow \lambda > 0$. When $\log f$ is twice differentiable almost everywhere (w.r.t. Lebesgue measure) on $\mathbb{R}^d$ and satisfies Cramér-type regularity conditions, the conditional limiting distribution of $(m+n)^{-1/2} W_{m,n}^{(\alpha)}$ under the sequence of contiguous alternatives $H_A^{(m,n)} : \boldsymbol{\delta} = (m+n)^{-1/2}\boldsymbol{\mu}$ such that $\boldsymbol{\mu} \in \mathbb{R}^d$ and $\boldsymbol{\mu} \neq \mathbf{0}$ is normal with the same variance covariance matrix Ψ_3 and a mean vector $\Lambda_3\{f, \Sigma^{-1/2}\boldsymbol{\mu}, \Sigma^{-1/2}\mathbf{X}(\alpha), \lambda\}$ that depends on $\lambda, f, \Sigma^{-1/2}\boldsymbol{\mu}$, and $\Sigma^{-1/2}\mathbf{X}(\alpha)$. Also, the limiting conditional power of the test under such a sequence of alternatives increases monotonically with the noncentrality parameter*

$$\begin{aligned}
&\Delta_3\{f, \Sigma^{-1/2}\boldsymbol{\mu}, \Sigma^{-1/2}\mathbf{X}(\alpha), \lambda\} \\
&= \left[\Lambda_3\{f, \Sigma^{-1/2}\boldsymbol{\mu}, \Sigma^{-1/2}\mathbf{X}(\alpha), \lambda\}\right]^{\mathrm{T}}\left[\Psi_3\{f, \Sigma^{-1/2}\mathbf{X}(\alpha), \lambda\}\right]^{-1} \\
&\quad \Lambda_3\{f, \Sigma^{-1/2}\boldsymbol{\mu}, \Sigma^{-1/2}\mathbf{X}(\alpha), \lambda\}.
\end{aligned} \tag{6}$$

Further, for any $\boldsymbol{\mu}$, Σ and any invertible matrix $\mathbf{A}$ such that $\mathbf{A}^{\mathrm{T}}\Sigma^{-1}\mathbf{A}$ is a diagonal matrix, we have $\Delta_3\{f, \Sigma^{-1/2}\boldsymbol{\mu}, \Sigma^{-1/2}\mathbf{B}, \lambda\} = \eta(f, \lambda)\boldsymbol{\mu}^{\mathrm{T}}\Sigma^{-1}\boldsymbol{\mu}$, where the scalar η depends only on f and λ, and for normal f and any invertible matrix $\mathbf{B}$, we have

$$\inf_{\mu:\mu^{\mathrm{T}}\Sigma^{-1}\mu=c} \Delta_3\{f, \Sigma^{-1/2}\boldsymbol{\mu}, \Sigma^{-1/2}\mathbf{B}, \lambda\} \leq c\eta(f, \lambda).$$

Once again for normal f, if one chooses the transformation matrix $\mathbf{X}(\alpha)$ in such a way that $\{\mathbf{X}(\alpha)\}^{\mathrm{T}}\Sigma^{-1}\mathbf{X}(\alpha)$ is as close as possible to a diagonal matrix, it maximizes the minimum power of the test.

OBSERVATION 3: With proper choice of $\mathbf{X}(\alpha)$, the Pitman efficiency of the test based on $W_{m,n}^{(\alpha)}$ can be made close to that of a coordinatewise two-sample rank-sum test when Σ is diagonal. Also, if the data follow normal distribution, it will outperform the coordinatewise two sample rank-sum test in a minimax sense when Σ is not diagonal.

We close this section by pointing out that using a similar approach one may define an affine invariant version of the multivariate median test in the two-sample problem. Define

$$V_{m,n}^{(\alpha)} = \sum_{\substack{i=1 \\ i \notin \alpha}}^{m} Sign\left(Z_i^{(\alpha)} - \hat{\phi}\right), \tag{7}$$

where $\hat{\phi}$ is the vector of coordinatewise medians based on the transformed observations of the combined sample $Z_i^{(\alpha)}$, $i = 1, \ldots, m+n$, $i \notin \alpha$. It is possible to establish the asymptotic normality of $(m+n)^{-1/2} V_{m,n}^{(\alpha)}$ following the results in Puri and Sen (1971). Here we may also adopt the strategy for selecting the transformation matrix $\mathbf{X}(\alpha)$ in such a way that $\{\mathbf{X}(\alpha)\}^{\mathrm{T}} \Sigma^{-1} \mathbf{X}(\alpha)$ becomes as close as possible to a diagonal matrix. In Section 3, we will explain in detail how to obtain P-values for the test and report simulation results comparing our test with other standard multivariate two-sample tests for different multivariate distributions.

2.3 Location Estimation Problem

As in Section 2.1, assume that the d-dimensional observations $X_1, \ldots, X_n$ are generated from a common distribution which is absolutely continuous w.r.t. Lebesgue measure on $\mathbb{R}^d$. Define the transformation matrix $\mathbf{X}(\alpha)$ and transformed observations $Y_j^{(\alpha)}$s, for $1 \leq j \leq n, j \notin \alpha$. One can then compute coordinatewise HL-estimate $\hat{\phi}_n^{(\alpha)}$ based on $Y_j^{(\alpha)}$'s by minimizing the sum $\sum_{i \notin \alpha} \sum_{\substack{j \notin \alpha \\ j \neq i}} \left| \frac{Y_j^{(\alpha)} + Y_i^{(\alpha)}}{2} - \Phi \right|$ (here for $x = (x_1, \ldots, x_d)$, $|x| = |x_1| + \cdots + |x_d|$).

Finally, in order to express the estimate back in terms of the original coordinate system, we need to retransform $\hat{\phi}_n^{(\alpha)}$ into $\hat{\theta}_n^{(\alpha)} = \mathbf{X}(\alpha)\hat{\phi}_n^{(\alpha)}$, which is our desired location estimate. Alternatively, one may also use the Euclidean norm-based procedures described by Chaudhuri (1992) for constructing $\hat{\phi}_n^{(\alpha)}$. However, we will not consider that here. In view of the construction, it is obvious that $\hat{\theta}_n^{(\alpha)}$ will be equivariant under any affine transformation of the data vectors whatever procedure is used to estimate $\hat{\phi}_n^{(\alpha)}$ (cf. Chakraborty and Chaudhuri, 1996).

THEOREM 4: *If $X_1, X_2, \ldots, X_n$ are independent and identically distributed with a common elliptically symmetric density* $\{\det(\Sigma)\}^{-1/2} f[(x-\theta)^{\mathrm{T}} \Sigma^{-1} (x-\theta)]$, *the conditional asymptotic distribution of $n^{1/2}(\hat{\theta}_n^{(\alpha)} - \theta)$ given the X_is with $i \in \alpha$, is d-variate normal with zero mean and a variance covariance matrix $\mathcal{V}\{f, \Sigma, \mathbf{X}(\alpha)\}$ that depends on f, Σ and the transformation matrix $\mathbf{X}(\alpha)$ as $n \to \infty$. Further, if the underlying distribution of X_is is d-dimensional normal, the positive definite matrix $\mathcal{V}$ is such that* $\det[\mathcal{V}\{f, \Sigma, \mathbf{A}\}] \geq \det[\mathcal{V}\{f, \Sigma, \mathbf{B}\}]$ *for any Σ and any two $d \times d$ invertible matrices $\mathbf{A}$ and $\mathbf{B}$ such that $\mathbf{B}^{\mathrm{T}} \Sigma^{-1} \mathbf{B}$ is a diagonal matrix.*

In fact, we do not need the assumption of elliptic symmetry for asymptotic normality of our transformation–retransformation HL-estimate. We made the assumption of elliptic symmetry for the asymptotic variance covariance matrix $\mathcal{V}$ to have the special structure

$$\mathcal{V}\{f, \Sigma, \mathbf{X}(\alpha)\} = \frac{1}{12}\left\{\int f_1^2(x)dx\right\}^{-2} \Sigma^{1/2}\{\mathbf{J}(\alpha)\}^{-1}\mathcal{U}(\alpha)\left[\{\mathbf{J}(\alpha)\}^{\mathrm{T}}\right]^{-1}\Sigma^{1/2}, \quad (8)$$

where f_1 is the univariate marginal density of the spherically symmetric density f, the matrix $\mathbf{J}(\alpha)$ is obtained by normalizing the rows of $\{\Sigma^{-1/2}\mathbf{X}(\alpha)\}^{-1}$, and $\mathcal{U}(\alpha)$ is the correlation matrix of $(F_1(Y_1), \ldots, F_d(Y_d))^{\mathrm{T}}$. Here F_i is the i-th marginal distribution function of $\mathbf{J}(\alpha)\boldsymbol{X}$ and Y_i is the i-th coordinate variable of $\mathbf{J}(\alpha)\boldsymbol{X}$, where $\boldsymbol{X}$ has a spherically symmetric density f. For normal f, the (i, j)-th element of the matrix $\mathcal{U}$ is $(6/\pi)\sin^{-1}(\rho_{ij}/2)$, where ρ_{ij} is the inner product of the i-th and the j-th rows of the matrix $\mathbf{J}(\alpha)$.

In Chakraborty et al. (1998), it was suggested that one may select optimal α by choosing $\mathbf{X}(\alpha)$ in such a way that $\{\mathbf{X}(\alpha)\}^{\mathrm{T}}\Sigma^{-1}\mathbf{X}(\alpha)$ becomes as close as possible to a diagonal matrix with all diagonal entries equal. Following the arguments used in Chakraborty and Chaudhuri (1998), it can be shown that for the above-mentioned procedure, the conditional asymptotic generalized variance of the proposed estimate is $[12n\{\int f_1^2(x)dx\}^2]^{-d}\det\{\Sigma\}$. Thus the efficiency, defined as the d-th root of the ratio of generalized variances (c.f. Bickel, 1964), of the proposed estimate $\hat{\boldsymbol{\theta}}_n^{(\alpha)}$ over the sample mean is the same as that of HL-estimate of location over mean in the univariate setup. It is also transparent from the above theorem that for normally distributed data and optimal transformation matrix $\mathbf{X}(\alpha)$, our affine equivariant estimate $\hat{\boldsymbol{\theta}}_n^{(\alpha)}$ is asymptotically more efficient than the nonequivariant coordinatewise HL-estimate when the scale matrix Σ is not a diagonal matrix.

In the next section, we will discuss some simulation studies in an attempt to see the performance of the transformation–retransformation HL-estimate $\hat{\boldsymbol{\theta}}_n^{(\alpha)}$ in finite samples for different elliptically symmetric distributions.

3. SIMULATION RESULTS AND DATA ANALYSIS

From the results and discussions in the previous section, it is clear that we need to choose the transformation matrix $\mathbf{X}(\alpha)$ in such a way that $\{\mathbf{X}(\alpha)\}^{\mathrm{T}}\Sigma^{-1}\mathbf{X}(\alpha)$ becomes as close as possible to a diagonal matrix. Since Σ will be unknown in practice, we have to estimate that from the data, and we will need a consistent estimate that will be invariant under location shift and equivariant under linear transformations of the data (say $\hat{\Sigma}$). If we can assume that the underlying distribution has finite second-order moments, we can use the usual variance covariance matrix for this purpose. To be in conformity with the selection procedure described by Chakraborty et al. (1998), we will try to select $\mathbf{X}(\alpha)$ in such a way that the eigenvalues of the positive definite matrix $\{\mathbf{X}(\alpha)\}^T \hat{\Sigma}^{-1}\mathbf{X}(\alpha)$ become as equal as possible. Rather than computing the eigenvalues of the matrix explicitly, we will minimize the ratio between the arithmetic mean and the geometric mean of the eigenvalues, which are given by the trace and the determinant of the matrix respectively. We have observed that instead of minimizing the ratio over all possible subsets α with size $d+1$ of $\{1, \ldots, n\}$, one can substantially reduce the amount of computation by stopping the search for optimal subset α as soon as the ratio becomes sufficiently close to one. Of course there are other different ways to achieve this goal of making $\{\mathbf{X}(\alpha)\}^T \hat{\Sigma}^{-1}\mathbf{X}(\alpha)$ as close as possible to a diagonal matrix. We have adopted a technique that is computationally convenient and has been observed to work fairly well in our numerical investigations.

3.1 Simulated Powers of Different Single-Sample Tests

We will present the results of a simulation study that was carried out to compare the finite sample powers of our affine invariant rank test and sign test, which are based on the statistics $R_n^{(\alpha)}$ and $T_n^{(\alpha)}$, respectively in the one-sample problem with the powers of the well-known Hotelling's T^2 test and the noninvariant sign test, which is based on the coordinatewise sign test statistic S_n. We have used sample size $n = 30$ and level of significance 5%. We estimated the powers in each case from 5000 Monte Carlo replications for $d = 2$ and from 3000 Monte Carlo replications for $d = 3$. We generated data from three different distributions, namely multivariate normal, multivariate Laplace (i.e., when $f(\boldsymbol{x}^T\boldsymbol{x}) = k\exp\{-(\boldsymbol{x}^T\boldsymbol{x})^{1/2}\}$) and multivariate t with 3 degrees of freedom, and for Σ we have used the matrix with each diagonal entry equal to 1 and each off-diagonal entry equal to ρ such that $\rho \in [0, 1)$. The results are presented in Tables 1 and 2. For Hotelling's T^2 test the critical value at 5% level of significance was determined from the F distribution table, and for the noninvariant coordinatewise sign test S_n and invariant sign test $T_n^{(\alpha)}$, we used χ^2 approximations (with 2 d.f. and 3 d.f. for $d = 2$ and 3, respectively) for the distributions of the test statistics. In the case of $R_n^{(\alpha)}$, we chose to simulate the permutation distribution of the test statistic as the use of a χ^2 approximation to $R_n^{(\alpha)}$ was observed not to be adequate for such a small sample size. It will be appropriate to note that in the case of the invariant sign test, it is not necessary to simulate the conditional null distribution because of the "distribution free" nature of the test, and χ^2 approximation was observed to lead to very good results. The matrix Σ was always estimated using the usual

Table 1. Finite sample power of affine invariant rank test and its competitors in the one-sample problem for $n = 30$, $d = 2$ and level of significance = 5%

			$(\theta^T \Sigma^{-1} \theta)^{1/2}$					
Distributions	Test	ρ	0.0	0.3	0.6	0.9	1.2	1.5
Laplace	$R_n^{(\alpha)}$	—	0.049	0.134	0.405	0.729	0.918	0.982
	$T_n^{(\alpha)}$	—	0.051	0.134	0.384	0.677	0.891	0.970
	T^2	—	0.049	0.121	0.367	0.692	0.902	0.982
	S_n	0.00	0.048	0.133	0.383	0.677	0.889	0.972
		0.75	0.047	0.134	0.413	0.708	0.905	0.971
		0.85	0.046	0.133	0.419	0.712	0.892	0.946
		0.95	0.035	0.123	0.392	0.647	0.796	0.818
Normal	$R_n^{(\alpha)}$	—	0.048	0.266	0.796	0.986	0.990	0.998
	$T_n^{(\alpha)}$	—	0.053	0.193	0.615	0.923	0.994	0.999
	T^2	—	0.049	0.267	0.805	0.989	1.000	1.000
	S_n	0.00	0.049	0.184	0.609	0.925	0.994	1.000
		0.75	0.046	0.196	0.645	0.938	0.984	0.969
		0.85	0.049	0.194	0.648	0.927	0.947	0.899
		0.95	0.033	0.175	0.597	0.831	0.782	0.667
t with 3 d.f.	$R_n^{(\alpha)}$	—	0.051	0.191	0.580	0.876	0.977	0.983
	$T_n^{(\alpha)}$	—	0.050	0.172	0.523	0.848	0.969	0.995
	T^2	—	0.043	0.157	0.489	0.799	0.938	0.981
	S_n	0.00	0.052	0.174	0.528	0.844	0.970	0.996
		0.75	0.048	0.184	0.559	0.868	0.973	0.980
		0.85	0.045	0.184	0.564	0.865	0.949	0.931
		0.95	0.035	0.169	0.524	0.778	0.819	0.749

sample variance covariance matrix. Since $R_n^{(\alpha)}$, $T_n^{(\alpha)}$, and Hotelling's T^2 are all invariant under affine transformations of the data, their powers do not depend on different values of ρ and depend only on the noncentrality parameter $(\theta^T \Sigma^{-1} \theta)^{1/2}$. It is quite clear from Tables 1 and 2 that our affine invariant modifications of sign and rank tests have a superior performance over noninvariant procedures for large values of ρ.

Table 2. Finite sample power of affine invariant rank test and its competitors in the one-sample problem for $n = 30$, $d = 3$ and level of significance = 5%

			$(\theta^T \Sigma^{-1} \theta)^{1/2}$					
Distributions	Test	ρ	0.0	0.3	0.6	0.9	1.2	1.5
Laplace	$R_n^{(\alpha)}$	—	0.052	0.103	0.272	0.530	0.775	0.922
	$T_n^{(\alpha)}$	—	0.051	0.103	0.247	0.471	0.694	0.857
	T^2	—	0.042	0.083	0.237	0.493	0.740	0.911
	S_n	0.00	0.046	0.080	0.204	0.415	0.662	0.851
		0.75	0.038	0.082	0.228	0.465	0.701	0.878
		0.85	0.035	0.073	0.223	0.463	0.691	0.847
		0.95	0.021	0.052	0.178	0.372	0.545	0.629
Normal	$R_n^{(\alpha)}$	—	0.051	0.243	0.741	0.976	0.987	0.992
	$T_n^{(\alpha)}$	—	0.056	0.197	0.561	0.893	0.991	0.998
	T^2	—	0.050	0.245	0.743	0.981	1.000	1.000
	S_n	0.00	0.041	0.156	0.542	0.884	0.988	0.999
		0.75	0.034	0.182	0.576	0.910	0.968	0.918
		0.85	0.030	0.181	0.572	0.878	0.878	0.743
		0.95	0.017	0.137	0.481	0.664	0.554	0.358
t with 3 d.f.	$R_n^{(\alpha)}$	—	0.050	0.215	0.667	0.923	0.981	0.994
	$T_n^{(\alpha)}$	—	0.054	0.195	0.605	0.893	0.984	0.996
	T^2	—	0.041	0.160	0.578	0.878	0.969	0.991
	S_n	0.00	0.047	0.171	0.575	0.901	0.987	0.997
		0.75	0.042	0.186	0.617	0.915	0.956	0.903
		0.85	0.037	0.185	0.614	0.870	0.852	0.723
		0.95	0.025	0.150	0.497	0.604	0.509	0.362

3.2 *P*-Value Computation with Real Data : The Single-Sample Case

Let us now consider a real data set and try to investigate the performance of our affine invariant versions of sign and rank tests $T_n^{(\alpha)}$ and $R_n^{(\alpha)}$ applied to it. In a study of cerebral metabolism in epileptic patients, Sperling et al. (1989) measured the metabolic rates of glucose at 10 cortical locations and 6 subcortical locations in the brain by positron emission tomography (PET) scans. The cortical locations were the *Frontal*, *Sensorimotor*, *Temporal*, *Parietal*, and *Occipital* locations of the right and left hemispheres of the brain. Similarly, subcortical locations were the *Caudate nucleus*, *Lenticular nucleus*, and *Thalamas* regions of the right and left hemispheres. The metabolic rates are measured in mg/100/g/min. We considered 18 patients forming a normal control group to see whether there is any difference in the metabolic rates of the right and left hemispheres of the brain. After fixing the transformation matrix $\mathbf{X}(\alpha)$, it is very easy to simulate the null permutation distribution of the invariant rank test statistic, and the P-value of the invariant rank test based on $R_n^{(\alpha)}$ was found to be 0.0033 in the case of cortical regions. The P-value of the invariant sign test based on the χ^2 (with 5 d.f.) approximation was 0.046, and those of Hotelling's T^2 and noninvariant sign tests were 0.006 and 0.169 respectively. In the case of subcortical regions, the P-value of the invariant rank test was found to be 0.0264, and that of invariant sign test was 0.014. Classical Hotelling's T^2 test and noninvariant sign test produced P-values that were 0.041 and 0.075, respectively. In both situations, we see that all the invariant tests conclude that there is a significant difference in the metabolic rates of right and left hemispheres of the brain at 5% level of significance, whereas the noninvariant sign test concludes the difference to be insignificant. This example amply highlights the necessity and importance of using invariant procedures over noninvariant procedures.

3.3 Simulated Powers of Different Two-Sample Tests

In the two-sample problem, we conducted a simulation study similar to that in Section 3.1 to judge the performance of our affine invariant two-sample rank test based on the statistic $W_{m,n}^{(\alpha)}$. Here we compared the power of the invariant rank test with that of the invariant two-sample median test based on the test statistic $V_{m,n}^{(\alpha)}$, the classical Hotelling's T^2 test for a two-sample location problem, and the noninvariant median test based on coordinatewise univariate two-sample median test statistics $S_{m,n}$. Again the powers of the tests were computed based on 5000 Monte Carlo replications in the case $d = 2$ and 3000 Monte Carlo replications for $d = 3$, and in both cases sizes of the samples were taken to be 30 (i.e., $m = n = 30$). The critical value corresponding to the nominal significance level of 5% for Hotelling's T^2 test is obtained from the F distribution, and those of the invariant and noninvariant median tests are obtained from χ^2 approximations (with d d.f.). The critical value for the invariant rank statistic is obtained by simulating the permutation distribution of the statistic $W_{m,n}^{(\alpha)}$. The results are presented in Tables 3 and 4 for dimensions $d = 2$ and $d = 3$, respectively, for the same three multivariate distributions used earlier. We have selected the optimal transformation matrix $\mathbf{X}(\alpha)$ in the same manner as discussed earlier, but this time we have restricted our search to the first sample only (i.e., we have considered only $1 \leq i_k \leq m$ for all $1 \leq k \leq d$) in order to reduce the size of the search problem substantially. From Tables 3 and 4, it is clear that the invariant two-sample rank test has a superior performance over Hotelling's T^2 for nonnormal distributions, and it is also better than noninvariant procedures for large values of ρ.

Table 3. Finite sample power of affine invariant rank test and its competitors in the two-sample problem for $n = 30$, $d = 2$ and level of significance = 5%

			$(\mu^T \Sigma^{-1} \mu)^{1/2}$					
Distributions	Test	ρ	0.0	0.3	0.6	0.9	1.2	1.5
Laplace	$W_{m,n}^{(\alpha)}$	—	0.052	0.082	0.226	0.444	0.692	0.864
	$V_{m,n}^{(\alpha)}$	—	0.049	0.085	0.217	0.418	0.651	0.837
	T^2	—	0.047	0.079	0.204	0.423	0.652	0.844
	$S_{m,n}$	0.00	0.041	0.082	0.199	0.410	0.638	0.819
		0.75	0.042	0.080	0.206	0.389	0.633	0.827
		0.85	0.040	0.075	0.181	0.383	0.630	0.829
		0.95	0.039	0.072	0.177	0.384	0.636	0.821
Normal	$W_{m,n}^{(\alpha)}$	—	0.049	0.141	0.467	0.832	0.973	0.999
	$V_{m,n}^{(\alpha)}$	—	0.052	0.109	0.333	0.664	0.894	0.984
	T^2	—	0.051	0.156	0.518	0.874	0.984	1.000
	$S_{m,n}$	0.00	0.046	0.109	0.292	0.661	0.878	0.983
		0.75	0.043	0.106	0.294	0.628	0.889	0.982
		0.85	0.040	0.089	0.286	0.634	0.892	0.982
		0.95	0.039	0.091	0.286	0.629	0.884	0.968
t with 3 d.f.	$W_{m,n}^{(\alpha)}$	—	0.049	0.111	0.317	0.606	0.852	0.956
	$V_{m,n}^{(\alpha)}$	—	0.046	0.095	0.281	0.561	0.804	0.942
	T^2	—	0.046	0.095	0.257	0.508	0.742	0.889
	$S_{m,n}$	0.00	0.050	0.089	0.276	0.503	0.763	0.930
		0.75	0.041	0.081	0.244	0.527	0.788	0.940
		0.85	0.043	0.087	0.243	0.529	0.785	0.944
		0.95	0.041	0.082	0.242	0.524	0.782	0.931

Table 4. Finite sample power of affine invariant rank test and its competitors in the two-sample problem for $n = 30$, $d = 3$ and level of significance = 5%

			$(\mu^{T}\Sigma^{-1}\mu)^{1/2}$					
Distributions	Test	ρ	0.0	0.3	0.6	0.9	1.2	1.5
Laplace	$W_{m,n}^{(\alpha)}$	—	0.052	0.074	0.145	0.296	0.506	0.665
	$V_{m,n}^{(\alpha)}$	—	0.052	0.082	0.156	0.265	0.456	0.635
	T^2	—	0.046	0.072	0.141	0.276	0.473	0.664
	$S_{m,n}$	0.00	0.052	0.072	0.145	0.217	0.444	0.632
		0.75	0.053	0.071	0.112	0.221	0.400	0.598
		0.85	0.054	0.078	0.109	0.226	0.408	0.599
		0.95	0.043	0.066	0.108	0.218	0.402	0.588
Normal	$W_{m,n}^{(\alpha)}$	—	0.052	0.129	0.391	0.760	0.955	0.998
	$V_{m,n}^{(\alpha)}$	—	0.054	0.120	0.314	0.633	0.870	0.977
	T^2	—	0.051	0.140	0.447	0.822	0.980	0.999
	$S_{m,n}$	0.00	0.049	0.105	0.313	0.624	0.885	0.959
		0.75	0.045	0.113	0.273	0.595	0.854	0.972
		0.85	0.046	0.113	0.284	0.593	0.856	0.977
		0.95	0.037	0.109	0.276	0.583	0.834	0.936
t with 3 d.f.	$W_{m,n}^{(\alpha)}$	—	0.051	0.098	0.268	0.514	0.772	0.952
	$V_{m,n}^{(\alpha)}$	—	0.054	0.108	0.254	0.504	0.750	0.904
	T^2	—	0.037	0.081	0.228	0.443	0.684	0.853
	$S_{m,n}$	0.00	0.047	0.092	0.211	0.453	0.718	0.891
		0.75	0.047	0.072	0.212	0.486	0.759	0.923
		0.85	0.047	0.073	0.210	0.489	0.764	0.930
		0.95	0.042	0.071	0.211	0.430	0.749	0.893

3.4 *P*-Value Computation for Real Data: The Two-Sample Case

In order to investigate the performance of the test statistic $W_{m,n}^{(\alpha)}$ when applied to real data, we analyzed data on the effect of a certain drug on three biochemical compounds found in the brain, which is reported by Morrison (1990, pp. 184–185). Twenty-four mice of the same strain were randomly divided into two equal groups with the second receiving periodic administrations of the drug. Both samples received the same care and diet, and two of the control-group mice died of natural causes during the experiment. Assays of the brains of the mice after they were killed revealed the amounts of the compounds in micrograms per gram of brain tissue. We estimated the P-value of the invariant rank statistic by simulating the null permutation distribution of the statistic with 2000 replications, and this was 0.0016. The P-value of the Hotelling's T^2 based on the F distribution was 0.00003, and those of the invariant median test and the noninvariant median test based on χ^2 approximation were 0.01376 and 0.02894 respectively. Here also there is a noticeable difference in the P-values of the invariant and noninvariant procedures.

3.5 Finite Sample Efficiency of the Multivariate Hodges–Lehmann Estimate

We conclude this section with a small simulation study on the efficiencies of our affine equivariant HL-estimate of location. To determine the efficiency of one multivariate location estimate over another, we have used the notion of efficiency introduced by Bickel (1964) that considers the d-th root of the ratio of the generalized variances of the competing estimates, where d is the dimension of the data. We carried out the simulations for three multivariate distributions, namely multivariate normal, multivariate Laplace, and multivariate t with 3 d.f. as before ,with sample size $n = 30$ and 5000 Monte Carlo replications. The matrix Σ was also chosen to have the same form, with each diagonal entry equal to one and each off-diagonal entry equal to ρ, as in Section 3.1. For different values of ρ, the efficiency of the affine equivariant HL-type estimate is computed over the usual sample mean, the vector of the coordinatewise median, the affine equivariant transformation–retransformation median (Chakraborty and Chaudhuri, 1996, 1998), and the affine equivariant transformation–retransformation spatial median (Chakraborty et al., 1998), which are denoted by e_1, e_2, e_3, and e_4, respectively. The results are summarized in Tables 5 and 6.

4. CONCLUDING REMARKS

REMARK 1: It is interesting to note that the procedure proposed here for selection of $\mathbf{X}(\alpha)$ does not require any knowledge of the form of the underlying density f. As has been observed by Chakraborty and Chaudhuri (1996, 1998) and Chakraborty et al. (1998), there is a nice and intuitively appealing geometric interpretation for the approach. The matrix $\{\mathbf{X}(\alpha)\}^{\mathrm{T}}\Sigma^{-1}\mathbf{X}(\alpha)$ becomes a diagonal matrix when the columns of $\Sigma^{-1/2}\mathbf{X}(\alpha)$ are orthogonal

Table 5. Finite sample efficiency of affine equivariant HL-estimate of location for $n = 30$ and $d = 2$

Distribution		ρ 0.00	0.75	0.80	0.85	0.90	0.95
Laplace	e_1	1.0919	0.0919	1.0919	1.0919	1.0919	1.0919
	e_2	0.9047	1.1356	1.2078	1.3086	1.4629	1.7579
	e_3	1.2854	1.2854	1.2854	1.2854	1.2854	1.2854
	e_4	0.8360	0.8360	0.8360	0.8360	0.8360	0.8360
Normal	e_1	0.8480	0.8480	0.8480	0.8480	0.8480	0.8480
	e_2	1.3534	1.7095	1.8224	1.9734	2.1929	2.6099
	e_3	1.7757	1.7757	1.7757	1.7757	1.7757	1.7757
	e_4	1.2328	1.2328	1.2328	1.2328	1.2328	1.2328
t with 3 d.f.	e_1	1.5883	1.5883	1.5883	1.5883	1.5883	1.5883
	e_2	0.9474	1.2261	1.2954	1.3992	1.5603	1.8810
	e_3	1.3971	1.3971	1.3971	1.3971	1.3971	1.3971
	e_4	0.8861	0.8861	0.8861	0.8861	0.8861	0.8861

Table 6. Finite sample efficiency of affine equivariant HL-estimate of location for $n = 30$ and $d = 3$

Distribution		ρ 0.00	0.75	0.80	0.85	0.90	0.95
Laplace	e_1	0.9926	0.9926	0.9926	0.9926	0.9926	0.9926
	e_2	1.0585	1.4729	1.5763	1.7326	1.9993	2.4915
	e_3	1.2425	1.2425	1.2425	1.2425	1.2425	1.2425
	e_4	0.8807	0.8807	0.8807	0.8807	0.8807	0.8807
Normal	e_1	0.8344	0.8344	0.8344	0.8344	0.8344	0.8344
	e_2	1.3382	1.8066	1.9327	2.1351	2.3848	2.9853
	e_3	1.5036	1.5036	1.5036	1.5036	1.5036	1.5036
	e_4	1.1299	1.1299	1.1299	1.1299	1.1299	1.1299
t with 3 d.f.	e_1	1.4273	1.4273	1.4273	1.4273	1.4273	1.4273
	e_2	0.9495	1.2946	1.3882	1.5395	1.7494	2.1956
	e_3	1.2362	1.2362	1.2362	1.2362	1.2362	1.2362
	e_4	0.8203	0.8203	0.8203	0.8203	0.8203	0.8203

to one another. In other words, our recommendation amounts to transforming the observation vectors using a new "data-driven coordinate system" determined by the transformation matrix $\mathbf{X}(\alpha)$ such that the coordinate system is as orthogonal as possible in a d-dimensional vector space, where the inner product and orthogonality are defined based on the positive definite scatter matrix Σ of the probability distribution associated with the data vectors.

REMARK 2: Chaudhuri (1996) and Möttönen and Oja (1995) gave detailed reviews of various notions of multivariate quantiles and ranks. An interesting alternative to our present approach is to use the rank vectors that are associated with the spatial median. Affine invariance can still be achieved through data-driven transformation–retransformation, as has been done in Chakraborty et al. (1998), where an affine equivariant version of the spatial median and an affine invariant version of the angle test were proposed and studied. Such geometric concepts of ranks and quantiles are very different in nature from the coordinatewise ranks and quantiles considered here.

REMARK 3: Chaudhuri (1992) proposed and studied a class of HL-type estimates in multidimensions that are equivariant under rotations (i.e., under orthogonal transformations) but they are not equivariant under arbitrary affine transformations of the data. We can employ our transformation–retransformation strategy in conjunction with those estimates to define a related class of affine equivariant versions of HL-type estimates retaining their good efficiency properties.

REMARK 4: It is also interesting to observe that the estimation of finite sample variation of our proposed transformation–retransformation estimate is quite simple. After selecting the optimal transformation matrix, one can use any resampling technique (e.g., bootstrap) on the transformed observations to estimate the variance covariance matrix of the proposed estimate. The simplicity of computing the finite-sample variation of the estimate is a major advantage of the transformation–retransformation estimate when it comes to practical applications (see also Chakraborty and Chaudhuri, 1998; Chakraborty et al., 1998).

ACKNOWLEDGMENT

This research was partially supported by a grant from the Indian Statistical Institute.

REFERENCES

Bennet, B.M. (1962). On multivariate sign tests. *J. R. Stat. Soc., Ser. B*, **24**, 159–161.

Bickel, P.J. (1964). On some alternative estimates for shift in the p-variate one sample problem. *Ann. Math. Stat.*, **35**, 1079–1090.

Bickel, P.J. (1965). On some asymptotically nonparametric competitors of Hotelling's T^2. *Ann. Math. Stat.*, **36**, 160–173.

Blumen, I. (1958). A new bivariate sign test. *J. Am. Stat. Assoc.*, **53**, 448–456.

Brown, B.M. and Hettmansperger, T.P. (1987). Affine invariant rank methods in the bivariate location model. *J. R. Stat. Soc., Ser. B*, **49**, 301–310.

Brown, B.M. and Hettmansperger, T.P. (1989). An affine invariant bivariate version of the sign test. *J. R. Stat. Soc., Ser. B*, **51**, 117–125.

Brown, B.M., Hettmansperger, T.P., Nyblom, J. and Oja, H. (1992). On certain bivariate sign tests and medians. *J. Am. Stat. Assoc.*, **87**, 127–135.

Chakraborty, B. and Chaudhuri, P. (1996). On a transformation and retransformation technique for constructing affine equivariant multivariate median. *Proceedings of the American Mathematical Society*, **124**, 2539–2547.

Chakraborty, B., and Chaudhuri, P. (1998) On an adaptive transformation–retransformation estimate of multivariate location. *J. R. Stat. Soc., Ser. B*, **60**, 145–157.

Chakraborty, B., Chaudhuri, P. and Oja, H. (1998). Operating transformation–retransformation on spatial median and angle test. *Stat. Sinica*, **8**, 767–784.

Chatterjee, S.K. (1966). A bivariate sign test for location. *Ann. Math. Stat.*, **37**, 1771–1782.

Chaudhuri, P. (1992). Multivariate location estimation using extension of R-estimates through U-statistics type approach. *Ann. Stat.*, **20**, 897–916.

Chaudhuri, P. (1996). On a geometric notion of quantiles for multivariate data. *J. Am. Stat. Assoc.*, **91**, 862–872.

Chaudhuri, P. and Sengupta, D. (1993). Sign tests in multidimension: Inference based on the geometry of the data cloud. *J. Am. Stat. Assoc.*, **88**, 1363–1370.

Hajek, J. and Sidak, Z. (1967). *Theory of Rank Tests*. New York: Academic Press.

Hettmansperger, T.P., Nyblom, J. and Oja, H. (1992). On multivariate notions of sign and rank. In Y. Dodge, ed. *L_1 Statistical Analysis and Related Methods*. Amsterdam: North Holland, pp. 267–278.

Hettmansperger, T.P., Nyblom, J. and Oja, H. (1994). Affine invariant multivariate one sample sign tests. *J. R. Stat. Soc., Ser. B*, **56**, 221–234.

Hettmansperger, T.P., Möttönen, J. and Oja, H. (1996a). Affine invariant multivariate one-sample signed-rank tests. Preprint.

Hettmansperger, T.P., Möttönen, J. and Oja, H. (1996b). Affine invariant multivariate two-sample rank tests. Preprint.

Hodges, J.L. (1955). A bivariate sign test. *Ann. Math. Stat.*, **26**, 523–527.

Hodges, J.L. and Lehmann, E.L. (1963). Estimates of location based on rank tests. *Ann. Math. Stat.*, **34**, 598–611.

Liu, R.Y. (1992). Data depth and multivariate rank tests. In Y. Dodge, ed., *L_1 Statistical Analysis and Related Methods*. Amsterdam: North Holland, pp. 279–302.

Liu, R.Y. and Singh, K. (1993). A quality index based on data depth and multivariate rank tests. *J. Am. Stat. Assoc.*, **88**, 252–259.

Morrison, D.F. (1990). *Multivariate Statistical Methods*. New York: McGraw-Hill.

Möttönen, J., and Oja, H. (1995). Multivariate spatial sign and rank methods. *J. Nonparametric Stat.*, **5**, 201–213.

Oja, H. and Nyblom, J. (1989). Bivariate sign tests. *J. Am. Stat. Assoc.*, **84**, 249–259.

Puri, M.L. and Sen, P.K. (1971). *Nonparametric Methods in Multivariate Analysis*. New York: Wiley.

Randles, R.H. (1989). A distribution-free multivariate sign test based on interdirections. *J. Am. Stat. Assoc.*, **84**, 1045–1050.

Sperling, M.R., Gur, R.C., Alavi, A., Gur, R.E., Resnick, S., O'Connor, M.J. and Reivich, M. (1989). Subcortical metabolic alterations in partial epilepsy. *Epilepsia*.

5. APPENDIX 1: PROOFS

Proof of theorem 1. First note that in view of the affine invariance of the test statistic $T_n^{(\alpha)}$ defined in equation (1), it is enough to prove the theorem only for $\Sigma = \mathbf{I}_d$. Now, since given $\mathbf{X}(\alpha)$ the transformed observations $\boldsymbol{Y}_i^{(\alpha)}$s are conditionally independent, and they are identically distributed with the elliptically symmetric density $\det\{\mathbf{X}(\alpha)\}f[\mathrm{y}^{\mathrm{T}}\{\mathbf{X}(\alpha)\}^{\mathrm{T}}\mathbf{X}(\alpha)\mathrm{y}]$, the conditional distribution of $T_n^{(\alpha)}$ does not depend on f and depends only on $\mathbf{X}(\alpha)$. This actually follows from the introductory discussion in Bickel (1965). The (i,j)-th element of the variance covariance matrix $\Psi_1\{\mathbf{X}(\alpha)\}$ is $r_i r_j (2/\pi)\sin^{-1}\rho_{ij}$, where $\{\mathbf{X}(\alpha)\}^{-1} = \mathbf{R}(\alpha)\mathbf{J}(\alpha)$, $\mathbf{R}(\alpha) = diag(r_1, \ldots, r_d)$ and each row of the matrix $\mathbf{J}(\alpha)$ is of unit length. ρ_{ij} is given by the inner product between the i-th and j-th rows of $\mathbf{J}(\alpha)$.

When $\log f$ is twice differentiable almost everywhere in $\mathbb{R}^d$ and satisfies Cramér-type regularity conditions, by Lemma 3.1 of Bickel (1965), the conditional limiting distribution of $n^{-1/2}T_n^{(\alpha)}$ given $\mathbf{X}(\alpha)$ under the sequence of contiguous alternatives $H_A^{(n)}$ is normal with mean equal to $\{\beta(f)\}^{1/2}\{\mathbf{X}(\alpha)\}^{-1}\boldsymbol{\delta} = \Lambda_1\{f, \boldsymbol{\delta}, \mathbf{X}(\alpha)\}$ and Ψ_1 as the variance covariance matrix. Here $\beta(f) = \{2f_1(0)\}^2$ is a scalar multiple that depends only on the univariate marginal f_1 of the spherically symmetric density f. This immediately implies that the conditional limiting distribution of $n^{-1}\{T_n^{(\alpha)}\}^{\mathrm{T}}[\Psi_1\{\mathbf{X}(\alpha)\}]^{-1}T_n^{(\alpha)}$ under $H_A^{(n)}$ is noncentral χ^2 with d d.f., and the noncentrality parameter is

$$\beta(f)\boldsymbol{\delta}^{\mathrm{T}}[\{\mathbf{J}(\alpha)\}^{-1}\mathbf{D}(\alpha)[\{\mathbf{J}(\alpha)\}^{\mathrm{T}}]^{-1}]^{-1}\delta, \tag{9}$$

where the (i,j)-th element of the matrix $\mathbf{D}(\alpha)$ is $(2/\pi)\sin^{-1}\rho_{ij}$. Consequently the limiting conditional power of the test under the sequence of contiguous alternatives will be a monotonically increasing function of this noncentrality parameter. Note now that when the rows of $\mathbf{J}(\alpha)$ are orthogonal, we have $\{\mathbf{J}(\alpha)\}^{-1}\mathbf{D}(\alpha)[\{\mathbf{J}(\alpha)\}^{\mathrm{T}}]^{-1} = \mathbf{I}_d$. Finally, the minimax ordering of the noncentrality parameter Δ_1 stated in the theorem follows from Theorem 3.2 of Chakraborty and Chaudhuri (1996), which implies the largest eigenvalue of $\{\mathbf{J}(\alpha)\}^{-1}\mathbf{D}(\alpha)[\{\mathbf{J}(\alpha)\}^{\mathrm{T}}]^{-1} \geq 1$. □

Proof of theorem 2. Here also we can assume that $\Sigma = \mathbf{I}_d$ in view of the affine invariance of the test statistic $R_n^{(\alpha)}$. The conditional limiting distribution of $n^{-1/2}R_n^{(\alpha)}$ under the null hypothesis and also under the sequence of contiguous alternatives follows from Bickel (1965). Let F_i be the marginal distribution of the i-th coordinate of the transformed observations $\boldsymbol{Y}_j^{(\alpha)}$s. Then the (i,j)-th element of the variance covariance matrix Ψ_2 is given by the covariance between $F_i(Y_{ki})$ and $F_j(Y_{kj})$, where Y_{ki} is the i-th coordinate of $Y_k^{(\alpha)}$, and the noncentrality parameter of the limiting distribution under alternative $H_A^{(n)}$ is given by $\boldsymbol{\delta}^{\mathrm{T}}[\mathcal{V}\{f, \mathbf{I}_d, \mathbf{X}(\alpha)\}]^{-1}\boldsymbol{\delta}$, where $\mathcal{V}\{f, \Sigma, \mathbf{X}(\alpha)\}$ is as defined in equation (8). Consequently, the minimax ordering of the noncentrality parameter under the assumption of normality of the density f follows from arguments similar to those used in the proof of Theorem 3.2 of Chakraborty and Chaudhuri (1996). □

Proof of theorem 3. In view of the affine invariance of the test statistic $W_{m,n}^{(\alpha)}$, we may assume that $\Sigma = \mathbf{I}_d$. Under H_0, $\boldsymbol{X}_1, \ldots, \boldsymbol{X}_m$ and $\boldsymbol{Y}_1, \ldots, \boldsymbol{Y}_n$ are assumed to have the same distribution with density function $h(\boldsymbol{x} - \boldsymbol{\theta})$. We can generalize the result of asymptotic normality of the univariate two-sample rank-sum test statistic to the d-dimensional rank statistic $W_{m,n}^{(\alpha)}$ through U-statistics-type representations (for a detailed proof, see Hodges and Lehmann, 1963; Puri and Sen, 1971). Let F_i be the marginal distribution of the i-th coordinate of the transformed observations $\boldsymbol{Z}_j^{(\alpha)}$s and $m/(m+n) \to \lambda$ as $m, n \to \infty$. Then the (i, j)-th element of the asymptotic variance covariance matrix of $(m+n)^{-1/2} W_{m,n}^{(\alpha)}$ is given by $\lambda(1-\lambda)$ times the covariance between $F_i(Z_{ki})$ and $F_j(Z_{kj})$, where Z_{ki} is the i-th coordinate of $\boldsymbol{Z}_k^{(\alpha)}$.

When $\log f$ is twice differentiable almost everywhere in $\mathbb{R}^d$ and satisfies Cramér type regularity conditions, the sequence of alternatives $H_A^{(m,n)} : \boldsymbol{\delta} = (m+n)^{-1/2}\boldsymbol{\mu}$ such that $\boldsymbol{\mu} \in \mathbb{R}^d$ and $\boldsymbol{\mu} \neq 0$ will form a contiguous sequence. Then we can apply results of Hajek and Sidak (1967) for the univariate two-sample linear rank statistic to have the asymptotic normality of $(m+n)^{-1/2} W_{m,n}^{(\alpha)}$ given $\mathbf{X}(\alpha)$ under the sequence of contiguous alternatives $H_A^{(m,n)}$ as $m, n \longrightarrow \infty$. The asymptotic mean of the nonnull distribution is given by $\lambda(1-\lambda)\{\int f_1^2(x)\,dx\}\mathbf{J}(\alpha)\boldsymbol{\mu}$, where the matrix $\mathbf{J}(\alpha)$ and the function f_1 are as defined earlier. Consequently, the noncentrality parameter of the limiting distribution under alternative $H_A^{(m,n)}$ is given by $\lambda(1-\lambda)\boldsymbol{\mu}^{\mathrm{T}}[\mathcal{V}\{f, \mathbf{I}_d, \mathbf{X}(\alpha)\}]^{-1}\boldsymbol{\mu}$ where $\mathcal{V}\{f, \Sigma, \mathbf{X}(\alpha)\}$ is as defined in equation (8). Again as before the minimax ordering of the noncentrality parameter under the assumption of normality of the density f follows from the arguments used in the proofs of Theorems 1 and 2, and Theorem 3.2 of Chakraborty and Chaudhuri (1996). □

Proof of theorem 4. As our estimate $\hat{\boldsymbol{\theta}}_n^{(\alpha)}$ is equivariant under arbitrary affine transformations of the observations, we can assume without loss of generality that $\Sigma = \mathbf{I}_d$. Note that $\hat{\boldsymbol{\phi}}_n^{(\alpha)}$ is the coordinatewise HL-estimate based on the transformed observations $\boldsymbol{Y}_j^{(\alpha)}$s, $j \notin \alpha$. Then by the results of Bickel (1964), $n^{1/2}(\hat{\boldsymbol{\phi}}_n^{(\alpha)} - \boldsymbol{\phi})$ is asymptotically normal given $\mathbf{X}(\alpha)$ with mean zero, where $\boldsymbol{\phi} = \{\mathbf{X}(\alpha)\}^{-1}\boldsymbol{\theta}$. Now $\hat{\boldsymbol{\theta}}_n^{(\alpha)}$ is defined as $\{\mathbf{X}(\alpha)\}\hat{\boldsymbol{\phi}}_n^{(\alpha)}$. Thus the limiting distribution of $n^{1/2}(\hat{\boldsymbol{\theta}}_n^{(\alpha)} - \boldsymbol{\theta})$ given $\mathbf{X}(\alpha)$ is d-dimensional normal with mean zero. Straightforward algebra shows the asymptotic variance covariance matrix to be $\mathcal{V}\{f, \mathbf{I}_d, \mathbf{X}(\alpha)\}$. The determinant ordering of the matrix $\mathcal{V}\{f, \mathbf{I}_d, \mathbf{X}(\alpha)\}$ when the underlying distribution is d-dimensional normal follows from the arguments used in the proof of Theorem 3.2 of Chakraborty and Chaudhuri (1996). □

18

Correspondence Analysis Techniques

JAN DE LEEUW and DEBORAH Y. WANG University of California, Los Angeles, California

GEORGE MICHAILIDIS University of Michigan, Ann Arbor, Michigan

1. CORRESPONDENCE ANALYSIS

Correspondence analysis can be introduced in many different ways, which is probably the reason why it was reinvented many times over the years. We do not repeat the various derivations in this chapter; instead we refer to the extensive discussions in the books by Greenacre (1984), Gifi (1990), and Benzécri (1992).

Usually, correspondence analysis is motivated in graphical language. It is often said, in this context, that "A picture is worth a thousand numbers." Complicated multivariate data are made more accessible by displaying the main regularities of the data in scatterplots. This graphical approach is outlined in considerable detail in the books mentioned above. We merely give a brief introduction, which differs in some important aspects from earlier ones because it emphasizes the *graph plot* and the *star plots* (defined below). This type of introduction, which discusses the data analysis technique as a graph-layout method, was first dicussed in the review articles by Hoffman and de Leeuw (1992) and Michailidis and de Leeuw (1998). We think it nicely captures the essential geometric characteristics of the technique.

We have to choose one of the many names the technique has had over the years (see de Leeuw (1973, 1983) for a historical overview). The most widely used name seems to be (multiple) correspondence analysis or MCA, and this is what we shall use in this review as well. By considering various generalizations, we actually review a very broad class of techniques under the MCA label.

1.1 Data

MCA starts with n observations on m categorical variables, where variable j has k_j categories (possible values). Using categorical variables causes no real loss of generality: so-called *continuous* variables are merely categorical variables with a large number of numerical categories. We use K for the total number of categories over all variables.

The data are coded as m *indicator matrices* or *dummies* G_j, where G_j is a binary $n \times k_j$ matrix with exactly one nonzero element in each row i (indicating in which category of variable j observation i falls). The $n \times K$ matrix $G = (G_1 | \dots | G_m)$ is called the *indicator supermatrix.*

1.2 Graph Layout

One can represent all information in the data by a bipartite graph with $n + K$ vertices and nm edges. Each edge connects an object and a category. Thus the n vertices corresponding to the objects all have degree m, and the K vertices corresponding to the categories have varying degrees, equal to the number of objects in the category. The indicator supermatrix G is the *adjacency matrix* of the graph. The idea of presenting categorical data as a graph is not new, of course, but the idea of interpreting a data analysis technique as a graph layout method does not seem to have been studied before.

We can make a drawing or layout of the graph by placing the vertices at $n + K$ locations in the plane, or, more generally, in R^p. If we then draw the nm edges, the resulting picture will generally be more informative and more aesthetically pleasing if the edges are short. In other words, if objects are close to the categories they fall into, and categories are close to the objects falling in them. Thus we want to make a *graph plot* that "minimizes the amount of ink," i.e. the total length of all edges.

There is a substantial literature in computer science about methods and criteria to draw graphs (di Battista et al., 1994). Graph drawing algorithms for bipartite graphs that emphasize minimizing edge crossing are discussed by Eades and Wormald (1994). MCA, i.e., our "minimum ink" criterion, is closely related to the force-directed or spring algorithms first introduced by Eades (1984). Many of the criteria discussed in the computer science literature lead to NP-complete problems, i.e., they are computationally infeasible even for fairly small problems. Our edge-length algorithm is designed to be practical even for very large bipartite graphs.

Actually, we will minimize the total *squared* length of the edges. The reasons for choosing the square are the classical ones.

> Of all the principles that can be proposed for this purpose, I think there is none more general, more exact, or easier to apply, that that which we have used in this work; it consists of making the sum of squares of the errors a *minimum*. By this method, a kind of equilibrium is established among the errors which, since it prevents the extremes from dominating, is appropriate for revealing the state of the system which most nearly approximates the truth.
>
> Legendre (1805), quoted by Stigler (1986, p. 13)

In order to implement a useful algorithm, we also need a *normalization constraint*. This is needed because we want distances between vertices that are connected to be small, but we do not require distances between edges that are not connected to be large. Merely minimizing the amount of ink, without requiring a normalization, can be done by collapsing the drawing into a single point.

1.3 Least-Squares Criterion

To formalize our "minimum ink" criterion in a convenient way, we use the indicator matrices G_j. If the $n \times p$ matrix X has the locations of the object vertices in $\mathbb{R}^p$, and Y_j has the location of the k_j category vertices of variable j, then the squared length of the n edges for variable j is

$$\sigma_j(X, Y_j) = \mathbf{SSQ}(X - G_j Y_j) \tag{1}$$

where $\mathbf{SSQ}()$ is short for the sum of squares. The corresponding graph drawing, with $n + k_j$ vertices and n edges, is known as the *star plot* for variable j. The *graph plot* is the union (overlay) of the m star plots.

The squared edge length over all variables is

$$\sigma(X, Y) = \sum_{j=1}^{m} \mathbf{SSQ}(X - G_j Y_j) \tag{2}$$

and this is the function we want to minimize. The book by Gifi (1990) is mainly about many different versions of this minimization problem, where the differences are a consequence of various *restrictions* imposed on the quantifications Y_j.

As we said earlier, minimizing equation (2) without any restrictions on the vertex locations is not possible. Or, more appropriately, it is too easy. We just collapse all vertices into a single point, and we use no ink at all. This means that in order to get a nontrivial solution, we have to impose some form of *normalization*. There are two obvious ways to normalize, defining, say, MCA_1 and MCA_2. In MCA_1 we require that the columns of X add up to zero, and are *orthonormal*, i.e., satisfy $mX'X = I$. In MCA_2 we normalize Y, i.e., we require that $u'DY = 0$ and $Y'DY = I$. Here u is a vector with all elements equal to $+1$, and D is the $K \times K$ diagonal matrix with the marginal frequencies of all m variables on the diagonal.

We emphasize that there are no compelling reasons, except for computational convenience, to choose these particular normalizations. Specifically, introducing additional dimensions by requiring orthonormality is in many respects not completely satisfactory. This was indicated in Guttman's classical 1941 MCA paper.

> we should be tempted to try a "multiple factor" analysis. But the present rationale was devised specifically for a "single factor" analysis and does not necessarily carry over to the other case. It may be quite a different task to devise a rationale for "multiple factor" analysis of attributes
> (Guttman, 1941, p. 332)

1.4 Eigenvalue Formulations

One of the reasons why squared edge lengths are so appealing is that the MCA_1 and MCA_2 problems we are trying to solve are basically eigenvalue problems. We discuss this in some detail, following Gifi (1990).

First we define some useful matrices. Define the $k_j \times k_\ell$ matrix $C_{j\ell} = G_j'G_\ell$. Matrix $C_{j\ell}$ is the *cross table* or *contingency table* of variables j and ℓ. Thus $D_j = C_{jj}$, where D_j is the diagonal matrix with the univariate marginals of variable j on the diagonal. The $K \times K$ supermatrix C is known in the correspondence analysis literature as the *Burt matrix*, after Burt (1950). Write $CY = mDY\Xi$ for the generalized eigenvalue problem associated with the Burt matrix.

We also define $P_j = G_jD_j^{-1}G_j'$, then P_j is the *between-category* projector, which transforms each vector in $\mathbb{R}^n$ into a vector in $\mathbb{R}^n$ with category means. Moreover $Q_j = I - P_j$ transforms each vector into a *within-category* vector of deviations from category means. Write $P_\star$ for the average of the P_j, and write Θ for the diagonal matrix of eigenvalues of $P_\star$.

THEOREM 1: *Suppose $(\hat{X}, \hat{Y})$ solves either the MCA_1 or the MCA_2 problem. Then*

$$P_\star \hat{X} = \hat{X}\Lambda \tag{3a}$$

$$C\hat{Y} = mD\hat{Y}\Lambda \tag{3b}$$

where $\Theta = \Xi = \Lambda$.

Proof. We first analyze MCA_1, in which X is normalized by $mX'X = I$, and Y is free. Define $\sigma(X, \bullet)$ as the minimum of $\sigma(X, Y)$ over all Y. Clearly the minimum is attained for

$$\hat{Y}_j = D_j^{-1} G_j' X \tag{4}$$

i.e., by locating a category quantification in the centroids of the objects in that category. We see that

$$\sigma(X, \bullet) = m \; \mathbf{tr} \; X'(I - P_\star)X \tag{5}$$

Clearly we minimize $\sigma(X, \bullet)$ over $mX'X = I$ by choosing $\hat{X}$ equal to the eigenvectors corresponding to the p largest eigenvalues of $P_\star$. Thus $P_\star \hat{X} = \hat{X}\Theta$ for MCA_1. Also, from equation (4), we see $G\hat{Y} = mP_\star \hat{X} = m\hat{X}\Theta$ and thus $C\hat{Y} = mG'\hat{X}\Theta = mD\hat{Y}\Theta$. This proves equation (3b), with $\Theta = \Lambda$.

We now travel the other route, and tackle MCA_2. Define $\sigma(\bullet, Y)$ as the minimum of $\sigma(X, Y)$ over all X. This minimum is attained by

$$\hat{X} = \frac{1}{m}\sum_{j=1}^{m} G_j Y_j \tag{6}$$

i.e., each object is located in the centroid of the m categories that it is in. Then

$$\sigma(\bullet, Y) = \mathbf{tr} \; Y'\left(D - \frac{1}{m}C\right)Y \tag{6}$$

and the minimum over $Y'DY = I$ is attained by finding $\hat{Y}$, the eigenvector corresponding with the largest eigenvalues of the eigen-problem $CY = mDY\Xi$. Now we use equation (6) to derive $mG'\hat{X} = C\hat{Y} = mD\hat{Y}\Xi$ and thus $mP_\star \hat{X} = G\hat{Y}\Xi = m\hat{X}\Xi$. This proves equation (3a), with $\Xi = \Lambda$.

□

There are several aspects of the proof which deserve some additional attention. Equation (4) is called the *first centroid principle*, and equation (6) is the *second centroid principle*. The first centroid principle shows clearly how the star plots get their name in MCA_1. Category vertices are in the centroid of the vertices of the objects in the category, and if we have a clear separation of the k_j categories, we see k_j stars in $\mathbb{R}^p$. This also shows that in MCA_1 the category vertices are in the convex hull of the object vertices; they form a more compact cloud. In MCA_2, it is the other way around.

Of course the theorem does not say the solutions to MCA_1 andMCA_2 are *identical*, they are in fact merely *proportional*. In MCA_1 we see from equation (4) that $\hat{Y}'D\hat{Y} = m\hat{X}'P_\star\hat{X} = \Lambda$. In the same way, in MCA_2, $m\hat{X}'\hat{X} = \Lambda$.

In fact, this leads to one last important construct in MCA_1. The matrix $\hat{Y}_j'D_j\hat{Y}_j = \hat{X}'P_j\hat{X}$ is known as the discrimination matrix. It is equal to the between-category dispersion matrix of variable j, i.e., to the size of the stars for that variable. The average *discrimination matrix* is equal to Λ, the diagonal matrix of eigenvalues. Since $P_\star$ is the average of m orthogonal projectors, we have that $\Lambda \leq I$. This can also be seen from the fact that each element of Λ is the average, over variables, of the ratio of the between-category variance and the total variance.

A more direct proof of the equivalence of the eigenvalue problems for MCA_1 and MCA_2 is possible by starting with the singular value decomposition problems $GY = mXM$ and $G'X = DYM$, which immediately gives $CY = mDYM^2$ and$P_\star X = XM^2$. In the French approach, this is expressed by saying that MCA is correspondence analysis applied to the "tableau disjonctif complet" G.

2. ASPECTS OF MULTIVARIABLES

We think the graphical or geometric approach to MCA outlined above is a valid and interesting way to introduce and discuss the technique, but in this paper we go back to the more analytical formulations first proposed by Guttman (1941). We unify and extend results in previous papers (de Leeuw, 1982, 1983, 1988, 1990, 1993).

One major reason for preferring the analytical approach is that it generalizes easily to different criteria and to more general (infinite-dimensional) situations. Another reason is that a more convincing treatment of multidimensional quantification becomes possible.

In the alternative (nongeometric) formulation we discuss here, the emphasis is on finding transformations of variables and on the construction of scales. This makes it easier to relate correspondence analysis to classical multivariate analysis techniques, such as principal components analysis.

2.1 Aspects and Feasible Transformations

In this paper, a variable is an element h of some Hilbert space $\mathcal{H}$. This could just be the space of vectors with n elements, but it could also be the space of random variables with finite variances. In fact, this is precisely the reason why we use $\mathcal{H}$; our subsequent formulas apply without modification to the "population" case in which our variables can be continuous random variables.

A *multivariable* is just a mapping of an index set $\mathcal{J}$ into $\mathcal{H}$, i.e., each $j \in \mathcal{J}$ corresponds with a variable h_j. A finite number of the h_j can be collected in a "matrix" $H = \{h_1, \ldots, h_J\}$. Observe that H, interpreted as a matrix, does not have a well-defined number of rows. Nevertheless it is straightforward to define the matrix operations we need in a consistent way. Matrix $H'H$ contains inner products of the elements of H, while $Hu \in \mathcal{H}$ is a linear combination of the h_j. If $f \in \mathcal{H}$, then $f'H$ has inner products of f with the elements of H.

Informally, an *aspect* of a multivariable is a well-defined function that is used to measure how well the multivariable satisfies some, presumably desirable, criterion. Because there are many such criteria, there are many different aspects.

DEFINITION 1: An *aspect* of a multivariable is a real-valued function ϕ defined on the set of multivariables on a given index set $\mathcal{J}$.

The idea of using aspects to define multivariate analysis techniques was introduced in de Leeuw (1990). The basic idea is simple. In many situations, especially in the social and behavioral sciences, we do not know precisely how to *express* our variables. Thus in a regression model the dependent variable "income," for instance, might be dollar-income, but it might also be log-dollar-income, or even some unknown monotonic *transformation* of dollar-income. For other variables there may be missing information on some of the observations, and we get different expressions for different *imputations*. In yet another scenario, there may be *latent variables*, which are completely unobserved and only defined by their place in the model. Finally, some variables may be *ordinal* or *nominal*, and they can be incorporated in a correlational analysis only after *quantification*.

Of course not all quantifications are *feasible*. If we impute missing data, we want the imputed variable to be equal to the data in the nonmissing part. If we transform or quantify an ordinal variable, we want the transformation to be monotone. If we quantify the nominal variable "religion," we want all protestants to get the same value, all buddhists to get the same value, and so on.

Thus the basic problem in this particular approach to multivariate analysis is to select an aspect, and to investigate how this aspect varies over all feasible transformations, quantifications, or imputations of the variables. One particular approach is to study what the maximum value of the aspect is. We will formalize this optimization problem for the case in which the feasible transformations are finite-dimensional subspaces of $\mathcal{H}$. This can easily be generalized to infinite-dimensional subspaces, and even to convex cones (see de Leeuw, 1990).

In the psychometric literature this approach is known as "optimal scaling," a term due to Darrell Bock. Thus optimality is defined in terms of the aspect, and each aspect defines its own form of optimal scaling.

2.2 Examples of Aspects

In a well-known paper, Kettenring (1971) uses a similar "aspect" approach to extend canonical correlation analysis to three or more sets of variables. Versions of these ideas were proposed even earlier by Steel (1951) and Horst (1961, 1965). Kettenring's contribution is also discussed in the book by Gnanadesikan (1977, pp. 69–81). From a slightly different (psychometric) angle, other related aspects were discussed by van de Geer (1984) and ten Berge (1988).

Proposer	Name	Description
Horst	SUMCOR	Sum of $r_{j\ell}$
Kettenring	SSQCOR	Sum of $r_{j\ell}^2$
Horst	MAXVAR	Largest eigenvalue of R
Kettenring	MINVAR	Smallest eigenvalue of R
Steel	GENVAR	Determinant of R

All of the correlational aspects in the table, except SUMCOR, are actually also *eigenvalue aspects*, i.e., they are functions of the eigenvalues of the correlation matrix R. SSQCOR is the sum of squares of the eigenvalues of the correlation matrix. Because the sum of the eigenvalues of the correlation matrix is a constant, using the SSQCOR aspect is also identical to looking at the variance of the eigenvalues. GENVAR, the determinant of the correlation matrix, is the product of the eigenvalues.

The correlational aspects in the table are all measures of "interdependence" of the variables, there is no notion of dependence or causal ordering in any of them. De Leeuw (1990) observed that we may as well include aspects such as the squared multiple correlation (SMC) coefficient between one variable and the rest, or we could use the sum or sum-of-squares of one or more canonical correlation coefficients for a given partitioning of the variables into sets. Also, generalizations such as the sum of the correlation coefficients to the power s, or the absolute value of the correlation coefficients to the power s could be considered. MAXVAR and MINVAR can be generalized by considering the sum of the p largest or smallest eigenvalues. The multinormal negative log-likelihood can also be analyzed as a correlational aspect. This is

$$\phi(R) = \min_{\Gamma} \log|\Gamma| + tr\ \Gamma^{-1}R \tag{8}$$

where the minimization is over a set of model-constrained correlation matrices (for instance, all matrices satisfying the Spearman two-factor model). We do not go into details here, but clearly the notion of a correlational aspect is very general.

In de Leeuw (1988) another aspect, which is noncorrelational, was studied in some detail. We form the difference of the sum of all correlation ratios and corresponding squared correlation coefficients. Thus the aspect is

$$\phi(H) = \sum_{j=1}^{m}\sum_{\ell=1}^{m}\{\eta_{j\ell}^2 - r_{j\ell}^2\} \tag{9}$$

Minimizing such an aspect means aiming for transformations that maximize linearity of the bivariate regressions.

Another noncorrelational aspect is the Box–Cox version of equation (8). This adds a penalty term to the log-likelihood equal to the logarithm of the transformation Jacobian. It penalizes for making the transformations too flat. Because of this there is no need to normalize, and we can use covariances instead of correlations. The aspect is given by

$$\phi(S) = \min_{\Sigma} \log|\Sigma| + tr\ \Sigma^{-1}S - 2\sum_{j=1}^{m}\log \mathcal{D}h_j \tag{10}$$

The original reference is Box and Cox (1964), and use of this aspect in multivariate analysis has been analyzed in detail in Meijerink (1996). One reason why it makes sense for us to look at aspects at this level of generality is because there is a simple algorithm that allows us to optimize many of them. We will discuss this below. The other reason is that in some theoretically and perhaps also practically interesting situations we can show that the transformations we find are independent of the choice of the aspect. This will be discussed in Section 6.

3. MAXIMIZING ASPECTS

We have seen that different multivariate analysis techniques are associated with different aspects. Canonical analysis looks at aspects defined in terms of the canonical correlations, principal component analysis looks at eigenvalue aspects. Multiple regression uses the SMC, path analysis uses the sum of a number of SMCs, and so on. Some of the aspects we have considered do not correspond with classical techniques at all. It is, of course, interesting to discuss the problem of how to choose an aspect, but this is not what this chapter is about. We deal with the situation in which the client arrives in our office with an aspect, and asks us for a method to optimize it.

We need some additional notation. Suppose G_j is a basis for the subspace $\mathcal{H}_j$ of feasible transformations of variable j. Thus G_j consists of a finite number of elements, say k_j elements, of $\mathcal{H}$, collected in the "matrix" H. Previously, G_j was the indicator matrix of variable j, now it is more general. Suppose D_j is the diagonal matrix of order k_j with the squared lengths of the elements of G_j on the diagonal. An element of $\mathcal{H}_j$ can obviously be written as a linear combination of the elements of G_j, i.e., in the form $h_j = G_j\theta_j$. We write $\mathbf{NORM}(\theta_j)$ for the normalization of θ_j that satisfies $\theta_j' D_j \theta_j = 1$.

3.1 Majorization

The algorithms proposed in this paper are all of the majorization type. In a majorization algorithm we want to maximize $\phi(\theta)$ over $\theta \in \Theta$. Suppose $\psi(\theta, \xi)$ on $\Theta \times \Theta$, which we call the *majorization function*, satisfies

$$\phi(\theta) \geq \psi(\theta, \xi) \qquad \text{for all} \qquad \theta, \xi \in \Theta \tag{11a}$$

$$\phi(\theta) = \psi(\theta, \theta) \qquad \text{for all} \qquad \theta \in \Theta \tag{11b}$$

Thus, for a fixed ξ, $\psi(\bullet, \xi)$ is below ϕ, and it touches ϕ at the point $(\xi, \phi(\xi))$.

There are two key theorems associated with these definitions.

THEOREM 2: *If ϕ attains its maximum on Θ at $\hat{\theta}$, then $\psi(\bullet, \hat{\theta})$ also attains its maximum on Θ at $\hat{\theta}$.*

Proof. Suppose $\psi(\tilde{\theta}, \hat{\theta}) > \psi(\hat{\theta}, \hat{\theta})$ for some $\tilde{\theta} \in \Theta$. Then, by equations (11a and 11b), $\phi(\tilde{\theta}) \geq \psi(\tilde{\theta}, \hat{\theta}) > \psi(\hat{\theta}, \hat{\theta}) = \phi(\hat{\theta})$, which contradicts the definition of $\hat{\theta}$ as the maximizer of ϕ on Θ. □

THEOREM 3: *If $\tilde{\theta} \in \Theta$ and $\hat{\theta}$ maximizes $\psi(\bullet, \tilde{\theta})$ over Θ, then $\phi(\hat{\theta}) \geq \phi(\tilde{\theta})$.*

Proof. By equation (11a) we have $\phi(\hat{\theta}) \geq \psi(\hat{\theta}, \tilde{\theta})$. By the definition of $\hat{\theta}$ we have $\psi(\hat{\theta}, \tilde{\theta}) \geq \psi(\tilde{\theta}, \tilde{\theta})$. And by equation (11b) we have $\psi(\tilde{\theta}, \tilde{\theta}) = \phi(\tilde{\theta})$. Combining the three results gives the desired conclusion. □

If ϕ is bounded above on Θ, then the algorithm generates a bounded increasing sequence of function values, and thus it converges. Some mild continuity considerations are needed to actually show that the sequence of θ values converges as well (see de Leeuw (1990), or for a general discussion the book by Zangwill (1969)).

3.2 General Aspects

We shall maximize the aspect

$$\phi(G_1\theta_1, \ldots, G_m\theta_m)$$

over the θ_j with the normalizations conditions $\theta_j' D_j \theta_j = 1$ for all j. The notion of a general aspect, natural as it is, has not been discussed before.

THEOREM 4: *If $(\theta_1, \ldots, \theta_m)$ maximizes the aspect $\phi(H)$ over the normalized θ_j then*

$$G_j' \frac{\partial \phi}{\partial h_j} = \lambda_j D_j \theta_j \tag{12}$$

Proof. This just applies the chain rule to the Langrangian of the optimization problem. □

It follows that the Lagrange multipliers λ_j are given by

$$\lambda_j = h_j' \frac{\partial \phi}{\partial h_j} \tag{13}$$

THEOREM 5: *Suppose $\phi(H)$ is a convex and differentiable function of H which is bounded above. Then the algorithm $\mathcal{A}$ defined by*

$$\bar{\theta}_j^{(k)} = D_j^{-1} G_j' \left.\frac{\partial \phi}{\partial h_j}\right|_{\theta=\theta^{(k)}} \tag{14a}$$

and

$$\theta_j^{(k+1)} = \mathbf{NORM}(\bar{\theta}_j^{(k)}) \tag{14b}$$

converges from any starting point.

Proof. Convexity implies that for all θ and $\tilde{\theta}$ we have the majorization

$$\phi(H(\theta)) \geq \phi(H(\tilde{\theta})) + \sum_{j=1}^{m}(\theta_j - \tilde{\theta}_j)' G_j' \left.\frac{\partial \phi}{\partial h_j}\right|_{\theta=\tilde{\theta}} \tag{15}$$

Maximizing the majorization function on the right-hand side gives the algorithm in the theorem. □

3.3 Correlational Aspects

For correlational aspects we use the fact that the covariance of feasible transformations of variables j and ℓ can be written as the simple bilinear-form $\theta_j' C_{j\ell} \theta_\ell$. The variances are given by $\theta_j' D_j \theta_j$ and $\theta_\ell' D_\ell \theta_\ell$. We now state two theorems very similar to those in the previous section. These results are formalizations of the discussion of correlational aspects by de Leeuw (1990).

THEOREM 6: *If* $(\theta_1, \ldots, \theta_m)$ *maximizes the aspect* $\phi(R)$ *over all normalized* θ, *then*

$$\sum_{\ell=1}^{m} \frac{\partial \phi}{\partial r_{j\ell}} C_{j\ell} \theta_\ell = \mu_j D_j \theta_j \tag{16}$$

Proof. Use the chain rule, just as before. □

The Lagrange multipliers μ_j are now given by

$$\mu_j = \sum_{\ell=1}^{m} \frac{\partial \phi}{\partial r_{j\ell}} r_{j\ell} \tag{17}$$

Unfortunately, the majorization algorithm for correlational aspects is somewhat less simple. The main difference is that we now have to update a single θ_j at a time, and then recompute the aspect and its derivatives before we update the next θ_j.

THEOREM 7: *Suppose* $\phi(R)$ *is a convex and differentiable function of* R *which is bounded above. Then the algorithm* $\mathcal{A}$ *defined by*

$$\overline{\theta}_j^{(k)} = D_j^{-1} \left\{ \sum_{\ell=1}^{j-1} \frac{\partial \phi}{\partial r_{j\ell}} C_{j\ell} \theta_\ell^{(k+1)} + \sum_{\ell=j+1}^{m} \frac{\partial \phi}{\partial r_{j\ell}} C_{j\ell} \theta_\ell^{(k)} \right\} \tag{18a}$$

and

$$\theta_j^{(k+1)} = \mathbf{NORM}\left(\overline{\theta}_j^{(k)}\right) \tag{18b}$$

converges from any starting point.

Proof. Suppose we update θ_j.The convexity of ϕ gives the majorization

$$\phi(R) \geq \phi(\tilde{R}) + \sum_{\substack{\ell=1 \\ \ell \neq j}}^{m} \frac{\partial \phi}{\partial r_{j\ell}} (r_{j\ell} - \tilde{r}_{j\ell})$$

Thus it obviously suffices to maximize

$$\sum_{\substack{\ell=1 \\ \ell \neq j}}^{m} \frac{\partial \phi}{\partial r_{j\ell}} \theta_j' C_{j\ell} \theta_\ell$$

over the normalized θ_j. This gives the update in the theorem. □

In de Leeuw (1990) several ways are discussed to simplify the above algorithm for correlational aspects. If we assume that $\partial \phi / \partial R$ is positive semi-definite for all R, then we can apply majorization a second time, and we find an algorithm that can update all θ_j in a single step, without having to recompute aspects and derivatives.

3.4 Convexity

The above theorems are useful if the aspects we study are convex, either in the transformed variables H or in the correlation matrix R. In de Leeuw (1990), it is shown that most interesting correlational and eigenvalue aspects are, indeed, convex. The key result used to prove convexity is the following lemma.

LEMMA 1: *Suppose $\psi(\theta, \xi)$ is convex in θ for every $\xi \in \Xi$. Then*

$$\phi(\theta) = \sup_{\xi \in \Xi} \psi(\theta, \xi)$$

is convex in θ as well.

Proof. See, for instance, Rockafellar, 1970, pp. 102–111. □

In particular, the first three aspects in Table 7 are convex in R, and the last two (which we usually want to minimize) are concave in R. Norms of R are convex, the SMC is convex, the sum of the p largest eigenvalues is convex, the multinormal log-likelihood in equation (8) is convex, and so on.

3.5 Canonical Correlation Aspects

It is observed in de Leeuw (1990) that the canonical correlations and most aspects based on them are not convex functions of the correlation coefficients. Thus, although they are correlational aspects, we cannot use the results based on convexity. Nevertheless successful algorithms based on canonical correlation aspects were tried out by Tijssen and de Leeuw (1989).

To some extent, however, reintroducing enough convexity is merely a matter of redefining the problem. Suppose we have three sets of variables. We then set up the analysis as if there are only three subspaces of $\mathcal{H}$, and we compute the correlations between elements of those three subspaces. Thus we do not transform each variable separately, we transform each set of variables with a feasible transformation. This brings us back to the correlational aspects, in particular those in the Kettenring table, applied to the 3×3 correlation matrix of the sets.

This is precisely the way in which the generalized canonical correlation program OVERALS, discussed in van der Burg et al. (1988), is fitted into the MCA loss function equation (2). We code the variables in the sets interactively, and then impose additivity restrictions on the quantifications. So far, applications of the generalized canonical correlation aspects are rare, and it is difficult to choose a natural set of invariants (and a corresponding aspect) if there are more than two multivariables.

4. THE LARGEST EIGENVALUE ASPECT

Suppose the aspect we want to maximize is the largest eigenvalue $\lambda_{max}(R)$. Thus, in a sense, we want to scale the variables in such a way that they are as one-dimensional as possible. It is of some interest that in recent numerical analysis and mathematical programming literature there is a great deal of interest in minimizing the largest eigenvalue of a parameter-dependent matrix. Compare, for example, various papers by Michael Overton and co-workers (Overton, 1988; Overton and Womersley, 1993; Haeberly and Overton, 1994). In fact, it might be interesting in general to look at the *range* of aspects, i.e., to compute both the minimum and the maximum over all monotone transformations.

To compute the derivatives of the aspect, we need a general result on the derivatives of eigenvalues. This result is classical. Background, and rigorous proofs, can be found in Kato (1976) or Baumgaertel (1985).

LEMMA 2: *Suppose $Rz = \lambda z$, where $z'z = 1$ and the eigenvalue λ is unique. Then*

$$\frac{\partial \lambda}{\partial R} = zz'$$

This result is powerful enough to implement the algorithm of the previous section, but actually this algorithm can be simplified considerably in this case.

THEOREM 8: *If (θ, μ) corresponds to a stationary value of the maximum eigenvalue aspect, then $y_j = z_j\theta_j$ and $\lambda = \sum_{j=1}^{m} \mu_j$ satisfy the generalized eigenvalue problem*

$$\sum_{\ell=1}^{m} C_{j\ell} y_\ell = \lambda D_j y_j \tag{19a}$$

where

$$\sum_{j=1}^{m} y_j' D_j y_j = 1 \tag{19b}$$

*Conversely, any eigenvalue–eigenvector pair (y, λ) of this generalized eigenvalue problem defines a stationary value of the maximum eigenvalue aspect with $\mu_j = \lambda z_j^2$ and $\theta_j =$ **NORM**(y_j).*

Proof. The stationary equations (3a and b) in this case are given by

$$\sum_{\ell=1}^{m} z_j z_\ell C_{j\ell} \theta_\ell = \mu_j D_j \theta_j \tag{20}$$

which implies that

$$\mu_j = \sum_{\ell=1}^{m} z_j z_\ell r_{j\ell} = \lambda z_j^2 \tag{21}$$

Substituting this and defining $y_j = z_j\theta_j$ gives the generalized eigenvalue problem. A similar substitution establishes the converse. □

This eigenvalue–eigenvector problem is, of course, precisely equation (3b). Thus finding quantifications which maximize the largest eigenvalue is the same thing as finding the dominant eigenvalue in MCA. Again this shows that the first dimension of MCA is special. The remaining dimensions, from this point of view, merely give the remaining stationary values of the maximum eigenvalue aspect.

5. NOTIONS OF MULTIDIMENSIONALITY

The technique we have discussed in the previous section can be extended, or "made multidimensional," in at least two different ways. The fact that there are two forms of multidimensionality in this context has created some confusion, but it also provides a framework in which the Guttman quotation from Subsection 1.3 and the "horseshoe" effect (in French, the "effect Guttman") can be discussed (see Section 7).

5.1 Multiple Quantifications

In our first multidimensional extension, we can compute additional solutions to the generalized eigenvalue problem $Cy = \lambda Dy$. Each one of these defines a stationary value of our maximum eigenvalue aspect, and a corresponding system of feasible transformations. This type of multidimensionality is used in *multiple correspondence analysis*. From the point of view of maximizing aspects it is not very natural to go this way, as Guttman already indicated a long time ago.

Also observe that each additional dimension produces a set of quantifications, which can be used to construct an "induced" correlation matrix. We have $m \times k$ nontrivial solutions with m variables coded with subspaces over dimension k. This produces a lot of correlation matrices (and each of these could be subjected to a principal component analysis, for instance). Gifi uses the expression "data production" in this context.

5.2 Multidimensional Aspects

In the second multidimensional extension, we use as a different aspect the sum of the first p largest eigenvalues. This is used in *nonlinear principal component analysis*, which is discussed in considerable detail in the multidimensional scaling literature. A classic reference is Young et al. (1978), and more recently the technique has been discussed in the ACE framework by Koyak (1987).

The aspect is a convex correlational aspect, but it does not now have a simple relationship with a single fixed generalized eigenvalue problem. Thus the computational problem is inherently more complicated. It is based on the obvious generalization of Lemma 2.

LEMMA 3: *Suppose $RZ = Z\Lambda$, where $Z'Z = 1$ and the p dominant eigenvalues are in the diagonal matrix Λ. If $\lambda_p > \lambda_{p+1}$ then*

$$\frac{\partial \sum_{s=1}^{p} \lambda_p}{\partial R} = ZZ'$$

The majorization algorithm alternates between finding the dominant eigenvalues and their eigenvectors with optimal transformation of the variables. Or, alternatively, we alternate a single simultaneous iteration for the eigenvectors with optimal scaling of the variables. Convergence of these algorithms follows from the general theory.

Observe that these multidimensional aspects give rise to stationary equations that in general also have more than one solution. Such additional solutions, corresponding with other stationary values, have not been studied, except in some very special cases.

6. LINEARIZING ALL BIVARIATE REGRESSIONS

In this section we discuss a very interesting robustness result which was first mentioned in de Leeuw (1982). It says that under some circumstances, the choice of the aspect does not matter. We find the same quantifications, no matter which aspect we maximize.

THEOREM 9: *Suppose we can find* $(\theta_1, \ldots, \theta_m)$ *that make all bivariate regressions linear. Then* $(\theta_1, \ldots, \theta_m)$ *satisfies the stationary equations* (3.5), *independently of the aspect.*

Proof. The bivariate regressions are linear if

$$C_{j\ell}\theta_\ell = r_{j\ell} D_j \theta_j \tag{22}$$

If we substitute this in equation (3.5) we find

$$\sum_{\ell=1}^{m} \frac{\partial \phi}{\partial r_{j\ell}} r_{j\ell} D_j \theta_j = \mu_j D_j \theta_j \tag{23}$$

which is an identity because of equation (17). □

Clearly, for such a bilinearizable multivariate distribution, nonlinear principal component analysis will give the same quantifications as the first multiple correspondence analysis dimension.

The question is, however, in how far we can expect to observe approximate bilinearizability in real data. It seems intuitively obvious that we may be able to find it in ordinal variables, such as attitude scales, but we are unlikely to observe it with purely nominal variables which do not have any obvious order on the categories. Also, more than one system of quantifications linearizing the regressions may exist.

7. SOME GAUGES

According to Gifi (1990), a gauge is a dataset whose structure we know. Thus we know what to expect from an analysis of such a gauge, and if a technique does not represent the essential features of the data, it fails the gauge. The notion is used in Gifi's book to broaden the relationship between models and techniques beyond the classical optimality relationship of mathematical statistics.

We shall discuss some gauges to show that bilinearizability occurs in at least three common situations.

7.1 A Single Bivariate Table

In a single bivariate table both regressions can be linearized by performing a correspondence analysis on the table. The row and column scores of the correspondence analysis linearize the regressions. In fact this is one way to define correspondence analysis: find row and column scores for a table which makes both regressions linear.

7.2 Binary Variables

If all variables are binary, then any set of scores linearizes the regressions, because two points are always on a line.

7.3 The Multivariate Normal

Suppose the multivariate distribution we analyze is a multivariate normal, with standard normal marginals and correlations $\rho_{j\ell}$. Of course if we *know* that the distribution is multivariate normal, we generally do not apply any "optimal scaling" technique. In that case classical multinormal multivariate analysis applies.

For the multinormal, the Hermite polynomials of degree s in $h_s(x_j)$ are a bilinearizing system. In fact

$$C_{j\ell}h_s(x_\ell) = \rho_{j\ell}^s D_j h_s(x_j) \tag{24}$$

Thus we have a denumerable system of bilinearizing systems, one for each polynomial degree. Each system induces a correlation matrix which is an element-wise (Hadamard) power of the correlation matrix R. The eigenvalues of MCA are the m eigenvalues of R, the m eigenvalues of $R^{(2)}$, and so on. Transformations corresponding to $R^{(s)}$ are all Hermite polynomials of degree s, weighted by the coefficient from the eigenvector.

This polynomial bilinearizibility, discussed for the bivariate case in great detail by Lancaster and his students, (Lancaster, 1969), is one possible explanation of the (in)famous "horseshoe" of correspondence analysis. If the largest eigenvalue of $R^{(2)}$ is larger than the second eigenvalue of $R = R^{(1)}$, which will happen for dominant first dimensions, then all second-dimension transformations are quadratic functions of the first-dimension transformations. Two dimensional transformation plots will look like horseshoes.

Observe that basically the same result applies to what Yule calls *strained multinormals*. In this case the variables we observe are smooth monotonic ("strained") transformations of a number of standard joint multinormals. Applying MCA (or any other aspect-based transformation technique) will "unstrain" the distribution by finding the inverse transformation, which linearizes all bivariate regressions. Some statistical consequences of bilinearizability are discussed by de Leeuw (1988).

7.4 KPL Diagonalization

The structure in our three gauges can be described nicely in terms of diagonalizing the Burt matrix. Let us look at the Burt matrix of a multinormal, for instance. We continue to use matrix notation, even though some of our operators are infinite-dimensional. Thus each $C_{j\ell}$ is a standard bivariate normal, with correlation $\rho_{j\ell}$. Collect the Hermite polynomials as columns in the "matrix" K. Then $K'C_{j\ell}K = \Lambda_{j\ell}$, where $\Lambda_{j\ell}$ is a diagonal matrix which contains the powers of $\rho_{j\ell}$. Thus, if we use the direct sum $\mathcal{K} = K \oplus \cdots \oplus K$ in $\mathcal{K}'C\mathcal{K}$, then all blocks will be diagonal. Thus there is a permutation matrix P, such that $P'\mathcal{K}'C\mathcal{K}P$ is the direct sum $R \oplus R^{(2)} \oplus \cdots$. Construct the direct sum $\mathcal{L} = L \oplus L_{(2)} \oplus \cdots$ which contains the eigenvalues of the $R^{(s)}$. Then $\mathcal{L}'P'\mathcal{K}'C\mathcal{K}P\mathcal{L}$ is diagonal, i.e., the matrix $\mathcal{K}P\mathcal{L}$ has the eigenvalues of the Burt matrix C.

This KPL structure for the eigenvectors also occurs (trivially) in our two other gauges (see de Leeuw (1982) and Bekker and de Leeuw (1988) for details). More importantly, however, there is an approximate KPL structure for the MCA eigenvalues in many actual examples with attitude or diagnostic scales. It provides a much more compact decomposition of the Burt matrix, and it shows in which respects an MCA of the matrix is redundant.

Of course assuming KPL is stricter than assuming bilinearizibility, because if KPL applies, each dimension is a bilinearizing system (with some of the bilinearizing systems inducing the same correlation matrix).

8. DISCUSSION

If we compare the geometrical approach to multivariate analysis with optimal scaling and the aspect-based approach, we see that using aspects gives us a great deal of additional generality, and that we stay relatively close to classical multivariate analysis. The same criteria are studied, and the notion of an aspect suggests many generalizations (and majorization can be used to produce simple algorithms).

Choice of a particular aspect is a major practical problem, similar to the problem of choosing a loss function for regression. We have seen that bilinearizability and KPL diagonalizability provide a partial solution to the dilemma, at least as far as the optimal transformations are concerned.

Bilinearizability is a very powerful property. It guarantees that different aspects will lead to the same transformations, and it guarantees a certain statistical robustness which entails that classical asymptotic tests and estimates can still be used (see de Leeuw, 1988). Although perfect bilinearizability only occurs in some of the standard gauges, it seems to occur approximately in many practical situations.

REFERENCES

Baumgaertel, H. (1985). *Analytic Perturbation Theory for Matrices and Operators*. Basel, Boston, Stuttgart: Birkhauser.

Bekker, P. and de Leeuw, J. (1988). Relation between variants of nonlinear principal component analysis. In: J. van Rijckevorsel and J. de Leeuw, eds. *Components and Correspondence Analysis*. Chichester: Wiley.

Benzécri, J. (1992). *Correspondence Analysis Handbook*. New York: Marcel Dekker.

Box, G. E. P. and Cox, D. R. (1964). An analysis of transformations (with discussion). *J. R. Stat. Soc.*, **B26**, 211–252.

Burt, C. (1950). The factorial analysis of qualitative data. *Br. J. Stat. Psychol.*, **3**, 166–185.

de Leeuw, J. (1973). Canonical analysis of categorical data. PhD Thesis, University of Leiden. Republished in 1985 by DSWO-Press, Leiden.

de Leeuw, J. (1982). Nonlinear principal component analysis. In: H. Caussinus et al., ed. *COMPSTAT 1982*, Vienna: Physika Verlag.

de Leeuw, J. (1983). On the prehistory of correspondence analysis. *Stat. Neerlandica*, **37**, 161–164.

de Leeuw, J. (1988). Multivariate analysis with linearizable regressions. *Psychometrika*, **53**, 437–454.

de Leeuw, J. (1990). Multivariate analysis with optimal scaling. In: S. D. Gupta and J. Sethuraman, eds. *Progress in Multivariate Analysis*. Calcutta: Indian Statistical Institute.

de Leeuw, J. (1993). Some generalizations of correspondence analysis. In: C. M. Cuadras and C. R. Rao, eds. *Multivariate Analysis: Future Directions 2*. Amsterdam, London, New York, Tokyo: North-Holland.

di Battista, G., Eades, P., Tamassia, R., and Tollis, I. (1994). Algorithms for automatic graph drawing: An annotated bibliography. *Comput. Geom.*, **4**, 235–282.

Eades, P. (1984). A heuristic for graph drawing. *Congressus Numerantium*, **42**, 149–160.

Eades, P. and Wormald, N. C. (1994). Edge crossings in drawings of bipartite graphs. *Algorithmica*, **11**, 379–403.

Gifi, A. (1990). *Nonlinear Multivariate Analysis*. Chichester: Wiley.

Gnanadesikan, R. (1977). *Methods for Statistical Data Analysis of Multivariate Observations*. New York: Wiley.

Greenacre, M. (1984). *Theory and Applications of Correspondence Analysis*. New York: Academic Press.

Guttman, L. (1941). The quantification of a class of attributes: A theory and method of scale construction. In: P. Horst, ed. *The Prediction of Personal Adjustment*. New York: Social Science Research Council.

Haeberly, J.-P. A. and Overton, M. (1994). A hybrid algorithm for optimizing eigenvalues of symmetric definite pencils. *SIAM J. Matrix Anal. Appl.*, **15**, 1141–1156.

Hoffman, D. L. and de Leeuw, J. (1992). Interpreting multiple correspondence analysis as a multidimensional scaling method. *Marketing Lett.*, **3**, 259–272.

Horst, P. (1961). Relations among *m* sets of measures. *Psychometrika*, **26**, 129–149.

Horst, P. (1965). *Factor Analysis of Data Matrices*. New York: Holt, Rinehart & Winston.

Kato, T. (1976). *Perturbation Theory for Linear Operators*, 2nd edn. Berlin, Heidenlberg, New York: Springer.

Kettenring, J. R. (1971). Canonical analysis of several sets of variables. *Biometrika*, **58**, 433–451.

Koyak, R. (1987). On measuring internal dependence in a set of random variables. *Ann. Stat.*, **15**, 1215–1228.

Lancaster, H. (1969). *The Chi-Squared Distribution*. New York: Wiley.

Legendre, A. M. (1805). *Nouvelles Méthodes pour la Détermination des Orbites des Comètes*. Paris: Courcier.

Meijerink, F. (1996). *A Nonlinear Structural Relations Model*. Leiden: DSWO Press.

Michailidis, G. and de Leeuw, J. (1998). The Gifi system for descriptive multivariate analysis. Statistical Science, 13.

Overton, M. (1988). On minimizing the maximum eigenvalue of a symmetric matrix. *SIAM J. Matrix Anal. Appl.*, **9**, 256–268.

Overton, M. and Womersley, R. (1993). Optimality conditions and duality theory for minimizing sums of the largest eigenvalues of symmetric matrices. *Math. Prog.*, **62**, 321–357.

Rockafellar, R. T. (1970). *Convex Analysis*. Princeton: Princeton University Press.

Steel, R. G. D. (1951). Minimum generalized variance for a set of linear functions. *Ann. Math. Stat.*, **22**, 456–460.

Stigler, S. M. (1986). *The History of Statistics*. Cambridge: Belknap Press.

ten Berge, J. M. F. (1988). Generalized approaches to the MAXBET problem and the MAXDIFF problem, with applications to canonical correlations. *Psychometrika*, **53**, 487–494.

Tijssen, R. and de Leeuw, J. (1989). Multi-set nonlinear canonical analysis via the burt-matrix. In: R. Coppi and S. Bolasko, eds. *Multiway Data Analysis*. Amsterdam, New York: North Holland.

van de Geer, J. P. (1984). Linear relations among K sets of variables. *Psychometrika*, **49**, 79–94.

van der Burg, E., de Leeuw, J., and Verdegaal, R. (1988). Homogeneity analysis with K sets of variables: An alternating least squares approach with optimal scaling features. *Psychometrika*, **53**, 177–197.

Young, F., Takane, Y., and de Leeuw, J. (1978). The principal components of mixed measurement level multivariate data: An alternating least squares method with optimal scaling features. *Psychometrika*, **45**, 279–281.

Zangwill, W. I. (1969). *Nonlinear Programming. A Unified Approach.* Englewood Cliffs: Prentice-Hall.

19

Growth Curve Models

MUNI S. SRIVASTAVA University of Toronto, Toronto, Ontario, Canada

DIETRICH VON ROSEN Uppsala University, Uppsala, Sweden

1. INTRODUCTION

In many experiments, the observations are taken over time on the same experimental unit to measure the effect of some drugs or, in more common terminology, to observe the growth of a subject. The usual null hypothesis of no change in growth or "no effect" of the drug can easily be carried out by a test proposed by Rao (1948) where several subjects chosen at random are subjected to the same treatment. In most cases the null hypothesis is expected to be rejected unless the drug happened to be a placebo. Thus, either after the rejection of the hypothesis or more realistically, right in the beginning, the growth, or the "drug effect," is assumed to depend on time, often to be a polynomial over time, but mathematically it may be assumed that the expected value of the observation vector $\mathbf{x}$ is given by $B\xi$, where $\mathbf{x}$ is the observation vector taken at p different time points, and B is a known $p \times q$ matrix and ξ is an unknown q-vector. Obviously, q should be less than p, otherwise, the p means of the observation vector are not related. As noted above, in growth curve models B is usually a polynomial. For example, for vector will be of order two, if there is a constant term in the model. Thus, for $p = 4$,

$$E(\mathbf{x}_i) = B\xi_1, \qquad i = 1, \ldots, N_1$$

where N_1 is the number of subjects (or observation vectors),

$$B = \begin{pmatrix} 1 & t_1 & t_1^2 \\ 1 & t_2 & t_2^2 \\ 1 & t_3 & t_3^2 \\ 1 & t_4 & t_4^2 \end{pmatrix}, \qquad \text{and} \qquad \xi_1 = \begin{pmatrix} \xi_{01} \\ \xi_{11} \\ \xi_{21} \end{pmatrix}$$

This model was introduced by Rao (1959) who also provided the statistical analysis. An alternative way of writing the above model is to write it as

$$E(X) = E(\mathbf{x}_1, \ldots, \mathbf{x}_{\mathbf{N}_1}) = B\xi_1 \mathbf{1}'_{N_1}$$

where $\mathbf{1}'_{N_1} = (1, \ldots, 1)$ is a row vector of ones of length N_1. An advantage of writing it in this manner is that if we have another observation matrix on a second treatment group, say $X_2 = (\mathbf{x}_1, \ldots, \mathbf{x}_{N_2})$, which has its expectation given by

$$E(X_2) = B\xi_2 \mathbf{1}'_{N_2}$$

where $\mathbf{1}'_{N_2} = (1, \ldots, 1)$ is again a row vector of ones but of length N_2, then we can write the two models together as

$$E(X) = E(X_1, X_2) = B(\xi_1, \xi_2)\begin{pmatrix} \mathbf{1}'_{N_1} & \mathbf{0}' \\ \mathbf{0}' & \mathbf{1}'_{N_2} \end{pmatrix}$$

and a comparison of ξ_1 with ξ_2 can be carried out, i.e., a comparison of two treatment groups. In the most general form, a growth curve model can be written as

$$X = B\xi A + E$$

where X is a $p \times N$ matrix of observations, $B : p \times q$ and $A : m \times N$ are known matrices, and $\xi : q \times m$ is a matrix of unknown parameters. It is assumed that the N columns of E are iid $N_p(\mathbf{0}, \Sigma)$.

This model was introduced by Potthoff and Roy (1964), and the maximum likelihood solution was given by Khatri (1966). The above model includes Rao's (1959) model for $m = 1$ and the two-group case considered above for $m = 2$. Further properties and analysis of the above model was carried out by Rao (1961, 1965, 1966, 1967), Grizzle and Allen (1969), Gleser and Olkin (1970), Khatri (1973), Khatri and Srivastava (1975, 1976), Srivastava (1997b), Srivastava and Khatri (1979), Srivastava and Carter (1977, 1983), Kariya (1978), Marden (1983), and Hooper (1983).

For some interesting applications of this model, see Potthoff and Roy (1964), Ware and Bowden (1977), Zerbe and Jones (1980), and Carter and Hubert (1984). Bayesian prediction and analysis were carried out by Lee and Geisser (1972, 1975). A review of the literature has appeared in von Rosen (1991a) and a book that considers this model exclusively is by Kshirsagar and Smith (1995). Robustness of test procedures were investigated by Khatri (1988). The missing observation situation was considered by Kleinbaum (1973), Srivastava and McDonald (1974), Liski (1985), Srivastava (1985) and Tsai and Koziol (1988).

The objective here is to cover the basic traditional estimation and testing problems to make it more accessible by practitioners of the growth curve models. The organization of this chapter is as follows. Section 2 gives the maximum likelihood estimates of the unknown parameters. Sections 3 and 4 discuss and propose tests for the adequacy of the growth curve model and whether a MANOVA model may be more appropriate. Section 5 gives the test for a general hypothesis in the growth curve model. Section 6 considers the sum of profiles or nested model introduced by Srivastava and Khatri (1979). Section 7 contains a generalization of Section 6, and Section 8 illustrates, by an example, the theory given in the other sections.

2. MAXIMUM LIKELIHOOD ESTIMATION

This section gives the maximum likelihood (ML) estimates of the unknown parameters in the growth curve model

$$X = B\xi A + E \tag{1}$$

where the columns of the $p \times N$ matrix E are iid $N_p(\mathbf{0}, \Sigma)$, the unknown covariance matrix Σ is positive definite, and ξ is a $q \times m$ matrix of unknown parameters. The matrices $B : p \times q, q \le p$ and $A : m \times N, m < N - p$ are known matrices of ranks q and m, respectively. In what follows, we shall write $(\mathbf{x} - \boldsymbol{\mu})(\ )'$ for $(\mathbf{x} - \boldsymbol{\mu})(\mathbf{x} - \boldsymbol{\mu})'$ or $(\mathbf{x} - \boldsymbol{\mu})A(\ )'$ for $(\mathbf{x} - \boldsymbol{\mu})A(\mathbf{x} - \boldsymbol{\mu})'$. Theorem 1 provides the ML estimates.

THEOREM 1: *The maximum likelihood estimator of ξ and Σ are given by*

$$\hat{\xi} = (B'V^{-1}B)^{-1}B'V^{-1}XA'(AA')^{-1} \tag{2}$$

and

$$N\hat{\Sigma} = V + (I - P)V_1(\ )' \tag{3}$$

respectively where

$$V = X[I - H]X' \qquad H = A'(AA')^{-1}A \tag{4}$$

$$V_1 = XHX' \tag{5}$$

$$P = B(B'V^{-1}B)^{-1}B'V^{-1} \tag{6}$$

Proof. The likelihood function is given by

$$L(\xi, \Sigma) = (2\pi)^{-\frac{1}{2}pN}|\Sigma|^{-\frac{1}{2}N}\text{etr} - \tfrac{1}{2}\Sigma^{-1}[(X - B\xi A)(X - B\xi A)']$$

where etr A denotes the exponential of the trace of the matrix A. For given ξ, the MLE of Σ is given by

$$N\hat{\Sigma} = (X - B\xi A)(X - B\xi A)'$$

To obtain the MLE of ξ, we need to minimize the determinant of $(X - B\xi A)(X - B\xi A)'$ with respect to ξ. By using Lemma A.2 in Appendix 1,

$$\hat{\xi} = (B'V^{-1}B)^{-1}B'V^{-1}XA'(AA')^{-1}$$

and

$$B\hat{\xi}A = PXH$$

Hence, the MLE of Σ is given by

$$\begin{aligned} N\hat{\Sigma} &= (X - PXH)(I - H)(X - PXH)' + (X - PXH)H(X - PXH)' \\ &= X(I - H)X' + (I - P)XHX'(I - P)' \\ &= V + (I - P)V_1(I - P)' \end{aligned}$$

The maximum likelihood estimators of ξ and Σ can also be obtained by directly differentiating the likelihood function, as has been done by von Rosen (1989) and Srivastava (1997a). In the next theorem, we state a property of the estimator $\hat{\xi}$.

THEOREM 2: *The maximum likelihood estimator $\hat{\xi}$ is an unbiased estimator of ξ, and the covariance matrix of $vec(\hat{\xi})$ is given by*

$$\text{cov}\left[vec(\hat{\xi})\right] = \frac{N - m - 1}{N - m - p + q - 1}(AA')^{-1} \otimes \left(B'\Sigma^{-1}B\right)^{-1}$$

where $vec(G) \equiv vec(\mathbf{g}_1, \ldots, \mathbf{g}_m) = (\mathbf{g}_1', \ldots, \mathbf{g}_m')'$, and $F \otimes G$ is the Kronecker product defined by $(f_{ij}G)$ where $F = (f_{ij})$. For brevity of notation, we shall write $\text{cov}(\hat{\xi})$ *for* $\text{cov}(vec(\hat{\xi}))$.

Proof. From normal theory, V and XA' are independently distributed. Hence,

$$\begin{aligned}E(\hat{\xi}) &= E\big[E\big((B'V^{-1}B)^{-1}B'V^{-1}XA'(AA')^{-1}|V\big)\big]\\ &= E\big[(B'V^{-1}B)^{-1}B'V^{-1}B\xi AA'(AA')^{-1}\big]\\ &= \xi\end{aligned}$$

Thus, $\hat{\xi}$ is an unbiased estimator of ξ.

From the definition of *vec* of a matrix given in the theorem, it follows that $vec(\hat{\xi}) \equiv vec(\hat{\xi}_1, \ldots, \hat{\xi}_m) \equiv (\hat{\xi}'_1, \ldots, \hat{\xi}'_m)'$; that is, $vec(\hat{\xi})$ is a qm vector since $\hat{\xi}_i$s are q-vectors. From the *vec* definition, it can also be verified that for any three matrices A, B, and C, such that the product ABC is defined,

$$vec(ABC) = (C' \otimes A)vec(B)$$

Since

$$\begin{aligned}\hat{\xi} &= (B'V^{-1}B)^{-1}B'V^{-1}XA'(AA')^{-1}\\ vec(\hat{\xi}) &= [(AA')^{-1} \otimes Q]vec(XA')\end{aligned}$$

where

$$Q = (B'V^{-1}B)^{-1}B'V^{-1}$$

Normal theory gives

$$\text{cov}[vec(XA')] = (AA') \otimes \Sigma$$

By letting

$$\boldsymbol{\theta} = E[vec\ XA']$$

it follows that

$$\begin{aligned}\text{cov}[vec(\hat{\xi})] &= E((AA')^{-1} \otimes Q)(vec(XA') - \boldsymbol{\theta})(vec(XA) - \boldsymbol{\theta})'((AA')^{-1} \otimes Q')\\ &= E\big[E\{(AA')^{-1} \otimes Q)(vec(XA') - \boldsymbol{\theta})(vec(XA') - \boldsymbol{\theta})'\\ &\quad ((AA')^{-1} \otimes Q')|V\}\big]\\ &= E[((AA')^{-1} \otimes Q)((AA') \otimes \Sigma)((AA')^{-1} \otimes Q')]\\ &= E[(AA')^{-1} \otimes (Q\Sigma Q')]\\ &= E[(AA')^{-1} \otimes (B'V^{-1}B)^{-1}B'V^{-1}\Sigma V^{-1}B(B'V^{-1}B)^{-1}]\end{aligned}$$

since under the normality assumption V and XA' are independently distributed, where $V \sim W_p(\Sigma, N-m)$. Let

$$B_1 = \Sigma^{-\frac{1}{2}}B, \qquad W_1 = \Sigma^{-\frac{1}{2}}V\Sigma^{-\frac{1}{2}}, \qquad W = \Gamma' W_1 \Gamma$$

and

$$W^{-1} = \begin{pmatrix} W^{11} & W^{12} \\ W^{12'} & W^{22} \end{pmatrix}$$

where $\Sigma^{-\frac{1}{2}}$ is a symmetric square root of $\Sigma^{-1} = \Sigma^{-\frac{1}{2}}\Sigma^{-\frac{1}{2}}$, and Γ is an orthogonal matrix given by $\Gamma' = (B_1(B_1'B_1)^{-\frac{1}{2}}, C)$ for some suitably chosen C such that $B_1'C = 0$. Then, $W \sim W_p(I, N-m)$, $B_1'\Gamma' = ((B_1'B_1)^{\frac{1}{2}}, 0)$, which implies that $B_1' = (B_1'B_1)^{\frac{1}{2}}(I : 0)\Gamma$ and

$$\begin{aligned}
&(B'V^{-1}B)^{-1}B'V^{-1}\Sigma V^{-1}B(B'V^{-1}B)^{-1} \\
&= (B_1'W_1^{-1}B_1)^{-1}B_1'W_1^{-2}B_1(B_1'W_1^{-1}B_1)^{-1} \\
&= (B_1'B_1)^{-\frac{1}{2}}(W_{11})^{-1}(W^{11} : W^{12})(W^{11} : W^{12})'(W^{11})^{-1}(B_1'B_1)^{-\frac{1}{2}} \\
&= (B_1'B_1)^{-\frac{1}{2}}[I + W_{12}W_{22}^{-2}W_{12}'](B_1'B_1)^{-\frac{1}{2}}
\end{aligned}$$

since $(W^{11})^{-1}W^{12} = -W_{12}W_{22}^{-1}$. From Srivastava and Khatri (1979, p. 79) we know that $W_{12}W_{22}^{-\frac{1}{2}}$ is $N_{q,p-q}(0, I, I)$ and is independently distributed of W_{22}. Hence, since $E(W_{22}^{-1}) = (N-m-p+q-1)^{-1}I$,

$$\begin{aligned}
E(\mathrm{I} + W_{12}W_{22}^{-2}W_{12}') &= [\mathrm{I} + (N-m-p+q-1)^{-1}E(W_{12}W_{22}^{-1}W_{12}')] \\
&= [1 + (p-q)(N-m-p+q-1)^{-1}]\ \mathrm{I}
\end{aligned}$$

This yields,

$$E[(B'V^{-1}B)^{-1}B'V^{-1}\Sigma V^{-1}B(B'V^{-1}B)^{-1}] = \frac{N-m-1}{N-m-p+q-1}(B'\Sigma^{-1}B)^{-1}$$

since $(B_1'B_1)^{-1} = (B'\Sigma^{-1}B)^{-1}$. Thus, $\mathrm{cov}[vec(\hat{\xi})]$ is as stated in the theorem.

Following the above steps, it can be shown that

$$E(\hat{\Sigma}) = \Sigma - \left(\frac{m}{N}\right)\left(1 - \frac{p-q}{N-m-p+q-1}\right)B(B'\Sigma^{-1}B)^{-1}B'$$

Hence, the maximum likelihood estimator of Σ is not an unbiased estimator. However, since (again following the above steps)

$$E(B'V^{-1}B)^{-1} = (N-m-p+q)(B'\Sigma^{-1}B)^{-1}$$

an unbiased estimator of Σ can be obtained. We can also obtain an unbiased estimator of the $\mathrm{cov}(\hat{\xi})$. These results are stated below. □

THEOREM 3: *An unbiased estimator of* Σ *and* $\mathrm{cov}(\hat{\xi})$, *respectively, are given by*

$$\hat{\Sigma} + \left(\frac{m}{N}\right)\frac{N-m-2p+2q-1}{(N-m-p+q)(N-m-p+q-1)}B(B'V^{-1}B)^{-1}B'$$

and

$$\widehat{\text{cov}}(\hat{\xi}) = \frac{N-m-1}{(N-m-p+q-1)(N-m-p+q)}(A'A)^{-1} \otimes (B'V^{-1}B)^{-1}$$

It should be noted that from equation (3), $NB'V^{-1}B = B'\hat{\Sigma}^{-1}B$, and thus estimators are based on the MLE of Σ.

3. ADEQUACY OF THE MODEL

Consider the multivariate regression model in which

$X = \beta A + E$

where β is a $p \times m$ matrix of unknown multivariate regression parameters, and the columns of E are iid $N_p(\mathbf{0}, \Sigma)$. The growth curve model is a special case when we assume that $\beta = B\xi$ for some known $p \times q$ matrix B, $p \geq q$; that is, the number of unknown parameters required to describe the model is fewer than pm. However, it may be desirable to check if this is indeed true before using this model. Thus, we wish to test the hypothesis

$$H_1 : \beta = B\xi \tag{7}$$

against the alternative $A_1 \neq H_1$.

If C be a $(p-q) \times p$ matrix of rank $(p-q)$ such that $CB = 0$, then the hypothesis that $\beta = B\xi$ is equivalent to testing the hypothesis that $C\beta = 0$. The likelihood ratio test for this hypothesis is given by

$$\lambda_1 \equiv U_{(p-q),m,n} = \frac{|CVC'|}{|CVC' + CV_1C'|}, \qquad n = N - m \tag{8}$$

where V and V_1 are given by equations (4) and (5), respectively. The asymptotic distribution of λ_1 is given by

$$\begin{aligned} &P\left\{-[n - \tfrac{1}{2}(p-q-m+1)]log\lambda_1 \geq z\right\} \\ &\quad = P\{\chi_f^2 \geq z\} + n^{-2}\gamma_2\left\{P(\chi_{f+4}^2 \geq z) - P(\chi_f^2 \geq z)\right\} + O(n^{-4}) \end{aligned} \tag{9}$$

where $f = (p-q)m$, $\gamma_2 = f[(p-q)^2 + m - 5]/48$, and χ_f^2 denotes a chi-square random variable with f df.

This test was given by Srivastava and Carter (1983), and later by Chinchilli and Elswick (1985). Observe that equation (9) is obtained from an expansion of (see Srivstava and Carter, 1983, Theorem 4.2.1)

$$U_{p,m,n} = \frac{|S_e|}{|S_e + S_H|}$$

where $S_e \sim W_p(\Sigma, n)$ is independent of $S_H \sim W_p(\Sigma, m)$.

In terms of the original variables, equation (8) is given by

$$\lambda_1 = \frac{|V|}{|V+V_1|}\frac{|B'V^{-1}B|}{|B'(V+V_1)^{-1}B|} \tag{10}$$

(see Srivastava, 1997b, for details). Thus, equation (10) does not require us to find a matrix C such that $CB = 0$, although C can be chosen as $M'[I - B(B'B)^{-1}B']$, where M is any $p \times (p-q)$ matrix such that (B, M) is nonsingular. We may chose M from $\Gamma = (L, M)$ such that $\Gamma'[I - B(B'B)^{-1}B']\Gamma = diag(0, \ldots, 0, 1, \ldots, 1)$, the last $(p-q)$ diagonal elements as one. Thus:

THEOREM 4: *The likelihood ratio test for the adequacy of the model, i.e., that the growth curve model is tenable, is given by equations (8) or (10). The asymptotic distribution is given in equation (9).*

4. GROWTH CURVE MODEL AS A CONDITIONAL MANOVA MODEL

Let $B_0 = (B_1, B_2)$ be a nonsingular $p \times p$ matrix such that $B_2'B_1 = 0$, $B_1 = B(B'B)^{-1} : p \times q$ and $B_2 : p \times (p-q)$. Then, $B_2'B = 0$ and $B_1'B = I_q$. Hence,

$$E(B_0'X) = E\begin{pmatrix} B_1'X \\ B_2'X \end{pmatrix} = \begin{pmatrix} \xi A \\ 0 \end{pmatrix} \tag{11}$$

since $E(X) = B\xi A$. If

$$Y = B_0'X, \qquad Y_1 = B_1'X \qquad \text{and} \qquad Y_2 = B_2'X \tag{12}$$

then the N columns of the random matrix Y are independently normally distributed with covariance matrix

$$\Lambda \equiv B_0'\Sigma B_0 = \begin{pmatrix} B_1'\Sigma B_1 & B_1'\Sigma B_2 \\ B_2'\Sigma B_1 & B_2'\Sigma B_2 \end{pmatrix} = \begin{pmatrix} \Lambda_{11} & \Lambda_{12} \\ \Lambda_{12}' & \Lambda_{22} \end{pmatrix} \tag{13}$$

Hence conditionally, given Y_2,

$$E[Y_1 | Y_2] = \xi A + \Lambda_{12}\Lambda_{22}^{-1} Y_2 = \beta A^* \tag{14}$$

where $\beta = (\xi, \Lambda_{12}\Lambda_{22}^{-1})$ and $A^* = (A', Y_2')'$. Also conditionally, given Y_2, the columns of Y_1 are independently normally distributed with covariance matrix

$$\begin{aligned}\Lambda_{11}-\Lambda_{12}\Lambda_{22}^{-1}\Lambda_{12}' &= B_1'\Sigma B_1 - B_1'\Sigma B_2(B_2'\Sigma B_2)^{-1}B_2'\Sigma B_1 \\ &= B_1'[\Sigma-\Sigma B_2(B_2'\Sigma B_2)^{-1}B_2'\Sigma]B_1 \\ &= B_1'B_1(B_1'\Sigma^{-1}B_1)^{-1}B_1'B_1 \qquad \text{from Lemma A.1} \\ &= (B'\Sigma^{-1}B)^{-1}\end{aligned} \tag{15}$$

Thus, conditionally given Y_2, the growth curve model is a MANOVA model (multivariate regression model).

We shall now consider the case when in equation (13) $\Lambda_{12}=0$. In this case, we get from equation (14)

$$E(Y_1\,|\,Y_2)=\xi A$$

and the columns of Y_1 are independently normally distributed with covariance matrix Λ_{11}. It is also distributed independently of Y_2. Thus, the least squares or maximum likelihood estimates of ξ will not depend on Y_2 and Y_1 follows unconditionally a MANOVA model. The least squares estimate of ξ when $\Lambda_{12}=0$ is given by

$$\hat{\xi}_{LS}=(B'B)^{-1}B'XA'(AA')^{-1}$$

This is also the least squares estimate of ξ in the full model when $\Sigma=\sigma^2 I$, without the assumption of normality.

Next we find the structure of Σ when $\Lambda_{12}=0$, i.e., $B_1'\Sigma B_2=0$. From Lemma A.1, it follows that

$$\begin{aligned}\Sigma &= (B(B'B)^{-1}B+B_2(B_2'B_2)^{-1}B)\Sigma(B(B'B)^{-1}B+B_2(B_2'B_2)^{-1}B_2) \\ &= B(B'B)^{-1}B\Sigma B(B'B)^{-1}B+B_2(B_2'B_2)^{-1}B\Sigma B_2(B_2'B_2)^{-1}B_2 \\ &= B\Lambda_{11}B'+B_2(B_2'B_2)^{-1}\Lambda_{22}(B_2'B_2)^{-1}B_2'\end{aligned}$$

Thus Σ is of the form

$$\Sigma=B\Gamma B'+B_2\Theta B_2' \tag{16}$$

where Γ and Θ are arbitrary unknown matrices such that Σ is positive definite and $B_2'B=0$, where (B, B_2) is a nonsingular matrix. On the other hand, if Σ is of the form of equation (16) $B_1'\Sigma B_2=0$, i.e., $\Lambda_{12}=0$. Thus the condition in equation (16) is necessary and sufficient for $\Lambda_{12}=0$.

It should be noted that we have assumed normality in obtaining the necessary and sufficient condition in equation (16). Rao (1967) has shown that equation (16) is necessary and sufficient in order that the least-square estimates of ξ obtained under the model $\Sigma=\sigma^2 I$ will be identical with the one obtained under general Σ, without the assumption of normality. However, he writes equation (16) in the form

$$\Sigma=B\Gamma_1 B'+B_2\Theta_2 B_2'+\sigma^2 I$$

where Γ_1 and Θ_2 are arbitrary matrices. It can easily be shown that this is equivalent to equation (16) since

$$B(B'B)^{-1}B' + B_2(B_2'B_2)^{-1}B_2' = I$$

In some sense, this representation is easier to apply. For example if $\Theta_2 = 0$, we have the representation

$$\Sigma = B\Gamma_1 B' + \sigma^2 I$$

Thus, when Σ can be represented in any of the three forms given above, the problem should be treated as a reduced MANOVA model with observation matrix Y_1, or simply obtain the ordinary least-square estimate of ξ from the whole model, that is, assuming $\Sigma = \sigma^2 I$. The models considered by Reinsel (1982, 1984) fall in this category.

Thus, before embarking on the analysis of growth curve models, it may be desirable to test the hypothesis that $\Lambda_{12} = 0$, or equivalently $H : \Sigma$ is of the form of equation (16) against the alternative A$\neq$ H. The likelihood ratio test for this hypothesis has been given by Lee and Geisser (1972). The likelihood ratio test is based on the statistic which is the ratio of the determinants of the two MLEs of Λ, one under the alternative and the other under the hypothesis.

From equation (3), as noted before, it can be shown that

$$B'\hat{\Sigma}^{-1}B = NB'V^{-1}B$$

and

$$B_2'\hat{\Sigma}B_2 = N^{-1}B_2'(V + V_1)B_2 = N^{-1}B_2'XX'B_2 = N^{-1}Y_2Y_2'$$

Hence, under the alternative hypothesis,

$$\begin{aligned}|\hat{\Lambda}| &= |B_0'\hat{\Sigma}B_0| = |\hat{\Lambda}_{11} - \hat{\Lambda}_{12}\hat{\Lambda}_{22}^{-1}\hat{\Lambda}_{12}'||\hat{\Lambda}_{22}| \\ &= |B'\hat{\Sigma}^{-1}B|^{-1}|B_2'\hat{\Sigma}B_2| \\ &= |NB'V^{-1}B|^{-1}|N^{-1}Y_2Y_2'|\end{aligned}$$

Under the null hypothesis, Y_1 and Y_2 are independently distributed. Hence, the MLE of Λ_{11} under H is

$$N\hat{\Lambda}_{11H} = Y_1[I - H]Y_1' = B_1'X[I - H]X'B_1 = B_1'VB_1$$

Thus,

$$\begin{aligned}|\hat{\Lambda}_H| &= |\hat{\Lambda}_{11H}||\hat{\Lambda}_{22H}| \\ &= |N^{-1}B_1'VB_1||B_2'\hat{\Sigma}B_2| \\ &= |B'B|^{-2}|N^{-1}B'VB||N^{-1}Y_2Y_2'|, \qquad n = N - m\end{aligned}$$

Hence, the likelihood ratio test equals

$$\tilde{\lambda} = \frac{|\hat{\Lambda}|}{|\hat{\Lambda}_H|} = U_{p-q,q,n-q-1}$$

$$= \frac{|B'B|^2}{|B'VB||B'V^{-1}B|}$$

The asymptotic null distribution of $\tilde{\lambda}$ is given by

$$P\{-[n-q-1-\tfrac{1}{2}(p-2q+1)log\tilde{\lambda} \geq z\}$$

$$= P\left\{\chi^2_{\tilde{f}} \geq z\right\} + n^{-2}\tilde{\gamma}_2\left\{P\left(\chi^2_{\tilde{f}+4} \geq z\right) - P\left\{\left(\chi^2_{\tilde{f}} \geq z\right)\right\} + O(n^{-4})\right.$$

where

$$\tilde{f} = q(p-q) \tag{17}$$

and

$$\tilde{\gamma}_2 = \tilde{f}[(p-q)^2 + q - 5]/48 \tag{18}$$

Thus, we have the following theorem.

THEOREM 5: *The growth curve model in equation (1) is a reduced dimension MANOVA model if Σ is of the form of equation (16). The likelihood ratio test for testing the hypothesis that Σ is of the form of equation (16) is based on the statistic $\tilde{\lambda}$. The MANOVA model is given by*

$$Y_1 \equiv B_1'X = \xi A + B_1'E$$

where the columns of $B_1'E$ are iid $N_p(\mathbf{0}, \Lambda_{11})$, $\Lambda_{11} = B_1'\Sigma B_1$, and $B_1 = B(B'B)^{-1}$.

5. TESTING GENERAL HYPOTHESES IN GCM

Sections 3 and 4 tested the adequacy of the growth curve model (GCM), given in equation (1), and the independence of the two transformed matrices, respectively. If the first hypothesis is accepted and the second hypothesis is rejected, the data should be analyzed as the growth curve model, but there is still a possibility of redundancy of parameters, that is, some of the parameters may be close to zero. A check for redundancy of the parameters is carried out by testing the general hypothesis

$$H_2 : C\xi D = 0 \qquad vs. \qquad A_2 \neq H_2$$

where $C : c \times q$ and $D : m \times d$ are known matrices of ranks $c \leq q$ and $d \leq m$, respectively. The matrices C and D are appropriately chosen to obtain the parameters of interest. Consider nonsingular matrices C_0 and D_0 defined by

$$C_0' = (C_1', C') : q \times q, \qquad CC_1' = 0$$

and

$$D_0 = (D_1, D) : m \times m \qquad D_1' D = 0$$

Then

$$\begin{aligned} E(X) &= BC_0^{-1} C_0 \xi D_0 D_0^{-1} A \\ &\equiv B_1 \eta A_1 \end{aligned} \tag{19}$$

where $B_1 = BC_0^{-1}$, $A_1 = D_0^{-1} A$ and

$$\eta = C_0 \xi D_0 = \begin{pmatrix} C_1 \xi D_1 & C_1 \xi D \\ C \xi D_1 & C \xi D \end{pmatrix}$$

Let

$$A_0' = [A_1'(A_3')^{-1},\ A_2']$$

be an orthogonal matrix such that $A_1 A_1' = D_0^{-1} A A' D_0^{-1} = A_3 A_3'$; $A_1 A_2' = 0$, where A_3 is a lower triangular matrix. Let

$$B_0 = [B_1 (B_1' B_1)^{-1},\ B_2]$$

be a nonsingular matrix with $B_2' B_1 = 0$. Note that

$$B_1' B_0 = (I_q, 0), \qquad A_1 A_0' = (A_3, 0)$$

Hence, with

$$Y = B_0' X A_0' = \begin{pmatrix} (B_1' B_1)^{-1} B_1' X A_0 \\ B_2' X A_0 \end{pmatrix}$$

$$\begin{aligned} E(Y) &= B_0' B_1 \eta A_1 A_0' \\ &= \begin{pmatrix} I_q \\ 0 \end{pmatrix} \eta (A_3, 0) \\ &= \begin{matrix} q \\ p-q \end{matrix} \overset{\begin{matrix} m & N-m \end{matrix}}{\begin{pmatrix} \eta A_3 & 0 \\ 0 & 0 \end{pmatrix}} \equiv \begin{matrix} q \\ p-q \end{matrix} \overset{\begin{matrix} m & N-m \end{matrix}}{\begin{pmatrix} \delta & 0 \\ 0 & 0 \end{pmatrix}} \end{aligned}$$

and under H_2 : $C \xi D = 0$

$$\delta = \eta A_3 = \begin{pmatrix} \eta_{11} & \eta_{12} \\ \eta_{21} & 0 \end{pmatrix} \begin{pmatrix} A_{11} & 0 \\ A_{12} & A_{22} \end{pmatrix} \equiv \begin{pmatrix} \delta_1 & \delta_3 \\ \delta_2 & 0 \end{pmatrix}$$

Then

$$E(Y) = \begin{pmatrix} \delta & 0 \\ 0 & 0 \end{pmatrix}; \qquad \text{and under } H_2, \qquad E(Y) = \begin{pmatrix} \delta_1 & \delta_3 & 0 \\ \delta_2 & 0 & 0 \\ 0 & 0 & 0 \end{pmatrix}$$

Let

$$Y = \begin{matrix} & m & N-m \\ q & \begin{pmatrix} Y_1 \\ Y_3 \end{pmatrix} & Y_4 \\ p-q & & \end{matrix}$$

$$Y = \begin{matrix} & m-d & d & N-m \\ q-c & Y_{11} & Y_{12} & \\ c & Y_{21} & Y_{22} & Y_4 \\ p-q & Y_{31} & Y_{32} & \end{matrix}$$

$$V = Y_4 Y_4' = \begin{pmatrix} P_{11} & P_{12} \\ P_{12}' & V_{33} \end{pmatrix} = \begin{pmatrix} V_{11} & V_{12} & V_{13} \\ V_{12}' & V_{22} & V_{23} \\ V_{13}' & V_{23}' & V_{33} \end{pmatrix}$$

Then, following the approach in Section 2, we need to minimize the following determinant d to obtain the maximum likelihood estimator of δ under A_2.

$$\begin{aligned} d &= \left| V + \begin{pmatrix} Y_1 - \delta \\ Y_3 \end{pmatrix} \begin{pmatrix} Y_1 - \delta \\ Y_3 \end{pmatrix}' \right| \\ &= \left| V \right| \left| I + (Y_1' - \delta', \ Y_3') V^{-1} \begin{pmatrix} Y_1 - \delta \\ Y_3 \end{pmatrix} \right| \\ &= |V||I + Y_3' V_{33}^{-1} Y_3 + (Y_1 - \delta - P_{12} V_{33}^{-1} Y_3)'(P_{11} - P_{12} V_{33}^{-1} P_{12}')^{-1}(\ \ )'| \\ &\geq |V||I + Y_3' V_{33}^{-1} Y_3| \\ &= \frac{|V|}{|V_{33}|} |V_{33} + Y_3 Y_3'| \end{aligned} \tag{20}$$

where for the third equality, Lemma A.3 has been used. The minimum occurs at

$$\hat{\eta} A_3 = \hat{\delta} = Y_1 - P_{12}' V_{33}^{-1} Y_3$$

This gives

$$\begin{aligned} C_0 \hat{\xi} D_0 A_3 &= \hat{\delta} \\ &= (B_1' B_1)^{-1} B_1' X A_0' - P_{12}' V_{33}' B_2' X A_0' \\ &= [(B_1' B_1)^{-1} B_1' - P_{12}' V_{33}^{-1} B_2'] X A_0' \end{aligned}$$

Consider H_2 and let

$$Q' = (Y_{12}' - \delta_3', \quad Y_{22}', \quad Y_{32}')$$

To obtain the MLE of $\delta_1, \delta_2, \delta_3$, we need to minimize

$$d \equiv \left| V + \begin{pmatrix} Y_{11} - \delta_1 \\ Y_{21} - \delta_2 \\ Y_{31} \end{pmatrix} (\ )' + \begin{pmatrix} Y_{12} - \delta_3 \\ Y_{22} \\ Y_{32} \end{pmatrix} (\ )' \right|$$

We shall minimize with respect to δ_2 and δ_3 using Lemma A.3

$$\begin{aligned} d &= |V + QQ'| \left| I + \begin{pmatrix} Y_{11} - \delta_1 \\ Y_{21} - \delta_2 \\ Y_{31} \end{pmatrix}' (V + QQ')^{-1} \begin{pmatrix} Y_{11} - \delta_1 \\ Y_{21} - \delta_2 \\ Y_{31} \end{pmatrix} \right| \\ &\geq |V + QQ'||I + Y_{31}'(V_{33} + Y_{32}Y_{32}')^{-1}Y_{31}| \\ &= \frac{|V + QQ'|}{|V_{33} + Y_{32}Y_{32}'|} |V_{33} + Y_{32}Y_{32}' + Y_{31}Y_{31}'| \\ &= \left| V + \begin{pmatrix} Y_{12} - \delta_3 \\ Y_{22} \\ Y_{33} \end{pmatrix} (\ )' \right| \frac{|V_{33} + Y_{32}Y_{32}' + Y_{31}Y_{31}'|}{|V_{33} + Y_{32}Y_{32}'|} \end{aligned}$$

Now, we shall minimize with respect to δ_3, again using Lemma A.3. This gives,

$$\begin{aligned} d &\geq |V| \left| I + \begin{pmatrix} Y_{22} \\ Y_{32} \end{pmatrix}' \begin{pmatrix} V_{22} & V_{23} \\ V_{23}' & V_{33} \end{pmatrix}^{-1} \begin{pmatrix} Y_{22} \\ Y_{32} \end{pmatrix} \right| \frac{|V_{33} + Y_{32}Y_{32}' + Y_{31}Y_{31}'|}{|V_{33} + Y_{32}Y_{32}'|} \\ &= \frac{|V| \left| \begin{pmatrix} V_{22} & V_{23} \\ V_{23}' & V_{33} \end{pmatrix} + \begin{pmatrix} Y_{22} \\ Y_{32} \end{pmatrix} \begin{pmatrix} Y_{22} \\ Y_{32} \end{pmatrix}' \right|}{\begin{vmatrix} V_{22} & V_{23} \\ V_{23}' & V_{33} \end{vmatrix}} \frac{|V_{33} + Y_{32}Y_{32}' + Y_{31}Y_{31}'|}{|V_{33} + Y_{32}Y_{32}'|} \end{aligned} \tag{21}$$

While obtaining the maximum likelihood estimator of δ under the alternative and δ_1, δ_2, and δ_3 under the null hypothesis, we also obtained the value of the determinant of the maximum likelihood estimators of the covariance matrix under the alternative and hypothesis. These are given by the expression on the right-hand side of equations (20) and (21), respectively. Using these expressions, we can easily obtain the likelihood ratio test for testing the hypothesis H_2 against A_2. This is based on the statistic

$$\lambda_2 = \frac{|\tilde{V}|}{|\tilde{V} + ZZ'|} \times \frac{|V_{33} + Z_2 Z_2'|}{|V_{33}|} \tag{22}$$

where

$$\tilde{V} \equiv \begin{pmatrix} V_{22} & V_{23} \\ V_{23}' & V_{33} \end{pmatrix} \quad \text{and} \quad Z = \begin{pmatrix} Z_1 \\ Z_2 \end{pmatrix} \equiv \begin{pmatrix} Y_{22} \\ Y_{32} \end{pmatrix} \tag{23}$$

In terms of the original variables, (see Khatri, 1966),

$$\lambda_2 = \frac{|Q|}{|P+Q|} \tag{24}$$

where

$$\begin{aligned} Q &= C(B'V^{-1}B)^{-1}C' \\ P &= (C\hat{\xi}D)(D'RD)^{-1}(C\hat{\xi}D)' \\ \hat{\xi} &= (B'V^{-1}B)^{-1}B'V^{-1}XA'(AA')^{-1} \\ R &= (AA')^{-1} + (AA')^{-1}AX'[V^{-1} - V^{-1}B(B'V^{-1}B)^{-1}B'V^{-1}]XA'(AA')^{-1} \\ V &= X[I - A(AA')^{-1}A']X' \end{aligned} \tag{25}$$

It should be noted that the statistic λ_2 is invariant under nonsingular linear transformations. Thus, we may assume, without any loss of generality, that $\Sigma = I$. Under this assumption, the joint distribution of $\tilde{V}$ and Z, defined in equation (23), under the null hypothesis is given by

$$Const.|\tilde{V}|^{(n-p+q-c-1)/2}[etr - \frac{1}{2}(\tilde{V} + ZZ')], \qquad n = N - m$$

Consider the transformations

$$\tilde{V} + ZZ' = TT' = \begin{pmatrix} T_1 & T_{12} \\ 0 & T_2 \end{pmatrix} \begin{pmatrix} T_1' & 0' \\ T_{12}' & T_2' \end{pmatrix}$$

$$Z = TW, \quad W' = (W_1', W_2')$$

Also, the Jacobian of the transformation

$$J(\tilde{V}, Z \to T, W) = J(\tilde{V} \to T)J(Z \to W) = 2^p \prod_{i=1}^{p-q+c} t_{ii}^{i}|T|^d = 2^p \prod_{i=1}^{p-q+c} t_{ii}^{i+d}$$

since Z is a matrix of order $(p - q + c) \times d$. It can be seen that T and W are independently distributed. The pdf of W is given by

$$Const.|I - WW'|^{(n-p+q-c-1)/2} \tag{26}$$

and

$$\lambda_2 = \frac{|I - WW'|}{|I - W_2W_2'|} = \frac{|I - W_1'W_1 - W_2'W_2|}{|I - W_2'W_2|} = |I - P'P|$$

where $P' = (I - W_2'W_2)^{-\frac{1}{2}}W_1'$. The pdf of P' follows from equation (26) and is given by

$$Const.|I - P'P|^{(n-p+q-c-1)/2}$$

Hence, under the null hypothesis H_2 (see Srivastava and Khatri, 1979, Chap. 6),

$$\lambda_2 = U_{c,d,n-p+q}, \quad n = N - m$$

The asymptotic distribution of λ_2 is given by

$$P\{-[n - p + q - \tfrac{1}{2}(c - d + 1)] \log U_{c,d,n-p+q} \geq z\}$$
$$= P(\chi_f^2 \geq z) n^{-2}\gamma_2\{P(\chi_{f+2}^2 \geq z) - P(\chi_f^2 \geq z)\} + O(n^{-4})$$

where $f = cd$ and $\gamma_2 = f(c^2 + d - 5)/48$.

THEOREM 6: *For the growth curve model in equation (1), the likelihood ratio test statistic for testing the hypothesis $C\xi D = 0$, $C : c \times q$, $D : m \times d$ is given by λ_2 given in equations (22) or (24).*

6. NESTED MODELS

Section 5 considered the problem of testing the hypothesis $H : C\xi D = 0$ in the GCM model $E(X) = B\xi A$, where the columns of X are independently normally distributed with common covariance Σ. This testing problem was rewritten in equation (19) as

$$E(X) = B_1 \eta A_1$$

where

$$\eta = \begin{pmatrix} \eta_1 & \eta_2 \\ \eta_3 & \eta_4 \end{pmatrix}$$

and testing the hypothesis was equivalent to testing the hypothesis that $\eta_4 = 0$. By writing $B_1 = (B_2, \tilde{B})$, $A_1' = (\tilde{A}_1', \tilde{A}_2')$, $\eta_{(1)} = (\eta_1', \eta_3')'$, and $\eta_{(2)} = \eta_2$, then $B_1\eta A_1$ can be written as

$$B_1\eta A_1 = B_1\eta_{(1)}\tilde{A}_1' + B_2\eta_{(2)}\tilde{A}_2' \tag{27}$$

where B_2 is a subset of B_1. Srivastava and Khatri (1979) proposed this model and gave an outline of the procedure to obtain tests and estimates in Problem 6.9, p. 196. The model of the type in equation (27) is also called the "sum of profiles." Banken (1984), Kariya (1985), Verbyla and Venables (1988) and von Rosen (1989) also independently considered this model. Andersson et al. (1993) have also discussed this model. For details of the estimation and testing of the hypothesis problems considered in Srivastava and Khatri (1979), see Srivastava (1997b).

Thus it has been demonstrated that the testing of the hypothesis problem of Section 5 can be written as a nested or sum of profiles models. We now give another example of a nested model. Consider two treatments applied to N_1 and N_2 subjects, respectively. Suppose the second treatment is a placebo. Then, if we represent the two responses by y_{ti} and z_{ti}, and if the response is linear in time for the first treatment (and constant for the second, placebo), we obtain

$$E(y_{ti}) = \beta_{01} + \beta_{11}t, \qquad i = 1, \ldots, N_1$$

and

$$E(z_{ti}) = \beta_{02}, \qquad i = 1, \ldots, N_2$$

$t = 1, 2, 3$. By writing

$$\mathbf{y}_i = \begin{pmatrix} y_{1i} \\ y_{2i} \\ y_{3i} \end{pmatrix}, \qquad i = 1, \ldots, N_1$$

and

$$\mathbf{z}_i = \begin{pmatrix} z_{1i} \\ z_{2i} \\ z_{3i} \end{pmatrix}, \qquad i = 1, \ldots, N_2$$

then

$$E(\mathbf{y}_i) = \begin{pmatrix} 1 & 1 \\ 1 & 2 \\ 1 & 3 \end{pmatrix} \begin{pmatrix} \beta_{01} \\ \beta_{11} \end{pmatrix}, \qquad i = 1, \ldots, N_1$$

and

$$E(z_i) = \begin{pmatrix} \beta_{02} \\ \beta_{02} \\ \beta_{02} \end{pmatrix} = \begin{pmatrix} 1 & 1 \\ 1 & 2 \\ 1 & 3 \end{pmatrix} \begin{pmatrix} \beta_{02} \\ 0 \end{pmatrix}, \qquad i = 1, \ldots, N_1$$

Thus, if

$$B_1 = \begin{pmatrix} 1 & 1 \\ 1 & 2 \\ 1 & 3 \end{pmatrix}$$

and

$$X = (\mathbf{y}_1, \ldots, \mathbf{y}_{N_1}, \mathbf{z}_1, \ldots, \mathbf{z}_{N_2})$$

then

$$E(X) = B_1 \begin{pmatrix} \beta_{01} & \beta_{02} \\ \beta_{11} & 0 \end{pmatrix} \begin{pmatrix} \mathbf{1}'_{N_1} & 0 \\ 0 & \mathbf{1}'_{N_2} \end{pmatrix}$$

This can be written as

$$E(X) \equiv B_1 \begin{pmatrix} \beta_{01} & \beta_{02} \\ \beta_{11} & 0 \end{pmatrix} A$$
$$= B_1 \xi_1 A_1 + B_2 \xi_2 A_2 \ = B_2 \tilde{\xi}_1 A + \tilde{B} \tilde{\xi}_2 A_1$$

where

$$B_1 = (B_2, \tilde{B}) \qquad \text{and} \qquad A = \begin{pmatrix} A_1 \\ A_2 \end{pmatrix}$$

6.1 Maximum Likelihood Estimators

This section considers the nested growth curve model in which

$$E(X) = B_1 \xi A$$

where

$$\xi = \begin{matrix} & m_1 & m_2 \\ q_1 & \xi_1 & \xi_2 \\ q_2 & \xi_3 & 0 \end{matrix}, \qquad m_1 + m_2 = m, \qquad q_1 + q_2 = q$$

By writing

$$\eta_1 = \begin{pmatrix} \xi_1 \\ \xi_3 \end{pmatrix}, \qquad \eta_2 \equiv \xi_2, \qquad B_1 = (B_2, \tilde{B}), \qquad A' = \begin{pmatrix} A_1' \\ A_2' \end{pmatrix}$$

we have

$$B_1 \xi A = B_1 \eta_1 A_1 + B_2 \eta_2 A_2 \tag{28}$$

We shall assume that the columns of X are independently normally distributed with covariance matrix Σ. It is also known that B_2 is a subset of B_1, and we assume that they are of full rank. To obtain the maximum likelihood estimators, following the approach in Section 2, note that for given η_1 and η_2, the MLE of Σ is given by

$$N\hat{\Sigma} = (X - B_1 \eta_1 A_1 - B_2 \eta_2 A_2)(\ \ )'$$

To obtain the values of η_1 and η_2, we minimize the determinant

$$d = |(X - B_1 \eta_1 A_1 - B_2 \eta_2 A_2)(\ \ )'|$$

with respect to η_1 and η_2. We minimize d first with respect to η_1 for fixed η_2. From Lemma A.2,

$$B_1 \hat{\eta}_1 A_1 = B_1 (B_1' S_\eta^{-1} B_1)^{-1} B_1' S_\eta^{-1} (X - B_2 \eta_2 A_2) P_1$$

where

$$P_1 = A_1'(A_1A_1')^{-1}A_1$$
$$S_\eta = (X - B_2\eta_2A_2)(\ \ )' - S_{1\eta}$$
$$S_{1\eta} = (X - B_2\eta_2A_2)P_1(X - B_2\eta_2A_2)'$$

Let B_0 be a matrix such that (B_1, B_0) is nonsingular and $B_0'B_1 = 0$. Then, at $\hat{\eta}_1$

$$\begin{aligned} d &= |S_\eta + [I - B_1(B_1'S_\eta^{-1}B_1)^{-1}B_1'S_\eta^{-1}]S_{1\eta}[\ \ ]'| \\ &= |S_\eta||I + S_{1\eta}[I - B_1(B_1'S_\eta^{-1}B_1)^{-1}B_1'S_\eta^{-1}]'S_\eta^{-1}[\ \ ]| \\ &= |S_\eta||I + S_{1\eta}[S_\eta^{-1} - S_\eta^{-1}B_1(B_1'S_\eta^{-1}B_1)'B_1'S_\eta^{-1}]| \\ &= |S_\eta||I + S_{1\eta}B_0(B_0'S_\eta B_0)^{-1}B_0'| \\ &= |S_\eta||(B_0'S_\eta B_0)^{-1}||B_0'(S_\eta + S_{1\eta})B_0| \end{aligned}$$

since $B_0'B_1 = 0$, $B_0'\tilde{B} = 0$, and $B_0'B_2 = 0$. Hence,

$$B_0'S_{1\eta}B_0 = B_0'XP_1X'B_0$$

and

$$\begin{aligned} B_0'S_\eta B_0 &= B_0'XX'B_0 - B_0'XP_1X'B_0 \\ &= B_0'XQ_1X'B_0, \qquad Q_1 = I - P_1 \end{aligned}$$

Thus, at $\hat{\eta}_1$,

$$\begin{aligned} d &= |S_\eta||B_0'XQ_1X'B_0|^{-1}|B_0'XX'B_0| \\ &= |(X - B_2\eta_2A_2)Q_1(X - B_2\eta_2A_2)'||B_0'XQ_1X'B_0|^{-1}|B_0'XX'B_0| \\ &= |(\tilde{X} - B_2\eta_2\tilde{A}_2)(\tilde{X} - B_2\eta_2\tilde{A}_2)'||B_0'XQ_1X'B_0|^{-1}|B_0'XX'B_0| \end{aligned}$$

where $\tilde{X} = XQ_1$ and $\tilde{A}_2 = A_2Q_1$. Again, using Lemma A.2 to minimize d with respect to η_2,

$$B_2\hat{\eta}_2\tilde{A}_2 = B_2(B_2'\tilde{S}^{-1}B_2)^{-1}B_2'\tilde{S}^{-1}\tilde{X}\tilde{P}_2$$

where

$$\tilde{S} = \tilde{X}\tilde{X}' - \tilde{S}_1, \qquad \tilde{S}_1 = \tilde{X}\tilde{P}_2\tilde{X}', \qquad \tilde{P}_2 = \tilde{A}_2'(\tilde{A}_2\tilde{A}_2')^{-}\tilde{A}_2$$

The minimum value of d at $\hat{\eta}_1$ and $\hat{\eta}_2$ equals

$$|\tilde{T}|\,|B_0'XQ_1X'B_0|^{-1}\,|B_0'XX'B_0|$$

where

$$\tilde{T} = \tilde{S} + [I - B_2(B_2'\tilde{S}^{-1}B_2)^{-1}B_2'\tilde{S}^{-1}]\tilde{S}_1[\ \ ]'$$

Note that

$$\begin{aligned}\tilde{S} &= XQ_1X' - XQ_1A_2'(A_2Q_1A_2)^-A_2Q_1X' \\ &= XQ_1[I - A_2'(A_2Q_1A_2')^-A_2]Q_1X'\end{aligned}$$

and

$$\begin{aligned}S_{\hat{\eta}} &= (X - B_2\hat{\eta}_2A_2)Q_1(X - B_2\hat{\eta}_2A_2)' \\ &= (\tilde{X} - B_2\hat{\eta}_2\tilde{A}_2)(\ )'\end{aligned}$$

Hence, the MLE of η_2 is

$$\hat{\eta}_2 = (B_2'\tilde{S}^{-1}B_2)^{-1}B_2'\tilde{S}^{-1}XQ_1A_2'(A_2Q_1A_2')^- \tag{29}$$

and the MLE of η_1 is

$$\hat{\eta}_1 = (B_1'S_{\hat{\eta}}^{-1}B_1)^{-1}B_1'S_{\hat{\eta}}^{-1}(X - B_2\hat{\eta}_2A_2)A_1'(A_1A_1')^{-1} \tag{30}$$

The MLE of Σ is given by

$$N\hat{\Sigma} = (X - B_1\hat{\eta}_1A_1 - B_2\hat{\eta}_2A_2)(\ )' \tag{31}$$

Thus:

THEOREM 7: *For the nested growth curve model in equation (27), the MLE of η_1, η_2 and Σ are given by equations (28), (29), and (30), respectively.*

6.2 Testing for Nested Model versus GCM

The nested growth curve model defined in equation (28) has fewer parameters than the general growth curve model defined in equation (1) with B_1 instead of B. Since fewer parameters are always preferable, it may be desirable to test the hypothesis that the model is as in equation (28) against the alternative that the model is as in equation (1) where B has been redefined as B_1. The likelihood procedure can be obtained on the lines of Section 5. However, it can also be obtained directly from the method of this section. It can be shown that the likelihood ratio test is based on the statistic

$$\lambda_3 = \frac{|S + (I - T_1S^{-1})S_1(I - S^{-1}T_1)|}{|\tilde{S} + (I - \tilde{T}_2\tilde{S}^{-1})\tilde{S}_1(I - \tilde{S}^{-1}\tilde{T}_2)|} \cdot \frac{|B_0'XQ_1X'B_0|}{|B_0'XX'B_0|}$$

where

$$T_1 = B_1(B_1'\tilde{S}^{-1}B_1)^{-1}B_1'$$

$$\tilde{T}_2 = B_2(B_2'\tilde{S}^{-1}B_2)^{-1}B_2'$$

$$S = X[I - A'(AA')^{-1}A]X'$$

$$S_1 = XA'(AA')^{-1}AX'$$

and $\tilde{S}_1$ and $\tilde{S}$ have been defined earlier. Following Srivastava (1997 b), it can be shown that

$$\frac{|B_0'XQ_1X'B_0|}{|B_0'XX'B_0|} = \frac{|B_1'(XQ_1X')^{-1}B_1|}{|B_1'(XX')^{-1}B_1|} \cdot \frac{|XQ_1X'|}{|XX'|}$$

since (B_1, B_0) is nonsingular and $B_1'B_0 = 0$. The distribution of λ_3 can be shown to be $U_{q_2,m_2,n-p+q}$.

7. GENERALIZED NESTED MODELS

With a slight change of notation, consider the GCM model in which

$$E(X) = B_1\xi A$$

where $B_1 : p \times q$, $\xi : q \times m$, $A : m \times N$, and the matrices A and B_1 are of full rank. Consider the case when

$$\xi = \begin{matrix} & m_1 & m_2 & m_3 \\ q_1 & \xi_1 & \xi_4 & \xi_7 \\ q_2 & \xi_2 & \xi_5 & \xi_8 \\ q_3 & \xi_3 & \xi_6 & \xi_9 \end{matrix}$$

where $\xi_6 = 0$, $\xi_8 = 0$, $\xi_9 = 0$. Let

$$\eta_1 = \begin{pmatrix} \xi_1 \\ \xi_2 \\ \xi_3 \end{pmatrix}, \qquad \eta_2 = \begin{pmatrix} \xi_4 \\ \xi_5 \end{pmatrix}, \qquad \eta_3 = (\xi_7)$$

$$B_1 = \left(\overset{q_1+q_2}{B_2}, \ \overset{q_3}{\tilde{B}} \right) = \left(\overset{q_1}{B_3}, \ \overset{q_2}{B_{22}}, \ \overset{q_3}{\tilde{B}} \right), \qquad A' = \left(\overset{m_1}{A_1'}, \ \overset{m_2}{A_2'}, \ \overset{m_3}{A_3'} \right)$$

where $q_1 + q_2 + q_3 = q$ and $m_1 + m_2 + m_3 = m$. Then,

$$B_1\xi A = B_1\eta_1 A_1 + B_2\eta_2 A_2 + B_3\eta_3 A_3$$

where $B_3 \subset B_2 \subset B_1$. This generalized nested model was considered by von Rosen (1989) (see also Banken, 1984; Andersson et al., 1993). We obtain the MLE of the parameters by following the approach of Srivastava and Khatri (1979). For testing and estimation problems, see Srivastava (1997 b). The MLE can also be found in von Rosen (1989). The first moments of the MLEs have been given in von Rosen (1990).

8. DATA ANALYSIS

In this final section, the results of the previous sections will be illustrated through a numerical example. In Wei and Lachin (1984) data material are presented where serum cholesterol on each patient has been measured five times; before the start (baseline) and at 6, 12, 20, and 24 months after the study start. The patients constitute two groups, of which one received an active drug and one a placebo. In our analysis, we consider only the cases on which complete information is available, that is, those cases which had missing values have been omitted. Mean values and some summary statistics are presented in Table 1, while the complete data set is given in Table 2.

A new data set was created by individually subtracting baseline values from the original observations. This data set is the one which is analysed below. The assumption of multivariate normality seems to hold: this was checked by plotting principal components on probability plot papers (see Srivastava, 1984).

We suppose that there exists a within-individuals mean structure and that the structure differs between the placebo and treatment groups. Moreover, the sample covariance matrix for the placebo group is

$$\frac{1}{N_{pl}-1}V_{pl} = \begin{pmatrix} 1150 & 840 & 725 & 774 \\ 840 & 1031 & 702 & 664 \\ 725 & 702 & 938 & 602 \\ 774 & 664 & 602 & 1346 \end{pmatrix}$$

and for the treatment group

$$\frac{1}{N_{tr}-1}V_{tr} = \begin{pmatrix} 1073 & 619 & 731 & 850 \\ 619 & 1215 & 556 & 1068 \\ 731 & 556 & 1206 & 867 \\ 850 & 1068 & 867 & 2490 \end{pmatrix}$$

These estimators are unbiased. The likelihood ratio test for the equality of two covariances were carried out (see Srivastava and Carter, 1983, p. 333). The p-value was found to be 0.06, which is somewhat low. However, we will still assume that the two covariances are equal. The differences arise mainly from the measurements at month 24.

Let the data be collected in $X : 4 \times 67$ where the first 31 columns comprise the data from the placebo group. Moreover, the between-individuals design matrix, $A : 2 \times 67$, is given by

$$A = \begin{pmatrix} \mathbf{1}'_{31} & 0 \\ 0 & \mathbf{1}'_{36} \end{pmatrix}$$

Table 1. Various descriptive statistics for the cholesterol data*

	Placebo group					Treatment group				
	Base	6	12	20	24	Base	6	12	20	24
Mean	237	243	245	262	257	227	250	253	253	257
Std. dev.	55.5	52.5	47.6	51.6	49.4	44.3	41.4	39.4	33.9	50.5
Std. err.	10.0	9.43	8.55	9.28	8.87	7.38	6.90	6.57	5.65	8.41
Skewness	1.03	0.222	0.151	0.451	0.629	0.230	0.0830	−0.276	0.558	0.575
Kurtosis	2.42	0.0631	0.0909	0.541	0.448	−0.529	−0.626	−0.939	0.0102	0.245

*Measurements taken at the start of the study and 6, 12, 20, and 24 months later.

Table 2. Cholesterol data from Wei and Lachin (1984)*

Placebo group					Treatment group				
Base	6	12	20	24	Base	6	12	20	24
251	262	239	234	248	178	246	295	228	274
233	218	230	251	273	254	260	278	245	340
250	258	258	286	240	185	232	215	220	292
141	143	157	162	169	219	268	241	260	320
418	371	363	384	387	205	232	265	242	230
229	218	228	244	179	182	213	173	200	193
271	289	270	296	346	310	334	290	286	248
312	323	318	383	310	191	204	227	228	196
194	220	214	256	204	245	270	209	255	213
211	232	189	230	231	229	200	238	259	221
205	299	278	259	266	245	293	261	297	231
191	248	283	268	233	240	313	251	307	291
249	217	236	266	235	234	281	277	235	210
301	270	282	287	268	210	252	275	235	237
201	214	247	274	224	275	231	285	238	251
277	242	249	293	306	269	332	300	320	335
294	313	295	295	271	148	180	184	231	184
212	236	235	272	287	181	194	212	217	205
230	315	300	305	341	165	242	250	249	312
246	205	249	225	236	293	276	276	278	306
245	192	215	214	242	195	190	205	217	238
179	202	194	239	234	210	230	249	240	194
165	142	188	192	200	212	224	246	271	256
262	274	245	275	278	243	271	304	273	318
212	216	228	221	223	259	279	296	262	283
285	292	300	319	277	202	214	192	239	172
166	171	166	186	220	184	192	205	253	217
179	206	214	189	250	238	272	297	282	251
298	280	280	328	318	263	283	248	334	271
238	267	269	268	280	144	226	261	227	283
191	208	162	218	206	220	272	222	246	253
					225	260	253	202	265
					307	252	316	258	283
					313	300	313	317	397
					206	177	194	194	212
					285	291	291	268	260

*See footnote to table 1.

where $\mathbf{1}_{31}$ is a vector of 31 ones and $\mathbf{1}_{36}$ is a vector of 36 ones.The mean structure is modelled with the help of the within-individuals design matrix, $B : 4 \times 3$, given by

$$B = \begin{pmatrix} 1 & 1 & 1 \\ 1 & 2 & 2^2 \\ 1 & \frac{10}{3} & \frac{10^2}{3} \\ 1 & 4 & 4^2 \end{pmatrix}$$

with the unit of 6 months. The following model is applied:

$$X = B\xi A + E$$

According to Section 2, the MLEs are

$$\hat{\xi} = \begin{pmatrix} 0.82 & 21.1 \\ 4.80 & 1.8 \\ 0.43 & 0.004 \end{pmatrix}$$

$$\hat{\Sigma} = \begin{pmatrix} 1076 & 698 & 708 & 789 \\ 698 & 1105 & 597 & 868 \\ 708 & 597 & 1056 & 712 \\ 789 & 868 & 712 & 1922 \end{pmatrix}$$

In $\hat{\xi}$, the first column gives the estimators of the parameters for the placebo group, whereas in the second column the estimators for the treatment group are given. There is a remarkable difference between the estimators of the mean parameters in respective columns but note that these estimators are not independent (see $\widehat{\text{cov}}(\hat{\xi})$, given below). Later some tests are performed where the differences are exploited.

In Theorem 3 an unbiased estimator of Σ was presented:

$$\hat{\Sigma}_u = \begin{pmatrix} 1109 & 721 & 729 & 815 \\ 721 & 1133 & 621 & 885 \\ 729 & 621 & 1084 & 741 \\ 815 & 885 & 741 & 1968 \end{pmatrix}$$

In particular, on the diagonal the elements of $\hat{\Sigma}_u$ are somewhat larger than those of $\hat{\Sigma}$, which is as expected. Moreover, Theorem 3 also gives an unbiased estimator of the dispersion of $\hat{\xi}$:

$$\hat{\text{cov}}(\hat{\xi}) = \begin{pmatrix} 148 & -113 & 22 & | & 0 & 0 & 0 \\ -113 & 110 & -23 & | & 0 & 0 & 0 \\ 22 & -23 & 5 & | & 0 & 0 & 0 \\ -- & -- & -- & | & -- & -- & -- \\ 0 & 0 & 0 & | & 127 & -97 & 19 \\ 0 & 0 & 0 & | & -97 & 95 & -20 \\ 0 & 0 & 0 & | & 19 & -20 & 4 \end{pmatrix}$$

Observe that the estimates from the treatment group and the estimates from the placebo group are uncorrelated. However, since $\hat{\xi}$ is not normally distributed, this would not imply that these estimators are independent. To strictly show non-independence one can use expressions for higher moments of $\hat{\xi}$ obtained in von Rosen (1991b).

It follows that the model via the matrix B seems reasonable to apply. However, we may strictly test for the adequacy of the model when following the results of Section 3. Thus, if the mean structure $\beta = B\xi$ is tested against an arbitrary

$$\beta = \begin{pmatrix} \beta_{11} & \beta_{12} \\ \beta_{21} & \beta_{22} \\ \beta_{31} & \beta_{32} \\ \beta_{41} & \beta_{42} \end{pmatrix}$$

it follows from Theorem 4 that the test statistic

$$\lambda_1 = 0.96$$

is well above the given approximate critical level, since $P(-65 ln\lambda_1 > 2.8) \approx 0.24$.

Next, we carry out a comparison of the placebo group with the treatment group. Thus, the hypothesis

$$H_2: \ \xi D = 0$$

where

$$D = \begin{pmatrix} 1 \\ -1 \end{pmatrix}$$

is set up versus the alternative $A_2: \ \xi D \neq 0$. According to Theorem 6, where $C = I$,

$$\lambda_2 = 0.91$$

and $P(\frac{62}{3}\frac{1-\lambda_2}{\lambda_2} > 2.1) \approx 0.10$. Thus, H_2 is not rejected and we may not conclude that the two groups differ.

Secondly, it is tested if the placebo and treatment groups each have constant cholesterol values over time. Hence, for the placebo group the following hypothesis is investigated:

$$H_3^{(1)}: \ C\xi D = 0, \qquad A_3^{(1)}: \ C\xi D \neq 0$$

where

$$C = \begin{pmatrix} 0 & 1 & 0 \\ 0 & 0 & 1 \end{pmatrix}, \qquad D = \begin{pmatrix} 1 \\ 0 \end{pmatrix}$$

Once again applying Theorem 6 gives

$$\lambda_3^{(1)} = 0.81$$

and $P(\frac{62}{3}\frac{1-\lambda_3^{(1)}}{\lambda_3^{(1)}} > 4.8) \approx 0.004$. Correspondingly, for the treatment group

$$H_3^{(2)}: \ C\xi D = 0 \ , \ A_3^{(2)}: \ C\xi D \neq 0$$

where

$$C = \begin{pmatrix} 0 & 1 & 0 \\ 0 & 0 & 1 \end{pmatrix}, \qquad D = \begin{pmatrix} 0 \\ 1 \end{pmatrix}$$

Now Theorem 6 gives

$$\lambda_3^{(2)} = 0.98$$

which is clearly nonsignificant.

Hence, a reasonable conclusion from the two last tests is that there exist some differences between the two groups, i.e., the treatment group is constant over time whereas the placebo group has increased cholesterol values. This is in correspondence with $\hat{\xi}$. The difference is mainly due to the fact that the placebo group has low values at month 6.

ACKNOWLEDGMENTS

We are grateful to the four referees for their very constructive suggestions. The research was supported by the Natural Science and Engineering Research Council of Canada and the Swedish Natural Research Council. Thanks are also due to David Emig and Boon Chew for their assistance in computing and typing of the manuscript.

REFERENCES

Andersson, S.A., Marden, J.I., and Perlman, M.D. (1993). Totally ordered multivariate linear models. *Sankhyā, Ser. A.*, **55**, 370–394.

Banken, L. (1984). Eine Verallgemeinerung des Gmanova Modells. Dissertation, University of Trier, Trier (in German).

Carter, E.M. and Hubert, J.J. (1984). A growth-curve model approach to multivariate quantal bioassay. *Biometrics*, **40,** 699–706.

Chinchilli, V.M. and Elswick, R.K. (1985). A mixture of the Manova and Gmanova models. *Commun. Stat.-Theor. Math.*, **14**, 3075–3089.

Gleser, L.J. and Olkin, I. (1970). Linear models in multivariate analysis. In: R.C. Bose, I.M. Chakravarti, P.C. Mahalanobis, C. Radhakrishna Rao, and K.J.C. Smith, eds. *Essays in Probability and Statistics*. pp. 267–292, Chapel Hill: University of North Carolina Press.

Grizzle, J.E. and Allen, D.M (1969). Analysis of growth and dose response curves. *Biometrics*, **25**, 357–381.

Hooper, P.M. (1983). Simultaneous interval estimation in the general multivariate analysis of variance model. *Ann. Stat.*, **11**, 666–673. Correction, *Ann. Stat.*, **12**, 785.

Kariya, T. (1978). The general Manova problem. *Ann. Stat.*, **6**, 200–214.

Kariya, T. (1985). *Testing in the Multivariate General Linear Model.* Tokyo: Kinokuniya.

Khatri, C.G. (1966). A note on a Manova model applied to problems in growth curve. *Ann. Inst. Stat. Math.*, **18**, 75–86.

Khatri, C.G. (1973). Testing some covariance structures under a growth curve model. *J. Multivariate Anal.*, **3**, 102–116.

Khatri, C.G. (1988). Robustness study for a linear growth model. *J. Multivariate Anal.*, **24**, 66–87.

Khatri, C.G. and Srivastava, M.S. (1975). On the likelihood ratio test for covariance matrix in growth curve model. In: R.P. Gupta, ed. *Applied Statistics*. pp. 187–198. New York: North-Holland.

Khatri, C.G. and Srivastava, M.S. (1976). Asymptotic expansions of the non-null distributions of the likelihood ratio critera for covariance matrices. II. *Metron*, **34**, 55–71.

Kleinbaum, D.G. (1973). A generalization of the growth curve model which allows missing data. *J. Multivariate Anal.*, **3**, 117–124.

Kshirsagar, A.M. and Smith, W.B. (1995). *Growth Curves*. New York: Marcel Dekker.

Lee, J.C. and Geisser, S. (1972). Growth curve prediction. *Sankyā Ser. A*, **34**, 393–412.

Lee, J.C. and Geisser, S (1975). Applications of growth curve prediction. *Sankyā Ser. A*, **37**, 239–256.

Liski, E.P. (1985). Estimation from incomplete data in growth curve models. *Commun. Stat. Comput.*, **14**, 13–27.

Marden, J.I. (1983). Admissibility of invariant tests in the general multivariate analysis of variance problem. *Ann. Stat.*, **11**, 1086–1099.

Potthoff, R.F. and Roy, S.N. (1964). A generalized multivariate analysis of variance model useful especially for growth curve problems. *Biometrika*, **51**, 313–326.

Rao, C. R. (1948). Tests of significance in multivariate analysis. *Biometrika*, **35**, 58–79.

Rao, C. R. (1959). Some problems involving linear hypotheses in multivariate analysis. *Biometrika*, **46**, 49–58.

Rao, C. R. (1961). Some observations on multivariate statistical methods in anthropological research. *Bull. Inst. Int. Stat.*, **38**, 99–109.

Rao, C. R. (1965). The theory of least squares when the parameters are stochastic and its application to the analysis of growth curves. *Biometrika*, **52**, 447–458.

Rao, C. R. (1966). Covariance adjustment and related problems in multivariate analysis. In: P.R. Krishnaiah, ed. *Multivariate Analysis*. pp. 87–103. NewYork: Academic Press.

Rao, C. R. (1967). Least squares theory using an estimated dispersion matrix and its application to measurement of signals. In: L.M. LeCam and J. Neuman, eds. *Proceedings of the Fifth Berkeley Symposium on Mathematical and Statistical Problems*, **1**, pp. 355–372. Berkeley and Los Angeles: University of California Press.

Reinsel, G.C. (1982). Multivariate repeated-measurement or growth curve models with multivariate random-effects covariance structure. *J. Am. Stat. Assoc.*, **77**, 190–195.

Reinsel, G.C. (1984). Estimation and prediction in a multivariate random effects generalized linear model. *J. Am. Stat. Assoc.*, **79**, 406–414.

Srivastava, J.N. and McDonald, L.L (1974). Analysis of growth curves under the hierarchical models. *Sankyā Ser. A*, **36**, 251–260.

Srivastava, M.S. (1984). A measure of skewness and kurtosis and a graphical method for assessing multivariate normality. *Statist. Probab. Lett.*, **2**, 263–267.

Srivastava, M.S. (1985). Multivariate data with missing observations. *Commun. Stat. Theory Methods*, **14**, 775–792.

Srivastava, M.S. (1997a). Reduced rank discrimination. *Scand. J. Stat.*, **24**, 115–124.

Srivastava, M.S. (1997b). Generalized multivariate analysis of variance models. Technical Report, University of Toronto.

Srivastava, M.S. and Carter, E.M. (1977). Asymptotic non-null distribution of a likelihood ratio criterion for sphericity in the growth curve model. *Sankyā Ser. B*, **39**, 160–165.

Srivastava, M.S. and Carter, E.M. (1983). An Introduction to Applied Multivariate Statistics. New York: North-Holland.

Srivastava, M.S. and Khatri, C.G. (1979). *An Introduction to Multivariate Statistics*. New York: North-Holland.

Tsai, K.T. and Koziol, J.A. (1988). Score and Wald tests for the multivariate growth curve model with missing data. *Ann. Inst. Stat. Math.*, **40**, 179–186.

Verbyla, A.P. and Venables, W.N. (1988). An extension of the growth curve model. *Biometrika*, **75**, 129–138.

von Rosen, D. (1989). Maximum likelihood estimators in multivariate linear normal models. *J. Multivariate Anal.*, **31**, 187–200.

von Rosen, D. (1990). Moments for a multivariate linear model with an application to the growth curve model. *J. Multivariate Anal.*, **35**, 243–259.

von Rosen, D. (1991 a). The growth curve model: A review. *Commun. Stat. Theory Methods*, **20**, 2791–2822.

von Rosen, D. (1991 b). Moments of maximum likelihood estimators in the growth curve model. *Statistics*, **22**, 111–131.

Ware, J.H. and Bowden, R.E. (1977). Circadian rhythm analysis when output is collected at intervals. *Biometrics*, **33**, 566–571.

Wei, L.J. and Lachin, J.M. (1984). Two-sample asymtotically distribution-free tests for incomplete multivariate observations. *J. Am. Stat. Assoc.*, **79**, 653–661.

Zerbe, G.O. and Jones, R.H. (1980). On application of growth curve techniques to time series data. *J. Am. Stat. Assoc.*, **75**, 507–508.

APPENDIX 1

These three Lemmas are used in obtaining the maximum likelihood tests and the estimators.

LEMMA A.1: Let (B, B_0) be a $p \times p$ nonsingular matrix such that $B'B_0 = 0$. Then for any $p \times p$ symmetric positive definite matrix S,

$$S^{-1} - S^{-1}B(B'S^{-1}B)^{-1}B'S^{-1} = B_0(B_0'SB_0)^{-1}B_0'$$

For a proof, see Srivastava and Khatri, 1979, p. 19.

LEMMA A.2: Let $X : p \times N$, $B : p \times q$, and $A : m \times N$ be matrices such that $(X - B\xi A)(X - B\xi A)'$ is positive definite for every $q \times m$ matrix ξ. Then

$$|(X - B\xi A)(X - B\xi A)'| \geq |T| \qquad \text{for all} \qquad \xi$$

where

$$T = S + [I - P]S_1[I - P]'$$
$$P = B(B'S^{-1}B)^- B'S^{-1}$$
$$S_1 = XHX', \qquad H = A'(AA')^- A$$
$$S = X[I - H]X' = XX' - S_1$$

The equality holds if and only if

$$B\xi A = B(B'S^{-1}B)^- B'S^{-1}XH$$

This is Lemma 1.10.3 of Srivaslava and Khatri (1979, p. 24).

Proof. Let $rank(B : B_1) = p$ and $B_1'B = 0$. Then,

$$
\begin{aligned}
|(X - B\xi A)(\ )'| = |S + (XH - B\xi A)(\ )'| &= |S||I + (XH - B\xi A)'S^{-1}(\ )'| \\
&= |S||I + (XH - B\xi A)'S^{-1}B(B'S^{-1}B)^{-}B'S^{-1}(\ )' \\
&\qquad + (XH - B\xi A)'B_1(B_1'SB_1)^{-}B_1'(\ )'| \\
&\geq |S||I + HX'B_1(B_1'SB_1)^{-}B_1'XH|
\end{aligned}
$$

which is independent of ξ and equality holds if and only if

$$B(B'S^{-1}B)^{-}B'(XH - B\xi A) = 0$$

LEMMA A.3: Let

$$X = \begin{pmatrix} X_1 \\ X_2 \end{pmatrix} \qquad \text{and} \qquad \Sigma = \begin{pmatrix} \Sigma_{11} & \Sigma_{12} \\ \Sigma_{21} & \Sigma_{22} \end{pmatrix}$$

Then,

$$X'\Sigma^{-1}X = (X_1 - \Sigma_{12}\Sigma_{22}^{-1}X_2)'(\Sigma_{11} - \Sigma_{12}\Sigma_{22}^{-1}\Sigma_{21})^{-1}(\ \ )' + X_2'\Sigma_{22}^{-1}X_2$$

20

Dealing with Uncertainties in Queues and Networks of Queues: A Bayesian Approach

C. ARMERO and M. J. BAYARRI University of Valencia, Valencia, Spain

1. INTRODUCTION

Queues are, unfortunately for us, much too familiar to need detailed definitions. In a queueing system, customers arrive at some facility requiring some type of service; if the server is busy, the customer waits in line to be served. The familiar notion of a queue thus parallels its mathematical meaning. "Customer" and "server" do not necessarily mean people, but refer to very general entities.

Since the economical and social impact of congestion can be considerable, the importance of analysing queueing systems has been clear ever since the pioneering work of Erlang on telecommunications at the beginning of the century. Queues possess for mathematicians an added bonus, namely that the mathematics involved can be hard and intriguing, and they have an enormous potential for varying the conditions and assumptions of the system. It does not come as a surprise that queues developed quickly into the huge area that it is today. Many papers, journals, meetings, and books are entirely devoted to the subject. We cannot possibly list here even all of the key references, and content ourselves with giving a small selection for the interested reader. Among the numerous books, it is worth mentioning the general books of Cox and Smith (1961), Takács (1962), Cooper (1981), Gross and Harris (1985), Medhi (1991), and Nelson (1995); Kleinrock (1975, 1976) and Newell (1982) are good sources of applications; Cooper (1990) is a nice survey with an extensive bibliography. Nevertheless, there is a plethora of survey books and papers on queueing, and Prabhu (1987) is a useful up-to-date survey.

Most of this vast effort, however, has been devoted to the probabilistic development of queueing models and to the study of its mathematical properties. That is, the parameters governing the models are, for the most part, assumed given. Statistical analyses, in which uncertainty is introduced, are comparatively very scarce. Good reviews are by Bhat and Rao (1987), Lehoczky (1990), and Bhat et al. (1997). While it could be argued that the general results of the (very developed) area of inference for stochastic processes could be applied to queues, these general results might not take advantage of the special structure of queueing systems nor address its specific questions (Bhat and Rao, 1987).

Inference in queueing systems is not easy. Development of the necessary sampling distributions can be very involved and often the analysis is restricted to asymptotic results. In this paper we try to argue that analyses can be much easier when approached from a Bayesian perspective, for which queueing systems seem to be nicely suited. We do not try to be dogmatic, nor are we trying to imply that frequentist analyses of queueing systems are worthless; we shall only try to show that Bayesian analyses can work nicely and easily, providing, perhaps, answers not easily obtained with other methodologies. From now on we adopt a Bayesian approach throughout.

Although still very rare, Bayesian analyses of queues are becoming more popular. To the best of our knowledge an up-to-date, comprehensive listing of Bayesian papers on queues is: Bagchi and Cunningham (1972), Muddapur (1972), Reynolds (1973), Morse (1979), Armero (1985, 1994), McGrath and Singpurwalla (1987), McGrath et al. (1987), Lehoczky (1990), Thiruvaiyaru and Basawa (1992), Armero and Bayarri (1994a,b, 1996, 1997a,b), Armero and Conesa (1997, 1998), Butler and Huzurbazar (1996), Rios Insua et al. (1996), Ruggeri et al. (1996), Sohn (1996), and Wiper (1996).

This paper (mostly based on previous work of the authors) will highlight the advantages of Bayesian analyses of queues, pointing out also some of its shortcomings. It is basically a review of previous results, and thus mathematical details are kept to a minimum. Throughtout this paper, results are demonstrated in simple examples using noninformative priors. The use of such priors allows for easy comparison with non-Bayesian results. Also, Bayesian answers with noninformative priors might be more appealing to non-die-hard Bayesians, since the controversial inclusion of influential prior information is avoided. (The interested reader can find details about informative analyses in the references mentioned in the paper.)

The paper contains seven sections, this Introduction being Section 1. In Section 2 we present the basic queueing models that we shall use in the rest of the paper. Sections 3 and 4 are devoted to a somewhat peculiar review of previous results developed through the study of some advantages (Section 3) and some difficulties (Section 4) of Bayesian analyses of queues. Section 5 introduces the very important area of queueing networks. Finally, in Section 6, we give a very succinct account of (Bayesian) inferences for queueing networks.

2. QUEUES

All queueing systems used in this paper can be described with the shorthand notation introduced by Kendall (1953), which consists of a series of symbols separated by slashes, e.g., $A/B/c$. "A" refers to the distribution of the interarrival times between customers; thus Er stands for Erlang, G for general, M for exponential (Markovian), etc. If customers arrive in batches of, say, X at a time, a superindex X is added. "B" characterises the service distribution for each server, with the same symbols as before, and "c" refers to the number of servers, who are usually assumed to be equally efficient. More symbols are added when needed, but we shall assume throughout the paper that there is no limitation either in the capacity of the "queueing room" or in the population that feeds the system.

Our basic queue will be the $M/M/c$ queue, in which it is assumed that customers arrive at the system according to a Poisson process of mean λ, so that the time between two consecutive arrivals to the queue, Y, has an exponential distribution with parameter λ (mean $1/\lambda$). The time required to serve a customer, X, is supposed to be independent of the arrivals and of the history of the queue, and it is given an exponential distribution with parameter μ. There are c identical servers, and the parameters are supposed to stay stable through time.

The very important $M/M/c$ queues are, perhaps, the most studied queueing systems, and include the fundamental $M/M/1$ queue. A limiting case is the $M/M/\infty$ queue, that, even though strictly speaking it is not a queue (there is no interaction between customers, no lines, no congestion), is nevertheless frequently used to model self-service systems and to study the approximate behaviour of $M/M/c$ queues as c grows (further arguments are given in Armero and Bayarri, 1997a).

$M/M/c$ queues are point processes whose probabilistic behavior is usually characterized in terms of its associated Markov (birth and death) process $\{N(t), t \geq 0\}$, where $N(t)$ denotes the number of customers in the system at time t. Since both the waiting room capacity and the size of the population are assumed unlimited, $N(t)$ can become arbitrarily large when t grows, and our common sense tells us that this will happen whenever arrivals come at a faster rate than the servers can handle (note the crucial assumptions here that μ and λ remain constant over time and that μ does not depend on the size of the queue). In mathematical terms, it can be shown that the queue (with probability one) either grows without limits when $\lambda \geq c\mu$ or reaches the steady state, which can be described by the equilibrium or steady-state distribution which is independent of t and of the initial state of the process. For most of this paper we shall assume that the queue is in equilibrium, and hence that the so-called *traffic intensity* $\rho = \lambda/(\mu c)$ is smaller than 1. This crucial condition is known as the *ergodic condition*, and it is of fundamental importance in queues, since it is required for the existence of steady-state distributions. Note that in queues this is a most natural assumption (for a discussion of stochastic models in equilibrium see Whittle, 1986). Indeed, queueing systems usually run for long periods of time and, in order for that to be possible without the system getting out of hand, either equilibrium is eventually reached or μ and λ have to vary with time (unless the system can be kept under control by the limited size of the waiting room and/or the population).

For further reference, we display the assumptions so far. We shall let Y denote an interarrival time, and assume $Y \sim \text{Ex}(\lambda)$. That is,

$$f(y \mid \lambda) = \lambda e^{-\lambda y}, \qquad y > 0 \tag{1}$$

X denotes service time, and it is assumed that $X \sim \text{Ex}(\mu)$, independently of Y, that is,

$$f(x \mid \mu) = \mu e^{-\mu x}, \qquad x > 0 \tag{2}$$

Also, whenever required, the queue will be in equilibrium and hence

$$\rho = \frac{\lambda}{\mu c} < 1 \tag{3}$$

3. WHY IS BAYESIAN ANALYSIS SO GOOD FOR QUEUES?

All our previous work on the statistical analysis of queues has been inside the Bayesian framework, and we have not attempted a frequentist analysis. We realised, however, that we could offer simple answers to problems that would have been very difficult to handle satisfactorily from a frequentist perspective, and that there were a number of reasons that made queueing systems ideally suited for Bayesian analyses. We will highlight some of these reasons.

3.1 Likelihood Principle

Bayesian inference (as well as maximum likelihood, ML, estimates) obeys the likelihood principle, and thus all that it is required from the data generating process is the likelihood of the observed data. This irrelevance of the sample space and other aspects of the sampling distribution can result in a substantial simplification not only of the analysis, but also of the experiment to be carried out.

Derivation of exact joint sampling distributions often requires observation of the system on $(0, T]$, where T can be fixed in advance or determined by some stopping rule. Many different quantities providing full information about the system can then be recorded. A review of different sampling schemes can be found in Armero and Bayarri (1996, 1997a). Some of these experiments can be very difficult, if not impossible, to implement; for instance, some of them might in practical terms require an experimenter who continuously follows each of the customers as they pass through the system. For queues with large λ, this might be prohibitively expensive to perform.

If all that is needed is a likelihood function, much easier and cheaper experiments can be performed, sometimes providing likelihood functions that are proportional to those provided by more involved experiments. The experiments we considered are very simple and consist of observing n_a interarrival times, $Y_1, \ldots, Y_{n_a}$, and, simultaneously or not, n_s service completions, $X_1, \ldots, X_{n_s}$. Note that, because of the likelihood principle, n_a and n_s can be taken to be fixed. This type of experiment has also been considered for maximum likelihood estimation (Thiruvaiyaru and Basawa, 1992; Sohn, 1996). From equations (1) and (2) it can be seen that the likelihood function is simply given by

$$l(\lambda, \mu) \propto \lambda^{n_a} e^{-\lambda t_a} \mu^{n_s} e^{-\mu t_s} \tag{4}$$

where $t_a = \sum_{i=1}^{n_a} y_i$, $t_s = \sum_{j=1}^{n_s} x_j$, are the totals of observed interarrival times and service times, respectively. We have also considered other possibilities in the context of $M/M/\infty$ queues, such as that of incorporating the observation of the initial number of customers in the queue; actually, when this quantity is observed, likelihood functions for (λ, μ) can be derived even if only the arrival process or only the service process are available for observation (Armero and Bayarri, 1997a).

3.2 Probabilities of Interest

From the Bayesian perspective, parameters are considered random variables with prior distribution $\pi(\lambda, \mu)$ which gets updated once data has been observed, producing the posterior distribution $\pi(\lambda, \mu \mid \mathbf{x}, \mathbf{y}) \propto l(\lambda, \mu)\pi(\lambda, \mu)$. From the posterior distribution all the probabilities of direct interest concerning λ, μ and the traffic intensity $\rho = \lambda/(c\mu)$ can be computed. One probability of particular importance is the probability that the ergodic condition holds,

$$\pi_e = Pr\left(\frac{\lambda}{\mu} < c \mid \mathbf{x}, \mathbf{y}\right) = \int_0^{\infty} \int_0^{c\mu} \pi(\lambda, \mu \mid \mathbf{x}, \mathbf{y})\, \mathrm{d}x\, \mathrm{d}y \tag{5}$$

and it can easily be computed, as shown above. In particular, with a Jeffrey's noninformative prior distribution, the posterior distribution of (λ, μ) is the product of two Gamma distributions $Ga(\lambda \mid n_a, t_a)\, Ga(\mu \mid n_s, t_s)$, and considering for simplicity the case in which $n_a = n_s = m$, it can be seen that the probability in equation (5) that the ergodic condition holds is given by (Armero and Bayarri, 1996)

$$\pi_e = \pi_e(\hat{\rho}, m) = \frac{1}{\hat{\rho}^m} \frac{F(2m, m; m+1; -1/\hat{\rho})}{mBe(m, m)} \tag{6}$$

where $\hat{\rho} = \hat{\lambda}/(c\hat{\mu})$ is the MLE of ρ, $Be(a, b)$ is the Beta function, and $F(a, b; c; z)$ the hypergeometric function (see Abramowitz and Stegun, 1964).

In Figure 1 we show equation (6) as a function of m for $\hat{\rho} = 0.3, 0.5, 0.8, 1.0, 1.2, 2.0$. The behaviour of π_e matches intuition. Indeed, the larger m, the more we trust $\hat{\rho}$ as an estimate of ρ, so that π_e should grow to 1 as m grows for values of $\hat{\rho} < 1$ and should decrease to 0 for $\hat{\rho} > 1$. Also, as would be expected, the increase (decrease) is faster for more extreme $\hat{\rho}$. Interestingly, $\pi_e = 0.5$ for all values of m when $\hat{\rho} = 1$.

Note that, while $\hat{\rho}$ and other estimates are easily available, there does not seem to be an easy parallel of equation (5) from the frequentist point of view. Note also that no subjective, prior information was used to derive equation (5), and that figures such as Figure 1 could be a very useful tool in the design of queues.

3.3 Restrictions on the Parameter Space

As we have said before, assuming equilibrium is very frequent in queueing theory, and derivation of the steady-state distributions remains the central focus in most queueing publications. However, the steady state requires $\rho < 1$, and this restriction is not easily incorporated into the frequentist

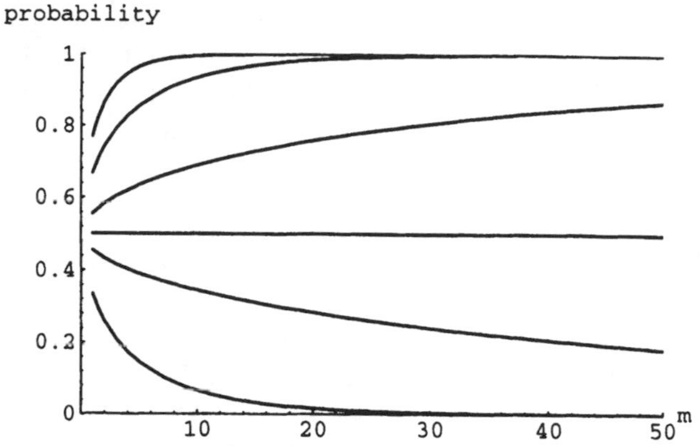

Figure 1. Probability π_e that the ergodic condition holds for values of $\hat{\rho}$, from top to bottom, equal to 0.3, 0.5, 0.8, 1, 1.2, and 2

statistical analysis; it is, however, trivial to incorporate into a Bayesian analysis. The usual approach consists of restricting the prior to be positive only for pairs (λ, μ) such that $\rho < 1$. This can be seen to be equivalent to performing the usual, unrestricted analysis and then restricting the posterior to be positive if $\rho < 1$. In other words;

$$\pi(\lambda, \mu \mid \mathbf{x}, \mathbf{y}, \rho < 1) = \frac{\pi(\lambda, \mu \mid \mathbf{x}, \mathbf{y})}{\pi_e} \qquad \text{for} \qquad \lambda < c\mu, \tag{7}$$

and 0 otherwise, where π_e is given in equation (5), and $\pi(\lambda, \mu \mid \mathbf{x}, \mathbf{y})$ is the posterior that would result from an unrestricted analysis.

Thus, for instance, for the simple situation considered in equation (5), if the steady state is *not* assumed we have (Armero and Bayarri, 1996)

$$\frac{\rho}{\hat{\rho}} \sim F(2m, 2m) \tag{8}$$

where F is the usual F distribution. Hence, without restrictions, the usual ML and the noninformative Bayesian analyses parallel each other (for ML analyses see Clarke, 1957, Cox 1965, Wolf, 1965, Basawa and Rao, 1980, Basawa and Prahbu, 1988, etc.). In fact, the usual Bayesian estimator $E(\rho \mid data) = \frac{m}{m-1}\hat{\rho}$ will be very close to $\hat{\rho}$ for moderate m, and interval estimates will numerically match.

On the other hand, however, suppose that we work under the assumption of stationarity, so that the restriction $\rho < 1$ applies. How should ρ be estimated? $\hat{\rho} = \hat{\lambda}/(c\hat{\mu})$ can easily result in values of $\hat{\rho}$ greater than 1, specially if m is not very large. The usual trick of taking $min\{1, \hat{\rho}\}$ is not very satisfactory, and can be very difficult to work with; in particular, confidence intervals might not be easy to derive. In contrast, introducing the restriction in equation (8) is very easy, and all of the analyses can be carried out just as well with $\pi(\rho \mid data, \rho < 1)$ instead. In particular, a Bayesian estimate is (Armero and Bayarri, 1996)

$$E(\rho \mid data, \rho < 1) = \frac{m}{m+1} \frac{F(2m, m+1; m+2; -1/\hat{\rho})}{F(2m, m; m+1; -1/\hat{\rho})} = \tilde{\rho} \tag{9}$$

In Figure 2 we show the factor by which we should multiply $\hat{\rho}$ to get the adjusted estimate $\tilde{\rho}$ as a function of $\hat{\rho}$ for $m = 5$, $m = 10$ and $m = 20$. It can be seen that the effect of introducing the restriction $\rho < 1$ in the estimate is to shrink large values of $\hat{\rho}$, the shrinkage being larger for smaller values of m. For large m ($m = 50$), $\hat{\rho}$ is taken basically at face value when $\hat{\rho}$ is moderately small ($\hat{\rho} < 0.7$). For small values of m, however, the restriction produces a larger estimate $\tilde{\rho}$ than the unrestricted $\hat{\rho}$, the correction factor being larger for smaller values of m.

Here again, equation (9) and Figure 2 are very useful tools, not requiring subjective inputs (if such are not desired; of course, a full Bayesian approach can be carried in an entirely similar manner) and not having easy frequentist counterparts. Also, from the truncated posterior distribution of ρ derived from equation (8), probabilities of interest and credible intervals can easily be derived.

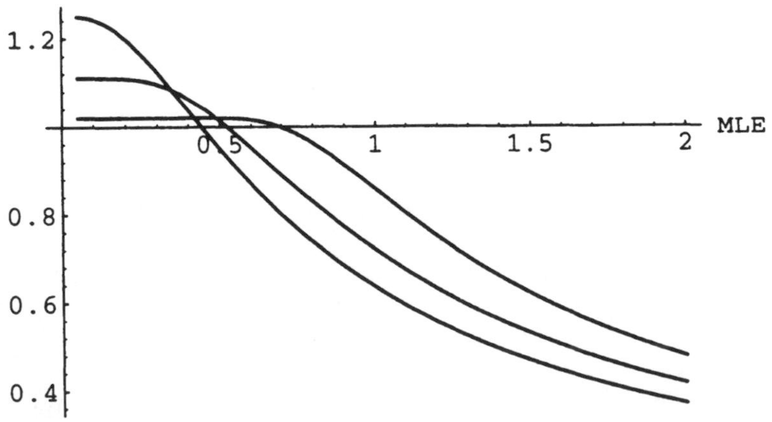

Figure 2. Correction factor for $\hat{\rho}$ to account for the restriction $\rho < 1$ when estimating ρ, as a function of $\hat{\rho}$, for values of $m = 5, 10, 50$

3.4 Accuracy of Estimators

A usual argument in favour of Bayesian analyses is the ease in deriving measures of performance of estimators, the posterior variance (or its square root) being one such natural measure (see, for instance, Berger, 1985). When dealing with queues, the argument reveals all of its potential, since while likelihood functions (and hence posterior distributions and posterior variances) are usually very easy to derive, this is not so for sampling distributions of estimators, and standard errors frequently have to be given in terms of asymptotic variances.

The same is true when a *standard error* has to be attached to a prediction, as we shall shortly see, or when restrictions in the parameter space are introduced. Indeed, posterior variances are basically as easy to compute as posterior means, without resorting to any asymptotic results. The simple scenario treated in the previous subsections leading to equation (8) is an exception in which frequentist and Bayesian measures of standard error of estimation agree, both giving the square root of

$$\mathrm{Var}(\rho \mid data) = \hat{\rho}^2 \frac{m(2m-1)}{(m-1)^2(m-2)} \tag{10}$$

However, if the restriction $\rho < 1$ is introduced, deriving a frequentist standard error for a suitable estimator can be very challenging. In contrast, the variance of the truncated version of equation (8) is readily derived, and, in fact

$$E(\rho^2 \mid data,\ \rho < 1) = \frac{m}{m+2} \frac{F(2m, m+2; m+3; -1/\hat{\rho})}{F(2m, m; m+1; -1/\hat{\rho})} \tag{11}$$

(see Armero and Bayarri, 1996, for details).

3.5 Prediction

So far we have addressed different issues arising in the estimation of the parameters governing the queue. As important as this inference might be, this is not the natural inferential aim when statistically analysing queues. In fact, interest usually lies in the prediction of observable quantities that describe the behavior of the queueing system, such as the number of customers in the system $N(t)$, or in the queue, $N_q(t)$, at time t, or its steady-state counterparts N and N_q. Other measures of interest are the (steady-state) time that a customer spends in the system, W, or in the line, W_q, the (steady-state) time that the servers are idle, . . . , etc.

Prediction (similar to the handling of nuisance parameters) is an area in which the operational advantages of the Bayesian approach become very clear. Indeed, there is not a unique, generally agreed-upon way to address prediction problems from a frequentist approach. If predictions are desired for the random variable Z with conditional density $f(z \mid \theta)$, maybe the most usual approach consists of plugging estimators $\hat{\theta}$ into the data-generating process and then using $f(z \mid \hat{\theta})$ as a predictive distribution. This procedure might produce very sensible point predictors; however, failing to take into account the error incurred in estimating θ usually produces inadequate standard errors, and sometimes even point predictors. This fact can be particularly noticeable when using $p(N \mid \hat{\lambda}, \hat{\mu})$ or $p(W \mid \hat{\lambda}, \hat{\mu})$ to make inferences (predictions) about the (steady-state) number of customers in the system, N, the waiting time of a customer, W, or other measures of performance. This is so because the behavior of queues is extremely sensitive to $\rho = \lambda/(c\mu)$ getting close to 1, and so ignoring the estimation error there can have dramatic influences, not only in assessing the standard error, but even when producing the point predictor itself. We have seen this extreme sensitivity in all of the examples we have analysed so far in our papers. Schruben and Kulkarni (1982) show the drastic effect that ignoring estimator variability can have in the distributions of measures of performance of $M/M/1$ queues.

In contrast, Bayesian predictive distributions do take into account the estimation uncertainties, and thus constitute a natural and powerful tool for prediction in queues. Thus, all of the inferences about the observable Z would be based on the posterior predictive distribution

$$p(z \mid data) = \int f(z \mid \theta)\, \pi(\theta \mid data)\, \mathrm{d}\theta \tag{12}$$

in particular, point predictors, standard errors of prediction, credible intervals, probabilities of interest concerning $Z, \ldots,$ etc.

Prediction in queues presents an attractive and interesting departure from the vast majority of prediction problems in that the quantity to be predicted is not the usual next observation, or the value of a sufficient statistics in a future sample, but one of the measures of performance whose relation with the parameters is more subtle and involves the derivation of entirely new conditional distributions.

As an example, let us take again the simplified situation we have been using as illustration so far ($M/M/c$ queue, noninformative prior distribution, and $n_a = n_s = m$). The predictive distribution of the (steady-state) number of customers, N, in the system is, for an $M/M/1$ queue (Armero and Bayarri, 1994a),

$$p_1(n \mid data,\ \rho < 1) = \frac{mF(2m, m+n; m+n+2; -1/\hat{\rho})}{(m+n)(m+n+1)F(2m, m; m+1; -1/\hat{\rho})} \tag{13}$$

and for an $M/M/\infty$ queue (Armero and Bayarri, 1997a),

$$p_\infty(n \mid data) = \frac{\Gamma(2m)}{\Gamma(m)\Gamma(m)} \frac{\Gamma(m+n)U(m+n, 1-m+n, 1/\hat{\rho})}{n!} \hat{\rho}^n \qquad (14)$$

where $U(a, b; c)$ is a Kummer function (see Abramowitz and Stegun, 1964). Both predictive distributions are shown in Figure 3 for $m = 10$ and $\hat{\rho} = 0.3, 0.7, 1.3$. These two systems correspond to the limiting cases of $M/M/c$ queues, and predictive distributions for $M/M/c$ systems would fall between the ones displayed here. Note that, as expected, the differences between the two predictive distributions are very small for small $\hat{\rho}$, and get progressively more important as $\hat{\rho}$ grows, and they are substantially different distributions for $\hat{\rho} = 1.3$. At first, the behavior shown in Figure 3 for $\hat{\rho} = 1.3$ might appear as counterintuitive, since the probability of n customers has a maximum at $n = 0$ for $M/M/1$ queues, while this maximum is attained at $n = 1$ for $M/M/\infty$ queues. Recall, however, that predictive distributions for $M/M/c$ queues are derived *under the assumption* that $\rho < 1$. This restriction is *not* imposed when computing predictive distributions for $M/M/\infty$ queues.

Predictive distributions for different models and/or under different priors are derived in many of the references listed in Section 1.

Note the conditioning requirement, $\rho < 1$, in equation (13). This is needed because the conditional distribution $p(N \mid \lambda, \mu)$ is not defined unless $\lambda < \mu c$. This is true for virtually every conditional distribution of measures of performance whose existence requires the ergodic condition to hold. Hence, the relevant posterior distribution to derive all of these predictive distributions is $\pi(\lambda, \mu \mid \mathbf{x}, \mathbf{y}, \lambda < c\mu)$.

There are a few measures of performance whose conditional distribution exists without this requirement, as, for example, the distribution of the time that the single server of an $M/M/1$ queue reminds idle. Of course, this requirement is not needed for systems that are guaranteed to reach the steady state, as for the $M/M/\infty$ queue.

3.6 Transient Analyses

The study of queues under the steady state is sensible when analysing queues that have been running for a long period of time, or when interest lies on average measures over long periods. It is thus particularly useful when designing queues. However, often of interest is the behavior of the queue under non-steady-state conditions. Usual (equilibrium) analyses provide no insight into the performance of the queue before it reaches the steady state and no indications of the length of time required to reach it.

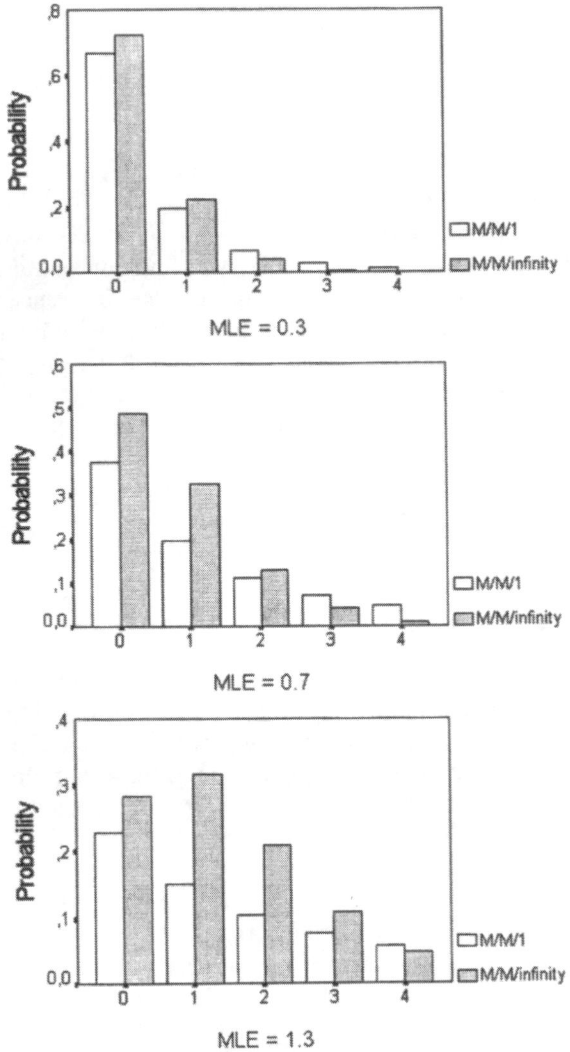

Figure 3. Number of customers in the system for $M/M/1$ and $M/M/\infty$ queues, when $m = 10$ and $\hat{\rho} = 0.3$ (top figure), $\hat{\rho} = 0.7$ (middle), and $\hat{\rho} = 1.3$ (bottom)

For Bayesian procedures, transient analysis poses no conceptual or methodological problem. To derive the likelihood, there is no need to assume the steady state, or even that it can ever be reached. Hence, posterior distributions have in fact been derived without ever assuming this condition.

Nevertheless, deriving the conditional distribution of the number of customers in the system at time t, $p(N(t) \mid \lambda, \mu)$, or those of other measures of performance, can be an extremely difficult task and only a few of them, for simple models, have been derived in closed form, usually in terms of the Laplace–Stieltjes transform. A particularly easy transient analysis occurs for the $M/M/\infty$ queue. In this model, $N(t) \mid \lambda, \mu \sim Po(\frac{\lambda}{\mu}(1 - e^{-\mu t}))$, and the predictive distribution is given in terms of a very simple unidimensional integral (see Armero and Bayarri, 1997a). Its moments can be derived in closed form, and in particular, for the simple illustration in this paper (non-informative prior, $n_a = n_s = m$) the expected value is

$$E(N(t) \mid data) = \frac{m}{m-1}\frac{\hat{\lambda}}{\hat{\mu}}\left[1 - \left(\frac{m}{m+\hat{\mu}t}\right)^{m-1}\right] \tag{15}$$

Note that $E(N(t) \mid \hat{\lambda}, \hat{\mu}) = \frac{\hat{\lambda}}{\hat{\mu}}(1 - e^{-\hat{\mu}t})$ would be the frequentist solution, and this is simply the limit of equation (15) as $m \to \infty$. This is highly intuitive, since when $m \to \infty$ we would know for sure that $\lambda = \hat{\lambda}$ and $\mu = \hat{\mu}$ and then there would be no need to take into account the effect of the uncertainty in the estimation of λ and μ. For small or moderate values of m, however, equation (15) seems a very sensible alternative to the classical estimator.

3.7 Design

Good design requires explicit incorporation of the opinions of the specialist into the analysis, and thus it is an ideal area for application of Bayesian methods. Besides, if, as it is the case in queues, the mere substitution for the parameters of their estimates results in totally inadequate predictions, the case for using Bayesian predictive distributions becomes even stronger.

In Armero and Bayarri (1997a) we used the predictive distribution of the number of customers in the system, $p(N(t) \mid data)$, to determine how long should we run an $M/M/\infty$ queue so as to achieve the steady state with some preassessed probability p.

Another kind of design was described in Armero and Bayarri (1996), in which the optimal number of servers c^* was found which would achieve different goals in an $M/M/c$ queue in equilibrium. The most elementary requirement is that the queue can run for a long period of time without exploding; this can be achieved by choosing c^* as the minimum c such that the ergodic condition holds with high enough probability. If the efficiency of the queue is the primary goal, then c^* could be chosen based on the (steady-state) distribution of the number of busy servers, N_b. On the other hand, in some queueing systems, a delay in service can be associated with an enormous loss. This is so in most emergency services, such as ambulances, polices cars, fire engines, ..., etc. In these cases, c^* could be chosen so that the (steady-state) probability that there are no customers waiting in line is high enough. However, this requirement is much too expensive to be used with more standard queues, even if they are designed with the comfort of the customers in mind. If such is the goal, a more realistic way to choose c^* would be to require that the number of people in line rarely exceeds a preassigned value; another possibility would be to choose c^* such that the (steady-state) probability is small that a customer has to queue up for service longer than some preassigned length of time. All these goals were addressed and explicit solutions given in Armero and Bayarri (1996).

An added bonus of Bayesian analyses for design purposes in the easy, natural way in which actual costs and losses can be introduced into the analysis in a full decision-theoretic approach (see, for instance, DeGroot, 1970; Berger, 1985). This possibility was explored by Bagchi and Cunningham (1972), where a loss function in terms of the cost of operating time was introduced to choose the optimal waiting room size and service rate in an $M/M/1$ queue in equilibrium with limited waiting room. The loss function involved the cost of maintaining service at level μ, renting the waiting space, the potential loss due to customers lost when the system is full, and a loss incurred which is proportional to the average waiting time of a customer. Of course, the full decision-theoretic approach to designing queues is the most appealing and the one that should be taken in designing queues with different, conflicting, objectives. Needless to say, its potential for applications to real-world problems is enormous.

4. CHALLENGING QUESTIONS IN BAYESIAN ANALYSIS

As attractive as the Bayesian approach to inference in queues is, it does have its drawbacks. The most challenging ones pertain (not surprisingly) to the choice and assessment of the prior distributions required in the Bayesian analysis. There is a middle-of-the road approach in which instead of fully specifying the prior distribution, its hyperparameters (or some features of the posterior, or of the risk function) are themselves estimated from the data. This approach, the so-called *empirical-Bayes*, gives results which are very appealing to many statisticians and have been explored for inference in queues by Lehoczky (1990), Thiruvaiyaru and Basawa (1992), and Sohn (1996). Its main drawbacks again, stem from the failure to take into account the estimation error when applied naively. We will not pursue the topic here, but instead concentrate on the challenges encountered when choosing a prior for statistically analysing queues.

Most of the analyses that we have performed so far in our papers have been carried out assuming either a "noninformative" type of prior, or a conjugate prior; often we have chosen natural conjugate priors, but none of these selections, however popular, is foolproof. In general terms, some difficulties with these standard priors are well known and discussed (see, for example, Berger, 1985), as for instance, the large number of different noninformative priors that can be produced, or the lack of adequacy of the natural conjugate prior to reflect actual subjective opinions. We shall not discuss these issues here, but will restrict our attention to pointing out the difficulties that can arise when using these (most standard and frequent) types of prior in the context of queueing models.

We first address the difficulties of choosing a noninformative, default or automatic prior. Since in all our analyses Jeffrey's prior produced sensible answers, and was very easy to derive, it was our default choice for a noninformative prior. Moreover, it did reproduce more sophisticated (and considerably more difficult to derive) priors in many situations (see, for example, the discussion and rejoinder of Armero and Bayarri, 1996). This said, we have to admit that we were not fully satisfied with this choice, and that considerably more research is needed to satisfactorily answer the many questions posed by a default statistical analysis of queues. For instance, what noninformative priors should be used when the steady state is assumed?; Should it be the same as the one for transient analysis (except for the restriction)?; What is the appropriate prior to be used for prediction of the different measures of performance? A typical queueing analysis would usually test for stationarity (see Armero, 1994) and then derive the predictive distributions of several measures of performance, and what is a good default prior for all these analyses is not clear.

Also, it should be kept in mind that queues can be observed in a huge variety of ways, so that the same arrival and service processes can result in many different "models" that generate "data." This poses a serious challenge to very heavily model-dependent (not merely likelihood-dependent) priors, in that the wide variety of different experiments that could be performed might produce an equally wide variety of priors which might be severely inappropriate if a slight modification of the planned experiment is the one ultimately performed (see Armero and Bayarri, 1997a). Thus, if a prescription of an "automatic," "default" type of analysis is desired, much research is still needed to apply this type of noninformative prior to queueing scenarios.

Most of the Bayesian analyses of queueing systems mentioned in the introduction result in posterior distributions belonging to the conjugate family, either because the prior itself was chosen to be conjugate, or because a noninformative analysis produced such a posterior. (Notable exceptions are Gaver and Lehoczky, 1987, McGrath et al., 1987, and Ruggeri et al,

1996, in which the prior independence, implied by conjugate priors, between arrivals and service rates when stationarity is *not* assumed is relaxed, and prior distributions which make λ and μ explicitly dependent a priori are used instead.)

Apart from the usual criticisms of conjugate priors, new questions arise when applied to the analysis of queueing systems which are assumed to be in equilibrium. How to deal with this assumption is not clear. In most of our analyses, we have followed the usual approach in which the conjugate, unrestricted analysis is performed and the resulting posterior distribution is truncated so that the ergodic condition holds. This, however, might not be appropriate, or might not truly reflect the prior opinions under stationarity.

A direct consequence of the use of these truncated conjugate posterior distributions can be seen in the form of the predictive distributions of measures of performance discussed in Section 3.5. A property common to many of those predictive distributions is their lack of moments. This is a direct consequence of averaging quantities which grow quickly to ∞ when ρ goes to 1 with respect to posterior distributions, $\pi(\lambda, \mu \mid data, \rho < 1)$, that, because of the truncation equation (7), are bounded away from zero in the neighbourhood of $\rho = 1$. This lack of Bayesian predictive moments has corresponding frequentist parallels, as, for instance, in the lack of moments of the distribution $p(N \mid \hat{\rho}, \hat{\rho} < 1)$ (see Schruben and Kulkarni, 1982). The lack of moments, per se, is not such an undesirable property, since a predictive distribution can most efficiently be described in terms of quantiles. However, it does indicate the existence of extremely long queues, that might clash with our intuition about the behavior of a queue in equilibrium. Besides, this procedure of truncating the natural conjugate posterior corresponds to using a truncated (at $\rho = 1$) prior for ρ, which again might not adequately represent prior opinions. As a matter of fact, under the assumption of equilibrium, a prior density whose upper tail goes to zero as ρ goes to 1 seems more appropriate. Thus, when assessing a joint prior distribution for the parameters of a queue, it may be desirable to enlarge the truncated natural conjugate prior so as to allow for different ways in which ρ can behave in the neighbourhood of $\rho = 1$. We did so for $M/M/1$ queues (and this can trivially be generalized to $M/M/c$ queues) in Armero and Bayarri (1994b). There we proposed using the same conditional for μ given ρ as the one derived from the natural conjugate prior (a Gamma density), but using as a marginal distribution for ρ a distribution that we called *Gauss hypergeometric* because it is derived from the Gauss hypergeometric function. Its density (for a, c, $z > 0$, $b \geq 1$) is

$$GH(\rho \mid a, b, c, z) = C\frac{\rho^{a-1}(1-\rho)^{b-1}}{(1+z\rho)^c}, \qquad 0 \leq \rho \leq 1 \tag{16}$$

The proportionality constant C is given by $C^{-1} = F(c, a; a + b; -z)/Be(a, b)$, where $Be(a, b)$ is the Beta function.

The distribution in equation (16) has a particular case in the Beta distribution (which is an intuitive choice for a prior for ρ under equilibrium), and also in the distribution derived from truncating the natural conjugate marginal for ρ; also, the corresponding joint density is conjugate, and thus the learning process and the description of distributions get notably simplified. The main advantage over the natural conjugate prior is in the parameter b, which controls the way that $\pi(\rho)$ goes to zero as ρ goes to 1; the larger b ($b > 1$) the faster $\pi(\rho)$ goes to zero. For $b = 1$ (truncated natural conjugate density), the density is bounded away from zero in the neighborhood of $\rho = 1$. Its effect on the moments of the predictive distributions of the measures of performance is crucial. We have already mentioned that the distribution of the number of customers in the queue, N_q, and that of the waiting time in the queue, T_q, have no moments for $b = 1$. It can be shown, however, that they both have mean for $b > 1$, variance for $b > 2$, and so on. The predictive distribution of idle periods always has moments. The existence of moments for the busy periods requires a faster decay to zero of $\pi(\rho)$; thus $\mathrm{E}(T_{busy})$ only exists for $b > 2$, and $\mathrm{E}(T_{busy}^2)$ for $b > 3$.

An entirely different extension of the natural conjugate prior occurred in an $M/M/\infty$ queueing scenario (Armero and Bayarri, 1997a). There, we analysed the four basic types of experiment that could be encountered, depending on whether or not the initial size of the queue was observed, and, if observed, which process (arrival and/or service) was recorded. Natural conjugate priors are, by construction, proportional to the likelihood function, and these four experiments gave rise to nonproportional likelihood functions, and hence to different families of prior distributions. We thought this to be highly nonintuitive in this queueing context, and preferred to use the same prior distribution for the same parameters no matter what experiment happened to be performed.

To accomplish this, we again enlarged the natural conjugate prior in the following way: we kept the same conditional distribution for $\mu \mid \rho$ as that derived from the natural conjugate, that is, a Gamma distribution. For the marginal distribution of $\rho = \lambda/\mu$ we introduced a new family of distributions which we called *Kummer distributions* because they were derived from a Kummer function (see Abramowitz and Stegun, 1964). The density is given by (for $a > 0$, $b > 0$ $d > 0$)

$$Ku(x \mid a, b, c, z) = C \, \frac{x^{a-1} \, e^{-bx}}{(1 + dx)^c}, \qquad x > 0 \tag{17}$$

where $C^{-1} = \Gamma(a)\, d^{-a}\, U(a,\ a+1-c,\ b/d$. Kummer distributions generalize both the Gamma and the F distributions.

A noteworthy property of the joint distribution derived in this way is that it is conjugate for *all* the four experiments described above, even though they result in nonproportional likelihood functions. Thus, the same family of distributions can conveniently be used to describe prior opinions, no matter which experiment ultimately gets performed. We find this to be an intuitive and desirable property.

Although the new families of distributions in equations (16) and (17) were derived from scratch, we have learnt that they also occur in different contexts (see Johnson and Kotz, 1995, for further details).

5. QUEUEING NETWORKS

So far we have considered statistical issues concerning isolated queues (single-stations queues). Quite often, however, customers require more than one service from different servers, and upon completion of service in one node, they have to proceed to another node. If the new server is busy, the customer has to queue for service again. This is a typical situation modelled by the so-called *networks of queues*. In these networks, congestion occurs not only because of the server serving this particular node, but also due to the interaction among the different nodes or stations.

Queueing networks are undoubtedly the main innovation in queueing theory, both in its theoretical developments and in its enormous potential for applications. It has become an extremely useful tool to evaluate (at least approximately) the performance of complex stochastic service systems.

Obvious applications of queueing networks occur in airport terminals and health care centres, but they have also traditionally been applied to biology (migration, population models, etc.), electrical engineering (flow models, ...), and chemistry (polymerisation, clustering, ...), among others. However, by far the most prevalent applications of queueing networks occur in the fields of manufacturing, computer networking, and telecommunications and broadcasting.

The analysis of networks of queues began in the 1950s, and received its start in the landmark paper by Jackson (Jackson, 1963), showing that, under a simple set of assumptions, the joint probability of the number of customers in each node can be factored into the product of the corresponding probabilities for each component node. This remarkable property is called a "product form result." Baskett et al. (1975) would later show that the "product form result" also occurs for several sets of assumptions, and not only for Jackson networks.

For the next 15 years or so, most of the advances in the field came from the performance analysis of computer systems and communication networks. The explosion of papers, both theoretical and applied, were in part motivated by two crucial developments. First, Gordon and Newell (1967) showed that the product-forms solution also holds, up to a normalizing constant, for closed queueing networks (which are much used to model computer systems). Then Buzen (1973) established a connection between Gordon and Newell's work and the problems of designing operating systems for optimal-level multiprogramming and other characteristics; most importantly, Buzen derived an easy implementable algorithm to compute the normalizing constant, which had been a formidable task (often impossible in a feasible computing time) for moderate, or even large, queueing networks. A plethora of papers with applications to computer systems and communications networks followed. Manufacturing applications joined later, by the end of the 1970s, after the influential work of Solberg (1977), who applied Buzen's work to the analysis of performance of flexible manufacturing systems (see Suri et al., 1995, for further details).

The literature on queueing networks is so vast that we hesitate to mention just a few references. Our intention in so doing is just to provide some useful references from which many others (perhaps more important) can be obtained. From the many books in networks of queues, we shall mention only three: Van Dijk (1993), which has many references, Robertazzi (1994), which is oriented toward computer applications, and Buzacott and Shanthikumar (1993), which considers manufacturing ones. Among the general review papers which the reader might find particularly useful are those of Asmussen (1993) and Lemoine (1977). More specialized review papers are those of Koenigsberg (1982) for closed networks, Suri et al. (1995) for manufacturing applications, and Kelly (1985) and Lavenberg (1988) for computer applications (the rest of this book of reviews is also very informative). As was the case with single queues, but is even more so in queueing networks, statistical inference is basically nonexistent. Notable exceptions are the papers by Thiruvaiyaru et al. (1991) and Thiruvaiyaru and Basawa (1992) on maximum likelihood and parametric empirical Bayes, respectively, of a couple of very simple queueing networks. Gaver and Lehoczky (1987) developed Bayes and empirical Bayes procedures for some simple Markov population processes which included some queueing systems. As far as we know, those are the only specific references to date.

In short, a network of queues is a network of *service centers* (also called *stations* or *nodes* in the queueing network jargon), each having its own service process, servers, queues, waiting room, ..., etc. Arrival processes occur to several (sometimes all) nodes in the system, and customers then proceed through the network according to some routing scheme, not necessarily the same for every customer (customers may return to nodes visited previously, skip some nodes, leave from any node, or remain in the system for ever). The networks which have received by far the most attention are the ones named after Jackson (1957), and we shall briefly treat them in the following section.

6. BAYESIAN INFERENCE AND PREDICTION IN JACKSON NETWORKS

Jackson networks are rather general systems, which include both *open* and *closed* (no external arrivals and departures) networks, but with a set of (restrictive) assumptions that make it possible for their steady-state probability distribution to exhibit the "product form." These systems are the natural extensions to networks of the most popular $M/M/c$ queues. The specific assumptions are as follows:

1. The network has K stations. Node i has c_i identical servers.
2. Outside customers arrive at node i according to a Poisson process with rate λ_i. Interarrival times are thus i.i.d. exponentials with mean $1/\lambda_i$.
3. Customers in node i are served FIFO (first-in-first-out) and service times are i.i.d. exponentials with rate μ_i (mean $1/\mu_i$).
4. After leaving node i, each customer either leaves the system, with probability p_{i0}, or goes to node j with probability p_{ij}, $j = 1, 2, \ldots, K$. That is, each customer chooses where to go next according to multinomial distribution.

It can be seen that 1, 2, and 3 just match the assumptions for $M/M/c$ queues. The probabilities p_{ij} ($\sum_{i=0}^{K} p_{ij} = 1$) are called *routing probabilities*, and a crucial assumption is that they are independent of the history and the state of the network. The matrix $\mathbf{P}$ formed by the p_{ij}, $i, j = 1, \ldots, k$ (excluding the p_{i0}s) is called the *routing matrix*. If at least one $\lambda_i > 0$, then the system is called an *open network*, otherwise (noexternal arrivals and $p_{i0} = 0$) it is a *closed network*, in which a fixed number of jobs M stays indefinitely in the system. If, in an open network, external arrivals only occur to one node (the "first node"), jobs then travel successively through all the nodes, one after another in a determined way, and all depart from the "last node," the system is called *tandem queues* or *series queues*. A closed network in series is called a *cyclic network*.

With an easy to perform experiment similar to that in Section 3.1, the likelihood function can easily be derived. Explicitly, for each node i to which external arrivals occur we observe n_{ia} external interarrival times, Y_{ij}, $j = 1, 2, \ldots, n_{ia}$. Also, for every node in the network, we observe n_{is} service completions, $X_{ij}, j = 1, 2, \ldots, n_{is}$. Finally, for each node i such that some of the routing probabilities p_{ij} are unknown, we observe the number of customers among the n_{is} who go to the node j, R_{ij}, for all such j. Then, the likelihood function is easily seen to be (for appropriate ranges of subindexes when not explicit)

$$l(\boldsymbol{\lambda}, \boldsymbol{\mu}, \mathbf{p}) \propto \left(\prod_{i=1}^{K} \lambda_i^{n_{ia}} e^{-\lambda_i t_{ia}}\right) \left(\prod_{i=1}^{K} \mu_i^{n_{is}} e^{-\mu_i t_{is}}\right) \left(\prod_i \prod_j p_{ij}^{R_{ij}}\right) \tag{18}$$

where $t_{ia} = \sum_{j=1}^{n_{ia}} y_{ij}$, $t_{is} = \sum_{j=1}^{n_{is}} x_{ij}$.

A much more difficult experiment to perform, that of observing the Markov process $\{N(s), 0 < s \leq t\}$, would produce a likelihood function asymptotically equivalent to equation (18). Our likelihood function in equation (18) ignores the initial and end pieces where nothing happens (no arrivals and no departures). Similar likelihood functions are used in Thiruvaiyaru et al. (1991) and Thiruvaiyaru and Basawa (1992).

The Bayesian conjugate analysis is very easy. The form of equation (18) suggests taking all the λ_is, μ_is and $\mathbf{p}_i$s independent, with the λ_is and μ_is having Gamma distributions and the $\mathbf{p}_i$s having Dirichlet distributions. The posterior $\pi(\boldsymbol{\lambda}, \boldsymbol{\mu}, \mathbf{p} \mid data)$ is then of the same form with the hyperparameters appropriately updated. A limiting (noninformative) case would produce as Bayes estimators the MLEs derived from equation (18). Note that we do not have to assume the steady state, nor even that the steady-state distribution exists in order to derive posterior distributions for the parameters governing the system.

If the system is in equilibrium, the flow into a given state has to equal the flow out of it, so that we can obtain the *effective arrival rate* at node i, γ_i (flow from outside as well as flows from the rest of the nodes) as the solution to the equations

$$\gamma_i = \lambda_i + \sum_{j=1}^{K} \gamma_j p_{ji} \tag{19}$$

which are known as *traffic equations*, *flow balance equations*, or *conservation equations*. The steady state then requires

$$\rho_i = \frac{\gamma_i}{c_i \mu_i} < 1 \tag{20}$$

Both equation (19) and (20) are highly intuitive conditions. It can be checked that under the steady state the solution to equation (19) is unique if $\mathbf{I} - \mathbf{P}$ is not singular, which is equivalent to the condition that every arriving customer eventually leaves the system.

Jackson's (1957) most amazing theorem stated that in an open network in equilibrium, if N_i is the number of customers at node i, then the conditional joint probability distribution exhibits the product form

$$p_{\mathbf{n}} = p(N_1 = n_1, \ldots, N_K = n_K \mid \boldsymbol{\theta}) = p_1(N_1 = n_1 \mid \boldsymbol{\theta}) \ldots p_K(N_K = n_K \mid \boldsymbol{\theta}) \tag{21}$$

(Here $\boldsymbol{\theta}$ denotes all unknown parameters.)

That is, each node behaves *as if* it was an independent $M/M/c$ queue, node i having c_i servers, arrival rate γ_i, and service rate μ_i. This result is all the more remarkable because the flow into each station is *not* Poisson as long as there is any kind of feedback. (Jackson (1963) and Gordon and Newell (1967) would extend this result to closed networks, showing that equation (21) holds except for a normalizing constant, but since this constant involves the n_is, the conditional independence is destroyed.)

From equation (19) and the joint posterior distribution, it is easy to derive the posterior distribution of effective arrival rates $\pi(\gamma \mid \textit{data, steady-state})$, requiring, at most, straight Monte-Carlo integration, where simulation is from (easy to simulate from) Gamma and Dirichlet distributions. In turn, this posterior distribution together with equation (21) can be used to derive the joint predictive distribution for the (steady-state) number of customers at each node. All of the details can be seen in Armero and Bayarri (1997b). We limit ourselves here to two simple examples (one open, one closed), of which we only give the outlines without full details.

EXAMPLE 1: *Open Jackson Network with two nodes*. This simple example appears in several places. We take the version in Bunday (1996). Figure 4 represents the production process for certain articles made in a factory. "Skeleton" parts arrive at random at station 1 at an average rate of λ and are processed at rate μ_1. A proportion p of this output is faulty and has to be "partially dismantled" at station 2 at rate μ_2 before being returned to station 1.

With an unknown p we observe:

n_a interarrival times of skeleton parts, $Y_1, Y_2, \ldots, Y_{n_a}$ with observed $t_a = \sum y_j$.

n_{1s} processing times at node 1, $X_{11}, \ldots, X_{1n_{1s}}$ with observed $t_{1s} = \sum x_{1j}$.

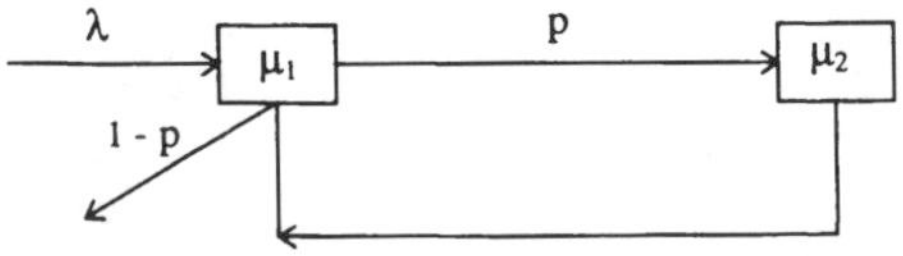

Figure 4. Open Jackson network with two nodes corresponding to Example 1

The number R_{12} among these n_{1s} which are faulty (and therefore proceed to node 2). Then $R_{10} = n_{1s} - R_{12}$ would be the number of items processed correctly and hence leaving the system.

n_{2s} "dismantling" times at node 2, $X_{21}, \ldots, X_{2n_{2s}}$ with observed $t_{2s} = \sum x_{2j}$.

A standard noninformative prior (limit of the natural conjugate one) combined with the likelihood function in equation (18) would produce the joint posterior distribution:

$$\pi(\lambda, \mu_1, \mu_2, p \mid data) = Ga(\lambda \mid n_a, t_a) Ga(\mu_1 \mid n_{1s}, t_{1s}) \\ Ga(\mu_2 \mid n_{2s}, t_{2s}) Be(p \mid r_{12}, r_{10}) \tag{22}$$

The flow equations here take the form

$$\gamma_1 = \lambda + \gamma_2, \qquad \gamma_2 = p\gamma_1 \tag{23}$$

so that the effective arrival rates are given by

$$\gamma_1 = \frac{1}{1-p}\lambda, \qquad \gamma_2 = \frac{p}{1-p}\lambda = p\gamma_1. \tag{24}$$

For any given p, the joint distribution of the γs is fully characterised by that of γ_1 (since γ_2 is simply $p\gamma_1$), which can easily be seen to be a $Ga(n_a, t_a(1-p))$ distribution. This would be the distribution to use in the frequently assumed case of a known proportion of faulty items p. (Note that some of the results in this analysis would be very similar to those obtained for single $M/M/1$ queues, since γ_1 has a Gamma distribution.) If p is unknown, then the marginal posterior distribution of γ_1 is easy to derive by integrating out p from the joint distribution $Ga(\gamma_1 \mid n_a, t_a(1-p))$ $Be(p \mid r_{12}, r_{10})$.

Under equilibrium $\gamma_1 < \mu_1$ and $\gamma_2 < \mu_2$, so that the relevant joint posterior distribution is

$$\pi(\lambda, \mu_1, \mu_2, p \mid data, steady\text{-}state) = \frac{\pi(\lambda, \mu_1, \mu_2, p \mid data)}{Pr(SS \mid data)} \\ \text{for} \quad \lambda < min\{(1-p)\mu_1, (1-p)\mu_2/p\} \tag{25}$$

where $Pr(SS \mid data) = Pr(\lambda < min\{(1-p)\mu_1, (1-p)\mu_2/p\} \mid data)$ is the probability that the ergodic condition holds and can easily be obtained via Monte-Carlo from equation (22). Equation (25) is the distribution that we use to integrate out λ, μ_1, μ_2, p to get the predictive distribution of the number of customers at each node. From Jackson's theorem

$$p(N_1 = n_1, N_2 = n_2 \mid \lambda, \mu_1, \mu_2, p)$$
$$= \rho_1^{n_1}(1 - \rho_1)\rho_2^{n_2}(1 - \rho_2)$$
$$= \frac{\lambda^{n_1+n_2} p^{n_2}}{\mu_1^{n_1+1} \mu_2^{n_2+1} (1 - p)^{n_1+n_2+2}} (\mu_1(1 - p) - \lambda)(\mu_2(1 - p) - \lambda p) \qquad (26)$$

and the predictive distribution $p(n_1, n_2 \mid \textit{data, steady-state})$ integrates out λ, μ_1, μ_2, p from equation (26) using the distribution in equation (25). Full details appear in Armero and Bayarri (1997b).

EXAMPLE 2: *Two-node closed network with feedback*. This example also appears in several places. We take here the version of Medhi (1991). Consider the computer system shown in Figure 5 in which node 1 corresponds to CPU and node 2 to an I/O device. Computer systems are frequently modeled as closed networks, with a fixed number of M circulating programs (as soon as a program is processed, it is immediately substituted by another one, so that M, known as the *level of multiprogramming*, is fixed). CPU execution bursts are assumed to be i.i.d. exponential with mean $1/\mu_1$, and I/O bursts are i.i.d. exponential with mean $1/\mu_2$. Assume further that at the end of a CPU burst a program requests an I/O operation with probability p.

With an unknown p we observe:

n_{1s} CPU execution times, $X_{11}, \ldots, X_{1n_{1s}}$ with observed $t_{1s} = \sum x_{1j}$.
The number R_{12} among the n_{1s} that require an I/O operation, so that $R_{11} = n_{1s} - R_{12}$ is the number directly fed back into the system. We denote by r_{12} and r_{11} the observed values.
n_{2s} I/O service times, $X_{21}, \ldots, X_{2n_{2s}}$ with observed $t_{2s} = \sum x_{2j}$.

Again, the likelihood function is of the form of equation (18) without the factors corresponding to external arrivals. A standard, noninformative prior (uniform for p) produces the joint posterior distribution

$$\pi(\mu_1, \mu_2, p \mid data) = Ga(\mu_1 \mid n_{1s}, t_{1s})Ga(\mu_2 \mid n_{2s}, t_{2s})Be(p \mid r_{12}, r_{11}) \qquad (27)$$

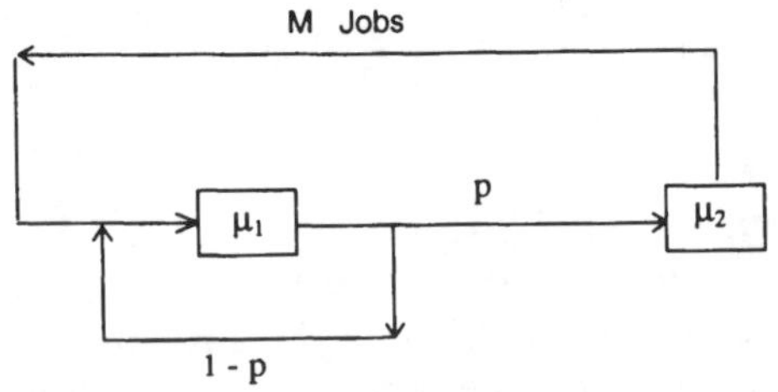

Figure 5. Two-node closed network with feedback of Example 2

In this case, the steady state is guaranteed to be reached eventually. Also, since flow equations take the form $\gamma' = \gamma'\mathbf{P}$ and the routing matrix $\mathbf{P}$ is stochastic, there are an infinite number of solutions of the form $\gamma_1 = a$, $\gamma_2 = pa$, with a arbitrary that we take, without loss of generality, to be equal to 1 (the proportionality constant takes care of the arbitrarity in the choice). Then again, according to the Jackson–Gordon–Newell result, the probability that there are n_1 jobs at the CPU station and $M - n_1$ jobs in the I/O facility (note that $n_1 - 1$ and $n_2 - 1$ have to be queueing up) is

$$p(N_1 = n_1, N_2 = M - n_1 \mid \mu_1, \mu_2, p) = C(M)\, \rho_1^{n_1} \rho_2^{M-n_1}, \qquad 0 \le n_1 \le M \tag{28}$$

where, as before, $\rho_i = \gamma_i/\mu_i$ and $C(M)$ is the appropriate normalizing constant. Note that for a closed network with K nodes and M circulating customers, the number of different state probabilities $p(\mathbf{N} = \mathbf{n} \mid \theta)$ is $\binom{K+M-1}{M}$, which is also the number of terms to be summed over in order to compute the normalizing constant C, so direct computation of C might be practically impossible for large or even moderate networks, and Buzen's work (and many other related works) reveals all its usefulness.

In our simple example, the proportionality constant can be computed in a closed form and equation (28) is given by

$$p(N_1 = n_1, N_2 = M - n_1 \mid \mu_1, \mu_2, p) = \frac{(1-\varphi)\varphi^{n_1}}{1-\varphi^{M+1}}, \qquad \varphi \ne 1 \tag{29}$$

and $= (M+1)^{-1}$ if $\varphi = 1$, where $\varphi = \mu_2/(p\mu_1)$. From equation (29) and the joint posterior in equation (28) the predictive marginal $p(N_1 = n_1, N_2 = M - n_1 \mid data)$ can easily be derived. Probabilities of interest are directly obtained from this predictive. Thus, for instance, the probability that the CPU is free is given by

$$Pr(CPU\,free) = p(0, M \mid data) = \int \frac{1-\varphi}{1-\varphi^{M+1}}\, \pi(\varphi \mid data)\, d\varphi \tag{30}$$

where $\pi(\varphi \mid data)$ can be obtained from equation (27) if desired. The utilization factor of the CPU node is precisely defined as $1 - Pr(CPU\,free)$. In a similar way, the utilization factor of the I/O node can be obtained as $1 - Pr(I/O\,free) = 1 - p(0, M \mid data)$.

Derivation of the conditional distribution of sojourn (or response or waiting) times at the different nodes as well as the that of the cycle time is very difficult for networks of queues, and very little is known with some generality. There exist two main problems. First, flows into the different nodes is Poisson only if the routing is arborescent (tree-like, with no feedback). Second, if there is any possibility that customers can bypass each other (as, for instance, when there is more than one server in each node, or when the routing of customers might differ), then sojourn times at the different nodes become dependent. Sometimes, partial answers (like the expected conditional times) have been derived for special networks. For tandem and cyclic queues (with one server) solutions have been derived and appear in many places (see Armero and Bayarri, 1997b).

ACKNOWLEDGMENT

Research supported in part by Generalitat Valenciana No. GV-1081/93.

REFERENCES

Abramowitz, M., Stegun, I.A. *Handbook of Mathematical Functions.* New York: Dover, 1964.

Armero, C. *Bayesian Analysis of M/M/1/FIFO/∞ queues. Bayesian Statistics 2* (J.M. Bernardo, M.H. DeGroot, D.V. Lindley, and A.F.M. Smith, eds.). Amsterdam: North-Holland, pp. 613–618, 1985.

Armero, C. Bayesian inference in Markovian queues. *Queueing Syst.*, 15: 419–426, 1994.

Armero, C., Bayarri, M.J. Bayesian prediction in *M/M/*1 queues. *Queueing Syst.*, 15: 401–417, 1994a.

Armero, C., Bayarri, M.J. Prior assessments for prediction in queues. *The Statistician*, 43: 139–153, 1994b.

Armero, C., Bayarri, M.J. *Bayesian Questions and Answers in Queues. Bayesian Statistics 5* (J.M. Bernardo, J.O. Berger, A.P. Dawid, and A.F.M. Smith, eds.). Oxford: Oxford University Press, 3–23, 1996.

Armero, C., Bayarri, M.J. A Bayesian analysis of a queueing system with unlimited service. *J. Stat. Plann. Inference*, 58: 241–261, 1997a.

Armero, C., Bayarri, M.J. Bayesian analysis of Jackson queueing networks. Technical Report 2-97, Department of Statistics and OR, Universitat de València, 1997b.

Armero, C., Conesa, D. Inference and prediction in bulk arrival queues and queues with service in stages. *Appl. Stoch. Models Data Anal.*, 14: 35–46, 1998.

Armero, C., Conesa, D. Prediction in Markovian bulk arrival queues. Technical Report 1-97, Department of Statistics and OR, Universitat de València, 1997.

Asmussen, S. A tutorial on queueing networks. In: *Network and Chaos-Statistical and Probabilistic Aspects* (O.E. Bardnoff-Nielsen, J.L. Jensen, and W.S. Kendall, eds.). London: Chapman and Hall, pp. 251–273, 1993.

Bagchi, T.P., Cunningham, A.A. Bayesian approach to the design of queueing systems. *INFOR*, 10: 36–46, 1972.

Basawa, I.V., Prahbu, N.U. Large sample inference from single server queues. *Queueing Syst.*, 3: 289–304, 1988.

Basawa, I.V., Prakasa Rao, B.L.S. *Statistical Inference for Stochastic Processes*. New York: Academic Press, 1980.

Baskett, F., Chandy, K.M., Muntz, R.R., Palacios, F. Open, closed, and mixed networks of queues with different classes of customers. *J. Assoc. Comput. Mach.*, 22: 248–260, 1975.

Berger, J.O. *Statistical Decision Theory and Bayesian Analysis*. 2nd edn. New York: Springer, 1985.

Bhat, U.N., Rao, S.S. Statistical analysis of queueing systems. *Queueing Syst.*, 1: 217–247, 1987.

Bhat, U.N., Miller, G.K., Subba Rao, S. Statistical analysis of queueing systems. In: *Frontiers in Queueing: Models and Applications in Science and Engineering* (J.H. Dshalalow, ed.). Boca Raton: CRC Press, pp. 351–394, 1997.

Bunday, B.D. *An Introduction to Queueing Theory*. London: Arnold, 1996.

Butler, R.W., Huzurbazar, A.V. Bayesian prediction of waiting times in stochastic models. Technical Report, Colorado State University and University of New Mexico, 1996.

Buzacott, J.A., Shanthikumar, J.G. *Stochastic Models of Manufacturing Systems*. New Jersey: Prentice Hall, 1993.

Buzen, J.P. Computational algorithms for closed queueing networks with exponential servers. *Commun. ACM*, 16: 527–531, 1973.

Clarke, A.B. Maximum likelihood estimates in a simple queue. *Ann. Math. Stat.*, 28: 1036–1040, 1957.

Cooper, R.B. *Introduction to Queueing Theory*. 2nd edn. New York: North-Holland, 1981.

Cooper, R.B. Queueing Theory. In: *Handbooks in OR & MS*, Vol 2 (D.P. Heyman, M.J. Sobel, eds.). Amsterdam: Elsevier North-Holland, pp 469–518, 1990.

Cox, D.R., Smith, W.L. *Queues*. London: Chapman and Hall, 1961.

Cox, D.R.. Some problems of statistical analysis connected with congestion. *Proceedings of the Symposium on Congestion Theory*. Chapel Hill: University of North Carolina Press, pp 289–316, 1965.

DeGroot, M.H. *Optimal Statistical Decisions*. New York: McGraw-Hill, 1970.

Gaver, D.P., Lehoczky, J.P. Statistical inference for random parameter Markov population process models. In: *Contributions to the Theory and Application of Statistics*. A Volume in Honor of Herbert Solomon. pp. 75–99, 1987.

Gordon, W.J., Newell, G.F. Closed queueing systems with exponential servers. *Oper. Res.*, 15: 254–265, 1967.

Gross, D., Harris, C.M. *Fundamentals of Queueing Theory*. 2nd edn. New York: Wiley, 1985.

Jackson, J.R. Networks of waiting lines. *Oper. Res.*, 5: 518–521, 1957.

Jackson, J.R. Jobshop-like queueing systems. *Manage. Sci.*, 10: 131–142, 1963.

Johnson, N.L., Kotz, S., Balakrishnan, N. *Continuous Univariate Distributions*. Vol. 2. 2nd edn. New York: Wiley, 1995.

Kelly, F.P. Stochastic models of computer communication systems. *J. R. Stat. Soc., Ser. B*, 4, 3: 379–395, 1985.

Kendall, D.G. Stochastic processes occurring in the theory of queues and their analysis by the method of imbedded Markov chains. *Ann. Math. Stat.*, 24: 338–354, 1953.

Kleinrock, L. *Queueing Systems. Vol. 1. Theory*. New York: Wiley, 1975.

Kleinrock, L. *Queueing Systems. Vol. 2. Computer Applications*. New York: Wiley, 1976.

Koenigsberg, E. Twenty-five years of cycle queues and closed queue networks: A review. *J. Oper. Res. Soc.*, 33: 605–619, 1982.

Lavenberg, S.S. A perspective on queueing models of computer performance. In: *Queueing Theory and its Applications*. Liber Amicorum for J.W. Cohen. (O.J. Boxma, R. Syski, ed.). Amsterdam: North-Holland, pp. 59–94, 1988.

Lehoczky, J. Statistical Methods. In: *Handbooks in OR & MS*, Vol 2 (D.P. Heyman, M.J. Sobel, eds.). Amsterdam: Elsevier North-Holland, pp. 469–518, 1990.

Lemoine, A.J. Networks of queues—a survey of equilibrium analysis. *Manage. Sci.*, 24: 464–481, 1977.

McGrath, M.F., Singpurwalla, N.D. A subjective Bayesian approach to the theory of queues. II. Inference and information. *Queueing Syst.*, 1: 335–353, 1987.

McGrath, M.F., Gross, D., Singpurwalla, N.D. A subjective Bayesian approach to the theory of queues. I. Modelling. *Queueing Syst.*, 1: 317–333, 1987.

Medhi, J. *Stochastic Models in Queueing Theory*. Boston: Academic Press, 1991.

Morse, P.M. A queueing theory, Bayesian model for the circulation of books in a library. *Oper. Res.*, 4: 693–716, 1979.

Muddapur, M.V. Bayesian estimates of parameters in some queueing models. *Ann. Inst. Math.*, 24: 327–331, 1972.

Nelson, R. *Probability, Stochastic Processes and Queueing Theory*. New York: Springer, 1995.

Newell, G.F. *Applications of Queueing Theory*. 2nd edn. London: Chapman and Hall, 1982.

Prabhu, N.U. A bibliography of books and survey papers on queueing systems: Theory and applications. *Queueing Syst.*, 2: 393–398, 1987.

Reynolds, J.F. On estimating the parameters in some queueing models. *Aust. J. Stat.*, 15: 35–43, 1973.

Rios Insua, D., Wiper, M., Ruggeri, F. Bayesian analysis of $M/Er/1$ and $M/Hk/1$ queues. Technical Report CNR IAMI, 1996.

Robertazzi, T.G. *Computer Networks and Systems: Queueing Theory and Performance Evaluation*. 2nd edn. New York: Springer, 1994.

Ruggeri, F., Wiper, M.P., Rios Insua, D. Bayesian analysis of dependence in $M/M/1$ models. Technical Report CNR IAMI, 1996.

Schruben, L., Kulkarni, R. Some consequences of estimating parameters for the *M/M/1* queue. *Oper. Res. Lett.*, 1: 75–78, 1982.

Sohn, S.Y. Empirical Bayesian analysis for traffic intensity: *M/M/1* queues with covariates. *Queueing Syst.*, 22: 383–401, 1996.

Solberg, J.J. A mathematical model of computerized manufacturing systems. *Proceedings of the 4th International Conference on Production Research*, Tokyo, pp. 1265–1275, 1977.

Suri, R., Diehl, G.W.W., de Treville, S., Tomsicek, M.J. From CAM-Q to MPX: Evolution of queueing software for manufacturing. *Interfaces*, 25: 128–150, 1995.

Takács, L. *Introduction to the Theory of Queues*. Oxford: Oxford University Press, 1962.

Thiruvaiyaru, D., Basawa, I.V. Empirical Bayes estimation for queueing systems and networks. *Queueing Syst.*, 11: 179–202, 1992

Thiruvaiyaru, D., Basawa, I.V., Bhat, U.N. Estimation for a class of simple queueing networks. *Queuing Syst.*, 9: 301–312, 1991.

Van Dijk, N.M. *Queueing Networks and Product Forms*. A Systems Approach. Chichester: Wiley, 1993.

Whittle, P. *Systems in Stochastic Equilibrium*. Chichester: Wiley, 1986.

Wiper, M.P. Bayesian analysis of *Er/M/1* and *Er/M/c* queues. Technical Report CNR IAMI, 1996.

Wolf, R.W. Problems of statistical inference for birth and death queueing models. *Oper. Res.*, 13: 343–357, 1965.

21

Optimal Bayesian Design for a Logistic Regression Model: Geometric and Algebraic Approaches

MARILYN AGIN Pfizer, Inc., Groton, Connecticut

KATHRYN CHALONER University of Minnesota, St. Paul, Minnesota

1. INTRODUCTION

Haines (1995) and Chaloner (1993) give closed-form results for Bayesian designs for nonlinear problems. Both papers use, among other examples, the logistic regression model with a known slope parameter. Both papers derive some of the same results for prior distributions with two points of support, but with very different methods. Haines uses a novel geometric approach and Chaloner uses a more traditional algebraic approach using an equivalence theorem. Prior distributions with a small number of support points are not of much practical use, but when closed-form solutions can be found they give an understanding of more general problems in which designs must be found numerically.

The two different approaches are compared and contrasted in Sections 2 and 3. A new result for a three-point prior distribution is given in Section 4, together with a discussion of the geometric approach. Some numerical results for multiple-point prior distributions are given in Section 5.

2. BAYESIAN DESIGN FOR NONLINEAR PROBLEMS

An extensive review of Bayesian approaches to design is given in Chaloner and Verdinelli (1995). A Bayesian approach to design in nonlinear problems is to think of design as a decision problem and to maximize a criterion representing an approximation to the expected utility. Two of the earliest implementations of this approach were by Chernoff (1953) and Tsutakawa (1972).

2.1 Notation

Let the design region $\mathcal{X}$ be a compact subset of $\Re^p$ for some p. For a fixed sample size n, the design problem is to choose the number of design points, k, the values of the explanatory variables x_i, $i = 1, \ldots, k$, and also the number of observations n_i to be taken at each value x_i. The n_i are integers summing to n. Define $w_i = n_i/n$ and $\mathcal{H}$ to be the set of probability measures over $\mathcal{X}$. When the values w_i are constrained to be nonnegative and to sum to one, but are not constrained so that nw_i are integers, the design problem becomes that of choosing a measure η from $\mathcal{H}$. Criteria $\phi(\eta)$ will be defined on $\mathcal{H}$ and η is chosen to maximize $\phi(\eta)$.

Let η denote both a probability measure and its density function. Also define $D(\eta_1, \eta_2)$ to be the directional derivative of $\phi(\eta)$ at η_1 in the direction of η_2. Specifically,

$$D(\eta_1, \eta_2) = \lim_{\epsilon \to 0^+} \frac{1}{\epsilon} [\phi((1-\epsilon)\eta_1 + \epsilon\eta_2) - \phi(\eta_1)]. \tag{1}$$

Further define η_x to be the probability measure which is point mass at x in $\mathcal{X}$. Then the directional derivative $D(\eta, \eta_x)$, as a function on $\mathcal{H} \times \mathcal{X}$, is denoted as $d(\eta, x)$.

If the criterion function $\phi(\eta)$ is concave on $\mathcal{H}$, the equivalence theorem of Whittle (1973) can be extended, as in Chaloner and Larntz (1989), to verify that a particular design is optimal.

EQUIVALENCE THEOREM:

(a) If ϕ is concave, then a ϕ-optimal design η_* can be equivalently characterized by any of the three conditions:
  (i) η_* maximizes $\phi(\cdot)$;
  (ii) η_* minimizes $\sup_{x\in\mathcal{X}} d(\eta, x)$;
  (iii) $\sup_{x\in\mathcal{X}} d(\eta_*, x) = 0$.

(b) The point (η_*, η_*) is a saddlepoint of D in that $D(\eta_*, \eta_1) \leq 0 = D(\eta_*, \eta_*) \leq D(\eta_2, \eta_*)$ for all $\eta_1, \eta_2 \in \mathcal{H}$.

(c) If ϕ is differentiable, then the support of η_* is contained in the set of x for which $d(\eta_*, x) = 0$ almost everywhere in η_* measure.

2.2 The Logistic Regression Problem

Independent Bernoulli responses are observed at a value x of the explanatory variable. Let the probability of success $p(x)$ be

$$p(x) = \frac{1}{(1 + e^{-(x-\theta)})}.$$

The design problem is to choose k values x_i of the explanatory variable, and corresponding weights $w_i, i = 1, \ldots, k$ subject to the constraint $\sum_{i=1}^{k} w_i = 1$.

Denote the expected Fisher information for the design putting one observation at x as $I(\theta, x)$, then

$$I(\theta, x) = \frac{e^{-(x-\theta)}}{(1 + e^{-(x-\theta)})^2}.$$

For the design, η, with weight w_i at x_i, $i = 1, \ldots, k$, define

$$\begin{aligned} I(\theta, \eta) &= \int I(\theta, x) \mathrm{d}\eta(x) \\ &= \sum_{i=1}^{k} w_i \frac{e^{-(x_i-\theta)}}{(1 + e^{-(x_i-\theta)})^2}. \end{aligned}$$

Then the Fisher information for this design is $nI(\theta, \eta)$.

As in Chaloner (1987) and Chaloner and Larntz (1989), under an asymptotic normal approximation to the posterior distribution of θ, the expected Shannon information gives

$$\phi_1(\eta) = \int \ln I(\theta, \eta) \mathrm{d}\pi(\theta)$$

as the Bayesian D-optimality criterion where π is a prior distribution for θ. Similarly, the criterion function ϕ_2 for Bayesian c-optimality approximates minus the squared error loss. That is

$$\phi_2(\eta) = -\int \frac{1}{I(\theta, \eta)} \mathrm{d}\pi(\theta)$$

and the ϕ_2-optimal design measure η in $\mathcal{H}$ maximizes $-E[I(\theta, \eta)]^{-1}$. Both $\phi_1(\eta)$ and $\phi_2(\eta)$ are concave functions on $\mathcal{H}$.

Consider a prior distribution with just two support points, at $\theta = \pm g$, with mass 1/2 at each point. The following result was proved algebraically in Chaloner (1993) for ϕ_1-optimality and proved geometrically and generalized in Haines (1995) for both ϕ_1-optimality and ϕ_2-optimality.

THEOREM 1 (Haines, 1995): For a prior distribution which puts probability 1/2 on each of $\theta = \pm g$, the ϕ_1-optimal and ϕ_2-optimal designs are as follows.

1. If $|g| \le \ln(2 + \sqrt{3})$, both optimal design measures put mass 1 at $x = 0$.
2. If $|g| > \ln(2 + \sqrt{3})$, define

$$B(g) = \ln \frac{e^{4g} - 6e^{2g} + 1 + (e^{2g} - 1)\sqrt{e^{4g} - 14e^{2g} + 1}}{2(e^{3g} + e^{g})}.$$

   Assuming that $\pm B(g) \in \mathcal{X}$, both optimal design measures put mass 1/2 at each of $x = \pm$ B(g).

2.3 The Algebraic Approach

Chaloner (1993) considered only ϕ_1-optimality, and so for completeness a new algebraic proof is given here for ϕ_2-optimality, using the equivalence theorem.

The derivative for ϕ_2-optimality has the simple expression

$$d_2(\eta, x) = E\left[\frac{I(\theta, x)}{I(\theta, \eta)^2} - \frac{1}{I(\theta, \eta)}\right].$$

For $|g| \le \ln(2 + \sqrt{3})$, define η_0 to be the one-point design putting mass 1 at $x = 0$. Define

$$Q1 = 2e^{2x+2g} - e^{x+4g} + 2e^{x+3g} + 2e^{x+2g} + 2e^{x+g} - e^{x} + 2e^{2g}.$$

Then

$$d_2(\eta_0, x) = -\frac{e^{-g}(e^{g} + 1)^2(e^{x} - 1)^2(Q1)}{2(e^{x} + e^{g})^2(e^{x+g} + 1)^2}.$$

$Q1$ is a quadratic in e^x and it is straightforward to show that if $|g| \le \ln(2 + \sqrt{3})$, the quadratic is positive and the derivative has a single root at $x = 0$, and so η_0 is the ϕ_2-optimal design.

Now consider $|g| > \ln(2 + \sqrt{3})$ and let η_2 be the symmetric design which puts mass 1/2 at each of $\pm B(g)$, where $\pm B(g)$ is defined in Theorem 1. Define

$$Q2 = e^{2x+3g} + e^{2x+g} - e^{x+4g} + 6e^{x+2g} - e^{x} + e^{3g} + e^{g}.$$

Then the directional derivative $d_2(\eta_2, x)$ at the design η_2 is

$$d_2(\eta_2, x) = -\frac{8(e^g - 1)^2(e^g + 1)^2(Q2)^2}{(e^{2g} + 1)^4(e^x + e^g)^2(e^{x+g} + 1)^2}$$

which is clearly nonpositive for all x. It can be shown that the derivative equals zero only at $x = \pm B(g)$, and so the design η_2 is ϕ_2-optimal. The algebraic proof is therefore complete for ϕ_2-optimality.

The algebraic approach requires a candidate optimal design for which the derivative can be calculated and optimality verified. The proof is straightforward although algebraically tedious. The geometric approach of Haines, in contrast, gives a more direct approach to finding an optimal design.

2.4 The Geometric Approach

In Haines' approach slightly different notation is required. Let the design space $\mathcal{X} = \Re$. For fixed g, and x in $\Re$ define the curve $C(x)$ in $\Re^2$ to have horizontal and vertical coordinates $(I(-g, x), I(g, x))$. Define S to be the closed convex hull of C(x). Then S can also be written as

$$S = \text{conv}(C) = \{(I(-g, \eta), I(g, \eta)) \mid \eta \in \mathcal{H}\}.$$

S is then a convex set over which the criterion function is maximized. C and S are compact.

As x varies from $-\infty$ to $+\infty$, the point $(I(-g, x), I(g, x))$ moves from the origin, along the curve $C(x)$, and then back to the origin. $C(x)$ lies within the first quadrant of $\Re^2$ and defines a closed and bounded parametric curve. Since $C(x)$ has a parametric representation, the formula for the signed curvature κ_x of $C(x)$ has a simple form. Let $c_1 = I(-g, x)$ and $c_2 = I(g, x)$, and for $i = 1, 2$ define $\dot{c}_i = \frac{\partial}{\partial x} c_i$ and $\ddot{c}_i = \frac{\partial^2}{\partial x^2} c_i$. Then also define

$$\kappa_x = \frac{\dot{c}_1\ddot{c}_2 - \dot{c}_2\ddot{c}_1}{(\dot{c}_1{}^2 + \dot{c}_2{}^2)^2}.$$

First consider $|g| \le \ln(2 + \sqrt{3})$. If κ_x does not change sign as x varies from $-\infty$ to ∞, then curve C will be the boundary of S, denoted ∂S. κ_x is a quadratic function of e^x and it can be shown that κ_x does not change sign if and only if the equation $\kappa_x = 0$ has no real roots or one real root. This condition holds if and only if $|g| \le \ln(2 + \sqrt{3})$. The solid line in Figure 1 shows the curve $C(x)$ for $g = 1$, and Figure 2 shows $C(x)$ for $g = 2$.

For the prior distribution with mass $\frac{1}{2}$ at $\theta = \pm g$, and the design with all mass at x, the ϕ_2-criterion function for a one-point design η_x is

$$\phi_2(\eta_x) = -\left[\tfrac{1}{2}I(-g, x)^{-1} + \tfrac{1}{2}I(g, x)^{-1}\right].$$

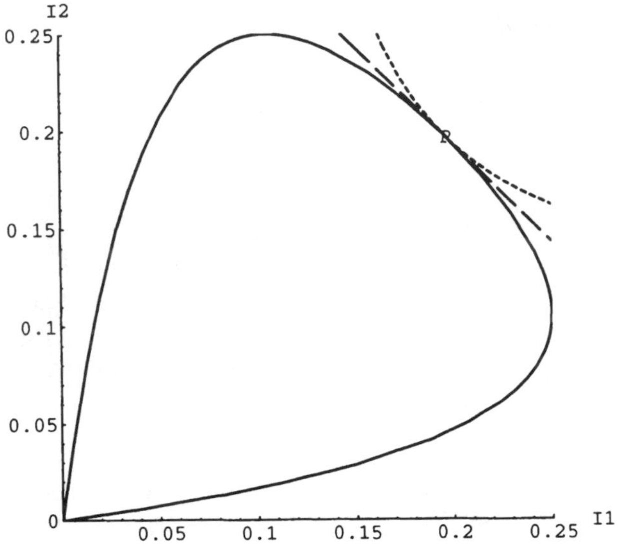

Figure 1. Symmetric prior distribution, $\theta = \pm 1$, and ϕ_2-optimality: C is the solid line, L the dashed line, and T_{z^*} the dotted line. For $z^* = -5.08616$, the maximum value of $\phi_2(\cdot)$, C, and T_{z^*} intersect at P

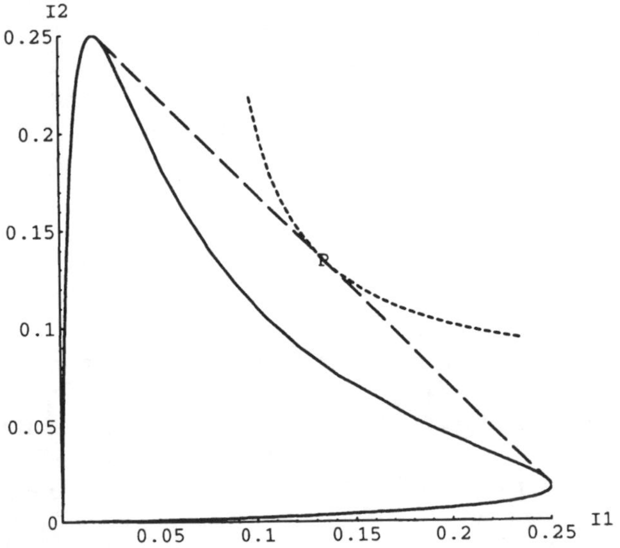

Figure 2. Symmetric prior distribution, $\theta = \pm 2$, and ϕ_2-optimality: C is the solid line, L the dashed line, and T_{z^*} the dotted line. For $z^* = -7.43479$, the maximum value of $\phi_2(\cdot)$, T_{z^*}, and ∂S intersect at P

Denote the horizontal and vertical coordinates as I_1 and I_2, respectively, and let

$$z = -\left[\frac{1}{2I_1} + \frac{1}{2I_2}\right]. \tag{2}$$

For $I_1 > 0$ and $I_2 > 0$ and a fixed value of $z \in \Re$, the set of points $\{(I_1, I_2)\}$ satisfying equation (2) describes a strictly convex contour curve. Let T_z be such a curve for a fixed value of z. From standard arguments in convex programming, in maximizing over S, when z is a maximum, say z^*, there is one T_{z^*} which intersects ∂S at a point P. This point represents the ϕ_2-optimal design. When $|g| \leq \ln(2+\sqrt{3})$, it is easy to see, by symmetry and from Figure 1, that the ϕ_2-optimal design is to put mass 1 at $x = 0$.

It is also easy to see that ϕ_1-optimality gives the same optimal design, as does any criterion $\phi(\cdot)$ with strictly convex contours which is also symmetric in the sense that $\phi(\eta_x) = \phi(\eta_{-x})$. Figure 1 shows the contour T_{z^*} and the curve $C(x)$ for ϕ_2-optimality and $g = 1$.

If $|g| > \ln(2+\sqrt{3})$, $C(x)$ is no longer the boundary of S. The solid line in Figure 2 is the curve $C(x)$ for $g = 2$. S can be constructed by first constructing the unique line segment L which joins two points $C(-x)$ and $C(x)$, and which is also tangent to the curve at those points. $C(x)$ is symmetric with respect to the line $I_1 = I_2$ which has slope equal to 1, so L must have a slope of -1.

Setting the slope of the line tangent to $C(x)$ at x equal to -1 and solving for x results in the two roots $x = \pm B(g)$, as defined earlier, so L is the line segment from the point $C(-B(g))$ to the point $C(B(g))$.

Using similar arguments as in the case when $|g| \leq \ln(2+\sqrt{3})$, it is clear that for any criterion with convex contours and the symmetry property that $\phi(\eta_x) = \phi(\eta_{-x})$, the point P where the optimal contour T_{z^*} intersects ∂S corresponds to a design with mass $1/2$ at each of $x = \pm B(g)$. Figure 2 shows the contour T_{z^*} for $g = 2$.

2.5 Discussion

In this special case of logistic regression with a symmetric two-point prior distribution on θ, the Bayesian ϕ_1- and ϕ_2-optimal designs are the same. This is not immediately clear using the algebraic approach. It is clear after noting that the algebraic forms of the directional derivatives at the optimal design are closely related. When $|g| \leq \ln(2+\sqrt{3})$, the quadratic term $Q1$ appears in the numerators of the directional derivatives for both ϕ_1- and ϕ_2-optimality. Specifically, denote d_2 to be the directional derivative for ϕ_2-optimality and d_1 to be the directional derivative for ϕ_1-optimality. Then

$$d_2(\eta_0, x) = e^{-g}(e^g + 1)^2 d_1(\eta_0, x)$$

and the directional derivatives must have the same roots. Similarly, when $|g| > \ln(2 + \sqrt{3})$, the quadratic term $Q2$ appears in the numerators of both directional derivatives, and

$$d_2(\eta_2, x) = \frac{8(e^g - 1)^2(e^g + 1)^2}{(e^{2g} + 1)^2} d_1(\eta_2, x).$$

In general, it is not the case that ϕ_1- and ϕ_2-optimal designs are the same. Haines (1995) gives conditions on the prior distribution for θ under which the ϕ_1- and ϕ_2-optimal designs are identical for this special case of logistic regression, and using the geometric approach these similarities are clearly apparent.

There is a connection between the geometric approach and the algebraic approach. In this example for ϕ_2-optimality, since the line L is tangent to T_{z^*} at P and is perpendicular to the gradient of T_{z^*} at P, the points (I_1, I_2) on L satisfy:

$$\frac{-1}{2}\frac{1}{I(-g, \eta^*)^2}(I_1 - I(-g, \eta^*)) + \frac{-1}{2}\frac{1}{I(g, \eta^*)^2}(I_2 - I(g, \eta^*)) = 0.$$

As the set S is contained in the lower half plane determined by L, substituting $I_1 = I(-g, x)$ and $I_2 = I(g, x)$ into the left-hand side will result in an expression which must be nonpositive. However, the resulting expression on the left-hand side is $d_2(\eta^*, x)$, and so we have $d_2(\eta^*, x) \leq 0$ as required by the equivalence theorem.

Similar connections can be shown for ϕ_1-optimality.

3. ASYMMETRIC PRIOR DISTRIBUTION

Consider the case when the two-point prior distribution for θ is asymmetric in the sense that θ takes values which are symmetric about zero, but the probabilities β_1 and $\beta_2 = 1 - \beta_1$ are not equal. Haines gave Theorem 2.

THEOREM 2 (Haines, 1995): Let the prior distribution $\pi(\theta)$ be such that $\theta = -g$ with probability β_1 and $\theta = +g$ with probability β_2.

(1) If $|g| \leq \ln(2 + \sqrt{3})$, define

$$R(g) = \ln\left[\frac{\sqrt{(\beta_2 - \beta_1)^2(e^{2g} - 1)^2 + 4e^{2g}}}{2e^g} + \frac{(e^{2g} - 1)(\beta_2 - \beta_1)}{2e^g}\right].$$

The ϕ_1-optimal design measure puts mass 1 at the point $R(g)$.

(2) If $|g| > \ln(2 + \sqrt{3})$, define $\rho(g)$ to be the slope of the line from the origin to the point C(B(g)) and $w = \frac{\beta_2\rho(g)-\beta_1}{\rho(g)-1}$. If $\frac{\beta_2}{\beta_1}$ satisfies

$$\frac{1}{\rho(g)} < \frac{\beta_2}{\beta_1} < \rho(g) \tag{3}$$

the ϕ_1-optimal design puts mass at $+B(g)$ and $-B(g)$ (with $\pm B(g)$ as in Theorem 1) with associated weights w and $(1 - w)$, respectively. If the inequality is not satisfied, then the optimal design puts mass 1 at the point $R(g)$ above.

Haines' geometric proof. A sketch of Haines' proof is given here. Note that $C(x)$ and S are as in Theorem 1.

First consider the case $|g| \leq \ln(2 + \sqrt{3})$. The optimal design must be a one point design, and it is straightforward to show algebraically that $\phi_1(\eta_x)$ is maximized at $x = R(g)$.

Second, let $|g| > \ln(2 + \sqrt{3})$. The only difference between the symmetric case is the shape of the contours T_z. The optimal contour curve T_{z^*} must touch ∂S at a point. Either the point corresponds to $x = R(g)$ or the optimal design has two support points $\pm B(g)$. All that remains is to find conditions on β_1 and β_2 which determine whether the optimal design is a one- or two-point design.

The slope of the line L joining $C(B(g))$ to $C(-B(g))$ is -1. So the slope of the line tangent to T_{z^*} at that point where it touches ∂S on L is also -1. The locus of all points on the family of curves $\{T_z \mid z \in \Re\}$ must be found at which the slope of the line tangent to T_z is -1. This locus is a line through the origin with slope $\frac{\beta_2}{\beta_1}$. Denote this line as L''.

Let $\rho(g)$ be the slope of a line from the origin to the point C(B(g)). Then the line L'' will intersect the line segment L' if and only if condition (3) holds. The point of intersection is a weighted combination of the points $C(+B(g))$ and $C(-B(g))$. The weight w is easily found to be $\frac{\beta_2\rho(g)-\beta_1}{\rho(g)-1}$. Therefore the ϕ_1-optimal design is to put weight w at $B(g)$ and weight $1 - w$ at $-B(g)$. If the condition in condition (3) does not hold, then T_{z^*} and ∂S intersect at the point $R(g)$ on C, so the ϕ_1-optimal design is to put mass 1 at $R(g)$. This completes the geometric proof.

An algebraic proof for an asymmetric prior distribution has been found to be intractable so far. The geometric approach has provided optimal designs when the algebraic approach has not. For any criterion with convex contours, Haines' geometric argument shows that the optimal design is either a one-point design or a two-point design with support at $\pm B(g)$.

The shape of the curve $C(x)$ depends only on the support points of the prior distribution and not on the probabilities at those points. The contour curves T_z, however, depend on the probabilities for θ. The curves will be moved closer to the axis which represents the Fisher information with the larger probability for θ.

For $|g| > \ln(2+\sqrt{3})$, consider β_2 getting larger. As β_2 gets larger, the curve T_{z^*} will tilt toward the I_2 axis, eventually touching ∂S at a point on C outside of the line segment L. In that case, a one-point design will be optimal, and it can be shown algebraically that this is the point $C(R(2))$ as defined in the statement of Theorem 2. It can also be shown that when $\frac{\beta_2}{\beta_1} = \rho(g)$, $R(g) = B(g)$, so the change from a one-point ϕ_1-optimal design to a two-point ϕ_1-optimal design occurs in a continuous manner.

This example is interesting because when a two-point design is optimal, the design is a weighted combination of the same two optimal design points $\pm B(g)$ found with a symmetric prior, a fact which is not apparent when using algebraic methods but is obvious with the geometric approach.

4. THREE-POINT PRIOR DISTRIBUTION

In this section, the optimal design problem for the special case of logistic regression is examined for a three-point, symmetric, prior distribution on θ. A new algebraic result is presented and the geometric argument discussed.

4.1 The Algebraic Approach

THEOREM 3: Let the prior distribution be such that $\theta = \{\pm g, 0\}$ each with probability 1/3. If $|g| \leq \ln(3+2\sqrt{2})$, the ϕ_1-optimal design measure puts mass 1 at 0.

Proof of theorem 3. By symmetry, the best one-point design puts mass 1 at $x = 0$. Define $Q3$ to be the following quartic in e^x:

$$\begin{aligned} Q3 = {} & 3e^{2g}e^{4x} + (-e^{4g} + 4e^{3g} + 6e^{2g} + 4e^{g} - 1)e^{3x} \\ & + (-e^{4g} + 4e^{3g} + 12e^{2g} + 4e^{g} - 1)e^{2x} \\ & + (-e^{4g} + 4e^{3g} + 6e^{2g} + 4e^{g} - 1)e^{x} + 3e^{2g}. \end{aligned}$$

Then the derivative at η_0 in the direction x is

$$d(\eta_0, x) = \frac{-(e^x - 1)^2(Q3)}{3(e^x + 1)^2(e^x + e^g)^2(e^{x+g} + 1)^2}.$$

If $Q3$ is strictly positive, the derivative will have one root at x = 0.

Let the standard form of a quartic in y be

$$ay^4 + 4by^3 + 6cy^2 + 4dy + f.$$

Further, let $y = e^x$ and write $Q3$ in standard form with

$$a = (3e^{2g})$$
$$4b = (-e^{4g} + 4e^{3g} + 6e^{2g} + 4e^g - 1)$$
$$6c = (-e^{4g} + 4e^{3g} + 12e^{2g} + 4e^g - 1)$$
$$4d = (-e^{4g} + 4e^{3g} + 6e^{2g} + 4e^g - 1)$$
$$f = 3e^{2g}.$$

Then from standard theory of equations (for example Barnard and Child, 1964, p. 186), let $H = ac - b^2$, $I = af - 4bd + 3c^2$, $J = acf + 2bcd - ad^2 - c - fb^2$, and $\Delta = I^3 - 27J^2$. Conditions on g must be satisfied so that the quartic is strictly positive; that is, all four roots are imaginary. This will happen if and only if $\Delta > 0$ and at least one of H and $2HI - 3aJ$ is positive.

Now

$$H = -\frac{(e^g - 1)^4(e^g + 1)^2(e^{2g} - 6e^g + 1)}{16}$$

so H has roots at $g = 0$ and at $g = \pm\ln(3 + 2\sqrt{2})$. When $|g| < \ln(3 + 2\sqrt{2})$, then $(e^{2g} - 6e^g + 1)$ will be negative and H will be positive. Similarly,

$$\Delta = -\frac{3(e^g - 1)^{12}(e^g + 1)^2(e^{2g} - 6e^g + 1)(e^{4g} - 4e^{3g} - 6e^{2g} - 4e^g + 1)^2}{256}$$

so Δ will also be positive when $|g| < \ln(3 + 2\sqrt{2})$. That is, when $|g| < \ln(3 + 2\sqrt{2})$, both Δ and H will be strictly positive, so the quartic will have four imaginary roots and thus be strictly positive.

Therefore, when $|g| \leq \ln(3 + 2\sqrt{2})$ the derivative $d(\eta_0, x)$ is nonpositive and has a single root at $x = 0$, thus verifying that the best one-point design is the ϕ_1-optimal design. This completes the proof of Theorem 3. □

Agin (1997) examines three-point prior distributions with $|g| > \ln(3 + 2\sqrt{2})$. She shows that for $|g| > \ln(3 + 2\sqrt{2})$ and $|g|$ simultaneously less than about 2.29, a two-point design is ϕ_1-optimal. She derives a lengthy closed-form algebraic expression for the optimal design points, denoted $\pm D(g)$, and shows that the transition from one to two points happens continuously. She also shows numerically that for $|g|$ greater than about 2.29, a three-point design is ϕ_1-optimal, with the weights not necessarily being equal. The transition from two to three points is continuous.

For ϕ_2-optimality, Agin (1997) also gives the following result and proves it using the algebraic approach.

THEOREM 4: Let the prior distribution be such that $\theta = \{\pm g, 0\}$, each with probability 1/3. If $|g| \leq \ln(\frac{5+\sqrt{21}}{2})$, the ϕ_2-optimal design measure puts mass 1 at 0.

The proof again involves finding the roots of a quartic polynomial in e^x and is a straightforward application of the equivalence theorem, but algebraically very cumbersome.

4.2 The Geometric Approach

For this prior distribution with three support points, at $\theta = 0$ and $\theta = \pm g$, a geometric approach seems initially promising. The curve $C(x)$ is a curve in three dimensions with coordinates $(I(-g, x), I(0, x), I(g, x))$. Figure 3 shows the curve $C(x)$ for $g = 1$, and it is immediately clear that for any prior distribution with equal prior mass at ± 1 and any positive mass at 0, a one-point design at $x = 0$ must be optimal. Figure 3 also shows the supporting hyperplane at $x = 0$ and the convex surface T_{z^*}. Figure 4 shows $C(x)$ for

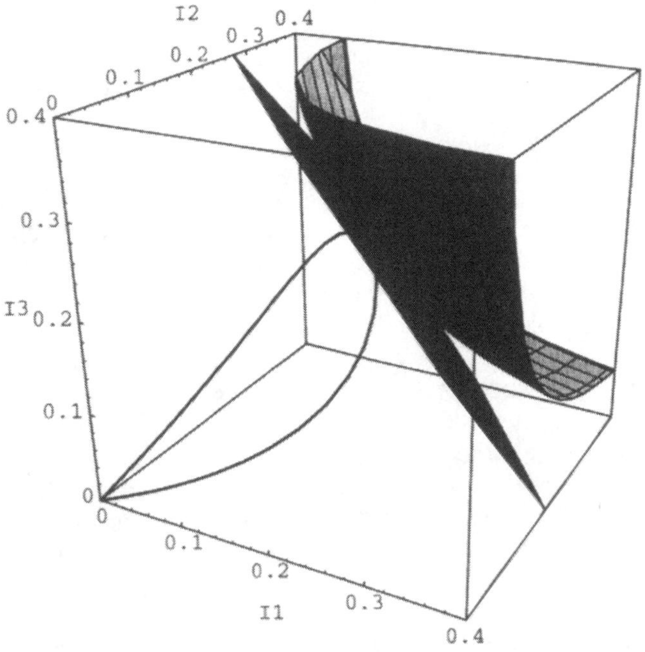

Figure 3. The curve C, supporting hyperplane H, and convex surface T_{z^*} for a prior distribution with probability 1/3 at each of 0 and ± 1 under ϕ_1-optimality

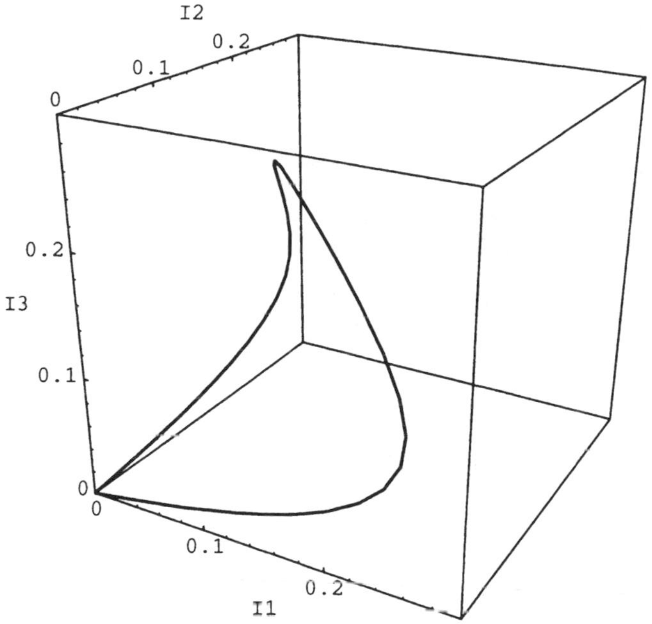

Figure 4. The curve $C(x)$ for a three-point prior distribution on 0 and ± 2

$g = 2$, and it is clear that more than one point may be needed. Figure 5 is the same curve from a different perspective showing the two points $\pm D(2) = \pm 0.89035$. These two points are the optimal design points for a three-point prior distribution with equal mass on $-2, 0, 2$. Figure 6 shows $C(x)$ for $g = 3.5$, and it is clear that three points may be required.

Deriving exact algebraic results geometrically, however, has so far proved intractable. For $|g| > \ln(3 + 2\sqrt{2})$, it is not easy to define the convex hull of C. Both the algebraic and geometric approach are difficult to use in three dimensions.

5. MULTIPLE-POINT PRIOR DISTRIBUTIONS

Denote $\pi_m(\theta)$ as the discrete uniform prior distribution, with mean zero, which puts mass $1/m$ at m equally spaced points. A result in Chaloner (1993) proves that for small enough support in the prior distribution a one-point design is ϕ_1-optimal. That one-point design must be at $x = 0$. For fixed m, let g_m be the largest value of $|\theta|$ with positive support in $\pi_m(\theta)$ such that a one-point design is ϕ_1-optimal. From Theorem 1, $g_2 = \ln(2 + \sqrt{3}) = 1.3169$,

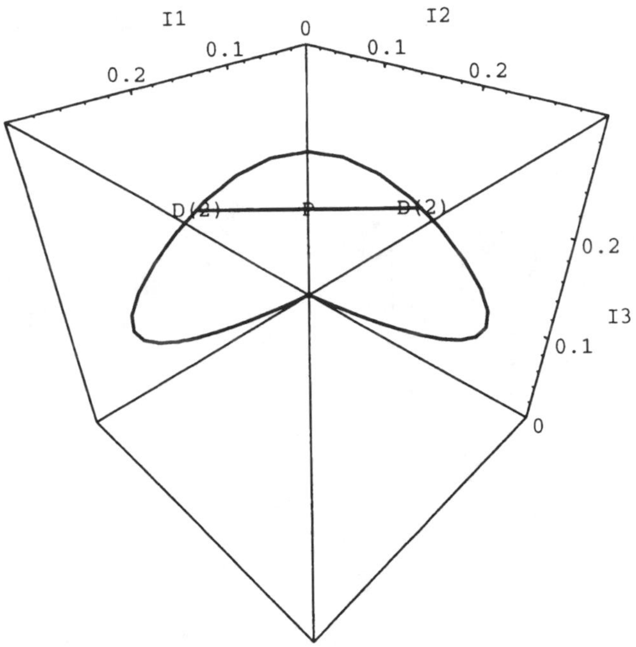

Figure 5. The curve $C(x)$ for a three-point prior distribution on 0 and ± 2, perspective from $(1, 1, 1)$. The points $C(D(2))$ and $C(D(-2))$ and the line joining them is indicated. The point P is $\frac{1}{2}C(D(-2)) + \frac{1}{2}C(D(2))$

and from Theorem 3, $g_3 = \ln(3 + 2\sqrt{2}) = 1.7627$. Numerical results show that g_m is increasing in m with $g_{29} = 2.4875$ and $g_{30} = 2.4904$.

6. CONCLUSION

In this paper, the geometric and algebraic approaches to finding closed form optimal designs for a one-parameter logistic regression model have been reviewed and discussed. Although closed-form solutions are important in understanding the problem in general, they are surprisingly difficult to find. The two different approaches complement each other and help to give an intuitive understanding of the problem.

ACKNOWLEDGMENT

This work was partly supported by a grant from the National Security Agency.

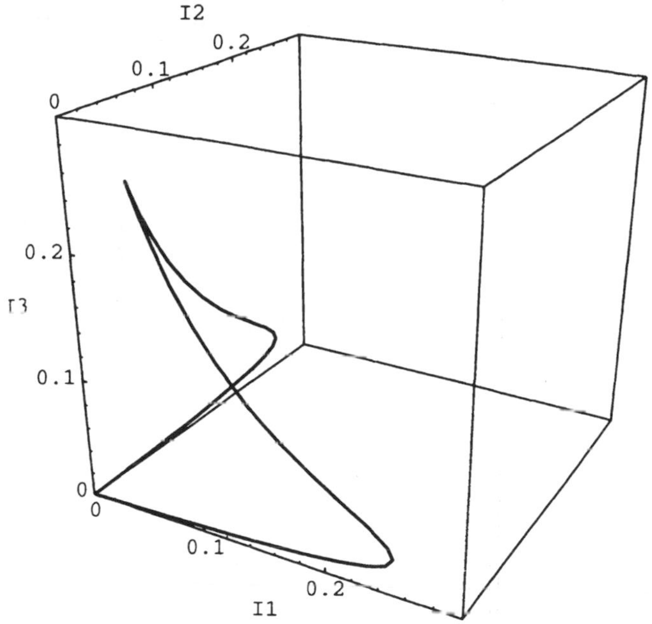

Figure 6. The curve $C(x)$ for any prior distribution with support on 0 and ± 3.5

REFERENCES

Agin, MA. Optimal Bayesian designs for nonlinear models. PhD Thesis, University of Minnesota, 1997.

Barnard, S and Child, JM. *Higher Algebra*. London: Macmillan, 1964.

Chaloner, K. An approach to design for generalized linear models. In: VV Fedorov and H Läuter, eds. *Model Oriented Data Analysis*. Lecture Notes in Economics and Mathematical Systems, Berlin: Springer, pp. 3–12, 1987.

Chaloner, K. A note on optimal Bayesian design for nonlinear problems. *J. Stat. Plann. Inference*, 37:229–235, 1993.

Chaloner, K and Larntz, K. Optimal Bayesian experimental design applied to logistic regression experiments. *J. Stat. Plann. Inference*, 21:191–208, 1989.

Chaloner, K and Verdinelli, I. Bayesian experimental design: A review. *Stat. Sci.*, 10:273–304, 1995.

Chernoff, H. Locally optimal designs for estimating parameters. *Ann. Math. Stat.*, 24:586–602, 1953.

Haines, LM. A geometric approach to optimal design for one-parameter non-linear models. *J. R. Stat. Soc., Ser. B*, 57:575–598, 1995.

Tsutakawa, RK. Design of an experiment for bioassay. *J. Am. Stat. Assoc.*, 67:584–590, 1972.

Whittle P. Some general points in the theory and construction of D-optimum experimental designs. *J. R. Stat. Soc., Ser. B*, 35:123–130, 1973.

22

Structure of Weighing Matrices of Small Order and Weight

HIROYUKI OHMORI Ehimc University, Matsuyama, Japan

TERUHIRO SHIRAKURA Kobe University, Kobe, Japan

1. INTRODUCTION

A weighing matrix $W = W(n, k)$ of order n and weight k is an $n \times n$ matrix with elements $+1$, -1, and 0 such that $W W^t = kI_n$, $k \leq n$, where I_n is the identity matrix of order n and W^t denotes the transpose of W. In particular, a $W(n, n)$ matrix is called a Hadamard matrix of order n. The following existence theorems for weighing matrices of order n and weight k are well known (cf. Geramita and Seberry, 1979):

E1. If n is odd then a $W(n, k)$ only exists if
   (i) k is a square
   (ii) $(n - k)^2 + (n - k) + 1 \geq n$.
E2. If $n \equiv 2$ (*mod* 4), then for a $W(n, k)$ to exist
   (i) $k \leq n - 1$
   (ii) k is the sum of two squares.

It was unknown whether a $W(17, 9)$ exists or not (see Craigen, 1996). Recently, Ohmori and Miyamoto (1997) have constructed a $W(17, 9)$. On the other hand, it is easy to show that a necessary condition for the existence of a Hadamard matrix of order n is $n = 1, 2$ or $n = 4t$, where t is a positive integer. Hadamard (1893) conjectured that the condition is also sufficient. The smallest order which is unknown whether Hadamard's conjecture is correct or not is 428 (Craigen, 1996).

The classification problem of weighing matrices is very interesting as well as the existence problem. Two weighing (Hadamard) matrices are said to be equivalent if one can be transformed into the other by using the following operations:

O1. multiply any row or column by -1;
O2. interchange two rows or two columns.

The classification of such matrices is related to several areas, for example, coding theory, cryptography, and graph theory. Therefore, it may be useful for obtaining a new series of matrices. It is known that the complete classification of Hadamard matrices whose orders are less than or equal to 28 has been accomplished. It is easy to show that the Hadamard matrices of orders up to 12 are unique up to equivalence. Hall (1961, 1965) showed that there are five and three inequivalent Hadamard matrices of order 16 and 20, respectively. Ito et al. (1981) showed that there are 59 inequivalent Hadamard matrices of order 24, but this classification was not complete. A new 60th matrix was found by Kimura (1989). Furthermore, Kimura and Ohmori (1987), and Kimura (1988, 1994) showed that there are 487 inequivalent matrices of order 28. The classification of Hadamard matrices of order 32 is not yet complete. On the other hand, the classification of weighing matrices has been completed in the following cases (see Chan et al., 1986; Ohmori, 1989, 1992, 1993):

C1. $W(n, k)$s, where $k < 6$ and $k \leq n$;
C2. $W(n, k)$s, where $1 \leq k \leq n \leq 13$;
C3. $W(14, k)$s, where $k \leq 8$. $W(14, 9)$s and $W(14, 10)$s are not yet completed.

The number of inequivalent weighing matrices for each order ≤ 14 are listed in Table 1. Chan et al. (1986) completed the classification of weighing matrices with weight $k \leq 5$ for every order n. Therefore those matrices may be deleted from the table.

As the next step of the investigation, it is appropriate to consider the classification problem of weighing matrices with weight six. As seen from Section 2, it is very difficult to classify weighing matrices with weight six for every order n. Therefore we shall deal with the classification of weighing matrices of the semi-biplane type. A set of matrices including all inequivalent weighing matrices of the semi-biplane type is constructed in Section 2.

Table 1. The number of inequivalent weight matrices $W(n, k)$*

n	k		References
8	5	1	Chan et al. (1986)
	6	1	Chan et al. (1986)
	7	1	Chan et al. (1986)
	8	1	Chan et al. (1986)
10	5	1	Chan et al. (1986)
	8	1	Chan et al. (1986)
	9	1	Chan et al. (1986)
12	5	2	Chan et al. (1986)
	6	7	Ohmori (1989)
	7	3	Ohmori (1989)
	8	6	Ohmori (1989)
	9	4	Ohmori (1989)
	10	4	Ohmori (1989)
	11	1	Chan et al. (1986)
	12	1	Chan et al. (1986)
13	9	8	Ohmori (1993)
14	5	3	Chan et al. (1986)
	8	65	Ohmori (1992)
	9	≥ 2	Chan et al. (1986)
	10	≥ 3	Chan et al. (1986)
	13	1	Chan et al. (1986)

*The numbers in the 3rd column denote the numbers of inequivalent weighing matrices $W(n, k)$.

Matrices of the semi-biplane type have been treated by Hughes (1977), and have certain interesting geometrical properties. In Section 3, we give the complete classification of weighing matrices with weight six of the semi-biplane type. In Section 4, weighing matrices with weight k and order 2^{k-1} of the semi-biplane type are constructed for $k \geq 2$.

2. CONSTRUCTION

Let W be an $n \times n$ weighing matrix with weight n ($n \geq 6$). Then it can be assumed, without loss of generality, that the first 6×6 submatrix of W is

$$\begin{bmatrix} 1 & \cdots & 1 \\ \vdots & a_{ij} & \\ 1 & & \end{bmatrix}, \qquad a_{ij} \in \{1, -1, 0\}$$

whose first row (column) is orthogonal to the other rows (columns), where

$$1 \le \Sigma_{i=2}^{6} |a_{ij}| \le 5 (2 \le j \le 6), \qquad 1 \le \Sigma_{j=2}^{6} |a_{ij}| \le 5 (2 \le i \le 6)$$

DEFINITION 1: *A* 6×6 *matrix satisfying the above conditions is called a feasible matrix, and a* 6×6 *submatrix of* W *such that the first row and column are of nonzero elements is called a* Γ*-submatrix of* W.

The following lemma can easily be proved.

LEMMA 1: *There are 25 feasible matrices, say* $F_1, F_2, \ldots, F_{25}$, *up to equivalence.*

These feasible matrices are listed in Figure 1.

REMARK 1: *A* Γ*-submatrix of* W *is equivalent to any one of* $F_1, F_2, \ldots, F_{25}$.

REMARK 2: *There is no weighing matrix with weight 6 including a* Γ*-submatrix* Γ_i, *which is equivalent to* F_i, *for* $i = 7$–$13, 15$–$17, 19$–20. *There is a unique weighing matrix, up to equivalence, based on* F_i, *for each* $i = 4$–$6, 14, 18, 23$. *For the other* F_i, *it is very difficult to obtain weighing matrices with weight 6 based on* F_i.

DEFINITION 2: *Let* $\mathbf{F} = \{F_{i_1}, \ldots, F_{i_l}\}$ *be a subset of* $\{F_1, \ldots, F_{25}\}$ *given as:*

(i) $1 \le i_1 < i_2 < \cdots < i_l \le 25$;
(ii) *for each* i_s, *there are* Γ*-submatrices of* W *which are equivalent to* F_{i_s}, *where* $1 \le s \le l$.

Then W *is called a matrix of type* F_{i_l} *if for any* Γ*-submatrix* Γ *of* W, *there is a* $F_{i_s} (\in \mathbf{F})$ *which is equivalent to* Γ.

In order to complete the classification of weighing matrices with weight six, irreducible weighing matrices of each type F_i may be constructed and classified in order.

$$F_1 = \begin{bmatrix} 1&1&1&1&1&1\\ 1&-&\cdot&\cdot&\cdot&\cdot\\ 1&\cdot&-&\cdot&\cdot&\cdot\\ 1&\cdot&\cdot&-&\cdot&\cdot\\ 1&\cdot&\cdot&\cdot&-&\cdot\\ 1&\cdot&\cdot&\cdot&\cdot&- \end{bmatrix} \quad F_2 = \begin{bmatrix} 1&1&1&1&1&1\\ 1&1&-&-&\cdot&\cdot\\ 1&-&\cdot&\cdot&\cdot&\cdot\\ 1&-&\cdot&\cdot&\cdot&\cdot\\ 1&\cdot&\cdot&\cdot&-&\cdot\\ 1&\cdot&\cdot&\cdot&\cdot&- \end{bmatrix} \quad F_3 = \begin{bmatrix} 1&1&1&1&1&1\\ 1&1&-&-&\cdot&\cdot\\ 1&-&\cdot&\cdot&\cdot&\cdot\\ 1&-&\cdot&\cdot&\cdot&\cdot\\ 1&\cdot&1&\cdot&-&-\\ 1&\cdot&-&\cdot&\cdot&\cdot \end{bmatrix} \quad F_4 = \begin{bmatrix} 1&1&1&1&1&1\\ 1&1&-&-&\cdot&\cdot\\ 1&-&1&\cdot&-&\cdot\\ 1&-&\cdot&\cdot&\cdot&\cdot\\ 1&\cdot&-&\cdot&\cdot&\cdot\\ 1&\cdot&\cdot&\cdot&\cdot&- \end{bmatrix} \quad F_5 = \begin{bmatrix} 1&1&1&1&1&1\\ 1&1&-&-&\cdot&\cdot\\ 1&-&1&-&\cdot&\cdot\\ 1&-&-&1&\cdot&\cdot\\ 1&\cdot&\cdot&\cdot&-&\cdot\\ 1&\cdot&\cdot&\cdot&\cdot&- \end{bmatrix}$$

$$F_6 = \begin{bmatrix} 1&1&1&1&1&1\\ 1&1&-&-&\cdot&\cdot\\ 1&-&1&\cdot&-&\cdot\\ 1&-&\cdot&\cdot&\cdot&\cdot\\ 1&\cdot&-&1&\cdot&-\\ 1&\cdot&\cdot&-&\cdot&\cdot \end{bmatrix} \quad F_7 = \begin{bmatrix} 1&1&1&1&1&1\\ 1&1&-&-&\cdot&\cdot\\ 1&-&\cdot&\cdot&\cdot&\cdot\\ 1&-&\cdot&\cdot&\cdot&\cdot\\ 1&\cdot&1&-&-&\cdot\\ 1&\cdot&-&1&\cdot&- \end{bmatrix} \quad F_8 = \begin{bmatrix} 1&1&1&1&1&1\\ 1&1&-&-&\cdot&\cdot\\ 1&-&1&-&\cdot&\cdot\\ 1&-&\cdot&\cdot&\cdot&\cdot\\ 1&\cdot&-&1&-&\cdot\\ 1&\cdot&\cdot&\cdot&\cdot&- \end{bmatrix} \quad F_9 = \begin{bmatrix} 1&1&1&1&1&1\\ 1&1&-&-&\cdot&\cdot\\ 1&-&1&\cdot&-&\cdot\\ 1&-&\cdot&\cdot&\cdot&\cdot\\ 1&\cdot&-&1&-&\cdot\\ 1&\cdot&\cdot&-&1&- \end{bmatrix} \quad F_{10} = \begin{bmatrix} 1&1&1&1&1&1\\ 1&1&-&-&\cdot&\cdot\\ 1&-&1&-&\cdot&\cdot\\ 1&-&-&\cdot&1&\cdot\\ 1&\cdot&\cdot&1&-&-\\ 1&\cdot&\cdot&\cdot&-&\cdot \end{bmatrix}$$

$$F_{11} = \begin{bmatrix} 1&1&1&1&1&1\\ 1&1&-&-&\cdot&\cdot\\ 1&-&\cdot&\cdot&1&-\\ 1&-&\cdot&\cdot&\cdot&\cdot\\ 1&\cdot&1&\cdot&-&-\\ 1&\cdot&-&\cdot&-&1 \end{bmatrix} \quad F_{12} = \begin{bmatrix} 1&1&1&1&1&1\\ 1&1&-&-&\cdot&\cdot\\ 1&-&\cdot&\cdot&1&-\\ 1&-&\cdot&\cdot&\cdot&\cdot\\ 1&\cdot&1&-&-&\cdot\\ 1&\cdot&-&1&-&\cdot \end{bmatrix} \quad F_{13} = \begin{bmatrix} 1&1&1&1&1&1\\ 1&1&-&-&\cdot&\cdot\\ 1&-&1&\cdot&-&\cdot\\ 1&-&\cdot&\cdot&1&-\\ 1&\cdot&-&1&-&\cdot\\ 1&\cdot&\cdot&-&\cdot&\cdot \end{bmatrix} \quad F_{14} = \begin{bmatrix} 1&1&1&1&1&1\\ 1&1&-&-&\cdot&\cdot\\ 1&-&1&\cdot&-&\cdot\\ 1&-&\cdot&-&1&\cdot\\ 1&\cdot&-&1&-&\cdot\\ 1&\cdot&\cdot&\cdot&\cdot&- \end{bmatrix} \quad F_{15} = \begin{bmatrix} 1&1&1&1&1&1\\ 1&1&-&-&\cdot&\cdot\\ 1&-&1&\cdot&-&\cdot\\ 1&-&\cdot&\cdot&\cdot&\cdot\\ 1&\cdot&-&\cdot&1&-\\ 1&\cdot&\cdot&\cdot&-&\cdot \end{bmatrix}$$

$$F_{16} = \begin{bmatrix} 1&1&1&1&1&1\\ 1&1&-&-&\cdot&\cdot\\ 1&-&\cdot&\cdot&1&-\\ 1&-&\cdot&\cdot&-&1\\ 1&\cdot&1&-&-&\cdot\\ 1&\cdot&-&1&\cdot&- \end{bmatrix} \quad F_{17} = \begin{bmatrix} 1&1&1&1&1&1\\ 1&1&-&-&\cdot&\cdot\\ 1&-&1&\cdot&-&\cdot\\ 1&-&\cdot&\cdot&1&-\\ 1&\cdot&-&1&\cdot&-\\ 1&\cdot&\cdot&-&-&1 \end{bmatrix} \quad F_{18} = \begin{bmatrix} 1&1&1&1&1&1\\ 1&1&-&-&\cdot&\cdot\\ 1&-&-&\cdot&1&\cdot\\ 1&-&\cdot&\cdot&-&1\\ 1&\cdot&1&-&\cdot&-\\ 1&\cdot&\cdot&1&-&- \end{bmatrix} \quad F_{19} = \begin{bmatrix} 1&1&1&1&1&1\\ 1&1&-&-&\cdot&\cdot\\ 1&-&1&-&\cdot&\cdot\\ 1&-&\cdot&\cdot&1&-\\ 1&\cdot&-&\cdot&-&1\\ 1&\cdot&\cdot&1&-&- \end{bmatrix} \quad F_{20} = \begin{bmatrix} 1&1&1&1&1&1\\ 1&1&-&-&\cdot&\cdot\\ 1&-&1&\cdot&-&\cdot\\ 1&-&\cdot&1&\cdot&-\\ 1&\cdot&-&\cdot&1&-\\ 1&\cdot&\cdot&-&-&1 \end{bmatrix}$$

$$F_{21} = \begin{bmatrix} 1&1&1&1&1&1\\ 1&1&1&-&-&-\\ 1&-&\cdot&\cdot&\cdot&\cdot\\ 1&-&\cdot&\cdot&\cdot&\cdot\\ 1&\cdot&-&\cdot&\cdot&\cdot\\ 1&\cdot&-&\cdot&\cdot&\cdot \end{bmatrix} \quad F_{22} = \begin{bmatrix} 1&1&1&1&1&1\\ 1&1&1&-&-&-\\ 1&-&\cdot&1&-&\cdot\\ 1&-&\cdot&-&1&\cdot\\ 1&\cdot&-&\cdot&\cdot&\cdot\\ 1&\cdot&-&\cdot&\cdot&\cdot \end{bmatrix} \quad F_{23} = \begin{bmatrix} 1&1&1&1&1&1\\ 1&1&1&-&-&-\\ 1&-&\cdot&1&-&\cdot\\ 1&-&\cdot&\cdot&\cdot&\cdot\\ 1&\cdot&-&-&1&\cdot\\ 1&\cdot&-&\cdot&\cdot&\cdot \end{bmatrix} \quad F_{24} = \begin{bmatrix} 1&1&1&1&1&1\\ 1&1&1&-&-&-\\ 1&-&\cdot&1&-&\cdot\\ 1&-&\cdot&-&\cdot&1\\ 1&\cdot&-&\cdot&1&-\\ 1&\cdot&-&\cdot&\cdot&\cdot \end{bmatrix} \quad F_{25} = \begin{bmatrix} 1&1&1&1&1&1\\ 1&-&1&1&-&-\\ 1&1&-&\cdot&-&\cdot\\ 1&1&\cdot&-&\cdot&-\\ 1&-&-&\cdot&1&\cdot\\ 1&-&\cdot&-&\cdot&1 \end{bmatrix}$$

Figure 1. The 25 feasible matrices

REMARK 3: *In particular, let* $W = W(n, 6)$ *be a weighing matrix of type* F_1 *and* $\overline{W}$ *be the incidence matrix obtained by ignoring the signs of elements of* W*. Then the number of columns (rows) having common nonzero elements for any two rows (columns) of* $\overline{W}$ *is 2 or 0. Such a matrix* $\overline{W}$ *is called the incidence matrix of a semi-biplane of parameters* $(n, 6)$ *(Hughes, 1977). Thus, a weighing matrix of type* F_1 *and order n may be called a matrix of the semi-biplane type of parameters* $(n, 6)$.

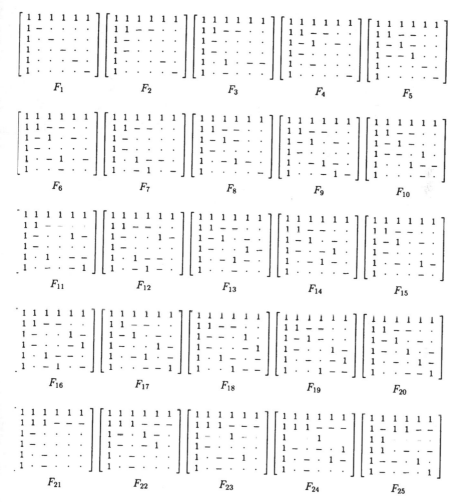

In this paper, we consider a weighing matrix of type F_1. Let W be a matrix of order n and weight six. Then without loss of generality, a 16×16 submatrix (say $\tilde{W}$) consisting of the first 16 rows and columns of W can be assumed to be

$$
\tilde{W} = \left[\begin{array}{cccccc|cccccccccc}
1 & 1 & 1 & 1 & 1 & 1 & \cdot & \cdot & \cdot & \cdot & \cdot & \cdot & \cdot & \cdot & \cdot & \cdot \\
1 & - & \cdot & \cdot & \cdot & \cdot & 1 & 1 & 1 & 1 & \cdot & \cdot & \cdot & \cdot & \cdot & \cdot \\
1 & \cdot & - & \cdot & \cdot & \cdot & - & \cdot & \cdot & \cdot & 1 & 1 & 1 & \cdot & \cdot & \cdot \\
1 & \cdot & \cdot & - & \cdot & \cdot & \cdot & - & \cdot & \cdot & - & \cdot & \cdot & 1 & 1 & \cdot \\
1 & \cdot & \cdot & \cdot & - & \cdot & \cdot & \cdot & - & \cdot & \cdot & - & \cdot & - & \cdot & 1 \\
1 & \cdot & \cdot & \cdot & \cdot & - & \cdot & \cdot & \cdot & - & \cdot & \cdot & - & \cdot & - & - \\
\hline
\cdot & 1 & - & \cdot & \cdot & \cdot & & & & & & & & & & \\
\cdot & 1 & \cdot & - & \cdot & \cdot & & & & & & & & & & \\
\cdot & 1 & \cdot & \cdot & - & \cdot & & & & & & & & & & \\
\cdot & 1 & \cdot & \cdot & \cdot & - & & & & & & & & & & \\
\cdot & \cdot & 1 & - & \cdot & \cdot & & & & & \mathbf{S} & & & & & \\
\cdot & \cdot & 1 & \cdot & - & \cdot & & & & & & & & & & \\
\cdot & \cdot & 1 & \cdot & \cdot & - & & & & & & & & & & \\
\cdot & \cdot & \cdot & 1 & - & \cdot & & & & & & & & & & \\
\cdot & \cdot & \cdot & 1 & \cdot & - & & & & & & & & & & \\
\cdot & \cdot & \cdot & \cdot & 1 & - & & & & & & & & & &
\end{array}\right]
$$

Here the notations $-$ and $\cdot$ stand for -1 and 0 in all vectors and matrices, respectively. For convenience, the symbols $\{\times, \ast, \diamond, \star, \circ, a_1, a_2, \ldots, a_{10}\}$ may be used for elements of S as follows:

$$
S = \begin{bmatrix}
a_1 & \times & \times & \times & \ast & \ast & \ast & & & \\
\times & a_2 & \times & \times & \diamond & & & \diamond & \diamond & \\
\times & \times & a_3 & \times & & \star & & \star & & \star \\
\times & \times & \times & a_4 & & & \circ & & \circ & \circ \\
\ast & \diamond & & & a_5 & \ast & \ast & \diamond & \diamond & \\
\ast & & \star & & \ast & a_6 & \ast & \star & & \star \\
\ast & & & \circ & \ast & \ast & a_7 & & \circ & \circ \\
 & \diamond & \star & & \diamond & \star & & a_8 & \diamond & \star \\
 & \diamond & & \circ & \diamond & & \circ & \diamond & a_9 & \circ \\
 & & \star & \circ & & \star & \circ & \star & \circ & a_{10}
\end{bmatrix}
$$

Note that a position with no mark is called a free position of S. Furthermore, the five submatrices C_i of S may be given as:

$$
C_1 = \begin{bmatrix}
a_1 & \times & \times & \times \\
\times & a_2 & \times & \times \\
\times & \times & a_3 & \times \\
\times & \times & \times & a_4
\end{bmatrix}, \qquad
C_2 = \begin{bmatrix}
a_1 & \ast & \ast & \ast \\
\ast & a_5 & \ast & \ast \\
\ast & \ast & a_6 & \ast \\
\ast & \ast & \ast & a_7
\end{bmatrix}
$$

$$C_3 = \begin{bmatrix} a_2 & \diamond & \diamond & \diamond \\ \diamond & a_5 & \diamond & \diamond \\ \diamond & \diamond & a_8 & \diamond \\ \diamond & \diamond & \diamond & a_9 \end{bmatrix}, \quad C_4 = \begin{bmatrix} a_3 & \star & \star & \star \\ \star & a_6 & \star & \star \\ \star & \star & a_8 & \star \\ \star & \star & \star & a_{10} \end{bmatrix}$$

$$C_5 = \begin{bmatrix} a_4 & \circ & \circ & \circ \\ \circ & a_7 & \circ & \circ \\ \circ & \circ & a_9 & \circ \\ \circ & \circ & \circ & a_{10} \end{bmatrix}$$

Clearly, each element of C_i is 1 or -1 or 0, and then we have the following lemma.

LEMMA 2: *The matrix C_i is a signed permutation matrix for $i = 1, 2, \ldots, 5$.*

To obtain weighing matrices with weight six of type F_1, all possibilities on S are calculated by the following algorithm.

Step I. Search all possibilities on signed permutation matrices $C_1, C_2, \ldots, C_5$, except for the cases reduced to one of type $F_i (i \geq 2)$. Moreover, for each column of S, determine the values of free positions to preserve the orthogonality with the first six columns of $\tilde{W}$.

Step II. Reduce the number of signed permutation matrices constructed in Step I by acting invariant permutations which are invariant the 2nd, 3rd, ..., 6th rows and columns of $\tilde{W}$.

Step III. Keep the orthogonality of rows of $\tilde{W}$ and the property of F_1 of $\tilde{W}$.

As a result, it turns out that there are only ten cases, say $S_1, S_2, \ldots, S_{10}$. They are called the base matrices with weight 6 of type F_1 and are listed in Figure 2. Let $\tilde{W}_i (1 \leq i \leq 10)$ be a matrix obtained from $\tilde{W}$ by using S_i instead of S. As the next step, we extend $\tilde{W}_i$ to weighing matrices of type F_1. Let W_i be an extended weighing matrix based on $\tilde{W}_i$. Then we put

$$W_i = \begin{bmatrix} \tilde{W}_i & Z_0 \\ Z_0 & K_i \end{bmatrix} = \begin{bmatrix} A_1 & A_2 & Z_0 \\ A_2^t & S_i & R_i \\ Z_0^t & U_i & K_i \end{bmatrix}$$

where A_1 and A_2 are the submatrices of the left upper and the right upper corners of $\tilde{W}_i$, respectively, and Z_0 is the zero matrix of an appropriate size.

DEFINITION 3: *The matrices R_i, U_i, and K_i are called the R-, U-, and K-matrix of W_i, respectively.*

THEOREM 1: *There is no weighing matrix based on $\tilde{W}_i$ for $i = 8, 9$.*

$$\begin{bmatrix} 1&\cdot&\cdot&\cdot&\cdot&\cdot&\cdot&\cdot&\cdot&\cdot\\ \cdot&1&\cdot&\cdot&\cdot&\cdot&\cdot&\cdot&\cdot&\cdot\\ \cdot&\cdot&1&\cdot&\cdot&\cdot&\cdot&\cdot&\cdot&\cdot\\ \cdot&\cdot&\cdot&1&\cdot&\cdot&\cdot&\cdot&\cdot&\cdot\\ \cdot&\cdot&\cdot&\cdot&1&\cdot&\cdot&\cdot&\cdot&\cdot\\ \cdot&\cdot&\cdot&\cdot&\cdot&1&\cdot&\cdot&\cdot&\cdot\\ \cdot&\cdot&\cdot&\cdot&\cdot&\cdot&1&\cdot&\cdot&\cdot\\ \cdot&\cdot&\cdot&\cdot&\cdot&\cdot&\cdot&1&\cdot&\cdot\\ \cdot&\cdot&\cdot&\cdot&\cdot&\cdot&\cdot&\cdot&1&\cdot\\ \cdot&\cdot&\cdot&\cdot&\cdot&\cdot&\cdot&\cdot&\cdot&1 \end{bmatrix}$$

S_1

$$\begin{bmatrix} 1&\cdot&\cdot&\cdot&\cdot&\cdot&\cdot&\cdot&\cdot&\cdot\\ \cdot&1&\cdot&\cdot&\cdot&\cdot&\cdot&\cdot&\cdot&\cdot\\ \cdot&\cdot&1&\cdot&\cdot&\cdot&\cdot&\cdot&\cdot&\cdot\\ \cdot&\cdot&\cdot&1&\cdot&\cdot&\cdot&\cdot&\cdot&\cdot\\ \cdot&\cdot&\cdot&\cdot&1&\cdot&\cdot&\cdot&\cdot&\cdot\\ \cdot&\cdot&\cdot&\cdot&\cdot&1&\cdot&\cdot&\cdot&\cdot\\ \cdot&\cdot&\cdot&\cdot&\cdot&\cdot&1&\cdot&\cdot&\cdot\\ \cdot&\cdot&\cdot&\cdot&\cdot&\cdot&\cdot&\cdot&1&-\\ \cdot&\cdot&\cdot&\cdot&\cdot&\cdot&\cdot&1&\cdot&1\\ \cdot&\cdot&\cdot&\cdot&\cdot&\cdot&\cdot&-&1&\cdot \end{bmatrix}$$

S_2

$$\begin{bmatrix} 1&\cdot&\cdot&\cdot&\cdot&\cdot&\cdot&\cdot&\cdot&\cdot\\ \cdot&1&\cdot&\cdot&\cdot&\cdot&\cdot&\cdot&\cdot&\cdot\\ \cdot&\cdot&1&\cdot&\cdot&\cdot&\cdot&\cdot&\cdot&\cdot\\ \cdot&\cdot&\cdot&1&\cdot&\cdot&\cdot&\cdot&\cdot&\cdot\\ \cdot&\cdot&\cdot&\cdot&\cdot&\cdot&1&\cdot&-&\cdot\\ \cdot&\cdot&\cdot&\cdot&1&\cdot&\cdot&\cdot&1&-\\ \cdot&\cdot&\cdot&\cdot&\cdot&1&\cdot&\cdot&\cdot&1\\ \cdot&\cdot&\cdot&\cdot&-&1&\cdot&\cdot&\cdot&\cdot\\ \cdot&\cdot&\cdot&\cdot&\cdot&-&1&1&\cdot&\cdot\\ \cdot&\cdot&\cdot&\cdot&\cdot&\cdot&\cdot&-&1&\cdot \end{bmatrix}$$

S_3

$$\begin{bmatrix} 1&\cdot&\cdot&\cdot&\cdot&\cdot&\cdot&\cdot&\cdot&\cdot\\ \cdot&1&\cdot&\cdot&\cdot&\cdot&\cdot&\cdot&\cdot&\cdot\\ \cdot&\cdot&\cdot&1&\cdot&\cdot&\cdot&\cdot&\cdot&-\\ \cdot&\cdot&1&\cdot&\cdot&\cdot&\cdot&\cdot&\cdot&1\\ \cdot&\cdot&\cdot&\cdot&\cdot&1&\cdot&-&\cdot&\cdot\\ \cdot&\cdot&\cdot&\cdot&1&\cdot&\cdot&1&\cdot&\cdot\\ \cdot&\cdot&\cdot&\cdot&\cdot&\cdot&1&\cdot&\cdot&\cdot\\ \cdot&\cdot&\cdot&\cdot&-&1&\cdot&\cdot&\cdot&\cdot\\ \cdot&\cdot&\cdot&\cdot&\cdot&\cdot&\cdot&\cdot&1&\cdot\\ \cdot&\cdot&-&1&\cdot&\cdot&\cdot&\cdot&\cdot&\cdot \end{bmatrix}$$

S_4

$$\begin{bmatrix} 1&\cdot&\cdot&\cdot&\cdot&\cdot&\cdot&\cdot&\cdot&\cdot\\ \cdot&\cdot&\cdot&1&\cdot&\cdot&\cdot&-&\cdot&-\\ \cdot&1&\cdot&\cdot&\cdot&\cdot&\cdot&1&\cdot&\cdot\\ \cdot&\cdot&1&\cdot&\cdot&\cdot&\cdot&\cdot&\cdot&1\\ \cdot&1&\cdot&-&\cdot&\cdot&1&\cdot&\cdot&\cdot\\ \cdot&-&1&\cdot&1&\cdot&\cdot&\cdot&\cdot&\cdot\\ \cdot&\cdot&-&1&\cdot&1&\cdot&\cdot&\cdot&\cdot\\ \cdot&\cdot&\cdot&\cdot&-&\cdot&1&\cdot&\cdot&-\\ \cdot&\cdot&\cdot&\cdot&\cdot&\cdot&\cdot&\cdot&1&\cdot\\ \cdot&\cdot&\cdot&\cdot&\cdot&-&1&\cdot&\cdot&\cdot \end{bmatrix}$$

S_5

$$\begin{bmatrix} 1&\cdot&\cdot&\cdot&\cdot&\cdot&\cdot&\cdot&\cdot&\cdot\\ \cdot&\cdot&\cdot&1&\cdot&\cdot&\cdot&\cdot&-&\cdot\\ \cdot&1&\cdot&\cdot&\cdot&\cdot&\cdot&\cdot&1&-\\ \cdot&\cdot&1&\cdot&\cdot&\cdot&\cdot&\cdot&\cdot&1\\ \cdot&1&\cdot&-&\cdot&\cdot&1&\cdot&\cdot&\cdot\\ \cdot&-&1&\cdot&1&\cdot&\cdot&\cdot&\cdot&\cdot\\ \cdot&\cdot&-&1&\cdot&1&\cdot&\cdot&\cdot&\cdot\\ \cdot&\cdot&\cdot&\cdot&-&1&\cdot&\cdot&\cdot&\cdot\\ \cdot&\cdot&\cdot&\cdot&\cdot&-&1&1&\cdot&\cdot\\ \cdot&\cdot&\cdot&\cdot&\cdot&\cdot&\cdot&-&1&\cdot \end{bmatrix}$$

S_6

$$\begin{bmatrix} 1&\cdot&\cdot&\cdot&\cdot&\cdot&\cdot&-&1&-\\ \cdot&\cdot&\cdot&1&\cdot&\cdot&\cdot&\cdot&-&\cdot\\ \cdot&1&\cdot&\cdot&\cdot&\cdot&\cdot&1&\cdot&\cdot\\ \cdot&\cdot&1&\cdot&\cdot&\cdot&\cdot&\cdot&\cdot&1\\ \cdot&\cdot&\cdot&\cdot&\cdot&1&\cdot&-&\cdot&\cdot\\ \cdot&\cdot&\cdot&\cdot&\cdot&\cdot&1&\cdot&\cdot&-\\ \cdot&\cdot&\cdot&\cdot&1&\cdot&\cdot&\cdot&1&\cdot\\ \cdot&-&1&\cdot&\cdot&\cdot&\cdot&\cdot&\cdot&\cdot\\ \cdot&\cdot&-&1&-&1&\cdot&\cdot&\cdot&\cdot\\ \cdot&\cdot&\cdot&\cdot&\cdot&-&1&\cdot&\cdot&\cdot \end{bmatrix}$$

S_7

$$\begin{bmatrix} 1&\cdot&\cdot&\cdot&\cdot&\cdot&\cdot&-&1&-\\ \cdot&\cdot&\cdot&1&\cdot&\cdot&\cdot&\cdot&-&\cdot\\ \cdot&1&\cdot&\cdot&\cdot&\cdot&\cdot&1&\cdot&\cdot\\ \cdot&\cdot&1&\cdot&\cdot&\cdot&\cdot&\cdot&\cdot&1\\ \cdot&\cdot&\cdot&\cdot&\cdot&1&\cdot&-&\cdot&\cdot\\ \cdot&\cdot&\cdot&\cdot&\cdot&\cdot&1&\cdot&\cdot&-\\ \cdot&\cdot&\cdot&\cdot&1&\cdot&\cdot&\cdot&1&\cdot\\ \cdot&-&\cdot&\cdot&\cdot&1&\cdot&\cdot&\cdot&\cdot\\ \cdot&\cdot&\cdot&1&-&\cdot&\cdot&\cdot&\cdot&\cdot\\ \cdot&\cdot&-&\cdot&\cdot&\cdot&1&\cdot&\cdot&\cdot \end{bmatrix}$$

S_8

$$\begin{bmatrix} \cdot&\cdot&\cdot&1&\cdot&\cdot&-&\cdot&\cdot&\cdot\\ 1&\cdot&\cdot&\cdot&\cdot&\cdot&1&\cdot&-&\cdot\\ \cdot&1&\cdot&\cdot&\cdot&\cdot&\cdot&\cdot&1&-\\ \cdot&\cdot&1&\cdot&\cdot&\cdot&\cdot&\cdot&\cdot&1\\ -&1&\cdot&\cdot&\cdot&\cdot&\cdot&\cdot&\cdot&\cdot\\ \cdot&-&1&\cdot&1&\cdot&\cdot&\cdot&\cdot&\cdot\\ \cdot&\cdot&-&1&\cdot&1&\cdot&\cdot&\cdot&\cdot\\ \cdot&\cdot&\cdot&\cdot&-&1&\cdot&\cdot&\cdot&\cdot\\ \cdot&\cdot&\cdot&\cdot&\cdot&-&1&1&\cdot&\cdot\\ \cdot&\cdot&\cdot&\cdot&\cdot&\cdot&\cdot&-&1&\cdot \end{bmatrix}$$

S_9

$$\begin{bmatrix} \cdot&\cdot&\cdot&1&\cdot&\cdot&-&\cdot&\cdot&\cdot\\ 1&\cdot&\cdot&\cdot&\cdot&\cdot&1&\cdot&-&\cdot\\ \cdot&1&\cdot&\cdot&\cdot&\cdot&\cdot&1&\cdot&\cdot\\ \cdot&\cdot&1&\cdot&\cdot&\cdot&\cdot&-&1&\cdot\\ -&1&\cdot&\cdot&\cdot&\cdot&\cdot&\cdot&\cdot&\cdot\\ \cdot&-&\cdot&1&1&\cdot&\cdot&\cdot&\cdot&-\\ \cdot&\cdot&\cdot&\cdot&\cdot&1&\cdot&\cdot&\cdot&1\\ \cdot&\cdot&\cdot&\cdot&-&1&\cdot&\cdot&\cdot&\cdot\\ \cdot&\cdot&\cdot&\cdot&\cdot&-&1&1&\cdot&\cdot\\ \cdot&\cdot&-&1&\cdot&\cdot&\cdot&\cdot&\cdot&\cdot \end{bmatrix}$$

S_{10}

Figure 2. The 10 base matrices with weight six of type F_1

Proof. Suppose that there exists a weighing matrix W_8 based on $\tilde{W}_8$ and let R_8 be an R-matrix of W_8. Then without loss of generality, the first 7×5 submatrices of R_8 can be written as

$$\begin{bmatrix} \cdot&\cdot&\cdot&\cdot\\ 1&1&\cdot&\cdot\\ -&\cdot&1&\cdot\\ \cdot&-&-&\cdot\\ -&\cdot&\cdot&1\\ \cdot&\cdot&-&-\\ 1&\cdot&1&\cdot \end{bmatrix}$$

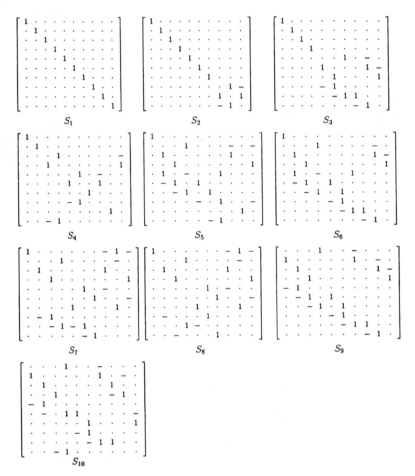

However, it is clear that we cannot construct a weighing matrix based on $\tilde{W}_8$ and R_8. This is similar to $\tilde{W}_9$. □

THEOREM 2: *There is a unique weighing matrix (say W_i) based on $\tilde{W}_i$, up to equivalence, for $i = 3, 4, 5, 6, 7, 10$.*

Proof. It is straightforward to construct matrices R_i, U_i, K_i based on $\tilde{W}_i$. □

Put $T_i = \begin{bmatrix} S_i & R_i \\ U_i & K_i \end{bmatrix}$. The matrices $\{T_i | i = 3, 4, 5, 6, 7, 10\}$ are listed in Figure 3.

$$\begin{bmatrix}
1 & \cdot & \cdot & \cdot & \cdot & \cdot & \cdot & \cdot & \cdot & \cdot & 1 & 1 & 1 & \cdot & \cdot & \cdot & \cdot & \cdot \\
\cdot & 1 & \cdot & \cdot & \cdot & \cdot & \cdot & \cdot & \cdot & \cdot & - & \cdot & \cdot & 1 & 1 & \cdot & \cdot & \cdot \\
\cdot & \cdot & 1 & \cdot & \cdot & \cdot & \cdot & \cdot & \cdot & \cdot & \cdot & - & \cdot & - & \cdot & 1 & \cdot & \cdot \\
\cdot & \cdot & \cdot & 1 & \cdot & \cdot & \cdot & \cdot & \cdot & \cdot & \cdot & \cdot & - & \cdot & - & - & \cdot & \cdot \\
\cdot & \cdot & \cdot & \cdot & \cdot & \cdot & 1 & \cdot & - & \cdot & 1 & \cdot & \cdot & \cdot & \cdot & \cdot & 1 & \cdot \\
\cdot & \cdot & \cdot & \cdot & 1 & \cdot & \cdot & \cdot & 1 & - & \cdot & 1 & \cdot & \cdot & \cdot & \cdot & \cdot & \cdot \\
\cdot & \cdot & \cdot & \cdot & \cdot & 1 & \cdot & \cdot & \cdot & 1 & \cdot & \cdot & 1 & \cdot & \cdot & \cdot & - & \cdot \\
\cdot & \cdot & \cdot & \cdot & - & 1 & \cdot & \cdot & \cdot & \cdot & \cdot & \cdot & \cdot & 1 & \cdot & \cdot & 1 & \cdot \\
\cdot & \cdot & \cdot & \cdot & \cdot & - & 1 & 1 & \cdot & \cdot & \cdot & \cdot & \cdot & \cdot & 1 & \cdot & \cdot & \cdot \\
\cdot & \cdot & \cdot & \cdot & \cdot & \cdot & \cdot & - & 1 & \cdot & \cdot & \cdot & \cdot & \cdot & \cdot & 1 & 1 & \cdot \\
1 & - & \cdot & \cdot & 1 & \cdot & \cdot & \cdot & \cdot & \cdot & \cdot & - & \cdot & 1 & \cdot & \cdot & \cdot & 1 \\
1 & \cdot & - & \cdot & \cdot & 1 & \cdot & \cdot & \cdot & \cdot & \cdot & \cdot & - & - & 1 & \cdot & \cdot & \cdot \\
1 & \cdot & \cdot & - & \cdot & \cdot & 1 & \cdot & \cdot & \cdot & - & \cdot & \cdot & \cdot & - & \cdot & \cdot & - \\
\cdot & 1 & - & \cdot & \cdot & \cdot & \cdot & 1 & \cdot & \cdot & \cdot & \cdot & \cdot & \cdot & - & 1 & \cdot & 1 \\
\cdot & 1 & \cdot & - & \cdot & \cdot & \cdot & \cdot & 1 & \cdot & 1 & - & \cdot & \cdot & \cdot & - & \cdot & \cdot \\
\cdot & \cdot & 1 & - & \cdot & \cdot & \cdot & \cdot & \cdot & 1 & \cdot & 1 & - & \cdot & \cdot & \cdot & \cdot & 1 \\
\cdot & \cdot & \cdot & \cdot & 1 & \cdot & - & 1 & \cdot & 1 & \cdot & \cdot & \cdot & \cdot & \cdot & \cdot & 1 & - \\
\cdot & \cdot & \cdot & \cdot & \cdot & \cdot & \cdot & \cdot & \cdot & \cdot & 1 & \cdot & - & 1 & \cdot & 1 & - & -
\end{bmatrix}$$

T_3

$$\begin{bmatrix}
1 & \cdot & \cdot & \cdot & \cdot & \cdot & \cdot & \cdot & \cdot & \cdot & \cdot & 1 & 1 & \cdot & \cdot & \cdot & 1 & \cdot \\
\cdot & 1 & \cdot & \cdot & \cdot & \cdot & \cdot & \cdot & \cdot & \cdot & - & \cdot & \cdot & \cdot & \cdot & 1 & - & \cdot \\
\cdot & \cdot & \cdot & 1 & \cdot & \cdot & \cdot & \cdot & \cdot & - & \cdot & \cdot & - & \cdot & \cdot & - & \cdot & \cdot \\
\cdot & \cdot & 1 & \cdot & \cdot & \cdot & \cdot & \cdot & \cdot & 1 & 1 & - & \cdot & \cdot & \cdot & \cdot & \cdot & \cdot \\
\cdot & \cdot & \cdot & \cdot & \cdot & 1 & \cdot & - & \cdot & \cdot & 1 & 1 & \cdot & \cdot & \cdot & \cdot & \cdot & \cdot \\
\cdot & \cdot & \cdot & \cdot & 1 & \cdot & \cdot & 1 & \cdot & \cdot & \cdot & \cdot & 1 & 1 & \cdot & \cdot & \cdot & \cdot \\
\cdot & \cdot & \cdot & \cdot & \cdot & \cdot & 1 & \cdot & \cdot & \cdot & - & \cdot & \cdot & - & \cdot & \cdot & 1 & \cdot \\
\cdot & \cdot & \cdot & \cdot & - & 1 & \cdot & \cdot & \cdot & \cdot & \cdot & \cdot & \cdot & \cdot & 1 & 1 & \cdot & \cdot \\
\cdot & \cdot & \cdot & \cdot & \cdot & \cdot & \cdot & \cdot & 1 & \cdot & \cdot & 1 & \cdot & \cdot & - & \cdot & - & \cdot \\
\cdot & \cdot & - & 1 & \cdot & \cdot & \cdot & \cdot & \cdot & \cdot & \cdot & \cdot & \cdot & 1 & 1 & \cdot & \cdot & \cdot \\
1 & \cdot & \cdot & - & 1 & \cdot & \cdot & \cdot & 1 & \cdot & \cdot & \cdot & - & \cdot & 1 & \cdot & \cdot & \cdot \\
\cdot & - & \cdot & 1 & 1 & \cdot & - & \cdot & \cdot & \cdot & \cdot & \cdot & \cdot & - & \cdot & 1 & \cdot & \cdot \\
\cdot & \cdot & \cdot & \cdot & \cdot & 1 & - & \cdot & \cdot & 1 & - & \cdot & \cdot & \cdot & \cdot & - & \cdot & 1 \\
1 & \cdot & - & \cdot & \cdot & 1 & \cdot & \cdot & \cdot & \cdot & \cdot & - & \cdot & \cdot & - & \cdot & \cdot & - \\
\cdot & 1 & - & \cdot & \cdot & \cdot & \cdot & 1 & \cdot & \cdot & 1 & \cdot & \cdot & - & \cdot & \cdot & \cdot & 1 \\
\cdot & \cdot & \cdot & \cdot & \cdot & \cdot & \cdot & 1 & - & 1 & \cdot & 1 & - & \cdot & \cdot & \cdot & \cdot & - \\
1 & - & \cdot & \cdot & \cdot & \cdot & 1 & \cdot & - & \cdot & \cdot & \cdot & \cdot & \cdot & \cdot & \cdot & - & 1 \\
\cdot & \cdot & \cdot & \cdot & \cdot & \cdot & \cdot & \cdot & \cdot & \cdot & \cdot & \cdot & 1 & - & 1 & - & - & -
\end{bmatrix}$$

T_4

Figure 3. Matrices $\{T_i | i = 3, 4, 5, 6, 7, 10\}$, $T_{1,1}$ and $T_{1,2}$, and $T_{2,1}$ and $T_{2,2}$

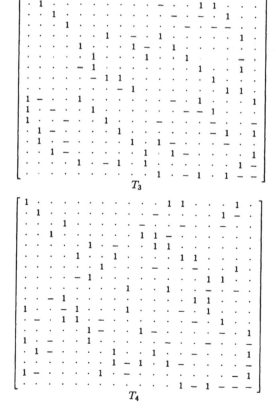

$$\begin{bmatrix} 1 & \cdot & \cdot & \cdot & \cdot & \cdot & \cdot & \cdot & \cdot & \cdot & 1 & 1 & 1 & \cdot \\ \cdot & \cdot & \cdot & 1 & \cdot & \cdot & \cdot & - & \cdot & - & - & \cdot & \cdot & \cdot \\ \cdot & 1 & \cdot & \cdot & \cdot & \cdot & \cdot & 1 & \cdot & \cdot & \cdot & - & \cdot & 1 \\ \cdot & \cdot & 1 & \cdot & \cdot & \cdot & \cdot & \cdot & \cdot & 1 & \cdot & \cdot & - & - \\ \cdot & 1 & \cdot & - & \cdot & \cdot & 1 & \cdot & \cdot & \cdot & \cdot & 1 & \cdot & \cdot \\ \cdot & - & 1 & \cdot & 1 & \cdot & \cdot & \cdot & \cdot & \cdot & \cdot & \cdot & 1 & \cdot \\ \cdot & \cdot & - & 1 & \cdot & 1 & \cdot & \cdot & \cdot & \cdot & 1 & \cdot & \cdot & \cdot \\ \cdot & \cdot & \cdot & \cdot & - & \cdot & 1 & \cdot & \cdot & - & \cdot & \cdot & \cdot & - \\ \cdot & \cdot & \cdot & \cdot & \cdot & \cdot & \cdot & \cdot & 1 & \cdot & - & 1 & \cdot & 1 \\ \cdot & \cdot & \cdot & \cdot & \cdot & - & 1 & \cdot & \cdot & \cdot & \cdot & - & 1 & \cdot \\ 1 & - & \cdot & \cdot & \cdot & \cdot & 1 & \cdot & - & \cdot & \cdot & \cdot & - & 1 \\ 1 & \cdot & - & \cdot & 1 & \cdot & \cdot & 1 & \cdot & \cdot & - & \cdot & \cdot & - \\ 1 & \cdot & \cdot & - & \cdot & 1 & \cdot & - & 1 & \cdot & \cdot & - & \cdot & \cdot \\ \cdot & \cdot & \cdot & \cdot & 1 & - & \cdot & \cdot & 1 & - & 1 & \cdot & - & \cdot \end{bmatrix}$$

T_5

$$\begin{bmatrix} 1 & \cdot & \cdot & \cdot & \cdot & \cdot & \cdot & \cdot & \cdot & \cdot & 1 & 1 & 1 & \cdot \\ \cdot & \cdot & \cdot & 1 & \cdot & \cdot & \cdot & \cdot & - & \cdot & - & \cdot & \cdot & 1 \\ \cdot & 1 & \cdot & \cdot & \cdot & \cdot & \cdot & \cdot & 1 & - & \cdot & - & \cdot & \cdot \\ \cdot & \cdot & 1 & \cdot & \cdot & \cdot & \cdot & \cdot & \cdot & 1 & \cdot & \cdot & - & - \\ \cdot & 1 & \cdot & - & \cdot & \cdot & 1 & \cdot & \cdot & \cdot & \cdot & 1 & \cdot & \cdot \\ \cdot & - & 1 & \cdot & 1 & \cdot & \cdot & \cdot & \cdot & \cdot & \cdot & \cdot & 1 & \cdot \\ \cdot & \cdot & - & 1 & \cdot & 1 & \cdot & \cdot & \cdot & \cdot & 1 & \cdot & \cdot & \cdot \\ \cdot & \cdot & \cdot & \cdot & - & 1 & \cdot & \cdot & \cdot & \cdot & - & 1 & \cdot & \cdot \\ \cdot & \cdot & \cdot & \cdot & \cdot & - & 1 & 1 & \cdot & \cdot & \cdot & \cdot & \cdot & 1 \\ \cdot & \cdot & \cdot & \cdot & \cdot & \cdot & \cdot & - & 1 & \cdot & - & \cdot & 1 & \cdot \\ 1 & - & \cdot & \cdot & \cdot & \cdot & 1 & - & \cdot & - & \cdot & \cdot & - & \cdot \\ 1 & \cdot & - & \cdot & 1 & \cdot & \cdot & 1 & \cdot & \cdot & - & \cdot & \cdot & - \\ 1 & \cdot & \cdot & - & \cdot & 1 & \cdot & \cdot & \cdot & 1 & \cdot & - & \cdot & 1 \\ \cdot & \cdot & \cdot & \cdot & 1 & \cdot & - & \cdot & 1 & \cdot & \cdot & 1 & - & 1 \end{bmatrix}$$

T_6

$$\begin{bmatrix} 1 & \cdot & \cdot & \cdot & \cdot & \cdot & \cdot & - & 1 & - & \cdot & \cdot & \cdot & \cdot \\ \cdot & \cdot & \cdot & 1 & \cdot & \cdot & \cdot & \cdot & - & \cdot & 1 & 1 & \cdot & \cdot \\ \cdot & 1 & \cdot & \cdot & \cdot & \cdot & \cdot & 1 & \cdot & \cdot & - & \cdot & 1 & \cdot \\ \cdot & \cdot & 1 & \cdot & \cdot & \cdot & \cdot & \cdot & \cdot & 1 & \cdot & - & - & \cdot \\ \cdot & \cdot & \cdot & \cdot & \cdot & 1 & \cdot & - & \cdot & \cdot & - & \cdot & \cdot & 1 \\ \cdot & \cdot & \cdot & \cdot & \cdot & \cdot & 1 & \cdot & \cdot & - & 1 & - & \cdot & \cdot \\ \cdot & \cdot & \cdot & \cdot & 1 & \cdot & \cdot & \cdot & 1 & \cdot & \cdot & 1 & \cdot & - \\ \cdot & - & 1 & \cdot & \cdot & \cdot & \cdot & \cdot & \cdot & \cdot & \cdot & 1 & \cdot & 1 \\ \cdot & \cdot & - & 1 & - & 1 & \cdot & \cdot & \cdot & \cdot & \cdot & \cdot & \cdot & \cdot \\ \cdot & \cdot & \cdot & \cdot & \cdot & - & 1 & \cdot & \cdot & \cdot & \cdot & \cdot & 1 & 1 \\ 1 & - & \cdot & \cdot & \cdot & \cdot & 1 & \cdot & - & \cdot & - & \cdot & \cdot & - \\ 1 & \cdot & - & \cdot & 1 & \cdot & \cdot & 1 & \cdot & \cdot & \cdot & \cdot & - & 1 \\ 1 & \cdot & \cdot & - & \cdot & 1 & \cdot & \cdot & \cdot & 1 & 1 & \cdot & 1 & \cdot \\ \cdot & 1 & \cdot & - & - & \cdot & 1 & \cdot & \cdot & \cdot & \cdot & 1 & - & \cdot \end{bmatrix}$$

T_7

$$\begin{bmatrix} \cdot & \cdot & \cdot & 1 & \cdot & \cdot & - & \cdot & \cdot & \cdot & 1 & 1 & \cdot & \cdot \\ 1 & \cdot & \cdot & \cdot & \cdot & \cdot & 1 & \cdot & - & \cdot & \cdot & \cdot & 1 & \cdot \\ \cdot & 1 & \cdot & \cdot & \cdot & \cdot & \cdot & 1 & \cdot & \cdot & - & \cdot & - & \cdot \\ \cdot & \cdot & 1 & \cdot & \cdot & \cdot & \cdot & - & 1 & \cdot & \cdot & - & \cdot & \cdot \\ - & 1 & \cdot & \cdot & \cdot & \cdot & \cdot & \cdot & \cdot & \cdot & 1 & \cdot & \cdot & 1 \\ \cdot & - & \cdot & 1 & 1 & \cdot & \cdot & \cdot & \cdot & - & \cdot & \cdot & \cdot & \cdot \\ \cdot & \cdot & \cdot & \cdot & \cdot & 1 & \cdot & \cdot & \cdot & 1 & \cdot & 1 & \cdot & - \\ \cdot & \cdot & \cdot & \cdot & - & 1 & \cdot & \cdot & \cdot & \cdot & \cdot & \cdot & 1 & 1 \\ \cdot & \cdot & \cdot & \cdot & \cdot & - & 1 & 1 & \cdot & \cdot & 1 & \cdot & \cdot & \cdot \\ \cdot & \cdot & - & 1 & \cdot & \cdot & \cdot & \cdot & \cdot & \cdot & - & \cdot & \cdot & 1 \\ 1 & \cdot & - & \cdot & \cdot & 1 & \cdot & \cdot & \cdot & \cdot & 1 & - & - & \cdot \\ 1 & \cdot & \cdot & - & 1 & \cdot & \cdot & \cdot & 1 & \cdot & \cdot & 1 & \cdot & 1 \\ \cdot & 1 & - & \cdot & \cdot & \cdot & \cdot & \cdot & 1 & - & \cdot & \cdot & 1 & - \\ \cdot & \cdot & \cdot & \cdot & 1 & \cdot & - & 1 & \cdot & 1 & \cdot & - & 1 & \cdot \end{bmatrix}$$

T_{10}

Figure 3. *(continued)*

THEOREM 3: *There are two weighing matrices (say $W_{1,1}$, $W_{1,2}$) based on $\tilde{W}_1$, up to equivalence.*

Proof. Up to equivalence, there are two R-matrices based on $\tilde{W}_1$, say $R_{1,1}$ and $R_{1,2}$, where

$$R_{1,1} = \begin{bmatrix} 1 & 1 & 1 & \cdot & \cdot & \cdot & \cdot & \cdot & \cdot & \cdot & \cdot & \cdot & \cdot & \cdot & \cdot & \cdot \\ - & \cdot & \cdot & 1 & 1 & \cdot & \cdot & \cdot & \cdot & \cdot & \cdot & \cdot & \cdot & \cdot & \cdot & \cdot \\ \cdot & - & \cdot & - & \cdot & 1 & \cdot & \cdot & \cdot & \cdot & \cdot & \cdot & \cdot & \cdot & \cdot & \cdot \\ \cdot & \cdot & - & \cdot & - & - & \cdot & \cdot & \cdot & \cdot & \cdot & \cdot & \cdot & \cdot & \cdot & \cdot \\ 1 & \cdot & \cdot & \cdot & \cdot & \cdot & 1 & 1 & \cdot & \cdot & \cdot & \cdot & \cdot & \cdot & \cdot & \cdot \\ \cdot & 1 & \cdot & \cdot & \cdot & \cdot & - & \cdot & 1 & \cdot & \cdot & \cdot & \cdot & \cdot & \cdot & \cdot \\ \cdot & \cdot & 1 & \cdot & \cdot & \cdot & \cdot & - & - & \cdot & \cdot & \cdot & \cdot & \cdot & \cdot & \cdot \\ \cdot & \cdot & \cdot & 1 & \cdot & \cdot & 1 & \cdot & \cdot & 1 & \cdot & \cdot & \cdot & \cdot & \cdot & \cdot \\ \cdot & \cdot & \cdot & \cdot & 1 & \cdot & \cdot & 1 & \cdot & - & \cdot & \cdot & \cdot & \cdot & \cdot & \cdot \\ \cdot & \cdot & \cdot & \cdot & \cdot & 1 & \cdot & \cdot & 1 & 1 & \cdot & \cdot & \cdot & \cdot & \cdot & \cdot \end{bmatrix}$$

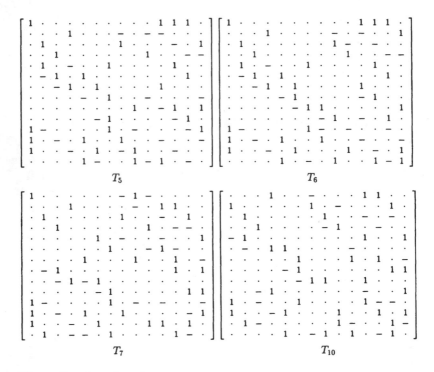

$$
\left[\begin{array}{cccccccccccccccccccccccccc}
1 & \cdot & \cdot & \cdot & \cdot & \cdot & \cdot & \cdot & \cdot & \cdot & 1 & 1 & 1 & \cdot & \cdot & \cdot & \cdot & \cdot & \cdot & \cdot & \cdot & \cdot & \cdot & \cdot & \cdot & \cdot \\
\cdot & 1 & \cdot & \cdot & \cdot & \cdot & \cdot & \cdot & \cdot & \cdot & - & \cdot & \cdot & 1 & 1 & \cdot & \cdot & \cdot & \cdot & \cdot & \cdot & \cdot & \cdot & \cdot & \cdot & \cdot \\
\cdot & \cdot & 1 & \cdot & \cdot & \cdot & \cdot & \cdot & \cdot & \cdot & \cdot & - & \cdot & - & \cdot & 1 & \cdot & \cdot & \cdot & \cdot & \cdot & \cdot & \cdot & \cdot & \cdot & \cdot \\
\cdot & \cdot & \cdot & 1 & \cdot & \cdot & \cdot & \cdot & \cdot & \cdot & \cdot & \cdot & - & \cdot & - & - & \cdot & \cdot & \cdot & \cdot & \cdot & \cdot & \cdot & \cdot & \cdot & \cdot \\
\cdot & \cdot & \cdot & \cdot & 1 & \cdot & \cdot & \cdot & \cdot & \cdot & 1 & \cdot & \cdot & \cdot & \cdot & \cdot & 1 & 1 & \cdot & \cdot & \cdot & \cdot & \cdot & \cdot & \cdot & \cdot \\
\cdot & \cdot & \cdot & \cdot & \cdot & 1 & \cdot & \cdot & \cdot & \cdot & \cdot & 1 & \cdot & \cdot & \cdot & \cdot & - & \cdot & 1 & \cdot & \cdot & \cdot & \cdot & \cdot & \cdot & \cdot \\
\cdot & \cdot & \cdot & \cdot & \cdot & \cdot & 1 & \cdot & \cdot & \cdot & \cdot & \cdot & 1 & \cdot & \cdot & \cdot & \cdot & - & - & \cdot & \cdot & \cdot & \cdot & \cdot & \cdot & \cdot \\
\cdot & \cdot & \cdot & \cdot & \cdot & \cdot & \cdot & 1 & \cdot & \cdot & \cdot & \cdot & \cdot & 1 & \cdot & \cdot & 1 & \cdot & \cdot & 1 & \cdot & \cdot & \cdot & \cdot & \cdot & \cdot \\
\cdot & \cdot & \cdot & \cdot & \cdot & \cdot & \cdot & \cdot & 1 & \cdot & \cdot & \cdot & \cdot & \cdot & 1 & \cdot & \cdot & 1 & \cdot & - & \cdot & \cdot & \cdot & \cdot & \cdot & \cdot \\
\cdot & \cdot & \cdot & \cdot & \cdot & \cdot & \cdot & \cdot & \cdot & 1 & \cdot & \cdot & \cdot & \cdot & \cdot & 1 & \cdot & \cdot & 1 & 1 & \cdot & \cdot & \cdot & \cdot & \cdot & \cdot \\
1 & - & \cdot & \cdot & 1 & \cdot & \cdot & \cdot & \cdot & \cdot & - & \cdot & \cdot & \cdot & \cdot & \cdot & \cdot & \cdot & \cdot & \cdot & 1 & 1 & \cdot & \cdot & \cdot & \cdot \\
1 & \cdot & - & \cdot & \cdot & 1 & \cdot & \cdot & \cdot & \cdot & \cdot & - & \cdot & \cdot & \cdot & \cdot & \cdot & \cdot & \cdot & \cdot & - & \cdot & 1 & \cdot & \cdot & \cdot \\
1 & \cdot & \cdot & - & \cdot & \cdot & 1 & \cdot & \cdot & \cdot & \cdot & \cdot & - & \cdot & \cdot & \cdot & \cdot & \cdot & \cdot & \cdot & \cdot & - & - & \cdot & \cdot & \cdot \\
\cdot & 1 & - & \cdot & \cdot & \cdot & \cdot & 1 & \cdot & \cdot & \cdot & \cdot & \cdot & - & \cdot & \cdot & \cdot & \cdot & \cdot & \cdot & 1 & \cdot & \cdot & 1 & \cdot & \cdot \\
\cdot & 1 & \cdot & - & \cdot & \cdot & \cdot & \cdot & 1 & \cdot & \cdot & \cdot & \cdot & \cdot & - & \cdot & \cdot & \cdot & \cdot & \cdot & \cdot & 1 & \cdot & - & \cdot & \cdot \\
\cdot & \cdot & 1 & - & \cdot & \cdot & \cdot & \cdot & \cdot & 1 & \cdot & \cdot & \cdot & \cdot & \cdot & - & \cdot & \cdot & \cdot & \cdot & \cdot & \cdot & 1 & 1 & \cdot & \cdot \\
\cdot & \cdot & \cdot & \cdot & 1 & - & \cdot & 1 & \cdot & \cdot & \cdot & \cdot & \cdot & \cdot & \cdot & \cdot & - & \cdot & \cdot & \cdot & - & \cdot & \cdot & \cdot & 1 & \cdot \\
\cdot & \cdot & \cdot & \cdot & 1 & \cdot & - & \cdot & 1 & \cdot & \cdot & \cdot & \cdot & \cdot & \cdot & \cdot & \cdot & - & \cdot & \cdot & \cdot & - & \cdot & \cdot & - & \cdot \\
\cdot & \cdot & \cdot & \cdot & \cdot & 1 & - & \cdot & \cdot & 1 & \cdot & \cdot & \cdot & \cdot & \cdot & \cdot & \cdot & \cdot & - & \cdot & \cdot & \cdot & - & \cdot & 1 & \cdot \\
\cdot & \cdot & \cdot & \cdot & \cdot & \cdot & \cdot & 1 & - & 1 & \cdot & \cdot & \cdot & \cdot & \cdot & \cdot & \cdot & \cdot & \cdot & - & \cdot & \cdot & \cdot & - & - & \cdot \\
\cdot & \cdot & \cdot & \cdot & \cdot & \cdot & \cdot & \cdot & \cdot & \cdot & 1 & - & \cdot & 1 & \cdot & \cdot & - & \cdot & \cdot & \cdot & 1 & \cdot & \cdot & \cdot & \cdot & 1 \\
\cdot & \cdot & \cdot & \cdot & \cdot & \cdot & \cdot & \cdot & \cdot & \cdot & 1 & \cdot & - & \cdot & 1 & \cdot & \cdot & - & \cdot & \cdot & \cdot & 1 & \cdot & \cdot & \cdot & - \\
\cdot & \cdot & \cdot & \cdot & \cdot & \cdot & \cdot & \cdot & \cdot & \cdot & \cdot & 1 & - & \cdot & \cdot & 1 & \cdot & \cdot & - & \cdot & \cdot & \cdot & 1 & \cdot & \cdot & 1 \\
\cdot & \cdot & \cdot & \cdot & \cdot & \cdot & \cdot & \cdot & \cdot & \cdot & \cdot & \cdot & \cdot & 1 & - & 1 & \cdot & \cdot & \cdot & - & \cdot & \cdot & \cdot & 1 & \cdot & - \\
\cdot & \cdot & \cdot & \cdot & \cdot & \cdot & \cdot & \cdot & \cdot & \cdot & \cdot & \cdot & \cdot & \cdot & \cdot & \cdot & 1 & - & 1 & - & \cdot & \cdot & \cdot & \cdot & 1 & 1 \\
\cdot & \cdot & \cdot & \cdot & \cdot & \cdot & \cdot & \cdot & \cdot & \cdot & \cdot & \cdot & \cdot & \cdot & \cdot & \cdot & \cdot & \cdot & \cdot & \cdot & 1 & - & 1 & - & 1 & -
\end{array}\right]
$$

$T_{1,1}$

$$
\left[\begin{array}{cccccccccccccccccc}
1 & \cdot & \cdot & \cdot & \cdot & \cdot & \cdot & \cdot & \cdot & \cdot & 1 & 1 & 1 & \cdot & \cdot & \cdot & \cdot & \cdot \\
\cdot & 1 & \cdot & \cdot & \cdot & \cdot & \cdot & \cdot & \cdot & \cdot & - & \cdot & \cdot & 1 & 1 & \cdot & \cdot & \cdot \\
\cdot & \cdot & 1 & \cdot & \cdot & \cdot & \cdot & \cdot & \cdot & \cdot & \cdot & - & \cdot & - & \cdot & 1 & \cdot & \cdot \\
\cdot & \cdot & \cdot & 1 & \cdot & \cdot & \cdot & \cdot & \cdot & \cdot & \cdot & \cdot & - & \cdot & - & - & \cdot & \cdot \\
\cdot & \cdot & \cdot & \cdot & 1 & \cdot & \cdot & \cdot & \cdot & \cdot & \cdot & 1 & \cdot & - & \cdot & \cdot & 1 & \cdot \\
\cdot & \cdot & \cdot & \cdot & \cdot & 1 & \cdot & \cdot & \cdot & \cdot & \cdot & \cdot & 1 & 1 & - & \cdot & \cdot & \cdot \\
\cdot & \cdot & \cdot & \cdot & \cdot & \cdot & 1 & \cdot & \cdot & \cdot & 1 & \cdot & \cdot & \cdot & 1 & \cdot & - & \cdot \\
\cdot & \cdot & \cdot & \cdot & \cdot & \cdot & \cdot & 1 & \cdot & \cdot & \cdot & \cdot & \cdot & \cdot & 1 & - & 1 & \cdot \\
\cdot & \cdot & \cdot & \cdot & \cdot & \cdot & \cdot & \cdot & 1 & \cdot & - & 1 & \cdot & \cdot & \cdot & 1 & \cdot & \cdot \\
\cdot & \cdot & \cdot & \cdot & \cdot & \cdot & \cdot & \cdot & \cdot & 1 & \cdot & - & 1 & \cdot & \cdot & \cdot & 1 & \cdot \\
1 & - & \cdot & \cdot & \cdot & \cdot & 1 & \cdot & - & \cdot & - & \cdot & \cdot & \cdot & \cdot & \cdot & \cdot & 1 \\
1 & \cdot & - & \cdot & 1 & \cdot & \cdot & \cdot & 1 & - & \cdot & - & \cdot & \cdot & \cdot & \cdot & \cdot & \cdot \\
1 & \cdot & \cdot & - & \cdot & 1 & \cdot & \cdot & \cdot & 1 & \cdot & \cdot & - & \cdot & \cdot & \cdot & \cdot & - \\
\cdot & 1 & - & \cdot & - & 1 & \cdot & \cdot & \cdot & \cdot & \cdot & \cdot & \cdot & - & \cdot & \cdot & \cdot & 1 \\
\cdot & 1 & \cdot & - & \cdot & - & 1 & 1 & \cdot & \cdot & \cdot & \cdot & \cdot & \cdot & - & \cdot & \cdot & \cdot \\
\cdot & \cdot & 1 & - & \cdot & \cdot & \cdot & - & 1 & \cdot & \cdot & \cdot & \cdot & \cdot & \cdot & - & \cdot & 1 \\
\cdot & \cdot & \cdot & \cdot & 1 & \cdot & - & 1 & \cdot & 1 & \cdot & \cdot & \cdot & \cdot & \cdot & \cdot & - & 1 \\
\cdot & \cdot & \cdot & \cdot & \cdot & \cdot & \cdot & \cdot & \cdot & \cdot & 1 & \cdot & - & 1 & \cdot & 1 & 1 & 1
\end{array}\right]
$$

$T_{1,2}$

Figure 3. (*continued*)

$$
R^{t}_{1,2} = \left[\begin{array}{cccccccccc}
1 & - & \cdot & \cdot & \cdot & \cdot & 1 & \cdot & - & \cdot \\
1 & \cdot & - & \cdot & 1 & \cdot & \cdot & \cdot & 1 & - \\
1 & \cdot & \cdot & - & \cdot & 1 & \cdot & \cdot & \cdot & 1 \\
\cdot & 1 & - & \cdot & - & 1 & \cdot & \cdot & \cdot & \cdot \\
\cdot & 1 & \cdot & - & \cdot & - & 1 & 1 & \cdot & \\
\cdot & \cdot & 1 & - & \cdot & \cdot & \cdot & - & 1 & \cdot \\
\cdot & \cdot & \cdot & \cdot & 1 & \cdot & - & 1 & \cdot & 1 \\
\cdot & \cdot & \cdot & \cdot & \cdot & \cdot & \cdot & \cdot & \cdot & \cdot
\end{array}\right]
$$

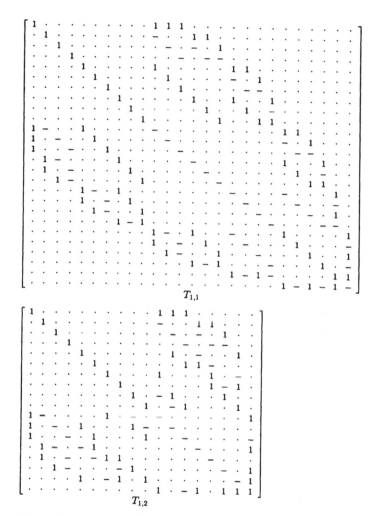

Then for each $R_{1,i}$, the U- and K-matrices are uniquely determined, say $U_{1,i}$, $K_{1,i}$ for $i = 1, 2$. □

Put $T_{1,i} = \begin{bmatrix} S_1 & R_{1,i} \\ U_{1,i} & K_{1,i} \end{bmatrix}$ for $i = 1, 2$. The matrices $T_{1,1}$ and $T_{1,2}$ are also listed in Figure 3.

THEOREM 4: *There are two weighing matrices (say $W_{2,1}$, $W_{2,2}$) based on $\tilde{W}_2$, up to equivalence.*

$$\left[\begin{array}{cccccccccccccccccccccc}
1&\cdot&\cdot&\cdot&\cdot&\cdot&\cdot&\cdot&\cdot&\cdot&1&1&1&\cdot&\cdot&\cdot&\cdot&\cdot&\cdot&\cdot&\cdot&\cdot\\
\cdot&1&\cdot&\cdot&\cdot&\cdot&\cdot&\cdot&\cdot&\cdot&-&\cdot&\cdot&1&1&\cdot&\cdot&\cdot&\cdot&\cdot&\cdot&\cdot\\
\cdot&\cdot&1&\cdot&\cdot&\cdot&\cdot&\cdot&\cdot&\cdot&\cdot&-&\cdot&-&\cdot&1&\cdot&\cdot&\cdot&\cdot&\cdot&\cdot\\
\cdot&\cdot&\cdot&1&\cdot&\cdot&\cdot&\cdot&\cdot&\cdot&\cdot&\cdot&-&\cdot&-&-&\cdot&\cdot&\cdot&\cdot&\cdot&\cdot\\
\cdot&\cdot&\cdot&\cdot&1&\cdot&\cdot&\cdot&\cdot&\cdot&1&\cdot&\cdot&\cdot&\cdot&\cdot&1&1&\cdot&\cdot&\cdot&\cdot\\
\cdot&\cdot&\cdot&\cdot&\cdot&1&\cdot&\cdot&\cdot&\cdot&\cdot&1&\cdot&\cdot&\cdot&\cdot&-&\cdot&1&\cdot&\cdot&\cdot\\
\cdot&\cdot&\cdot&\cdot&\cdot&\cdot&1&\cdot&\cdot&\cdot&\cdot&\cdot&1&\cdot&\cdot&\cdot&\cdot&-&-&\cdot&\cdot&\cdot\\
\cdot&\cdot&\cdot&\cdot&\cdot&\cdot&\cdot&\cdot&1&-&\cdot&\cdot&\cdot&1&\cdot&\cdot&1&\cdot&\cdot&\cdot&\cdot&\cdot\\
\cdot&\cdot&\cdot&\cdot&\cdot&\cdot&\cdot&1&\cdot&1&\cdot&\cdot&\cdot&\cdot&1&\cdot&\cdot&1&\cdot&\cdot&\cdot&\cdot\\
\cdot&\cdot&\cdot&\cdot&\cdot&\cdot&\cdot&-&1&\cdot&\cdot&\cdot&\cdot&\cdot&\cdot&1&\cdot&\cdot&1&\cdot&\cdot&\cdot\\
1&-&\cdot&\cdot&1&\cdot&\cdot&\cdot&\cdot&\cdot&-&\cdot&\cdot&\cdot&\cdot&\cdot&\cdot&\cdot&\cdot&1&1&\cdot\\
1&\cdot&-&\cdot&\cdot&1&\cdot&\cdot&\cdot&\cdot&\cdot&-&\cdot&\cdot&\cdot&\cdot&\cdot&\cdot&\cdot&-&\cdot&1\\
1&\cdot&\cdot&-&\cdot&\cdot&1&\cdot&\cdot&\cdot&\cdot&\cdot&-&\cdot&\cdot&\cdot&\cdot&\cdot&\cdot&\cdot&-&-\\
\cdot&1&-&\cdot&\cdot&\cdot&\cdot&1&\cdot&\cdot&\cdot&\cdot&\cdot&\cdot&-&1&\cdot&\cdot&\cdot&1&\cdot&\cdot\\
\cdot&1&\cdot&-&\cdot&\cdot&\cdot&\cdot&1&\cdot&\cdot&\cdot&\cdot&-&\cdot&-&\cdot&\cdot&\cdot&\cdot&1&\cdot\\
\cdot&\cdot&1&-&\cdot&\cdot&\cdot&\cdot&\cdot&1&\cdot&\cdot&\cdot&1&-&\cdot&\cdot&\cdot&\cdot&\cdot&\cdot&1\\
\cdot&\cdot&\cdot&\cdot&1&-&\cdot&1&\cdot&\cdot&\cdot&\cdot&\cdot&\cdot&\cdot&\cdot&\cdot&-&1&-&\cdot&\cdot\\
\cdot&\cdot&\cdot&\cdot&1&\cdot&-&\cdot&1&\cdot&\cdot&\cdot&\cdot&\cdot&\cdot&\cdot&-&\cdot&-&\cdot&-&\cdot\\
\cdot&\cdot&\cdot&\cdot&\cdot&1&-&\cdot&\cdot&1&\cdot&\cdot&\cdot&\cdot&\cdot&\cdot&1&-&\cdot&\cdot&\cdot&-\\
\cdot&\cdot&\cdot&\cdot&\cdot&\cdot&\cdot&\cdot&\cdot&\cdot&1&-&\cdot&1&\cdot&\cdot&-&\cdot&\cdot&\cdot&1&-\\
\cdot&\cdot&\cdot&\cdot&\cdot&\cdot&\cdot&\cdot&\cdot&\cdot&1&\cdot&-&\cdot&1&\cdot&\cdot&-&\cdot&1&\cdot&1\\
\cdot&\cdot&\cdot&\cdot&\cdot&\cdot&\cdot&\cdot&\cdot&\cdot&\cdot&1&-&\cdot&\cdot&1&\cdot&\cdot&-&-&1&\cdot
\end{array}\right]$$

$T_{2,1}$

$$\left[\begin{array}{cccccccccccccccccc}
1&\cdot&\cdot&\cdot&\cdot&\cdot&\cdot&\cdot&\cdot&\cdot&1&1&1&\cdot&\cdot&\cdot&\cdot&\cdot\\
\cdot&1&\cdot&\cdot&\cdot&\cdot&\cdot&\cdot&\cdot&\cdot&-&\cdot&\cdot&1&1&\cdot&\cdot&\cdot\\
\cdot&\cdot&1&\cdot&\cdot&\cdot&\cdot&\cdot&\cdot&\cdot&\cdot&-&\cdot&-&\cdot&1&\cdot&\cdot\\
\cdot&\cdot&\cdot&1&\cdot&\cdot&\cdot&\cdot&\cdot&\cdot&\cdot&\cdot&-&\cdot&-&-&\cdot&\cdot\\
\cdot&\cdot&\cdot&\cdot&1&\cdot&\cdot&\cdot&\cdot&\cdot&\cdot&1&\cdot&-&\cdot&\cdot&1&\cdot\\
\cdot&\cdot&\cdot&\cdot&\cdot&1&\cdot&\cdot&\cdot&\cdot&1&\cdot&\cdot&1&\cdot&\cdot&\cdot&1\\
\cdot&\cdot&\cdot&\cdot&\cdot&\cdot&1&\cdot&\cdot&\cdot&\cdot&\cdot&1&\cdot&\cdot&\cdot&-&-\\
\cdot&\cdot&\cdot&\cdot&\cdot&\cdot&\cdot&\cdot&1&-&-&1&\cdot&\cdot&\cdot&\cdot&\cdot&\cdot\\
\cdot&\cdot&\cdot&\cdot&\cdot&\cdot&\cdot&1&\cdot&1&\cdot&\cdot&\cdot&\cdot&1&\cdot&1&\cdot\\
\cdot&\cdot&\cdot&\cdot&\cdot&\cdot&\cdot&-&1&\cdot&\cdot&\cdot&\cdot&\cdot&\cdot&1&\cdot&1\\
1&-&\cdot&\cdot&\cdot&\cdot&1&\cdot&-&\cdot&-&\cdot&\cdot&\cdot&\cdot&\cdot&\cdot&1\\
1&\cdot&-&\cdot&\cdot&1&\cdot&\cdot&\cdot&\cdot&\cdot&\cdot&-&\cdot&\cdot&1&\cdot&-\\
1&\cdot&\cdot&-&1&\cdot&\cdot&\cdot&1&\cdot&\cdot&-&\cdot&\cdot&\cdot&-&\cdot&\cdot\\
\cdot&1&-&\cdot&\cdot&\cdot&\cdot&1&\cdot&\cdot&\cdot&\cdot&\cdot&-&\cdot&\cdot&-&1\\
\cdot&1&\cdot&-&-&\cdot&1&\cdot&\cdot&\cdot&\cdot&\cdot&\cdot&\cdot&-&\cdot&1&\cdot\\
\cdot&\cdot&1&-&\cdot&\cdot&\cdot&\cdot&\cdot&1&\cdot&1&-&\cdot&\cdot&\cdot&-&\cdot\\
\cdot&\cdot&\cdot&\cdot&1&-&\cdot&1&\cdot&\cdot&\cdot&\cdot&\cdot&1&-&1&\cdot&\cdot\\
\cdot&\cdot&\cdot&\cdot&\cdot&1&-&\cdot&\cdot&1&-&\cdot&1&\cdot&-&\cdot&\cdot&\cdot
\end{array}\right]$$

$T_{2,2}$

Figure 3. (*continued*)

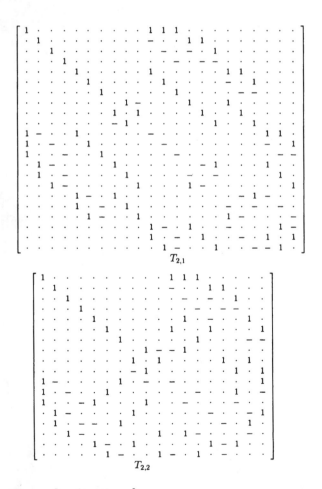

Proof. Up to equivalence, there are two R-matrices based on $\tilde{W}_2$, say $R_{2,1}$ and $R_{2,2}$, where

$$R_{2,1}^t = \begin{bmatrix} 1 & - & \cdot & \cdot & 1 & \cdot & \cdot & \cdot & \cdot & \cdot \\ 1 & \cdot & - & \cdot & \cdot & 1 & \cdot & \cdot & \cdot & \cdot \\ 1 & \cdot & \cdot & - & \cdot & \cdot & 1 & \cdot & \cdot & \cdot \\ \cdot & 1 & - & \cdot & \cdot & \cdot & \cdot & 1 & \cdot & \cdot \\ \cdot & 1 & \cdot & - & \cdot & \cdot & \cdot & \cdot & 1 & \cdot \\ \cdot & \cdot & 1 & - & \cdot & \cdot & \cdot & \cdot & \cdot & 1 \\ \cdot & \cdot & \cdot & \cdot & 1 & - & \cdot & 1 & \cdot & \cdot \\ \cdot & \cdot & \cdot & \cdot & 1 & \cdot & - & \cdot & 1 & \cdot \\ \cdot & \cdot & \cdot & \cdot & \cdot & 1 & - & \cdot & \cdot & 1 \\ \cdot & \cdot & \cdot & \cdot & \cdot & \cdot & \cdot & \cdot & \cdot & \cdot \\ \cdot & \cdot & \cdot & \cdot & \cdot & \cdot & \cdot & \cdot & \cdot & \cdot \\ \cdot & \cdot & \cdot & \cdot & \cdot & \cdot & \cdot & \cdot & \cdot & \cdot \end{bmatrix}$$

$$R_{2,2}^t = \begin{bmatrix} 1 & - & \cdot & \cdot & \cdot & 1 & \cdot & - & \cdot & \cdot \\ 1 & \cdot & \cdot & \cdot & 1 & \cdot & \cdot & 1 & \cdot & \cdot \\ 1 & \cdot & \cdot & - & \cdot & \cdot & 1 & \cdot & \cdot & \cdot \\ \cdot & 1 & - & \cdot & - & 1 & \cdot & \cdot & \cdot & \cdot \\ \cdot & 1 & \cdot & - & \cdot & \cdot & \cdot & \cdot & 1 & \cdot \\ \cdot & \cdot & 1 & - & \cdot & \cdot & \cdot & \cdot & \cdot & 1 \\ \cdot & \cdot & \cdot & \cdot & 1 & \cdot & - & \cdot & 1 & \cdot \\ \cdot & \cdot & \cdot & \cdot & \cdot & 1 & - & \cdot & \cdot & 1 \end{bmatrix}$$

Then for each $R_{2,i}$, the U- and K-matrices are uniquely determined, say $U_{2,i}$, $K_{2,i}$, for $i = 1, 2$. □

Put $T_{2,i} = \begin{bmatrix} S_2 & R_{2,i} \\ U_{2,i} & K_{2,i} \end{bmatrix}$ for $i = 1, 2$. The matrices $T_{2,1}$ and $T_{2,2}$ are also listed in Figure 3. By the above theorems, the set of weighing matrices which contains all inequivalent matrices of type F_1 is constructed.

3. CLASSIFICATION

In Section 2, the set of weighing matrices with weight six, including all inequivalent matrices of the semi-biplane type, is constructed. In this section, those ten matrices are classified to five inequivalent matrices. Let W_i and $W_{k,j}$ be weighing matrices based on T_i and $T_{k,j}$, respectively, where $i \in \{3, 4, 5, 6, 7, 10\}$ and $k, j \in \{1, 2\}$. The following lemma is useful to determine whether two weighing matrices are equivalent or not.

LEMMA 3: *Let W be a weighing matrix of type F_1 with weight 6 and order n. For a fixed row of W, consider a $3 \times n$ submatrix of W, say K, whose the first row corresponds to the row. Then the matrix ignoring the signs of K is equivalent to one of the following seven matrices.*

$$A_1 = \begin{bmatrix} 1 & 1 & 1 & 1 & 1 & 1 & \cdot & \cdot & \cdot & \cdot & \cdot & \cdot & \cdots & \cdot \\ 1 & 1 & \cdot & \cdot & \cdot & \cdot & 1 & 1 & 1 & \cdot & \cdot & \cdot & \cdots & \cdot \\ 1 & \cdot & 1 & \cdot & \cdot & \cdot & 1 & \cdot & \cdot & 1 & 1 & 1 & \cdots & \cdot \end{bmatrix}$$

$$A_2 = \begin{bmatrix} 1 & 1 & 1 & 1 & \cdot & \cdot & 1 & 1 & \cdot & \cdot & \cdot & \cdot & \cdots & \cdot \\ 1 & 1 & \cdot & \cdot & 1 & 1 & \cdot & \cdot & 1 & 1 & \cdot & \cdot & \cdots & \cdot \\ \cdot & \cdot & 1 & 1 & 1 & 1 & \cdot & \cdot & \cdot & \cdot & 1 & 1 & \cdots & \cdot \end{bmatrix}$$

$$A_3 = \begin{bmatrix} 1 & 1 & 1 & 1 & 1 & 1 & \cdot & \cdot & \cdot & \cdot & \cdot & \cdot & \cdot & \cdot & \cdots & \cdot \\ 1 & 1 & \cdot & \cdot & \cdot & \cdot & 1 & 1 & 1 & 1 & \cdot & \cdot & \cdot & \cdot & \cdots & \cdot \\ \cdot & \cdot & 1 & 1 & \cdot & \cdot & \cdot & \cdot & \cdot & \cdot & 1 & 1 & 1 & 1 & \cdots & \cdot \end{bmatrix}$$

$$A_4 = \begin{bmatrix} 1 & 1 & 1 & 1 & 1 & 1 & \cdot & \cdot & \cdot & \cdot & \cdot & \cdot & \cdot & \cdot & \cdots & \cdot \\ 1 & 1 & \cdot & \cdot & \cdot & \cdot & 1 & 1 & 1 & 1 & \cdot & \cdot & \cdot & \cdot & \cdots & \cdot \\ \cdot & \cdot & \cdot & \cdot & \cdot & \cdot & 1 & 1 & \cdot & \cdot & 1 & 1 & 1 & 1 & \cdots & \cdot \end{bmatrix}$$

$$A_5 = \begin{bmatrix} 1 & 1 & 1 & 1 & 1 & 1 & \cdot & \cdot & \cdot & \cdot & \cdot & \cdot & \cdot & \cdot & \cdot & \cdot & \cdots & \cdot \\ 1 & 1 & \cdot & \cdot & \cdot & \cdot & 1 & 1 & 1 & 1 & \cdot & \cdot & \cdot & \cdot & \cdot & \cdot & \cdots & \cdot \\ \cdot & \cdot & \cdot & \cdot & \cdot & \cdot & \cdot & \cdot & \cdot & \cdot & 1 & 1 & 1 & 1 & 1 & 1 & \cdots & \cdot \end{bmatrix}$$

$$A_6 = \begin{bmatrix} 1 & 1 & 1 & 1 & 1 & 1 & \cdot & \cdot & \cdot & \cdot & \cdot & \cdot & \cdot & \cdot & \cdot & \cdot & \cdots & \cdot \\ \cdot & \cdot & \cdot & \cdot & \cdot & \cdot & 1 & 1 & 1 & 1 & 1 & 1 & \cdot & \cdot & \cdot & \cdot & \cdots & \cdot \\ \cdot & \cdot & \cdot & \cdot & \cdot & \cdot & 1 & 1 & \cdot & \cdot & \cdot & \cdot & 1 & 1 & 1 & 1 & \cdots & \cdot \end{bmatrix}$$

$$A_7 = \begin{bmatrix} 1 & 1 & 1 & 1 & 1 & 1 & \cdot & \cdot & \cdot & \cdot & \cdot & \cdot & \cdot & \cdot & \cdot & \cdot & \cdot & \cdot & \cdots & \cdot \\ \cdot & \cdot & \cdot & \cdot & \cdot & \cdot & 1 & 1 & 1 & 1 & 1 & 1 & \cdot & \cdot & \cdot & \cdot & \cdot & \cdot & \cdots & \cdot \\ \cdot & \cdot & \cdot & \cdot & \cdot & \cdot & \cdot & \cdot & \cdot & \cdot & \cdot & \cdot & 1 & 1 & 1 & 1 & 1 & 1 & \cdots & \cdot \end{bmatrix}$$

Consider all $3 \times n$ submatrices of W for a fixed row vector ($\mathbf{r}$, say) and the distribution on $A_1, \ldots, A_7$ associated with $\mathbf{r}$. Let n_i be the number of submatrices which are equivalent to A_i, where $i = 1, 2, \ldots, 7$. Then $(n_1, \ldots, n_7)$ is called the $\mathbf{r}$-distribution. By considering all $\mathbf{r}$-distributions of W, a new distribution can be obtained. The distribution is called an R-distribution of W. Let $\{(n_1, \ldots, n_7), \ldots, (m_1, \ldots, m_7)\}$ be the set of different $\mathbf{r}$-distributions of W, and $\{\alpha, \ldots, \beta\}$ be the set of multiplicities of $\mathbf{r}$-distributions of W. Then the R-distribution is denoted by $(n_1, \ldots, n_7)^\alpha \ldots (m_1, \ldots, m_7)^\beta$. The C-distribution for columns of W can be defined similarly. The following lemma can easily be proved.

LEMMA 4: *Let W and W' be equivalent weighing matrices of type F_1. Then their R-distributions and C-distributions, respectively, are the same.*

For W_i and $W_{j,k}$, where $i \in \{3, 4, 5, 6, 7, 10\}$ and $j, k \in \{1, 2\}$, the R-distribution and C-distribution are calculated. The result is expressed in the following table.

	R-distribution	C-distribution
W_3	$(60, 10, 35, 70, 50, 25, 3)^{24}$	
W_4	$(60, 9, 36, 72, 48, 24, 4)^{8}$	
	$(60, 6, 39, 78, 42, 21, 7)^{16}$	
W_5	$(60, 21, 24, 48, 12, 6, 0)^{20}$	
W_6	$(60, 21, 24, 48, 12, 6, 0)^{20}$	
W_7	$(60, 21, 24, 48, 12, 6, 0)^{20}$	
W_{10}	$(60, 21, 24, 48, 12, 6, 10)^{20}$	Same as the left-hand column
$W_{1,1}$	$(60, 0, 45, 90, 150, 75, 45)^{32}$	
$W_{1,2}$	$(60, 10, 35, 70, 50, 25, 3)^{24}$	
$W_{2,1}$	$(60, 3, 42, 84, 96, 48, 18)^{28}$	
$W_{2,2}$	$(60, 9, 36, 72, 48, 24, 4)^{8}$	
	$(60, 6, 39, 78, 42, 21, 7)^{16}$	

Now, five new matrices $\{\overset{*}{W}_1, \overset{*}{W}_2, \overset{*}{W}_3, \overset{*}{W}_4, \overset{*}{W}_5\}$ are proposed, and they are listed in Figure 4.

The R-distributions of $\{\overset{*}{W}_1, \overset{*}{W}_2, \overset{*}{W}_3, \overset{*}{W}_4, \overset{*}{W}_5\}$ are $\{(60,0,45,90,150,75,45)^{32}$, $(60,3,42,84,96,48,18)^{28}$, $(60,10,35,70,50,25,3)^{24}$, $(60,9,36,72,48,24,4)^{8}$, $(60,6,39,78,42,21,7)^{16}$, $(60,21,24,48,12,6,0)^{20}\}$ in order.

REMARK 4: *Each $\overset{*}{W}_i$ is a symmetric matrix . So, the C-distribution of $\overset{*}{W}_i$ equals an R-distribution.*

REMARK 5: *The R-distribution and C-distribution of the first eight rows and columns of $\overset{*}{W}_4$ is* (60, 9, 36, 72, 48, 24, 4).

By summing up the above discussion, the following theorem is established.

THEOREM 5: *A weighing matrix with weight 6 of the semi-biplane type is equivalent to one of matrices $\{\overset{*}{W}_1, \overset{*}{W}_2, \ldots, \overset{*}{W}_5\}$.*

Proof. By Lemma 3, $\overset{*}{W} = \{\overset{*}{W}_1, \overset{*}{W}_2, \ldots, \overset{*}{W}_5\}$ are inequivalent matrices. On the other hand, from Section 2, any weighing matrix with six of the semi-biplane type is equivalent to one of $\mathbf{W} = \{W_3, W_4, W_5, W_6, W_7, W_{10}, W_{1,1}, W_{1,2}, W_{2,1}, W_{2,2}\}$. Therefore, it is sufficient that a matrix in $\mathbf{W}$ is equivalent to one in $\overset{*}{W}$. The corresponding table is given below.

W_3	Row	2 1 3 4 5 6 $\underline{7}$ $\underline{8}$ $\underline{9}$ $\underline{10}$ 17 18 19 20 21 22 11 12 13 14 15 16 23 $\underline{24}$	$\overset{*}{W}_3$
	Column	1 $\underline{2}$ 7 8 9 10 3 4 5 6 11 12 13 14 15 16 $\underline{17}$ $\underline{18}$ $\underline{19}$ $\underline{20}$ $\underline{21}$ $\underline{22}$ 24 23	
W_4	Row	1 5 $\underline{17}$ 18 10 11 $\underline{23}$ $\underline{24}$ 2 6 4 3 16 $\underline{9}$ $\underline{12}$ $\underline{14}$ $\underline{8}$ $\underline{13}$ $\underline{19}$ 21 $\underline{15}$ $\underline{7}$ 20 $\underline{22}$	$\overset{*}{W}_4$
	Column	1 5 17 $\underline{18}$ 10 $\underline{11}$ 23 $\underline{24}$ 2 6 3 4 16 $\underline{9}$ $\underline{14}$ $\underline{12}$ $\underline{7}$ $\underline{15}$ $\underline{21}$ 19 $\underline{13}$ $\underline{8}$ 22 $\underline{20}$	
W_5	Row	5 19 9 $\underline{16}$ 2 6 $\underline{10}$ 17 $\underline{1}$ 14 12 $\underline{11}$ $\underline{15}$ $\underline{7}$ $\underline{4}$ $\underline{18}$ 8 $\underline{3}$ $\underline{13}$ 20	$\overset{*}{W}_5$
	Column	$\underline{5}$ $\underline{18}$ $\underline{14}$ $\underline{12}$ $\underline{4}$ $\underline{3}$ $\underline{11}$ 17 1 $\underline{9}$ 16 $\underline{10}$ 7 15 2 20 8 6 $\underline{13}$ $\underline{19}$	
W_6	Row	5 9 $\underline{14}$ $\underline{19}$ 4 2 $\underline{18}$ $\underline{8}$ $\underline{3}$ $\underline{13}$ 15 $\underline{11}$ $\underline{7}$ $\underline{20}$ $\underline{1}$ $\underline{16}$ 12 $\underline{6}$ $\underline{17}$ 10	$\overset{*}{W}_5$
	Column	$\underline{12}$ $\underline{18}$ $\underline{5}$ 16 3 $\underline{13}$ 6 $\underline{19}$ 1 $\underline{14}$ $\underline{9}$ 8 15 2 11 17 $\underline{4}$ $\underline{7}$ 10 $\underline{20}$	
W_7	Row	8 11 15 $\underline{19}$ $\underline{13}$ $\underline{7}$ $\underline{10}$ 18 $\underline{6}$ 2 $\underline{20}$ 3 $\underline{5}$ $\underline{16}$ 4 $\underline{14}$ $\underline{1}$ $\underline{17}$ 12 $\underline{9}$	$\overset{*}{W}_5$
	Column	10 12 $\underline{4}$ 17 $\underline{1}$ 13 8 5 $\underline{15}$ 18 2 3 20 $\underline{14}$ $\underline{6}$ $\underline{9}$ $\underline{11}$ $\underline{16}$ $\underline{7}$ $\underline{19}$	
W_{10}	Row	$\underline{5}$ $\underline{2}$ 3 17 $\underline{12}$ $\underline{9}$ $\underline{20}$ $\underline{7}$ 14 $\underline{15}$ 13 11 $\underline{8}$ $\underline{18}$ $\underline{1}$ $\underline{6}$ $\underline{4}$ $\underline{16}$ $\underline{19}$ 10	$\overset{*}{W}_5$
	Column	12 $\underline{7}$ $\underline{1}$ 9 4 20 6 15 5 $\underline{16}$ 14 $\underline{8}$ $\underline{10}$ 2 11 13 $\underline{3}$ $\underline{19}$ 17 $\underline{18}$	
$W_{1,1}$	Row	Identity	$\overset{*}{W}_1$
	Column		
$W_{1,2}$	Row	Identity	$\overset{*}{W}_3$
	Column		
$W_{2,1}$	Row	Identity	$\overset{*}{W}_2$
	Column		
$W_{2,2}$	Row	15 4 $\underline{12}$ $\underline{9}$ $\underline{6}$ $\underline{21}$ $\underline{18}$ 7 $\underline{5}$ $\underline{16}$ 20 23 1 14 $\underline{8}$ $\underline{11}$ $\underline{22}$ 10 $\underline{19}$ 2 13 $\underline{24}$ 3 17	$\overset{*}{W}_4$
	Column	14 4 10 13 5 $\underline{20}$ 19 $\underline{7}$ 16 $\underline{6}$ 21 23 15 1 $\underline{8}$ $\underline{11}$ $\underline{12}$ $\underline{24}$ $\underline{3}$ $\underline{17}$ $\underline{22}$ $\underline{9}$ 18 $\underline{2}$	

Here it means that, for example, W_5 is transformed to $\overset{*}{W}_5$ by arranging rows and columns of W_5 in order (5, 19, 9, $\underline{16}$, 2, ...) and ($\underline{5}$, $\underline{18}$, $\underline{14}$, $\underline{12}$, ...), respectively, and the notation $\underline{n}$ means the nth row (column) of W_5 is multiplied by -1. □

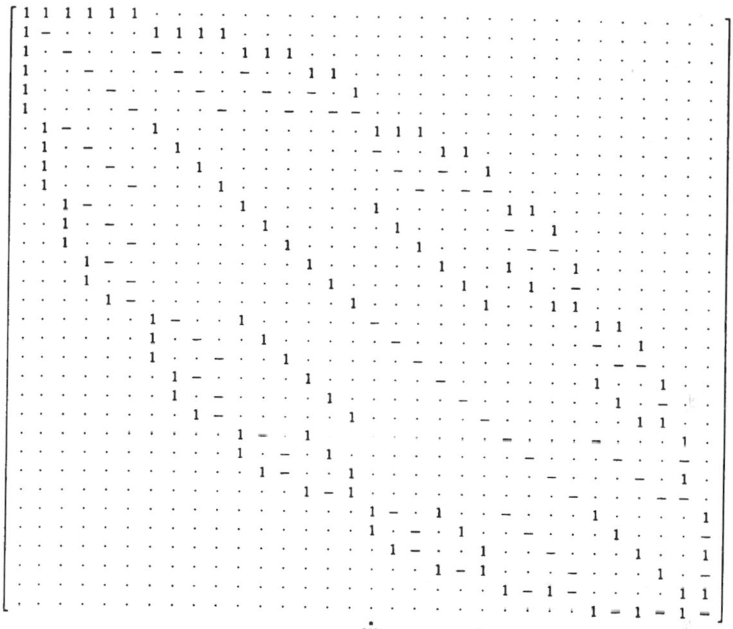

$\overset{*}{W}_1$

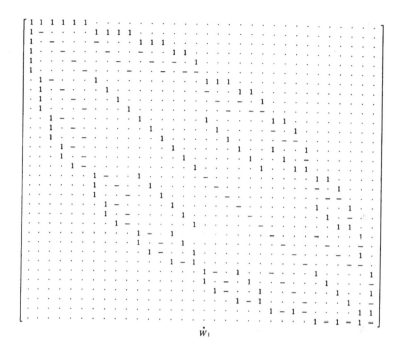

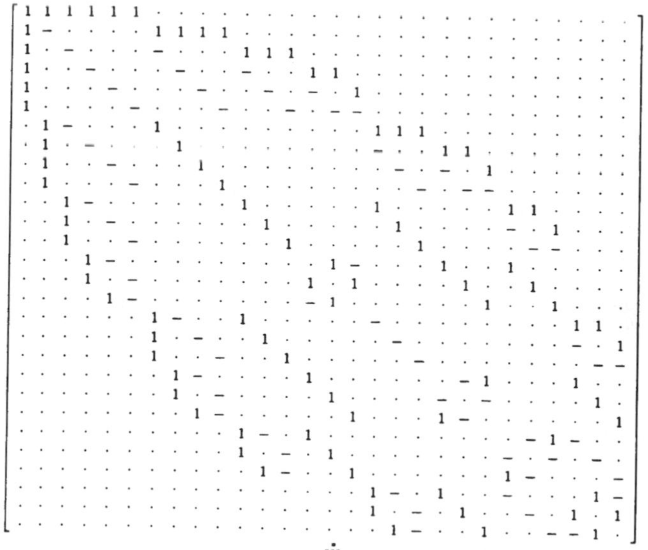

$\overset{*}{W}_2$

Figure 4. Five new matrices $\{\overset{*}{W}_1, \overset{*}{W}_2, \overset{*}{W}_3, \overset{*}{W}_4, \overset{*}{W}_5\}$

$$\begin{bmatrix}
1 & 1 & 1 & 1 & 1 & 1 & \cdot & \cdot & \cdot & \cdot & \cdot & \cdot & \cdot & \cdot & \cdot & \cdot & \cdot & \cdot & \cdot & \cdot & \cdot & \cdot & \cdot & \cdot \\
1 & - & \cdot & \cdot & \cdot & \cdot & 1 & 1 & 1 & 1 & \cdot & \cdot & \cdot & \cdot & \cdot & \cdot & \cdot & \cdot & \cdot & \cdot & \cdot & \cdot & \cdot & \cdot \\
1 & \cdot & - & \cdot & \cdot & \cdot & - & \cdot & \cdot & \cdot & 1 & 1 & 1 & \cdot & \cdot & \cdot & \cdot & \cdot & \cdot & \cdot & \cdot & \cdot & \cdot & \cdot \\
1 & \cdot & \cdot & - & \cdot & \cdot & \cdot & - & \cdot & \cdot & - & \cdot & \cdot & 1 & 1 & \cdot & \cdot & \cdot & \cdot & \cdot & \cdot & \cdot & \cdot & \cdot \\
1 & \cdot & \cdot & \cdot & - & \cdot & \cdot & \cdot & - & \cdot & \cdot & - & \cdot & - & \cdot & 1 & \cdot & \cdot & \cdot & \cdot & \cdot & \cdot & \cdot & \cdot \\
1 & \cdot & \cdot & \cdot & \cdot & - & \cdot & \cdot & \cdot & - & \cdot & \cdot & - & \cdot & - & - & \cdot & \cdot & \cdot & \cdot & \cdot & \cdot & \cdot & \cdot \\
\cdot & 1 & - & \cdot & \cdot & \cdot & 1 & \cdot & \cdot & \cdot & \cdot & \cdot & \cdot & \cdot & \cdot & \cdot & 1 & 1 & 1 & \cdot & \cdot & \cdot & \cdot & \cdot \\
\cdot & 1 & \cdot & - & \cdot & \cdot & \cdot & 1 & \cdot & \cdot & \cdot & \cdot & \cdot & \cdot & \cdot & \cdot & - & \cdot & \cdot & 1 & 1 & \cdot & \cdot & \cdot \\
\cdot & 1 & \cdot & \cdot & - & \cdot & \cdot & \cdot & 1 & \cdot & \cdot & \cdot & \cdot & \cdot & \cdot & \cdot & \cdot & - & \cdot & - & \cdot & 1 & \cdot & \cdot \\
\cdot & 1 & \cdot & \cdot & \cdot & - & \cdot & \cdot & \cdot & 1 & \cdot & \cdot & \cdot & \cdot & \cdot & \cdot & \cdot & \cdot & - & \cdot & - & - & \cdot & \cdot \\
\cdot & \cdot & 1 & - & \cdot & \cdot & \cdot & \cdot & \cdot & \cdot & 1 & \cdot & \cdot & \cdot & \cdot & \cdot & \cdot & 1 & \cdot & - & \cdot & \cdot & 1 & \cdot \\
\cdot & \cdot & 1 & \cdot & - & \cdot & \cdot & \cdot & \cdot & \cdot & \cdot & 1 & \cdot & \cdot & \cdot & \cdot & \cdot & \cdot & 1 & 1 & - & \cdot & \cdot & \cdot \\
\cdot & \cdot & 1 & \cdot & \cdot & - & \cdot & \cdot & \cdot & \cdot & \cdot & \cdot & 1 & \cdot & \cdot & \cdot & 1 & \cdot & \cdot & \cdot & 1 & \cdot & - & \cdot \\
\cdot & \cdot & \cdot & 1 & - & \cdot & \cdot & \cdot & \cdot & \cdot & \cdot & \cdot & \cdot & 1 & \cdot & \cdot & \cdot & \cdot & \cdot & \cdot & 1 & - & 1 & \cdot \\
\cdot & \cdot & \cdot & 1 & \cdot & - & \cdot & \cdot & \cdot & \cdot & \cdot & \cdot & \cdot & \cdot & 1 & \cdot & - & 1 & \cdot & \cdot & \cdot & 1 & \cdot & \cdot \\
\cdot & \cdot & \cdot & \cdot & 1 & - & \cdot & \cdot & \cdot & \cdot & \cdot & \cdot & \cdot & \cdot & \cdot & 1 & \cdot & - & 1 & \cdot & \cdot & \cdot & 1 & \cdot \\
\cdot & \cdot & \cdot & \cdot & \cdot & \cdot & 1 & - & \cdot & \cdot & \cdot & \cdot & 1 & \cdot & - & \cdot & - & \cdot & \cdot & \cdot & \cdot & \cdot & \cdot & 1 \\
\cdot & \cdot & \cdot & \cdot & \cdot & \cdot & 1 & \cdot & - & \cdot & 1 & \cdot & \cdot & \cdot & 1 & - & \cdot & - & \cdot & \cdot & \cdot & \cdot & \cdot & \cdot \\
\cdot & \cdot & \cdot & \cdot & \cdot & \cdot & 1 & \cdot & \cdot & - & \cdot & 1 & \cdot & \cdot & \cdot & 1 & \cdot & \cdot & - & \cdot & \cdot & \cdot & \cdot & - \\
\cdot & \cdot & \cdot & \cdot & \cdot & \cdot & \cdot & 1 & - & \cdot & - & 1 & \cdot & \cdot & \cdot & \cdot & \cdot & \cdot & \cdot & - & \cdot & \cdot & \cdot & 1 \\
\cdot & \cdot & \cdot & \cdot & \cdot & \cdot & \cdot & 1 & \cdot & - & \cdot & - & 1 & 1 & \cdot & \cdot & \cdot & \cdot & \cdot & \cdot & - & \cdot & \cdot & \cdot \\
\cdot & \cdot & \cdot & \cdot & \cdot & \cdot & \cdot & \cdot & 1 & - & \cdot & \cdot & \cdot & - & 1 & \cdot & \cdot & \cdot & \cdot & \cdot & \cdot & - & \cdot & 1 \\
\cdot & \cdot & \cdot & \cdot & \cdot & \cdot & \cdot & \cdot & \cdot & \cdot & 1 & \cdot & - & 1 & \cdot & 1 & \cdot & \cdot & \cdot & \cdot & \cdot & \cdot & - & 1 \\
\cdot & \cdot & \cdot & \cdot & \cdot & \cdot & \cdot & \cdot & \cdot & \cdot & \cdot & \cdot & \cdot & \cdot & \cdot & \cdot & 1 & \cdot & - & 1 & \cdot & 1 & 1 & 1
\end{bmatrix}$$

$\dot{W}_3$

$$\begin{bmatrix}
1 & 1 & \cdot & \cdot & \cdot & \cdot & \cdot & \cdot & 1 & 1 & 1 & 1 & \cdot & \cdot & \cdot & \cdot & \cdot & \cdot & \cdot & \cdot & \cdot & \cdot & \cdot & \cdot \\
1 & - & \cdot & \cdot & \cdot & \cdot & \cdot & \cdot & \cdot & \cdot & \cdot & \cdot & 1 & 1 & 1 & 1 & \cdot & \cdot & \cdot & \cdot & \cdot & \cdot & \cdot & \cdot \\
\cdot & \cdot & \cdot & \cdot & 1 & 1 & \cdot & \cdot & \cdot & \cdot & \cdot & \cdot & \cdot & \cdot & \cdot & \cdot & 1 & 1 & 1 & 1 & \cdot & \cdot & \cdot & \cdot \\
\cdot & \cdot & \cdot & \cdot & 1 & - & \cdot & \cdot & \cdot & \cdot & \cdot & \cdot & \cdot & \cdot & \cdot & \cdot & \cdot & \cdot & \cdot & \cdot & 1 & 1 & 1 & 1 \\
\cdot & \cdot & 1 & 1 & \cdot & \cdot & \cdot & \cdot & 1 & - & \cdot & \cdot & 1 & - & \cdot & \cdot & \cdot & \cdot & \cdot & \cdot & \cdot & \cdot & \cdot & \cdot \\
\cdot & \cdot & 1 & - & \cdot & \cdot & \cdot & \cdot & \cdot & \cdot & 1 & - & \cdot & \cdot & 1 & - & \cdot & \cdot & \cdot & \cdot & \cdot & \cdot & \cdot & \cdot \\
\cdot & \cdot & \cdot & \cdot & \cdot & \cdot & 1 & 1 & \cdot & \cdot & \cdot & \cdot & \cdot & \cdot & \cdot & \cdot & 1 & - & \cdot & \cdot & 1 & - & \cdot & \cdot \\
\cdot & \cdot & \cdot & \cdot & \cdot & \cdot & 1 & - & \cdot & \cdot & \cdot & \cdot & \cdot & \cdot & \cdot & \cdot & \cdot & \cdot & 1 & - & \cdot & \cdot & 1 & - \\
1 & \cdot & \cdot & \cdot & 1 & \cdot & \cdot & \cdot & - & \cdot & \cdot & \cdot & \cdot & - & \cdot & \cdot & - & \cdot & \cdot & \cdot & \cdot & - & \cdot & \cdot \\
1 & \cdot & \cdot & \cdot & - & \cdot & \cdot & \cdot & \cdot & - & \cdot & \cdot & - & \cdot & \cdot & \cdot & \cdot & 1 & \cdot & \cdot & 1 & \cdot & \cdot & \cdot \\
1 & \cdot & \cdot & \cdot & \cdot & 1 & \cdot & \cdot & \cdot & \cdot & \cdot & - & \cdot & \cdot & - & \cdot & \cdot & - & \cdot & \cdot & \cdot & 1 & \cdot & \cdot \\
1 & \cdot & \cdot & \cdot & \cdot & - & \cdot & \cdot & \cdot & \cdot & - & \cdot & \cdot & \cdot & \cdot & - & 1 & \cdot & \cdot & \cdot & - & \cdot & \cdot & \cdot \\
\cdot & 1 & \cdot & \cdot & 1 & \cdot & \cdot & \cdot & \cdot & - & \cdot & \cdot & \cdot & 1 & \cdot & \cdot & \cdot & \cdot & - & \cdot & \cdot & \cdot & \cdot & - \\
\cdot & 1 & \cdot & \cdot & - & \cdot & \cdot & \cdot & - & \cdot & \cdot & \cdot & 1 & \cdot & \cdot & \cdot & \cdot & \cdot & \cdot & 1 & \cdot & \cdot & 1 & \cdot \\
\cdot & 1 & \cdot & \cdot & \cdot & 1 & \cdot & \cdot & \cdot & \cdot & - & \cdot & \cdot & \cdot & 1 & \cdot & \cdot & \cdot & \cdot & - & \cdot & \cdot & \cdot & 1 \\
\cdot & 1 & \cdot & \cdot & \cdot & - & \cdot & \cdot & \cdot & \cdot & \cdot & - & \cdot & \cdot & \cdot & 1 & \cdot & \cdot & 1 & \cdot & \cdot & \cdot & - & \cdot \\
\cdot & \cdot & 1 & \cdot & \cdot & \cdot & 1 & \cdot & - & \cdot & \cdot & 1 & \cdot & \cdot & \cdot & \cdot & \cdot & \cdot & \cdot & \cdot & \cdot & 1 & - & \cdot \\
\cdot & \cdot & 1 & \cdot & \cdot & \cdot & - & \cdot & \cdot & 1 & - & \cdot & \cdot & \cdot & \cdot & \cdot & \cdot & \cdot & \cdot & \cdot & 1 & \cdot & \cdot & - \\
\cdot & \cdot & 1 & \cdot & \cdot & \cdot & \cdot & 1 & \cdot & \cdot & \cdot & \cdot & - & \cdot & \cdot & 1 & \cdot & \cdot & \cdot & \cdot & - & \cdot & 1 & \cdot \\
\cdot & \cdot & 1 & \cdot & \cdot & \cdot & \cdot & - & \cdot & \cdot & \cdot & \cdot & \cdot & 1 & - & \cdot & \cdot & \cdot & \cdot & \cdot & \cdot & - & \cdot & 1 \\
\cdot & \cdot & \cdot & 1 & \cdot & \cdot & 1 & \cdot & \cdot & 1 & \cdot & - & \cdot & \cdot & \cdot & \cdot & \cdot & 1 & - & \cdot & \cdot & \cdot & \cdot & \cdot \\
\cdot & \cdot & \cdot & 1 & \cdot & \cdot & - & \cdot & - & \cdot & 1 & \cdot & \cdot & \cdot & \cdot & \cdot & 1 & \cdot & \cdot & - & \cdot & \cdot & \cdot & \cdot \\
\cdot & \cdot & \cdot & 1 & \cdot & \cdot & \cdot & 1 & \cdot & \cdot & \cdot & \cdot & \cdot & 1 & \cdot & - & - & \cdot & 1 & \cdot & \cdot & \cdot & \cdot & \cdot \\
\cdot & \cdot & \cdot & 1 & \cdot & \cdot & \cdot & - & \cdot & \cdot & \cdot & \cdot & - & \cdot & 1 & \cdot & \cdot & - & \cdot & 1 & \cdot & \cdot & \cdot & \cdot
\end{bmatrix}$$

$\dot{W}_4$

Figure 4. (*continued*)

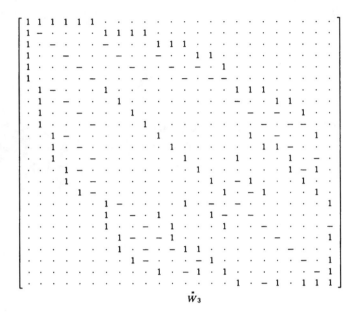

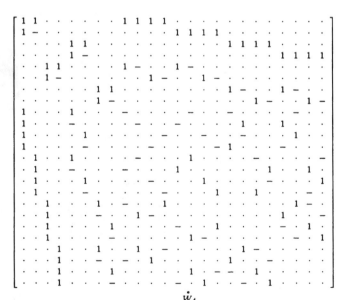

$$\begin{bmatrix}
1 & \cdot & 1 & 1 & \cdot & \cdot & \cdot & \cdot & 1 & 1 & 1 & \cdot & \cdot & \cdot & \cdot & \cdot & \cdot & \cdot & \cdot & \cdot \\
\cdot & 1 & 1 & - & \cdot & \cdot & \cdot & \cdot & \cdot & \cdot & \cdot & 1 & 1 & 1 & \cdot & \cdot & \cdot & \cdot & \cdot & \cdot \\
1 & 1 & - & \cdot & \cdot & \cdot & \cdot & \cdot & \cdot & \cdot & \cdot & \cdot & \cdot & \cdot & 1 & 1 & 1 & \cdot & \cdot & \cdot \\
1 & - & \cdot & - & \cdot & \cdot & \cdot & \cdot & \cdot & \cdot & \cdot & \cdot & \cdot & \cdot & \cdot & \cdot & \cdot & 1 & 1 & 1 \\
\cdot & \cdot & \cdot & \cdot & \cdot & \cdot & \cdot & \cdot & 1 & - & \cdot & - & 1 & \cdot & - & \cdot & 1 & \cdot & \cdot & \cdot \\
\cdot & \cdot & \cdot & \cdot & \cdot & \cdot & \cdot & \cdot & 1 & \cdot & - & 1 & \cdot & - & \cdot & \cdot & \cdot & - & 1 & \cdot \\
\cdot & \cdot & \cdot & \cdot & \cdot & \cdot & \cdot & \cdot & \cdot & 1 & - & \cdot & \cdot & \cdot & - & 1 & \cdot & 1 & \cdot & - \\
\cdot & \cdot & \cdot & \cdot & \cdot & \cdot & \cdot & \cdot & \cdot & \cdot & \cdot & \cdot & 1 & - & \cdot & 1 & - & \cdot & - & 1 \\
1 & \cdot & \cdot & \cdot & 1 & 1 & \cdot & \cdot & - & \cdot & \cdot & \cdot & \cdot & \cdot & - & \cdot & \cdot & - & \cdot & \cdot \\
1 & \cdot & \cdot & \cdot & - & \cdot & 1 & \cdot & \cdot & \cdot & - & \cdot & \cdot & \cdot & \cdot & - & \cdot & \cdot & - & \cdot \\
1 & \cdot & \cdot & \cdot & \cdot & - & - & \cdot & \cdot & - & \cdot & \cdot & \cdot & \cdot & \cdot & \cdot & - & \cdot & \cdot & - \\
\cdot & 1 & \cdot & \cdot & - & 1 & \cdot & \cdot & \cdot & \cdot & \cdot & - & \cdot & \cdot & \cdot & \cdot & - & \cdot & 1 & \cdot \\
\cdot & 1 & \cdot & \cdot & 1 & \cdot & \cdot & 1 & \cdot & \cdot & \cdot & \cdot & \cdot & - & \cdot & - & \cdot & 1 & \cdot & \cdot \\
\cdot & 1 & \cdot & \cdot & \cdot & - & \cdot & - & \cdot & \cdot & \cdot & \cdot & - & \cdot & - & \cdot & \cdot & \cdot & \cdot & 1 \\
\cdot & \cdot & 1 & \cdot & - & \cdot & - & \cdot & - & \cdot & \cdot & \cdot & \cdot & - & \cdot & \cdot & 1 & \cdot & \cdot & \cdot \\
\cdot & \cdot & 1 & \cdot & \cdot & \cdot & 1 & 1 & \cdot & - & \cdot & \cdot & - & \cdot & \cdot & 1 & \cdot & \cdot & \cdot & \cdot \\
\cdot & \cdot & 1 & \cdot & 1 & \cdot & \cdot & - & \cdot & \cdot & - & - & \cdot & \cdot & 1 & \cdot & \cdot & \cdot & \cdot & \cdot \\
\cdot & \cdot & \cdot & 1 & \cdot & - & 1 & \cdot & - & \cdot & \cdot & \cdot & 1 & \cdot & \cdot & \cdot & \cdot & \cdot & 1 & \cdot \\
\cdot & \cdot & \cdot & 1 & \cdot & 1 & \cdot & - & \cdot & - & \cdot & 1 & \cdot & \cdot & \cdot & \cdot & \cdot & 1 & \cdot & \cdot \\
\cdot & \cdot & \cdot & 1 & \cdot & \cdot & - & 1 & \cdot & \cdot & - & \cdot & \cdot & 1 & \cdot & \cdot & \cdot & \cdot & \cdot & 1
\end{bmatrix}$$

$\dot{W}_5$

Figure 4. (*continued*)

4. CONSTRUCTION OF $W(2^{k-1}, k)$ OF THE SEMI-BIPLANE TYPE

In order to construct of $W(2^{k-1}, k)$ of the semi-biplane type, we introduce several definitions. The following shaped matrix is called a Γ–hooked matrix of size(a, a'), where $a' \geq 0$.

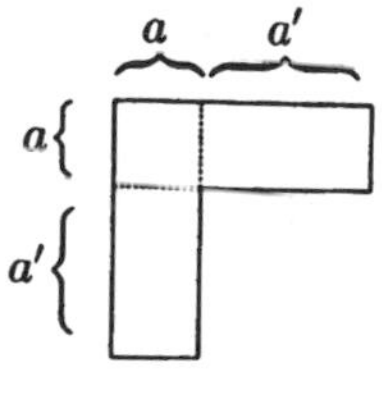

Let A, B be Γ-hooked matrices of size (a, a'), (b, b'), respectively. When $a' = b$, we define $A \dot{+} B$ as

$A \dot{+} B =$

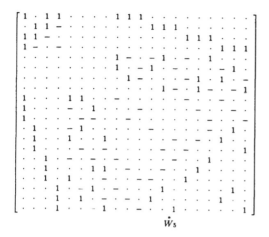

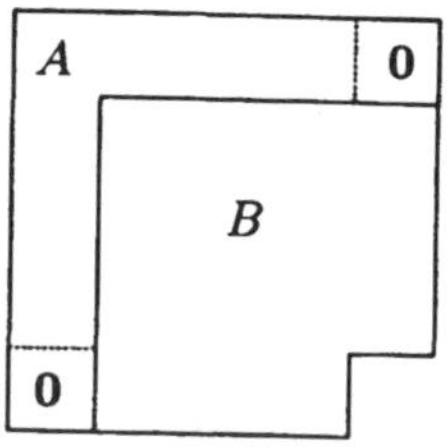

,

where a zero vector or zero matrix is denoted by **0** in this section. It is obvious that the operation $\dot{+}$ satisfies the associative law. Note that the left lower and right upper corners of $A\dot{+}B$ are zero matrices, and $A\dot{+}B$ is also a Γ-hooked matrix of size $(a+b, b')$.

Let X and Y be $\alpha \times \beta$ and $\gamma \times \delta$ matrices, respectively. When $\alpha \geq \gamma$, we define the $\alpha \times (\beta + \delta)$ matrix $X \oplus Y$ as

$$X \oplus Y = \left[\begin{array}{c:c} X & \begin{array}{c} \mathbf{0} \\ \hdashline Y \end{array} \end{array}\right]$$

It is obvious that the operation $\oplus$ satisfies the associative law.

Let $S_\alpha^{(1)} = [1, 1, \ldots, 1]$ be a $1 \times \alpha$ matrix which is called the S-matrix of level 1 and size α. Inductively, the S-matrix of level $i+1$ and size α, say $S_\alpha^{(i+1)}$, is defined as

$$S_\alpha^{(i+1)} = \begin{bmatrix} S_{\alpha-1}^{(i)} \\ \\ (-1)^i I_{c(\alpha-1,i)} \end{bmatrix} \oplus \cdots \oplus \begin{bmatrix} S_{\alpha-u}^{(i)} \\ \\ (-1)^i I_{c(\alpha-u,i)} \end{bmatrix} \oplus \cdots \oplus \begin{bmatrix} S_{\alpha-(\alpha-i)}^{(i)} \\ \\ (-1)^i I_{c(i,i)} \end{bmatrix}$$

where $\alpha \geq i+1 \geq 2$, $c(\gamma, i) = \binom{\gamma}{i}$, and I_β is the identity matrix of order β.

LEMMA 5: *The matrix $S_\alpha^{(i+1)}$ is of size $c(\alpha, i) \times c(\alpha, i+1)$. Also, there are $(\alpha - i)$'s and $(i+1)$'s nonzero elements for each row and column of $S_\alpha^{(i+1)}$, respectively.*

Proof. The first statement is obvious by induction on α and i, and the relations

$$\binom{\alpha-1}{i} + \binom{\alpha-1}{i-1} = \binom{\alpha}{i}$$

$$\sum_{j=1}^{\alpha-i} \binom{\alpha-j}{i} = \binom{\alpha}{i+1}$$

Note that the relation between the uth and $(u-1)$th column components of $S_\alpha^{(i+1)}$ is as

$$\left[\begin{array}{c:c} \begin{array}{c} \mathbf{0} \\ \hdashline S_{\alpha-(u-1)}^{(i)} \end{array} & \mathbf{0} \\ \hdashline (-1)^i I_{c(\alpha-u+1,i)} & \begin{array}{c} S_{\alpha-u}^{(i)} \\ \hdashline (-1)^i I_{c(\alpha-u,i)} \end{array} \end{array}\right]$$

Let the mth row of $S_\alpha^{(i+1)}$ belong to the uth row component of $S_\alpha^{(i+1)}$. Then by induction, the number of nonzero elements of the mth row is $(u-1)+(\alpha-u)-(i-1)=\alpha-i$. The number of nonzero elements in a column of $S_\alpha^{(i+1)}$ is $(i+1)$ by the induction hypothesis and from the above figure. □

The $c(\alpha, i) \times c(\alpha-u, i)$ submatrix of $S_\alpha^{(i+1)}$ containing $S_{\alpha-u}^{(i)}$ is called the uth column component of $S_\alpha^{(i+1)}$. Similarly, the $c(\alpha-u, i-1) \times c(\alpha, i+1)$ submatrix of $S_\alpha^{(i+1)}$ containing $S_{\alpha-u}^{(i)}$ is called the uth row component of $S_\alpha^{(i+1)}$, where $1 \le u \le \alpha - i$.

REMARK 6: *To make the matrices presentable, the transpose matrix of matrix A is denoted by ${}^tA$ in this section.*

LEMMA 6: *The following relation holds for ${}^tS_\alpha^{(i)}$ and $S_\alpha^{(i+1)}$:*

$$[{}^tS_\alpha^{(i)} \vdots S_\alpha^{(i+1)}] \cdot {}^t[{}^tS_\alpha^{(i)} \vdots S_\alpha^{(i+1)}] \;=\; \alpha I_{c(\alpha,i)}$$

Proof. It is proved by induction on α and i, where $\alpha > i \ge 1$.

$$[{}^tS_\alpha^{(1)} \vdots S_\alpha^{(2)}] = \left[\begin{array}{cccccccccccc} 1 & 1 & \cdots & 1 & 0 & \cdots & 0 & & & & & \\ 1 & - & & & 1 & \cdots & 1 & & & & \mathbf{0} & \\ \vdots & & & \mathbf{0} & - & & & & & & & \\ \vdots & & \ddots & & & & 0 & \cdots & & 1 & 1 & 0 \\ \vdots & \mathbf{0} & & & \mathbf{0} & & & \ddots & & - & 0 & 1 \\ 1 & & & - & & & - & & & 0 & - & - \end{array}\right]$$

Thus, the lemma holds for the case of $i = 1$ and any $\alpha > 1$. Suppose that the lemma holds for any $i'(< i)$ and $\alpha'(< \alpha)$, where $i' < \alpha'$. Let $T_\alpha^{(i+1)}(u, v)$ be the common submatrix of the uth row and vth column components of $S_\alpha^{(i+1)}$, where $1 \le u, v \le \alpha - i$. Then

$$T_\alpha^{(i+1)}(u, v) \;=\; \begin{cases} S_{\alpha-u}^{(i)} & u = v \\ \mathbf{0} & u < v \\ R_{u,v}^{(i)} & u > v \end{cases}$$

where $R_{u,v}^{(i)} \;=\; [\mathbf{0} \;\vdots\; (-1)^i I_{c(\alpha-u,i-1)} \;\vdots\; \mathbf{0}]$. Note that

$$(-1)^i I_{c(\alpha-v,i)} \;=\; \begin{bmatrix} R_{u,v+1}^{(i)} \\ \vdots \\ R_{u,u}^{(i)} \end{bmatrix}$$

and $R^{(i)}_{u,v} \cdot {}^tR^{(i)}_{u,v'} = \mathbf{0}$ when $v \neq v'$. Let X_u and X_v be the uth and vth row components of $[{}^tS^{(i)}_\alpha \;\vdots\; S^{(i+1)}_\alpha]$ for $u > v$, respectively. Then

$$\begin{aligned} \mathrm{X}_u &= \left[\mathbf{0} \vdots {}^tS^{(i-1)}_{\alpha-u} \vdots (-1)^{i-1} I_{c(\alpha-u,i-1)} \vdots T^{(i)}_\alpha(u,1),\ldots,T^{(i)}_\alpha(u,\alpha-i)\right] \\ &= \left[\mathbf{0} \vdots {}^tS^{(i-1)}_{\alpha-u} \vdots (-1)^{i-1} I_{c(\alpha-u,i-1)} \vdots \; R^{(i)}_{u,1},\ldots,R^{(i)}_{u,u-1} \vdots S^{(i)}_{\alpha-u} \vdots \mathbf{0}\right] \end{aligned}$$

Thus,

$$\begin{aligned} X_u \cdot {}^tX_u &= \left[{}^tS^{(i-1)}_{\alpha-u} \vdots S^{(i)}_{\alpha-u}\right] \cdot {}^t\left[{}^tS^{(i-1)}_{\alpha-u} \vdots S^{(i)}_{\alpha-u}\right] + uI_{c(\alpha-u,i-1)} \\ &= (\alpha-u)I_{c(\alpha-u,i-1)} + uI_{c(\alpha-u,i-1)} = \alpha I_{c(\alpha-u,i-1)} \end{aligned}$$

by the hypothesis of induction. Next, it follows that

$$\begin{aligned} & X_v \cdot {}^t X_u \\ &= \left[\mathbf{0} \vdots {}^tS^{(i-1)}_{\alpha-v} \vdots \quad (-1)^{i-2} I_{c(\alpha-v,i-1)} \quad \vdots R^{(i)}_{v,1} \cdots\cdots R^{(i)}_{v,v-1} \quad S^{(i)}_{\alpha-v} \vdots \quad \mathbf{0}\right] \\ &\cdot {}^t\left[\quad \mathbf{0} \quad \vdots {}^tS^{(i-1)}_{\alpha-u} \vdots (-1)^{i-2} I_{c(\alpha-u,i-1)} \vdots R^{(i)}_{u,1} \cdots\cdots R^{(i)}_{u,v-1} \quad R^{(i)}_{u,v} \cdots S^{(i)}_{\alpha-u} \vdots \mathbf{0}\right] \\ &= (-1)^{i-2} S^{(i-1)}_{\alpha-u} + I_{c(\alpha-u,i-1)} + S^{(i)}_{\alpha-v} \cdot {}^tR^{(i)}_{u,v} \\ &= (-1)^{i-2} S^{(i-1)}_{\alpha-u} + I_{c(\alpha-u,i-1)} + (-1)^{i-1} S^{(i-1)}_{\alpha-u} - I_{c(\alpha-u,i-1)} = \mathbf{0} \end{aligned}$$

because of

$$\begin{aligned} & S^{(i)}_{\alpha-v} = \\ & \begin{bmatrix} S^{(i-1)}_{\alpha-v-1} \\ \\ (-1)^{i-1} I_{c(\alpha-v-1,i-1)} \end{bmatrix} \oplus \cdots \oplus \begin{bmatrix} S^{(i-1)}_{\alpha-u} \\ \\ (-1)^{(i-1)} I_{c(\alpha-u,i-1)} \end{bmatrix} \oplus \cdots \oplus \\ & \begin{bmatrix} S^{(i-1)}_{i-1} \\ \\ (-1)^{(i-1)} I_{c(i-1,i-1)} \end{bmatrix} \end{aligned}$$

and by the definition of $R^{(i)}_{u,v}$. □

Let $\Gamma^{(i)}_\alpha$ be a Γ-hooked matrix of size $(c(\alpha, i-1), c(\alpha, i))$ for $1 \leq i \leq \alpha$, and

$$\Gamma_\alpha^{(i)} = \begin{array}{|c|c|}\hline (-1)^{i-1} I_{e(\alpha,i-1)} & S_\alpha^{(i)} \\ \hline {}^t S_\alpha^{(i)} & \\ \cline{1-1} \end{array}$$

For convenience, we write $\Gamma_\alpha^{(\alpha+1)} = (-1)^\alpha$. Note that $\Gamma \dot{+} \Gamma_\alpha^{(\alpha+1)}$ is well defined for any Γ-hooked matrix of size $(\alpha, 1)$.

THEOREM 6: *There exists a $W(2^{k-1}, k)$ of type F_1.*

Proof. Put $W = \Gamma_{k-1}^{(1)} \dot{+} \Gamma_{k-1}^{(2)} \dot{+} \cdots \dot{+} \Gamma_{k-1}^{(k)}$. Then W is a weighing matrix with weight k and order 2^{k-1} from Lemmas 5 and 6. Also, it is clear that W is a matrix of the semi-biplane type from the method of construction of W and $\Gamma_{k-1}^{(i)}$. □

Note that $\overset{*}{W}_1$ in Section 3 is the matrix of the case $k = 6$. By the construction of $W(2^{k-1}, k)$ of the semi-biplane type, we may state the following conjecture.

CONJECTURE: *For irreducible weighing matrices $W(n, k)$ of the F_1 type, the following inequality holds:*

$$\frac{k(k-1)}{2} + 1 \leq n \leq 2^{k-1}$$

ACKNOWLEDGMENTS

The authors wish to thank the referees for their valuable comments and suggestions for improvement of the paper. The author also wish to thank Takashi Miyamoto for his assistance in computational works.

REFERENCES

Chan, H. C., Rodger, C. A., and Seberry, J. On inequivalent weighing matrices, *ARS Combinatoria*, A 21, 299–333, 1986.

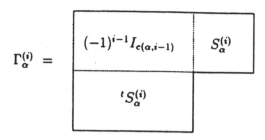

Craigen, R. Weighing matrices and conference matrices. In: C. J. Colbourn and J. H. Dinitz, eds. *The CRC Handbook of Combinatorial Designs*. Boca Raton, New York, London, Tokyo: CRC Press, pp. 492–498, 1996.

Geramita, A. V., and Seberry, J. *Orthogonal Designs: Quadratic forms and Hadamard Matrices*. New York: Marcel Dekker, 1979.

Hadamard, J. Résolution d'une question aux déterminants. *Bull. Sci. Math.*, 17, 240–248, 1893.

Hall, M., Hadamard matrices of order 16. J.P.L.Research Summary 32–10, Vol. 1, pp. 21–26, 1961.

Hall, M. Hadamard matrices of order 20. J.P.L. Technical Report 32–761, Pasadena, 1965.

Hughes, D. Biplanes and semi-biplanes. Lecture Notes in Mathematics, No. 686, pp. 55–58, Springer, Berlin, 1977.

Ito, N., Leon, J. S., Longyear, J. Q. Classification of 3-(24,12,5) designs and 24-dimensinal Hadamard matrices. *J. Combin. Theory*, **A**, 31, 66–93, 1981.

Kimura, H. On equivalence of Hadamard matrices. *Hokkaido Math. J.*, 17, 139–146, 1988.

Kimura, H. New Hadamard matrix of order 24. *Graphs Combin.*, 5, 235–242, 1989.

Kimura, H. Classification of Hadamard matrices of order 28 with Hall sets. *Discrete Math.*, 128, 257–268, 1994.

Kimura, H. and Ohmori, H. Hadamard matrices of order 28. *Mem. Fac. Educ. Ehime Univ. Nat. Sci.*, 7, 7–57, 1987.

Ohmori, H. On the classifications of weighing matrices of order 12. *J. Combin. Math. Combin. Comput.*, 5, 161–216, 1989.

Ohmori, H. Classification of weighing matrices. *Hiroshima Math. J.*, 22, 129–176, 1992.

Ohmori, H. Classification of weighing matrices of order 13 and weight 9. *Discrete Math.*, 112, 35–77, 1993.

Ohmori, H., and Miyamoto, T. Construction of weighing matrices $W(17, 9)$ having the intersection number 8. In press.

Appendix: THE PUBLICATIONS OF JAGDISH N. SRIVASTAVA

Books and Monographs

Written

1. Roy, S. N., Gnanadesikan, R. and Srivastava, J.N. (1970). *Analysis and Design of Certain Quantitative Multiresponse Experiments.* Pergamon Press, Oxford, England, 1970.

Edited

1. *A Survey of Combinatorial Theory*, (in cooperation with Professors Frank Harary, C. R. Rao, G. C. Rota, and S. S. Shrikhande); North-Holland Publishing Company, Amsterdam, 1973.
2. *A Survey of Statistical Design and Linear Models.* North-Holland Publishing Company, Amsterdam, 1975.
3. Combinatorial Mathematics, Optimal Designs and Their Applications, North-Holland Publishing Company, Amsterdam, 1980.
4. *Probability and Statistics*; North Holland, 1988.

Papers in Statistics and Mathematics

1. Srivastava, J. N. (1957). Number of red-rot lesions on the midrib of sugarcane leaves. *Jour. Ind. Soc. Agri. Stat.*, **9**, 52-60.
2. Kishen, K. and Srivastava, J. N. (1958). Confounding in asymmetrical factorial designs in relation to finite geometries. *Current Science*, **11**.
3. Kishen, K. and Srivastava, J. N. (1959). Mathematical theory of confounding in asymmetrical and symmetrical factorial designs. *Journ. Ind. Soc. Agri. Stat.*, **11**, 73-100.
4. Bose, R. C. and Srivastava, J. N. (1963). Mathematical theory of factorial designs; Part I. Analysis. *Bull. of International Stat. Inst.*, 34th Session, 780-787.

5. Bose, R. C. and Srivastava, J. N. (1963). Mathematical theory of factorial designs; Part II. Construction. *Bull. of International Stat. Inst.*, 34th Session, 788-794.
6. Bose, R. C. and Srivastava, J. N. (1964). Analysis of irregular factorial fractions. *Sankhya Ser. A*, **26**, 117-144.
7. Roy, S. N. and Srivastava, J. N. (1964). Hierarchical and p-block multiresponse designs and their analysis. *Contributions to Statistics, Mahalanobis Dedicatory Volume* 419-428, Pargamon Press.
8. Srivastava, J. N. (1964). On the monotonicity of property of the three main tests for multivariate analysis of variance. *Jour. Roy. Stat. Soc. Series B.*, **26**, 77-81.
9. Srivastava, J. N. (1964). A multivariate extension of Gauss-Markov theorem. *Ann. Inst. Stat. Math.*, **17**, 63-66.
10. Bose, R. C. and Srivastava, J. N. (1964). Multidimensional partially balanced designs and their analysis, with applications to partially balanced factorial fraction. *Sankhya Ser. A*, **26**, 145-168.
11. Bose, R. C. and Srivastava, J. N. (1964). On a bound useful in the theory of factorial designs and error-correction codes. *Ann. Math. Stat.*, **35**, 408-414.
12. Bose, R. C. and Srivastava, J. N. (1965). Economic partially balanced 2^n factorial fractions. *Ann. Inst. Stat. Math.*, **18**, 55-73.
13. Srivastava, J. N. (1966). Incomplete multiresponse designs. *Sankhya, Ser. A.*, **28**, 277-388.
14. Srivastava, J. N. (1966). On testing hypotheses regarding a class of covariance structures. *Psychometrika*, **31**, 147-164.
15. Srivastava, J. N. (1966). Some generalizations of multivariate analysis of variance. In *MULTIVARIATE ANALYSIS* (edited by P. R. Krishnaiah), 129-145, Academic Press, New York.
16. Srivastava, J. N. (1967). On a general class of designs for multiresponse experiments. *Ann. Math. Stat.*, **39**, 1825-1843.
17. Srivastava, J. N. (1967). On the extension of Gauss-Markov theorem to complex multivariate linear models. *Ann. Inst. Stat. Math.*, **19**, 417-437.
18. Maik, R. L. and Srivastava, J. N. (1967). On a new property of partially balanced association schemes useful in psychometric structural analysis. *Psychometrika*, **32**, 279-289.
19. Srivastava, J. N. (1967). Discussion on papers entitled "On applications of half-norms to cyclic difference sets" (by Kiochi Yamamota), and "Note on symmetric block designs with $\gamma = 2$" (by Marshall Hall, Jr.). *Proceedings of the Symposium of Combinatorial Mathematics*, Chapel Hill, North Carolina, 186, 254-255.

20. Srivastava, J. N. (1967). A new class of error-correcting and error-detecting codes. Remarks (unpublished) made at the Combinatorial Symposium, University of North Carolina, Chapel Hill, April 1967. These remarks are, however, incorporated in the book em ALGEBRAIC CODING THEORY by E. R. Berlekamp (McGraw-Hill, 1969) in Section 15.1 entitled, "Srivastava codes: noncyclic codes with an algebraic decoding algorithm", 350-352.
21. Roy, S. N. and Srivastava, J. N. (1968). Inference on treatment effects in incomplete block designs. *Review of International Stat. Inst.*, **36**, 1-6.
22. Srivastava, J. N. (1969). Some studies on intersection tests in multivariate analysis of variance. In *MULTIVARIATE ANALYSIS II*, (edited by P. R. Krishnaiah), Academic Press, New York, 145-168.
23. Srivastava, J. N. (1968, 1971). Optimal balanced 2^m factorial designs. In *S. N. ROY MEMORIAL VOLUME*, Univ. of North Carolina, 689-706.
24. Srivastava, J. N. (1968). Mathematical theory of confounded and fractionally replicated factorial designs with emphasis on the symmetrical case. In *V. G. PANSE VOLUME, INDIAN SOCIETY OF AGRICULTURE STATISTICS*, 227-241.
25. Srivastava, J. N. and McDonald, L. L. (1969). On the costwise optimality of hierarchical multiresponse randomized block designs under the trace criterion. *Ann. Inst. Stat. Math.*, **21**, 507-514.
26. Srivastava, J. N. and Anderson, D. A. (1970). Optimal fractional factorial plans for main effects orthogonal to two-factor interactions: 2^m series. *Jour. Amer. Stat. Assoc.*, **65**, 828-843.
27. Srivastava, J. N. and Anderson, D. A. (1970). Some basic properties of multidimensional partially balanced designs *Ann. Math. Stat.*, **41**, 1438-1445.
28. Srivastava, J. N. and McDonald, L. L. (1970). On the hierarchical two-response (cyclic PBIB) designs, costwise optimal under the trace criterion. *Ann. Inst. Stat. Math.*, **22**, 507-518.
29. Srivastava, J. N. and McDonald, L. L. (1970). On a large class of incomplete multivariate models, which can be transformed to make MANOVA applicable. *Metron.* **28**, 241-252.
30. Srivastava, J. N. and Chopra, D. V. (1971). Optimal balanced 2^m factorial designs of resolution V, $m \leq 6$ *Technometrics*, **13**, 257-269.
31. Srivastava, J. N. and McDonald, L. L. (1971). On the costwise optimality of certain hierarchical and standard multiresponse models under the determinant criterion. *Jour. of Multivariate Anal.*, **1**, 118-128.
32. Srivastava, J. N. and Chopra, D. V. (1971). On the characteristic roots of the information matrix of balanced fractional 2^m factorial designs of resolution V, with applications. *Ann Math. Stat.*, **42**, 722-734.

33. Srivastava, J. N. and Chopra, D. V. (1971). On the comparison of certain classes of balanced 2^m fractional actorial designs of resolution V with respect to the trace criterion. *Jour. Ind. Soc. Agr. Stat.* (Silver Jubilee Volume), 124-131.
34. Srivastava, J. N. and Anderson, D. A. (1971). Factorial subassembly association scheme and multidimensional partially balanced designs. *Ann. Math. Stat.*, **42**, 1167-1181.
35. Srivastava, J. N. (1972). Some general existence conditions for balanced arrays of strength *t* and 2 symbols. Jour. Comb. Th., Ser. A, **13**, 198-2
36. Srivastava, J. N. and Zaatar, M. K. (1972). On the maximum likelihood classification rule for incomplete multivariate samples and its admissibility. *Jour. Multi. Anal.*, **1**, 115-126.
37. Srivastava, J. N. and Anderson, D. A. (1972). Reduced-sized fractional factorial designs of resolution IV: $2^m \times 3$ series. *Jour. Roy. Stat. Soc., Series B*, **34**, 377-384.
38. Srivastava, J. N., Williams, M. C., and Rowland, W. F. (1972). On the choice of an experimental unit and a block system for weather modification experiments. *Proceedings of the Third National Conference on Weather Modification*, 231-234.
39. Srivastava, J. N. and McDonald, L. L. (1973). On the extension of Gauss-Markov theory to a subset of a parameter space, under complex multivariate linear models. *Ann. Inst. Stat. Math.*, **25**, 383-393.
40. Srivastava, J. N. and Chopra, D. V. (1973). Balanced arrays and orthogonal arrays, Chapter 35. In *A SURVEY OF COMBINATORIAL THEORY*, North-Holland Publishing Co., Amsterdam; (J. N. Srivastava, ed. in Cooperation with F. Harary, C. R. Rao, G. C. Rota, and S. S. Shrikhande).
41. Srivastava, J. N. and Zaatar, M. K. (1973). A Monte-Carlo comparison of four estimators of the dispersion matrix of a bivariate normal population, using incomplete data. *Jour. Amer. Stat. Assoc.*, **68**, 180-183.
42. Srivastava, J. N. (1973). Application of the information function to dimensionality analysis and curved manifold clustering. *Proceedings of the Third International Symposium on Multivariate Analysis*, (P. R. Krishnaiah, ed.), 369-381.
43. Srivastava, J. N. and Chopra, D. V. (1973). Optimal balanced resolution V factorial designs of the 2^n series $N \leq 42$. *Ann. Inst. Statist. Math.*, **25**, 587-604.
44. Srivastava, J. N. and Chopra, D. V. (1973). Balanced fractional factorial designs of resolution V for 3^m series. *Proceedings of International Stat. Inst.*, **39**, 196-201.

45. Srivastava, J. N. and Rasmuson, D. M. (1973). A new class of confounded 2-factor designs with small number of replications. *Proceedings of International Stat. Inst.*, **39**, 449-546.
46. Srivastava, J. N. and Chopra, D. V. (1974). Optimal balanced 2^7 factorial design of resolution V, $49 \leq N \leq 55$. *Comm. Stat.*, **2**, 59-84.
47. Srivastava, J. N. and Chopra, D. V. (1974). Optimal balanced 2^8 factorial designs of resolution V, $37 \leq N \leq 51$. *Sankhya, Ser. A.*, **36**, 41-52.
48. Srivastava, J. N. and McDonald, L. L. (1974). Analysis of growth curves under hierarchical models and spline regression. *Sankhya, Ser. A.*, **36**, 251-260.
49. Srivastava, J. N. and Zaatar, M. K. (1974). Incomplete multivariate designs, optimal with respect to Fisher's information matrix. *Ann. Inst. Stat. Math.*, **26**, 299-313.
50. Srivastava, J. N. and Chopra, D. V. (1974). Trace-optimal 2^7 fractional factorial designs of resolution V, with 56 to 68 runs. *Utilitas Mathematicas*, **5**, 263-279.
51. Srivastava, J. N. and Anderson, D. A. (1974). A comparison of the determinant, trace, and largest root optimality criteria. *Comm. Stat.*, **3**, 933-940.
52. Srivastava, J. N. (1975). Designs for searching non-negligible effects; A SURVEY OF STATISTICAL DESIGN AND LINEAR MODELS. (J. N. Srivastava, ed.), North-Holland Publishing Co., Amsterdam, 507-519.
53. Srivastava, J. N. and Chopra, D. V. (1975). Optimal balanced 2^7 factorial designs of resolution V, $43 \leq N \leq 48$. *Sankhya, Ser. B*, **37**, 429-447.
54. Srivastava, J. N. (1975). On the physical applicability of the methods of probability and statistics, I. Randomization theory. *Proceedings of the International Stat. Inst. 40* **46**, 349-356.
55. Srivastava, J. N., Raktoe, B. L. and Pesotan, H. (1976). On invariance and randomization in fractional replications. *Annals of Statistics*, **4**, 423-430.
56. Srivastava, J. N. and Rasmuson, D. M. (1976). A class of good confounded factorial designs with two or three factors. In: *ESSAYS IN PROBABILITY AND STATISTICS*, (S. Ikeda, ed.), 95-103.
57. Srivastava, J. N. (1976). Some further theory of search linear models. *Contributions to Statistics*, Swiss-Austrian Region of Biometric Society, 249-256.
58. Srivastava, J. N. (1976). Smaller sized factor screening designs through the use of search linear models. *Proceedings of the 9th International Conference*, **1**, 139-162.

59. Srivastava, J. N. (1976). An infinite series of balanced 2^m factorial designs of resolution V which allow search and estimation of one extra unknown effect. *Sankhya, Ser. B*, **38**, 280-289.
60. Srivastava, J. N. and Ghosh, S. (1977). Balanced 2^m factorial designs of resolution *V* which allow search and estimation of one extra unknown effect $4 \leq m \leq 8$. *Comm. Stat.*, **A6(2)**, 141-166.
61. Srivastava, J. N. and Ghosh, S. (1977). On the existence of search designs with continuous factors (with S. Ghosh); *Annals of the Institute of Statistical Mathematics*, **29**, 301-306.
62. Srivastava, J. N. (1977). Optimal search designs, or designs optimal under biased-free optimality criteria. *Proceedings of the International Symposium on Decision Theory*, Purdue University, 375-409.
63. Srivastava, J. N. (1977). Some recent work on optimal discrete factorial designs. *Developments in Statistics*, 267-328.
64. Srivastava, J. N. (1977). On the linear independence of sets of 2^q columns with certain $(1, -1)$ matrices with a group structure, and its connection with finite geometries. *Proceedings of the International Symposium on Combinatorial Mathematics*, Canberra, Australia, 79-88.
65. Srivastava, J. N. (1977). An application of search linear models to the diagnosis of disease of patients from observable symptoms. *Proceedings of the International Statistics Institute*, New Delhi, India, 539-542.
66. Srivastava, J. N. and Ariyaratna, W. M. (1977). Theory of optimal balanced 3^m factorial designs. *Proceedings of the International Symposium on Optimization in Statistics*, New Delhi, India, 443-460.
67. Srivastava, J. N. (1978). Presidential Address. *Indian Society of Agricultural Statistics, 1977 Session*, 1-10.
68. Srivastava, J. N. and Caprihan, A. (1978). Some new results on Srivastava Codes. *J. Combinatorics*, Information, and Systems Science, **3**, 11-20.
69. Srivastava, J. N. and Gupta, B. C. (1979). Main effect plan for 2^m factorials which allow search and estimation of one extra unknown effect. *JSPI* **3**, 259-265.
70. Srivastava, J. N. (1979). On the characteristic polynomial of the information matrix of balanced 2^m factorials, approached through 2^m factorials. *Proceedings of ISI/Manila Conference.*
71. Srivastava, J. N. (1980). Some basic results on search linear models with nuisance paramters. *Contributions in Statistics*, Banach Center of Mathematics, Warsaw Poland, **6**, 309-313.

72. Srivastava, J. N. and Ghosh, S. (1980). On the enumeration and representation of certain non-isomorphic bipartite graphs arising in factorial search designs. *Annals of Discrete Mathematics*, **6**, 315-332.
73. Srivastava, J. N. and Ariyaratna, W. M. (1982). Inversion of the information matrices of balanced 3^m factorial designs of resolution V, and optimal designs, 671-688. In *Essays in Honor of C.R. Rao, STATISTICS AND PROBABILITY*. (G. Kallianpur, P. R. Krishnaiah, and J. K. Ghosh, eds.)
74. Srivastava, J. N. and Ariyaratna, W. M. (1981). Balanced arrays of strength t with three symbols and $(t+l)$ rows. *J. Combinatorics, Information and System Sciences*.
75. Srivastava, J. N. (1981). Some problems in experiments with nested nuisance factors. *Proceedings of the ISI*, Buenos Aires. Invited Paper.
76. J. Srivastava, Open Problems 8, in M. Rosenfeld and J. Zaks, Eds., Convexity and Graph Theory, *Annals of Discrete Mathematics* **(20)**, North Holland, Amsterdam, 1981, 336-337.
77. Srivastava, J. N. and Katona, G. (1983). Minimal 2-coverings of a finite Euclidean space based on *GF*(2). *JSPI*, **8**, 375-388.
78. Srivastava, J. N. and Wijetunga, A.M. (1983). *D*-optimal cyclic nested 2-dimensional block designs. *Trabajos de Statistica*, **34**, 95-108.
79. Srivastava, J. N. (1984). Sensitivity and revealing power: Two fundamental statistical criteria other than optimality arising in discrete experimentation. In *EXPERIMENTAL DESIGN, STATISTICAL MODELS, AND GENETIC STATISTICS*, (K. Hinkelman, ed.), Marcel Dekker, 95-118.
80. Srivastava, J. N. (1984). Characteristic polynomial of certain second order 3^n factorials approaced throuth 2^n factorials. *J. Inf. Opt. Sci.*, **5**, 119-146.
81. Srivastava, J. N. Anderson, D. A., and Mardekian, J. (1984). Theory of factorial designs of the parallel flats type-I. *JSPI*, **9**, 229-252.
82. Srivastava, J. N. (1984). Structural richness: An important requirement for a good statistical experimental design. *Indian Agric. Stat. Research Institute*, **Souvenir Volume**, 267-271.
83. Srivastava, J. N. and Mallenby, D. W. (1985). On a decision rule using dichotomies for identifying the nonnegligible parameter in certain linear models. *J. Multivariate Analysis*, **16**, 318-334.
84. Srivastava, J. N. (1985). On a general theory of sampling using experimental design concepts, I: Estimation. *Bull. Intl. Stat. Inst.*, **51**, 10.3 (1-16).
85. Srivastava, J. N. and Saleh, F. (1985). On the need of t-designs in sampling theory. *Utilitas Mathematics*, **28**, 5-17.

86. Srivastava, J. N. and Beaver, R. J. (1986). On the superiority of the nested multidimensional block designs, relative to the classical incomplete block designs. *JSPI*, **13**, 133-150.
87. Srivastava, J. N. (1985). Design of experiments for the assessment of reliability. In *RELIABILITY AND QUALITY CONTROL*, (A. P. Basu, ed.), North Holland, Amsterdam, 369-394.
88. Srivastava, J. N. and Shirakura, T. (1987). Characteristic polynomial of 4^n factorial approached through 2^n factorials. *J. Inf. Opt. Sci.*, **8**, 335-357.
89. Srivastava, J. N. (1987). More efficient statistical experimental designs for reliability assessment. *JSPI*, **16**, 389-413.
90. Srivastava, J. N. (1987). Advances in the general theory of factorial designs based on partial pencils in Euclidean n-space. *Utilitas Mathematics*, **32**, 75-94.
91. Srivastava, J. N. and Siddiqui, M.M. (1987). When A- and D-optimality criteria conflict. *Comm. Stat.*, **16**, 1675-1682.
92. Srivastava, J. N. and Arora, S. (1987). On a minimal resolution 3.2 design for the 2^4 factorial experiment. *Ind. J. Math*, **29**, 309-320.
93. Srivastava, J. N. (1987). On the inadequacy of the customary orthogonal arrays in quality control and general scientific experimentation, and the need of probing designs of higher revealing power. *Comm. Stat.*, **16**, 2901-2941.
94. Srivastava, J. N. (1988). Advances in the statistical theory of comparison of lifetimes of machines under the generalized Weibull distribution. *Comm. Stat.*, **18**, 1031-1045.
95. Srivastava, J. N. (1988). On a general theory of sampling using experimental design concepts, II. Relation with Arrays. In *ESSAYS IN PROBABILITY AND STATISTICS*, North Holland, Amsterdam, 267-283.
96. Srivastava, J. N. and Arora, S. (1988). Minimal search designs of resolution 3.1 and 3.2 for the 2^4 experiment. In *CODING THEORY AND DESIGN THEORY, II.* (Raychaudhuri, D., ed.), Springer Verlag, 336-361.
97. Srivastava, J. N. (1990). Some aspects of response surface theory and the philosophy of statistical experimental designs. *JSPI*, **25**, 415-435.
98. Srivastava, J. N. and Throop, D. (1990). Orthogonal arrays obtained from partial pencils in Euclidean n-space. *Linear Algebra and Its Applications*, **127**, 283-300.
99. Srivastava, J. N. and Hveberg, R. (1992). Sequential factorial probing designs for identifying and estimating nonnegligible parameters. *JSPI*, **30**, 141,-162.

100. Srivastava, J. N. and Arora, S. (1991). An infinite class of 2^m factorial designs of resolution 3.2 for general *m*. *Disc. Math.*, **98**, 35-56.
101. Srivastava, J. N. and Ouyang, Z. (1992). Sampling theory and experimental design concepts. In *CURRENT ISSUES IN STATISTICAL INFERENCE*, (ed: M. Ghosh and P.K. Pathak) IMS Lecture Note Series, **17**, 241-264.
102. Srivastava, J. N. (1990). Modern factorial design theory for experimenters. In *STATISTICAL DESIGN AND ANALYSIS OF INDUSTRIAL EXPERIMENTS*, (ed: S. Ghosh) Marcel Dekker, 311-406.
103. Srivastava, J. N. (1992). A 2^8 factorial search design with good revealing power. *Sankhya*, **54**, 461-474.
104. Ouyang, Z., Srivastava, J. N., and Schreuder, H. T. (1993). A general ratio estimator, and its application to regression model sampling in forestry. *Annals. Inst. Stat. Math.* **45**, 113-127.
105. Srivastava, J. N. and Ouyang, Z. (1992). Studies on the general estimator in sampling theory, based on the sample weight function. *JSPI*, **31**, 199-218.
106. Srivastava, J. N. and Ouyang, Z. (1992). Some properties of a certain general estimator in finite population sampling. *Sankhya* **54**. 261-264.
107. Srivastava, J. N. (1993). Multivariate analysis with few or incomplete obervations. In *FUTURE DIRECTIONS FOR MULTIVARIATE ANALYSIS*, (ed., C.R. Rao) North-Holland.
108. Srivastava, J. N. (1993). Some basic contributions to the theory of comparative life testing experiments. In *ADVANCES IN RELIABILITY*, (ed., A.P. Basu) North-Holland, 365-378.
109. Srivastava, J. N. (1993) Nonadditivity in row-column designs. *Jour. Comb. Inf. Sys. Sc.* **18**, 85-96.
110. Srivastava, J. N. (1994) A new mathematical space with applications to statistical experimental design. JSPI, **40**, 113-126.
111. Srivastava, J.N. and Li, J.F. (1996). Orthogonal designs of parallel flat type. JSPI **53**, 261-283.
112. Srivastava, J.N, and Ghosh S. On nonorthogonality and nonoptimality of Addelman's main-effect plans satisfying the condition of proportional frequencies. *Statistics and Probability Letters*, **26**, 1996, 51-60
113. A critique of some aspects of experimental designs. *Handbook of Statistics* **13** (S. Ghosh and C.R. Rao, ed) 1996, 309-341.
114. T. Shirakura, K. Takahashi and J.N. Srivastava Searching probabilities for nonzero effects in search designs for the noisy case. *Annals of Statistics* **24**, 1997, 2560-2568.
115. Li, J.F. and Srivastava, J.N. Optimal 2^n Factorial Designs of Parallel Flats Type,. *Comm. Stat. Theory Meth.* **26** (10), 1997, 2473-2488.

116. Srivastava J.N. and Wang Y.C. Row-column designs. Non-additivity makes them hazardous to use, 1998. (to appear in JSPI)
117. Srivastava, J.N. and Chu, J.Y. Multistage designs procedures for identifying two-factor interactions, when higher effects are negligible, 1997. (to appear)
118. Srivastava, J.N. Some important classes of problems in statistical experimental design, multivariate analysis, and sampling theory, 1998. (to appear)
119. Srivastava, J.N. On the usage in agricultural statistics of the new class of estimators, pre-eminent in sampling theory, 1998. (to appear)
120. Srivastava, J.N. Further results on the new class of estimators, pre-eminent in sampling theory. (under processing)
121. Srivastava, J.N. On the lack of relevance of and support for medical experiments from the statistical design viewpoint. (under processing)

Index

Printed in the United States
by Baker & Taylor Publisher Services